ELECTRIC POWER SYSTEMS

BOOKS IN THE IEEE PRESS SERIES ON POWER ENGINEERING

Analysis of Faulted Power Systems
P. M. Anderson
1995 Hardcover 536pp 0-7803-1145-0

Subsynchronous Resonance in Power Systems
P. M. Anderson, B. L. Agrawal, J. E. Van Ness
1990 Softcover 282pp 0-7803-5350-1

Power System Protection
P. M. Anderson
1999 Hardcover 1344pp 0-7803-3427-2

Power and Communication Cables: Theory and Applications
Edited by R. Bartnikas and K. D. Srivastava
2000 Hardcover 896pp 0-7803-1196-5

Understanding Power Quality Problems: Voltage Sags and Interruptions
Math H. J. Bollen
2000 Hardcover 576pp 0-7803-4713-7

Electric Power Applications of Fuzzy Systems
Edited by M. E. El-Hawary
1998 Hardcover 384pp 0-7803-1197-3

Inspection of Large Synchronous Machines: Checklists, Failure Identification, and Troubleshooting
Kerszenbaum
1996 Hardcover 200pp 0-7803-1148-5

Power System Stability, Volumes I, II, and III
An IEEE Press Classic Reissue Set
E. W. Kimbark
1995 Hardcover 1008pp 0-7803-1135-3

Analysis of Electric Machinery and Drive Systems, Second Edition
Paul C. Krause, Oleg Wasynczuk, and Scott D. Sudhoff
2002 Hardcover 624pp 0-471-14326-X

Power System Control and Stability, Second Edition
P. M. Anderson, A. A. Fouad
2002 Hardcover 672pp 0-471-23862-7

ELECTRIC POWER SYSTEMS

Analysis and Control

Fabio Saccomanno

IEEE PRESS SERIES ON POWER ENGINEERING

IEEE Press Series on Power Engineering
Mohamed E. El-Hawary, *Series Editor*

IEEE

IEEE PRESS

WILEY-INTERSCIENCE

A JOHN WILEY & SONS PUBLICATION

IEEE Press
445 Hoes Lane
Piscataway, NJ 08854

Published simultaneously in Canada.

Library of Congress Cataloging-in-Publication Data is available.

ISBN 0-471-23439-7

Printed in the United States of America

10 9 8 7 6 5 4 3 2 1

CONTENTS

FOREWORD

It gives me great pleasure to write the Foreword for this book because it covers a subject very dear to me and is written by someone whose work I have followed with interest for many years.

Research and development in electric power systems analysis and control has been an area of significant activity for decades. However, because of the increasing complexity of present-day power systems, this activity has increased in recent years and continues to do so because of the great economic significance of this field in the evolving scenario of a restructured electric power industry. I cannot think of a more qualified person than Professor Fabio Saccomanno to write on this subject. He has worked at the leading edge of developments in power system analysis and control for more than three decades. In addition to his extensive industrial and academic experience, he has made significant contributions to this area through his participation in the activities of international technical organizations, such as CIGRE, IEEE, IFAC, and PSCC.

This book covers a wide range of topics related to the design, operation, and control of power systems that are usually treated separately. Various issues are treated in depth with analytical rigor and practical insight. The subject matter is presented in a very interesting and unique perspective. It combines, in a structured way, control theory, characteristics and modeling of individual elements, and analysis of different aspects of power system performance.

While the book naturally covers topics presented in many other books on the subject, it includes many important original contributions based on pioneering work by the author, in particular, in analysis and control of electromechanical oscillations, nonlinear stability analysis, dynamic modeling and experimental identification, reactive compensation, emergency control, and generation scheduling. The comprehensive and rigorous coverage of all aspects of the subject was accompanied by the search for simplification and practical applications using intuition and common sense. The original Italian edition of this book was published in 1992 by UTET, Turin (Italy). It is not surprising that the original edition received the Galileo Ferraris Award from the Associazione Elettrotecnica ed Elettronica Italiana (AEI) in 1994. I am pleased to have been involved in editing the English translation of the original publication, along with Professor Stefano Massucco, Dr. Lei Wang, and Mr. G. K. Morison.

This book will be an invaluable source of reference for teachers and students of power engineering courses as well as practicing engineers in an area of major significance to the electric power industry.

Dr. Prabha Kundur
President & CEO
Powertech Labs, Inc.

Surrey, British Columbia, Canada
December, 2002

PREFACE

Some years ago, before I started to write this book, I already had an idea about how to structure the Preface. It was to be organized essentially as a "Preface-diary," to assist and encourage me in my work by recording the difficulties encountered along the way, the choices made and the subsequent changes, which often required rearranging the entire order of the subjects and rewriting complete parts of the book.

Now that the book is written, such a Preface would make no sense, not even to me. With the work finished, a strange feeling is aroused in me, a mixture of pride and perplexity, mainly when thinking about the courage and the tenacity I have had.

Drawing from the results of a long personal experience–technical, scientific, and teaching–which has been intensively matured in industrial and academic fields, I have tried to put together whatever I thought was necessary to achieve an up-to-date, organized, and coherent treatise; quite a challenging project which I have been thinking about for a long time.

The typical topics of electric power area (and related areas such as hydro and thermal power plants) are discussed here according to a "system approach" in order to allow, according to the most recent theories and methods, a global and right vision of the different problems involved. Special attention is dedicated to operational scheduling, control, and modelling of phenomena (essential, by the way, for simulations), and to the interpretation of the phenomena themselves, to make them more understandable to the reader and to ensure a sufficient mastery of the problems.

On the whole, the aim of this book is to be critical and constructive, not only for the ability to "do" but, before that, for "knowledge." It expresses the constant desire to clarify concepts and justify simplifications, so to maintain the *human being* at the core of problems. Therefore, this book is intended as a basic and up-to-date text, both for students and for anyone concerned, working in universities, industries or consulting. I hope that the presence of commonly separated topics will offer, for the homogenity of the treatment, interesting comparisons, useful correlations, and a deeper and wider knowledge for specialists in different branches.

I cited at the beginning, pride and perplexity; it seems to me that what has been said could justify such feelings, at least for the variety of contents and ambitious intention. The attempt of assembling (consistently) so many topics may appear successful, but such an effort has required, apart from a sort of cultural "challenge," a continuous and tiring research of the most effective way of presentation, to avoid lack of uniformity in style and emphasis; it is difficult for me to evaluate the result.

I am also aware of having, perhaps, "invaded" too many fields, with the risk of appearing sometimes superficial to specialists, but I hope they can forgive me by considering this invasion an obliged (and possibly discreet) choice and by still appreciating the outcome of the work and the intention that has driven me.

Just before closing, I wish to thank everybody who contributed to my preparation, and the ones who shared, in different ways, more or less directly, the discomforts originating from my work.

FABIO SACCOMANNO

Genoa, Italy
December, 2002

ACKNOWLEDGMENTS

Professor Stefano Massucco played an important role in the preparation of the English version of this book. He has also been precious for constant assistance in the project and a spirit of deep participation which has characterized his contribution. Dr. Prabha Kundur reviewed and edited the English translation, and provided valuable comments and helpful suggestions. He was ably assisted by Dr. Lei Wang and Mr. G. K. Morison in this effort. The large and tiring work of editing the equations has been carried out with competence and great care by Mr. Paolo Scalera.

F.S.

CHAPTER 1

INTRODUCTION TO THE PROBLEMS OF ANALYSIS AND CONTROL OF ELECTRIC POWER SYSTEMS

1.1. PRELIMINARIES

1.1.1 Electric power can be easily and efficiently *transported* to locations far from production centers and *converted* into desired forms (e.g., mechanical, thermal, light, or chemical).

Therefore, electric power can satisfy the requirements of a variety of users (e.g., factories, houses, offices, public lighting, traction, agriculture), widely spread around the intended territory.

On the other hand, it is generally convenient to concentrate electric power generation into a few appropriately sized generating plants. Moreover, generating plants must be located according to both technical and economic considerations. For example, the availability of water is obviously of primary concern to hydroelectric power plants as well as the availability of fuels and cooling water to thermoelectric power plants. General requirements — about primary energy sources to be used, area development planning, and other constraints, e.g., of ecological type — must also be considered.

Consequently, the network for electric power transportation must present a branched configuration, and it can be required to cover large distances between generation and end-users. Moreover, the possible unavailability of some generating units or interconnection lines could force electric power flows to be routed through longer paths, possibly causing current overloads on interconnection lines.

These considerations make it preferable to have a network configuration sufficiently meshed to allow greater flexibility in system operation (as an adequate rerouting when encountering partial outages) thus avoiding excessive current flows in each line and limiting voltage dips and power losses to acceptable levels.

1.1.2 As it is widely known, electric power is produced, almost entirely, by means of synchronous three-phase generators (i.e., alternators) driven by steam or water turbines. Power is transported through a three-phase alternating current (ac) system operated by transformers at different voltage levels.

More precisely:

- Transportation that involves larger amounts of power and/or longer distances is carried out by the "transmission" system, which consists of a meshed network and operates at a very high voltage level (relative to generator and end-user voltages). This system ensures that at the same transmitted powers the corresponding currents are reduced, thereby reducing voltage dips and power losses(1).
- Power transportation is accomplished through the "distribution" system, which also includes small networks of radial configuration and voltages stepped down to end-user levels.

The use of ac, when compared with direct current (dc), offers several advantages, including:

- transformers that permit high-voltage transmission and drastically reduces losses;
- ac electrical machines that do not require rotating commutators;
- interruption of ac currents that can be accomplished in an easier way.

Moreover, the three-phase system is preferable when compared with the single-phase system because of its superior operating characteristics (rotating field) and possible savings of conductive materials at the same power and voltage levels.

For an ac three-phase system, reactive power flows become particularly important. Consequently, it is also important that transmission and distribution networks be equipped with devices to generate or absorb (predominantly) reactive power. These devices enable networks to adequately equalize the reactive power absorbed or generated by lines, transformers, and loads to a larger degree than synchronous machines are able.

These devices can be static (e.g., inductive reactors, capacitors, static compensators) or rotating (synchronous compensators, which can be viewed as

(1) Moreover, an improvement in stability can be obtained, at the same transmitted powers, due to reduced angular shifts between synchronous machine emfs, resulting in a smoother synchronism between machines.

synchronous generators without their turbines or as synchronous motors without mechanical loads).

Furthermore, interconnection between different systems — each taking advantage of coordinated operation — is another important factor. The electrical network of the resulting system can become very extensive, possibly covering an entire continent.

1.1.3 The basic elements of a power system are shown in Figure 1.1. Each of the elements is equipped with devices for maneuvering, measurement, protection, and control.

The nominal frequency value is typically 50 Hz (in Europe) or 60 Hz (in the United States); the maximum nominal voltage ranges 20–25 kV (line-to-line voltage) at synchronous machine terminals; other voltage levels present much larger values (up to 1000 kV) for transmission networks, then decrease for distribution networks as depicted in Figure 1.1.

Generation is predominantly accomplished by thermal power plants equipped with steam turbines using "traditional" fuel (coal, oil, gas, etc.) or nuclear fuel, and/or hydroelectric plants (with reservoir or basin, or fluent-water type). Generation also can be accomplished by thermal plants with gas turbines or diesel engines, geothermal power plants (equipped with steam turbines), and other sources (e.g., wind, solar, tidal, chemical plants, etc.) whose actual capabilities are still under study or experimentation.

The *transmission* system includes an extensive, relatively meshed network. A single generic line can, for example, carry hundreds or even thousands of megawatts (possibly in both directions, according to its operating conditions),

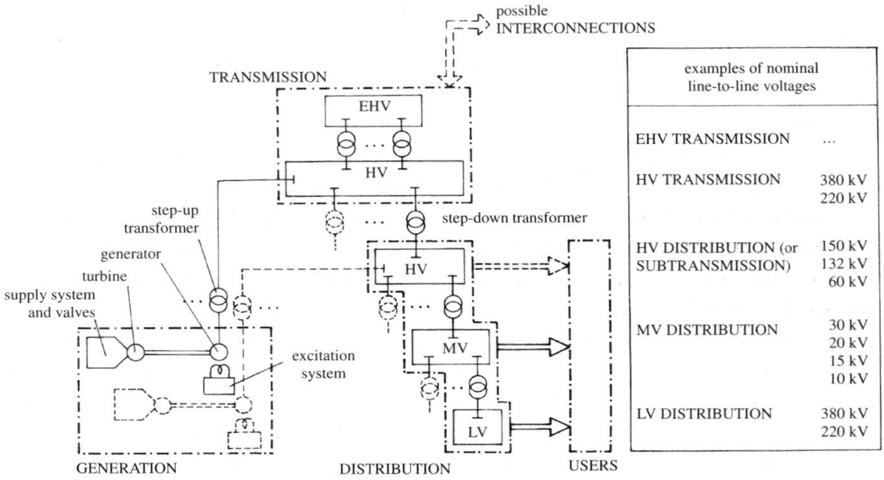

Figure 1.1. Basic elements of an electric power system (EHV, HV, MV, LV mean, respectively, extra-high, high, medium, and low voltage).

covering a more or less great distance, e.g., from 10 km to 1500 km and over. The long lines might present large values of shunt capacitance and series inductance, which can be, at least partially, compensated by adding respectively shunt (inductive) reactors and series capacitors.

The task of each generic *distribution* network at high voltage (HV), often called a "subtransmission" network, is to carry power toward a single load area, more or less geographically extended according to its user density (e.g., a whole region or a large urban and/or industrial area). The power transmitted by each line may range from a few megawatts to tens of megawatts.

Electric power is then carried to each user by means of medium voltage (MV) distribution networks, each line capable of carrying, for example, about one megawatt of power, and by low voltage (LV) distribution networks. To reduce the total amount of reactive power absorbed, the addition of shunt capacitors might be helpful ("power factor correction").

Reactor and *capacitor* types can be fixed or adjustable (through the use of switching devices); the adjustment increases the networks' operation flexibility and may be realized before ("no-load") or even during operation ("under-load", or "on-load").

To further improve system behavior, *controlled compensators* (*synchronous* and/or *static* ones) may be added in a shunt configuration at proper busbars of HV (transmission and subtransmission) networks. *Tap-changing transformers*, which are controllable under load, are also adopted, mostly at the HV to MV transformation, sometimes between HV transformations. While at the MV/LV transformation, the use of tap-changing transformers, set up at no load, can be sufficient.

Moreover, some transmission lines are equipped with series *"regulating" transformers*, by which a range of voltage variations (both in magnitude and phase)—particularly useful to control line power flows—can be achieved.

More recently, the so-called FACTS (Flexible AC Transmission Systems) have also emerged; these equipments recall and integrate the above-cited functions, providing controlled injections of active and reactive powers, through the use of high-performance electronic devices.

The possibility of adopting *direct current links*, by using controlled converters (i.e., rectifiers and inverters) at line terminals, also must be discussed. This is particularly helpful with very long distances and/or with cable connections (e.g., sea-crossing connections); that is, when the ac option would prevent voltage variations within given ranges at the different locations or the synchronism between connected networks.

Finally, the *interconnections* between very large systems (e.g., neighboring countries) are generally developed between their transmission networks. Similar situations involving a smaller amount of power can occur, even at the distribution level, in the case of "self-generating users" (e.g., traction systems, large chemical or steel processing plants, etc.), which include not only loads in the strict sense but also generators and networks.

1.2. THE EQUILIBRIUM OPERATION

1.2.1 A proper definition of the generic steady-state (or equilibrium) operating condition (i.e., the "working point" at which the system may be required to operate) refers to a well-defined mathematical model of the system itself, as discussed in detail in the following chapters. The present section is limited to a general definition at this preliminary stage.

Let us assume that the "configuration"(2) and the system parameters are constant, as well as the external variables which define, together with parameters concerning users, each load requirement (e.g., braking torques externally applied to electromechanical users). Let us also assume that the three-phase electrical part of the system is "physically symmetrical." Moreover, we may assume that the electrical part of the system is linear with regard to the relationships between *phase* voltages and currents, thus allowing sinusoidal operations of *phase* variables without waveform distortions or production of harmonics.

Note however that, in this concern, the presence of nonlinearities also may be assumed, provided they can be simply translated into nonlinear *time-invariant* relationships between (voltage and current) *Park's vectors*, as specified in Section 5.6.1.

We will say that the system is in equilibrium operation if (and only if):

- excitation voltages of synchronous machines are constant;
- all synchronous machine shafts rotate at the same electrical speed (*"synchronous" operation*), so that electrical angular shifts among rotors are constant;
- such speed is constant.

Under the above-mentioned conditions, each three-phase set of the emfs applied to the electrical part of the system results in a positive sequence sinusoidal set, at a frequency equal to the electrical speed of the synchronous machines; the same applies for voltages and currents at any generic point inside the electrical network. More precisely, the frequency of these sets, which comes from the synchronous motion of the machines, can be given the name of *"network" frequency* because of its common value at every point of the network.

The following important consequences apply:

- by using the Park's transformation (see Appendix 2) with a "synchronous" reference (i.e., rotating at synchronous speed), both voltages and currents at any generic point of the network are represented by constant vectors;

(2) By the term *configuration* we imply both the system "composition" (i.e., the whole set of operating components) and its "structure" (i.e., the connection among such components).

- active and reactive powers at any point of the network are constant, as well as active powers generated by alternators; consequently, the driving powers also must be constant, otherwise a variation in machines' speeds would result[(3)].

The definition of the steady-state condition is both useful and appropriate, as it can be transformed, by means of the Park's transformation, into an operating condition characterized by constant values. The definition also has practical aspects, as the synchronous operation at a given speed can be viewed (at ideal operating conditions and once stability conditions[(4)] are satisfied) as a result of the "synchronizing" actions between the machines and the frequency regulation (see also Sections 1.3 and 1.6).

The generic equilibrium operation is determined by:

- system configuration and parameters,
- load requirements,
- network frequency,
- synchronous machine excitation voltages,
- synchronous machine (electrical) angular shifts.

Note that, once all the N excitation voltages and the $(N-1)$ angular shifts are known (where N is the number of synchronous machines), the N vectors corresponding, through the Park's transformation, to the synchronous machine emfs in equilibrium conditions, can be directly deduced, both in magnitude and phase, by assuming an arbitrary reference phase.

However, for a better characterization of the steady-state, one could specify the value of other $(2N-1)$ scalar variables, as detailed in Chapter 2.

For example, instead of excitation voltages, it is usually preferable to specify the terminal voltage values (magnitude) of all synchronous machines, as these values are of paramount importance for the system operation (and are under the so-called "v/Q control"; see Section 1.3).

[(3)] The driving power of each generating unit is obviously limited between minimum and maximum values, which are dependent (at the given speed) upon the characteristics of the supply system and the turbine. At the steady-state, each generated active power matches the corresponding net driving power and is subjected to the same limitations. The maximum real power made available by all the operating plants at the steady-state, which is called "*rotating power*," must be large enough to supply — with an adequate margin, named "rotating reserve" or "*spinning reserve*" — the total active load and network losses (we obviously imply that possible powers generated by nonmechanical sources have been previously subtracted from the total load).

[(4)] The stability properties can vary according to the considered operating point, due to nonlinearities in the equations relating active and reactive powers, magnitude and phase of voltage vectors, etc. Moreover, stability is particularly related to synchronizing phenomena which govern the relative motion between the machines, and to actions (possibly having stabilizing effects) through the control devices; as a consequence, the stability analysis may use some schematic approaches, such as those presented in Section 1.8.

Similarly, it is preferable to specify, instead of angular shifts:

- all active powers of generating plants except one, that is, the active power generation dispatching: this distribution is, in fact, important for system operation (and is related, with frequency regulation, to the "f/P control"; see Section 1.3);
- mechanical powers generated by synchronous motors and compensators; powers can be considered known for motors based on actual loading conditions, whereas powers for compensators can be neglected, as their value is only equal to mechanical losses at the given speed.

1.2.2 Nevertheless, the equilibrium operation previously defined corresponds to, with regard to voltage and current behavior, an ideal situation which in practice can be only approximately achieved.

Regarding the above-mentioned hypotheses (and assuming that stability holds), the most important reasons for deviation from the ideal behavior are:

- *network configuration variations, in proximity to loads*: for example, frequent inserting and disconnecting operations of loads, or opening and closing operations of distribution networks due to local requirements or operation of protection systems (e.g., with stormy weather);
- *load variations*: for example, those caused by intermittent operating cycles (traction systems, rolling mills, tooling machines, excavators, welding machines, etc.);
- the *physical dissymmetries* of the electrical part of the system: for example, in lines, transformers, and mostly in loads (as single-phase loads), which can be amplified by anomalous connections (e.g., the disconnection of a phase or an unsymmetrical short-circuit);
- the *nonlinearities* of the electrical part, with reference to the instantaneous values of voltages, currents, magnetic fluxes, etc.: for example, saturations and magnetic hystereses, and "granular" effects due to winding distribution and slots in the machines; electrical characteristics of arc furnaces, fluorescent lights, thyristor controlled converters, static compensators, etc.

As far as network configuration variations and/or load variations are concerned, they can be treated, in terms of a real quasi-steady-state operating condition, similar to small, random "load fluctuations" (both active and reactive) with a zero mean value, whose fastest variations can only be partially compensated by control devices(5). In practice, these fluctuations can significantly affect voltage and current behavior, particularly in proximity of loads, where filtering actions

(5) Here, we are not considering significant and typically deterministic perturbations (e.g., the opening of a major connection in the transmission network, the outage of a generator or a significant load rejection, etc.), in which case the role of control actions becomes essential; see Section 1.7.

might be recommended. On the contrary, the effects on machines' speeds and network frequency are generally modest, because of the filtering actions of the machines' inertias.

Physical dissymmetries generate voltage and current components of negative or zero sequence; however, such components usually can be kept within acceptable limits by properly equalizing loads connected at each phase (see Section 6.1) and by avoiding (with the help of protective devices) permanent anomalous connections. Furthermore, the presence of zero-sequence components can be limited to a particular section of the network near the element that caused them, by proper transformer winding connections (delta or wye) and neutral conductor connection of the wye windings.

Nonlinearities, instead, are responsible for current and voltage waveform distortion and can generate harmonic components that might produce undesired disturbances to telephonic and data transmission systems. Harmonic effects can be reduced by introducing filtering actions close to those components responsible for harmonic generation. Often, a significant filtering is already provided by the same reactive elements adopted to equalize reactive power flows in the network.

In the following — except when differently specified — all previously mentioned phenomena will be considered within acceptable limits. Consequently, at the considered operating condition (synchronous and at constant speed), both voltage and current Park's vectors and active and reactive powers will be considered constant as above specified.

1.3. OPERATING REQUIREMENTS

1.3.1 Different operating requirements can be classified according to the following fundamental aspects: quality, economy, and security.

Quality of operation must be evaluated by considering:

- load supply conditions, which should not be much different from contractual ones;
- operating conditions of each system's equipment, which should not deviate much from optimal design conditions, in both performances or life duration.

Economy implies evaluation of the overall operating cost necessary to provide service to users, with specific reference to:

- availability and costs of energy sources;
- maintenance costs, personnel costs, and so forth, which are relatively dependent on the "operational scheduling" of each equipment.

Security of operation[6] implies the warranty, from a probabilistic point of view, of continuity in system operation (particularly of continuity in supplying

(6) Obviously, here reference is not made to equipment or human safety, which is rather demanded of protection devices, according to considerations developed in Section 1.4.

load), when faced with significant perturbations. The equilibrium stability, for "small" variations, can be viewed as requirement for both quality and security aspects.

Fundamental requirements concerning the quality of the generic equilibrium operation are(7):

- network *frequency* should be at its "nominal" value (the choice of the nominal value is a technical and economic compromise among design and operating characteristics of main components, with specific regard to generators, transformers, lines, and motors);
- *voltage magnitudes* (positive sequence) should match their nominal values, within a range, e.g., of ±5% or ±10% at each network busbar, particularly at some given load busbars.

One should note that, in a pure transmission line, the *voltage support* at values near nominal voltage also may be important to guarantee satisfactory voltages at the line terminals and avoid a reduction in transmittable active power (see Section 1.5.).

The fulfillment of these requirements should comply with "admissibility" limits of each equipment piece (see Section 1.4): for example, it is necessary to avoid, at any network location, excessive *current amplitudes* which may cause tripping of protective devices.

Moreover, the agreed *power supply to users* should be respected as well as the agreed *exchanges of power* (or energy) with other utilities, in the case of interconnected systems.

As far as the quasi–steady-state operating condition described in Section 1.2 is concerned, voltage waveform deviations, nonpositive sequence components, and effects of small and unavoidable zero-mean random load fluctuations are required to be negligible. For instance, voltage *flicker* on lighting and on television apparatus must be limited to avoid disturbances to human eyes (e.g., for voltage variations greater than approximately 1.5% at a frequency of 10 Hz).

Problems related to system operation economy will be discussed in Chapter 2. One can anticipate that, once the system configuration and load demand are given(8) (as well as possible interconnection power exchanges), economy

---

(7) There are exceptions to these requirements, such as the case of a small system temporarily operating in island conditions, for which out-of-range frequency deviations may be accepted, or a system with lack of sufficient generating capacity (for technical, human, or other reasons) for which the requirements of spinning reserve may suggest reduction of active power absorbed by loads by lowering the voltage profile of the network.

(8) One should note that if load voltages are imposed, load currents — and consequently active and reactive powers — are only related to parameters and external variables which define the loads themselves. For instance, the knowledge of the resistance and reactance values of a generic user which can be represented by an equivalent shunt branch allows the definition of load demand directly in terms of absorbed active and reactive powers.

requirements dictate the most adequate dispatching of total power generated among plants in steady-state conditions.

Finally, security requirements have a strong effect (detailed later in Chapter 2) on the system configuration choice and can suggest further limitations on electrical line currents. If, for instance, the spinning reserve is increased and adequately distributed throughout the system, and if power flows and network voltages are adjusted, there can be a reduced risk that perturbations might cause (see Section 1.7):

- instability conditions;
- unacceptable current redistributions that might cause line tripping (due to overcurrent protective devices) leading to a nonsecure network configuration.

1.3.2 To match all operating requirements, the system configuration and working point must be adequately scheduled.

Scheduling is performed by considering situations preevaluated (*"previsional" scheduling*) or measured during system operation (*"real-time" scheduling*), with emphasis on load demand and equipment availability.

According to Section 1.2.1 and Chapter 2, we can assume that the degrees of freedom in choosing the working point for any system configuration are given by:

- the excitation voltages (or the terminal voltages) of the synchronous machines;
- the dispatching of generated active power (whose amount matches the total load demand and system losses);
- the values of adjustable parameters of system devices, such as reactors, capacitors, static compensators, tap-changing transformers, regulating transformers (some of these values are actually adjusted by control systems, whereas the other ones are chosen before the device operation and kept constant).

It should be noted that actual ranges for the preceding degrees of freedom are limited.

Furthermore, facing the effects of perturbations, particularly of those lasting longer, and keeping the system at satisfactory steady-state conditions can be done with two fundamental controls:

- frequency and active power control (in short named *f/P control*), which acts on control valves of prime movers (except for plants generating power at fixed rate), to regulate frequency (and exchanged active powers in case of interconnected operation) and dispatch active powers generated by each plant[9].

- voltages and reactive power control (v/Q *control*), which acts on the excitation circuit of synchronous machines and on adjustable devices (e.g., reactors, capacitors, static compensators, underload tap-changing transformers), to achieve acceptable voltage profiles with adequate power flows in the network.

It should be noted that f/P and v/Q control problems substantially differ for the following reasons:

- Regulated frequency is common to the whole system and can be affected by all the driving powers. Therefore, the f/P control must be considered with respect to the whole system, as the result of different contributions (to be suitably shared between generating plants). In other words, the f/P control must present a "hierarchical" structure in which local controls (also named "primary" controls) on each turbine are coordinated through a control at the system level (named "secondary" control).
- Regulated voltages are instead distinct from each other (as they are related to different network points), and each control predominantly acts on voltages of the nearest nodes. Consequently, the v/Q control problem can be divided into more primary control problems (of the local type), which may be coordinated by a secondary control (at the system level) or simply coordinated at the scheduling stage.

The control systems should also be provided (see Section 1.5) with sufficient margin for actions. This can be accomplished during real-time scheduling by performing "adaptive"-type actions on system configuration, adjustable parameters, parameters and "set-points" of the f/P and v/Q controls, etc. Adaptive actions

[9] Frequency regulation implies the modulation of driving powers which must match, at steady-state conditions, the total active load (apart from some deviations due to mechanical and electrical losses, or contributions from nonmechanical energy sources). One should note that, after a perturbation, the task of frequency regulation is not only to make net driving powers and generated active powers coincide but, moreover, to return frequency to the desired value. Therefore, even the regulation itself must cause transient unbalances between the powers until the frequency error returns to zero.

As a final remark, transient frequency errors, integrated over the time, cause a "phase error" which affects time keeping by electric clocks operating on the basis of network frequency; such an error can be reset to zero by forcing the system — using the f/P control, for instance, at night — to operate with frequency errors of the opposite sign for an adequate time duration (*"phase" regulation*). In an analogous way, one may compensate the transient errors which arise in the exchanged power regulation, thus returning to agreed values of the energy exchange at interconnections (*"energy" regulation*).

can also be suggested by a timely "diagnosis" of the perturbed system operation (see Section 1.7.2.).

1.3.3 Before concluding, it is worth emphasizing the advantages that may be offered by *interconnections*, with reference to quality, economy, and security. The following observations can be made:

- *Quality*. The voltage profile in the transmission network is better supported and distribution networks benefit because more generators provide their contribution to it, with an increased total capability (more specifically, an increased "short-circuit power" is obtained, at the busbars which are influenced by interconnections; see also Section 5.7.2). The same considerations apply to improved frequency behavior, with respect to any deterministic active power perturbation of given amplitude and to random perturbations. Random perturbations increase but, due to a partial statistical compensation, to an extent less than proportional to the total active power (i.e., in practice, to the total inertia of units and to the total driving power available for regulation purposes), so their relative influence is reduced.
- *Economy*. Different from isolated systems, it is possible to reduce the total set of generating plants and, consequently, operational and investment costs. This can result from the diminished influence of load perturbations and (for analogous reasons) errors on total load forecasting (thus allowing the reduction of the total spinning reserve), and time "compensation" of individual system load diagrams. Moreover, operational (including plant start-ups and shutdowns) and generation scheduling of units can be more economically coordinated by exploiting the flexibility offered by interconnections and by optimizing the scheduling of exchanged powers.
- *Security*. The chances of rerouting transmitted power flows in response to perturbations are increased and, more specifically, each system can benefit from the help of others even when spinning reserves were not sufficient for isolated operation.

1.4. ADMISSIBILITY LIMITS FOR SINGLE COMPONENTS

1.4.1 Each equipment of the power system (including generation, transmission, distribution, and utilization) is required to operate within limits expressed in terms of related variable ranges.

Some limits are *intrinsic*, as they are directly derived from the physical characteristics of the equipment. Examples are limits on excitation voltages of synchronous machines or the maximum allowable opening limit of turbine valves, which may be translated (at given conditions of the motive fluid and of speed) into a limit for the available motive power (see Section 1.2, footnote[(3)]).

On the contrary, other limits are related to operating ranges[(10)], according to different requirements, specifically:

(1) necessity of avoiding, for each equipment, *anomalous or nonacceptable operations* from the technical and economical points of view (then also considering efficiencies, duty, etc.); corresponding limits are: minimum and maximum values of voltage (and frequency) for the electrical auxiliary systems of power plants and, more generally, for users; the minimum technical value of generated power for steam units; maximum load currents related to contractual agreements; and so on;
(2) necessity of avoiding *equipment damage and any possible consequent damage*: for example, insulation damage due to excessive voltage; mechanical damage due to overspeed or to electrodynamic stresses between conductors by overcurrents; damage to insulating and conducting materials due to overtemperatures (which are, in turn, related to overcurrents);
(3) necessity of avoiding *situations which do not respect quality and security requirements*, also reducing instability risks (by limiting, for instance, the generator underexcitation to avoid unstable operating points)[(11)].

The fulfillment of requirements (1), (2), and (3) can be made easier through an appropriate system configuration and steady-state operating point. Control and, if necessary, protection actions are then integrative during operation.

1.4.2 Without much detail about functional requirements of *protection devices* (e.g., response speed, reliability, selectivity, etc.), the following remarks can be made:

- Many protective devices refer to local variables that are not directly under control or on which no significant effect may be expected through control actions (especially in short times). The case of short-circuit currents and fast overvoltages caused by external perturbations at a generic network location (usually with dissymmetric effects) is one such example. In fact, these situations are particularly vulnerable, and protection actions may be the only way to address them. On the other hand, the role of control actions becomes

---

[(10)] Such limits may depend on particular conditions and on the duration of the considered phenomena. For instance, limits on currents set to avoid excessive overtemperatures can vary according to local temperature and must account for overcurrent duration.

[(11)] For given values of voltage and frequency, the operating limits of a generator (see Section 2.2.1) can be expressed, in terms of delivered active power P and reactive power Q, by the maximum and minimum values for P, and by two curves that define the over- and underexcitation limits (i.e., the maximum and minimum Q values, at each value of P). Curiously, these four limits can be viewed as excellent examples of the four above-mentioned motivations, respectively; in fact, the maximum limit on P can be considered an intrinsic limit, and the minimum limit on P can be considered as a type (1) limit, whereas the overexcitation limit is type (2) and the underexcitation limit is often related to stability constraints, and hence a type (3).

essential to overcome even more severe phenomena, such as relatively slow voltages and frequency variations, due to large generator tripping or other causes. In these cases, the protection system is required to operate only "in extremis", i.e., once the control has not succeeded in its action.

- Some protective equipment (e.g., surge arresters) consists of limiting devices that address external cause by eliminating its undesired effect through a nonlinear behavior which does not alter the system structure. Many other protections, as in the typical case of short-circuit current protection, disconnect the faulted equipment (connecting it again in case of temporary fault), causing a structural change. Therefore, in this occasion (and apart from cases of extreme intervention, when control action is too late), the control also must address the subsequent (and sometimes severe) effects of structural changes caused by the protection system itself.

Setting protection devices is usually done according to "local" criteria, which are not strictly related to control requirements of the whole system(12).

Therefore, the interaction between control and protection systems can be seen as quite a critical problem (which can be made worse, for example, by protection out-of-settings); some examples will be given (see Sections 1.7 and 7.3), with reference to typical situations in the dynamic behavior of the system.

1.5. TYPICAL EXAMPLES OF NONEXISTENCE OF THE EQUILIBRIUM OPERATION

The system configuration must particularly:

(1) allow the set-up of the desired steady-state condition or, at least, of a condition meeting the operating requirements;

(2) guarantee a sufficient margin for control actions.

Even to fulfill condition (1) at the adopted configuration, it may be not enough to have several variables available. In fact, their limitations in variation ranges, and the nonlinearities of the system equations may be prohibitive.

More precisely, let us assume that some system variables properly chosen (e.g., voltages at specific busbars of the network), are set at the desired values, and their number is as large as possible. If the system configuration is not properly chosen, it may result that the "static" model of the system, and particularly the equations relating active and reactive powers, voltages, etc.:

(12) On the other hand, network protection coordination implies specific difficulties, in both selectivity requirements (in order to correctly identify the location of the original fault) and convenience of arranging the most appropriate intervention sequences, even if device settings are choosen independent of the system's general operating conditions.

- admit one solution (or more than one, because of nonlinearities), but with some variables assuming undesired (e.g., voltage far from nominal value at some location) or unacceptable values, i.e., so as to cause protection device intervention (e.g., excessive current flow in a line);
- admit no solution at all.

These considerations also involve condition (2), which states that control systems must work with a sufficient margin of action. In fact, if control systems were programmed to work around a steady-state point close to the limits of solution existence (or to the limits of protection intervention), they might be ineffective or even cause instability, even in response to relatively small perturbations.

The following are meaningful, although simple[(13)], examples that illustrate:

- typical operating limits that are dependent on system configuration and can be responsible for the nonexistence of the desired solution (such limits can be considered intrinsic, in addition to those related to each equipment, as mentioned in Section 1.4.1);
- different instability situations that may occur when conditions cannot produce the desired solution.

Example 1

Let us consider the system illustrated by obvious notations in Figure 1.2a,b, where[(14)]:

- the block (G) includes the generating system, constituted by more plants and represented by a single equivalent unit, with emf $\overline{e}_a$ (vector) behind a series reactance X_a; more precisely, the amplitude e_a of the emf depends on the excitation voltage of the equivalent generating unit, whereas its phase is the (electrical) angular position of the rotor, apart from a difference by a constant value depending on the Park's transformation angular reference;
- the block (M) includes a static compensator of adjustable susceptance B;
- the block (L) represents a connection link of reactance X, given by more parallel lines; therefore, the value of X depends on the actual set of operating lines;
- finally, the block (U) represents the utilization system, defined as a resistance R, the value of which depends on the actual set of users.

(13) Generally, to keep the examples related to practical cases, the block (U) in Figure 1.2a can include a reactance in series to the resistance R; nevertheless, the resulting conclusions are analogous to those presented afterward (see also Chapter 6).

(14) Vectors are defined according to the Park's transformation (see Appendix 2), and impedances are intended as evaluated at nominal frequency. (See Chapters 4 and 5 for more details and to evaluate the adopted approximations.)

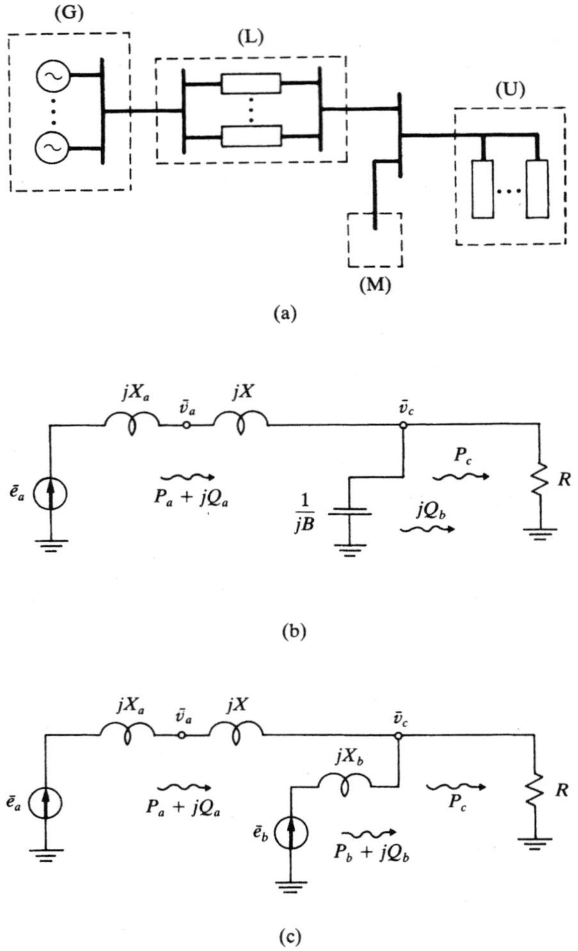

Figure 1.2. Some elementary examples (see text): (a) Reference diagram; (b) Example 1, with static compensator; (c) Example 2, with synchronous compensator (or generator).

(For the sake of simplicity, no transformer is included, although one could account for it in an obvious way).

To achieve the desired values of the voltages v_a and v_c (amplitudes), let us assume that the control of v_a is performed by means of e_a, and the control of v_c through the adjustment of B.

It should be specifically noted that, by imposing $v_a = v_c$, it results $Q_a = Q_b$, which is to say that a reactive power equal to half of what is absorbed by the link (L) (and varying in accordance to the P_a to be transmitted) must be injected at each end of the line. This condition is obviously not related to the rest of the system and therefore can be extended to the following Example 2.

It should be noted that:

- at each given B, both voltages v_a and v_c are proportional to e_a;
- on the other hand, the ratio v_c/v_a depends only on B, according to the equation:

$$\left(\frac{v_c}{v_a}\right)^2 = \frac{R^2}{R^2(1 - BX)^2 + X^2}$$

and cannot be larger than R/X (see Fig. 1.3); consequently, the desired voltage profile can be obtained only if the ratio R/X is large enough.

In other words (and not considering the variation limits for e_a and B) the desired value for v_a can always be achieved by acting on e_a, while the desired value of v_c can be achieved only if it is not larger than Rv_a/X, which can then be defined as *"supportability limit" of the voltage* v_c; the considered case also implies a limitation on the power P_c that can be transmitted to the load, as given by:

$$P_c = \frac{v_c^2}{R} \leq \frac{Rv_a^2}{X^2}$$

By assuming that the regulation of v_c is of the "integral" type, with $\mathrm{d}B/\mathrm{d}t$ proportional to the regulation error ($v_{c\text{des}} - v_c$) (so that B and v_c may vary according to the arrows reported in Fig. 1.3), the value $v_c = v_{c\text{des}}$ can be achieved only if $v_{c\text{des}}$ is below the supportability limit. Otherwise, the regulation error remains

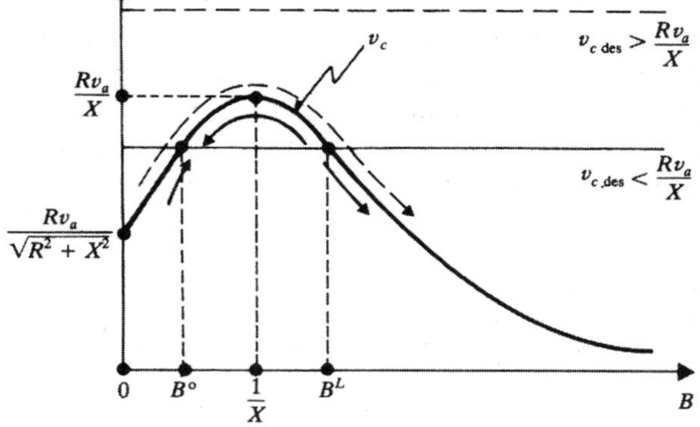

Figure 1.3. Dependence of voltage v_c (amplitude) on susceptance B, and variations due to control actions as per Figure 1.2b.

positive and B continues to increase, leading to an instability phenomenon, *voltage instability*, which results in the "collapse" of voltage v_c[(15)].

It is therefore necessary to guarantee relatively large values of R/X, which is equivalent to maintaining a sufficient set of lines in the link (L), so to have a reactance X small enough, and to reduce the risk of R/X below its critical value, in case of line tripping (resulting in larger X) and/or load increase (resulting in smaller R)[(16)].

Moreover, if the desired voltage profile can be achieved ($v_{c\text{des}} < Rv_a/X$), it is also necessary that the turbines can supply the driving power required to operate at the given constant value of frequency. Specifically, it is necessary to maintain an adequate number of units to guarantee a sufficient "spinning reserve", reducing the risk that a tripped generating unit would cause *frequency instability* and collapse of the frequency itself.

Example 2

Consider the system of Figure 1.2a and c, which differ from that in the previous example only in block (M), which now includes a synchronous compensator (or other generators, still considered an equivalent unit) with an emf $\overline{e}_b$ (vector) behind a series reactance X_b. The magnitude and phase of $\overline{e}_b$ have analogous meanings as per $\overline{e}_a$; in particular, the phase difference between vectors $\overline{e}_a$ and $\overline{e}_b$ is the (electrical) angular shift between the rotors of the equivalent machines in (G) and (M).

By not limiting the variability of e_a and e_b, both the desired values of v_a and v_c can be achieved (by just acting on e_a and e_b), so that no limitation on P_c applies (contrary to the previous example).

However, opposite to this, a limit on P_a arises since, if α is the angular shift between $\overline{v}_a$ and $\overline{v}_c$:

$$P_a = \frac{v_a v_c}{X} \sin\alpha$$

so that P_a cannot be larger than $v_a v_c/X$, which is the *"transmissibility limit" of the active power* through the link (L); while the power P_b, given by $P_b = P_c - P_a$, cannot be lower than $P_c - v_a v_c/X$.

Therefore, if the reactance X is not small enough, the link (L) might become an unacceptable "bottleneck" for active power transmission.

To run the operation at the desired frequency, it is not enough that the total rotating power matches the load demand P_c with an acceptable margin. In fact, it is now necessary to consider the above-mentioned limits on P_a and P_b:

---

(15) The regulation of v_c is then coresponsible for the described phenomenon. In real cases, B is increased up to its maximum value, i.e., the regulation is upper limited, and the voltage collapse may be avoided. Similar situations can be reached even with nonintegral regulation, provided the "static gain" of the regulator is large enough, as generally required by the v_c regulation. The phenomenon can be more complex because of interactions with the regulation of v_a (more specifically: v_a can also decrease if e_a reaches its maximum limit and/or its regulation is not fast enough), protection intervention (due to low voltage or increased line and generator currents), and other factors.

(16) Similar results are obtained in a system with no capacitor and in which block (L) presents at its terminal, at the load side, a tap-changing transformer used to regulate v_c.

- by properly sharing the rotating power between (G) and (M);
- by maintaining an acceptable line set in the link (L), thus providing a sufficiently high transmissibility limit, also considering the risk of line outages.

This last measure can become essential when the units in (M) are of small or even zero rotating power, with the latter being the case of a synchronous compensator.

In fact, in the opposite case (i.e., if the total rotating power is enough but the rotating power in (M) is below $P_c - v_a v_c/X$), the generating set (M) lacks driving power and slows down, so that the desired steady-state cannot be achieved. Particularly, it can be seen that at specific conditions concerning driving powers the synchronism between (G) and (M) might be achieved but at a frequency progressively lower (collapse); otherwise, if the total driving powers in (G) and (M) were equal to P_c, according to the power balance necessary for frequency regulation, the lack of power for the unit (M) would result in a surplus of power for the unit (G), causing the latter to accelerate. The final consequence would be another instability, known as *loss of synchronism* between (G) and (M) (see also Section 1.6).

1.6. SYNCHRONIZING ACTIONS BETWEEN MACHINES

Synchronous machines have the property, which is fundamentally important to the steady-state operation of the system, of spontaneously synchronizing with one another under proper operating conditions. In other words, if these machines present initial (electrical) speeds different from one another, the variations in their reciprocal angular shifts cause subsequent variations in active generated powers. These variations usually slow down the faster rotors and speed up the slower rotors, until — obviously if no further perturbation is applied — the speed deviations are reduced to zero.

This synchronizing phenomenon is generally characterized by damped oscillations (called *"electromechanical" oscillations*). However, the oscillations might degenerate into the so-called "loss of synchronism" between one or more machines and the remaining ones, following particular perturbations of relative severity.

To qualitatively ascertain these phenomena, consider the simple system in Figure 1.2c, which includes only two machines.

By denoting P_{ma} and P_{mb} as the driving powers of the two units and Ω_a and Ω_b as the electrical speeds of their rotors, the motion of the units (considering only their inertias) can be estimated with the following equations:

$$\begin{cases} P_{ma} - P_a = M_a \dfrac{\mathrm{d}\Omega_a}{\mathrm{d}t} \\[2ex] P_{mb} - P_b = M_b \dfrac{\mathrm{d}\Omega_b}{\mathrm{d}t} \end{cases}$$

with M_a and M_b constant.

If a "static" model for the electrical part[(17)] of the system is assumed, the electrical powers P_a and P_b are only dependent on the magnitudes e_a and e_b of the emfs and on the angular shift δ_{ab} between the emfs themselves.

Specifically, by developing the relations between emfs and currents, the following equations can be derived:

$$\begin{cases} P_a = A \sin \delta_{ab} + B \cos \delta_{ab} + C_a \\ P_b = -A \sin \delta_{ab} + B \cos \delta_{ab} + C_b \end{cases}$$

where A, B, C_a and C_b are functions of e_a and e_b, according to[(18)]:

$$\begin{cases} A \triangleq \dfrac{R^2}{R^2 + X'^2} \dfrac{e_a e_b}{X_a + X + X_b} \\ B \triangleq \dfrac{RX'}{R^2 + X'^2} \dfrac{e_a e_b}{X_a + X + X_b} \\ C_a \triangleq \dfrac{R}{R^2 + X'^2} \left(\dfrac{e_a X_b}{X_a + X + X_b} \right)^2 \\ C_b \triangleq \dfrac{R}{R^2 + X'^2} \left(\dfrac{e_b (X_a + X)}{X_a + X + X_b} \right)^2 \end{cases}$$

where, for the sake of simplicity,

$$X' \triangleq \frac{(X_a + X) X_b}{X_a + X + X_b}$$

Moreover, the angular shift δ_{ab} is equal to the electrical angular shift between the rotors of the two units, so that the following equation holds:

$$\frac{\mathrm{d}\delta_{ab}}{\mathrm{d}t} = \Omega_a - \Omega_b \triangleq \Omega_{ab}$$

The previous equations can be depicted by Figure 1.4a. Specifically, the relative motion of the rotors is defined, according to the block diagram of

(17) More precisely (see Section 3.1.1), M_a and M_b should be considered respectively proportional to Ω_a and Ω_b, but the assumption of constant M_a and M_b may be accepted due to the negligible speed variations that can occur. Besides that, a more rigorous description of the system should account for the dependence of e_a and e_b on speeds and more generally for the actual dynamic behavior of the electrical part of the machines and network. In particular, the reactances X_a, X, X_b are calculated at the nominal frequency as if the speeds Ω_a and Ω_b were equal to each other and to the nominal network frequency; for more details, see Chapters 4 and 5.

(18) Other network variables, particularly v_a and v_b, are related to e_a, e_b, δ_{ab}. If "ideal" voltage regulations are assumed (i.e., with e_a and e_b to keep v_a and v_b exactly constant), e_a and e_b would become functions of δ_{ab}. The same would happen for A, B, C_a, C_b, and the dependence of P_a, P_b on δ_{ab} would no longer be sinusoidal.

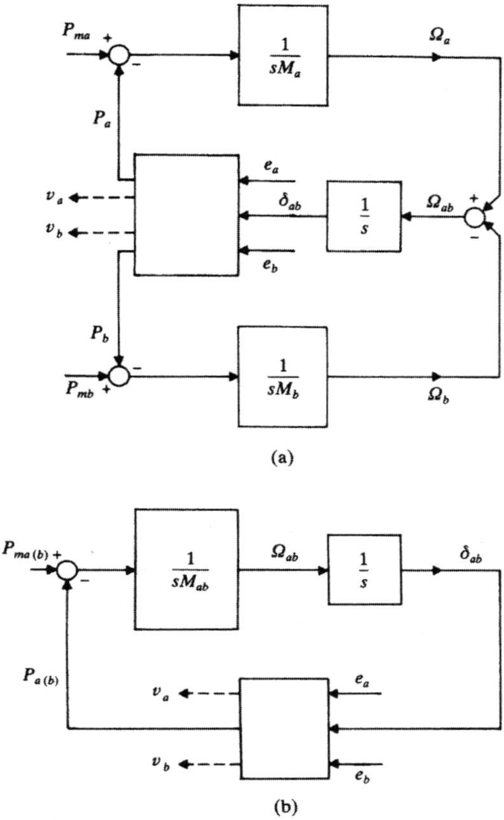

Figure 1.4. Block diagram for the system of Figure 1.2c: (a) motion of the two machines; (b) relative motion between the two machines.

Figure 1.4b, by:

$$\begin{cases} \dfrac{\mathrm{d}\delta_{ab}}{\mathrm{d}t} = \Omega_{ab} \\ \dfrac{\mathrm{d}\Omega_{ab}}{\mathrm{d}t} = \dfrac{P_{ma} - P_a}{M_a} - \dfrac{P_{mb} - P_b}{M_b} = \dfrac{P_{ma(b)} - P_{a(b)}}{M_{ab}} \end{cases} \qquad [1.6.1]$$

in which:

$$\begin{cases} P_{ma(b)} \triangleq P_{ma} - \dfrac{M_a(P_{ma} + P_{mb})}{M_a + M_b} \\ P_{a(b)} \triangleq P_a - \dfrac{M_a(P_a + P_b)}{M_a + M_b} \\ M_{ab} \triangleq \dfrac{M_a M_b}{M_a + M_b} \end{cases}$$

where, in accordance to the above:

$$P_{a(b)} = A \sin \delta_{ab} + \frac{M_b - M_a}{M_a + M_b} B \cos \delta_{ab} + \frac{M_b C_a - M_a C_b}{M_a + M_b}$$

Let us now assume, for simplicity, that $P_{ma(b)}$ remains constant (because, for instance, P_{ma} and P_{mb} are constant, or even variable due to the f/P control but proportionally to M_a and M_b, respectively). Let us also assume that e_a and e_b are constant (not considering the regulation of v_a and v_c), so that the power $P_{a(b)}$ is dependent only on δ_{ab} as shown in Figure 1.5[19].

The equilibrium of the relative motion (i.e., the synchronous operation) is defined by the following conditions:

$$\begin{cases} 0 = \dfrac{\mathrm{d}\delta_{ab}}{\mathrm{d}t} \\ 0 = \dfrac{\mathrm{d}\Omega_{ab}}{\mathrm{d}t} \end{cases}$$

from which the steady-state values of δ_{ab} and Ω_{ab} can be deduced.

Specifically, $\Omega_{ab} = 0$, i.e. $\Omega_a = \Omega_b$, can be deduced from the first condition, while the second condition gives $P_{a(b)} = P_{ma(b)}$ from which the two solutions δ^o_{ab}, δ^L_{ab} result, as generically shown in Figure 1.5, apart from the periodical repetition every 360° with respect to δ_{ab}[20].

The solution $\delta_{ab} = \delta^o_{ab}$, where P_{ab} is an increasing function of δ_{ab}, corresponds to a stable equilibrium point around which electromechanical oscillations may occur. The solution $\delta_{ab} = \delta^L_{ab}$ corresponds to an unstable equilibrium point.

To verify this, let us assume that the shift δ_{ab} and the slip Ω_{ab} had the initial values $\delta^i_{ab} = \delta'_{ab}$, $\Omega^i_{ab} = 0$ (the superscript "i" stands for initial value), with

---

[19] One should note that if one of the two units had an infinite inertia (e.g., $M_b = \infty$), it would simply result $\Omega_b =$ constant. Moreover, $P_{ma(b)} = P_{ma}$, $P_{a(b)} = P_a$, $M_{ab} = M_a$. Furthermore, if $M_b = \infty$ and $X_b = 0$, which is equivalent to considering the node at voltage v_c as an "infinite busbar," the following (simpler) result would be obtained:

$$B = Ca = 0, \qquad P_{a(b)} = P_a = A \sin \delta_{ab} = \frac{e_a e_b}{X_a + X} \sin \delta_{ab}$$

The treatment can be viewed as a generalization of such simple cases, nevertheless with no formal difference. It could be also extended in an analogous way to the case of "ideal" voltage regulations, by considering the dependence (again of the "static" type) of $P_{a(b)}$ on δ_{ab}, according to what is stated in footnote[18]. However, this is not recommended for practical purposes, because the actual behavior of regulations increases the system dynamic order and might lead to instabilities not revealed by this analysis (see Section 7.2.2).

[20] It is assumed that $P_{ma(b)}$ is within the minimum and maximum values of $P_{a(b)}$, otherwise no solution would exist and the "loss of synchronism" between the two units would certainly result (see Section 1.5, Example 2).

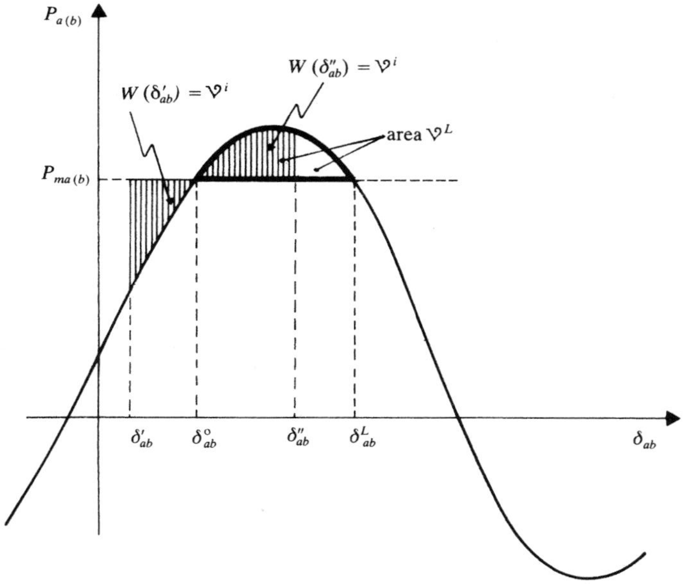

Figure 1.5. Active power versus angular shift curve, used to analyze the relative motion between machines, as in Figure 1.2c.

$\delta'_{ab} < \delta^o_{ab}$ as in Figure 1.5[(21)]. Under such assumptions, it follows that $P^i_{a(b)} < P_{ma(b)}$ and $(\mathrm{d}\Omega_{ab}/\mathrm{d}t)^i > 0$, so that the slip $\Omega_{ab} = \mathrm{d}\delta_{ab}/\mathrm{d}t$ becomes positive and the shift δ_{ab} increases. Then $P_{a(b)}$ becomes larger, $\mathrm{d}\Omega_{ab}/\mathrm{d}t$ diminishes (it goes to zero at $\delta_{ab} = \delta^o_{ab}$ and becomes negative for $\delta_{ab} > \delta^o_{ab}$), and Ω_{ab} reaches its maximum value (Ω_{abM}) at $\delta_{ab} = \delta^o_{ab}$ and subsequently decreases by again crossing zero (under the following conditions) at a given value $\delta_{ab} = \delta''_{ab}$.

Specifically, from Equations [1.6.1] it follows that:

$$(P_{a(b)} - P_{ma(b)})\mathrm{d}\delta_{ab} + M_{ab}\Omega_{ab}\mathrm{d}\Omega_{ab} = 0$$

and consequently:

$$\mathcal{V}(\delta_{ab}, \Omega_{ab}) \triangleq W(\delta_{ab}) + \frac{M_{ab}\Omega^2_{ab}}{2} = \text{constant} = W(\delta^i_{ab}) + \frac{M_{ab}(\Omega^i_{ab})^2}{2} = \mathcal{V}^i \qquad [1.6.2]$$

(21) Such a situation may occur if, for example, the system is initially in equilibrium condition with $P_{ma(b)} = P_{a(b)}$, $\Omega_{ab} = 0$ and one of the lines connecting the two units is tripped, causing a sudden variation of the curve $(P_{a(b)}, \delta_{ab})$ to that in Figure 1.5. On the contrary, generic values of δ^i_{ab}, $\Omega^i_{ab} \neq 0$ can be the consequence of a multiple perturbation, such as an opening-closing after a short-circuit fault or other situations.

in which[22]:

$$W(\delta_{ab}) \triangleq \int_{\delta_{ab}^o}^{\delta_{ab}} (P_{a(b)} - P_{ma(b)}) \mathrm{d}\delta_{ab}$$

Therefore, the values Ω_{abM}, δ''_{ab} must satisfy the following conditions:

$$\begin{cases} \dfrac{M_{ab}\Omega_{abM}^2}{2} = \mathcal{V}^i \\ W(\delta''_{ab}) = \mathcal{V}^i \end{cases}$$

where $\mathcal{V}^i = W(\delta'_{ab})$ is known.

The equation $\Omega_{abM} = \sqrt{2\mathcal{V}^i/M_{ab}}$ can be derived, while δ''_{ab} can be obtained according to Figure 1.5, where the two dotted areas—respectively equal to $W(\delta'_{ab})$ and $W(\delta''_{ab})$—are both equal to $\mathcal{V}^i$ with $\delta''_{ab} \in (\delta^0_{ab}, \delta^L_{ab})$ (*"equal area" criterion*). It should be noted that $W(\delta''_{ab})$ cannot be larger than the "limiting" value $\mathcal{V}^L$ as defined in Figure 1.6 (the area limited by the bold line). Consequently, the solution at δ''_{ab} exists if and only if $\mathcal{V}^i \le \mathcal{V}^L$, where $\mathcal{V}^L$ is the value of the function $\mathcal{V}(\delta_{ab}, \Omega_{ab})$ at the second equilibrium point ($\delta_{ab} = \delta^L_{ab}$, $\Omega_{ab} = 0$), i.e., $\mathcal{V}^L \triangleq \mathcal{V}(\delta^L_{ab}, 0)$.

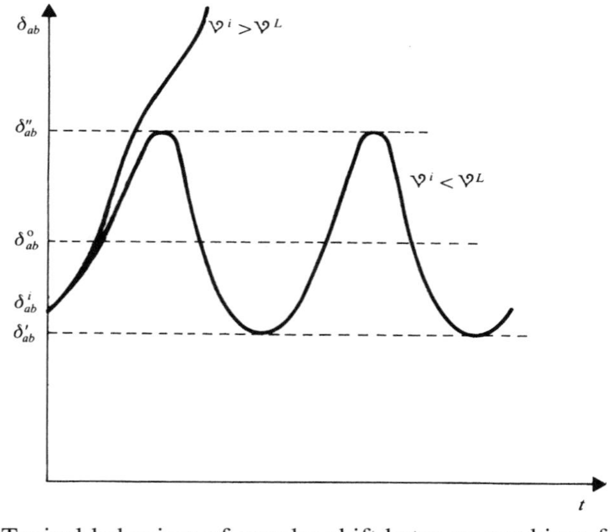

Figure 1.6. Typical behaviors of angular shift between machines for the system of Figure 1.2c.

[22] Actually, in the definition of $W(\delta_{ab})$, the lower integration limit may be arbitrarily chosen or even different from δ^o_{ab}. One should note that the function $\mathcal{V}(\delta_{ab}, \Omega_{ab})$ is (independent from the possibility of assigning to it any particular physical meaning in terms of energy) a *Lyapunov function*, used to analyze stability properties of the system (see Section 8.3).

At the limit case $\mathcal{V}^i = \mathcal{V}^L$, it is easy to see that the system would settle itself at the second equilibrium point, indefinitely.

When $\mathcal{V}^i < \mathcal{V}^L$ (obviously is the most interesting case from the practical point of view), the transient of Ω_{ab}, δ_{ab} continues to oscillate. Starting from the point at which $\delta_{ab} = \delta''_{ab}$, the system behaves similarly to what is described above, but with Ω_{ab} negative and δ_{ab} decreasing, until Ω_{ab} (after having reached its minimum $-\Omega_{abM}$ at $\delta_{ab} = \delta^o_{ab}$) again crosses zero at $\delta_{ab} = \delta'_{ab}$. Then Ω_{ab} and δ_{ab} again vary as described above, through persistent oscillations around the equilibrium point $\delta_{ab} = \delta^o_{ab}$, $\Omega_{ab} = 0$. One should note that, contrary to possible expectations, the resistance R in Figure 1.2c has effect only on the function $P_{a(b)}(\delta_{ab})$, without leading to any damping of oscillations.[23]

On the other hand, if $\mathcal{V}^i > \mathcal{V}^L$, Ω_{ab} remains positive and δ_{ab} always increases, leading to a loss of synchronism because of the lack of synchronizing actions. *The loss of synchronism then can happen not only because of the lack of equilibrium points* (see footnote[20] and Section 1.5), *but also in the presence of a stable equilibrium point*, if the initial point is "too far," i.e., with $\mathcal{V}^i > \mathcal{V}^L$, from it.

During the increase of δ_{ab} at the loss of synchronism, currents and voltages in Figure 1.2c experience unacceptable transients, which actually cause the intervention of protective actions, with the disconnection of the two units. By simply assuming, for instance, $X_b = 0$, $\overline{v}_c = \overline{e}_b$, one can say that, for $\delta_{ab} = 180°$:

- the current in the link between the two units reaches its maximum value $(e_a + e_b)/(X_a + X)$;
- the voltage v_a (amplitude) reaches its minimum value $(X_a e_b - X e_a)/(X_a + X)$;
- the voltage becomes zero (as if a short-circuit occurred) at an intermediate point of the branch connecting $\overline{e}_a$ and $\overline{e}_b$ (this point, also named "electrical center," is defined by the pair of reactances X_1 (from the $\overline{e}_a$ side) and X_2 (from the $\overline{e}_b$ side), with $X_1 + X_2 = X_a + X$, $e_a/X_1 = e_b/X_2$).

In conclusion, the oscillations discussed above depend only on the value $\mathcal{V}^i$ of the function $\mathcal{V}(\delta_{ab}, \Omega_{ab})$ at the initial instant. Therefore, they also may occur starting from $\delta^i_{ab} = \delta''_{ab}$, $\Omega^i_{ab} = 0$ or, more generally, from any pair of initial values $\delta^i_{ab} \in (\delta'_{ab}, \delta''_{ab})$, $\Omega^i_{ab} \neq 0$ that provides the same value of $\mathcal{V}^i$.[24]

---

[23] Under the adopted assumptions, the oscillations are persistent and the equilibrium is "weakly" stable. Nevertheless, in a real two-machine system, oscillations are generally damped because of the dynamic behavior of different components (particularly because of the effect of rotor circuits of machines). However, such *damping* may be negative (e.g., because of dynamic interactions with voltage regulators; see also footnote[19]), if proper stabilizing actions are not provided through control systems (see Section 7.2.2).

[24] For "small" variations (i.e., when $\delta^i_{ab} \to \delta^o_{ab}$, $\Omega^i_{ab} \to 0$) the oscillations of δ_{ab}, Ω_{ab} tend to become sinusoidal, with amplitudes $(\delta''_{ab} - \delta'_{ab})/2 \to \sqrt{2\mathcal{V}^i/K}$ (where $K \triangleq (dP_{a(b)}/d\delta_{ab})^o$) and $\Omega_{abM} = \sqrt{2\mathcal{V}^i/M_{ab}}$ respectively, and at frequency $\Omega_{abM}/((\delta''_{ab} - \delta'_{ab})/2) \to \sqrt{K/M_{ab}}$, as it also can be deduced by "linearizing" the system of Figure 1.4b.

The only condition for the existence of oscillations (for generic initial conditions δ^i_{ab}, Ω^i_{ab}) is then:

$$\mathcal{V}^i < \mathcal{V}^L$$

whereas, when $\mathcal{V}^i > \mathcal{V}^L$, the loss of synchronism results, as shown in Figure 1.6. Additionally, it is easy to determine, under analogous considerations, that the equilibrium point $(\delta^L_{ab}, 0)$ is unstable, because any generic deviation from it would cause oscillations around $(\delta^o_{ab}, 0)$ or loss of synchronism.

All the phenomena described here can be accounted for more concisely by considering (see Fig. 1.7) the possible "trajectories" on the $(\delta_{ab}, \Omega_{ab})$ plane. Each trajectory is, as shown, characterized by a constant value of $\mathcal{V}(\delta_{ab}, \Omega_{ab})$ and is described in the direction indicated by arrows. Then, the knowledge of the initial point $(\delta^i_{ab}, \Omega^i_{ab})$ immediately permits the subsequent evolution of the angular shift δ_{ab} and slip Ω_{ab}, thus confirming what is stated above.

Finally, similar phenomena occur in the more general (and more realistic) case of n machines, with $n > 2$. In such situations, $(n-1)$ angular shifts and $(n-1)$ slips must be considered, i.e., $(n-1)$ possible oscillatory modes, which interact according to system nonlinearities. Unstable situations may occur both "in the small" (e.g., due to negative damping) and "in the large" (with loss of synchronism), or because of the lack of equilibrium points (and subsequent loss of synchronism).[25]

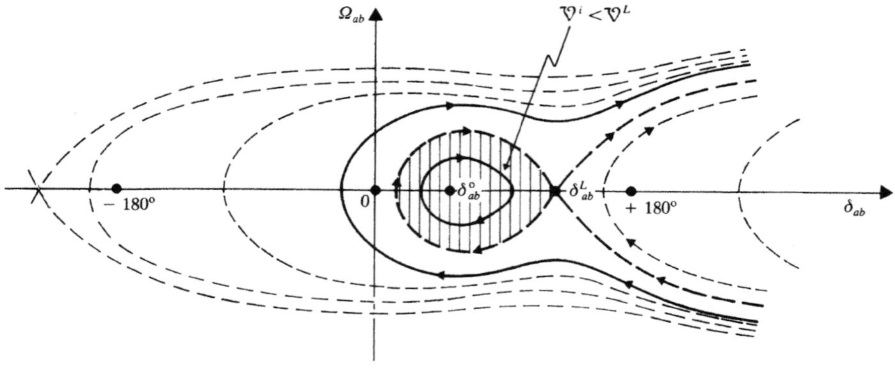

Figure 1.7. Trajectories on the $(\delta_{ab}, \Omega_{ab})$ plane, for the system of Figure 1.2c.

[25] In the usual practice, with a more or less "formal" correctness, these three types of instability are named, respectively, "*dynamic*," "*transient*," and "*static*" *instability*. The transient instability is also called "first swing" instability, with a clear reference to the undamped two-machine case considered above (on the contrary, when considering more general cases, loss of synchronism also may occur after some oscillations).

The analysis of such phenomena generally requires the use of simulations, except with indicative analyses (see Section 8.3) based upon a simplified dynamic model similar to that considered here[(26)].

1.7. THE PERTURBED OPERATION

1.7.1 As in Section 1.2, the steady-state is actually a limit situation with superimposed unavoidable small, zero average, "load fluctuations," whose effects can be considered predominantly local and, consequently, moderately important to the whole system. In this section, such fluctuations will not be addressed; instead, more relevant perturbations (see also Section 1.2, footnote[(5)]) will be considered because of their influence on the whole system with possible risk for its operation.

With these problems, the combination of cases is complex but can be summarized as follows (under the hypothesis of significant perturbations):

(1) perturbations altering only the system configuration and only in a transient way: for instance, a short-circuit from a nonpersistent cause, cleared by protective devices temporarily disconnecting the faulty equipment;

(2) perturbations not altering the system configuration (with regard to generation, transmission, and the most important distribution links), but forcing the system to leave its original steady-state: for instance, a "gradual" load variation;

(3) perturbations permanently altering the system configuration: for instance, a short-circuit from a permanent cause and subsequent disconnection of the affected equipment by protective devices; with more detail, it can happen that:

  (3a) the resulting system is still connected: for instance, because of a tripped network with alternative routings (Fig. 1.8a), a generating plant outage (Fig. 1.8b), or the loss of a user group (Fig. 1.8c)[(27)];

  (3b) the system is separated into two parts, each including generators, transmission, and distribution systems, and loads: for instance, as a result of a tripped (single) interconnection line (Fig. 1.8d).

[(26)] In the above example, due to the adopted assumptions, the loss of synchronism phenomenon occurs in a definitive way, i.e., without any chance to restore synchronism.

In real systems, the combined effects of the different damping actions — due, for instance, to machine rotor circuits and control systems — also might permit this restoration (cases of temporary loss of synchronism), provided that the links between the units are not disconnected by tripping from overcurrent protections.

[(27)] Actually, the cases of Figures 1.8b,c also imply a disconnection of the original system and might be treated as limit cases of type (3b).

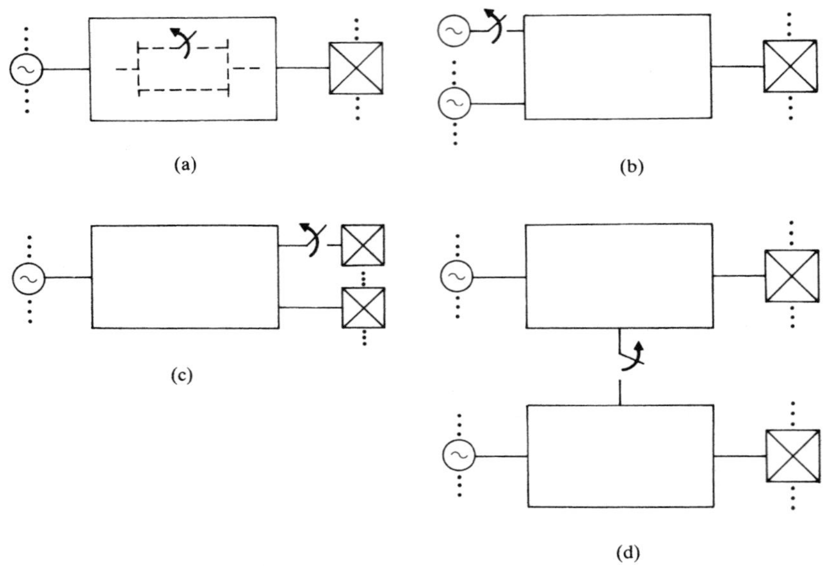

Figure 1.8. Examples of perturbations that permanently alter the system configuration: (a) internal line trip; (b) generating plant outage; (c) loss of a user group; (d) (single) interconnection line trip.

In addition, other perturbations, such as the outage of control system parts, cannot be excluded.

In case (1), the system is required to return to its original (stable) steady-state. In the other cases, above all, the system—or each of the two resulting parts in the case (3b)—is required to settle in a stable steady-state. Furthermore, when this occurs, it should meet all operating requirements. To verify all requirements are satisfied, a telemetering data report must be collected about the system's steady-state.

An accurate estimation of the operating state can be evaluated by using system equations (termed "*state estimation*"; see Section 2.5).

In fact, based on collected data and verification, the steady-state could be modified using available variables and acting on control parameters and "setpoints."

Additionally, it might be worth modifying the system configuration: adjusting it to the new load conditions in case (2), and restoring the original configuration, or achieving a new one, in cases (3a) and (3b). However, *"restoration" of the original configuration* is not possible if the outaged equipment is affected by permanent faults. The change to a new configuration must involve rectifying any critical situations; for example, adding generators or connecting new links.

Nevertheless, the operations required to put into service an equipment also may require special attention and/or long times, particularly with long lines, specific users, and steam-turbine power plants (the case is much easier for hydroelectric or gas-turbine power plants, which can be seen as a type of "quick reserve"; see also Section 2.4.2a).

1.7.2 The perturbed operation may become more complex and critical because of one of the following reasons:

(1) during the transient itself (approaching the stable steady-state condition), protection system action (e.g., due to excessive transient overcurrent in some line) causes new perturbations;
(2) instability situations arise, specifically:
- equilibrium points exist but are unstable[(28)]: for instance, with unstable electromechanical oscillations;
- the final configuration (as in cases (3a), (3b)) does not permit equilibrium points, thus leading, e.g., to voltage or frequency instability or loss of synchronism (see Section 1.5);
- stable equilibrium points exist but are not reachable from the initial system "state": see the simple system in Figure 1.2b, when (even assuming $v_{c\text{des}} < Rv_a/X$) the susceptance B is initially larger than the value B^L reported in Figure 1.3, so that the result is again voltage instability; alternatively, refer to cases of loss of synchronism illustrated in Section 1.6 with $\mathcal{V}^i > \mathcal{V}^L$.

Instability situations also may result in protective actions of generator or load disconnections, or line tripping, which could be due to excessive currents and/or voltage or frequency dips.

Therefore, the subsequent behavior of the system, eventually split into more parts, must be examined in its new operating conditions.

In most real cases, system operation might not be disrupted, with the exception of local outages: for example, the trip of a load with a voltage instability (to remove the instability itself) or the disconnection of a relatively small unit with a loss of synchronism.

Nevertheless, much more severe situations may occur, i.e., *emergency* situations capable of leading the system to a total outage (*blackout*). This may happen in case of a cascade tripping caused by overcurrent protections because of the progressive weakening of the network configuration. It also can occur when the rotating power becomes insufficient, causing a frequency collapse; this can be

(28) This might happen even when considering the case (2), for which the configuration is not altered. One should not be surprised because, due to the nonlinearities of the system, the stability properties can change with the equilibrium point.

worsened by further power plant disconnections because of protective actions in auxiliary systems.

To some extent, all these risks may be considered in operation scheduling, by imposing security checks on the generic steady-state. However, it is important to conduct an up-to-date *diagnosis* of the system during the perturbed operation itself, considering the results of preventive analyses, possibly synthesized in terms of stability "indices" or similar measures which can be quickly evaluated in real-time.(29)

If the diagnosis reveals an emergency situation as described above, then it is necessary to modify the system with new controls to avoid possible outages. An example would be operating the forced disconnection of some loads (*load-shedding*) to eliminate rotating power deficiencies (see Section 3.5) or prevent, or stop, cascade protection interventions ("cascading outages"). When outages are unavoidable it is also important to initiate actions that make easier and quicker the subsequent restoring operations: for instance, "isolating" thermal power plants from the network before a unit trip occurs, allowing operation via their auxiliary systems (*load-rejection*) and local loads, to be ready for reconnection to the network.

1.8. DYNAMIC PHENOMENA AND THEIR CLASSIFICATION

1.8.1 Dynamic relations among variables that characterize the generic system can be summarized as in the block diagram of Figure 1.9, where it can be seen:

- subsystem (a) of a predominantly mechanical type, consisting of generating unit rotating parts (specifically, inertias) and supply systems (thermal, hydraulic, etc.);
- subsystem (b) of a predominantly electrical type(30), consisting of the remaining parts, i.e., generator electrical circuits, transmission, and distribution systems, and users (and possible energy sources of the nonmechanical type), with the latter possibly assimilated with electrical equivalent circuits(31).

(29) In particular, the diagnosis may be organized by considering the above-described cases, specifically the type of perturbations and the possible phenomena of the type (a) or (b).

(30) Subsystem (b) includes mechanical rotating parts of synchronous compensators and electromechanical loads. The mechanical parts of synchronous compensators and of synchronous motors — the latter including their loads — can be considered, if worthy, in subsystem (a) without any particular difficulties (but see footnote(31)).

(31) The equivalence must account for the dynamic behavior of loads, as "seen" from the network. However, with regard to the overall system behavior, strong approximations, which are unavoidable during the analysis stage, can be accepted (above all in the case of loads composed, in an aggregation difficult to determine, by a number of users different in type and with modest unitary power). Equivalent circuits may be used for whole load areas, including in them the MV and LV distribution networks or even subtransmission networks (see Fig. 1.1).

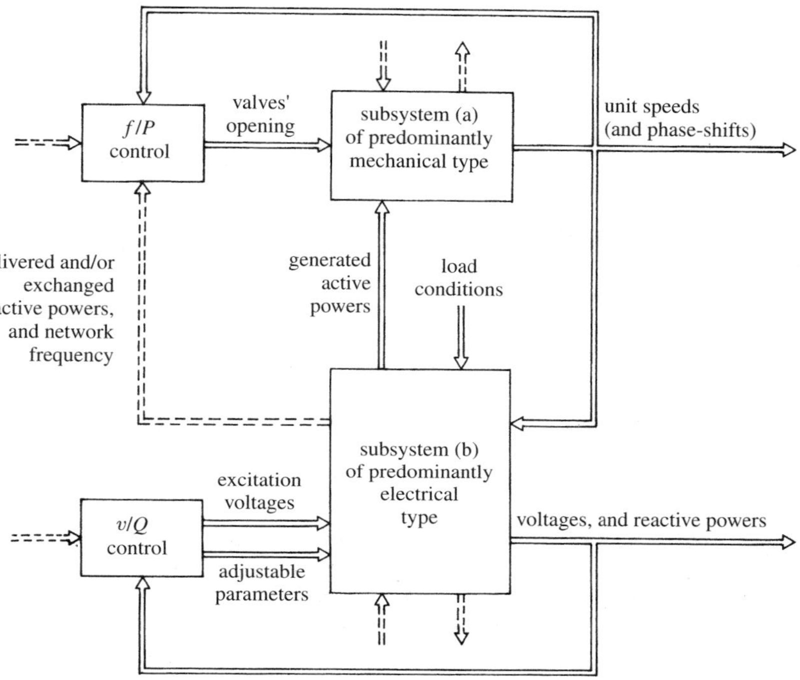

Figure 1.9. Broad block diagram of a generic system.

The input variables of the system composed by (a) and (b) are essentially (apart from structural perturbations):

- openings of prime mover valves, which "enter" into subsystem (a), affecting driving powers (at given operating conditions of the supply systems, e.g., set points of the boiler controls, water stored in reservoirs);
- excitation voltages of synchronous machines, which "enter" into subsystem (b), affecting the amplitude of emfs applied to the three-phase electrical system;
- different parameters that can be adjusted for control purposes (specifically, for the v/Q control): capacitances and inductances of reactive components (of the static type), transformer ratios of underload tap-changing transformers, etc.;
- load conditions dictated by users, which are further inputs for the subsystem (b), in terms of equivalent resistances (and inductances) or in terms of absorbed mechanical powers, etc.[32].

[32] The power produced by nonmechanical sources can be accounted for in an analogous way as a further input to subsystem (b). For simplicity, the modulation of these powers for the f/P control is not considered here.

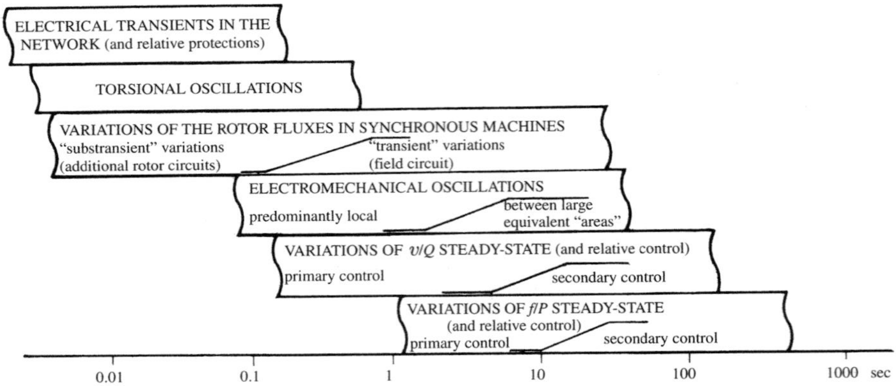

Figure 1.10. Typical time intervals for analysis and control of the most important dynamic phenomena.

According to Figure 1.9, the f/P control is achieved by acting on valves' opening, while the v/Q control is achieved by acting on excitation voltages and the adjustable parameters mentioned above. The load conditions instead constitute "disturbance" inputs for both types of control.

Subsystems (a) and (b) interact with each other, specifically through:

- generated active powers;
- electrical speeds of generating units (or, more generally, of synchronous machines; see footnote[(30)]) and (electrical) shifts between their rotors.

In fact, generated active powers clearly behave as resistant powers (opposed to driving powers) on each unit shaft. Consequently, they affect rotor speeds and relative angular shifts; the speed and angular shifts, vice versa, influence the emf vectors and thus the active powers produced by generating units (besides the other variables of the three-phase electrical system). These phenomena cause, under normal operating conditions, the previously mentioned synchronizing actions. Through electromechanically damped oscillations these actions usually permit the recovery of synchronism; only in the presence of large disturbances, might they be unsuccessful in preventing loss of synchronism.

1.8.2 Regarding response times (see Fig. 1.10), it is important to emphasize that subsystem (a) generally presents much slower "dynamics" than subsystem (b)[(33)], primarily because of the effects of rotor inertias, limits on driving power rate of change, and delay times by which (because of the dynamic characteristics of supply systems) driving powers match opening variations of the valves.

[(33)] Except with torsional phenomena on turbine-generator shafts, which are quite fast (see Section 4.1.4).

This fact supports many simplifications, which are particularly useful in identifying the most significant and characterizing factors of phenomena, performing dynamic analyses with reasonable approximation, and selecting the criteria and implementing the significant variables on which the real-time system operation (control, protection, supervision, etc.) should be based.

According to this order of reasoning, dynamic phenomena can be structured into one of the following categories:

- *"predominantly mechanical" phenomena*, caused by perturbations in subsystem (a) and in f/P control, which are slow enough to allow rough estimates on the transient response of subsystem (b), up to the adoption of a purely "static" model (an example is the case of phenomena related to frequency regulation);
- *"predominantly electrical" phenomena*, caused by perturbations in subsystem (b) and in v/Q control, which are fast enough that machine speeds can be assumed constant (for instance, the initial part of voltage and current transients following a sudden perturbation in the network) or which are such to produce negligible variations in active powers, again without involving the response of subsystem (a) (for instance, phenomena related to voltage regulation, in case of almost purely reactive load);
- *"strictly electromechanical" phenomena*, for which interaction between subsystems (a) and (b) is essential, but it looks acceptable to simplify the dynamic models of components according to the frequencies of the most important electromechanical oscillations (e.g., phenomena related to a single-machine oscillation against the rest of the system, when the latter may be represented as an equivalent connection line and an "infinite power" network; see Chapter 7).

However, when analyzing more complex cases for which simplifications may not seem acceptable, computer simulations can become necessary.

ANNOTATED REFERENCES

Among the works of more general interest, the following may be quoted: 5, 11, 21, 25, 30, 37, 46, 50, 53, 54, 231, 337.

Moreover, as far as dynamic and control problems are concerned: 6, 28, 32, 38, 39, 40, 48, 57, 104, 210, 234, 246, 323, 330, and more specifically:

- with reference to terminology: 227, 263;
- with reference to voltage instability: 55, 59, 229, 259, 286;
- with reference to the perturbed (and emergency) operation: 199, 207, 226, 250, 308.

As far as the most peculiar aspects of power engineering are concerned: 24 (especially for what concerns harmonics), 60.

CHAPTER 2

CONFIGURATION AND WORKING POINT

2.1. PRELIMINARIES

2.1.1. Basic Assumptions

The definition of the generic steady-state working point primarily requires, according to Section 1.2.1, that:

- the system configuration and parameters are constant;
- load demands are constant;
- the three-phase electrical part is physically symmetrical with linear behavior.

If one would account for what actually happens in distribution and utilization systems,

- an enormous amount of data, difficult (or practically impossible) to be collected and subjected to significant uncertainties, would have to be known;
- the overall system model would be excessively overburdened possibly in an unjustifiable way, because many details may actually have effects that are predominantly local;
- the previously mentioned hypotheses could appear unrealistic (remember what was already pointed out in Section 1.2.2).

However, such inconveniences are of minor importance in the overall behavior of the system; so it appears reasonable (and convenient) that distribution and utilization systems be considered only for behavior as "seen" from the transmission network, by treating them as equivalent circuits (more or less approximately) consisting of:

- "equivalent loads," directly fed by the transmission network through nodes (called "load nodes") to which distribution networks are connected;
- connections among these nodes, to account for interactions (e.g., at the subtransmission level) between distribution networks.

(In a more detailed analysis, some subtransmission networks can be kept with the transmission system.)

Each "equivalent load" may be defined, more simply, by absorbed powers (active and reactive) that are constant or slowly varying, apart from discontinuities — such as due to the disconnection of important loads — which actually are significant at the corresponding load node. On the other hand, prearranged actions on the network (e.g., voltage regulations and local filtering) and statistical user compensation can practically allow such approximation.

The (quasi) constancy of absorbed powers enables their identification during the operation, which can prove a valuable asset in real-time system operation. Additionally, these approximations can be particularly reasonable in previsional scheduling, where load forecasting is more reliable when the user sets they refer to are wider.

Figure 2.1 may be helpful in representing a generic system; this figure shows the block named "network" defined within types of "terminal" nodes, which can be classified as follows:

- *generation nodes*, which correspond to the synchronous generator terminals (i.e., at the primary side terminals of step-up transformers);

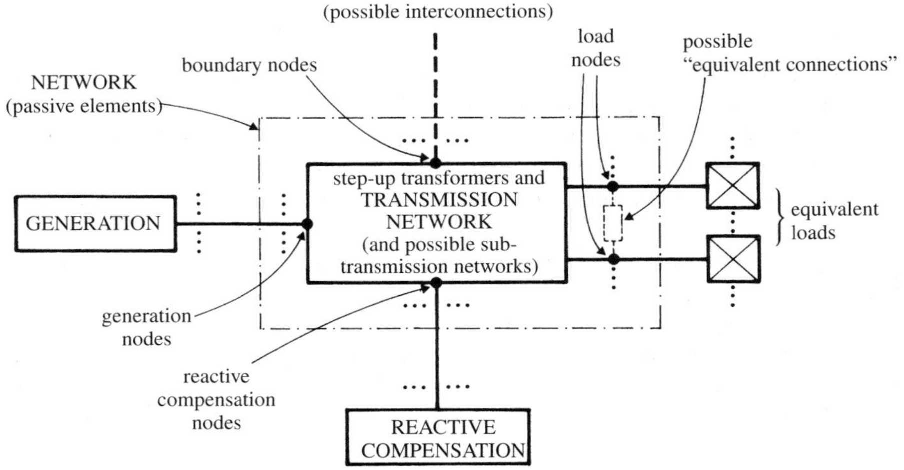

Figure 2.1. Schematic representation of a generic system.

- *reactive compensation nodes*, which correspond in an analogous way to synchronous and static compensator terminals;
- *load nodes*, i.e., nodes supplying the equivalent loads;
- possible *boundary nodes*, for connection with external systems.

Adjustable shunt reactors and condensers may be considered similar to compensators. Additionally, adjustable series equipment (e.g., tap-changing transformers, regulating transformers) may be viewed as external components, considering the relationships between voltages and currents (or powers) at their terminals; the same may be said for dc links (see Section 5.5), as seen by the ac side of the converters.

Moreover, within block "network" are only "passive" elements, i.e., lines and transformers of the transmission network (and possibly subtransmission networks), step-up transformers, reactors, condensers, etc., in addition to equivalent connections between load nodes, as above specified. These elements are predominantly "reactive," and this fact leads to important properties which will be discussed later in this book.

2.1.2. Network Representation

In the generic steady-state, under conditions specified in Section 1.2, phase voltages and currents are, at any given network point, sinusoidal, of positive sequence, and at a "network" frequency equal to the electrical speed of the synchronous machines. Also, at any given point, active and reactive powers are constant. Specifically, by applying the Park's transformation with a "synchronous" reference (i.e., rotating at the same electrical speed as the synchronous machines), the following holds:

- each set of phase voltages or currents transforms into a constant vector;
- the characteristics of the (passive) elements of the network and the relationships between the mentioned vectors are defined by the corresponding positive sequence equivalent circuits, both passive and linear, with impedances (or admittances) evaluated at the network frequency, and "phase-shifters" in the case of transformers with complex ratio (see Sections 5.2, 5.3, and 5.4).

(For simplicity, we could also assume as negligible the inductive coupling between lines close to each other, the variation of parameters, e.g., resistances, with the temperature, that is with local temperature and current, etc.)

Therefore, the whole network is represented by a passive and linear circuit, with "nodes" connected through "branches". More precisely, apart from the "reference" node for voltage vectors, the following node types can be identified:

- *terminal nodes* (see Section 2.1.1), through which an outside "injection" of current or power is generally performed;

- *internal nodes*, which refer only to network elements and do not allow any outside injection.

The set of internal nodes may include particular nodes to be retained (e.g., intermediate nodes of a given line, the sections of which are seen as different elements connected in series); furthermore, it may depend on the type of equivalent circuits adopted for network elements (e.g., a T equivalent circuit leads to a further intermediate node).

A generic branch may be:

- a *series branch*, if it connects a pair of the above-mentioned nodes (terminal and/or internal);
- a *shunt branch*, if it connects one of the above-mentioned nodes with the voltage reference node.

Both the set of branches and internal nodes depend on the detail of representation of the network as well as the single equivalent circuits adopted (e.g., capacitive shunt branches must be considered in the equivalent circuit of a relatively long line).

Furthermore, a shunt branch corresponding to an adjustable reactive element, which is connected at an internal node, must not be considered part of the network if, as already specified, it is to be retained (and the corresponding node becomes a "terminal," reactive compensation node).

Note that, for a generic transformer with turns proportionate to nominal voltages, it is possible to identify a simple (with no shunt branch) equivalent circuit, if magnetizing current and iron losses are neglected (as usually acceptable), and provided that the quantities are expressed in "per unit" values (see Section 5.3).

2.1.3. Network Equations Between Node Voltages and Branch Currents

With N equal to the total number of nodes (terminal and internal), the currents (and powers) of the network in Figure 2.1 can be easily determined if the values, constant in magnitude and phase, of node voltage vectors $\overline{v}_1, \ldots, \overline{v}_N$ are known.

In fact, when no "phase-shifter" is present, let us assume (Fig. 2.2) that the generic node h is connected:

- to the reference node: through a branch or parallel branches having a combined admittance $\overline{y}_{ho}$;
- to the generic node k, with $k \neq h$: through a branch or parallel branches having a combined admittance $\overline{y}_{hk}$.

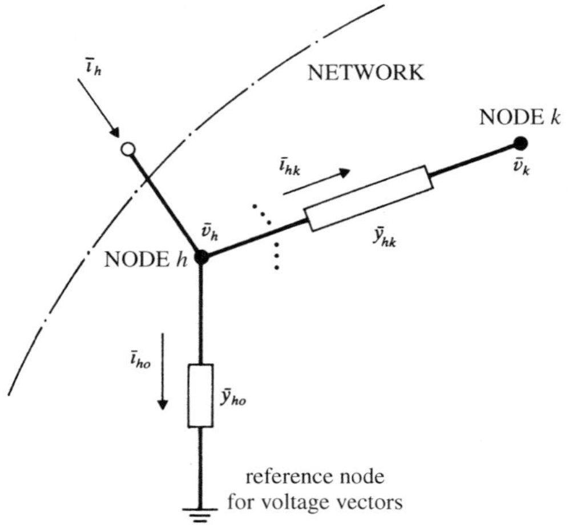

Figure 2.2. Admittances corresponding to a generic node (with no phase-shifters).

Knowing the values of $\overline{v}_h$, $\overline{v}_k$ enables immediate identification of the currents $\overline{\imath}_{ho}$, $\overline{\imath}_{hk}$ which are defined in Figure 2.1, as it results:

$$\left.\begin{aligned} \overline{\imath}_{ho} &= \overline{y}_{ho}\overline{v}_h \\ \overline{\imath}_{hk} &= \overline{y}_{hk}(\overline{v}_h - \overline{v}_k) \end{aligned}\right\} \qquad [2.1.1]$$

from which active and reactive powers can also be determined. Similarly, it is possible to obtain the current in each branch and the power at its terminals.

The following sequence can be used to determine the currents in each branch:

- branches are numbered (from 1 to L, if L is the number of branches) and the diagonal matrix $\overline{Y}^L$, named "*branch admittance*" *matrix*, is determined by putting the admittances of the branches on the diagonal, in the assumed order;
- nodes are numbered (from 1 to N) and the matrix C, named "*connection*" *matrix* (with L rows and N columns) is determined, where the nonzero elements are defined as follows:
  - if the generic branch $l(l = 1, \ldots, L)$ is of the series type and connects nodes h and k, with $h < k$: $C_{lh} = -1$, $C_{lk} = +1$;
  - if, instead, the branch is a shunt branch between the reference node and the node k: $C_{lk} = +1$;
- the column matrix $(L,1)$ of the "branch voltages" $\overline{u}_1, \ldots, \overline{u}_L$ — assuming, with previous notations, that $\overline{u}_l = \overline{v}_k - \overline{v}_h$ if the branch l is of the series type, and $\overline{u}_l = \overline{v}_k$ if it is a shunt branch — is then given by:

$$\overline{u} = C\overline{\mathcal{V}}$$

  where $\overline{\mathcal{V}}$ is the column matrix $(N,1)$ of node voltages;

- finally, the column matrix $(L,1)$ of the "branch currents" $\overline{f}_1, \ldots, \overline{f}_L$—evaluated, with previous notations, from node k (of higher index) to node h for a series branch, and from node k to the reference node for a shunt branch—is given by:

$$\overline{f} = \overline{Y}^L \overline{u} = \overline{Y}^L C \overline{v}$$

Terminal currents for any generic network component can be easily identified. For instance, in a component represented by an equivalent Π circuit for which the terminals are the generic nodes h and k, the following equations can be developed, using Figure 2.3 as a model:

$$\overline{\imath}' = \overline{y}'_o \overline{v}_h + \overline{y}_s(\overline{v}_h - \overline{v}_k), \qquad \overline{\imath}'' = \overline{y}_s(\overline{v}_h - \overline{v}_k) - \overline{y}''_o \overline{v}_k$$

Note that $\overline{y}'_o$ differs from the admittance $\overline{y}_{ho}$ defined in Figure 2.2 if other shunt branches are present at node h; similarly, $\overline{y}_s$ may be different from $\overline{y}_{hk}$, and so on. Moreover, currents $\overline{\imath}'$ and $\overline{\imath}''$ are usually different from each other and from the $\overline{\imath}_{hk}$ of Figure 2.2.

This treatment must be modified if series branches include *phase-shifters*, incorporated because of the presence of complex ratio transformers in the network (i.e., see Section 5.3, transformers with different type of connections at the different windings, such as wye-delta or vice versa; or "quadrature-regulating" transformers). The effect of phase-shifters can be determined by observing that, for the generic series branch reported in Figure 2.4, it results in:

$$\overline{\imath}'_s = \overline{\imath}''_s \epsilon^{j\beta} = \overline{y}_s(\overline{v}_h - \overline{v}_k \epsilon^{j\beta})$$

and consequently:

$$\left.\begin{aligned} \overline{\imath}'_s &= \overline{y}_s \overline{v}_h - (\overline{y}_s \epsilon^{j\beta})\overline{v}_k \\ \overline{\imath}''_s &= (\overline{y}_s \epsilon^{-j\beta})\overline{v}_h - \overline{y}_s \overline{v}_k \end{aligned}\right\} \qquad [2.1.2]$$

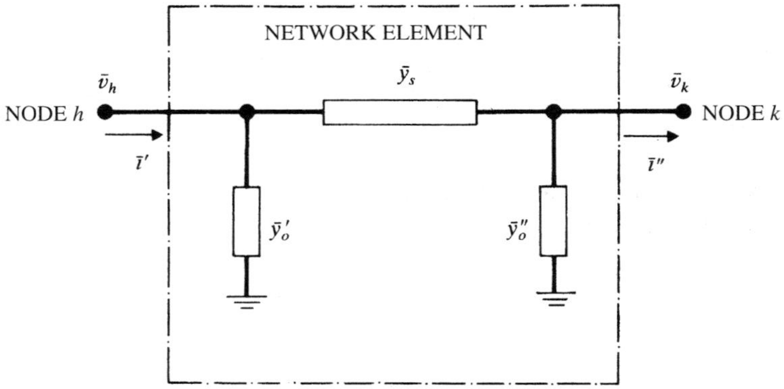

Figure 2.3. Representation for a generic network element (in the absence of phase-shifters).

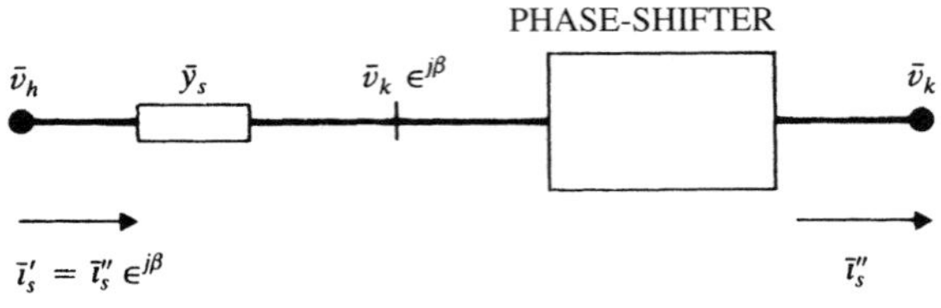

Figure 2.4. Series branch with phase-shifter.

Without phase-shifter it would be, more simply:

$$\bar{\imath}'_s = \bar{\imath}''_s = \bar{y}_s(\bar{v}_h - \bar{v}_k)$$

2.1.4. Network Equations Between Node Voltages and Injected Currents

Knowing the values of currents in different branches enables the immediate evaluation of the currents $\bar{\imath}_1, \ldots, \bar{\imath}_N$ injected (by the external systems) at different nodes as well as the corresponding powers, because the generic current $\bar{\imath}_h$ injected at the node h is the sum of the currents that flow into the branches connected to it. Actually, the currents injected at the internal nodes must be null; this will be accounted for at the end of this section.

Since the network is represented by a passive linear circuit, it is easy to see that the node-injected currents and voltages are related through a matrix equation:

$$\bar{\mathcal{I}} = \bar{\mathcal{Y}}\bar{\mathcal{V}} \qquad [2.1.3]$$

where:

- $\bar{\mathcal{I}}$, $\bar{\mathcal{V}}$ are column matrices (N,1), respectively, constituted by vectors $\bar{\imath}_h$, $\bar{v}_h$ that correspond to all the N nodes (both terminal and internal) of the network ($h = 1, \ldots, N$);
- $\bar{\mathcal{Y}}$ is a square matrix (N,N), called "nodal admittance matrix" or simply "*admittance*" *matrix.*

Diagonal elements $\bar{\mathcal{Y}}_{hh}$ are called "self-admittances" or "driving admittances," and off-diagonal elements are called "mutual admittances" or "transfer admittances."

When no phase-shifter is present, it holds (see Fig. 2.2 and Equations [2.1.1]) that:

$$\bar{\imath}_h = \bar{\imath}_{ho} + \sum_1^N {}_{k \neq h}\bar{\imath}_{hk} = \bar{y}_{ho}\bar{v}_h + \sum_1^N {}_{k \neq h}\bar{y}_{hk}(\bar{v}_h - \bar{v}_k) \quad (h = 1, \ldots, N) \quad [2.1.4]$$

Consequently, the elements of the $\overline{\mathcal{Y}}$ matrix are defined by:

$$\left.\begin{aligned} \overline{\mathcal{Y}}_{hh} &= \overline{y}_{ho} + \sum_{1}^{N}{}_{k\neq h}\overline{y}_{hk} \\ \overline{\mathcal{Y}}_{hk} &= -\overline{y}_{hk} \qquad\qquad (k \neq h) \end{aligned}\right\} \qquad [2.1.5]$$

$(h, k = 1, \ldots, N)$.

From this, $\overline{\mathcal{Y}}_{hk} = \overline{\mathcal{Y}}_{kh}$, so that (when no phase-shifter is present) the $\overline{\mathcal{Y}}$ matrix is symmetrical.

However, in a general case that includes phase-shifters, the symmetry of $\overline{\mathcal{Y}}$ is not verified. To determine this, one must refer to Equations [2.1.2], noting that the contribution to the mutual admittance $\overline{\mathcal{Y}}_{hk}$ due to the generic branch of Figure 2.4 is $(-\overline{y}_s e^{j\beta})$, while the contribution to $\overline{\mathcal{Y}}_{kh}$ is $(-\overline{y}_s e^{-j\beta})$, and they are different.

However, a *symmetrical matrix-based treatment* may usually be adopted, because the effect of phase-shifters can be accounted for separately. This is the case for transformers equipped with windings connected differently (as in the case of step-up transformers having their primary windings, from the generator side, delta connected and with their secondary windings wye connected) and linking two "subnetworks". Such a link systematically introduces (see Section 5.3) a 30° lead shift in all voltage and current vectors pertaining to one of the two subnetworks. It is possible to treat the system as if no phase-shifter were present, by applying a variable transformation (of the type $\overline{v}'_h = \overline{v}_h e^{-j30^\circ}$, $\overline{\imath}'_h = \overline{\imath}_h e^{-j30^\circ}$) to these vectors.

The situation differs when quadrature-regulating transformers are present, because they produce angular shifts on single branches (only some degrees, however not negligible with respect to angular shifts between voltages). Therefore, the mesh equations are modified, with effects on the network steady-state and, more specifically (see Section 2.1.5a), on active power flows. However, these angular shifts are usually small compared with the branch admittances' phases, so the resulting dissymmetry in the admittance matrix may be disregarded.

The admittance matrix $\overline{\mathcal{Y}}$ can be easily determined, starting from the network configuration and from each branch parameters. For instance, assuming there are no phase-shifters or they can be disregarded as explained above, the matrix $\overline{\mathcal{Y}}$ can be evaluated according to:

$$\overline{\mathcal{Y}} = C^T \overline{Y}^L C$$

where $\overline{Y}^L$ and C are the matrices already defined in Section 2.1.3, while C^T is the transposed matrix of C. In fact, the column matrix $\overline{\jmath}$ (of the injected currents) is related to the column matrix $\overline{f}$ of the branch currents according to the equation $\overline{\jmath} = C^T\overline{f}$, where $\overline{f} = \overline{Y}^L C\overline{v}$. It then follows $\overline{\jmath} = \overline{\mathcal{Y}}\overline{v}$, with $\overline{\mathcal{Y}} = C^T\overline{Y}^L C$.

From the matrix Equations [2.1.3] one may also derive:

$$\overline{\mathcal{V}} = \overline{\mathcal{Z}}\,\overline{\mathcal{I}} \qquad [2.1.6]$$

where $\overline{\mathcal{Z}}$ is the square matrix (N,N), called "nodal impedance matrix" or simply "*impedance*" *matrix*, defined by $\overline{\mathcal{Z}} \triangleq \overline{\mathcal{Y}}^{-1}$. Diagonal elements $\overline{\mathcal{Z}}_{hh}$ are called "self-impedances" and off-diagonal elements are called "mutual impedances". With no phase-shifters, $\overline{\mathcal{Z}}$ is obviously symmetrical, like $\overline{\mathcal{Y}}$.

The admittance matrix is usually very "sparse," i.e., with many zero elements, while the impedance matrix is usually "full". As a rule of thumb, the number of nonzero elements in $\overline{\mathcal{Y}}$ is in the order of $(4–5)N$ — instead of N^2 — of which N are diagonal and $(3–4)N$ are off-diagonal. This corresponds to the case for which each node is, on average, directly connected only to other 3 to 4 nodes.

The determination of the matrix $\overline{\mathcal{Z}}$ is not as easy as that of $\overline{\mathcal{Y}}$. For instance (differently from the mutual admittances; see the last of Equations [2.1.5]) the generic impedance $\overline{\mathcal{Z}}_{hk}$ generally does not depend only on the parameters of the connection between nodes h and k.

Nevertheless, it is not difficult to account for the modification of such two matrices when, for instance, a branch of impedance $\overline{z}$ is added (or a branch of impedance $-\overline{z}$ is disconnected; the two actions are obviously equivalent).

The effect on $\overline{\mathcal{Y}}$ is immediately seen, according to previous notes. As far as $\overline{\mathcal{Z}}$ is concerned, denoting by $\overline{\mathcal{Z}}^0$ the original matrix, i.e., the one before branch addition, the following can be observed.

- If a branch of impedance $\overline{z}$ is added in a shunt connection at the generic node r (see Fig. 2.5a), it follows that:

$$\overline{v}_h = \sum_{k \neq r} \overline{\mathcal{Z}}^0_{hk}\overline{\imath}_k + \overline{\mathcal{Z}}^0_{hr}\left(\overline{\imath}_r - \frac{\overline{v}_r}{\overline{z}}\right) = \sum_k \overline{\mathcal{Z}}^0_{hk}\overline{\imath}_k - \overline{\mathcal{Z}}^0_{hr}\frac{\overline{v}_r}{\overline{z}} \qquad (h = 1, \ldots, N)$$

  For $h = r$, an equation is then obtained which identifies $\overline{v}_r$ as a function of $\overline{\imath}_1, \ldots, \overline{\imath}_N$; by substituting it in the previous equation $(h = 1, \ldots, N)$, the following can be finally deduced:

$$\overline{v}_h = \sum_k \left(\overline{\mathcal{Z}}^0_{hk} - \frac{\overline{\mathcal{Z}}^0_{hr}\overline{\mathcal{Z}}^0_{rk}}{\overline{z} + \overline{\mathcal{Z}}^0_{rr}}\right) \cdot \overline{\imath}_k \qquad [2.1.7]$$

  where the coefficient of the generic $\overline{\imath}_k$ defines the element $\overline{\mathcal{Z}}_{hk}$ of the resulting matrix $\overline{\mathcal{Z}}$.

- If the branch of impedance $\overline{z}$ is instead added between the generic nodes r and s (see Fig. 2.5b), the following can be written:

$$\begin{aligned}\overline{v}_h &= \sum_{k \neq r,s} \overline{\mathcal{Z}}^0_{hk}\overline{\imath}_k + \overline{\mathcal{Z}}^0_{hr}\left(\overline{\imath}_r - \frac{\overline{v}_r - \overline{v}_s}{\overline{z}}\right) + \overline{\mathcal{Z}}^0_{hs}\left(\overline{\imath}_s + \frac{\overline{v}_r - \overline{v}_s}{\overline{z}}\right)\\ &= \sum_k \overline{\mathcal{Z}}^0_{hk}\overline{\imath}_k - (\overline{\mathcal{Z}}^0_{hr} - \overline{\mathcal{Z}}^0_{hs})\frac{\overline{v}_r - \overline{v}_s}{\overline{z}} \qquad (h = 1, \ldots, N)\end{aligned}$$

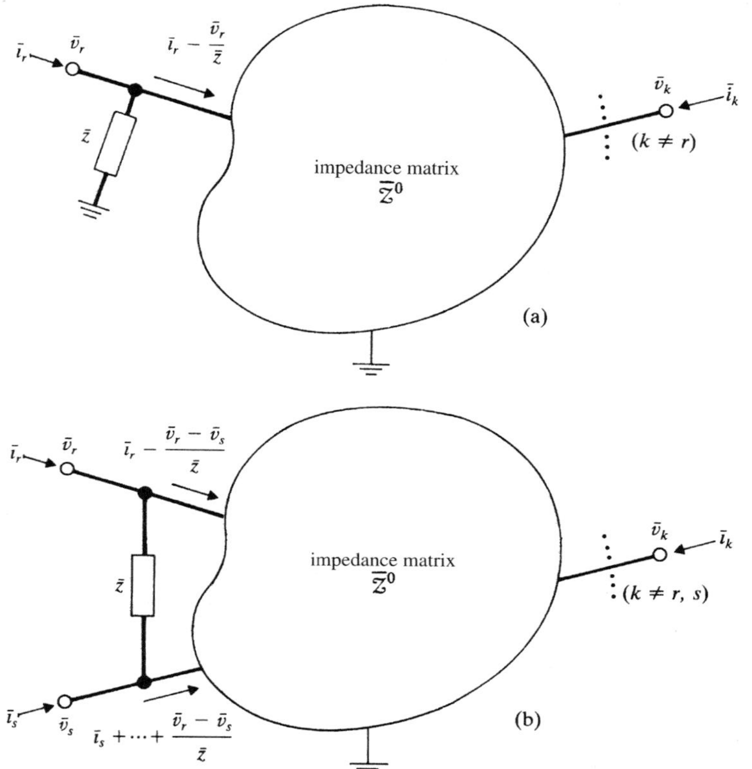

Figure 2.5. Addition of a branch: (a) shunt branch; (b) series branch.

By subtracting the expressions of $\overline{v}_r$ and $\overline{v}_s$ (respectively obtained for $h = r$ and $h = s$), an equation is obtained which evaluates $(\overline{v}_r - \overline{v}_s)$ as a function of $\overline{\imath}_1, \ldots, \overline{\imath}_N$; by substituting it in the previous equation ($h = 1, \ldots, N$), the following can be finally deduced:

$$\overline{v}_h = \sum_k \left(\overline{Z}^0_{hk} - \frac{(\overline{Z}^0_{hr} - \overline{Z}^0_{hs})(\overline{Z}^0_{rk} - \overline{Z}^0_{sk})}{\overline{z} + \overline{Z}^0_{rr} + \overline{Z}^0_{ss} - \overline{Z}^0_{rs} - \overline{Z}^0_{sr}} \right) \cdot \overline{\imath}_k \qquad [2.1.8]$$

where the coefficient of the generic $\overline{\imath}_k$ still defines (as desired) the element $\overline{Z}_{hk}$ of the resulting matrix.

The matrices $\overline{\mathcal{Y}}$ and $\overline{\mathcal{Z}}$, constituted by complex elements, can be put in the form of $\overline{\mathcal{Y}} = \mathcal{G} + j\mathcal{B}$, $\overline{\mathcal{Z}} = \mathcal{R} + j\mathcal{X}$. $\mathcal{G}$ is a real element square matrix (N, N), called "nodal conductance matrix" or simply "conductance matrix," constituted by "self conductances" (on the diagonal) and "mutual conductances" (off the diagonal). The same can be said for $\mathcal{B}$, $\mathcal{R}$, $\mathcal{X}$, with analogous denominations, by respectively substituting the word "conductance" with the words "susceptance," "resistance," and "reactance."

In particular, if the connection between the generic nodes h, k is through a single, purely reactive branch of reactance x, it follows (with known notation):

$$\overline{y}_{hk} = \frac{1}{jx} = -\frac{j}{x}, \quad \overline{\mathcal{Y}}_{hk} = \overline{\mathcal{Y}}_{kh} = -\overline{y}_{hk} = +\frac{j}{x}$$

(see the second part of Equations [2.1.5]), and then $\mathcal{G}_{hk} = \mathcal{G}_{kh} = 0$, $\mathcal{B}_{hk} = \mathcal{B}_{kh} = +1/x$. Instead, if the branch also includes a phase-shifter, as per Figure 2.4, it then results:

$$\overline{\mathcal{Y}}_{hk} = -\overline{y}_{hk}e^{j\beta} = j\frac{e^{j\beta}}{x}, \quad \overline{\mathcal{Y}}_{kh} = -\overline{y}_{hk}e^{-j\beta} = j\frac{e^{-j\beta}}{x}$$

and thus:

$$\mathcal{G}_{hk} = -\mathcal{G}_{kh} = -\frac{\sin\beta}{x}, \quad \mathcal{B}_{hk} = \mathcal{B}_{kh} = \frac{\cos\beta}{x}$$

i.e. the mutual susceptances between the nodes h, k are still equal to each other, whereas the mutual conductances are different from zero and have opposite values.

However, *usually, the network elements are predominantly reactive, whereas the effect of possible phase-shifters can be, as seen, considered separately or disregarded.* Then, as a first-level approximation, it may be assumed $\overline{\mathcal{Y}} = j\mathcal{B}$ and similarly $\overline{\mathcal{Z}} = j\mathcal{X}$ (where $\mathcal{X} = -\mathcal{B}^{-1}$) with $\mathcal{B}$ and $\mathcal{X}$ symmetrical matrices.

In the equations developed up to now, it was not considered that currents injected at the internal nodes of the network are necessarily null. If, when numbering network nodes, the terminal nodes are listed *before* the internal nodes, Equation [2.1.3] can be rewritten as:

$$\begin{pmatrix} \overline{\imath} \\ 0 \end{pmatrix} = \begin{pmatrix} \overline{Y}^{(tt)} & \overline{Y}^{(ti)} \\ \overline{Y}^{(it)} & \overline{Y}^{(ii)} \end{pmatrix} \begin{pmatrix} \overline{v} \\ \overline{v}^{(i)} \end{pmatrix}$$

where:

- $\overline{v}$, $\overline{v}^{(i)}$ are, respectively, column matrices of the terminal node voltages and internal node voltages;
- $\overline{\imath}$ is the column matrix of currents injected at the terminal nodes;
- $\overline{Y}^{(tt)}$, $\overline{Y}^{(ti)}$, $\overline{Y}^{(it)}$, and $\overline{Y}^{(ii)}$ are submatrices obtained by partitioning the matrix $\overline{\mathcal{Y}}$.

From the above, it can be deduced:

$$\overline{v}^{(i)} = -(\overline{Y}^{(ii)})^{-1}\overline{Y}^{(it)}\overline{v} \qquad [2.1.9]$$

and furthermore, by defining $\overline{Y} \triangleq \overline{Y}^{(tt)} - \overline{Y}^{(ti)}(\overline{Y}^{(ii)})^{-1}\overline{Y}^{(it)}$,

$$\overline{\imath} = \overline{Y}\,\overline{v} \qquad [2.1.10]$$

Actually, the steady-state behavior of the network is completely determined once the *voltages (vectors) at only the terminal nodes* are given, whereas:

- voltages at the remaining nodes (and, consequently, currents in the different branches, etc.) can be derived from Equation [2.1.9];
- currents injected at the terminal nodes are expressed by Equation [2.1.10].

The matrix $\overline{Y}$ in Equation [2.1.10] is the admittance matrix corresponding only to the terminal nodes. It is a square matrix having as many rows and columns as terminal nodes, and is less sparse than the $\overline{\mathcal{Y}}$ matrix, and even full. Furthermore, it is symmetrical and/or purely imaginary, if the matrix $\overline{\mathcal{Y}}$ is so.

A passive and linear equivalent circuit, for which the nodes are terminal nodes, can be associated with Equation [2.1.10]. If $\overline{Y}$ is symmetrical, the equivalent branch connecting the generic terminal nodes h, k must have an admittance $(-\overline{Y}_{hk})$, whereas the equivalent shunt branch at node h has an admittance $(\overline{Y}_{hh} + \sum_{k\neq h}\overline{Y}_{hk})$, i.e., $\sum_h \overline{Y}_{hk}$.

Note that, by eliminating (internal) node N, a $(N-1, N-1)$ admittance matrix $\overline{\mathcal{Y}}^{(N-1)}$ would result, with the generic element:

$$\overline{\mathcal{Y}}_{hk}^{(N-1)} = \overline{\mathcal{Y}}_{hk} - \frac{\overline{\mathcal{Y}}_{hN}\overline{\mathcal{Y}}_{Nk}}{\overline{\mathcal{Y}}_{NN}}$$

A relatively simple way to deduce the matrix $\overline{Y}$ can be based on the elimination of the internal nodes, one at a time, by repetitively using expressions analogous to the previous one.

Finally, by partitioning the matrix $\overline{\mathcal{Z}}$, Equation [2.1.6] can be rewritten as:

$$\begin{pmatrix} \overline{v} \\ \overline{v}^{(i)} \end{pmatrix} = \begin{pmatrix} \overline{Z}^{(tt)} & \overline{Z}^{(ti)} \\ \overline{Z}^{(it)} & \overline{Z}^{(ii)} \end{pmatrix} \begin{pmatrix} \overline{\imath} \\ 0 \end{pmatrix}$$

thus resulting in equations $\overline{v} = \overline{Z}^{(tt)}\overline{\imath}$, $\overline{v}^{(i)} = \overline{Z}^{(it)}\overline{\imath}$; i.e., $\overline{\imath} = (\overline{Z}^{(tt)})^{-1}\overline{v}$, $\overline{v}^{(i)} = \overline{Z}^{(it)}(\overline{Z}^{(tt)})^{-1}\overline{v}$, which are equivalent to Equations [2.1.10] and [2.1.9] respectively, as — due to $\overline{\mathcal{Y}}\overline{\mathcal{Z}} = I_{(N)}$ — it results $(\overline{Z}^{(tt)})^{-1} = \overline{Y}^{(tt)} - \overline{Y}^{(ti)}(\overline{Y}^{(ii)})^{-1}\overline{Y}^{(it)} = \overline{Y}$, $\overline{Z}^{(it)}(\overline{Z}^{(tt)})^{-1} = -(\overline{Y}^{(ii)})^{-1}\overline{Y}^{(it)}$. More importantly, the impedance matrix relative to the terminal nodes is directly given by $\overline{Z}^{(tt)}$.

2.1.5. Network Equations Between Node Voltages and Powers

(a) Generalities

As already discussed in Sections 2.1.3 and 2.1.4, knowing the values of node voltages and different currents entering and leaving the nodes allows an easy estimation of the corresponding values of the active and reactive powers.

With reference, for instance, to the simple circuit of Figure 2.3, it is easy to determine the complex power entering node h:

$$P' + jQ' = \overline{v}_h \overline{\imath}'^* = v_h^2 \overline{y}_0'^* + (v_h^2 - \overline{v}_h \overline{v}_k^*)\overline{y}_s^*$$

where the term $v_h^2 \overline{y}_0'^*$ is the complex power adsorbed by the shunt branch. By letting $\overline{v}_h \triangleq v_h e^{j\alpha_h}$, $\overline{v}_k \triangleq v_k e^{j\alpha_k}$, $\alpha_{hk} \triangleq \alpha_h - \alpha_k$ and $\overline{y} \triangleq g + jb$, the following equations may be deduced:

$$\begin{cases} P' = v_h^2 g_o' + v_h(v_h - v_k \cos\alpha_{hk})g_s - v_h v_k \sin\alpha_{hk} b_s \\ Q' = -v_h^2 b_o' - v_h(v_h - v_k \cos\alpha_{hk})b_s - v_h v_k \sin\alpha_{hk} g_s \end{cases}$$

which define the dependence of P' (active power) and Q' (reactive power) on the magnitudes of the voltages $\overline{v}_h$, $\overline{v}_k$ and on the phase shift between them. Similar relationships may be deduced for node k.

In particular, if the circuit is purely reactive, it follows that, by letting $\overline{y}_s = jb_s = -j/x$ (x being the reactance of the series branch):

$$\begin{cases} P' = \dfrac{v_h v_k}{x} \sin\alpha_{hk} \\ Q' = -v_h^2 b_o' + \dfrac{v_h(v_h - v_k \cos\alpha_{hk})}{x} \end{cases}$$

where P' is the active power leaving node k, i.e., it is the active power transmitted (from h to k) through the circuit.

Furthermore, if such transmission occurs at small phase shift α_{hk} (e.g., 10° or less), so that it may be assumed $\cos\alpha_{hk} \cong 1$, $\sin\alpha_{hk} \cong \alpha_{hk}$, the following holds:

$$\left.\begin{aligned} P' &\cong \frac{v_h v_k}{x}\alpha_{hk} \\ Q' &\cong -v_h^2 b_o' + \frac{v_h(v_h - v_k)}{x} \end{aligned}\right\} \qquad [2.1.11]$$

from which it is evident that the phase shift α_{hk} practically influences only the active power flow, whereas the reactive power Q' (and similarly, the reactive power relative to node k) is essentially dependent on the voltage magnitudes v_h and v_k.

Under the adopted assumptions, these last conclusions still hold in the presence of phase-shifting due to quadrature-regulating transformers (see also Section 5.3), because of the modest value of the considered phase shifts.

Now consider the complex powers $P_1 + jQ_1, \ldots, P_N + jQ_N$ injected (from the external) at nodes. From Equation [2.1.3] it can be deduced:

$$\begin{aligned} P_h + jQ_h = \overline{v}_h \overline{\imath}_h^* &= \overline{v}_h \sum_k \overline{\mathcal{Y}}_{hk}^* \overline{v}_k^* \\ &= v_h^2 \overline{\mathcal{Y}}_{hh}^* + v_h \sum_1^N {}_{k \neq h} \overline{\mathcal{Y}}_{hk}^* v_k e^{j\alpha_{hk}} \end{aligned}$$

from which the following equations can be obtained:

$$\left.\begin{aligned} P_h &= v_h^2 \mathcal{G}_{hh} + v_h \sum_{1}^{N}{}_{k \neq h} v_k (\mathcal{G}_{hk} \cos\alpha_{hk} + \mathcal{B}_{hk} \sin\alpha_{hk}) \\ Q_h &= -v_h^2 \mathcal{B}_{hh} + v_h \sum_{1}^{N}{}_{k \neq h} v_k (\mathcal{G}_{hk} \sin\alpha_{hk} - \mathcal{B}_{hk} \cos\alpha_{hk}) \end{aligned}\right\} \quad (h = 1, \ldots, N)$$

[2.1.12]

The above equations define injected active and reactive powers as functions of the node voltage magnitudes and phase shifts.

Actually, powers injected at the internal nodes must be zero (as well as currents), whereas internal node voltages depend on terminal node voltages according to Equation [2.1.9].

To deduce powers injected at terminal nodes as functions of voltages at the same nodes, one can refer to Equation [2.1.10] (again by assuming that, when ordering nodes, the terminal nodes are considered first), by directly obtaining a similar equation:

$$\left.\begin{aligned} P_h &= v_h^2 G_{hh} + v_h \sum_{1}^{n}{}_{k \neq h} v_k (G_{hk} \cos\alpha_{hk} + B_{hk} \sin\alpha_{hk}) \\ Q_h &= -v_h^2 B_{hh} + v_h \sum_{1}^{n}{}_{k \neq h} v_k (G_{hk} \sin\alpha_{hk} - B_{hk} \cos\alpha_{hk}) \end{aligned}\right\} \quad (h = 1, \ldots, n)$$

[2.1.13]

where n is the number of terminal nodes, and $\overline{Y}_{hk} \triangleq G_{hk} + jB_{hk}$ is a generic element of the matrix $\overline{Y}$.

It is evident that the generic regime of (both injected and internal) powers can be determined by assigning the n magnitudes $v_1, \ldots, v_n$ and the $(n-1)$ phase shifts $\alpha_{21}, \ldots, \alpha_{n1}$ (from which $\alpha_{hk} = \alpha_{h1} - \alpha_{k1}$). Generally, this means that, *in terms of active and reactive powers, voltage magnitudes, and phase shifts, the number of independent variables is* $(2n-1)$ *and not* $2n$.

In fact, if the phases of all terminal node voltages $\overline{v}_h$ were changed by the same arbitrary quantity, the same would occur to voltages and currents at any point of the network. Then, the magnitudes of vectors and their respective phase shifts would remain unchanged, and the same would happen to active and reactive powers. However, during the system operation, the value of phases may be significant if "phase" regulation is of concern (see Section 1.3.2, footnote[(9)]).

(b) Possible Simplifications

To introduce simplifications acceptable to many applications, it may be useful to note that, in practical situations:

- *the network is predominantly reactive,* so that in Equations [2.1.12] and [2.1.13] the terms including conductances may be disregarded as a first approximation;

- *the phase shifts α_{hk} are generally small*, so that it may be assumed — again as a first approximation — $\cos\alpha_{hk} = 1$, $\sin\alpha_{hk} = \alpha_{hk}$ [(1)].

Therefore, from Equations [2.1.12], the following approximate equations may be developed:

$$\left.\begin{aligned} P_h &= v_h \sum_1^N {}_{k\neq h} v_k \mathcal{B}_{hk}\alpha_{hk} = \sum_1^N {}_k \mathcal{B}'_{hk}\alpha_{hk} \\ Q_h &= -v_h \sum_1^N {}_k \mathcal{B}_{hk} v_k \end{aligned}\right\} \qquad [2.1.14]$$

or, alternatively:

$$\left.\begin{aligned} P_h &= -Q_h\alpha_h - \sum_1^N {}_k \mathcal{B}'_{hk}\alpha_k = \sum_1^N {}_k \mathcal{B}'_{hk}\alpha_{hk} \\ Q_h &= -\sum_1^N {}_k \mathcal{B}'_{hk} = -v_h \sum_1^N {}_k \mathcal{B}_{hk} v_k \end{aligned}\right\} \quad (h = 1, \ldots, N) \qquad [2.1.15]$$

with:

$$\mathcal{B}'_{hk} \triangleq v_h v_k \mathcal{B}_{hk} \qquad [2.1.16]$$

(Note that as seen, $\mathcal{B}$ must be considered symmetrical, so that $\mathcal{B}'_{hk} = \mathcal{B}'_{kh}$).

Similarly, (again assuming that nodes $1, \ldots, n$ are the terminal nodes, and $n+1, \ldots, N$ are the internal nodes) it follows from Equations [2.1.13]:

$$\left.\begin{aligned} P_h &= -Q_h\alpha_h - \sum_1^n {}_k B'_{hk}\alpha_k = \sum_1^n {}_k B'_{hk}\alpha_{hk} \\ Q_h &= -\sum_1^n {}_k B'_{hk} = -v_h \sum_1^n {}_k B_{hk} v_k \end{aligned}\right\} \qquad [2.1.17]$$

with:

$$B'_{hk} \triangleq v_h v_k B_{hk} \qquad [2.1.18]$$

Equations [2.1.17] define powers injected at the terminal nodes as function of voltages (magnitudes and phase shifts) at such nodes.

---

(1) Here, we obviously assume that the geographical extension of the network is not too much wide. On the other hand, even in the case of interconnected systems, the network considered here corresponds to only a generic member of the pool, as limited by its boundary nodes. Moreover, we assume that, as already pointed out, the shifts due to possible quadrature regulating transformers have negligible effects.

Obviously, these equations also can be obtained starting from Equations [2.1.15] by setting powers injected at the internal nodes equal to zero. Correspondingly, by partitioning matrices $\mathcal{B}$ and $\mathcal{B}'$ (with elements $\mathcal{B}_{hk}$ and $\mathcal{B}'_{hk}$, respectively), it follows with obvious notation:

- from the first part of Equations [2.1.15]:

$$\alpha^{(i)} = -(B'^{(ii)})^{-1} B'^{(it)} \alpha \qquad [2.1.19]$$

- from the second part of Equations [2.1.15]:

$$v^{(i)} = -(B^{(ii)})^{-1} B^{(it)} v \qquad [2.1.20]$$

where v, $v^{(i)}$ (and α, $\alpha^{(i)}$) are the column matrices constituted by the magnitudes (and phases) of the terminal node voltages and of the internal node voltages, respectively. With the adopted approximations, these last equations also can be directly derived from Equation [2.1.9]. Note that, in the matrix $(-(B'^{(ii)})^{-1} B'^{(it)})$, the sum of the elements of each row is 1; this means that, as already stressed, the same arbitrary variation for all phases $\alpha_1, \ldots, \alpha_n$ results in an identical variation for $\alpha_{(n+1)}, \ldots, \alpha_N$.

From the first part of Equations [2.1.17], for $h = 2, \ldots, n$, the following matrix equation can be finally developed:

$$P' = C'\alpha' \qquad [2.1.21]$$

where:

- P', α' are the column matrices $(n-1,1)$ constituted by $P_2, \ldots, P_n$ and $\alpha_{21}, \ldots, \alpha_{n1}$, respectively;
- C' is the matrix $(n-1, n-1)$ defined by:

$$C'_{hh} = -Q_h - B'_{hh} = \sum_{1}^{n}{}_{r \neq h} B'_{hr}, \qquad C'_{hk} = -B'_{hk} \quad (k \neq h)$$

whereas $P_1 = -(P_2 + \cdots + P_n)$.

Equation [2.1.21] also can be used to determine phase shifts, starting from active powers injected at nodes $2, \ldots, n$; in fact, the following equation can be obtained:

$$\alpha' = (C')^{-1} P' \qquad [2.1.22]$$

from which, remembering Equation [2.1.19]:

$$\alpha^{(i)'} = -(B'^{(ii)})^{-1} B'^{(it)} (C')^{-1} P' \qquad [2.1.23]$$

where $\alpha^{(i)'}$ is the column matrix constituted by the shifts $\alpha_{(n+1)1}, \ldots, \alpha_{N1}$ (note that, at reactive compensation nodes, the injected active powers are zero).

Similarly, the second part of Equations [2.1.17] also can be rewritten as:

$$(Q/v) = -Bv \qquad [2.1.24]$$

where B, v have clear meaning, whereas (Q/v) is the column matrix $(n,1)$ constituted by $Q_1/v_1, \ldots, Q_n/v_n$.

The (approximate) Equations [2.1.14] define the so-called "*direct current*" *model* of the network, as:

- active powers $P_1, \ldots, P_N$ injected in the network and node voltage phases $\alpha_1, \ldots, \alpha_N$ may be respectively interpreted as currents and voltages applied to the ("direct current") circuit reported in Figure 2.6a, where the generic $\mathcal{B}'_{hk}$ is the conductance of the branch connecting nodes h and k;
- similarly, currents $Q_1/v_1, \ldots, Q_N/v_N$ and voltage magnitudes $v_1, \ldots, v_N$ may be respectively interpreted as currents and voltages applied to the ("direct current") circuit of Figure 2.7, where the branch connecting generic nodes h, k has a conductance $\mathcal{B}_{hk}$, whereas the shunt branch at node h has a conductance $(-\mathcal{B}_{hh} - \sum_1^N {}_{k \neq h} \mathcal{B}_{hk})$.

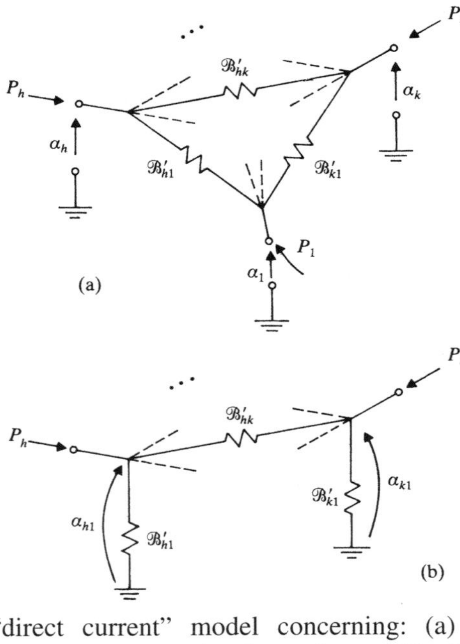

Figure 2.6. The "direct current" model concerning: (a) active powers and voltage phases; (b) active powers and voltage phase shifts. In the above circuits, $h, k = 2, \ldots, N$; moreover, $\mathcal{B}'_{hk}$ etc. are the conductances of the respective branches.

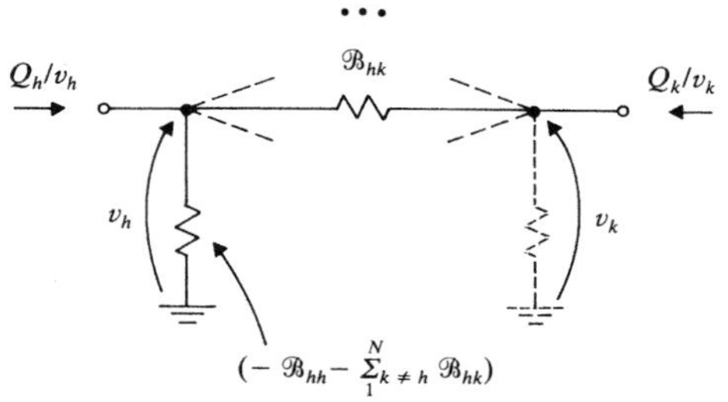

Figure 2.7. The "direct current" model concerning reactive powers and voltage magnitudes. In the above circuit, $h, k = 1, \ldots, N$; moreover, $\mathcal{B}_{hk}$ etc. are the conductances of the respective branches.

Note that Q_h/v_h is the reactive current—lagging with respect to voltage—absorbed by the network through node h; the second part of Equations [2.1.14] and the equivalent circuit of Figure 2.7, are directly deducible from Equation [2.1.3], by assuming $\overline{\mathcal{Y}} = j\mathcal{B}$ and disregarding the voltage phase shifts.

This treatment confirms the strict relationship between reactive powers and voltage magnitudes, whereas voltage phase shifts are especially effective on the active power regime.

Note that the circuit in Figure 2.6a has no shunt branches, so that:

- its structure is simpler than that of the circuit in Figure 2.7 (and of the network itself);
- it satisfies the condition $\sum_1^N {}_h P_h = 0$, for which (active power) losses are zero, in agreement with the assumption of a reactive network.

By assuming, for instance, node 1 as the phase-reference node (we assume that it is a terminal node, with P_1 not necessarily zero), the circuit in Figure 2.6a can be translated into that of Figure 2.6b, which corresponds to the following equations (see the first part of Equations [2.1.14]):

$$P_h = \left(\sum_1^N {}_{k\neq h}\mathcal{B}'_{hk}\right)\alpha_{h1} - \sum_2^N {}_{k\neq h}\mathcal{B}'_{hk}\alpha_{k1} \quad (h = 2, \ldots, N) \qquad [2.1.25]$$

(while $P_1 = -\sum_2^N {}_h P_h$).

Since the powers injected at the internal nodes are zero, additional equivalent circuits also may be determined, in which only the terminal nodes are retained ("reduced" direct current model). These circuits correspond to Equations [2.1.17].

Specifically, matrices C', $-B$ are the "conductance" matrices of the circuits obtained, respectively, from Figures 2.6b and 2.7, by eliminating the internal nodes.

With the circuits in Figure 2.6, a further reduction can be obtained, by remembering that active powers injected at the reactive compensation nodes equal zero.

As explained in Section 2.3.1, it is important to determine, using the parameters of Figure 2.6b circuit, active power flow variations in the branches, caused by:

(1) a variation of the injected power at one node $2, \ldots, n$ (and more generally $2, \ldots, N$), accompanied by an opposite variation of P_1;

(2) a branch opening for given injected powers.

For this purpose, Equation [2.1.25] may be translated into the matrix equation:

$$\begin{bmatrix} \alpha_{21} \\ \vdots \\ \alpha_{N1} \end{bmatrix} = \begin{bmatrix} T_{22} & \cdots & T_{2N} \\ \vdots & & \vdots \\ T_{N2} & \cdots & T_{NN} \end{bmatrix} \begin{bmatrix} P_2 \\ \vdots \\ P_N \end{bmatrix}$$

where the matrix, with elements T_{hk} (h, $k = 2, \ldots, N$), is the (symmetrical) "resistance matrix" of the circuit in Figure 2.6b; additionally, it holds $P_1 = -(P_2 + \cdots + P_N)$. Moreover, for the generic branch l that connects nodes r and s, the active power flow from r to s is given by:

$$F_l = \mathcal{B}'_{rs}(\alpha_{r1} - \alpha_{s1})$$

In case (1), assuming a variation ΔP_i ($i = 2, \ldots, N$) with $\Delta P_1 = -\Delta P_i$, it can be immediately determined:

$$\Delta F_l = \mathcal{B}'_{rs}(T_{ri} - T_{si})\Delta P_i \triangleq a_{li}\Delta P_i \qquad [2.1.26]$$

where the coefficient a_{li} may be called *sensitivity coefficient to an "injection shifting" from node i to node 1*, because the variation ΔP_i at node i is, by assumption, balanced by an opposite variation at node 1. If $r = 1$ or $s = 1$ (i.e., if the branch l is connected to node 1), Equation [2.1.26] holds with $T_{ri} = 0$ or $T_{si} = 0$, respectively.

In case (2), assuming the opened branch is branch a between nodes p, q (see Fig. 2.8a), it is evident that for the rest of the circuit this opening is equivalent to two opposite injections—$\pm F$ at p and q—to force the flow in branch a (from p to q) to vary from the initial value F_a to F, according to Figure 2.8b. Therefore the following equations can be developed:

$$\begin{cases} \Delta\alpha_{p1} = T_{pp}F + T_{pq}(-F) = (T_{pp} - T_{pq})F \\ \qquad\qquad (\text{assuming } T_{pp} = T_{pq} = 0 \text{ if } p = 1) \\ \Delta\alpha_{q1} = T_{qp}F + T_{qq}(-F) = (T_{qp} - T_{qq})F \\ \qquad\qquad (\text{assuming } T_{qp} = T_{qq} = 0 \text{ if } q = 1) \end{cases}$$

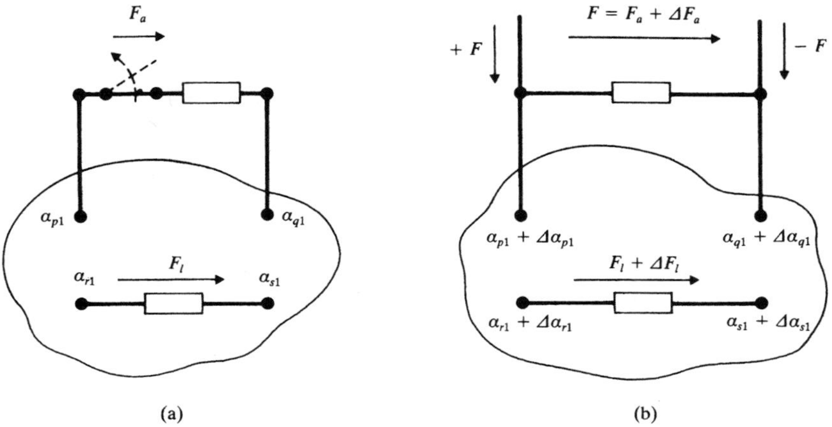

Figure 2.8. Opening of a branch between two given nodes.

with the condition:

$$F - F_a = \Delta F_a = \mathcal{B}'_{pq}(\Delta\alpha_{p1} - \Delta\alpha_{q1})$$

from which it follows:

$$F = \frac{1}{1 - \mathcal{B}'_{pq}(T_{pp} + T_{qq} - 2T_{pq})} F_a$$

and finally:

$$\Delta F_l = a_{lp} F + a_{lq}(-F) = \frac{a_{lp} - a_{lq}}{1 - \mathcal{B}'_{pq}(T_{pp} + T_{qq} - 2T_{pq})} F_a \triangleq d_{la} F_a \qquad [2.1.27]$$

where the coefficient d_{la} may be called "*branch-to-branch redistribution*" *coefficient of the active power flow, from branch a to branch l.*

Actually, *active power losses* do not equal zero and, because of the first part of Equations [2.1.12], they amount to:

$$p = \sum_1^N{}_h P_h = \sum_1^N{}_h v_h^2 \mathcal{G}_{hh} + \sum_1^N{}_h v_h \sum_1^N{}_{k \neq h} v_k \mathcal{G}_{hk} \cos\alpha_{hk}$$

where, because of Equations [2.1.5]:

$$\mathcal{G}_{hh} = g_{ho} - \sum_1^N{}_{k \neq h} \mathcal{G}_{hk}$$

by assuming that g_{ho} is the conductance of the shunt branch at node h.

Here, the network is assumed without phase-shifters, so that matrices $\mathcal{G}$, $\mathcal{B}$ are symmetrical; in the general case, a similar somewhat more complex treatment can be followed, which leads to similar results.

With the above, the active power losses can be expressed as:

$$p = p_o + p_u + p_\alpha \qquad [2.1.28]$$

with:

$$\begin{cases} p_o \triangleq \sum_1^N {}_h v_h^2 g_{ho} \\ p_u \triangleq \sum_1^N {}_h v_h \sum_1^N {}_{k \neq h} (v_k - v_h) \mathcal{G}_{hk} \\ p_\alpha \triangleq \sum_1^N {}_h v_h \sum_1^N {}_{k \neq h} v_k (\cos \alpha_{hk} - 1) \mathcal{G}_{hk} \end{cases}$$

where p_o represents the (generally negligible) losses in shunt branches, whereas $(p_u + p_\alpha)$ represents the series branch losses. Specifically, p_u is the contribution caused by differences in voltage magnitudes with zero phase shifts, whereas p_α is the further contribution caused by phase shifts.

The term p_u may be determined through the following equation:

$$p_u = u^T G_u u$$

where, by comparing voltage magnitudes with that of node 1:

- u is the column matrix $(N - 1, 1)$ constituted by the differences $(v_h - v_1)$;
- G_u is the matrix $(N - 1, N - 1)$ defined by:

$$(G_u)_{hh} \triangleq -\sum_1^N {}_{r \neq h} \mathcal{G}_{hr} = \mathcal{G}_{hh} - g_{ho}, \quad (G_u)_{hk} \triangleq \mathcal{G}_{hk} \quad (k \neq h)$$

$(h, k = 2, \ldots, N)$.

Moreover, with an accurate approximation, one may assume $\cos \alpha_{hk} \cong 1 - \alpha_{hk}^2/2$ in the expression of p_α, so that the following equation can be similarly developed:

$$p_\alpha = \alpha''^T G'_u \alpha''$$

where, by again comparing voltage phases with that at node 1:

- α'' is the column matrix $(N - 1, 1)$ constituted by phase shifts α_{h1}, i.e., with already adopted symbols $\alpha''^T \triangleq \left[\alpha'^T \alpha^{(i)'T}\right]$;

- G'_u is the matrix $(N-1, N-1)$ defined by:

$$(G'_u)_{hh} \triangleq -\sum_{1}^{N}{}_{r \neq h} v_h v_r \mathcal{g}_{hr}, \qquad (G'_u)_{hk} \triangleq v_h v_k \mathcal{g}_{hk} \quad (k \neq h)$$

$(h,\ k = 2, \ldots, N)$. By recalling Equations [2.1.20], [2.1.22], and [2.1.23], the following equation can be finally written:

$$p_\alpha = P'^T W P' \qquad [2.1.29]$$

where W is an easily identified matrix $(n-1, n-1)$, which is dependent on network parameters and magnitudes of terminal node voltages.

By adding up p_o and p_u, losses p can be identified, usually with an absolutely negligible error, starting from the n magnitudes of the terminal node voltages and active powers injected at nodes $2, \ldots, n$. This result will be considered in Section 2.3.

2.2. CHOICE OF THE WORKING POINT

2.2.1. Constraints on the Working Point

At the scheduling stage, every future working point should be chosen by considering not only the network equations, but also:

- conditions at the terminal nodes, which are defined by operating characteristics and admissibility limits of the external equipment connected to the network itself (e.g., generators, equivalent loads, etc.);
- operating requirements (quality, economy, security);
- admissibility limits for each network equipment.

For more details, see Sections 1.3 and 1.4.

To meet quality requirements in this connection, the *network frequency* should be kept at the nominal, e.g., $\omega = \omega_{nom}$, whereas *node voltages* (particularly terminal node voltages) must have magnitudes not far from the nominal values, according to "inequality" constraints:

$$v \in [v_{min}, v_{max}] \qquad [2.2.1]$$

with, for instance, $v_{min} = 0.90$–$0.95\ v_{nom}$ and $v_{max} = 1.05$–$1.10\ v_{nom}$. Furthermore, phase shifts between terminal node voltages, particularly those between voltages at the generation nodes, should be small, to avoid excessive electrical shifts between the synchronous machines and, consequently, weak synchronizing actions (see Section 1.6) which could even result in the instability of the working point.

Moreover, at the terminal nodes, further constraints on *injected powers* must be evaluated, concerning the characteristics of the equipment external to the network.

With *load nodes*, the active power P_c and the reactive power Q_c absorbed by a generic "equivalent load" may be assumed as input data to the problem, so that "equality" constraints:

$$\left.\begin{aligned} P &= -P_c \\ Q &= -Q_c \end{aligned}\right\} \qquad [2.2.2]$$

must be set, in which the negative sign is the result of assuming P and Q as the generic injected powers entering into the network.

Denoting by v_c the voltage magnitude at the generic load node, one can assume, for greater generality, $Q_c = Q_c(v_c)$. Particularly, the relationship $Q_c = Q_c^0 + av_c^2$, with a, Q_c^0 given, can be easily evaluated by adding (at the internal side of the network) a shunt branch with a reactance $1/a$ at the load node, and considering the load defined by adsorbed powers P, Q_c^0 as an equivalent load.

With *generation nodes*, the capability limits of each generating unit should be evaluated, with inequality constraints (see Fig. 2.9):

$$\left.\begin{aligned} P &\in [P_{\min}, P_{\max}] \\ Q &\in [Q_{\min}, Q_{\max}] \end{aligned}\right\} \qquad [2.2.3]$$

In the above, P and Q are net active and reactive powers injected into the network, not including powers absorbed by plant auxiliaries (nor electrical losses inside the plant, which are generally negligible).

At the given operating frequency, the values of $P_{\min}$ and $P_{\max}$ for a given generating unit, depend on the characteristics of the supply system and the turbine (or, generally, of the prime mover, as assumed in the following). These values

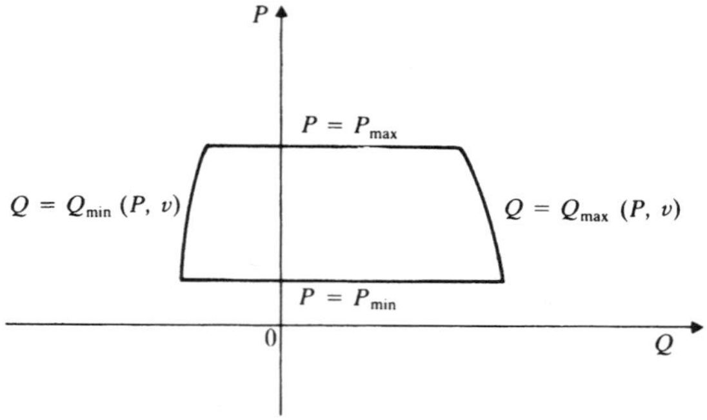

Figure 2.9. Capability limits of a generating unit.

can be considered as assigned. Instead, $Q_{\min}$ and $Q_{\max}$ depend on the behavior of the electrical side and are functions of P, v. More precisely:

- $P_{\min}$ is the so-called "technical minimum," which, in the case of thermal generating plants, is essentially conditioned by good operation of the boiler and auxiliary systems, and may be approximately 10%–30% of the nominal active power; with hydroelectric plants, this value may be assumed as zero, with some exceptions related to plants with Kaplan turbines, for which some minimum limit may be imposed to avoid cavitation phenomena;
- $P_{\max}$ is determined by the maximum limitation on mechanical power produced by the turbine;
- $Q_{\min}(P, v)$ is the so-called "underexcitation limit," imposed to avoid excessive overheating (caused by eddy currents in stator ends) or to avoid unstable working points with loss of synchronism between the generator and network;
- $Q_{\max}(P, v)$ is the "overexcitation limit," caused by heating reasons and related to the maximum tolerable excitation current, apart from further limitations related to iron losses and stator copper losses.

At some generating nodes, the active power P may be considered an input data because of particular requirements (e.g., flowing water hydroelectric plants, geothermal plants) or, at least as a first assumption, because of previous scheduling (e.g., hydroelectric plants with storage reservoir or basin), so that the corresponding equality constraint should be considered, in addition to the inequality constraints on Q and v.

Such a situation occurs at *reactive compensation nodes*, where $P = 0$. The active power absorbed by the generic synchronous compensator is only the result of electrical and mechanical losses, and may be disregarded.

Analogous situations can be encountered when considering *boundary nodes*, where active power P is assigned according to an agreed exchange program.

Finally, admissibility limits for each network equipment must be evaluated, specifically by means of inequality constraints on currents, like:

$$i \le i_{\max} \qquad [2.2.4]$$

as well as economic requirements, which may be translated (as shown in Sections 2.2.5 and 2.3) into further constraints on nonassigned variables at the terminal nodes and, more specifically, on active power dispatching for the different generating nodes. Security requirements are mostly related to the amount (and allocation) of rotating power and available reactive compensation, and to the possibility of equipment overload. Consequently, they may imply further constraints on the mentioned variables (in addition to possible modifications in system configuration and/or parameters, to avoid precarious situations).

Before concluding, note that:

(1) In previsional scheduling, the working point must be chosen based on forecasted load conditions.

(2) In real-time scheduling, the working point to be achieved must be chosen based on the actual load conditions estimated from field measurements (typically: P, Q, v values). Thus, more generally, the choice of the desired working point may be preceded by the estimation of the actual one.

The working point determination also may become necessary when considering other problems (different from operational scheduling). Examples are:

(3) "Planning" of future system development, for which a solution (e.g., adding of power plants, lines, etc.) must be found. This solution must allow acceptable working points, even for new load conditions foreseen in the long term (the problem is still of the previsional type, but data are particularly uncertain and the number of degrees of freedom—i.e., of solutions to be compared—is very large; then, it is reasonable to use suitably simplified models).
(4) Dynamic analyses and simulations, based on hypothetical situations (as to perturbations, configurations, parameter values, working points, etc.), for their validation, a deeper knowledge of the dynamic phenomena, the derivation of suitable dynamic models, etc.
(5) Reconstruction, via simulation (starting from a working point estimated from available experimental data), of an event that actually occurred (including possible outages, etc.) to ultimately identify causes and develop solutions to be implemented into the existing system.

In the cases (2) and (5), which refer to real situations, the working point "estimation" does not imply any choice, and the constraints to be considered are given only by measured data (see Section 2.5.).

2.2.2. Typical Solution Procedure

According to the previous section, the working point constraints are particularly related to terminal node voltages and (active and reactive) powers injected at such nodes. Therefore, it is suitable to explicitely consider the relationships between these quantities.

Generally, active and reactive power flows in the network define, for given powers injected at load nodes, the so-called "load-flow," or "power flow;" whereas the choice of active and reactive powers injected at the remaining nodes constitutes the so-called "dispatching."

Equations [2.1.13] must be recalled:

$$\left.\begin{aligned} P_h &= v_h^2 G_{hh} + v_h \sum_1^n {}_{k \neq h} v_k (G_{hk} \cos\alpha_{hk} + B_{hk} \sin\alpha_{hk}) \\ Q_h &= -v_h^2 B_{hh} + v_h \sum_1^n {}_{k \neq h} v_k (G_{hk} \sin\alpha_{hk} - B_{hk} \cos\alpha_{hk}) \end{aligned}\right\} \quad (h = 1, \ldots, n)$$

[2.2.5]

where n is the number of terminal nodes and:

- P_h, Q_h are, respectively, the active and reactive powers injected at node h;
- v_h is the magnitude of the voltage at node h;
- α_{hk} is the phase shift between voltage vectors at nodes h, k (i.e., $\alpha_{hk} \triangleq \alpha_h - \alpha_k$, where α_h, α_k are respective phases).

The number of independent variables (for given admittances G_{hk} and susceptances B_{hk}) is equal to $(2n - 1)$.

The *equality constraints* defined in the previous section only concern active and reactive powers at load nodes, active power at reactive compensation nodes (where $P = 0$) and possibly other nodes, so that the number of such constraints is less than $(2n - 1)$.

Consequently, there is a certain number of degrees of freedom, which can be used to determine a working point which:

- satisfies the inequality constraints already described (related to "admissibility" and "quality," at terminal nodes and inside the network);
- is sufficiently "secure";
- is optimal or at least satisfactory from the economical point of view.

One can use further degrees of freedom, as the values of adjustable parameters and, if necessary, the system configuration itself.

A procedure like that in Figure 2.10 can be adopted. It includes:

(1) *"nonrigid" assignments* of further voltages and/or powers at the terminal nodes, *up to a total of* $(2n - 1)$ *equality constraints*;
(2) *solution of Equations* [2.2.5] to obtain all the $(4n - 1)$ variables $v_1, \ldots, v_n$, $\alpha_{21}, \ldots, \alpha_{n1}$, $P_1, \ldots, P_n$, $Q_1, \ldots, Q_n$ (the resulting working point corresponds to the so-called *"nonlimited" load-flow*);
(3) *checks* on such working point, more precisely:
   - quality and admissibility checks at the terminal nodes,
   - similar checks inside the network,
   - security checks,
   - economic checks;

with *possible corrections* of assignments cited at item (1) and consequent iteration of (2), (3), until the working point checks are satisfactory (see also Section 2.2.6.)

2.2.3. Nonrigid Assignments at the Terminal Nodes

It is suitable that variables to be assigned at the terminal nodes (in addition to those already rigidly assigned) are active powers and/or voltage magnitudes. In fact, the dispatching of all active powers is important from the economic point of

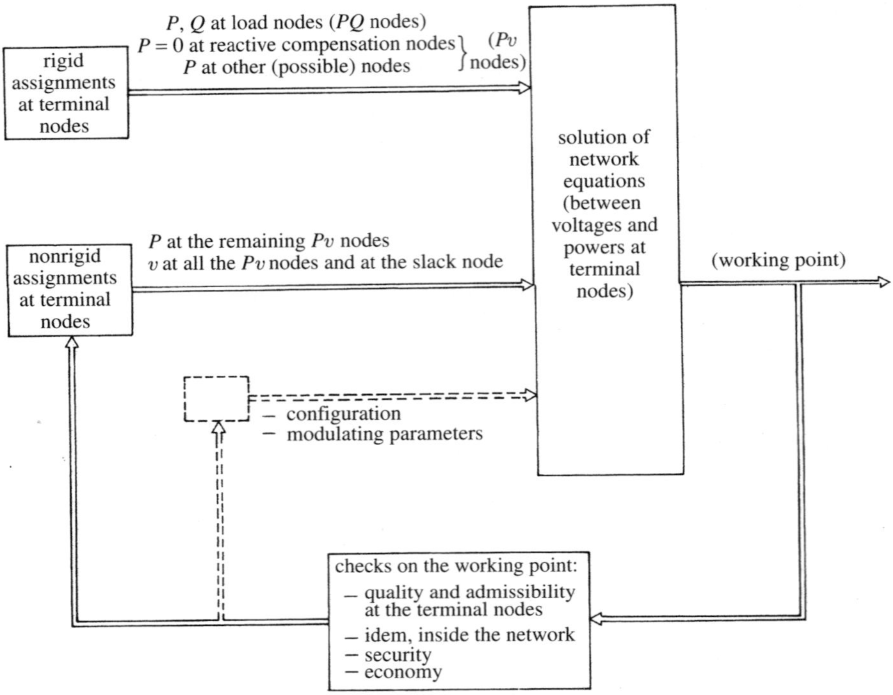

Figure 2.10. Schematic diagram for the choice of the working point.

view, as it determines the overall generation cost, whereas the voltage magnitudes are subject to relatively narrow constraints (see condition [2.2.1]).

On the other hand, Equations [2.2.5] are well suited for a "direct" solution starting from the values of all v_h, α_{h1}. Then, an iterative procedure may be adopted, by modifying the values v_h (not assigned) and α_{h1} at each step until they are consistent with the desired values of active and reactive powers (see Section 2.2.4).

The efficiency of such an iterative procedure is essentially dependent on the degree of sensitivity to variations Δv_h, $\Delta \alpha_{hk}$ of the constrained active and reactive powers. By inspection of the respective partial derivatives, one may easily determine that:

- variations in voltage phase shifts predominantly result in active power variations;
- variations in voltage magnitudes predominantly result in reactive power variations,

so that there is approximately a "decoupling" between the subsystems ($\Delta P_1, \ldots, \Delta P_n$, $\Delta \alpha_{21}, \ldots, \Delta \alpha_{n1}$) and ($\Delta Q_1, \ldots, \Delta Q_n$, $\Delta v_1, \ldots, \Delta v_n$), respectively characterized by $(n - 1)$ and n degrees of freedom. (One should recall constraints [2.2.1] on voltages and approximations (network predominantly reactive and small phase shifts) already outlined in Section 2.1.5. Again, it is assumed that node 1 is the phase reference node.)

To determine the so-called nonlimited load-flow, it is optimal to assume $(n - 1)$ *equality constraints on active powers*[(2)], *and the remaining n on reactive powers and/or voltage magnitudes.*

Usually:

- the generic phase shift α_{h1} mostly affects P_h, so that active powers are assigned at nodes $2, \ldots, n$ (leaving P_1 unknown, at the phase reference node) and the phase shifts $\alpha_{21}, \ldots, \alpha_{n1}$ are adapted;
- the generic v_h mostly affects Q_h, so that voltage magnitudes at all nodes are assigned, except for load nodes (possibly changing these assignments if they result in nonadmissible reactive powers; see Section 2.2.5) and the v_h values at the load nodes are changed until the assigned values of Q_h at such nodes are achieved.

As a consequence of the above, terminal nodes may be classified as follows (see Fig. 2.10):

- "*PQ*" *nodes:* nodes for which active and reactive powers are assigned;
- "*Pv*" *nodes:* nodes for which active powers and voltage magnitudes are assigned;
- node "v," i.e., node 1 (the phase reference node), for which only the voltage magnitude is assigned.

Node "v" is also called the "*slack node*" or "balance node," as its role is to achieve the due balance of active powers, considering active powers injected at other nodes and losses. For this reason, this node should be a generation node of significant power.

Additionally, the following typically holds[(3)]:

- load nodes are *PQ* nodes;
- reactive compensation nodes may be considered *Pv* nodes (or possibly *PQ* nodes) with $P = 0$;
- generation nodes are also *Pv* nodes, apart from the slack node.

---

(2) The assignment of all the n active powers $P_1, \ldots, P_n$, which would determine active power losses in the network, would not be theoretically impossible. However, it would easily result in unacceptable v_h values at one or more nodes, because of the small sensitivity of active powers with respect to voltage magnitudes.

(3) A *PQ* node and a *Pv* node coincident to each other would act as a *Pv* node. This assumption seems plausible, even with distinct but close nodes.

Boundary nodes may be considered Pv nodes, provided the static characteristics of the connected external systems are considered.

Finally, the "nonrigid assignments" are referred to:

- voltage magnitudes at all Pv nodes and at the slack node;
- active powers at some of the Pv nodes (typically at generation nodes).

At the boundary nodes, the constraints caused by static characteristics of the external systems may actually rearrange the P, v assignments at such nodes, based on values Q, α correspondingly evaluated.

Usually, it is reasonable to assume that such rearrangements are small, and that they may be automatically considered by representing the external systems with a simplified equivalent, such as that indicated in Figure 2.11 (see also Section 2.3.1c).

2.2.4. Solution of Network Equations Between Voltages and Powers at the Terminal Nodes

According to Section 2.2.3, it is necessary to develop an iterative solution for (see Fig. 2.12):

- the first part of Equations [2.2.5] (with P_h known) at all PQ and Pv nodes (i.e., all the terminal nodes with the exception of the slack node, $h = 2, \ldots, n$);
- the second part of Equations [2.2.5] (with Q_h known) only at the PQ nodes,

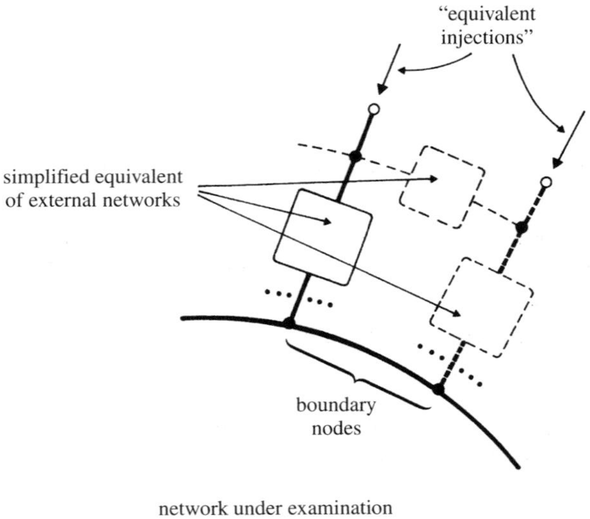

Figure 2.11. A representation of possible external systems.

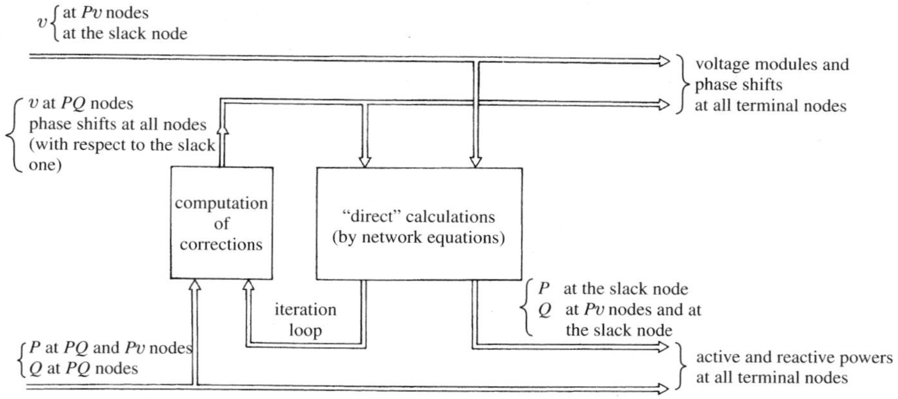

Figure 2.12. Schematic diagram for the solution of network equations.

so that one can obtain all the phase shifts $\alpha_{21}, \ldots, \alpha_{n1}$ and nonassigned voltage magnitudes v_h (i.e., those at the PQ nodes).

If, at a generic PQ node, an equality constraint $Q_h = Q_h(v_h)$ is actually held, the assignment of Q_h also would be iterated until this constraint is met.

The remaining parts of Equations [2.2.5] are used to derive the active and reactive powers at the slack node, and the reactive powers at the Pv nodes.

The *choice of the initial values* $v_h^{(o)}$, $\alpha_{h1}^{(o)}$ (related to nonassigned v_h and to α_{h1}, as the starting values for the iterative procedure) is obviously important for its effects on procedure convergence and final results. In fact, due to the nonlinearity of the equations, one could also obtain solutions of no practical interest, or even no solution at all. One may assume that $v_h^{(o)}$ is equal to nominal voltage (obviously at the PQ nodes where v_h is not already assigned) and $\alpha_{h1}^{(o)} = 0$. Alternatively, these values can be determined by using the simplified "dc" model (see Equations [2.1.17]).

From the values mentioned above, one can obtain the values $v_{h1}^{(1)}$, $\alpha_{h1}^{(1)}$ at the end of the first iteration step and so on, until the variations for the unknown variables are small enough.

The commonly used iterative methods refer, through possible adaptations and simplifications, to the Gauss-Seidel or Newton-Raphson methods.

By the *Gauss-Seidel method*, the equations to be solved (in the generic unknown variables $x_1, \ldots, x_N$) are arranged in the form:

$$x_h = f_h(x_1, \ldots, x_N) \quad (h = 1, \ldots, N) \qquad [2.2.6]$$

and the unknown values at the i-th iteration step are computed in sequence (from 1 to N) according to:

$$x_h^{(i)} = f_h(x_1^{(i)}, \ldots, x_{(h-1)}^{(i)}, x_h^{(i-1)}, \ldots, x_N^{(i-1)})$$

Such iterations are known as "chained," because the most updated values of the unknown variables are used in the sequence; the choice of the order $1, \ldots, N$ may, of course, be important to convergence.

By using the procedure suggested by Glimm-Stagg, Equation [2.2.6] is obtained observing that:

- for the PQ nodes (where P_h, Q_h are assigned, while the unknowns are v_h, α_{h1}): in vector terms and with known notation, it results:

$$\sum_{1}^{N} {}_k\overline{Y}_{hk}\overline{v}_k = \overline{\imath}_h = \frac{P_h - jQ_h}{\overline{v}_h^*}$$

  so that Equations [2.2.5] are equivalent to:

$$\overline{v}_h = \left(\frac{P_h - jQ_h}{\overline{v}_h^*} - \sum_{1}^{n} {}_{k \neq h}\overline{Y}_{hk}\overline{v}_k \right) \bigg/ \overline{Y}_{hh}$$

  from which v_h, α_{h1} can be deduced in the desired form;
- for Pv nodes (where P_h, v_h are assigned, while α_{h1} is unknown): by letting

$$i'_h e^{j\beta'_h} \triangleq \overline{\imath}_h - \overline{Y}_{hh}\overline{v}_h = \sum_{1}^{n} {}_{k \neq h}\overline{Y}_{hk} v_k e^{j\alpha_k}$$

  the first part of Equations [2.2.5] can be rewritten as:

$$P_h = v_h^2 G_{hh} + v_h i'_h \cos(\alpha_h - \beta'_h)$$

  so that:

$$\alpha_h = \beta'_h + \cos^{-1}\left(\frac{P_h - v_h^2 G_{hh}}{v_h i'_h} \right)$$

  and consequently α_{h1} can be deduced in the desired form.

With phase shifts, one may assume for simplicity that $\alpha_1 = 0$ and consequently $\alpha_{21} = \alpha_2, \ldots, \alpha_{n1} = \alpha_n$.

To improve the convergence of iterations, the computation can be started from the unknown variables relative to the most meshed nodes, i.e., those nodes into which more branches are connected.

By the *Newton-Raphson method*, the equations are instead arranged in the form:

$$c_h = W_h(x_1, \ldots, x_N) \quad (h = 1, \ldots, N) \qquad [2.2.7]$$

with $c_1, \ldots, c_N$ assigned, that is, with matrix notation:

$$c = W(x)$$

By generically imposing:

$$c = W(x^{(i)}) \cong W(x^{(i-1)}) + \frac{\mathrm{d}W}{\mathrm{d}x}(x^{(i-1)})(x^{(i)} - x^{(i-1)})$$

and by disregarding the approximation symbol, an equation in $x^{(i)}$ is derived, that is:

$$x^{(i)} = x^{(i-1)} + \left(\frac{\mathrm{d}W}{\mathrm{d}x}(x^{(i-1)})\right)^{-1}(c - W(x^{(i-1)}))$$

From the above, the values of the unknowns $x_h^{(i)}$ can be computed at the i-th iteration step.

The matrix $\mathrm{d}W/\mathrm{d}x$ is the so-called "Jacobian matrix." The matrix equation in $x^{(i)}$ is equivalent to a system of N linear equations in $x_1^{(i)}, \ldots, x_N^{(i)}$, which, for large N, may be more conveniently solved iteratively; for instance, by the Gauss-Seidel method itself, as previously mentioned, so that the inversion of the Jacobian matrix can be avoided.

Additionally, the iteration convergence may be improved by adopting possible "acceleration factors," i.e., by overestimating (e.g., by a multiplication factor 1.5) the generic correction $(x^{(i)} - x^{(i-1)})$ with respect to the above equation. Finally, the Jacobian matrix should be computed at each iteration, but this may be avoided—thus alleviating the computation process, at the cost of slower convergence speed—particularly when close to the final solution, where the corrections at each iteration would be small.

It is clear that Equations [2.2.5] are already in the form of Equation [2.2.7], so the application of the Newton-Raphson method requires the computation of the right sides of these equations, and of the Jacobian submatrices $\partial P'/\partial\alpha'$, $\partial P'/\partial v'$, $\partial Q'/\partial\alpha'$, $\partial Q'/\partial v'$ obtainable from Equations [2.2.5], where:

- P', α' are column matrices constituted by $P_2, \ldots, P_n$ and by $\alpha_{21}, \ldots, \alpha_{n1}$ respectively;
- Q', v' are column matrices constituted by Q_h and v_h, respectively, at the *PQ* nodes only.

To avoid the relatively burdensome computation of trigonometric functions (sine and cosine) that appear at the right side of Equations [2.2.5] and in above submatrices, the polar coordinates v_h, α_{h1} can be abandoned and a model based on Cartesian coordinates adopted:

$$\begin{cases} v_{hR} \triangleq v_h \cos\alpha_{h1} \\ v_{hI} \triangleq v_h \sin\alpha_{h1} \end{cases}$$

Equations [2.2.5] can be then replaced with:

- for all PQ and Pv nodes (i.e., for $h = 2, \ldots, n$):

$$P_h = G_{hh}(v_{hR}^2 + v_{hI}^2) + \sum_{1}^{n}{}_{k \neq h}(G_{hk}(v_{hR}v_{kR} + v_{hI}v_{kI})$$
$$- B_{hk}(v_{hR}v_{kI} - v_{hI}v_{kR}))$$

and moreover:

- for Pv nodes only:

$$v_h^2 = v_{hR}^2 + v_{hI}^2$$

- for PQ nodes only:

$$Q_h = -B_{hh}(v_{hR}^2 + v_{hI}^2) + \sum_{1}^{n}{}_{k \neq h}(-G_{hk}(v_{hR}v_{kI} - v_{hI}v_{kR})$$
$$- B_{hk}(v_{hR}v_{kR} + v_{hI}v_{kI}))$$

where all the quantities at the left sides are assigned (as well as $v_{1R} = v_1$, $v_{1I} = 0$), so that equations follow Equation [2.2.7] format, with v_{hR}, v_{hI} $(h = 2, \ldots, n)$ the unknown variables. However, by so doing, the number of equations and unknown variables becomes $2(n-1)$ and is therefore larger than with polar coordinates. (The difference is equal to the number of Pv nodes.)

Among the *simplified versions* of the Newton-Raphson method, which are based on estimates of the Jacobian matrix (apart from avoiding the updating of this matrix) the following can be mentioned:

- *With Cartesian coordinates*: the Ward-Hale method, in which the Jacobian matrix is considered, after a proper ordering of the indexes, of the "block diagonalized form," with $(n-1)$ (2,2) submatrices, by considering:
  - only terms $\partial P_h/\partial v_{hR}$, $\partial P_h/\partial v_{hI}$, $\partial(v_h^2)/\partial v_{hR}$, $\partial(v_h^2)/\partial v_{hI}$ for Pv nodes;
  - only terms $\partial P_h/\partial v_{hR}$, $\partial P_h/\partial v_{hI}$, $\partial Q_h/\partial v_{hR}$, $\partial Q_h/\partial v_{hI}$ for PQ nodes.
- *With polar coordinates:* the Carpentier method (or "*decoupled*" *load-flow* method) in which — by remembering the approximate decoupling between the subsystems ($\Delta P'$, $\Delta\alpha'$) and ($\Delta Q'$, $\Delta v'$) (see Section 2.2.3) — the Jacobian matrix is again considered in the block diagonalized form, by only accounting for the submatrices $\partial P'/\partial\alpha'$, $\partial Q'/\partial v'$. Finally, the Stott method (or "*fast decoupled*" *load-flow* method), which is particularly useful for very large networks and in which these submatrices are evaluated by using

Equations [2.1.17] ("dc" model, that makes use of no trigonometric functions), by further assuming:

$$\partial P_h/\partial \alpha_{k1} \cong -v_h v_k B_{hk}, \quad \partial Q_h/\partial v_k \cong -v_h B_{hk}$$

also for $k = h$ (actually, for $k = h$, it instead results:

$$\partial P_h/\partial \alpha_{h1} = -v_h^2 B_{hh} - Q_h \cong \sum_1^n {}_{k\neq h} v_h v_k B_{hk}$$

$$\partial Q_h/\partial v_h = (-v_h^2 B_{hh} + Q_h)/v_h \cong -2v_h B_{hh} - \sum_1^n {}_{k\neq h} v_k B_{hk}$$

but the terms $\pm Q_h$ can be usually disregarded with respect to $-v_h^2 B_{hh}$).

2.2.5. Checks on the Working Point and Possible Corrections

The working point obtained starting from the $(2n - 1)$ assignments at the terminal nodes (nonlimited load flow) must actually meet different conditions, related to:

- quality and admissibility at the terminal nodes and inside the network;
- security;
- economy.

If these conditions are not met, the nonrigid assignments must be changed (by accounting for their respective admissibility limits; see Equations [2.2.1], [2.2.3]), or the value of possible adjustable parameters must be modified (or even, if necessary, the system configuration itself).

The aim of this section is limited to broad indications on the most typical situations and approximations that are usually accepted.

(a) Quality and Admissibility at the Terminal Nodes

- The *active power* P_1 *at the slack node* must be within the admissible range, according to the constraints:

$$P_1 \in [P_{1\,\min}, P_{1\,\max}]$$

(see the first part of conditions [2.2.3]). Actually, such constraints can be considered when assigning the active powers at all the other nodes, based on an estimated value of losses. If this condition is not met, active power assignments at Pv nodes can be changed, and particularly at nodes near the slack node.

- Also the *nonassigned reactive powers* Q_h (i.e., those at Pv nodes and the slack node) must be within admissible ranges, according to constraints:

$$Q_h \in [Q_{h\,\min}, Q_{h\,\max}]$$

(see the second part of conditions [2.2.3]). If this condition is violated at one or more nodes, the voltage magnitudes (nonrigidly assigned) at such nodes can be rearranged; an increment of the generic v_h usually results in an increment in the respective Q_h and, at a smaller amount, in a reduction of the reactive power injected at the other nodes. At nodes where the above-mentioned condition is not met, Q_h could be assumed to equal the violated limit, leaving the value of v_h unknown, as with the operation of over- and underexcitation limiters (see Section 6.2).

For a boundary node with reactive compensation, it can be written $Q_h = Q'_h + Q''_h$, by denoting Q'_h as the reactive power supplied from the external network, and Q''_h as the reactive power delivered by the compensator, with $Q''_h \in [Q''_{h\,\min}, Q''_{h\,\max}]$.
The limits:

$$Q_{h\,\min} = Q'_h + Q''_{h\,\min}, \quad Q_{h\,\max} = Q'_h + Q''_{h\,\max}$$

are consequently variable with Q'_h, which itself is dependent on the assigned P_h,v_h and on the characteristics of the external system.

- The *nonassigned voltage magnitudes* v_h (i.e., those at PQ nodes) must satisfy constraints:

$$v_h \in [v_{h\,\min}, v_{h\,\max}]$$

(see condition [2.2.1]). If this is not true, the voltage magnitude assignments at the other nodes (Pv and slack node) again may be modified, by considering that their increment usually results in a voltage (magnitude) increment at PQ nodes.

The possibility of obtaining a better voltage profile by correcting the adjustable parameter values should be further considered, as described later.

(b) Quality and Admissibility Inside the Network

Typical conditions to be checked are:

- the voltage magnitudes at the "internal" nodes must meet constraints:

$$v \in [v_{\min}, v_{\max}]$$

- the current magnitudes (through various equipments) must meet constraints:

$$i \leq i_{\max}$$

Therefore, it is necessary to determine voltages at the internal nodes by using Equation [2.1.9] (or by estimation with Equation [2.1.20]), then to determine branch currents and current flowing through each component, according to Section 2.1.3.

If the check on *voltage magnitudes* is not successful, the assignments v_h (at Pv nodes and at the slack node) may be rearranged, particularly those having larger influence on the voltages that are out of limit. These assignments are constrained enough because of conditions discussed above, so availability of further degrees of freedom is important to achieve voltage magnitudes (and reactive power flows) acceptable for all the network. This is possible with the presence of adjustable parameter elements, such as shunt reactors and capacitors, tap-changing transformers, and "in-phase" regulating transformers (see Section 5.3).

Note that, because of previous scenarios, the solution of such a problem is not very sensitive to the value of phase shifts between voltages; on the other hand, the phase shifts may be important for the value of currents in series branches (and transmitted active powers), and consequently of *currents* in different components.

A significant example is a series branch between nodes h and k, having a reactance x, for which the current is given by:

$$i^2 = (v_h^2 + v_k^2 - 2v_h v_k \cos\alpha_{hk})/x^2$$

while the transmitted active power is:

$$P = \frac{v_h v_k}{x} \sin\alpha_{hk}$$

so that, by assuming $\cos\alpha_{hk} \cong 1 - \alpha_{hk}^2/2$, $\sin\alpha_{hk} \cong \alpha_{hk}$, one can develop this equation:

$$i^2 \cong \left(\frac{v_h - v_k}{x}\right)^2 + \frac{P^2}{v_h v_k}$$

where the first term vanishes if $v_h = v_k$.

If some currents do not meet the above-mentioned constraint, the active power assignments among Pv nodes, and more specifically among generators, must be redistributed to modify the phase shift values (both at the terminal and internal nodes; refer to Equations [2.1.22] and [2.1.23]). This causes active power flows in the network to be redistributed. To meet the same goal, "quadrature"-regulating transformers may be used as adjustable devices (see Section 5.3).

(c) Security

Security checks should require a broad but detailed plan of dynamic simulations, which should reflect the real behavior of different components — generators, network, loads, protection and control systems, etc. — in response to significant perturbations (*contingency analysis*). The stability "in the small," i.e., for small perturbations, can be checked in a more direct way by the dynamic analysis of the linearized system.

The goal of such simulations is to reveal "risk" situations, not only for single components (e.g., calculation of short-circuit currents, to be compared with breaker duties), but also for the whole system, according to Section 1.7.

Without getting into many details, it may be observed that:

- the simulation plan may be limited to events which appear the most dangerous, according to the characteristics of the specific system under examination and to the previous operating experience;
- each simulation can often use simplifications, with relation to the most interesting phenomena (see Section 1.8 and all the following chapters).

Much stronger simplifications usually are needed in real-time scheduling, for which the information concerning security must be quickly available (e.g., analyzing hundreds of events in a few minutes; similar conclusions may also apply in planning, although for very different reasons, i.e., related to the uncertainty of data and the large number of future configurations to be compared).

Quite often, security checks are limited to the so-called "*static*" *security*, evaluating only the possible static effects of perturbations without considering transients caused by them and by checking if such effects correspond to a steady-state that:

- exists (more specifically, the rotating power and the reactive compensation must still be sufficient and well spread, to avoid frequency or voltage instability, etc.);
- is tolerable (i.e., it does not cause the intervention of protections, particularly for currents which should not be too large in each network branch).

For an unacceptable case, one should again:

- correct the voltage profile (i.e., the corresponding assignments) to increase the *reactive power margins* at generation and compensation nodes; for instance, by setting the generic Q_h in the middle position within its limits $Q_{h\,\min}$, $Q_{h\,\max}$ (this criterion also may be adopted to share the reactive power generation between different units of a same power plant; see Section 6.2.2);
- correct the active power regime — specifically the active dispatching between the Pv nodes — to increase the *current margins* (with respect to $i_{\max}$) for each branch, similar to what has been discussed in (b).

The adjustable parameters may be acted upon or the system configuration reinforced (e.g., *addition of units or lines*).

For a more comprehensive evaluation of the new steady-state, at a frequency which may be different than the nominal frequency, the following should be considered:

- the static characteristics of control systems;
- the static dependency of loads on voltage and frequency;

without neglecting possible external systems.

For simplicity, the "dc" model may be assumed, for evaluating the new (magnitude) voltage and reactive power steady-state and then the new active power and phase shift steady-state, and for deducing possible corrective actions, considering the opportunity of using the full model for final validations.

Finally, the assumed events generally involve "single" perturbations of particular impact, such as:

- a line outage, which results in a reduction in transmittable active powers or supportable voltages (see Section 1.5), and causes an increment of current flow in the surrounding lines;
- a power plant disconnection, which causes a reduction in rotating power and overcurrents in some lines (particularly those connected to power plants which are mostly required to supply lost power).

(d) Economy

For a given set of thermal power units, the *cost of generation* can be evaluated based on fuel consumption and unit costs (at generation site, i.e., including transportation costs, etc.) of adopted fuels.

For each operating plant, the fuel consumption per unit of time depends on the generated driving power or (as internal losses may be neglected) on the delivered active power. By subtracting from this power the active power absorbed by auxiliary systems (which may be approximately 5% of the total delivered power), and incorporating the unit cost of fuel, the cost per unit of time (C) can be determined starting from the active power (P) supplied to the grid by the generic unit.

The characteristic (C, P) may be, for instance, similar to that in Figure 2.13, with multiple pieces caused by the sequential opening of the turbine valves. (In Section 2.3, a function $C(P)$, like the dashed one reported in Figure 2.13, will be adopted with a continuous and increasing derivative $\mathrm{d}C/\mathrm{d}P$.)

This characteristic can be determined based on design data and/or experimental tests. The values of C may be increased to account for further costs related to unit operation (e.g., accessory materials) and for possible charges related to the emission of polluting materials. Additionally, the characteristic may vary with the operating conditions of the thermal cycle (e.g., temperature of the cooling water at the condenser or external air) and with the efficiency of the plant (with

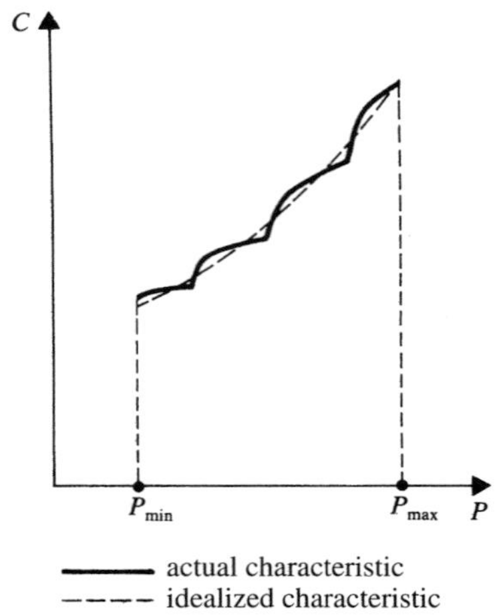

Figure 2.13. Example of the characteristic (C, P) for a thermal unit (C = cost per unit of time, P = active power).

respect to the time elapsed from the last maintenance). Further complications are encountered with multiple interacting boilers and turbines because the resulting characteristic depends on the coordination of the operation of different parts.

Even for a power plant with more than one operating unit, the minimization (for any given value of total active power delivered by the plant, disregarding internal losses) of the corresponding cost per unit of time is clearly possible. In fact, this cost generally varies according to the way the total power is shared between different units, and may be minimized by adopting the most suitable sharing option. If this minimization is systematically carried out at local sites by coordinating the use of the operating units, then it is possible to define (based only on cost characteristics of single units) a univocal relationship between the total power delivered by the plant and the corresponding cost per unit of time.

When the cost characteristics are not particularly suitable for analytic manipulation (see Section 2.3.1a), the optimal solution may be achieved by a procedure based on "dynamic programming."

For this purpose:

- let us assume that the set of units $1, \ldots, m$ operating at the given plant is known (and numbered in arbitrary order from 1 to m);
- let us consider discrete values (multiple of an elementary step ΔP) for active powers $P_{t1}, \ldots, P_{tm}$ delivered by units, and for their sum:

$$P_{\text{tot}} = \sum_1^m {}_i P_{ti} \in \left[\sum_1^m {}_i P_{ti\,\min}, \sum_1^m {}_i P_{ti\,\max} \right]$$

- let us denote by $C^0_{\text{tot}(k)}(P_{\text{tot}})$ the minimum total cost per unit of time, obtained — for any given P_{tot} — by properly choosing $P_{t1}, \ldots, P_{tk}$ and by keeping $P_{t(k+1)}, \ldots, P_{tm}$ at their respective technical minima (obviously, for $P_{\text{tot}} \leq \sum_1^k {}_i P_{ti\,\max} + \sum_{k+1}^m {}_i P_{ti\,\min}$). The following formula may then be used:

$$C^0_{\text{tot}(k)}(P_{\text{tot}}) = \min_{P_{tk}} \big(C^0_{\text{tot}(k-1)}(P_{\text{tot}} - P_{tk} + P_{tk\,\min}) + C_k(P_{tk}) - C_k(P_{tk\,\min})\big) \quad (k = 1, \ldots, m)$$

where:

- for $k = 1$, it simply results:

$$C^0_{\text{tot}(1)}(P_{\text{tot}}) = C_1\left(P_{\text{tot}} - \sum_2^m {}_i P_{ti\,\min}\right) + \sum_2^m {}_i C_i(P_{ti\,\min})$$

(obviously, for $P_{\text{tot}} \leq P_{t1\,\max} + \sum_2^m i\, P_{ti\,\min}$);
- for $k = 2, \ldots, m$ the function $C^0_{\text{tot}(k-1)}(P_{\text{tot}} - P_{tk} + P_{tk\,\min})$ must be evaluated for all possible values $(P_{\text{tot}} - P_{tk} + P_{tk\,\min})$, each of which leads to a solution $P_{t1}, \ldots, P_{t(k-1)}$.

The searched solution is finally defined, for any given P_{tot}, by the value $C^0_{\text{tot}(m)}(P_{\text{tot}})$ and by the corresponding values $P_{t1}, \ldots P_{tm}$.

The same problem would exist for the whole system, if the total (active) power delivered by the thermal power plants could be considered assigned. In this case, the optimal power share between the different plants (and units) could be determined, again, only based on individual cost characteristics.

However, the above hypothesis requires that:

- in addition to active powers absorbed by loads, powers delivered by hydroelectric plants and those exchanged through (possible) interconnections are assigned;
- network losses are known;

so that the total amount of power required from thermal power plants can be determined. Note that only thermal and hydroelectric generations are considered here to simplify the presentation. Other types of generation may be treated similarly to one of these two kinds of generation.

The former of the above conditions may be assumed as satisfied, as a consequence of previous scheduling of hydraulic resources and interconnection exchanges (see Sections 2.3.3 and 2.3.4).

For a generic hydroelectric unit, the used water flow q (disregarding losses) depends on the active power P supplied to the grid, according to a characteristic shown in Figure 2.14a. Such a characteristic may vary with the operating conditions of the plant (e.g., water level in the reservoir) and its efficiency status.

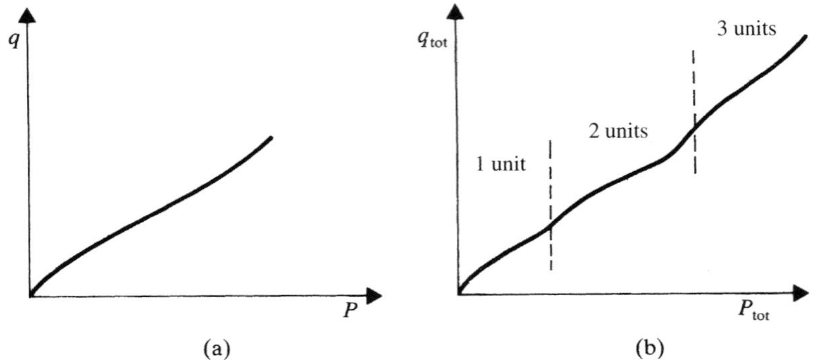

Figure 2.14. (a) An example of the characteristic (q, P) for a hydroelectric unit (q = used water flow, P = active power). (b) An example of the optimal characteristic (q_{tot}, P_{tot}) for a hydroelectric plant with several units (q_{tot} = used water flow, P_{tot} = active power).

When the hydroplant has more than one operating unit, it is convenient to maximize—at any given total water flow q_{tot}—the total delivered power P_{tot}, to reduce the generation demand to thermal power plants and the total thermal generation cost. Based on the above-mentioned characteristics, it is possible to determine the optimal sharing of the total water flow (and power) between the different units, by adopting a dynamic programming procedure like the one previously illustrated. This allows the definition of an univocal relationship between P_{tot} and q_{tot} (see Fig. 2.14b).

However, it is not possible to determine *network losses* in advance because they depend on the power sharing between different plants or all thermal units. Therefore network characteristics, which determine such dependency, must be considered.

According to the stated hypotheses, the slack node (1) must correspond to a thermal unit ($P_1 = P_{t1}$) or even a whole thermal plant, as P_1 cannot be assigned. Network losses can be expressed as a function of active powers $P_2, \ldots, P_n$, besides reactive powers Q_{cj} at load nodes (PQ nodes) and voltage magnitudes v_h at remaining terminal nodes, i.e., in the form:

$$p = \phi(P_2, \ldots, P_n; \ldots, Q_{cj}, \ldots; \ldots, v_h, \ldots) \qquad [2.2.8]$$

A small variation $\Delta P_i = \Delta P_{ti}$ in one of the thermal units ($i \neq 1$) causes not only a cost variation per unit of time ΔC_i in the thermal unit itself, but also a variation:

$$\Delta P_1 = -\Delta P_i + \Delta p = -\Delta P_i \left(1 - \frac{\partial \phi}{\partial P_i}\right)$$

and, consequently, a variation ΔC_1 in the thermal unit (or plant) connected at the slack bus. Instead, possible variations Δv_h, within the limits imposed by previously mentioned requirements, would only modify network losses, with usually small effects on the resulting cost.

By numerical computation, the ideal can be reached by modifying (according to respective constraints) each $P_i = P_{ti}$, to obtain a negative $\Delta C_i + \Delta C_1$, and then by iterating until, with some tolerance:

$$\Delta C_i + \Delta C_1 = 0$$

Such a condition may also be expressed in the form:

$$\frac{\Delta C_i / \Delta P_i}{1 - \partial\phi / \partial P_i} = \Delta C_1 / \Delta P_1 \qquad [2.2.9]$$

which is analogous to Equation [2.3.5] reported in Section 2.3.

2.2.6. Active and Reactive Dispatching

According to what has been shown up to now, the choices of the values P and v (nonrigid assignments; see Fig. 2.10) are based on very different criteria. By considering the constraints (in particular those on voltages), the effects of v assignments on the choice of P values are quite modest and vice versa. Thus, the determination of the working point may be practically carried out (at the possible price of some iterations) by solving two distinct problems, the so-called "active dispatching" (choice of P, with v assigned) and "reactive dispatching" (choice of v, with P assigned). Analogous considerations apply to the choice of adjustable parameter values, as specified below.

It may be stated that the goal of *active dispatching* is the determination of the active power share at different generation nodes and, possibly, boundary nodes, which minimizes the overall generation cost, accounting for constraints on generated active powers and currents (with adequate margins, for security requirements)[(4)]. Active dispatching also can take advantage of adjustable parameters, if the system is equipped with "quadrature"-regulating transformers to modify the different branch currents in the network. To obtain a satisfactory solution, the system configuration, and more specifically the set of operating generators, may require correction to meet the requirements of spinning reserve.

On the other hand, *reactive dispatching* implies the choice of voltages (except at load nodes) by considering constraints on all voltages (at terminal and internal nodes) and generated reactive powers (at generation and reactive compensation nodes), and providing sufficient reactive power margins. In reactive dispatching, the role of adjustable parameters—corresponding to tap-changing transformers, adjustable condensers and reactors, and "in-phase" regulating transformers—is particularly important. To obtain a satisfactory solution, the system configuration

---

[(4)] In some cases (see Sections 2.3.2 and 2.3.3) active dispatching must be formulated as an "over time" problem, instead of the purely "instantaneous" problem considered here.

and, more specifically, the whole set of operating compensators (and/or generators themselves) may need to be corrected so that at the corresponding nodes the required reactive power margins can be achieved. Finally, the solution of reactive dispatching also may be univocally found, by imposing the minimization of a particular "objective function," for instance $\sum_h a_h Q_h^2$ (with proper a_h), so that the reactive power margins are shared among the generation and/or compensation nodes (h), according to security requirements. Alternatively, if these requirements do not look critical, the minimization of network losses may be imposed, to somewhat reduce the total generated active power and consequently its cost, according to economical requirements.

Based on the above information, the working point may be determined by an *iterative procedure*, by solving each of the two problems starting from the solution obtained for the other. Each solution may be numerically obtained using the gradient method, by minimizing the objective function through successive corrections, consistently with constraints. By doing so, the scheme of Figure 2.10 may be translated into that of Figure 2.15.

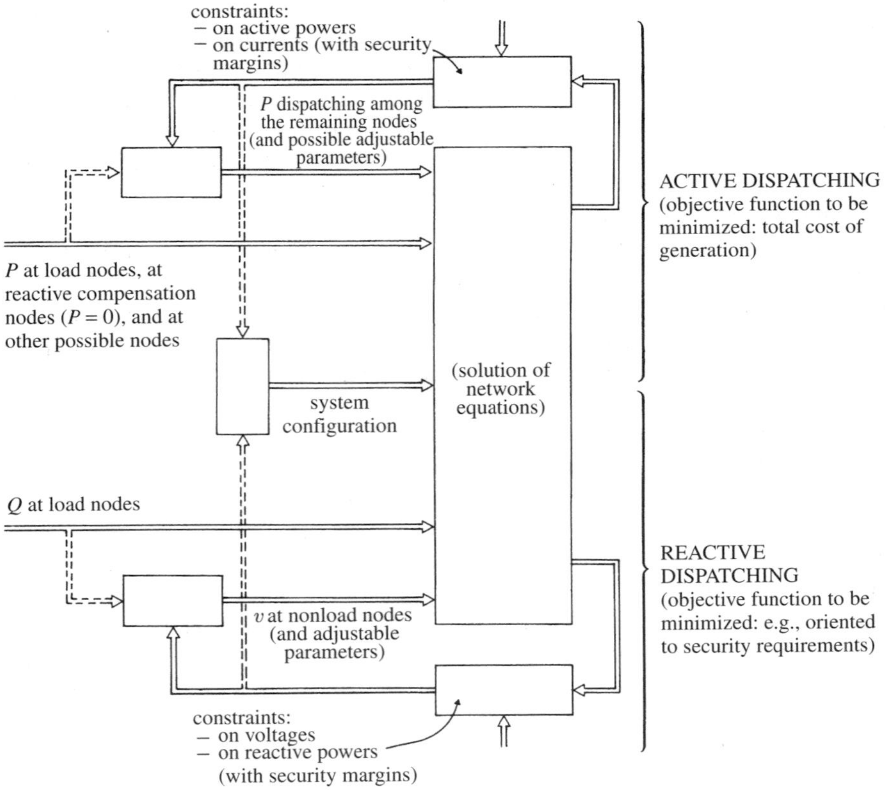

Figure 2.15. Choice of the working point, through active and reactive dispatching.

To start the procedure, *the active dispatching problem may be solved first by accounting for the generation costs and ignoring the presence of the network* (see also Section 2.3.1a). From the solution obtained and considering the network equations, the reactive dispatching problem may be solved, so as to find a first-step "preliminary" working point, and so on.

However, with an often acceptable approximation, active dispatching also may be solved by using a simplified model of the network (as, for instance, the "dc" model), and the iterations themselves may be avoided by assuming, as parameters of such a model, those values corresponding to the "preliminary" working point (see Sections 2.3.1b,c,d).

2.3. ANALYTICAL CRITERIA FOR ECONOMIC OPTIMIZATION

2.3.1. Economic Optimization of Thermoelectric Generation with Constraints on Variations over Time

This section states in analytical terms — even at the expense of some simplifications — the economic optimization problem and the criteria for its solution, besides what is generically recalled in Section 2.2.5d. The analytical expressions (adopted here and in the following) can be helpful, not only for the interpretation of optimal conditions but also for setting useful guidelines in the choice and implementation of numerical procedures, thus avoiding a purely empirical search of optimal solutions.

It has already been pointed out that the economic optimization problem may be essentially translated into the choice of the nonrigidly assigned active powers ("active dispatching"), whereas the choice of the voltage and reactive power steady-state may be substantially considered as a separate problem ("reactive dispatching"), in which the solution may not significantly influence the economic result. If the system is represented by a static model, with no dynamic constraints on variations over time, the mentioned active powers must be chosen in such a way as to minimize the total cost per unit of time, by accounting both for maximum and minimum active power values and for other conditions that may affect the choice (more precisely, limits on currents, and security requirements related to active power availability and current margins inside the network).

For ease of explanation, in the following we will assume that the generation is only thermal and hydroelectric, as any other type of generation may be formally treated in analogy to one of such two types.

Moreover, the following will be assumed to be preassigned:

- the values of adjustable parameters, which substantially affect only the v/Q steady-state (voltage magnitudes and reactive powers)[(5)];

(5) As an exception, we may recall the case of possible "quadrature"-regulating transformers, the adjustment of which corrects the active power flows and consequently currents in the different branches. This fact should be considered, in addition to what is expressed in the following Sections (c) and (d).

- the reactive powers Q_{cj} absorbed by loads, and the voltage magnitudes v_h at the remaining nodes (Pv nodes, and the slack node), or more generally (according to the "dc" approximation) the whole v/Q steady-state;
- the active powers P_{cj} absorbed by loads, as well as the active powers delivered by the hydroelectric plants and the active power "equivalent injections" from external systems (Fig. 2.11) (as to criteria on which their choice is based, see Sections 2.3.3 and 2.3.4, respectively), possibly in addition to the active powers delivered by specific thermal plants.

The n terminal nodes of the network may then be divided into (with $n = n_t + n_o + n_c$):

- n_t nodes with nonassigned thermal generation ($P_{t1}, \ldots, P_{tn_t}$);
- n_o nodes with preassigned injection ($P_{o1}, \ldots, P_{on_o}$) (hydroelectric and possibly thermal generation, reactive compensation and possible equivalent injections);
- n_c load nodes ($P_{c1}, \ldots, P_{cn_c}$).

In the above node-numbering scheme, the n_t nodes are to be considered first, i.e., they are the nodes $1, \ldots, n_t$, with node 1 being the slack node (in fact, it must be a nonassigned injection node).

With such a premise, the network steady-state operating condition will depend on the choice of the active powers $P_i = P_{ti}, i = 2, \ldots, n_t$.

(a) Optimization Based Only on Cost Characteristics

Let us assume at first that all the above-defined n_t nodes may be considered one unique node, as if all the thermal units with nonassigned generation would supply the grid through the same node. In such conditions, this node is just the slack node (1) and the steady-state operating condition of the network is already completely determined by the assignments.

In particular, network losses (p) come to be known, as well as the total active power $\sum_i P_{ti}$ demand for thermal plants with nonassigned generation (by P_{ti}, the single powers supplied to the grid by such units are denoted). The goal is to determine the most economical share of $\sum_i P_{ti}$.

To simplify the analytical treatment, a cost C_i per unit of time is assumed to correspond to each P_{ti} (see Section 2.2.5d) according to an ideal characteristic as shown in Figure 2.13, i.e., with a *continuous and increasing derivative* $\mathrm{d}C_i/\mathrm{d}P_{ti}$ (with $P_{ti} \in [P_{ti\,\min}, P_{ti\,\max}]$). (This characteristic also could be assumed to be piece-wise linear, with $\mathrm{d}C_i/\mathrm{d}P_{ti}$ constant at intervals. In such a case, different optimization procedures than those described below should be adopted.)

By temporarily ignoring the limits $P_{ti\,\min}, P_{ti\,\max}$, the following must be imposed:

$$\min \sum_i C_i(P_{ti})$$

(where $\sum_i C_i(P_{ti})$ is the so-called "objective function" to be minimized), with the constraint:

$$0 = P_c + p - P_o - \sum_i P_{ti}$$

which defines the active power balance, with $P_c \triangleq \sum_j P_{cj}$, $P_o \triangleq \sum_k P_{ok}$, so that $(P_c + p - P_o)$ has an assigned value.

To solve this problem, the well-known Lagrange method leads to the following "Lagrangian function:"

$$\mathcal{L} \triangleq \sum_i C_i(P_{ti}) + \lambda \left(P_c + p - P_o - \sum_i P_{ti} \right)$$

The solution must satisfy the conditions:

$$0 = \frac{\partial \mathcal{L}}{\partial P_{ti}} = \frac{dC_i}{dP_{ti}} - \lambda$$

or:

$$\frac{dC_i}{dP_{ti}} = \lambda \qquad \forall i = 1, \ldots, n_t$$

This indicates that the derivatives dC_i/dP_{ti}, also called *incremental costs "at generation,"* must be equal to one another; see for example, for the case of two units, the points OO in Figure 2.16[(6)].

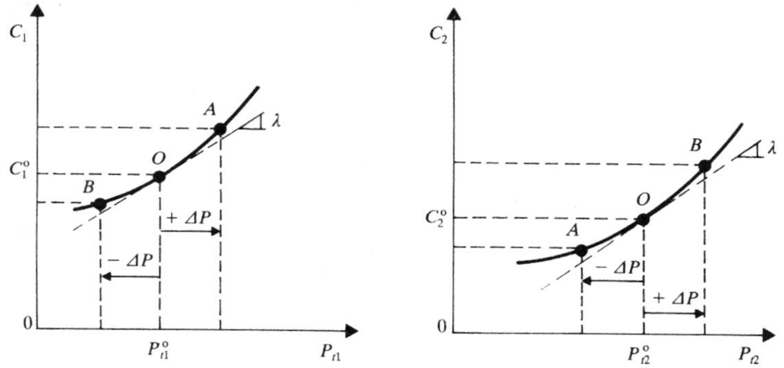

Figure 2.16. Optimization only based on cost characteristics, in case of two thermal units. For the given $P_{t1} + P_{t2} = P_{t1}^o + P_{t2}^o$, both solutions AA, BB lead to $C_1 + C_2 > C_1^o + C_2^o$. The solution OO, for which $(dC_1/dP_{t1})^o = (dC_2/dP_{t2})^o = \lambda$, is the most economical one.

(6) In general, these conditions are necessary to minimize the objective function. However, they also are sufficient in the present case, because of the hypothesis that dC_i/dP_{ti} increases with the respective P_{ti} (see also Fig. 2.16). The condition sufficiency will be similarly assumed in the following sections.

For any given value of λ, the single P_{ti} and sum $\sum_i P_{ti}$ can then be calculated, and vice versa. Once $\sum_i P_{ti} = P_c + p - P_o$ is known, the corresponding value of λ, and consequently the optimal dispatching defined by the P_{ti} values, can be derived.

By differentiating, it follows:

$$\mathrm{d}\lambda = \mathrm{d}\left(\frac{\mathrm{d}C_i}{\mathrm{d}P_{ti}}\right) = \frac{\mathrm{d}^2 C_i}{\mathrm{d}P_{ti}^2} \cdot \mathrm{d}P_{ti} \qquad \forall i$$

and then:

$$\mathrm{d}\sum_i P_{ti} = \mathrm{d}\lambda \sum_i \frac{1}{\mathrm{d}^2 C_i/\mathrm{d}P_{ti}^2} \triangleq \frac{\mathrm{d}\lambda}{h}$$

where h is the generic slope of the characteristic $(\lambda, \sum_i P_{ti})$.
It then results:

$$\mathrm{d}P_{ti} = \frac{h}{\mathrm{d}^2 C_i/\mathrm{d}P_{ti}^2}\,\mathrm{d}\sum_k P_{tk} \triangleq g_i\,\mathrm{d}\sum_k P_{tk}$$

where g_i (with $\sum_i g_i = 1$) defines the optimal participation of $\mathrm{d}P_{ti}$ to the overall variation $\mathrm{d}\sum_k P_{tk}$; thus, it is also called "*participation factor*" for the i-th generator.
Often, relations of the following can be assumed:

$$C_i = C_{io} + a_i P_{ti} + b_i \frac{P_{ti}^2}{2}$$

with C_{io}, a_i, b_i positive constants; the incremental costs at generation are then $\mathrm{d}C_i/\mathrm{d}P_{ti} = a_i + b_i P_{ti}$; i.e., linear functions of respective P_{ti}. At the optimal conditions, it holds:

$$P_{ti} = \frac{\lambda - a_i}{b_i}, \qquad \sum_i P_{ti} = \left(\sum_i \frac{1}{b_i}\right)\lambda - \sum_i \frac{a_i}{b_i}$$

so that the single P_{ti}'s and their sum are linearly increasing with λ. It then follows $C_i = C_{io} - a_i^2/2b_i + \lambda^2/2b_i$, and:

$$\lambda = \frac{\displaystyle\sum_i P_{ti} + \sum_i \frac{a_i}{b_i}}{\displaystyle\sum_i \frac{1}{b_i}}$$

so one can finally deduce the optimal values C_i, $\sum_i C_i$ for any given $\sum_i P_{ti}$.
With two generators and generic conditions, it results (by indicating with the superscript "o" the optimal values; see Fig. 2.16):

$$\frac{\mathrm{d}C_1}{\mathrm{d}P_{t1}} = \lambda + b_1(P_{t1} - P_{t1}^o), \qquad \frac{\mathrm{d}C_2}{\mathrm{d}P_{t2}} = \lambda - b_2(P_{t1} - P_{t1}^o)$$

$$C_1 = C_1^o + \lambda(P_{t1} - P_{t1}^o) + b_1 \frac{(P_{t1} - P_{t1}^o)^2}{2}, \qquad C_2 = C_2^o - \lambda(P_{t1} - P_{t1}^o) + b_2 \frac{(P_{t1} - P_{t1}^o)^2}{2}$$

to which the following *overcost per unit of time* (with respect to the optimal solution $P_{t1} = P_{t1}^o$, $dC_1/dP_{t1} = dC_2/dP_{t2} = \lambda$) can be derived:

$$(C_1 + C_2) - (C_1 + C_2)^o = (b_1 + b_2)\frac{(P_{t1} - P_{t1}^o)^2}{2} = \frac{\left(\frac{dC_1}{dP_{t1}} - \frac{dC_2}{dP_{t2}}\right)^2}{2(b_1 + b_2)}$$

If the equality of the mentioned incremental costs holds, it follows that $d\sum_i C_i = \lambda\, d\sum_i P_{ti} = \lambda\, d(P_c + p - P_o)$, which is equivalent to say that λ represents, for a small variation in total demanded power, the ratio between the resulting variation of cost per unit of time and the total power variation (or even, for all the time for which the considered steady-state holds, the ratio between the variation of cost and that of total energy demanded).

By expressing the constraint on powers in the form of $P_{t1} = P_c + p - P_o - \sum_{k\neq 1} P_{tk}$, the objective function $C_{tot} \triangleq \sum_i C_i$ can be considered as a function of P_{ti}'s, $i \neq 1$, with:

$$\frac{\partial C_{tot}}{\partial P_{ti}} = \frac{dC_i}{dP_{ti}} - \frac{dC_1}{dP_{t1}} \qquad (i \neq 1)$$

By numerical means, the minimization of C_{tot} may be obtained, for instance, by the gradient method, by assuming at each step:

$$\Delta P_{ti} = -k\left(\frac{dC_i}{dP_{ti}} - \frac{dC_1}{dP_{t1}}\right) \qquad (i \neq 1)$$

with a positive and appropriate k; or by varying the only value P_{ti} corresponding to the largest absolute value of $(dC_i/dP_{ti} - dC_1/dP_{t1})$, and so on. As an alternative, the Newton-Raphson method (Section 2.2.4) may be used, deducing ΔP_{ti}'s at each step starting from the equations:

$$0 = \frac{dC_i}{dP_{ti}} - \frac{dC_1}{dP_{t1}} + \frac{d^2C_i}{dP_{ti}^2}\Delta P_{ti} - \frac{d^2C_1}{dP_{t1}^2}\cdot\left(-\sum_{k\neq 1}\Delta P_{tk}\right) \qquad (i \neq 1)$$

Such numerical procedures (with obvious adaptations and possible modifications) can be useful also in solving problems that are presented later on.

Now consider the *limits on the* P_{ti}'s, by imposing:

$$\left.\begin{aligned} &\min \sum_i C_i(P_{ti}) \\ &0 = P_c + p - P_o - \sum_i P_{ti} \\ &P_{ti} \in [P_{ti\,\min}, P_{ti\,\max}] \qquad \forall i \end{aligned}\right\} \qquad [2.3.1]$$

The Kuhn-Tucker method (an extension of the Lagrange method, in the presence of inequality constraints) may be used to solve the problem, leading to the Lagrangian function:

$$\mathcal{L} \triangleq \sum_i C_i(P_{ti}) + \lambda\left(P_c + p - P_o - \sum_i P_{ti}\right) + \sum_i \lambda_i'(P_{ti\,\min} - P_{ti}) + \sum_i \lambda_i''(P_{ti} - P_{ti\,\max})$$

with the usual conditions:

$$0 = \frac{\partial \mathcal{L}}{\partial P_{ti}}$$

to which the following conditions (called "excluding" conditions) must be added:

$$\begin{cases} \lambda_i' = 0 \quad \text{if} \quad P_{ti} > P_{ti\,\min}, & \lambda_i' \geq 0 \quad \text{if} \quad P_{ti} = P_{ti\,\min} \\ \lambda_i'' = 0 \quad \text{if} \quad P_{ti} < P_{ti\,\max}, & \lambda_i'' \geq 0 \quad \text{if} \quad P_{ti} = P_{ti\,\max} \end{cases}$$

Note that, by generically imposing $\min F(x_1, \ldots, x_n)$ in the presence of the inequality constraint $0 \geq g(x_1, \ldots, x_n)$, such a constraint can be changed into an equality constraint by writing $0 = g(x_1, \ldots, x_n) + a^2$, with a as an unknown variable. If the constraint is of the type $0 \leq h(x_1, \ldots, x_n)$, h can be substituted with $-h$. By the Lagrange method with $\mathcal{L} \triangleq F + \lambda(g + a^2)$, the following conditions then hold:

$$\begin{cases} 0 = \dfrac{\partial \mathcal{L}}{\partial x_i} & (i = 1, \ldots, n) \\ 0 = \dfrac{\partial \mathcal{L}}{\partial a}, & \text{i.e., } 0 = \lambda a \end{cases}$$

and furthermore (to obtain $\min F$):

$$\frac{\partial^2 \mathcal{L}}{\partial a^2} \geq 0, \quad \text{i.e., } \lambda \geq 0$$

so that, from the last two conditions, by recalling the constraint $0 = g + a^2$, the following ("excluding") conditions can be deduced:

$$\lambda = 0 \quad \text{if } a \neq 0 \text{ (i.e., } 0 > g\text{)}, \qquad \lambda \geq 0 \quad \text{if } a = 0 \text{ (i.e., } 0 = g\text{)}$$

which are analogous to the above-mentioned ones.

It then follows $0 = \mathrm{d}C_i/\mathrm{d}P_{ti} - \lambda - \lambda_i' + \lambda_i''$, or equivalently:

$$c_i = \lambda \quad \forall i = 1, \ldots, n_t \qquad [2.3.2]$$

by denoting $c_i \triangleq \mathrm{d}C_i/\mathrm{d}P_{ti} - \lambda_i' + \lambda_i''$; that is:

$$c_i \begin{cases} \leq \dfrac{\mathrm{d}C_i}{\mathrm{d}P_{ti}}(P_{ti\,\min}^{+}) & \text{if } P_{ti} = P_{ti\,\min} \\ = \dfrac{\mathrm{d}C_i}{\mathrm{d}P_{ti}} & \text{if } P_{ti} \in (P_{ti\,\min}, P_{ti\,\max}) \\ \geq \dfrac{\mathrm{d}C_i}{\mathrm{d}P_{ti}}(P_{ti\,\max}^{-}) & \text{if } P_{ti} = P_{ti\,\max} \end{cases} \qquad [2.3.3]$$

As a consequence of Equation [2.3.2], it still holds that the incremental costs at generation must equal each other, provided this denomination is extended to the c_i's defined by Equations [2.3.3], which depend on their respective P_{ti} according to Figure 2.17a.

For any given value of λ, each P_{ti} can be evaluated as well as the sum $\sum_i P_{ti}$ (see the example with three generators, Fig. 2.17); then, the knowledge of the characteristic $(\lambda, \sum_i P_{ti})$ determines, through the value of λ, the optimal dispatching for any given $\sum_i P_{ti} = P_c + p - P_o$ (obviously included within the limits $\sum_i P_{ti\,\min}, \sum_i P_{ti\,\max}$).

The characteristic $(\lambda, \sum_i P_{ti})$ is generally constituted by several pieces with positive slopes, separated by discontinuity points. At each discontinuity point, (at least) one of the P_{ti}'s leaves—for an increasing λ—its lower limit $P_{ti\,\min}$, or

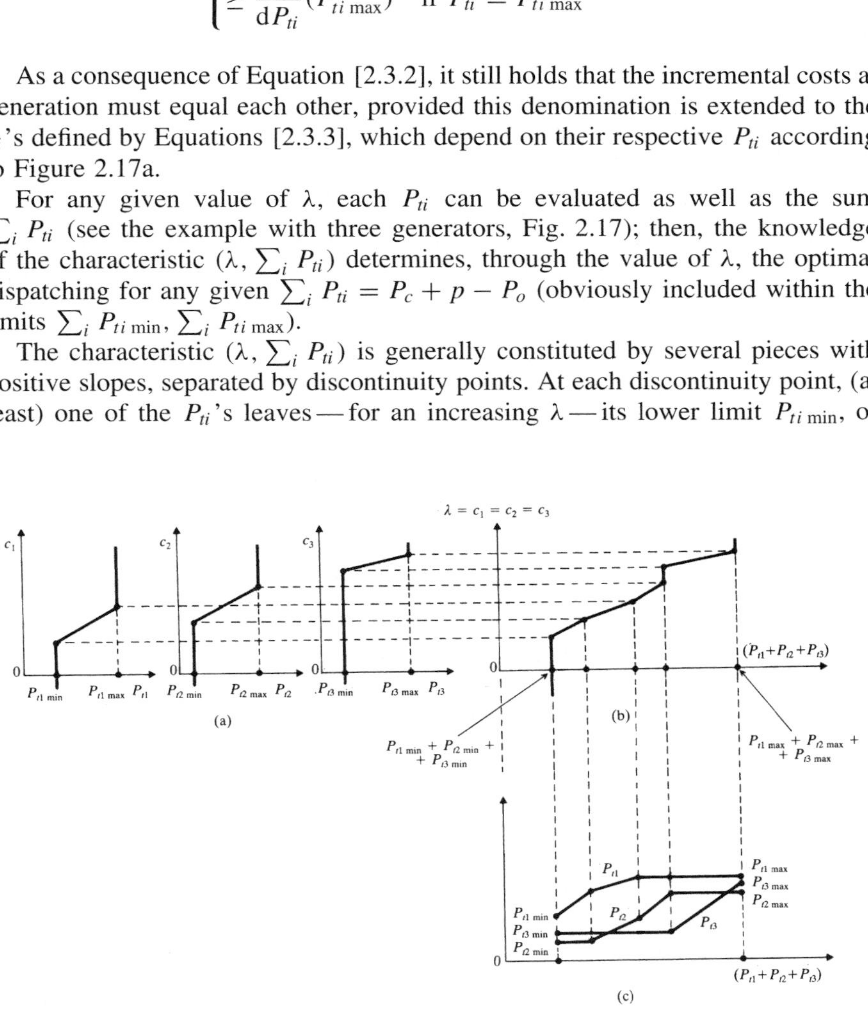

Figure 2.17. Equality of incremental costs at generation (example with three thermal units): (a) characteristics of individual units; (b) dependence of total generated power, on incremental cost at generation; (c) dispatching of total generated power, between different units.

reaches its upper limit $P_{ti\,\max}$. Each piece is linear if it can be assumed that costs per unit of time are quadratic functions of the respective P_{ti}'s, as indicated above. In any case, it still holds that $\mathrm{d}\sum_i C_{ti} = \lambda\,\mathrm{d}\sum_i P_{ti} = \lambda\,\mathrm{d}(P_c + p - P_o)$, so that λ retains the previously stated meaning.

If the operating units of a single power plant are represented by their ideal characteristics, and plant losses are disregarded, the condition of equal incremental costs at generation can be used to determine the optimal share of any possible value of the total delivered active power (by obviously disregarding possible generators operating at assigned power). The whole power plant can then be represented (for the considered units) by means of only one characteristic (C_{tot}, P_{tot}), with $P_{\text{tot}} \in [P_{\text{tot min}}, P_{\text{tot max}}]$, generally consisting of several pieces with $\mathrm{d}C_{\text{tot}}/\mathrm{d}P_{\text{tot}}$ increasing with P_{tot}.

According to Section 2.2.5d, for a more realistic solution it is possible to optimize the share between units of the same power plant, starting from the actual characteristics, not from ideal ones. We will assume that the resulting characteristic (C, P_t) of the generic power plant is with one or more pieces, in which $\mathrm{d}C/\mathrm{d}P_t$ increases with P_t, or that it may be likewise approximated.

Instead, with several thermal power plants (each having an unassigned total generation), the steady-state of the network depends on the active power dispatching between the power plants (Section 2.2.5d), particularly for losses. Thus, the condition of equal incremental costs at generation may be adopted with close estimation for the whole set of power plants, only if they are close to each other or if the dependence of network losses on active dispatching may be disregarded. Otherwise, it is worth considering such dependency, according to (b) below.

(b) Account for Variations of Network Losses

To account for network losses, as dependent on active power dispatching between units, Equations [2.3.1] must be replaced by:

$$\min \sum_i C_i(P_{ti}), \quad 0 = P_c + p - P_o - \sum_i P_{ti}, \quad P_{ti} \in [P_{ti\,\min}, P_{ti\,\max}]$$
$$(i = 1, \ldots, n_t)$$

where $P_c \triangleq \sum_j P_{cj}$, $P_o \triangleq \sum_k P_{ok}$ and:

$$p = \phi(P_{t2}, \ldots, P_{tn_t}; \ldots, P_{ok}, \ldots; \ldots, P_{cj}, \ldots; \ldots, Q_{cj}, \ldots; \ldots, v_h, \ldots) \tag{2.3.4}$$

according to Equation [2.2.8]. By recalling Equations [2.1.24] and [2.1.28], the dependence of p on active powers can be assumed to be the result of the term p_α, which can be approximated by Equation [2.1.29].

Once the Lagrangian function is assumed:

$$\mathcal{L} \triangleq \sum_i C_i(P_{ti}) + \lambda \left(P_c + p - P_o - \sum_i P_{ti} \right) + \sum_i \lambda_i'(P_{ti\,\min} - P_{ti}) + \sum_i \lambda_i''(P_{ti} - P_{ti\,\max})$$

with P_{ok}'s, P_{cj}'s, Q_{cj}'s, v_h's assigned, the following conditions can be determined:

$$c_1 = \frac{c_i}{1 - \dfrac{\partial \phi}{\partial P_{ti}}} = \lambda \qquad (i = 2, \ldots, n_t) \qquad [2.3.5]$$

(similarly to Equation [2.2.9]), which are, for any given value of λ, one equation in P_{t1} and $(n_t - 1)$ equations in $P_{t2}, \ldots, P_{tn_t}$. From Equation [2.3.5] each $P_{ti}\,(i = 1, \ldots, n_t)$, as well as p, $\sum_i P_{ti}$ and $(P_c - P_o)$, can be determined (see Fig. 2.18a). The knowledge of the $(\lambda, (P_c - P_o))$ characteristic allows then to derive, through the value of λ, the optimal values of P_{ti}'s for any assigned $(P_c - P_o)$.

Note that:

$$\mathrm{d}\sum_i C_i = \mathrm{d}C_1 + \sum_{i \neq 1} \mathrm{d}C_i = \lambda \left(\mathrm{d}P_{ti} + \sum_{i \neq 1} \left(1 - \frac{\partial \phi}{\partial P_{ti}}\right) \mathrm{d}P_{ti} \right)$$

$$= \lambda \left(\mathrm{d}\sum_i P_{ti} - \mathrm{d}\phi_t \right) = \lambda(\mathrm{d}(P_c - P_o) + \mathrm{d}\phi_o + \mathrm{d}\phi_c + \mathrm{d}\phi_{vQ})$$

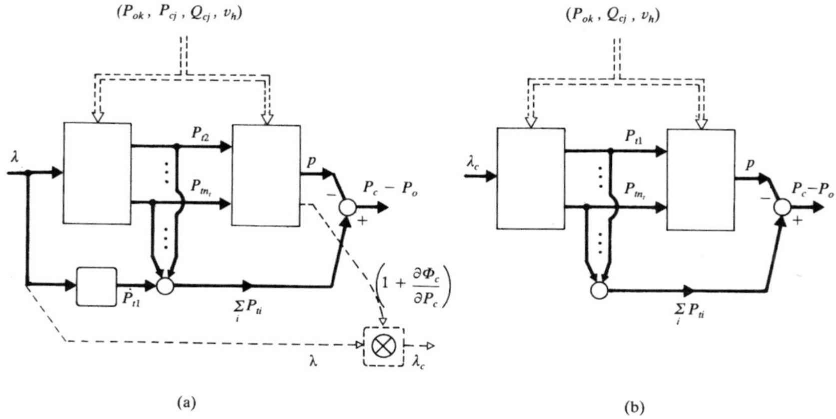

Figure 2.18. Block diagram showing the relationship between the total preassigned active power $(P_c - P_o)$ and: (a) the incremental cost at generation for unit 1 $(\lambda = c_1)$; (b) the incremental cost at load (λ_c) under the hypothesis of "conform" loads.

where:

$$d\phi_t \triangleq \sum_{i\neq 1} \frac{\partial \phi}{\partial P_{ti}} dP_{ti}$$

$$d\phi_o \triangleq \sum_{k} \frac{\partial \phi}{\partial P_{ok}} dP_{ok}$$

$$d\phi_c \triangleq \sum_{j} \frac{\partial \phi}{\partial P_{cj}} dP_{cj}$$

$$d\phi_{vQ} \triangleq \sum_{j} \frac{\partial \phi}{\partial Q_{cj}} dQ_{cj} + \sum_{h} \frac{\partial \phi}{\partial v_h} dv_h$$

(while $dp = d\phi_t + d\phi_o + d\phi_c + d\phi_{vQ}$, $d\sum_i P_{ti} - dp = d(P_c - P_o)$), from which the meaning of λ with respect to possible variations of the assigned values P_{ok}, P_{cj}, Q_{cj}, v_h can be determined. In particular, if each P_{ok}, Q_{cj}, v_h does not change, it results $d\sum_i C_i = \lambda(dP_c + d\phi_c)$, where $d\phi_c$ is the variation of losses caused by the variation of the assigned P_{cj}s only.

If v_h's were neither assigned nor constrained, then following further conditions would hold:

$$0 = \frac{\partial \phi}{\partial v_h}$$

(for all hs that do not refer to load nodes), which would correspond intuitively to a choice of v_h's so as to minimize network losses (for any given set of $P_{t2}, \ldots, P_{tn_t}$). However, these conditions do not usually have practical interest, as voltage magnitudes are already (as mentioned before) constrained enough by their limits and by requirements related to the steady-state reactive power (network flows, and reactive power margins at generation and compensation nodes).

In practice — especially if the generic network considered covers an area which, under different aspects, looks homogeneous and not too wide (see footnote[(1)], Section 2.1.5) — single load demands P_{cj}'s, may be assumed to partially depend on common causes (e.g., meteorological conditions, working time-tables, television programs etc.), according to:

$$P_{cj} = P_{cj}(P_c) \qquad [2.3.6]$$

("*conform*" *load* hypothesis). For instance $P_{cj} = P^*_{cj} + b_j(P_c - \sum P^*_{ci})$, with P^*_{cj}, b_j known constants (and $\sum b_j = 1$), where the P^*_{cj}'s are unrelated base values, whereas the b_j coefficients define the way any variation of P_c from the base value is shared between load nodes.

Because of $0 = P_c + p - P_o - \sum_i P_{ti}$, the P_{cj}'s depend on $(P_o + \sum_i P_{ti} - p)$, and by Equation [2.3.4] the following can be derived:

$$p = \theta(P_{t1}, \ldots, P_{tn_t}; \ldots, P_{ok}, \ldots; \ldots, Q_{cj}, \ldots; \ldots, v_h, \ldots; p) \qquad [2.3.7]$$

from which (at the price of approximations) losses p can be derived as a function of all powers $P_{t1}, \ldots, P_{tn_t}$, besides P_{ok}, Q_{cj}, v_h. This leads to the so-called *"loss" formula*(7):

$$p = p(P_{t1}, \ldots, P_{tn_t}; \ldots, P_{ok}, \ldots; \ldots, Q_{cj}, \ldots; \ldots, v_h, \ldots) \qquad [2.3.8]$$

(If it were $n_c = 1$, this formula would be directly determined similarly to Equation [2.2.8], by formally assuming as node 1 the unique load node.)

Therefore, conditions of the following type can be deduced

$$\frac{c_i}{1 - \dfrac{\partial p}{\partial P_{ti}}} = \lambda_c \quad \forall i = 1, \ldots, n_t \qquad [2.3.9]$$

This is a formulation more "symmetrical" than the previous one (as all nodes of thermoelectric generation are now treated in a similar way). Furthermore:

- the active powers that must be known to compute the derivatives $\partial p/\partial P_{ti}$ (also called "incremental losses at generation") are only the single P_{ti}, P_{ok}, which are operating variables that can be easily determined even during operation itself, differently from what usually happens for P_c and each P_{cj};
- the constant λ_c has a more interesting meaning than that of the previous constant λ, per the information specified below.

Equation [2.3.9] corresponds, at each given value of λ_c, to n_t equations in $P_{t1}, \ldots, P_{tn_t}$. Each P_{ti} can be determined, as well as p, $\sum_i P_{ti}$, and $(P_c - P_o)$ (see Fig. 2.18b). This leads to the characteristic $(\lambda_c, (P_c - P_o))$. Knowing this characteristic enables, for any given value of $(P_c - P_o)$, the determination of λ_c and thus the optimal values of P_{ti}'s.

Equation [2.3.9] is equivalent to Equations [2.3.5], as it results:

$$\frac{\partial \theta}{\partial p} = -\sum_j \frac{\partial \phi}{\partial P_{cj}} \frac{\mathrm{d}P_{cj}}{\mathrm{d}P_c} = -\frac{\mathrm{d}\phi_c}{\mathrm{d}P_c}$$

(7) For general purposes it might also be assumed—with obvious notation—that $v_{t1}, \ldots, v_{tn_t}$ vary with $P_{t1}, \ldots, P_{tn_t}$ in such a way to satisfy constraints of the type $Q_{ti} = k_i' + k_i'' P_{ti}$, with k_i', k_i'' known constants.

additionally,

$$\frac{\partial \theta}{\partial P_{t1}} = -\frac{\partial \theta}{\partial p}$$

$$\frac{\partial \theta}{\partial P_{ti}} = \frac{\partial \phi}{\partial P_{ti}} - \frac{\partial \theta}{\partial p} \quad \text{for } i = 2, \ldots, n_t$$

$$\frac{\partial p}{\partial P_{ti}} = \frac{\dfrac{\partial \theta}{\partial P_{ti}}}{1 - \dfrac{\partial \theta}{\partial p}} \quad \text{for } i = 1, \ldots, n_t \qquad [2.3.10]$$

so that, finally:

$$1 - \frac{\partial \phi}{\partial P_{ti}} = \frac{1 - \dfrac{\partial p}{\partial P_{ti}}}{1 - \dfrac{\partial p}{\partial P_{t1}}} \quad \text{for } i = 2, \ldots, n_t$$

while the constants λ, λ_c differ from each other according to:

$$\lambda_c = \frac{\lambda}{1 - \dfrac{\partial p}{\partial P_{t1}}} = \lambda \left(1 + \frac{\partial \phi_c}{\partial P_c}\right) \qquad [2.3.11]$$

If it is assumed that the costs per unit of time C_i are quadratic functions of the corresponding P_{ti}'s and that ϕ is a quadratic function in $P_{t2}, \ldots, P_{tn_t}$ (which can be easily obtained by using Equations [2.1.24], [2.1.28], and [2.1.29]), the conditions [2.3.5] may be translated, for each given value of λ, into a system of linear equations in the P_{ti}'s (or, better, in those P_{ti}'s which are within their limits, while for the other P_{ti}'s it holds either $P_{ti} = P_{ti\,\min}$ or $P_{ti} = P_{ti\,\max}$). This can significantly simplify the deduction of each P_{ti} from λ. By accounting for Equation [2.3.11], this situation also can be used for determining the correspondence between λ_c and P_{ti}'s, and between λ_c and $(P_c - P_o)$ (see Fig. 2.18a).

Additionally:

$$\mathrm{d}\sum_i C_i = \lambda_c \sum_i \left(1 - \frac{\partial p}{\partial P_{ti}}\right) \mathrm{d}P_{ti}$$

from which, by adopting symbols already defined, it may be determined that:

$$\mathrm{d}\sum_i C_i = \lambda_c \left(\mathrm{d}P_c + \frac{-\mathrm{d}P_o + \mathrm{d}\phi_o + \mathrm{d}\phi_{vQ}}{1 + \dfrac{\mathrm{d}\phi_c}{\mathrm{d}P_c}}\right)$$

If each assigned P_{ok}, Q_{cj}, v_h is kept unchanged, it then follows in a much simpler way $\mathrm{d}\sum_i C_i = \lambda_c\,\mathrm{d}P_c$, so that *the value of λ_c defines—for a small variation*

$\mathrm{d}P_c$ — *the ratio between the variation of cost per unit of time and the total load variation which causes it.* For this reason, the generic quantity $c_i/(1 - \partial p/\partial P_{ti})$, which is in the left side of Equation [2.3.9], is called the *incremental cost "at load."* Thus, such conditions imply the equality of the incremental costs "at load" instead of those "at generation" considered in section (a).

It is to be noted that, for each i, the incremental cost at load is somewhat different from that at generation, due to the *"penalty factor"* $1/(1 - \partial p/\partial P_{ti})$. As a practical consequence, the more penalized the generic generating unit (as if its fuel had a higher cost) the larger the increase of losses with its delivered power P_{ti}. With two generating units having the same characteristics (C_i, P_{ti}), more power should be supplied from the unit having the smaller $\partial p/\partial P_{ti}$, i.e. (obviously, assuming a single load node), from the unit "closer" to the load.

Practically, deducting the loss formula from the implicit form of Equation [2.3.7] requires some simplification. As an initial (and acceptable) estimation, the dependence of function θ on p may be disregarded (as if the P_{cj}'s would depend on $P_0 + \sum_i P_{ti}$), assuming the approximate equation:

$$p \cong \theta(P_{t1}, \ldots, P_{tn_t}; \ldots, P_{ok}, \ldots; \ldots, Q_{cj}, \ldots; \ldots, v_h, \ldots;\ 0)$$
$$\triangleq \theta^*(P_{t1}, \ldots, P_{tn_t}; \ldots, P_{ok}, \ldots; \ldots, Q_{cj}, \ldots; \ldots, v_h, \ldots) \qquad [2.3.12]$$

which is already in the desired form of Equation [2.3.8].

However, deducting the partial derivatives $\partial p/\partial P_{ti}$ should not be based on the above formula; in fact, in order not to worsen the degree of approximation, Equation [2.3.10] must be recalled, assuming:

$$\frac{\partial p}{\partial P_{ti}} \cong \frac{\dfrac{\partial \theta}{\partial P_{ti}}}{1 - \dfrac{\partial \theta}{\partial p}}(P_{t1}, \ldots, P_{tn_t}; \ldots, P_{ok}, \ldots; \ldots, Q_{cj}, \ldots; \ldots, v_h, \ldots; 0)$$

instead of $\partial p/\partial P_{ti} \cong \partial \theta^*/\partial P_{ti}$, as it would result from Equation [2.3.12].

(c) Influence of Limits on Currents

With reference to the current flowing into a generic branch of the network, now assume the constraint (see condition [2.2.4]):

$$i \le i_{\max}$$

The Lagrangian function can be written as:

$$\mathcal{L} \triangleq \sum_i C_i(P_{ti}) + \lambda\left(P_c + p - P_o - \sum_i P_{ti}\right) + \sum_i \lambda_i'(P_{ti\min} - P_{ti})$$
$$+ \sum_i \lambda_i''(P_{ti} - P_{ti\max}) + \mu(i - i_{\max})$$

where losses p (see Equation [2.3.4]) and current i depend on active powers at all terminal nodes, except that at the slack node (node 1), i.e., on:

$$P_{t2}, \ldots, P_{tn_t}; \quad P_{o1}, \ldots, P_{on_o}; \quad P_{c1}, \ldots, P_{cn_c}$$

In addition, they also depend on reactive powers $Q_{c1}, \ldots, Q_{cn_c}$ at load nodes and voltage magnitudes at the remaining terminal nodes. (If Equation [2.3.6] could be accepted, i.e., if the loads may be assumed to "conform," the active powers $P_{t1}, \ldots, P_{tn_t}; P_{o1}, \ldots, P_{on_o}$ may still be assumed as independent variables, instead of those listed previously.)

The following optimality conditions are then obtained:

$$c_1 = \frac{c_i + \mu \dfrac{\partial i}{\partial P_{ti}}}{1 - \dfrac{\partial \phi}{\partial P_{ti}}} = \lambda \quad (i = 2, \ldots, n_t) \tag{2.3.13}$$

where $\mu = 0$ if $i < i_{\max}$, $\mu \geq 0$ if $i = i_{\max}$.

By comparison with Equations [2.3.5] the effect of the constraint on i can be found to be formally equivalent (when $i = i_{\max}$) to an increase in the incremental cost c_i if $\partial i/\partial P_{ti} > 0$, and to a decrease if $\partial i/\partial P_{ti} < 0$ $(i = 2, \ldots, n_t)$. This can lead to a redistribution of $P_{t1}, \ldots, P_{tn_t}$. It is here assumed, as in (d), that the problem can actually have a solution, consistent with the constraints, by acting on the P_{ti}'s. Generally, one can also take advantage of adjustment of "quadrature"-regulating transformers and re-adjust the assigned values P_{ok}, v_h.

These considerations can be extended to constraints on more than one current. However, the treatment requires the (not so easy) analytical definition of the dependence of currents on the above-defined independent variables. Therefore, the use of simplifications becomes a must.

If the approximations of the "dc" model are accepted and the v/Q regime is already preassigned (and it remains unchanged in spite of corrections on active powers), the constraints on currents may be conveniently replaced (Section 2.2.5b) by the constraints:

$$F_l \in [-F_{l\max}, +F_{l\max}]$$

where F_l is the *active power flow* in the generic branch l (series branch) between nodes r, s (terminal or internal nodes). F_l is assumed to be positive when current flows from r to s.

With such hypotheses it holds $F_l = (v_r v_s / x_{rs})\alpha_{rs}$ (see the first part of Equations [2.1.11]). The phase shift α_{rs} linearly depends (per Equations [2.1.22] and [2.1.23]) on active powers injected at the terminal nodes except for the slack node, i.e., on active powers $P_{ti} (i \neq 1)$, P_{ok}, P_{cj}. This leads to an approximate equation:

$$F_l = \sum_{i \neq 1} a_{lti} P_{ti} + \sum_k a_{lok} P_{ok} + \sum_j a_{lcj}(-P_{cj}) \tag{2.3.14}$$

where coefficients $a_{lti}, \ldots$ are the sensitivity coefficients defined by Equation [2.1.26].

For interconnected operation, approximations of the dc model may further simplify the representation of external networks, in addition to what can be obtained with the usual simplified equivalents (see Fig. 2.11). Active powers exchanged at the boundary nodes can be expressed as linear functions of:

- active powers of generators and loads within external systems;
- phase shifts between voltages at boundary nodes.

The former of such contributions may be expressed—with reference to the equivalent circuit of Figure 2.6a or 2.6b—by (active power) "*equivalent injections*" directly applied at the boundary nodes, without any dependence on phase shifts; whereas the latter contribution can be accounted for by adding proper "equivalent branches" between nodes themselves (if more than one).

By assuming again that P_{ok}'s, P_{cj}'s are assigned (although the treatment can be obviously extended to the case in which the P_{ok}'s are not assigned) the following can be derived:

$$F_l^{(t)} \in [F_{l\,\min}^{(t)}, F_{l\,\max}^{(t)}]$$

with:

$$F_l^{(t)} \triangleq \sum_{i \neq 1} a_{lti} P_{ti} \qquad [2.3.15]$$

$$\begin{cases} F_{l\,\min}^{(t)} \triangleq -F_{l\,\max} - \sum_k a_{lok} P_{ok} + \sum_j a_{lcj} P_{cj} \\ F_{l\,\max}^{(t)} \triangleq +F_{l\,\max} - \sum_k a_{lok} P_{ok} + \sum_j a_{lcj} P_{cj} \end{cases}$$

With reference to a given branch it can be written (with obvious extension to constraints on more than one branch):

$$\mathcal{L} \triangleq \sum_i C_i(P_{ti}) + \lambda \left(P_c + p - P_o - \sum_i P_{ti} \right) + \sum_i \lambda_i'(P_{ti\,\min} - P_{ti})$$
$$+ \sum_i \lambda_i''(P_{ti} - P_{ti\,\max}) + \mu'(F_{l\,\min}^{(t)} - F_l^{(t)}) + \mu''(F_l^{(t)} - F_{l\,\max}^{(t)})$$

from which it can be derived:

$$c_1 = \frac{c_i + (\mu'' - \mu')a_{lti}}{1 - \dfrac{\partial \phi}{\partial P_{ti}}} = \lambda \quad (i = 2, \ldots, n_t) \qquad [2.3.16]$$

where:

$$(\mu'' - \mu') \begin{cases} = -\mu' \leq 0 & \text{if } F_l^{(t)} = F_{l\min}^{(t)} \quad (\text{i.e., } F_l = -F_{l\max}) \\ = 0 & \text{if } F_l^{(t)} \in (F_{l\min}^{(t)}, F_{l\max}^{(t)}) \ (\text{i.e., } F_l \in (-F_{l\max}, +F_{l\max})) \\ = \mu'' \geq 0 & \text{if } F_l^{(t)} = F_{l\max}^{(t)} \quad (\text{i.e., } F_l = +F_{l\max}) \end{cases}$$

Therefore, *if the power F_l is at one of its limits, the quantity $(\mu'' - \mu')a_{lti}$* (the sign of which depends also on the sign of the respective sensitivity coefficient a_{lti}) *has to be added to the generic incremental cost c_i*, thus modifying the active powers $P_{t1}, \ldots, P_{tn_t}$.

As a significant example, assume $C_i = C_{io} + a_i P_{ti} + b_i(P_{ti}^2/2)$ (with $C_{io}, a_i, b_i > 0$) and suppose that all the P_{ti}'s are within their respective limits $P_{ti\min}$, $P_{ti\max}$, so that:

$$\lambda_i' = \lambda_i'' = 0, \quad c_i = \frac{dC_i}{dP_{ti}} = a_i + b_i P_{ti}$$

Furthermore, losses are disregarded.

When $F = F_{l\max}$, it can be derived, from Equations [2.3.16]:

$$P_{t1} = \frac{\lambda - a_1}{b_1}, \quad P_{ti} = \frac{\lambda - a_i - \mu'' a_{lti}}{b_i} \quad (i = 2, \ldots, n_t)$$

(with $\mu'' \geq 0$). By imposing $\sum_i P_{ti} = P_c - P_o$, $F_l^{(t)} = F_{l\max}^{(t)}$, two linear equations in the unknowns λ, μ'' are obtained. Once these unknowns are determined, all the active powers $P_{t1}, \ldots, P_{tn_t}$ corresponding to the minimum $\sum_i C_i$ can be derived. If $\mu'' < 0$, the assumption $F_l = F_{l\max}$ should be removed.

In the above-expressed conditions, it is easy to check that, by denoting with the superscript "o" the values obtainable at the same $(P_c - P_o)$ in the absence of the constraint $F_l \leq F_{l\max}$, it results in particular:

$$\lambda - \lambda^o = \mu'' a_l^*$$

$$P_{ti} - P_{ti}^o = \frac{\lambda - \lambda^o - \mu'' a_{lti}}{b_i} = \mu'' \frac{a_l^* - a_{lti}}{b_i} \quad (i = 2, \ldots, n_t)$$

where:

$$a_l^* \triangleq \frac{\displaystyle\sum_{k \neq 1} \frac{a_{ltk}}{b_k}}{\displaystyle\sum_k \frac{1}{b_k}}$$

so that the constraint has the effect of increasing or decreasing the generic P_{ti}, according to the fact that the respective coefficient a_{lti} is smaller or larger than the critical value a_l^* defined above (while $P_{t1} - P_{t1}^o = (\lambda - \lambda^o)/b_1 = \mu''(a_l^*/b_1)$ is obviously equal to $-\sum_{i \neq 1}(P_{ti} - P_{ti}^o)$).

Similar considerations apply when considering the lower limit $F_l = -F_{l\max}$.

When two subnetworks are connected only through the branch under consideration, the coefficients a_{lti}'s would be:

- all zero, for the nodes in the subnetwork that includes the slack node;
- all equal to each other, for the nodes in the other subnetwork;

so that the considered constraint would increase the generation in one subnetwork while decreasing generation in the other; consequently, the units of the former subnetwork would be set at a higher incremental cost than that corresponding to the units in the latter.

To obtain a better approximation, the above-described simplifications may be adopted only when readjusting the working point ("*redispatching*"), starting from the "base" solution obtained by neglecting the limits on currents (by adopting the full model, and even resorting only to a numerical procedure).

Of course, redispatching becomes necessary only if in the base working point one or more limits are violated; in this case, the correction will increase the total cost per unit of time ($\sum_i C_i$). Therefore, increase of the total cost may be assumed as the objective function to be minimized, imposing:

$$\min \sum_i \Delta C_i, \qquad \sum_i \Delta C_i \triangleq \sum_i (C_i(P_{ti}^o + \Delta P_{ti}) - C_i(P_{ti}^o))$$

(the superscript "o" denotes the values corresponding to the base working point, and it is generically intended that $\Delta P_{ti} \triangleq P_{ti} - P_{ti}^o$, etc.)

By further assuming that the steady-state values of voltage magnitudes (and of reactive powers) remain unchanged with respect to the base working point, as well as the set of values P_{ok}, P_{cj}:

- the constraint caused by the balance of active powers can be written as:

$$0 = \Delta p - \Delta \sum_i P_{ti}$$

- the limits on P_{ti}'s may be expressed by the following constraints:

$$\Delta P_{ti} \in [P_{ti\,\min} - P_{ti}^o, P_{ti\,\max} - P_{ti}^o] \quad (i = 1, \ldots, n_t)$$

- the limits on currents may be expressed as constraints on transmitted powers:

$$\Delta F_l \in [-F_{l\,\max} - F_l^o, +F_{l\,\max} - F_l^o]$$

- finally, it may be assumed (by recalling Equation [2.1.26]):

$$\Delta F_l = \sum_{i \neq 1} a_{lti} \Delta P_{ti} + \sum_k a_{lok} \Delta P_{ok} + \sum_j a_{lcj}(-\Delta P_{cj}) \qquad [2.3.17]$$

or, with P_{ok}'s, P_{cj}'s assigned:

$$\Delta F_l = \sum_{i \neq 1} a_{lti} \Delta P_{ti}$$

where the coefficients a_{lti} retain the meaning defined earlier.

(The possible extension to the case when P_{ok}'s are not assigned is then obvious.)

By considering, for simplicity, the constraints on one branch for which the current base value has been found to be in excess of the admitted value, it may be written:

$$\mathcal{L} \triangleq \sum_i \Delta C_i(\Delta P_{ti}) + \lambda \left(\Delta p - \sum_i \Delta P_{ti} \right) + \sum_i \lambda_i'(\Delta P_{ti\,\min} - \Delta P_{ti})$$
$$+ \sum_i \lambda_i''(\Delta P_{ti} - \Delta P_{ti\,\max}) + \mu'(\Delta F_{l\,\min} - \Delta F_l) + \mu''(\Delta F_l - \Delta F_{l\,\max})$$

so that conditions as in Equations [2.3.16] may still be derived, where:

$$(\mu'' - \mu') \begin{cases} = -\mu' \leq 0 & \text{if } \Delta F_l = -F_{l\,\max} - F_l^o \\ = 0 & \text{if } \Delta F_l \in (-F_{l\,\max} - F_l^o, +F_{l\,\max} - F_l^o) \\ = \mu'' \geq 0 & \text{if } \Delta F_l = +F_{l\,\max} - F_l^o \end{cases}$$

with consequences similar to those illustrated above.

Note that with a simplified, likely acceptable approach we may also assume, if P_{ti}^o and P_{ti} are both within the limits $P_{ti\,\min}$, $P_{ti\,\max}$:

$$\Delta C_i = \left(\frac{\mathrm{d}C_i}{\mathrm{d}P_{ti}} \right)^o \Delta P_{ti} + \left(\frac{\mathrm{d}^2 C_i}{\mathrm{d}P_{ti}^2} \right)^o \frac{(\Delta P_{ti})^2}{2}$$

By assuming $\Delta P_{ti} = 0$ at the other possible nodes (for which $P_{ti} = P_{ti}^o = P_{ti\,\min}$ or $P_{ti} = P_{ti}^o = P_{ti\,\max}$ holds), we may then derive:

$$\sum_i \Delta C_i = \sum_i c_i^o \Delta P_{ti} + \sum_i \left(\frac{\mathrm{d}^2 C_i}{\mathrm{d}P_{ti}^2} \right)^o \frac{(\Delta P_{ti})^2}{2}$$

where the first term in the right side is:

$$\lambda^o \left(\Delta p - \sum_{i \neq 1} \frac{\partial \phi}{\partial P_{ti}} \Delta P_{ti} \right)$$

(remember Equations [2.3.5] and $0 = \Delta p - \Delta \sum_i P_{ti}$) and then it may be disregarded, or approximated by a quadratic function of $\Delta P_{t2}, \ldots, \Delta P_{tn_t}$.

No matter what simplifying procedure is adopted, the results should be checked by evaluating, on the basis of the full model, the working point which actually corresponds to the above found $P_{t2}, \ldots, P_{tn_t}$, and correcting this last working point if constraints on currents are not satisfied.

(d) Influence of Security Constraints

The analytical formulation illustrated in the previous sections also may be properly adapted to account for security requirements. However, it may seem unavoidable to use strong simplifications. First, the attention can be limited only to the static behavior of the system, in response to considered perturbations ("*static*" *security*); additionally, the static model itself can be simplified by accepting the approximations described in (c) (i.e., "dc" model and assigned v/Q steady-state), and by assuming that:

- network losses are negligible;
- the v/Q steady-state remains unchanged even after perturbations[(8)].

By doing so, the relationships between active powers (injected at the terminal nodes and flowing through the branches) and voltage phase shifts (at all nodes) are simply defined by the equivalent circuit of Figure 2.6a or b, where:

- conductances have assigned values;
- active power injections (depending on generation and load, in the examined system and any external systems connected to it) have a zero sum, because of the hypothesis on losses.

With interconnected operations, the "equivalent injections" at the boundary nodes must be considered, and the possible "equivalent branches" between these nodes should be included in the circuit according to Section (c).

With reference to such an equivalent circuit, perturbations may be schematically classified into:

- perturbations of injected active power;
- structural perturbations inside the equivalent circuit;

according to the specifications below.

A *perturbation of injected active power* may occur at:

- a generation node (thermal or hydroelectric generation), as resulting from the casual trip of a generating unit or a whole power plant;

[(8)] Additionally, reactance variations are disregard, even if the perturbation leads to a frequency value different from the initial one. When the network configuration is changed, the unavoidable variations of voltage magnitudes at the internal nodes should be accounted for (see Fig. 2.7), with consequences on the limits $\pm F_{l\,\max}$ and various coefficients considered below.

- a load node, caused by load connection or disconnection (or to a continuous variation of demand);
- a boundary node, caused by perturbations concerning generation and/or load in the external systems (which cause a variation in the equivalent injection)(9).

A perturbation like the one considered here results in a redistribution of all injected powers (as their sum must still be zero), which is determined by the overall system characteristics as will be better specified in the following chapters. Such a redistribution causes a transient unbalance between the driving power and the generated power on each generating unit, with consequent speed variations followed by the intervention of the f/P control system (see Chapter 3) until a new steady-state is reached.

However, for the existence of the new steady-state, it is necessary that units under control are capable of generating the overall demanded variation of active power, staying respectively within driving power minimum and maximum limits (refer to the first part of conditions [2.2.3]). The constraint concerning the minimum limits is not of particular concern, as it could be activated as with a large, total or almost total, load rejection (and/or loss of the exported power), which can indeed be managed by intentionally disconnecting some generating units. Instead, the situation involving the maximum limits may be more critical, such as when a power plant disconnection or load connection (or loss of imported power) occurs, because the total set of operating units should not be excessive (due to economical requirements; see also Section 2.4.2b), and the start-up of new units can be difficult, if not impossible, in short times. Therefore, it is necessary to guarantee sufficient *spinning reserve*, with particular reference to units under control, so as to reduce — from the probabilistic point of view — the risk of power deficiency. More simply, one may impose a spinning reserve sufficiently larger than:

- the maximum power of the largest unit (and/or the maximum power imported through a single link);
- a given percentage (e.g., 2–5%) of the total load.

Additionally, the spinning reserve also must exhibit a proper geographical spread, not only for the considerations illustrated below, but also for operation when the system is split into two or more electrical islands (Section 1.7.1).

Actually, the new steady-state may not even exist because of the nonlinearities disregarded by the "dc" model. In particular, the transmissibility limits of active power (Section 1.5) are not considered by such a model.

---

(9) With a single boundary node, the opening of the interconnection has similar consequences to those caused by a power plant trip or a load rejection, according to the sense of the exchanged power before the opening itself. This also holds when other boundary nodes are present, provided they are not connected (through equivalent branches, as explained above) to the considered node; otherwise, the opening may be seen as a structural perturbation.

By assuming that the generic perturbation does not prevent the system from reaching a new steady-state, let us now examine the effects of a *trip* of a thermal generating unit with a *nonassigned generation*, with specific concern to power flows in different branches.

By adopting the notation P_{ti}, P_{ok}, P_{cj}, F_l previously defined for the values prior to the trip, and by denoting with an apex the values corresponding to the final steady-state, let us assume that the s-th generating unit is disconnected, so that $P'_{ts} = 0$.

The corresponding lack of the injection P_{ts} is balanced in the new steady-state by proper variations of injections at the remaining nodes (the sum of injected active powers is zero by hypothesis, and each single variation is determined by the f/P control).

Equations such as the following can be assumed:

$$\left.\begin{aligned} P'_{ti} - P_{ti} &\triangleq g_{is} P_{ts} \qquad (i = 1, \ldots, n_t; i \neq s) \\ P'_{ok} - P_{ok} &\triangleq g_{oks} P_{ts} \qquad (k = 1, \ldots, n_o) \\ -(P'_{cj} - P_{cj}) &\triangleq g_{cjs} P_{ts} \qquad (j = 1, \ldots, n_c) \end{aligned}\right\} \qquad [2.3.18]$$

with:

$$\sum_{i \neq s} g_{is} + \sum_k g_{oks} + \sum_j g_{cjs} = 1$$

where the values of the coefficients g_{is}, g_{oks}, g_{cjs} depend on the "static" model assumed for the system. If only the static effects of the primary f/P control are considered, which generally result in a network frequency variation (Section 3.3.1), such coefficients are determined by the primary regulators' parameters (at those nodes for which the primary control is activated) and dependence of loads on frequency. Instead, if the static effects of the secondary f/P control are considered (Section 3.3.2), with network frequency equal to the initial value, only the coefficients g_{is}, g_{oks} corresponding to powers under secondary control are not zero.

These coefficients may sometimes change depending on the initial steady-state (particularly on disconnected power), e.g., when, after the disconnection, some unit reaches its maximum power limit. The meaning of the above-mentioned coefficients is evident from the preceding equations. They may be called "*node-to-node redistribution*" *coefficients of the injected active power, from node s to the remaining nodes i, ok, cj*.

As a consequence of Equations [2.3.17] and [2.3.18], the following equation can be derived:

$$F'_l = F_l + f_{ls} P_{ts} \qquad [2.3.19]$$

with:

$$f_{ls} \triangleq \sum_{i \neq 1} a_{lti} g_{is} + \sum_k a_{lok} g_{oks} + \sum_j a_{lcj} g_{cjs} \qquad [2.3.20]$$

(assuming $g_{ss} = -1$ if $s \neq 1$), where F_l, F'_l are, respectively, the initial and final values of the active power flow in generic branch l.

For security reasons, it is necessary that the following new constraints (concerning the final conditions after the possible disconnection):

$$F'_l \in [-F_{l\,\max}, +F_{l\,\max}]$$

are added to the constraints already considered in section (c):

$$F_l \in [-F_{l\,\max}, +F_{l\,\max}]$$

The values corresponding to the "base" working point, found without model approximations but disregarding the above-mentioned constraints, are again denoted by a superscript "o" (as in Section (c)). The resulting active power flow in the branch l, at the final steady-state following the disconnection, is denoted by $F_l'^o$, so that $F_l'^o = F_l^o + f_{ls} P_{ts}^o$.

If F_l^o and/or $F_l'^o$ do not match their constraints, a "*redispatching*" procedure may be used, which is similar to that described in Section (c), by again adopting corrections $\Delta P_{ti} = P_{ti} - P_{ti}^o$ for $i = 2, \ldots, n_t$, and considering the additional constraints:

$$\Delta F'_l \in [-F_{l\,\max} - F_l'^o, +F_{l\,\max} - F_l'^o]$$

where, according to Equation [2.3.19], with f_{ls} constant:

$$F_l'^o = F_l^o + f_{ls} P_{ts}^o \qquad [2.3.21]$$

$$\Delta F'_l = \Delta F_l + f_{ls} \Delta P_{ts} \qquad [2.3.22]$$

with $\Delta F_l = \sum_{i \neq 1} a_{lti} \Delta P_{ti}$.

It further holds, because of the hypothesis of zero losses, that $\Delta P_{t1} = -\sum_{i \neq 1} \Delta P_{ti}$. Additionally, it must be $\Delta F_l \in [-F_{l\,\max} - F_l^o, +F_{l\,\max} - F_l^o]$.

By considering the expression for ΔF_l, Equation [2.3.22] also may be written as:

$$\Delta F'_l = \sum_{i \neq 1} b_{li} \Delta P_{ti} \qquad [2.3.23]$$

where:

- if the disconnection takes place at a node $s \neq 1$:

$$\begin{cases} b_{li} \triangleq a_{lti} & \text{for } i \neq s \\ b_{ls} \triangleq a_{lts} + f_{ls} = \sum_{h \neq 1,s} a_{lth} g_{hs} + \sum_k a_{lok} g_{oks} + \sum_j a_{lcj} g_{cjs} \end{cases}$$

- if the disconnection takes place at the node $s = 1$ (i.e., at the slack node):

$$b_{li} \triangleq a_{lti} - f_{ls}$$

However, Equation [2.3.22] is based on the assumption that f_{ls}, as well as the coefficients g_{is} etc., are invariant. Indeed, such coefficients also might vary, depending on the initial steady-state obtained by the redispatching.

The treatment appears to be easier in other cases, for which the perturbation concerns an *injected power of assigned value*, which does not undergo changes caused by possible redispatching (assigned generations, loads, equivalent injections). In fact, for such cases, an equation similar to Equation [2.3.19] again results, where the last term of the right side may now be considered assigned (comments on the invariance of f_{ls} should however be reminded), so that Equation [2.3.22] may be replaced by the simpler expression:

$$\Delta F_l' = \Delta F_l$$

which is Equation [2.3.23] with $b_{li} = a_{lti}$.

A *structural perturbation* inside the equivalent circuit is typically constituted by a branch opening between two nodes (terminal and/or internal nodes). In a real network, this may correspond to a line and/or transformer opening.

By assuming that the a-th branch is opened, the corresponding active power flow then changes from its initial value F_a to the final value $F_a' = 0$, whereas the final steady-state of injected active powers may be considered unchanged with respect to the initial one, i.e., $P_{ti}' = P_{ti}$, etc.

Actually, this perturbation generally causes variations in each injected power, according to the dynamic characteristics of generators, loads etc. However, these variations may be modest, and purely transient due to f/P control. In present conditions, the existence of the new steady-state—even if not compromised by minimum and maximum limits on driving power—should be still checked by using the full model, with particular regard to active power transmissibility limits and voltage supportability limits (Section 1.5).

By recalling Equation [2.1.27], the final value of the active power flow in the generic branch l is now given by:

$$F_l' = F_l + d_{la} F_a$$

where the coefficient d_{la} is the "*branch-to-branch redistribution*" *coefficient of the active power flow* (*from branch a to branch l*), that can be deduced by the equivalent circuit.

The treatment is similar to that of the previous cases, provided that Equations [2.3.21] and [2.3.22] are substituted by:

$$F_l'^o = F_l^o + d_{la} F_a^o, \quad \Delta F_l' = \Delta F_l + d_{la} \Delta F_a$$

where $\Delta F_l = \sum_{i \neq 1} a_{lti} \Delta P_{ti}$, $\Delta F_a = \sum_{i \neq 1} a_{ati} \Delta P_{ti}$ (it then follows Equation [2.3.23], with $b_{li} = a_{lti} + d_{la} a_{ati}$).

2.3.2. Economic Optimization of Thermoelectric Generation with Constraints on Variations Over Time

(a) Influence of the Limits on the Rate of Change for Active Power in Thermal Units

In the previous sections, it has been assumed that the system is represented by a purely static model, as if the system could be kept indefinitely at the generic steady-state. Generally, as powers absorbed by loads vary with time, the assumption of the static model implies that the steady-state can be instantaneously adapted to different situations, without any dynamic limitation.

However, active powers delivered by thermal plants cannot vary too fast, because of functional and/or safety reasons (such as thermal stresses). Thus it may seem reasonable to resort to a "dynamic" model that includes at least such limitations, if fast enough load variations occur.

The resulting dispatching is called "dynamic dispatching," in opposition to the "static dispatching" considered so far. However, to avoid misunderstanding, in both cases the dispatching is the previsional type, starting from load forecasts covering a given time interval, and the difference is only related to the model type. More precisely, with the static model, the dispatching problem becomes "instantaneous" unrelated problems, whereas by using the dynamic model, only a single *dispatching "over time"* problem must be solved, the final solution of which depends, at each time, also on past and future load demands.

To formulate the effect of the considered limitations, let us assume:

$$\frac{\mathrm{d}P_{ti}}{\mathrm{d}t} = u_i \in [u_{i\,\min}, u_{i\,\max}] \quad (i = 1, \ldots, n_t)$$

where $u_{i\,\min} < 0$, $u_{i\,\max} > 0$ (for instance $|u_{i\,\min}| = u_{i\,\max} = kP_{ti\,\max}$, with k in the order of a few percent per minute); and assume, for simplicity, that limits $u_{i\,\min}, u_{i\,\max}$ may be actually activated only for the l-th unit.

Denoting by $[0, T]$ the considered time interval, and using the notation defined in Section 2.3.1, the problem may be formulated as follows:

$$\min \int_0^T \sum_i C_i(P_{ti}(t))\,\mathrm{d}t$$

with the following constraints ($\forall t \in [0, T]$):

$$0 = P_c(t) + p(t) - P_o(t) - \sum_i P_{ti}(t)$$

$$0 = \frac{\mathrm{d}P_{tl}}{\mathrm{d}t}(t) - u_l(t)$$

$$P_{ti}(t) \in [P_{ti\,\min}, P_{ti\,\max}] \quad (i = 1, \ldots, n_t)$$

$$u_l(t) \in [u_{l\,\min}, u_{l\,\max}]$$

where the initial value $P_{tl}(0)$ is assumed to be known, whereas losses p are expressed, at each time, by Equation [2.3.4]. (For simplicity, limits on currents and security constraints are not considered here; moreover *the system configuration* — operating generators etc. — *is assumed not to change* for the whole considered time interval.)

By subdividing the time interval into a sufficiently large number R of elementary intervals of duration $\tau \triangleq T/R$, the following "discrete" formulation may be deduced (which may be compared in an easier way with what is previously seen, and includes the total load forecasts in their discrete form $P_c^{(1)}, \ldots, P_c^{(R)}$, which is the form usually available):

$$\min \sum_r \sum_i C_i(P_{ti}^{(r)})\tau$$

with the constraints ($\forall r = 1, \ldots, R$):

$$0 = P_c^{(r)} + p^{(r)} - P_o^{(r)} - \sum_i P_{ti}^{(r)}$$

$$0 = P_{tl}^{(r)} - P_{tl}^{(r-1)} - u_l^{(r)}\tau$$

$$P_{ti}^{(r)} \in [P_{ti\,\min}, P_{ti\,\max}] \quad (i = 1, \ldots, n_t)$$

$$u_l^{(r)} \in [u_{l\,\min}, u_{l\,\max}]$$

where $P_{tl}^{(0)} \triangleq P_{tl}(0)$ is known (whereas $p^{(r)}$ depends on $P_{t2}^{(r)}, \ldots, P_{tn_t}^{(r)}, \ldots$ according to Equation [2.3.4]).

With the previously stated assumptions, a Lagrangian function of the type $\mathcal{L} = \sum_r \mathcal{L}^{(r)}\tau$ may then be assumed, with:

$$\mathcal{L}^{(r)} \triangleq \sum_i C_i(P_{ti}^{(r)}) + \lambda^{(r)}\Big(P_c^{(r)} + p^{(r)} - P_o^{(r)} - \sum_i P_{ti}^{(r)}\Big) + \sum_i \lambda_i'^{(r)}(P_{ti\,\min} - P_{ti}^{(r)})$$
$$+ \sum_i \lambda_i''^{(r)}(P_{ti}^{(r)} - P_{ti\,\max}) + \beta_l^{(r)}(P_{tl}^{(r)} - P_{tl}^{(r-1)} - u_l^{(r)}\tau)$$
$$+ [\beta_l'^{(r)}(u_{l\,\min} - u_l^{(r)}) + \beta_l''^{(r)}(u_l^{(r)} - u_{l\,\max})]\tau$$

Such Lagrangian function — by assuming assigned (besides $P_{tl}^{(0)} = P_{tl}(0)$) $P_{ok}, P_{cj}, Q_{cj}, v_h$ at the different time intervals — depends on $P_{t1}^{(r)}, \ldots, P_{tn_t}^{(r)}, u_l^{(r)}$ $(r = 1, \ldots, R)$.

By setting to zero the partial derivatives of $\mathcal{L}$ with respect to $P_{t1}^{(r)}, \ldots, P_{tn_t}^{(r)}$, the following conditions may be derived:

- for $i \neq l$:

$$0 = \frac{\partial \mathcal{L}^{(r)}}{\partial P_{ti}^{(r)}} = \left(\frac{\mathrm{d}C_i}{\mathrm{d}P_{ti}}\right)^{(r)} - \lambda^{(r)}\left(1 - \left(\frac{\partial \phi}{\partial P_{ti}}\right)^{(r)}\right) - \lambda_i'^{(r)} + \lambda_i''^{(r)}$$

$$= c_i^{(r)} - \lambda^{(r)}\left(1 - \left(\frac{\partial \phi}{\partial P_{ti}}\right)^{(r)}\right) \quad (r = 1, \ldots, R)$$

- for $i = l$:

$$\begin{cases} 0 = \dfrac{\partial \mathcal{L}}{\tau \partial P_{tl}^{(r)}} = \left(\dfrac{\mathrm{d}C_l}{\mathrm{d}P_{tl}}\right)^{(r)} - \lambda^{(r)}\left(1 - \left(\dfrac{\partial \phi}{\partial P_{tl}}\right)^{(r)}\right) \\ \quad -\lambda_l'^{(r)} + \lambda_l''^{(r)} + \beta_l^{(r)} - \beta_l^{(r+1)} \\ = c_l^{(r)} - \lambda^{(r)}\left(1 - \left(\dfrac{\partial \phi}{\partial P_{tl}}\right)^{(r)}\right) + \beta_l^{(r)} - \beta_l^{(r+1)} \quad (r = 1, \ldots, R-1) \\ 0 = \dfrac{\partial \mathcal{L}^{(R)}}{\partial P_{tl}^{(R)}} = \left(\dfrac{\mathrm{d}C_l}{\mathrm{d}P_{tl}}\right)^{(R)} - \lambda^{(R)}\left(1 - \left(\dfrac{\partial \phi}{\partial P_{tl}}\right)^{(R)}\right) - \lambda_l'^{(R)} + \lambda_l''^{(R)} + \beta_l^{(R)} \\ = c_l^{(R)} - \lambda^{(R)}\left(1 - \left(\dfrac{\partial \phi}{\partial P_{tl}}\right)^{(R)}\right) + \beta_l^{(R)} \end{cases}$$

or equivalently, with $i \neq l$:

$$\begin{cases} \dfrac{c_i^{(r)}}{1 - \left(\dfrac{\partial \phi}{\partial P_{ti}}\right)^{(r)}} = \dfrac{c_l^{(r)} + \beta_l^{(r)} - \beta_l^{(r+1)}}{1 - \left(\dfrac{\partial \phi}{\partial P_{tl}}\right)^{(r)}} = \lambda^{(r)} \quad (r = 1, \ldots, R-1) \\ \dfrac{c_i^{(R)}}{1 - \left(\dfrac{\partial \phi}{\partial P_{ti}}\right)^{(R)}} = \dfrac{c_l^{(R)} + \beta_l^{(R)}}{1 - \left(\dfrac{\partial \phi}{\partial P_{tl}}\right)^{(R)}} = \lambda^{(R)} \end{cases} \qquad [2.3.24]$$

where $c_1^{(r)}, \ldots, c_{n_t}^{(r)}$ are the incremental costs at generation defined by Equations [2.3.3], whereas $(\partial \phi / \partial P_{t1})^{(r)} = 0$.

Additionally, by setting to zero the partial derivatives of $\mathcal{L}$ with respect to $u_l^{(r)}$, it may be derived:

$$0 = \frac{\partial \mathcal{L}^{(r)}}{\partial u_l^{(r)}} = (-\beta_l^{(r)} - \beta_l'^{(r)} + \beta_l''^{(r)})\tau \quad (r = 1, \ldots, R)$$

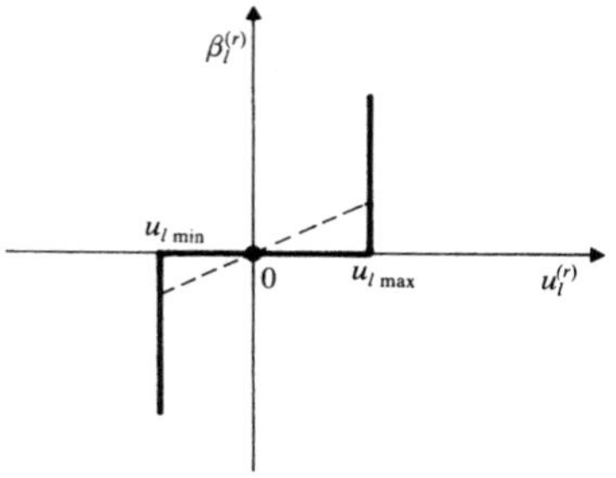

Figure 2.19. Dependence of the Lagrangian multiplier $\beta_l^{(r)}$ on rate of change $u_l^{(r)}$ for power of the l-th unit.

from which, by considering the "excluding" conditions on $\beta_l'^{(r)}$, $\beta_l''^{(r)}$:

$$\beta_l^{(r)} = -\beta_l'^{(r)} + \beta_l''^{(r)} \begin{cases} = -\beta_l'^{(r)} \leq 0 & \text{if } u_l^{(r)} = u_{l\,\min} \\ = 0 & \text{if } u_l^{(r)} \in (u_{l\,\min}, u_{l\,\max}) \\ = \beta_l''^{(r)} \geq 0 & \text{if } u_l^{(r)} = u_{l\,\max} \end{cases} \qquad [2.3.25]$$

according to Figure 2.19.

For dynamic conditions, the generation costs per unit of time can be considered slightly higher (for given P_{ti}'s) than those considered up to now. For example, we may assume the expression $C_i(P_{ti}) + C_i'(u_i)$, where the generic term C_i' is an additional cost (which is zero for $u_i = 0$).

By considering this, the expression of $\mathcal{L}^{(r)}$ becomes:

$$\begin{aligned} \mathcal{L}^{(r)} = & \sum_i (C_i(P_{ti}^{(r)}) + C_i'(u_i^{(r)})) + \lambda^{(r)} \left(P_c^{(r)} + p^{(r)} - P_o^{(r)} - \sum_i P_{ti}^{(r)} \right) \\ & + \sum_i \lambda_i'^{(r)} (P_{ti\,\min} - P_{ti}^{(r)}) + \sum_i \lambda_i''^{(r)} (P_{ti}^{(r)} - P_{ti\,\max}) \\ & + \sum_i \beta_i^{(r)} (P_{ti}^{(r)} - P_{ti}^{(r-1)} - u_i^{(r)} \tau) + \left[\beta_l'^{(r)} (u_{l\,\min} - u_l^{(r)}) + \beta_l''^{(r)} (u_l^{(r)} - u_{l\,\max}) \right] \tau \end{aligned}$$

so that $\mathcal{L}^{(r)}$ also depends on all the $u_i^{(r)}$ with $i \neq l$. Then, for all $i = 1, \ldots, n_t$, it can be derived:

$$\begin{cases} \dfrac{c_i^{(r)} + \beta_i^{(r)} - \beta_i^{(r+1)}}{1 - \left(\dfrac{\partial \phi}{\partial P_{ti}} \right)^{(r)}} = \lambda^{(r)} \quad (r = 1, \ldots, R-1) \\ \dfrac{c_i^{(R)} + \beta_i^{(R)}}{1 - \left(\dfrac{\partial \phi}{\partial P_{ti}} \right)^{(R)}} = \lambda^{(R)} \end{cases}$$

additionally:

$$\begin{cases} \beta_i^{(r)} = \left(\dfrac{dC_i'}{du_i}\right)^{(r)} \cdot \dfrac{1}{\tau} & (i \neq l, r = 1, \ldots, R) \\ \beta_l^{(r)} = \left(\dfrac{dC_l'}{du_l}\right)^{(r)} \cdot \dfrac{1}{\tau} - \beta_l'^{(r)} + \beta_l''^{(r)} & (r = 1, \ldots, R) \end{cases}$$

More specifically, by assuming C_i' proportional to u_i^2, the generic $\beta_i^{(r)}$ is proportional to $u_i^{(r)}$; and this holds true also for $i = l$, within the limits $u_{l\min}$, $u_{l\max}$; see the dashed line in Figure 2.19.

By neglecting, for simplicity, loss variations, conditions [2.3.24] can be rewritten as follows ($r = 1, \ldots, R$):

- $c_i^{(r)} = \lambda^{(r)}$ for all $i \neq l$: this means the incremental costs at generation for all units with $i \neq l$, are equal, thus allowing the definition of a single "equivalent characteristic" (λ, $\sum_{i \neq l} P_{ti}$) for the whole set of such units, according to Section 2.3.1a. Additionally, since $\sum_{i \neq l} P_{ti}^{(r)} = P_c^{(r)} + p - P_o^{(r)} - P_{tl}^{(r)}$, with $P_c^{(r)}$, p, $P_o^{(r)}$ known, the generic value $\lambda^{(r)}$ can be determined from $P_{tl}^{(r)}$ or vice versa.
- $\lambda^{(r)} - c_l^{(r)} = \beta_l^{(r)} - \beta_l^{(r+1)}$ (with the condition $\beta_l^{(R+1)} = 0$), where the left-side term can be deduced from $P_{tl}^{(r)}$ or vice versa.

Furthermore, it must be considered that $P_{tl}^{(r)} = P_{tl}^{(r-1)} + u_l^{(r)}\tau$, where $P_{tl}^{(0)}$ is assigned, whereas $\beta_l^{(r)}$ depends on $u_l^{(r)}$ according to Equations [2.3.25], i.e., according to the characteristic of Figure 2.19.

The block diagram of Figure 2.20 (with $r = 1, \ldots, R$) then applies, which shows that all variables under consideration finally depend on $\beta_l^{(R+1)} = 0$ and on $P_{tl}^{(0)}$.

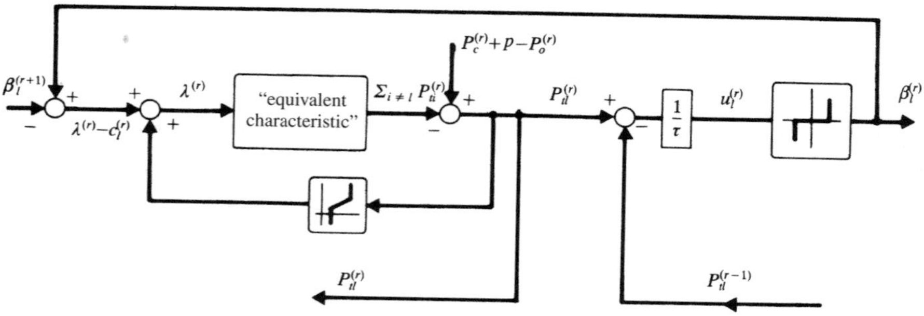

Figure 2.20. A computational block diagram in the presence of rate of change limits for the power of the l-th unit (under the assumption of constant losses).

For example, if the costs per unit of time are quadratic functions of delivered powers (of the type $C_i = C_{io} + a_i P_{ti} + b_i(P_{ti}^2/2)$ as already seen), and if it is assumed that limits $P_{ti\,\min}$, $P_{ti\,\max}$ are not reached, it follows:

$$\lambda = a + b\sum_{i \neq l} P_{ti} = a + b(P_c + p - P_o - P_{tl})$$

$$c_l = a_l + b_l P_{tl}$$

with a, b, a_l, b_l as proper (positive) constants; it then results:

$$\lambda^{(r)} - c_l^{(r)} = a - a_l + b(P_c^{(r)} + p - P_o^{(r)}) - (b + b_l)P_{tl}^{(r)}$$

Without constraints on u_l, the optimal solution would be defined, for each $r = 1, \ldots, R$, by $\beta_l^{(r)} = 0$, $\lambda^{(r)} = c_l^{(r)}$, and consequently, it would be:

$$P_{tl}^{(r)} = \frac{a - a_l + b(P_c^{(r)} + p - P_o^{(r)})}{b + b_l} \triangleq P_{tl}^{(r)o} \qquad [2.3.26]$$

as with R purely instantaneous dispatchings.

Now assume, for simplicity, that the number of elementary intervals is $R = 3$; furthermore:

$$P_c^{(1)} + p - P_o^{(1)} = P', \quad P_c^{(2)} + p - P_o^{(2)} = P_c^{(3)} + p - P_o^{(3)} = P' + \alpha\tau \quad (\alpha > o)$$

(see Fig. 2.21a), with the initial condition $P_{tl}^{(0)} = (a - a_l + bP')/(b + b_l)$ corresponding, for the given P', to the optimal value when disregarding constraints on u_l.

Equation [2.3.26] leads to:

$$P_{tl}^{(1)} = P_{tl}^{(0)} = P_{tl}^{(1)o}, \quad P_{tl}^{(2)} = P_{tl}^{(3)} = P_{tl}^{(0)} + \frac{b\alpha\tau}{b + b_l} = P_{tl}^{(2)o} = P_{tl}^{(3)o}$$

(Fig. 2.21b, case 1), but such a solution, which corresponds to three instantaneous dispatchings, is seen as optimal if and only if $b\alpha/(b + b_l) = u_l^{(2)} \leq u_{l\,\max}$; otherwise, it cannot be achieved because it requires too large a value of $u_l^{(2)}$.

More precisely:

- If $b\alpha/(b + b_l) \in (u_{l\,\max}, 3u_{l\,\max})$, the optimal solution is that shown in Figure 2.21b, case 2. The results of the instantaneous dispatching are valid only at the final point $r = 3$ (with $P_{tl}^{(3)} = P_{tl}^{(0)} + b\alpha\tau/(b + b_l) = P_{tl}^{(3)o}$), but not at the intermediate points $r = 1, 2$, where $P_{tl}^{(1)} > P_{tl}^{(1)o}$ and $P_{tl}^{(2)} < P_{tl}^{(2)o}$, with $u_l^{(2)} = u_{l\,\max}$, $u_l^{(1)} = u_l^{(3)} = (b\alpha/(b + b_l) - u_{l\,\max})/2$. Specifically, note the "*forward type*" *correction on* $P_{tl}^{(1)}$, *caused by the forecast of the following increase of* $(P_c + p - P_o)$.

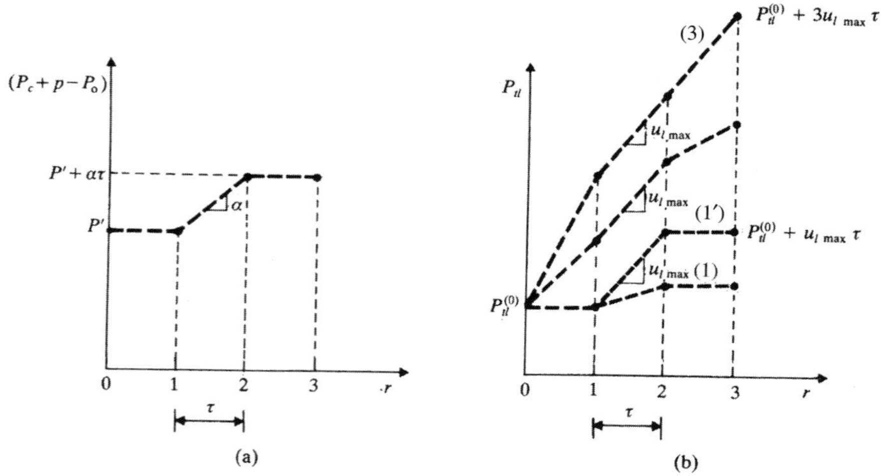

Figure 2.21. An example of dispatching "over time" ($R = 3$ elementary intervals), as influenced by limits on rate of change for the power of the l-th unit: (a) total active power demanded; (b) power of the l-th unit. (1) $b\alpha/(b + b_l) < u_{l\,\max}$, (1′) $b\alpha/(b + b_l) = u_{l\,\max}$, (2) $b\alpha/(b + b_l) \in (u_{l\,\max}, 3u_{l\,\max})$, (3) $b\alpha/(b + b_l) \geq 3u_{l\,\max}$.

- If $b\alpha/(b + b_l) \geq 3u_{l\,\max}$, the optimal solution is that in the same figure, case 3, with $u_l^{(1)} = u_l^{(2)} = u_l^{(3)} = u_{l\,\max}$. That is equivalent to saying that P_{tl} grows at its maximum rate in all the elementary intervals, until reaching the value $P_{tl}^{(3)} = P_{tl}^{(0)} + 3u_{l\,\max}\tau$ (which is no longer consistent with the instantaneous dispatching, unless it were $b\alpha/(b + b_l) = 3u_{l\,\max}$).

The dependence of $u_l^{(1)}, u_l^{(2)}, u_l^{(3)}$ on $b\alpha/(b + b_l)$ is illustrated in Figure 2.22a. The above-reported equations enable easy identification of each value $P_{tl}^{(r)}, \lambda^{(r)} - c_l^{(r)}, \beta_l^{(r)}$ for the different cases.

Recalling Section 2.3.1a for the case of two units, the *overcost* caused by constraint $u_l \leq u_{l\,\max}$ is given by:

$$S = \sum_r \left[\sum_i C_i\left(P_{ti}^{(r)}\right) - \sum_i C_i\left(P_{ti}^{(r)o}\right) \right] \tau = \frac{b + b_l}{2}\tau \sum_r \left(P_{tl}^{(r)} - P_{tl}^{(r)o}\right)^2$$

$$= \frac{1}{2(b + b_l)}\tau \sum_r \left(\lambda^{(r)} - c_l^{(r)}\right)^2$$

which depends on $b\alpha/(b + b_l)$, according to Figure 2.22b.

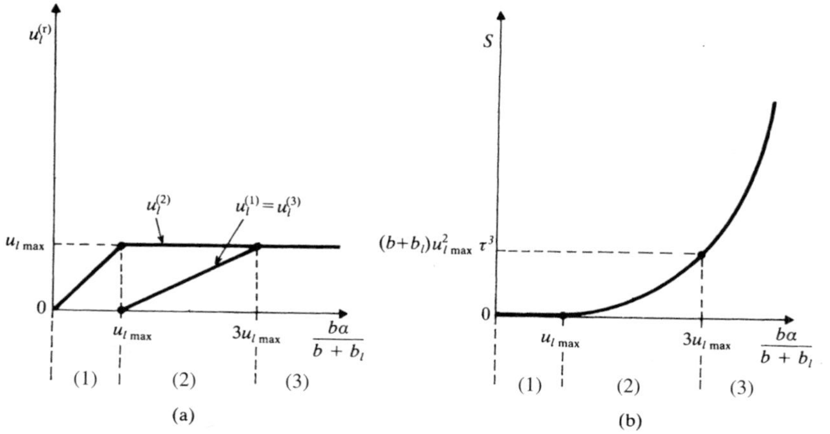

Figure 2.22. Example of Figure 2.21; effect of $b\alpha/(b + b_l)$ on: (a) rate of change u_l for the power of the l-th unit; (b) over-cost S due to the limit on the rate of change.

(b) Influence of Constraints Concerning the Fuel Consumption

In the previous treatment, it has been assumed that fuel flows used by the different units can be chosen, at any instant, with no limitation. Actually, such flows are related to other variables according to a model, which may result in further constraints on variations over time, accounting for the conditions of fuel acquisition and storage, the type of supplying contracts, environmental requirements (limits on pollution, etc.), and so on.

Consider an interval $[0, T]$ of proper duration (which may vary from case to case, and may also be quite different from that considered in Section (a)) and assume, for simplicity, that such constraints arise just for the l-th power plant and concern a single fuel.

The following problems are particularly meaningful (although they are defined very schematically), with possible generalizations.

Problem 1 (Fig. 2.23a):

- the law $q_{al}(t)$, according to which the fuel flows to the power plant in the interval $[0, T]$, is known (q_{al} is just the incoming fuel flow);
- it is possible to store the fuel, according to the law:

$$\frac{\mathrm{d}V_l}{\mathrm{d}t} = q_{al} - q_l$$

with $V_l(0)$ known (V_l denotes the quantity of stored fuel, and q_l the used fuel flow, which depends on P_{tl});

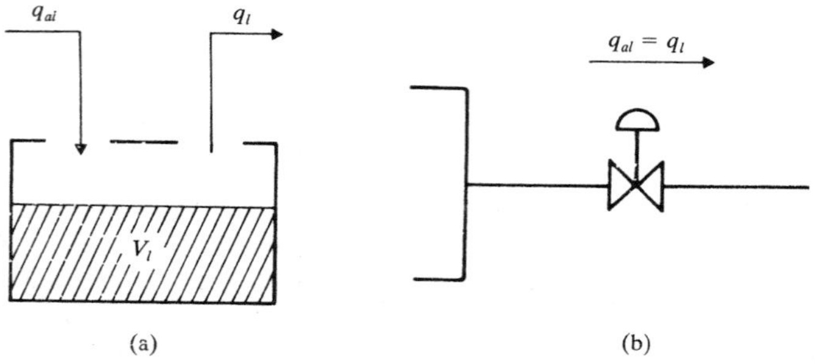

Figure 2.23. Examples of dispatching "over time" caused by: (a) limits on fuel storage (problem 1); (b) minimum amount of fuel to be used (problem 2).

- the quantity V_l must satisfy, at each instant, the constraints:

$$V_l \in [V_{l\,\min}, V_{l\,\max}]$$

Problem 2 (Fig. 2.23b):

- there is no storage capability, so that $q_l(t) = q_{al}(t)$;
- the supply contract implies, for the interval $[0, T]$, a minimum cost $G_{l\,\min}$ which corresponds to the acquisition of a given quantity of fuel $W_{l\,\min}$, so that at least this quantity should be used:

$$W_l \triangleq \int_o^T q_l \,\mathrm{d}t = \int_o^T q_{al} \,\mathrm{d}t \geq W_{l\,\min}$$

- furthermore, any possible surplus $(W_l - W_{l\,\min})$ implies an additional cost $k'(W_l - W_{l\,\min})$.

Both of the above-defined problems may be translated into dispatching "over time" problems, even if the power system is represented by a static model (as we will assume, disregarding the limits on $\mathrm{d}P_{ti}/\mathrm{d}t$; additionally, the system configuration will be assumed unchanged over the entire interval $[0, T]$).

In *Problem 1*, with notation similar to those used in Section (a), a Lagrangian function $\mathcal{L} = \sum_r \mathcal{L}^{(r)}\tau$ may be adopted, with:

$$\mathcal{L}^{(r)} \triangleq \sum_i C_i(P_{ti}^{(r)}) + \lambda^{(r)}\left(P_c^{(r)} + p^{(r)} - P_o^{(r)} - \sum_i P_{ti}^{(r)}\right) + \sum_i \lambda_i'^{(r)}(P_{ti\,\min} - P_{ti}^{(r)}) + \sum_i \lambda_i''^{(r)}(P_{ti}^{(r)} - P_{ti\,\max})$$

$$+ \gamma_l^{(r)} \left[\frac{V_l^{(r)} - V_l^{(r-1)}}{\tau} - q_{al}^{(r)} + q_l(P_{tl}^{(r)}) \right]$$

$$+ [\gamma_l'^{(r)}(V_{l\,\min} - V_l^{(r)}) + \gamma_l''^{(r)}(V_l^{(r)} - V_{l\,\max})]/\tau$$

which depends on $P_{t1}^{(r)}, \ldots, P_{tn_t}^{(r)}, V_l^{(r)} (r = 1, \ldots, R)$. ($V_l^{(0)} \triangleq V_l(0)$ is known, as well as all the values of $q_{al}, P_{ok}, P_{cj}, Q_{cj}, v_h$; whereas losses p can be derived by using Equation [2.3.4]).

By setting to zero the partial derivatives of $\mathcal{L}$ with respect to $P_{t1}^{(r)}, \ldots, P_{tn_t}^{(r)}$, the following conditions are then derived ($r = 1, \ldots, R$):

- for $i \neq l$:

$$0 = \frac{\partial \mathcal{L}^{(r)}}{\partial P_{ti}^{(r)}} = \left(\frac{\mathrm{d}C_i}{\mathrm{d}P_{ti}}\right)^{(r)} - \lambda^{(r)} \left(1 - \left(\frac{\partial \phi}{\partial P_{ti}}\right)^{(r)}\right) - \lambda_i'^{(r)} + \lambda_i''^{(r)}$$

$$= c_i^{(r)} - \lambda^{(r)} \left(1 - \left(\frac{\partial \phi}{\partial P_{ti}}\right)^{(r)}\right)$$

- for $i = l$:

$$0 = \frac{\partial \mathcal{L}^{(r)}}{\partial P_{tl}^{(r)}} = \left(\frac{\mathrm{d}C_l}{\mathrm{d}P_{tl}}\right)^{(r)} - \lambda^{(r)} \left(1 - \left(\frac{\partial \phi}{\partial P_{tl}}\right)^{(r)}\right) - \lambda_l'^{(r)} + \lambda_l''^{(r)}$$

$$+ \gamma_l^{(r)} \left(\frac{\mathrm{d}q_l}{\mathrm{d}P_{tl}}\right)^{(r)} = c_l^{(r)} - \lambda^{(r)} \left(1 - \left(\frac{\partial \phi}{\partial P_{tl}}\right)^{(r)}\right) + \gamma_l^{(r)} \left(\frac{\mathrm{d}q_l}{\mathrm{d}P_{tl}}\right)^{(r)}$$

that is to say, with $i \neq l$:

$$\frac{c_i^{(r)}}{1 - \left(\dfrac{\partial \phi}{\partial P_{ti}}\right)^{(r)}} = \frac{c_l^{(r)} + \gamma_l^{(r)} \left(\dfrac{\mathrm{d}q_l}{\mathrm{d}P_{tl}}\right)^{(r)}}{1 - \left(\dfrac{\partial \phi}{\partial P_{tl}}\right)^{(r)}} = \lambda^{(r)} \quad (r = 1, \ldots, R) \qquad [2.3.27]$$

where $c_1^{(r)}, \ldots, c_{n_t}^{(r)}$ are the incremental costs at generation defined by Equations [2.3.3], whereas $(\partial \phi / \partial P_{t1})^{(r)} = 0$.

By assuming $C_l = kq_l + h$, with k, h as proper constants (k is the unit cost of fuel at the l-th power plant), it holds:

$$\left(\frac{dC_l}{dP_{tl}}\right)^{(r)} + \gamma_l^{(r)}\left(\frac{dq_l}{dP_{tl}}\right)^{(r)} = \left(k + \gamma_l^{(r)}\right)\left(\frac{dq_l}{dP_{tl}}\right)^{(r)}$$

so that the generic value $\gamma_l^{(r)}$ may be viewed as an *"equivalent" variation of the unit cost of fuel*, due to the possible activation of constraints on V_l (see the following text).

Furthermore, by setting to zero the partial derivatives of $\mathcal{L}$ with respect to $V_l^{(r)}$, it follows:

$$\begin{cases} 0 = \dfrac{\partial \mathcal{L}}{\tau \partial V_l^{(r)}} = (\gamma_l^{(r)} - \gamma_l^{(r+1)} - \gamma_l'^{(r)} + \gamma_l''^{(r)})/\tau & (r = 1, \ldots, R-1) \\ 0 = \dfrac{\partial \mathcal{L}^{(R)}}{\partial V_l^{(R)}} = (\gamma_l^{(R)} - \gamma_l'^{(R)} + \gamma_l''^{(R)})/\tau \end{cases}$$

(this last condition would disappear if $V_l^{(R)}$ is assigned), from which, by considering the "excluding" conditions on $\gamma_l'^{(r)}$, $\gamma_l''^{(r)}$:

$$\gamma_l^{(r+1)} \begin{cases} = \gamma_l^{(r)} - \gamma_l'^{(r)} \leq \gamma_l^{(r)} & \text{if } V_l^{(r)} = V_{l\,\min} \\ = \gamma_l^{(r)} & \text{if } V_l^{(r)} \in (V_{l\,\min}, V_{l\,\max}) \\ = \gamma_l^{(r)} + \gamma_l''^{(r)} \geq \gamma_l^{(r)} & \text{if } V_l^{(r)} = V_{l\,\max} \end{cases} \qquad [2.3.28]$$

($r = 1, \ldots, R$), according to Figure 2.24, with $\gamma_l^{(R+1)} = 0$.

If constraints on V_l are not activated, it can be deduced $\gamma_l^{(1)} = \cdots = \gamma_l^{(R)} = 0$, and the optimal solution is constituted by R purely instantaneous dispatchings.

Otherwise, the optimal solution implies, generally, the definition of m subintervals. Within each of these subintervals the value $\gamma_l^{(r)}$ remains constant

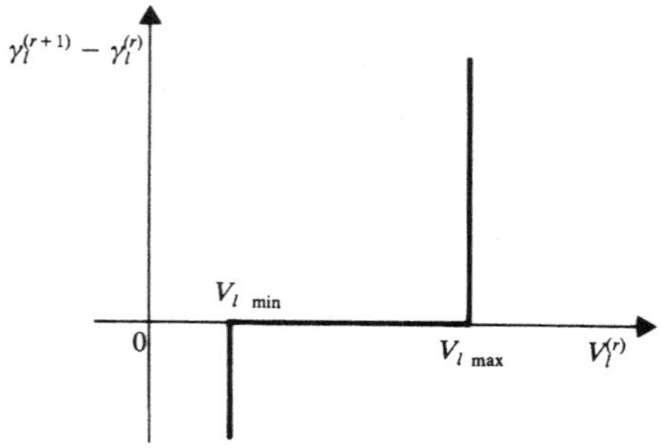

Figure 2.24. Dependence of the variation of the Lagrange multiplier γ_l on the quantity V_l of stored fuel (problem 1).

($\gamma_{l(1)}, \ldots, \gamma_{l(m)}$, respectively). By indicating with $V_{l(0)}, V_{l(1)}, \ldots, V_{l(m)}$ the values assumed by V_l at the extremes of such subintervals, it results $V_{l(s)} = V_{l\,\min}$ or $V_{l(s)} = V_{l\,\max}$ for $s = 1, \ldots, m-1$, whereas $V_{l(0)} = V_l^{(0)}$ is assigned and $V_{l(m)} = V_l^{(R)} \in [V_{l\,\min}, V_{l\,\max}]$ is the value reached at $t = T$.

The treatment must be modified if, besides $V_l(0)$, the final value $V_l^{(R)} = V_l(T)$ is also assigned (or even, since $\int_0^T q_{al}\,dt$ is known, if the quantity $W_l \triangleq \int_0^T q_l\,dt$ of fuel to be used is assigned). In this case, similar to that discussed in Section 2.3.3b:

- the condition $\gamma_l^{(R+1)} = 0$ (which is a consequence of $0 = \partial\mathcal{L}^{(R)}/\partial V_l^{(R)}$) needs no longer to be imposed;
- the value of $\gamma_l^{(R)}$ can be determined starting from $P_{tl}^{(R)}$, or from $(V_l^{(R)} - V_l^{(R-1)})$ (for instance, the diagram of the following Figure 2.25 changes in an obvious way, for $r = R$).

In a generic subinterval, the final quantity of stored fuel $V_{l(s)}$ increases with $\gamma_{l(s)}$ (see Fig. 2.25), i.e., with the "equivalent" unit fuel cost. Thus $\gamma_{l(s)}$ must be increased — as if fuel were more expensive — if the fuel consumption for the given subinterval is excessive (accounting for limits on V_l), and must be decreased in the opposite case.

By disregarding, for simplicity, the dependence of losses on generated power dispatching, from conditions [2.3.27] it can be derived ($r = 1, \ldots, R$):

- $c_i^{(r)} = \lambda^{(r)}$ for all $i \neq l$. Consequently, it is possible to reference the "equivalent characteristic" ($\lambda, \sum_{i \neq l} P_{ti}$), then derive $\lambda^{(r)}$ from $P_{tl}^{(r)}$ (with $P_c^{(r)}$, p, $P_o^{(r)}$ assigned) or vice versa.

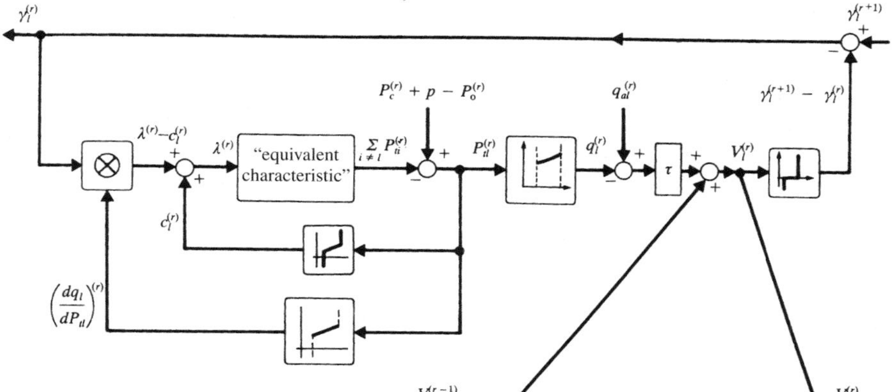

Figure 2.25. A computational block diagram for problem 1 (under the assumption of constant losses).

- $(\lambda^{(r)} - c_l^{(r)})/(\mathrm{d}q_l/\mathrm{d}P_{tl})^{(r)} = \gamma_l^{(r)}$, where the left side can be derived from $P_{tl}^{(r)}$ or vice versa.

Furthermore, it must be considered that $V_l^{(r)} = V_l^{(r-1)} + [q_{al}^{(r)} - q_l(P_{tl}^{(r)})]\tau$, where $V_l^{(0)}$ is assigned, whereas $(\gamma_l^{(r+1)} - \gamma_l^{(r)})$ (with the condition $\gamma_l^{(R+1)} = 0$) depends on $V_l^{(r)}$ according to Equations [2.3.28], i.e., according to the characteristic shown in Figure 2.24.

The block diagram in Figure 2.25 (with $r = 1, \ldots, R$) then holds, which shows that all the considered variables become dependent on $\gamma_l^{(R+1)} = 0$ and $V_l^{(0)}$.

As in Section (a), the subsequent values $(\lambda^{(r)} - c_l^{(r)})$ allow (under the adopted assumptions) the evaluation of *overcost* caused by constraints.

The problem becomes much simpler if, in the considered time interval, no fuel supply is scheduled, so that $q_{al}(t) = 0\ \forall t \in [0, T]$. In such a case, in fact, it holds $(\mathrm{d}V_l/\mathrm{d}t)(t) = -q_l(t) < 0$, and consequently only the limit $V_{l\,\min}$ may be considered, and only at $t = T$. With $W_l \triangleq \int_0^T q_l\,\mathrm{d}t$ as the consumed quantity of fuel, the constraint $V_l(T) = V_l(0) - W_l \geq V_{l\,\min}$ also can be written as $W_l \leq W_{l\,\max}$, with $W_{l\,\max} \triangleq V_{l(0)} - V_{l\,\min}$. As an alternative, by assuming $C_l = kq_l + h$, it also can be written $G_l \triangleq \int_0^T C_l\,\mathrm{d}t \leq k(V_l(0) - V_{l\,\min}) + hT$. It certainly follows:

$$\gamma_l'^{(1)} = \cdots = \gamma_l'^{(R-1)} = 0$$

$$\gamma_l''^{(1)} = \cdots = \gamma_l''^{(R)} = 0$$

and the solution implies a single subinterval ($m = 1$), with:

$$\gamma_l^{(1)} = \cdots = \gamma_l^{(R)} = \gamma_l'^{(R)} \begin{cases} = 0 & \text{if } V_l^{(R)} > V_{l\,\min} \\ \geq 0 & \text{if } V_l^{(R)} = V_{l\,\min} \end{cases}$$

where $V_l^{(R)} = V_l(T) = V_l(0) - \sum_r q_l(P_{tl}^{(r)})\tau$, whereas the Lagrangian function is reduced to:

$$\mathcal{L} = \sum_r \left[\sum_i C_i(P_{ti}^{(r)}) + \lambda^{(r)} \left(P_c^{(r)} + p^{(r)} - P_o^{(r)} - \sum_i P_{ti}^{(r)} \right) + \sum_i \lambda_i'^{(r)} (P_{ti\,\min} - P_{ti}^{(r)}) \right.$$
$$\left. + \sum_i \lambda_i''^{(r)} (P_{ti}^{(r)} - P_{ti\,\max}) \right] \tau + \gamma_l'^{(R)} \left(V_{l\,\min} - V_l^{(0)} + \sum_r q_l(P_{tl}^{(r)})\tau \right) \qquad [2.3.29]$$

If the assumption $\gamma_l'^{(R)} = 0$ leads to $V_l^{(R)} < V_{l\,\min}$, it must be imposed $V_l^{(R)} = V_{l\,\min}$, by deriving $\gamma_l'^{(R)}$. This last value is equivalent to an increase—constant all over the interval $[0, T]$—of the unit cost of fuel.

If, instead, $V_l(T)$ is assigned (as well as the quantity W_l), it follows:

$$\gamma_l'^{(1)} = \cdots = \gamma_l'^{(R)} = 0$$

$$\gamma_l''^{(1)} = \cdots = \gamma_l''^{(R)} = 0$$

again with a single subinterval ($m = 1$), and:

$$\gamma_l^{(1)} = \cdots = \gamma_l^{(R)}$$

where this last value (to be determined on the basis of the condition $V_l^{(R)} = V_l(T)$) is equivalent to a variation—constant within $[0, T]$—of the unit cost.

Finally, in the case of the *problem 2*, the cost G_l of generation at the l-th power plant for the whole interval $[0, T]$ can be assumed:

$$G_l = G_{l\,\min} + k'(W_l - W_{l\,\min}) + hT$$

(where h is a proper constant), with the constraint:

$$W_l \triangleq \int_0^T q_l \, \mathrm{d}t \geq W_{l\,\min}$$

which is equivalent to $G_l \geq G_{l\,\min} + hT$; then it again may be written;

$$G_l = \int_0^T C_l \, \mathrm{d}t$$

by assuming $C_l \triangleq k'q_l + h - (k'W_{\min} - G_{l\,\min})/T$.

The problem is similar to the previous one, leading to Equation [2.3.29], with the only difference that the quantity W_l of consumed fuel is now minimally bounded, instead of maximally bounded. As for the Lagrangian function, it may be then assumed as a function similar to Equation [2.3.29], obtained by replacing the last term $\gamma_l'^{(R)}\left(V_{l\,\min} - V_l^{(0)} + \sum_r q_l(P_{tl}^{(r)})\tau\right)$ by a term $\gamma_l''^{(R)}\left(W_{l\,\min} - \sum_r q_l(P_{tl}^{(r)})\tau\right)$, where:

$$\gamma_l''^{(R)} \begin{cases} = 0 & \text{if } W_l > W_{l\,\min} \\ \geq 0 & \text{if } W_l = W_{l\,\min} \end{cases}$$

with $W_l = \sum_r q_l(P_{tl}^{(r)})\tau$.

If the assumption $\gamma_l''^{(R)} = 0$ leads to $W_l < W_{l\,\min}$, it must be imposed $W_l = W_{l\,\min}$, by deriving $\gamma_l''^{(R)}$. This last value is equivalent to a decrease—constant within $[0, T]$—of the unit cost of fuel.

2.3.3. Choice of the Hydroelectric Generation Schedule

(a) Preliminaries

The hydroelectric generation schedule has been assumed to be preassigned up to now. Actually it must be properly coordinated with the thermal generation schedule, so that the most economical overall solution may be obtained.

Such coordination must be performed "over time," based on:

- forecasting hydraulic inflows, and spillages for uses different from generation;
- scheduling water storages in reservoirs and basins;

which determine the amount of water to be used for generation purposes, in each given time interval $[0, T]$ (e.g., a day or a week, according also to the capability of reservoirs and basins). The water storages must be kept within their minimum and maximum limits at each instant.

Even from these points of view the typology of hydraulic plants is very wide, since water inflows and storage capabilities may be small or large. With a pumping-generating plant, the inflow can be artificially increased by pumping as specified later. Additionally, the type of plants may differ significantly for the available water "head," which can vary from a few meters to several hundreds of meters. The water head determines the type of turbine to be adopted, which may be a "reaction" type for "low head" (Kaplan turbines or similar ones) or "medium head" (Francis turbines), and "action" type for "high head" (Pelton turbines).

The scheduling problem becomes more complicated with the presence, in the same valley, of *hydraulically coupled plants*, for which the water used by a plant contributes to the inflow of the downstream plants (with water travel times, from one plant to the other, which may not be disregarded in long distances). Finally, possible environmental requirements resulting from the use of water for agriculture, ship transportation, fishing, or use of the surrounding area, etc. may introduce further constraints on water flows, their rate of change, level variations in reservoirs, and so on.

The amounts of water to be used in the assigned interval $[0, T]$ may be translated, by considering the water heads and by properly estimating losses in hydraulic supply systems and turbines, into *available energy for hydroelectric generation*, obviously within the power limits of respective units. We will assume that this energy is *insufficient* with respect to total demand in $[0, T]$, so that generation cannot be only hydroelectric.

With such assumptions, the coordination of hydroelectric and thermoelectric generations may be established to minimize the thermoelectric generation cost, with the following known information:

- the load power demand as a function of time;
- the energy for hydroelectric generation and thus the amount of energy required from thermoelectric generation (apart from network losses);
- the power limits of the hydroelectric power plants (besides minimum and maximum storage limits).

It is assumed that within each hydroelectric plant, the use of operating units is coordinated, according to Section 2.2.5d (see also Fig. 2.14b). Furthermore it is assumed that the set of operating thermal units can be properly chosen, so that they can supply the demanded power, accounting for their respective limits.

The most frequent situation is case (1), in which the overall hydroelectric generation power is not enough to meet the total demanded power, at each instant within $[0, T]$. In this case, the thermal generation must be used for all the interval $[0, T]$, and the time behavior of the corresponding power must be optimized (specifically, as it will be seen, "leveling" it as much as possible).

A different problem arises in case (2), in which the hydroelectric power is sufficient at each instant within $[0, T]$ and the thermal generation, needed only to compensate for the insufficiency of available hydroenergy, may be used even for a subinterval within $[0, T]$. Thus, it is possible to optimize the choice of this subinterval (and specifically its duration), starting from the knowledge of the energy required from thermal generation.

Problems similar to the above mentioned may occur in the intermediate cases, when the hydropower is sufficient only for a part of $[0, T]$.

(b) Case with Hydropower Always Insufficient

(b1) Generalities

To qualitatively illustrate the above-defined case (1) (i.e., always insufficient hydropower), assume for simplicity that the system includes a single hydroelectric plant, with reservoir or basin. (For a flowing water plant, the generation schedule is simply based on the requirement to use the inflows at its disposal.)

With P_w denoting the active power delivered by this plant, and using the notation defined in Section 2.3.1, the problem of economical optimization along $[0, T]$ may be formulated as follows[(10)]:

$$\min \int_0^T \sum_i C_i\,(P_{ti}(t))\,\mathrm{d}t$$

with the constraints ($\forall t \in [0, T]$):

$$0 = P_c(t) + p(t) - P_o'(t) - \sum_i P_{ti}(t) - P_w(t)$$

$$P_{ti}(t) \in [P_{ti\,\min}, P_{ti\,\max}] \quad (i = 1, \ldots, n_t)$$

$$P_w(t) \in [P_{w\,\min}, P_{w\,\max}]$$

where it is assumed $P_o' \triangleq P_o - P_w$ (assigned value), while losses p are at each instant $p = \phi(P_{t2}, \ldots, P_{tn_t}, P_w, \ldots)$ (see Equation [2.3.4]). In addition:

$$\frac{\mathrm{d}V}{\mathrm{d}t} = q_a - q$$

$V(0)$, $V(T)$ both assigned

(10) Limits on currents and security constraints, and limits on variations over time considered in Section 2.3.2 are not included here for simplicity. Additionally, it is still assumed that the *system configuration* is kept *constant* along the whole $[0, T]$.

with the constraints ($\forall t \in [0, T]$):

$$V(t) \in [V_{\min}, V_{\max}]$$

where:

- V is the quantity of stored water (in the reservoir or in the basin);
- q_a is the total incoming water flow (difference between inflows and spillages), with $q_a(t)$ assigned within all $[0, T]$;
- q is the used water flow, depending on P_w (the head will be assumed constant).

The quantity of water to be used is:

$$W \triangleq \int_0^T q \, \mathrm{d}t = \int_0^T q_a \, \mathrm{d}t - (V(T) - V(0))$$

Due to constraints on V, it is assumed that:

- $V = V_{\min}$ implies $q \leq q_a$, and thus $q_a \geq q(P_{w\min})$. Otherwise, the plant should be considered out of service.
- $V = V_{\max}$ implies $q \geq q_a$. Actually, it might also be:

$$q < q_a, \quad 0 = \frac{\mathrm{d}V}{\mathrm{d}t} = q_a - q - q_s$$

where $q_s = q_a - q (>0)$ is an "overflow" which should be avoided as it cannot be reused unless other plants are present downstream.

It is easy to understand the strict analogy between the considered problem and problem 1 presented in Section 2.3.2b, with $V_l(0)$, $V_l(T)$ both assigned.

By assuming that the interval $[0, T]$ is divided into a large number R of elementary subintervals (each lasting $\tau \triangleq T/R$), a Lagrangian function of the type $\mathcal{L} = \sum_r \mathcal{L}^{(r)} \tau$ may be assumed, with:

$$\begin{aligned}
\mathcal{L}^{(r)} \triangleq & \sum_i C_i(P_{ti}^{(r)}) + \lambda^{(r)} \left(P_c^{(r)} + p^{(r)} - P_o'^{(r)} - \sum_i P_{ti}^{(r)} - P_w^{(r)} \right) \\
& + \sum_i \lambda_i'^{(r)} (P_{ti\min} - P_{ti}^{(r)}) + \sum_i \lambda_i''^{(r)} (P_{ti}^{(r)} - P_{ti\max}) \\
& + \lambda_w'^{(r)} (P_{w\min} - P_w^{(r)}) + \lambda_w''^{(r)} (P_w^{(r)} - P_{w\max}) \\
& + \gamma^{(r)} \left[\frac{V^{(r)} - V^{(r-1)}}{\tau} - q_a^{(r)} + q(P_w^{(r)}) \right] \\
& + [\gamma'^{(r)} (V_{\min} - V^{(r)}) + \gamma''^{(r)} (V^{(r)} - V_{\max})]/\tau
\end{aligned}$$

which depends on $P_{t1}^{(r)}, \ldots, P_{tn_t}^{(r)}, P_w^{(r)}$ with $r = 1, \ldots, R$, and on $V^{(r)}$ with $r = 1, \ldots, R-1$ (while $V^{(0)} = V(0)$ and $V^{(R)} = V(T)$ are assigned).

By setting to zero the corresponding partial derivatives, the following conditions result:

$$\frac{c_i^{(r)}}{1 - \left(\dfrac{\partial \phi}{\partial P_{ti}}\right)^{(r)}} = \frac{c_w^{(r)}}{1 - \left(\dfrac{\partial \phi}{\partial P_w}\right)^{(r)}} = \lambda^{(r)} \quad (i = 1, \ldots, n_t;\ r = 1, \ldots, R)$$

$$\gamma^{(r+1)} \begin{cases} = \gamma^{(r)} - \gamma'^{(r)} \le \gamma^{(r)} & \text{if } V^{(r)} = V_{\min} \\ = \gamma^{(r)} & \text{if } V^{(r)} \in (V_{\min}, V_{\max}) \ (r = 1, \ldots, R-1) \\ = \gamma^{(r)} + \gamma''^{(r)} \ge \gamma^{(r)} & \text{if } V^{(r)} = V_{\max} \end{cases}$$

similar to Equations [2.3.27] and [2.3.28], where $c_1^{(r)}, \ldots, c_{n_t}^{(r)}$ are the incremental costs at generation defined by Equations [2.3.3], with $(\partial\phi/\partial P_{t1})^{(r)} = 0$, whereas:

$$c_w^{(r)} \triangleq \gamma^{(r)} \left(\frac{\mathrm{d}q}{\mathrm{d}P_w}\right)^{(r)} - \lambda_w'^{(r)} + \lambda_w''^{(r)} \begin{cases} \le \gamma^{(r)} \dfrac{\mathrm{d}q}{\mathrm{d}P_w}(P_{w\min}^+) & \text{if } P_w^{(r)} = P_{w\min} \\ = \gamma^{(r)} \left(\dfrac{\mathrm{d}q}{\mathrm{d}P_w}\right)^{(r)} & \text{if } P_w^{(r)} \in (P_{w\min}, P_{w\max}) \\ \ge \gamma^{(r)} \dfrac{\mathrm{d}q}{\mathrm{d}P_w}(P_{w\max}^-) & \text{if } P_w^{(r)} = P_{w\max} \end{cases}$$

The above may be interpreted as the incremental cost at generation for the hydroelectric plant, in the generic r-th subinterval ($r = 1, \ldots, R$), as if water were a fuel with an "*equivalent*" *unit cost* equal to $\gamma^{(r)}$.

The optimal solution "over time" can be formally seen as R purely instantaneous dispatchings (with the hydroelectric plant considered as a thermal plant) but on the basis of the values $\gamma^{(1)}, \ldots, \gamma^{(R)}$ which must be determined by imposing the conditions $V^{(0)} = V(0)$, $V^{(R)} = V(T)$ and the constraints $V^{(1)}, \ldots, V^{(R-1)} \in [V_{\min}, V_{\max}]$.

By disregarding, for simplicity, the dependence of losses on generated power dispatching, the block diagram of Figure 2.26 may be derived, from which it is evident that all the variables involved depend on $V^{(0)}$ and on $V^{(R)}$. (Such a scheme is derived similarly to that of Figure 2.25, from which it differs only in the final part ($r = R$).)

Furthermore, by recalling Section 2.3.2b, it can be understood that, if constraints on V are never activated, the "equivalent" unit cost of water is constant along the whole interval $[0, T]$. In the opposite case, there are more subintervals separated by the condition $V = V_{\min}$ or $V = V_{\max}$, and in each of such subintervals, $\gamma^{(r)}$ remains constant (at a proper value, as already seen).

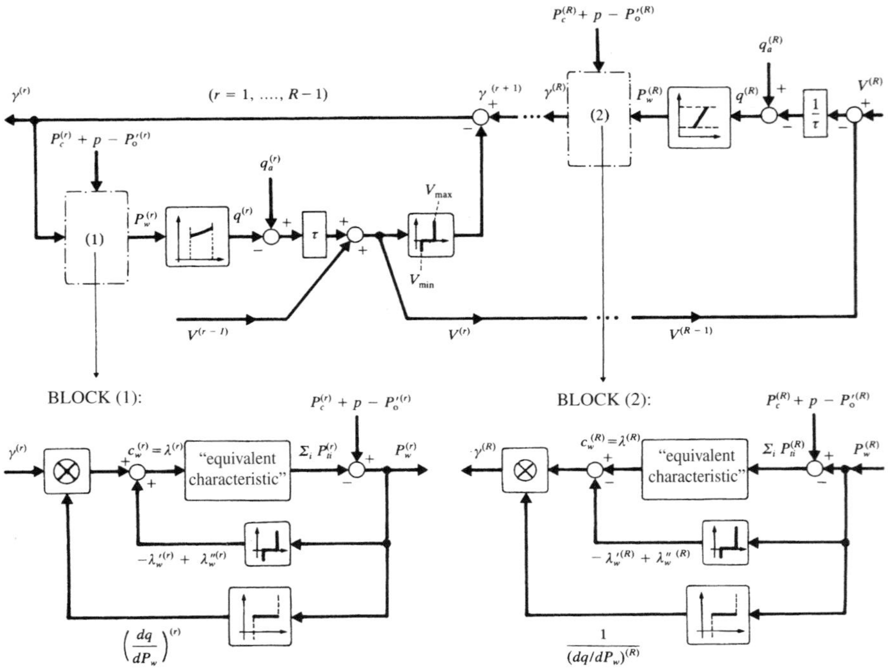

Figure 2.26. A computational block diagram in the presence of a single hydroelectric plant (under the assumption of constant losses).

The treatment becomes more complicated when:

- the head may not be considered constant (specifically, for low and medium head plants equipped with reaction turbines, it also depends on the water level at the water outflow);
- the dependence of q on P_w (or also that of the generic C_i on P_{ti}) is not suitable for analytical treatment;
- other constraints must be considered, such as constraints on dq/dt within certain subintervals, caused by use of water for agriculture or other requirements.

For such cases, a procedure based on dynamic programming may be useful, e.g., according to the following formula:

$$\mathcal{F}^o(r, X_k) = \min_{X_j} \left[\mathcal{F}^o(r-1, X_j) + \sum_i C_i(r-1, X_j; r, X_k)\tau \right]$$

where, for discrete values X_j, X_k:

- $\mathcal{F}^o(r, X_k)$ is the minimum total cost that can be obtained by going from $V = V^{(0)}$ at $t = 0$, to $V = X_k$ at $t = r\tau$ (for $r = 1$ it is $\mathcal{F}^o(1, X_k) = \sum_i C_i(0, V_o; 1, X_k)\tau$

- $\sum_i C_i(r-1, X_j; r, X_k)\tau$ is the thermal generation cost that results when going from $V = X_j$ at $t = (r-1)\tau$, to $V = X_k$ at $t = r\tau$ (in fact, under this condition, the values of q, P_w, $\sum_i P_{ti}$, $\sum_i C_i$ can be sequentially evaluated for the given $t = r\tau$)

with the assumption of directly accounting for all constraints at any step. The optimal solution is defined by $\mathcal{F}^o(R, V^{(R)})$ and the corresponding values of V etc. for $t = \tau, \ldots, R\tau$.

The extension to the case of more hydroelectric plants, hydraulically noninteracting, is obvious; more complications may arise in the presence of hydraulically coupled plants in the same valley, for which it is necessary to properly coordinate the generation to avoid overflows, etc. (the equivalent unit costs of water at each plant also may vary because of storage limits on downstream plants).

(b2) Simplified Analysis

The above-illustrated problem may be treated with some important simplifications, based on the following (usually acceptable) hypotheses:

(1) the water flow q is proportional to P_w (see also Fig. 2.14b), i.e., $q = bP_w$, $\mathrm{d}q/\mathrm{d}P_w = b$, with b a known constant;
(2) the storage constraints are not active (i.e. $V(t) \in (V_{\min}, V_{\max})$) along all $[0, T]$, so that the "equivalent" unit cost of water has a constant value γ;
(3) network losses are independent of the generated power dispatching, thus allowing to assume $\partial\phi/\partial P_{ti} = \partial\phi/\partial P_w = 0$;
(4) the minimum value of P_w is zero, i.e., $P_{w\min} = 0$ (whereas, as already specified, it is assumed $P_{w\max} < P_c(t) + p - P_o'(t)$ $\forall t \in [0, T]$, which means insufficient hydroelectric power at each instant).

As a consequence of hypotheses (1), (2), and (4), it results:

$$c_w \begin{cases} \leq \gamma b & \text{if } P_w = 0 \\ = \gamma b & \text{if } P_w \in (0, P_{w\max}) \\ \geq \gamma b & \text{if } P_w = P_{w\max} \end{cases}$$

where γb is constant. Whereas, due to hypothesis (3), it must be at any instant:

$$c_i = c_w = \lambda \quad (i = 1, \ldots, n_t)$$

which means the equality of incremental costs at generation, for both the thermal plants and the hydroelectric plant.

For any given value of γ, everything is then known; thus, the characteristic $\left(\lambda, \sum_i P_{ti} + P_w\right)$ can be evaluated in the usual way and, for any given total demanded power $\sum_i P_{ti} + P_w = P_c + p - P_o' \triangleq P_c'$, the values $\lambda, P_{t1}, \ldots, P_{tn_t}$, P_w can be derived; see Figure 2.27 (in Fig. 2.27a, the characteristic $\left(c_t, \sum_i P_{ti}\right)$, i.e., the "equivalent" characteristic of the set of thermal power plants, is represented as a piecewise line of only three segments for reasons of graphical simplicity; however, the diagrams in Fig. 2.27c remain valid independently of this simplification).

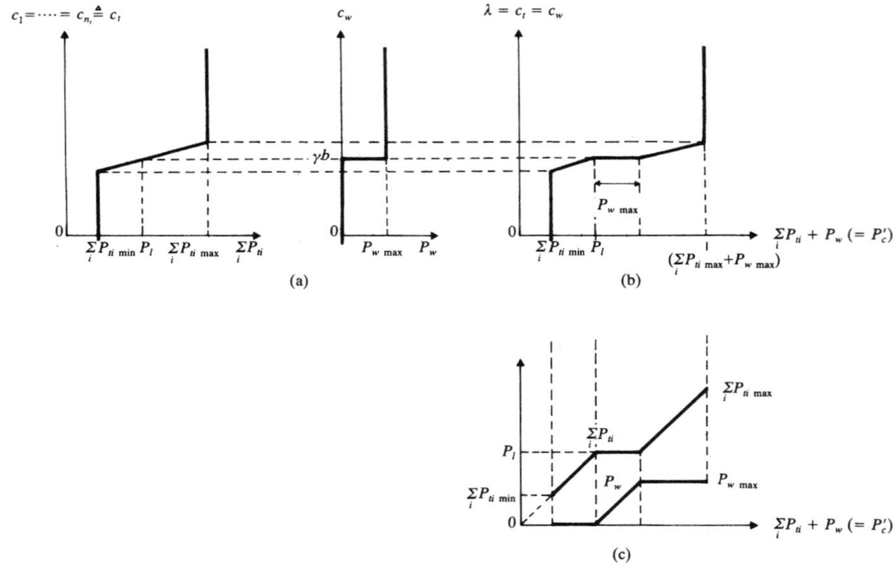

Figure 2.27. Equality of incremental costs at generation in the presence of a single hydroelectric plant (not a pumping station), for a given "equivalent" unit cost γ of the water: (a) equivalent characteristic of thermal units and characteristic of the hydroelectric plant; (b) dependence of total generated power on incremental cost at generation; (c) generated power dispatching, between thermal units and the hydroelectric plant.

It can be derived that $P_w = 0$ when $\lambda < \gamma b$ (i.e., P'_c smaller than the value P_l reported in the Fig. 2.27), whereas for increasing values of λ, or equivalently of P'_c, the hydroelectric plant is required to supply power when $\lambda = \gamma b$ (which corresponds to $\sum_i P_{ti}$ constant and equal to P_l), with $P_w = P_{w\,\max}$ when $\lambda > \gamma b$.

However, the value of γ is not known at the start of the procedure, and it must be determined based on assigned values $V(0)$ and $V(T)$, i.e., based on water quantity to be used $W \triangleq \int_0^T q(t)\,dt$. Again, the hypothesis (1), which states that $W = b\int_0^T P_w(t)$, is useful, since the energy $E_w \triangleq \int_0^T P_w(t)$ supplied by hydroelectric generation turns out to be assigned[(11)].

The optimal solution may then be directly obtained without involving the value of λ, starting from the law $P'_c(t)$ and from the values $P_{w\,\max}$, E_w, according to examples of Figure 2.28.

All the above analysis leads to the criterion of using hydroelectric generation to "*level*" *as much as possible the diagram* $\sum_i P_{ti}(t)$, by considering that at each

[(11)] This conclusion also can be reached by generally assuming that q is not proportional to P_w, but linearly depends on it.

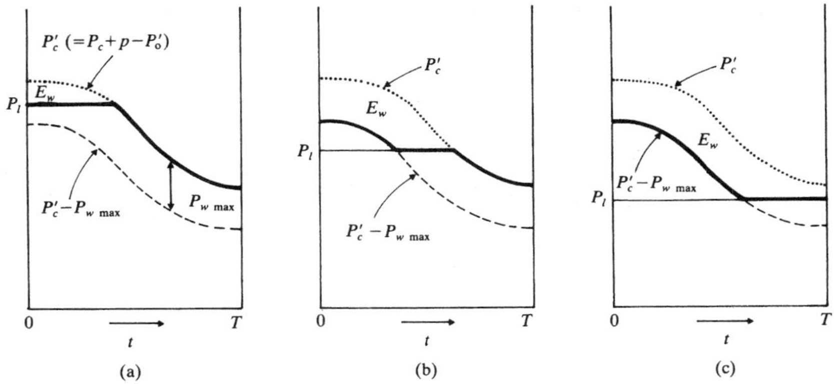

Figure 2.28. Leveling the diagram of total thermal generation power $\sum_i P_{ti}(t)$, for three different values of the hydrogeneration energy E_w (the smallest value of E_w is in case (1), and the largest is in case (3); the opposite happens for the values of P_l and γ reported in Fig. 2.27).
For each of the three cases:

- the bold line diagram represents the total thermal generation power $\sum_i P_{ti}(t)$;
- the hydrogeneration power is given by $P_w = P'_c - \sum_i P_{ti}$;
- the hydrogeneration energy corresponds to the area indicated with E_w.

instant it must be $\sum_i P_{ti}(t) \in [P'_c(t) - P_{w\max}, P'_c(t)]$[12]. (In Fig. 2.28, it has been assumed $P_{w\max} < P'_{c\max} - P'_{c\min}$, thus the complete leveling ($\sum_i P_{ti}$ constant) is not possible for any value of E_w. If instead it is $P_{w\max} > P'_{c\max} - P'_{c\min}$, the total leveling can be obtained, at intermediate values of E_w.)

The convenience of leveling $\sum_i P_{ti}(t)$ may be more directly justified by observing that:

- under the adopted assumptions, the following energy is assigned:

$$E_t \triangleq \int_0^T \sum_i P_{ti}(t)\, dt = \int_0^T P'_c(t)\, dt - E_w$$

[12] Once P_l is known, powers $\sum_i P_{ti}$, P_w will depend only on P'_c (independently of t), according to Figure 2.27c. Additionally, as it can be easily checked, the graphic deduction of P_l (see Fig. 2.28) also can be performed starting from the "*duration*" *diagram* (P'_c, θ) instead of the temporal diagram (P'_c, t). (The "duration" $\theta(a)$, corresponding to a generic value $P'_c = a$, is defined as the total duration for which $P'_c(t) \geq a$; if $P'_c(t) = a$ for one or more intervals of duration different from zero, it is assumed that $\theta(a)$ includes all values within $[\theta(a^+), \theta(a^-)]$.)

More specifically, the diagrams of Figure 2.28 (and those of Fig. 2.30) also might be "duration" diagrams, with t replaced by θ.

- the total cost per unit of time $\sum_i C_i$ is, for a given constant set of operating units and with optimal dispatching of P_{ti}'s, a continuous function of $\sum_i P_{ti}$, with increasing derivative (and possible discontinuities in the derivative itself),

so that, if E_t and T are assigned without further constraints, the total cost $\int_0^T \sum_i C_i \, dt$ will be minimum for $\sum_i P_{ti} = \text{constant} = E_t/T$.

It is also interesting that, to level $\sum_i P_{ti}(t)$, the objective function $\int_0^T \sum_i C_i \, dt$ also may be replaced by $\int_0^T (\sum_i P_{ti})^2 \, dt$. This assumption may allow clear simplifications in more general cases (e.g., when storage constraints are considered).

The *extension to the case of more hydroelectric plants* is obvious (even in the case of valleys, again under the assumption that storage constraints are not active). Specifically, the optimal solution can be obtained through subsequent leveling — performed by starting from highest values of P_c' — provided that the different plants are used according to a proper order, which may be determined on the basis of the law $P_c'(t)$ (or of the corresponding duration diagram) and of the values $P_{w(j)\max}$, $E_{w(j)}$ for each plant ($j = 1, 2, \ldots$).

However, it must be considered that all results described up to now have been obtained by assuming a constant system configuration, and thus assuming that all the considered units are in operation for the whole interval $[0, T]$. Actually, the set of operating units within the hydroelectric plant also might vary according to the value of P_w (Fig. 2.14b), and the whole plant might be set out of operation at those subintervals in which $P_w(t) = 0$ (Fig. 2.28a,b). Similarly, with an insufficiently leveled $\sum_i P_{ti}(t)$, it may be more economical — or even necessary, when this power is below the sum of "technical minimum" limits — to shut down some thermal units in the low-load subintervals. In this case, the problem must be reformulated, because the cost characteristics in the different subintervals are correspondingly modified (in other words, the hydroelectric generation scheduling interacts with the *operational scheduling of thermal plants*); see Section 2.4.2b.

By only assigning the demanded energy, the economic optimum can be obtained by using the thermal units at full power (or nearly), and possibly not all of them for the whole interval $[0, T]$, that is with a varying configuration and an unleveled $\sum_i P_{ti}(t)$; see the example reported in Section (c). Such solutions indeed appear to be of scarce interest (particularly for predominantly thermal generation), as they could easily affect the spinning reserve requirements.

Up to now, the behavior of powers injected through possible boundary nodes (see the term $P_o'(t)$) was considered as assigned; on the contrary, the behavior of such injections might be adjusted, according to the law by which the incremental cost varies within $[0, T]$ (see Section 2.3.4).

(b3) Convenience of Possible Pump-Storage

If thermal generation is required to supply, within $[0, T]$, significantly different values of $\sum_i P_{ti}$, the use of pump-storage hydroelectric plants may allow a further leveling of $\sum_i P_{ti}(t)$ and, consequently, a reduction in generation cost. To obtain this result, it is assumed that pumping actions are performed at lower loads (when the thermal generation cost is relatively small) and that the water corresponding to pumping is used to reduce the thermal generation when it is more expensive (i.e., at higher loads).

Depending on the characteristics of plants, the generation-pumping cycle may be *daily*, with pumping at night, or *weekly*, with (further) pumping during weekends, etc.

The new results obtained may suggest proper modifications to the operational scheduling of units and/or interchange scheduling through interconnections (see Section 2.4.2b).

By confining our attention, for simplicity, to the case of a single hydroelectric plant, and by assuming that it can work in the whole interval $P_w \in [-P_{p\,\max}, +P_{w\,\max}]$, with:

$$\begin{cases} P_w > 0 \quad (P_w \in (0, P_{w\,\max}]), \quad q = q_g(P_w) > 0 \quad \text{during generation} \\ P_w < 0 \quad (P_w \in [-P_{p\,\max}, 0)), \quad q = q_p(P_w) < 0 \quad \text{during pumping} \end{cases}$$

the previous treatment may be extended without difficulties, by again defining a proper value for the "equivalent" unit cost of water (or more values, if constraints on V are activated), and so on, and by considering that the dependence of q on P_w is generally different in the generating phase and in the pumping phase.

If, for simplicity:

- it is assumed that:

$$\begin{cases} q = b_g P_w, \quad \dfrac{dq}{dP_w} = b_g \quad \text{during generation} \\ q = b_p P_w, \quad \dfrac{dq}{dP_w} = b_p \quad \text{during pumping} \end{cases}$$

  with b_g, b_p known constants, and, more precisely, with $\eta \triangleq b_p/b_g < 1$ (where η can be interpreted as the overall efficiency of the generating-pumping cycle);
- hypotheses (2) and (3) defined earlier are accepted (constraints on V not activated, and losses independent of the generated power dispatching), whereas the hypothesis (4) must be removed, as it now holds $P_{w\,\min} = -P_{p\,\max}$;

then, diagrams like those in Figure 2.29 can be obtained (instead of those in Fig. 2.27).

Moreover, the quantity of water to be used (which is known) is now:

$$W \triangleq \int_0^T q(t)\,dt = b_g E_g - b_p E_p = b_g(E_g - \eta E_p)$$

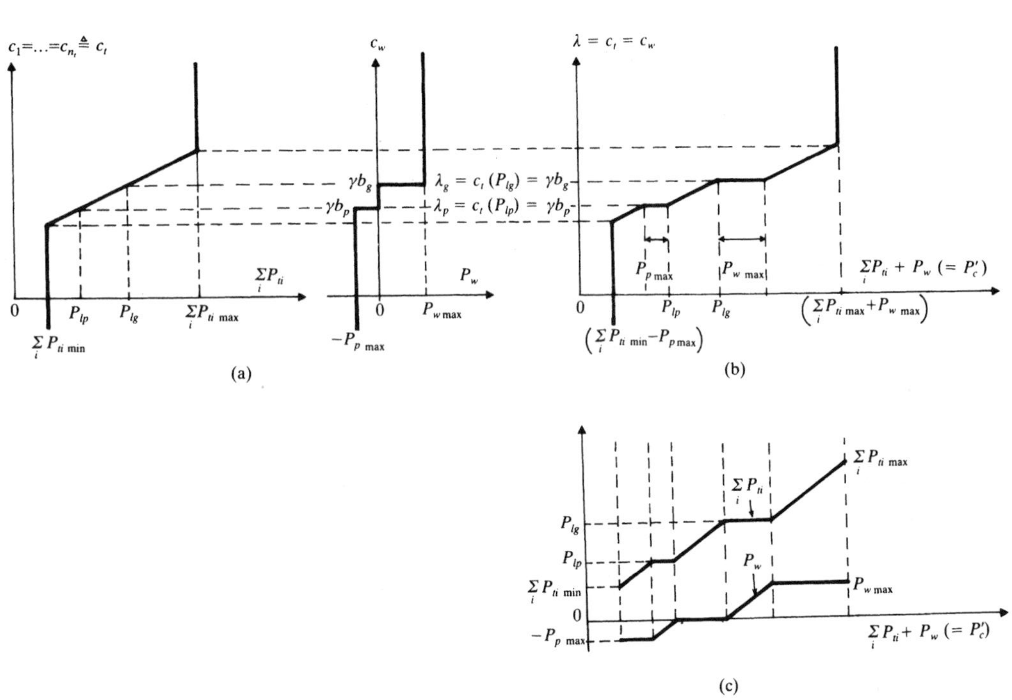

Figure 2.29. Equality of incremental costs at generation in the presence of a pump-storage hydroelectric plant, for a given "equivalent" unit cost γ of the water: (a) equivalent characteristic of thermal units and characteristic of the hydroelectric plant; (b) dependence of total generated power on incremental cost at generation; (c) generated power dispatching, between thermal units and the hydroelectric plant.

by assuming that E_g is the energy delivered during the generation phase, and E_p is the energy absorbed during the pumping phase. If $E_w = W/b_g$ is the delivered energy in the absence of pumping (as in the previously treated case; see Fig. 2.28), it then follows that:

$$\Delta E_g \triangleq E_g - E_w = \frac{b_p}{b_g} E_p = \eta E_p$$

Consequently, the optimal solution can again be directly derived, without involving the value γ of the "equivalent" unit cost, from $P_c'(t)$ and from $P_{w\max}$, $P_{p\max}$, E_w, η, according to examples of Figure 2.30.

The leveling of the diagram $\sum_i P_{ti}(t)$ can be then improved. On the other hand, the constraints to be considered are $\sum_i P_{ti}(t) \in [P_c'(t) - P_{w\max}, P_c'(t) + P_{p\max}]$, and are thus less stringent than those previously considered ($P_{p\max} = 0$).

Obvious variations must be integrated into the treatment when it is convenient, for efficiency reasons, that the pumping phase be operated only at P_w values near $-P_{p\max}$.

(c) Case with Hydropower Always Sufficient

Now examine case (2) defined in Section (a) (always sufficient hydroelectric power), for which the necessity of thermal generation is only caused by lack of hydroenergy at disposal.

More precisely, assume that the most economic law $\sum_i P_{ti}(t)$ ($t \in [0, T]$) must be found, based on the total energy $E_t \triangleq \int_0^T \sum_i P_{ti}(t)\,dt$ demanded from thermal plants, and without concern for load diagram and spinning reserve requirements (we assume that such problems can be solved by the contribution of hydroelectric generation).

For greater generality, assume that the generic i-th unit is kept into operation along the subinterval $[t_{oi}, t_{oi} + T_i]$ (which may possibly coincide with $[0, T]$). The energy supplied by this unit is then $E_{ti} = \int_{t_{oi}}^{t_{oi}+T_i} P_{ti}(t)\,dt$, while the corresponding cost:

- is reduced to only the *generation cost* $G_i = \int_{t_{oi}}^{t_{oi}+T_i} C_i(P_{ti}(t))\,dt$, if the considered unit is always in operation ($t_{oi} = 0$, $T_i = T$);
- includes, on the contrary, the *cost* A_i *for startup and/or shutdown.*

In the last case, it may be thought that the problem arises periodically (or nearly so) with period T. The cost A_i may be considered an increasing function of the no-operation duration $(T - T_i)$ or, equivalently, a decreasing function of T_i, with $T_i \in (0, T)$ (see also Section 2.4.2b).

With such assumptions, the aim is to minimize the total cost $\sum_i (G_i + A_i)$ for the given total energy $\sum_i E_{ti} = E_t$.

To solve this problem, it is useful to first determine the operating schedule which, for the generic unit, minimizes the cost $(G_i + A_i)$ at each given energy E_{ti}.

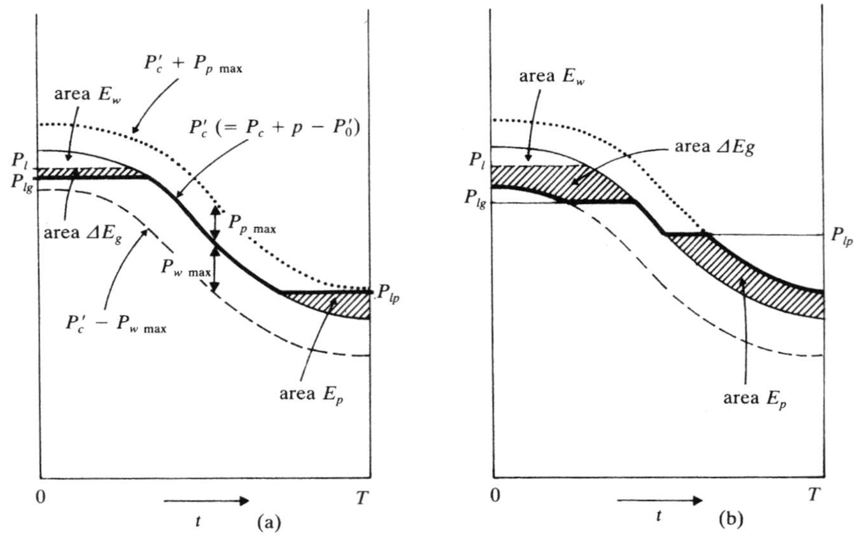

Figure 2.30. Leveling the diagram of total thermal generation power $\sum_i P_{ti}(t)$, in the presence of a pump-storage hydroelectric plant, and for two different values of the efficiency η [small in case (a), and large in case (b)].
It is assumed that, in the absence of pumping, the solution is similar to that indicated in Figure 2.28a (i.e., $\sum_i P_{ti} = \min(P_c', P_l)$).
For each of the two cases:

- the bold line diagram represents the total thermal generation power $\sum_i P_{ti}$ in the presence of pumping;
- the hydrogeneration power (negative in the pumping phase) is given by $P_w = P_c' - \sum_i P_{ti}$;
- the hydroelectric plant supplies the energy $E_w + \Delta E_g$ during the generation phase, and absorbs the energy E_p during the pumping phase.

Values P_{lp}, P_{lg}, ΔE_g, E_p are subject to the conditions: $\lambda_p/\lambda_g = c_t(P_{lp})/c_t(P_{lg}) = b_p/b_g = \eta$ (see also Fig. 2.29), $\Delta E_g/E_p = \eta$.

In this concern, according to previous information, the function $C_i(P_{ti})$ is continuous and with positive slope. Thus, for a generic pair of values E_{ti}, T_i, the cost G_i is minimum at $P_{ti}(t) = \text{constant} = E_{ti}/T_i$ (provided that $E_{ti}/T_i \in [P_{ti\,\min}, P_{ti\,\max}]$), i.e., it is convenient to operate the unit at constant power. In the following, we then assume that:

$$P_{ti} = \frac{E_{ti}}{T_i}$$

$$G_i = C_i(P_{ti})T_i = \frac{C_i(P_{ti})}{P_{ti}} E_{ti}$$

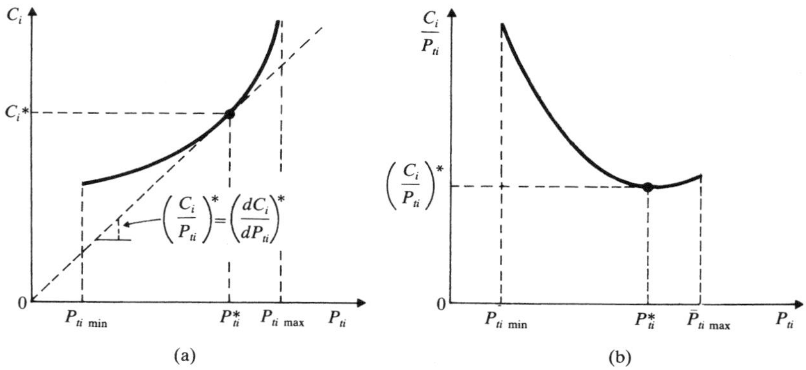

Figure 2.31. An example of characteristics for the i-th thermal unit: (a) (C_i, P_{ti}); (b) $(C_i/P_{ti}, P_{ti})$; (C_i = cost per unit of time, P_{ti} = generated power, C_i/P_{ti} = specific cost).

where the ratio $C_i(P_{ti})/P_{ti}$ (also called "*specific*" *cost*) can be viewed as the energy unit cost when operating at constant power. Such a cost depends on P_{ti} according to a characteristic like the one indicated in Figure 2.31b, and is minimum for a value $P_{ti} = P_{ti}^*$ usually close to or coincident with $P_{ti\,\max}$. In the former case it is evident that $(C_i/P_{ti})^* = (\mathrm{d}C_i/\mathrm{d}P_{ti})^*$, i.e., the specific cost and the incremental cost are equal to each other at $P_{ti} = P_{ti}^*$ (see also Fig. 2.31a.)

For any given E_{ti} (not larger than $P_{ti\,\max}T$) G_i may be minimized by imposing $C_i/P_{ti} = (C_i/P_{ti})^*$ and thus $P_{ti} = P_{ti}^*$; however, this is possible only if the consequent value $T_i = E_{ti}/P_{ti}^*$ is not larger than T, i.e., $E_{ti} \leq P_{ti}^* T$. On the contrary, it must be simply set $T_i = T$, $P_{ti} = E_{ti}/T$.

It then follows:

- if $E_{ti} < P_{ti}^* T : T_i = E_{ti}/P_{ti}^* < T$, $P_{ti} = P_{ti}^*$, $G_i = (C_i/P_{ti})^* E_{ti}$ (with $\mathrm{d}G_i/\mathrm{d}E_{ti} = (C_i/P_{ti})^*$);
- if $E_{ti} \geq P_{ti}^* T : T_i = T$, $P_{ti} = E_{ti}/T$, $G_i = C_i(E_{ti}/T) \cdot T$ (with $\mathrm{d}G_i/\mathrm{d}E_{ti} = \mathrm{d}C_i/\mathrm{d}P_{ti}$);

according to Figure 2.32a. However, such results correspond to the minimized generation cost G_i only; consequently they must be adopted only if the cost $A_i(T_i)$ (different from zero if $T_i \in (0, T)$) may be disregarded.

To account for the startup and/or shutdown cost, the following may be observed, for any given energy $E_{ti} \in [0, P_{ti\,\max}T]$.

- If $T_i = T$ (unit always in operation), it simply holds:

$$P_{ti} = \frac{E_{ti}}{T}, \quad G_i + A_i = G_i = C_i\left(\frac{E_{ti}}{T}\right) \cdot T$$

(from which, more specifically: $\mathrm{d}(G_i + A_i)/\mathrm{d}E_{ti} = \mathrm{d}C_i/\mathrm{d}P_{ti}$, with $P_{ti} = E_{ti}/T$).

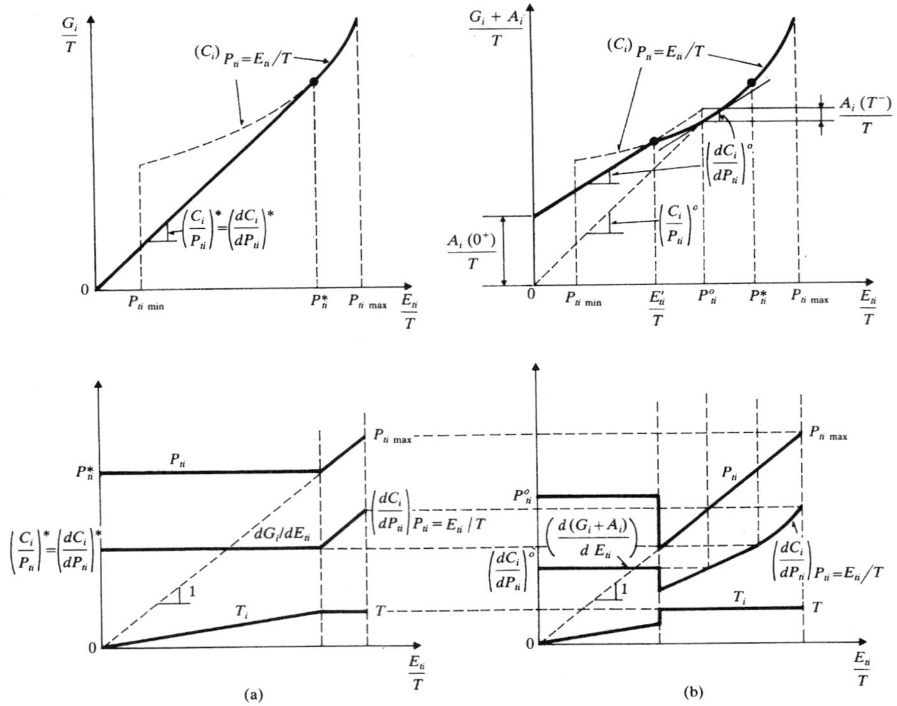

Figure 2.32. Minimization (starting from the knowledge of the energy E_{ti} supplied by the i-th unit) of: (a) only the generation cost G_i; (b) the total cost $(G_i + A_i)$, including also the cost (A_i) for startup and/or shutdown.

- If instead $T_i \in (0, T)$, it holds $T_i = E_{ti}/P_{ti}$ and, generically, the total cost is:

$$G_i + A_i = \frac{C_i(P_{ti})}{P_{ti}} E_{ti} + A_i\left(\frac{E_{ti}}{P_{ti}}\right)$$

In the last case, the total cost can be minimized by adequately choosing P_{ti}; more precisely, by setting to zero its partial derivative with respect to P_{ti}, the following condition is obtained:

$$\frac{\mathrm{d}C_i}{\mathrm{d}P_{ti}} = \frac{C_i + A_i'}{P_{ti}} \quad \left(\text{where } A_i' \triangleq \frac{\mathrm{d}A_i}{\mathrm{d}T_i}, \text{ with } T_i = \frac{E_{ti}}{P_{ti}}\right)$$

From the above equation, the solution $P_{ti} = P_{ti}^o(E_{ti})$ can be derived (with $P_{ti}^o < P_{ti}^*$, because $A_i' < 0$ and so $(\mathrm{d}C_i/\mathrm{d}P_{ti})^o < (C_i/P_{ti})^o$); it then follows that, for $T_i \in (0, T)$ and, consequently, for $E_{ti} \in (0, P_{ti}^o T)$:

$$T_i = \frac{E_{ti}}{P_{ti}^o}, \quad G_i + A_i = \left(\frac{C_i}{P_{ti}}\right)^o E_{ti} + A_i\left(\frac{E_{ti}}{P_{ti}^o}\right)$$

(it has been assumed that $P_{ti}^o \geq P_{ti\,\min}$; otherwise P_{ti}^o must be replaced by $P_{ti\,\min}$).

In particular, if A_i were linearly dependent on T_i (i.e., if A'_i were constant) it can be derived for any E_{ti}:

$$P^o_{ti} = \text{constant}$$

$$\frac{d(G_i + A_i)}{dE_{ti}} = \left(\frac{C_i}{P_{ti}}\right)^o + \frac{A'_i}{P^o_{ti}} = \left(\frac{dC_i}{dP_{ti}}\right)^o = \text{constant}$$

Then, for any assigned value of E_{ti}, the situation at minimum cost must be chosen. Referring to notation used in Figure 2.32b (in which A'_i is assumed to be a constant, and $P^o_{ti} > P_{ti\,\min}$), it can be derived that the unit must be kept into operation:

- for a duration $T_i = E_{ti}/P^o_{ti} < T$ (and with $P_{ti} = P^o_{ti}$), if $E_{ti} < E'_{ti}$;
- for the whole time T (and with $P_{ti} = E_{ti}/T$), if $E_{ti} \geq E'_{ti}$;

where E'_{ti} is the value of E_{ti} for which the two reported situations result in the same cost, or:

$$C_i\left(\frac{E_{ti}}{T}\right)T = \left(\frac{C_i}{P_{ti}}\right)^o E_{ti} + A_i\left(\frac{E_{ti}}{P^o_{ti}}\right)$$

(The solution must be searched within the interval $E_{ti} \in [P_{ti\,\min}T, P^o_{ti}T]$; if it does not exist, E'_{ti} should be replaced with $P_{ti\,\min}T$.)
If it is assumed that $C_i = C_{io} + a_i P_{ti} + b_i(P^2_{ti}/2)$, $A_i = A_{io} - k_i T_i$, with all coefficients being positive, it can be derived:

$$\frac{dC_i}{dP_{ti}} = a_i + b_i P_{ti}$$

$$P^*_{ti} = \sqrt{\frac{2C_{io}}{b_i}} \qquad \text{(assuming } P^*_{ti} < P_{ti\,\max}\text{)}$$

$$P^o_{ti} = \sqrt{\frac{2(C_{io} - k_i)}{b_i}} \qquad \text{(assuming } k_i < C_{io},\ P^o_{ti} > P_{ti\,\min}\text{)}$$

$$E'_{ti} = P^o_{ti}T - \sqrt{\frac{2(A_{io} - k_i T)T}{b_i}}$$

(obviously assuming that $A_i(T^-) = A_{io} - k_i T > 0$ and furthermore $E'_{ti} > P_{ti\,\min}T$).

By again examining the problem with more than one thermal unit, it must be imposed:

$$\min \sum_i (G_i + A_i)(E_{ti})$$

with the constraints:

$$0 = E_t - \sum_i E_{ti}, \qquad E_{ti} \in [0, P_{ti\,\max}T]$$

where each cost $(G_i + A_i)$ can be considered a function of the respective energy E_{ti}, according to the above.

By assuming, for simplicity, that possible startup and/or shutdown have a negligible cost A_i, the following Lagrangian function may be assumed:

$$\mathcal{L} \triangleq \sum_i G_i(E_{ti}) + \mu\left(E_t - \sum_i E_{ti}\right) + \sum_i \mu'_i \cdot (-E_{ti}) + \sum_i \mu''_i \cdot (E_{ti} - P_{ti\max}T)$$

and the following conditions may be derived:

$$0 = \frac{\partial \mathcal{L}}{\partial E_{ti}} = \frac{\mathrm{d}G_i}{\mathrm{d}E_{ti}} - \mu - \mu'_i + \mu''_i$$

or equivalently:

$$g_i = \mu \quad \forall i$$

by assuming:

$$g_i \triangleq \frac{\mathrm{d}G_i}{\mathrm{d}E_{ti}} - \mu'_i + \mu''_i \begin{cases} \leq \dfrac{\mathrm{d}G_i}{\mathrm{d}E_{ti}}(0^+) & \text{if } E_{ti} = 0 \\ = \dfrac{\mathrm{d}G_i}{\mathrm{d}E_{ti}} & \text{if } E_{ti} \in (0, P_{ti\max}T) \\ \geq \dfrac{\mathrm{d}G_i}{\mathrm{d}E_{ti}}(P_{ti\max}T^-) & \text{if } E_{ti} = P_{ti\max}T \end{cases}$$

according to what is reported in Figure 2.33a.

For any given value of μ, each energy E_{ti} may be derived, as well as the sum $\sum_i E_{ti} = E_t$ (see the example for two units, in Figure 2.33). Knowing the characteristic (μ, E_t) enables the deduction, through μ, of the optimal values E_{ti} for any assigned E_t. Note the strict analogy with the problem of the most economical dispatching of thermal generation powers (Section 2.3.1a). Also, for E_t values that require the use of more than one unit, the power dispatching meets the condition of equal incremental costs (see also Fig. 2.33).

By knowing energy E_{ti}, the values T_i, P_{ti} can be finally derived, which define the operational scheduling of each unit.

The problem can be *significantly simplified in the case—not too far from reality—that* $P^*_{ti} = P_{ti\max}$ for all units (i.e., the specific cost is at its minimum at maximum power). In fact, under such assumption, $g_i = \text{constant} = (C_i/P_{ti})^*$ for the whole interval $E_{ti} \in (0, P_{ti\max}T)$; if, with the adopted numbering, it is assumed that:

$$\left(\frac{C_1}{P_{t1}}\right)^* < \left(\frac{C_2}{P_{t2}}\right)^* < \ldots$$

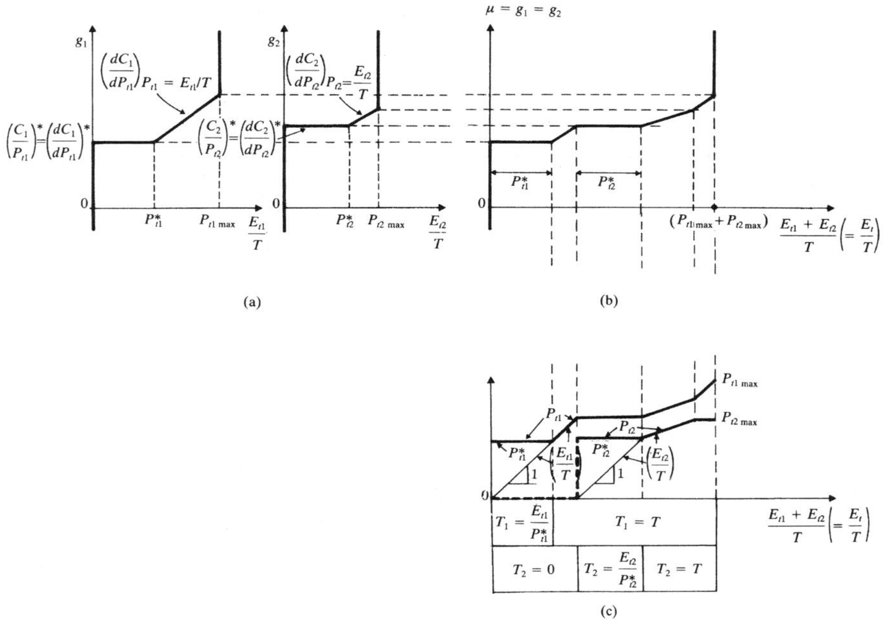

Figure 2.33. Minimization of total generation cost $\sum_i G_i$ (see text).

the solution is then achieved by simply using units at full power, and more precisely:

- only unit 1, for a duration $T_1 = E_t/P_{t1\,\max}$, if $E_t \leq P_{t1\,\max}T$;
- unit 1 for the whole duration T, and unit 2 for a duration $T_2 = E_{t2}/P_{t2\,\max}$ (with $E_{t2} = E_t - P_{t1\,\max}T$), if $E_t \in (P_{t1\,\max}T, (P_{t1\,\max} + P_{t2\,\max})T)$;

and so on. (The same result may be reached more directly by observing that the least expensive energy is the one supplied by unit 1, then by unit 2, etc.)

Things are more complicated if costs A_i (caused by possible startup and/or shutdown) also must be considered. In fact, the behavior of the generic cost $(G_i + A_i)$ as a function of E_{ti} (with a discontinuity at $E_{ti} = 0$ and an abrupt reduction of its slope at $E_{ti} = E'_{ti}$; see Fig. 2.32b) may pose prejudices on the conditions $0 = \partial\mathcal{L}/\partial E_{ti}$. To avoid such shortcomings, a proper numerical procedure may be adopted to perform the optimal search by using, for instance, dynamic programming. Qualitatively speaking, because of costs A_i, the priority sequence $1, 2, \ldots$ may undergo some modifications with respect to the one previously defined (based only on values $(C_i/P_{ti})^*$), as it is not convenient to schedule the startup and/or the shutdown of those units exhibiting rather high A_i costs.

2.3.4. Choice of Powers Exchanged Through Interconnections

In the treatment developed up to now, the possible presence of external systems — interconnected to the one under examination — has been considered by means of proper simplified equivalents with preassigned active power injections. Actually, the optimal solution should instead be determined jointly, based on characteristics (operational and economical) of all single systems, with a uniform approximation of respective models. To simplify the terminology, the single systems interconnected with each other will be named "*areas*," whereas the entire system may be called multiarea "composite" system.

If, for simplicity, the variations of network losses are disregarded, the optimal solution for the whole composite system implies the equality of incremental costs in the different areas. If, for instance, the system includes two areas and their respective incremental costs $\lambda_{(1)}$, $\lambda_{(2)}$ are — for a given value of the exchange power P_{12} from area 1 to area 2 — different from each other with $\lambda_{(1)} < \lambda_{(2)}$, it is easy to understand that the mentioned exchange should be increased. The incremental cost $\lambda_{(1)}$ can be relatively small when, for instance, the load in area 1 is modest or, in this area, it is necessary to use a lot of water, stored or flowing, for hydroelectric generation.
A "small" increment $\mathrm{d}P_{12} > 0$ in fact results into:

$$\mathrm{d}C_{(1)} = \lambda_{(1)}\,\mathrm{d}P_{12}, \qquad \mathrm{d}C_{(2)} = -\lambda_{(2)}\,\mathrm{d}P_{12}$$

so that the additional cost $\mathrm{d}C_{(1)}$ (per unit of time) in area 1 is smaller than the saving $(-\mathrm{d}C_{(2)})$ in area 2; thus, indicating by $h\,\mathrm{d}P_{12}$ the price (per unit of time) paid by 2 to 1, the cost variations become:

$$\mathrm{d}C_{(1)} - h\,\mathrm{d}P_{12} = (\lambda_{(1)} - h)\,\mathrm{d}P_{12} \quad \text{in area 1}$$

$$\mathrm{d}C_{(2)} + h\,\mathrm{d}P_{12} = (h - \lambda_{(2)})\,\mathrm{d}P_{12} \quad \text{in area 2}$$

Consequently, they can be both negative (that means there can be a saving for both areas) if the unit price h is assumed within $\lambda_{(1)}$ and $\lambda_{(2)}$. Note that the overall saving is:

$$-\mathrm{d}C_{(1)} - \mathrm{d}C_{(2)} = (\lambda_{(2)} - \lambda_{(1)})\,\mathrm{d}P_{12} > 0$$

and can be divided into equal parts by assuming $h = (\lambda_{(1)} + \lambda_{(1)}/2)$. Similar considerations hold for not "small" variations ΔP_{12}, $\Delta C_{(1)}$, $\Delta C_{(2)}$, as long as incremental costs are different from each other. However, the existence of constraints on P_{12} also can call for different incremental costs, according to Section 2.3.1c.

However, a unique global dispatching, committed to a proper *centralized "coordination" office*, may be generally accepted only as a general guideline.

In fact, accounting for all details in each area would easily lead to an excessively burdensome global model and significant difficulties in the centralized (and updated) collection of necessary data. At least for analyzing such details, it is usually convenient to resort to better approximated dispatchings, performed within each area.

Furthermore, the autonomy in the management of different areas (which may belong to different states) may result, because of local reasons of different nature, in specific strategies that cannot be easily committed to others. In such situations, power exchanges through interconnections become the object of contracting between the involved utilities. Meanwhile, the centralized office may offer a general overview of the situation, evaluate the obtainable savings, start the negotiations, perform intermediate actions, and so on.

With different conditions to be considered, it is useful to qualitatively identify the following:

- it is efficient to sell power (or equivalently energy, in the considered time interval) particularly to those areas having higher incremental costs, and to buy from those areas having lower incremental costs; in case of separate negotiations, the agreed prices also may depend on the sequence of the negotiations themselves;
- the efficiency of modifying an exchanged power should be evident, accounting for the uncertainty on load forecasting in the interested areas (and errors in exchanged power regulation during operation; see Section 3.4); with respect to this matter, prices may be readjusted, with reference to what is estimated at the previsional stage, according to actual values of loads and exchanged powers;
- when modifying the exchange power between two areas, a variation in the losses of other areas (e.g., interposed between the two considered areas; see also the simplified treatment below) may be induced, thus affecting their generation costs; this should be considered for possible refunding;
- because of the exchange program between two areas for a given time, the unit sets to be kept in operation may be reduced in one area and increased in the other (with corresponding shift of rotating power from one area to the other); this fact implies further variations in respective costs;
- negotiations can be carried out in advance, by guaranteeing supply of the agreed amount (in which case, the price may be rather high); however, there may be different situations, for instance, when the amount of the generic exchange is not guaranteed, subject to actual supply area availabilities;
- the exchange program between two areas may imply long-term compensations, if the annual load peaks in the two areas are shifted with each other, for instance, with winter peak (electric heating) in one area and summer peak (air conditioning) in the other; or if there is a significant amount of water used for hydroelectric generation, in one area or in the other, according to seasons; and so on; similar compensations may occur even in shorter-term periods (e.g., during the same day, in case of shifted load diagrams due to an hourly fuse difference).

The global dispatching for an N-area system may resort to simplifying assumptions. As a first-level approximation, and with possible refinements in the model, it may be assumed that:

(1) the configuration is assigned (actually, as cited, the exchange program itself may suggest modifications to the configuration; for instance, as far as the sets of operating units in the different areas are concerned);
(2) the v/Q steady-state is preassigned (and not sensitive to active power corrections);
(3) loads "conform" (see Equation [2.3.6]) within each area (note that such an hypothesis might be unacceptable, if referred to different areas);
(4) in each area a "generation-load subnetwork" can be defined, which is connected to boundary nodes of the area itself by means of a single "area node"; see Figure 2.34 (the subnetwork that links all area nodes, including the boundary nodes, will be called "interconnection subnetwork").

Note that assumptions (1), (2), and (3) have already been largely used in the previous sections. Assumption (4), which substantially recalls simplified equivalents like those reported in Figure 2.11, may allow consideration of effects of each exchanged power on losses in the different areas, according to what is already mentioned.

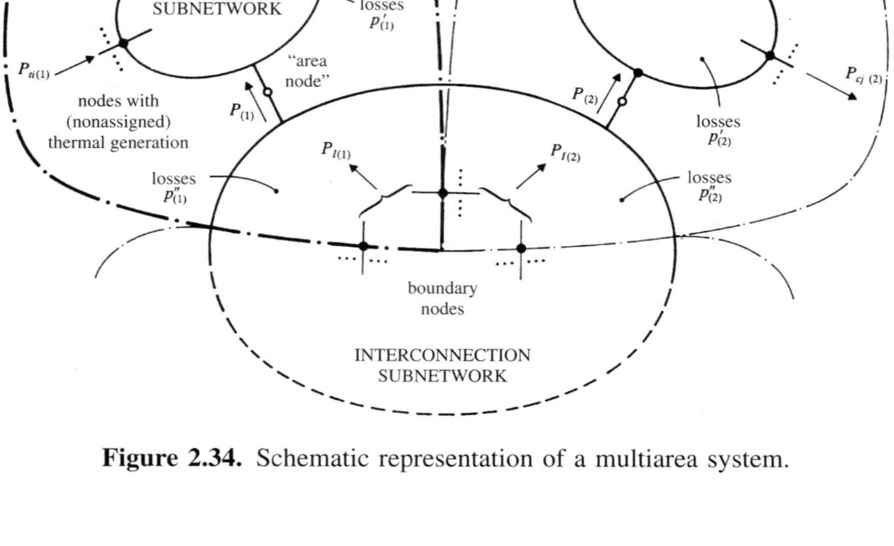

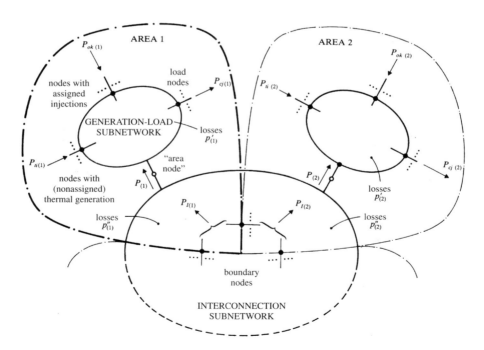

Figure 2.34. Schematic representation of a multiarea system.

The generic *generation-load subnetwork* may be represented by an equation like:

$$\sum_i P_{ti(r)} + P_{o(r)} + P_{(r)} = P_{c(r)} + p'_{(r)} \quad (r = 1, \ldots, N)$$

with constraints, for each thermoelectric unit:

$$P_{ti(r)} \in \left[P_{ti(r)\min}, P_{ti(r)\max}\right]$$

where it is assumed that $P_{o(r)} \triangleq \sum_k P_{ok(r)}$ and $P_{c(r)} \triangleq \sum_j P_{cj(r)}$ are assigned, whereas $P_{(r)}$ is the active power injected through the area node, and losses $p'_{(r)}$ depend—because of the hypothesis of load conformity—on $P_{ti(r)}$, $P_{(r)}$.

For any given value of the power $P_{(r)}$, it is possible to derive conditions of the following type, by means of separate dispatching (see Equation [2.3.9]):

$$\frac{c_{i(r)}}{1 - \dfrac{\partial p'_{(r)}}{\partial P_{ti(r)}}} = \lambda_{c(r)}$$

where the value $\lambda_{c(r)}$ is the incremental cost "at load," in common to all the units of the considered subnetwork.

For small variations of $P_{(r)}$, with $P_{o(r)}$ and $P_{c(r)}$ assigned, it can be derived:

$$\begin{aligned} 0 &= \mathrm{d}\sum_i P_{ti(r)} + \mathrm{d}P_{(r)} - \mathrm{d}p'_{(r)} \\ &= \sum_i \left(1 - \frac{\partial p'_{(r)}}{\partial P_{ti(r)}}\right) \mathrm{d}P_{ti(r)} + \left(1 - \frac{\partial p'_{(r)}}{\partial P_{(r)}}\right) \mathrm{d}P_{(r)} \end{aligned}$$

and, consequently, the overall cost variation (per unit of time) in the r-th area:

$$\mathrm{d}C_{(r)} \triangleq \mathrm{d}\sum_i C_{i(r)} = \lambda_{c(r)} \sum_i \left(1 - \frac{\partial p'_{(r)}}{\partial P_{ti(r)}}\right) \mathrm{d}P_{ti(r)} = -\lambda_{c(r)} \left(1 - \frac{\partial p'_{(r)}}{\partial P_{(r)}}\right) \mathrm{d}P_{(r)}$$

or equivalently:

$$\mathrm{d}C_{(r)} = -\mu_{(r)}\, \mathrm{d}P_{(r)} \quad (r = 1, \ldots, N) \qquad [2.3.30]$$

with:

$$\mu_{(r)} \triangleq \lambda_{c(r)} \left(1 - \frac{\partial p'_{(r)}}{\partial P_{(r)}}\right)$$

The quantity $\mu_{(r)}$ can be interpreted in terms of *incremental cost at the r-th area node*. Alternatively, it is also possible to assimilate the rest of the system, as viewed from such a node, to an equivalent generator, supplying the power $P_{(r)}$ at an incremental cost (at generation) equal to $\mu_{(r)}$; by so doing, it is possible to account for constraints $P_{(r)} \in [P_{(r)\min}, P_{(r)\max}]$ by means of $\mu_{(r)}$, similarly to that proved for one generator in Equations [2.3.3].

With the *interconnection subnetwork*, it must first be underlined that powers "imported" from different areas through their respective boundary nodes (see Fig. 2.34) sum up to zero, so that $\sum_r P_{I(r)} = 0$. This implies that only $(N-1)$ of the N imported powers can be chosen independently.

On the other hand, because of the assumed hypotheses, the steady-state of the interconnection subnetwork is determined by $(N-1)$ independent variables, e.g., constituted by phase shifts between voltages at "area nodes." Since the imported powers are the base for possible negotiations, it is convenient to assume that the $(N-1)$ independent variables are just powers $P_{I(2)}, \ldots, P_{I(N)}$ (whereas $P_{I(1)} = -P_{I(2)} - \ldots - P_{I(N)}$). For any given choice of these powers, it is possible to derive:

- powers exchanged between any pair of areas (and, specifically, through any single boundary node);
- losses $p''_{(1)}, \ldots, p''_{(N)}$ in the interconnection subnetwork, and pertaining to the different areas;
- powers $P_{(1)}, \ldots, P_{(N)}$ flowing through area nodes, expressed by;

$$P_{(r)} = P_{I(r)} - p''_{(r)}(P_{I(2)}, \ldots, P_{I(N)}) \quad (r = 1, \ldots, N)$$

and so on.

Therefore, it is possible to obtain:

$$\mathrm{d}P_{(1)} = -\sum_{2}^{N}{}_{s}\left(1 + \frac{\partial p''_{(1)}}{\partial P_{I(s)}}\right)\mathrm{d}P_{I(s)}$$

$$\mathrm{d}P_{(r)} = \left(1 - \frac{\partial p''_{(r)}}{\partial P_{I(r)}}\right)\mathrm{d}P_{I(r)} - \sum_{2}^{N}{}_{s\neq r}\frac{\partial p''_{(r)}}{\partial P_{I(s)}}\mathrm{d}P_{I(s)}$$

$(r = 2, \ldots, N)$; and by recalling Equation [2.3.30] it is finally possible to determine the cost variations $\mathrm{d}C_{(1)}, \ldots, \mathrm{d}C_{(N)}$ caused, in different areas, by small variations $\mathrm{d}P_{I(2)}, \ldots, \mathrm{d}P_{I(N)}$ of imported powers.

More precisely, if the power exchange P_{ab} from area a to area b is increased by the (small) amount:

$$\mathrm{d}P_{ab} = -\mathrm{d}P_{I(a)} = +\mathrm{d}P_{I(b)}$$

with powers imported by other areas being unchanged (i.e., with $\mathrm{d}P_{I(r)} = 0$ for $r \neq a, b$), it is possible to determine the following cost variations:

$$\mathrm{d}C_{(a)} = -\mu_{(a)}\,\mathrm{d}P_{(a)} = +\mu_{(a)}\left(1 - \frac{\partial p''_{(a)}}{\partial P_{I(a)}} + \frac{\partial p''_{(a)}}{\partial P_{I(b)}}\right)\mathrm{d}P_{ab}$$

$$\mathrm{d}C_{(b)} = -\mu_{(b)}\,\mathrm{d}P_{(b)} = -\mu_{(b)}\left(1 - \frac{\partial p''_{(b)}}{\partial P_{I(b)}} + \frac{\partial p''_{(b)}}{\partial P_{I(a)}}\right)\mathrm{d}P_{ab}$$

Furthermore, in the other areas ($r \neq a, b$):

$$\mathrm{d}C_{(r)} = -\mu_{(r)}\,\mathrm{d}P_{(r)} = +\mu_{(r)}\left(\frac{\partial p''_{(r)}}{\partial P_{I(b)}} - \frac{\partial p''_{(r)}}{\partial P_{I(a)}}\right)\mathrm{d}P_{ab}$$

(such expressions also can be used when $a = 1$ or $b = 1$, provided the partial derivatives with respect to $P_{I(1)}$ are assumed to be zero). This can provide important suggestions to arrange exchange programs between different areas.

The *optimal solution for the whole system* then corresponds to the $(N-1)$ conditions:

$$0 = \frac{\partial \sum_{1}^{N}{}_{s} C_{(s)}}{\partial P_{I(r)}}, \quad \text{or equivalently} \quad 0 = \sum_{1}^{N}{}_{s}\, \mu_{(s)} \frac{\partial P_{(s)}}{\partial P_{I(r)}} \qquad (r = 2, \ldots, N)$$

which, by considering what has been previously established, may be translated in the matrix equation:

$$A\begin{bmatrix}\mu_{(2)}\\ \vdots \\ \mu_{(N)}\end{bmatrix} = \begin{bmatrix}1\\ \vdots \\ 1\end{bmatrix}\mu_{(1)} \qquad [2.3.31]$$

in $\mu_{(1)}, \ldots, \mu_{(N)}$, where the elements of the matrix A are defined by:

$$A_{rr} \triangleq \frac{1 - \dfrac{\partial p''_{(r)}}{\partial P_{I(r)}}}{1 + \dfrac{\partial p''_{(1)}}{\partial P_{I(r)}}}$$

$$A_{rs} \triangleq \frac{-\dfrac{\partial p''_{(s)}}{\partial P_{I(r)}}}{1 + \dfrac{\partial p''_{(1)}}{\partial P_{I(r)}}} \qquad (s \neq r)$$

($r, s = 2, \ldots, N$). Note that the A matrix depends on powers $P_{I(2)}, \ldots, P_{I(N)}$. For any values of these last powers, Equation [2.3.31] allows, starting from $\mu_{(1)}$,

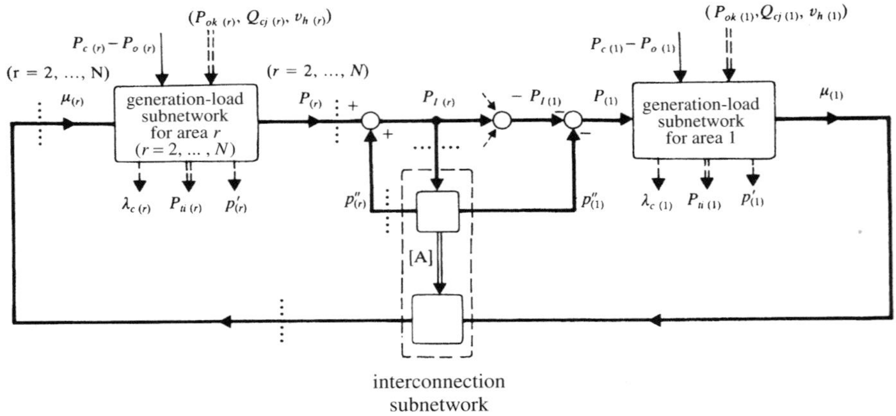

Figure 2.35. A computational block diagram for a multiarea system.

determination of $\mu_{(2)}, \ldots, \mu_{(N)}$; the final solution can be derived according to the diagram of Figure 2.35, by considering further links between involved variables.

The $(N-1)$ conditions expressed by the matrix Equation [2.3.31] also may be set into other equivalent forms such as, for instance, Equation [2.3.32] or [2.3.33] reported below. More precisely:

- if each of powers $P_{(1)}, \ldots, P_{(N)}$ is expressed as a function of the phases $\alpha_1, \ldots, \alpha_N$ of area node voltages, by setting:

$$0 = \frac{\partial \sum_1^N {}_s C_{(s)}}{\partial \alpha_r} \quad (r = 1, \ldots, N)$$

and by recalling Equation [2.3.30] the following $(N-1)$ conditions are derived:

$$0 = \sum_1^N {}_s \mu_{(s)} K_{sr} \quad (r = 2, \ldots, N) \tag{2.3.32}$$

with $K_{sr} \triangleq \partial P_{(s)}/\partial \alpha_r$ (the similar condition for $r = 1$ can be disregarded, as powers $P_{(s)}$ actually depend only on phase shifts, and therefore it holds $\sum_1^N {}_r K_{sr} = 0$);

- if total losses in the interconnection subnetwork are assumed to be of the form $\sum_1^N {}_r p''_{(r)} = \phi''(P_{(2)}, \ldots, P_{(N)})$, it follows:

$$P_{(1)} = -(\phi'' + P_{(2)} + \cdots + P_{(N)})$$

$$\mathrm{d} \sum_1^N {}_s C_s = -\sum_1^N {}_s \mu_{(s)} \, \mathrm{d}P_{(s)} = \sum_2^N {}_s \left(\left(1 + \frac{\partial \phi''}{\partial P_{(s)}} \right) \mu_{(1)} - \mu_{(s)} \right) \mathrm{d}P_{(s)}$$

and then the following $(N-1)$ conditions:

$$\mu_{(s)} = \left(1 + \frac{\partial \phi''}{\partial P_{(s)}}\right) \mu_{(1)} \quad (s = 2, \ldots, N) \qquad [2.3.33]$$

from which $\mu_{(2)}, \ldots, \mu_{(N)}$ starting from $\mu_{(1)}$.

If variations of losses in the interconnection subnetwork are disregarded, the results simplify to:

$$\mu_{(1)} = \mu_{(2)} = \cdots = \mu_{(N)}$$

while the matrix A becomes the $(N-1, N-1)$ identity matrix, and in Equations [2.3.32] and [2.3.33] it respectively holds $\sum_1^N {}_s K_{sr} = 0$, $\partial \phi''/\partial P_{(2)} = \cdots = \partial \phi''/\partial P_{(N)} = 0$. In this case, therefore, *the optimal solution implies (further than the equality of the incremental costs at load, within each generation-load subnetwork) the equality of the incremental costs at the different area nodes*, that is:

$$\lambda_{c(1)}\left(1 - \frac{\partial p'_{(1)}}{\partial P_{(1)}}\right) = \cdots = \lambda_{c(N)}\left(1 - \frac{\partial p'_{(N)}}{\partial P_{(N)}}\right)$$

2.4. PREVISIONAL SCHEDULING

2.4.1. Generalities

Data for the previsional scheduling basically concern:

- system components;
- load demands;
- different inflows available for generation.

Such data are obviously affected by uncertainties, for what respectively concerns (as specified later on) the risks of forced unavailability of components and the difficulties in forecasting load demands and inflows.

Data about *system components* concern not only already existing equipment, but also those on the way into service. Thus problems relevant for the operational scheduling also can overlap—particularly in the long term, e.g., from 6 months to several years—with those concerning the system development planning.

Each component is intended to be specified by its main operating characteristics, in addition to its operating constraints. In particular, for generating units it is necessary to consider the maximum active powers that they can generate (and other capability limits; see Section 2.2.1), generation costs (by assuming as known the unit costs of fuels; see Section 2.2.5d), and further constraints and costs related to startups and shutdowns (Section 2.4.2b).

Furthermore, for each component, ordinary maintenance requirements must be specified in terms of duration and frequency.

Finally, the risk of *forced unavailabilities* caused by damages (more or less casual in nature) must be taken into consideration. Apart from damages resulting from specific deterministic causes, such as errors in design, manufacturing, or maintenance, random damage events should be considered and treated by probabilistic models. These events can be characterized by experimentally derived parameters; for instance, the mean "outage" duration τ_a (continuous unavailability, with repair time included) may be in the range of at least 1–5 days for hydroelectric or gas turbine units, and 3–10 days or more for thermal units equipped with steam turbines. For the most usual values of the outage frequency f_a, the "availability factor" for a unit (defined as the ratio between the availability duration and the total duration, and thus equal to $(1 - f_a\tau_a)$) may be 0.80–0.95 for hydroelectric units, and even less for thermal units equipped with steam or gas turbines. Outage frequency may actually be higher for both new and old units (such as for 40–50 year old hydrounits, or 20–25 year old thermal units), respectively, caused by settlement and wear reasons. Based on such data, it is possible to determine probabilistic indications on the actual availability, along the time, of each single component. Once we know the units scheduled to be in operation and, for each of them, the date of the most recent outage, we also may evaluate the probability that the total available power, at any generic instant, is not lower than a given value, and so on.

Load demands are defined, in detail, by the variations with time ("load diagram"), as active and reactive powers absorbed at each load node. However, because of uncertainties in forecasting, it may be preferable to accept simpler more easily predictable specifications, by grouping loads at a single or a few "equivalent" nodes and/or assuming step-varying load diagrams defined by mean values within each time interval, etc., according to the following (risks resulting from forecasting errors should be accounted for).

Also, load demands—apart from the small, rather quick fluctuations with zero mean value, already considered in Section 1.2.2—are related to requirements of different nature which are typically periodical or almost so. The typical periodicity (see Fig. 2.36) is:

- daily, with higher demands in daytime hours (mostly due to industrial loads at working hours, and loads for "lighting" in the early evening) with respect to night hours;
- weekly, with higher demands (and of different nature) in working days with respect to weekends;
- yearly, with different demands depending on months and weeks (e.g., results from winter heating, holiday weeks, etc.),

and with different amounts in the subsequent years, according to variations of the demand law in the long term.

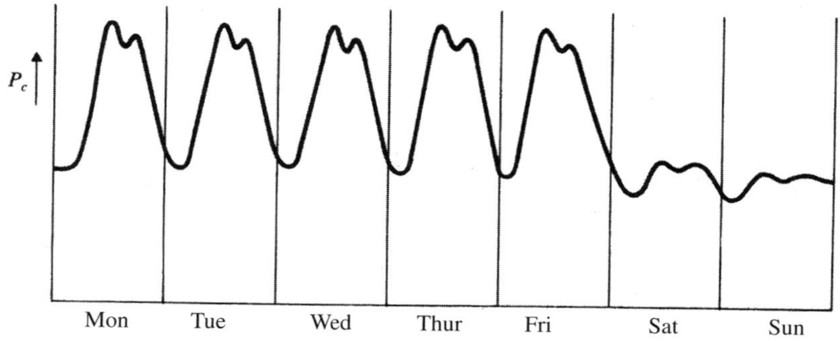

Figure 2.36. Example of a weekly load diagram (P_c = total active load power).

Load demand *forecasting* is essentially based on:

- past information (e.g., obtained by recording at load nodes, or at generation and boundary nodes), concerning, for instance, the shape of the load diagram and the energy absorbed in days similar to the considered one;
- previsions about parameters affecting load demands themselves, among which the industrial development and (particularly in the short term) the meteorological conditions, further than other contingent causes of different nature (modifications in working times, television programs, etc.).

Probabilistic indications on the reliability of load forecasts may be obtained by detecting past errors for any given day and hour (e.g., peak hour), and so on.

The *different inflows available for generation* are typically constituted by water inflows (having subtracted possible spillages) in the hydroplants. Similarly, we may consider natural inflows of fuel or of motive fluid to possible geothermal power plants, and so on. Hydraulic inflows are basically the *natural* ones, caused by rain and snow and ice melting, and then depending on meteorological conditions and water travel times. Moreover, in the case of snow and ice melting, the inflows depend on the state of snow fields and glaciers, and thus on the meteorological conditions of the preceding winter. Further inflows may be added to natural ones, such as those *due to hydraulic coupling*, i.e., caused by the outflow from hydroplants located upstream and belonging to the same valley, and those *due to pumping*.

It also may be thought that the causes which determine the hydraulic inflows arise in a rather similar way year by year. However, it would be unreasonable to estimate future inflows as functions of time, within the span of a whole year. For long intervals (e.g., 1 year, 1 month, or even 1 week, as per interpolation on each month), previsions can be formulated only on water volumes due to inflows, based on previous multiyear experience (and for instance in terms of probability that the total energy at disposal from the totality of inflows is not below a given value). The inflows as functions of time are instead predicted only for shorter intervals, not too far in the future (e.g., the next day).

The basic problems related to previsional scheduling are qualitatively summarized in Figure 2.37. For some problems the solution may be required, broadly, with a significant advance and with reference to long periods of operation. Reference can be made to the necessity of in-time agreements with suppliers of fuel (its acquisition imply may also delays in transportation etc.) and with other utilities (for what concerns exchanges through interconnections), and furthermore to the convenience of scheduling—even for long-time horizons—fuel storages, water storages in reservoirs, maintenance periods of units, etc. Thus it is necessary to carry out, at first, a "*long-term*" *scheduling* (which, for instance, covers 1 year and is updated, for instance, monthly with reference to the subsequent 12 months), based on data relatively global and therefore less affected by forecast uncertainties, and sufficient to provide a quick response to the above-cited requirements. In this context, the search for detailed solutions would be scarcely significant (apart from the enormous computing complications) because necessary data would be too uncertain, and many results would not be necessary so much in advance.

On the other hand, more detailed, useful, and significant solutions can be obtained only with reference to more limited future intervals. A "*medium-term*" *scheduling* (concerning, for instance, the next week) is then performed, to better define the amount of water to be used and possibly pumped, in hydroelectric plants within each day, the operating intervals of thermal units, and so on. The detailed determination of configurations and working points to be achieved, for instance, during the next day, can be then obtained by a "*short-term*" *scheduling*, based on the most updated forecast. Results so obtained may be adjusted by the real-time scheduling, accounting for actual operating conditions.

2.4.2. Typical Formulations and Procedures

For each of the three scheduling levels mentioned (long-, medium-, and short-term) the *total interval* of time under consideration (lasting T) is subdivided into several *elementary intervals* (lasting τ). The results of scheduling are referred to these elementary intervals, in accordance with a "discrete" formulation. The different pairs of values T, τ, must be chosen according to the nature of the problems considered; in fact, the choice of the duration τ must account for a reasonable level of detail to be achieved in the definition of results, based on enough credible data along the interval T.

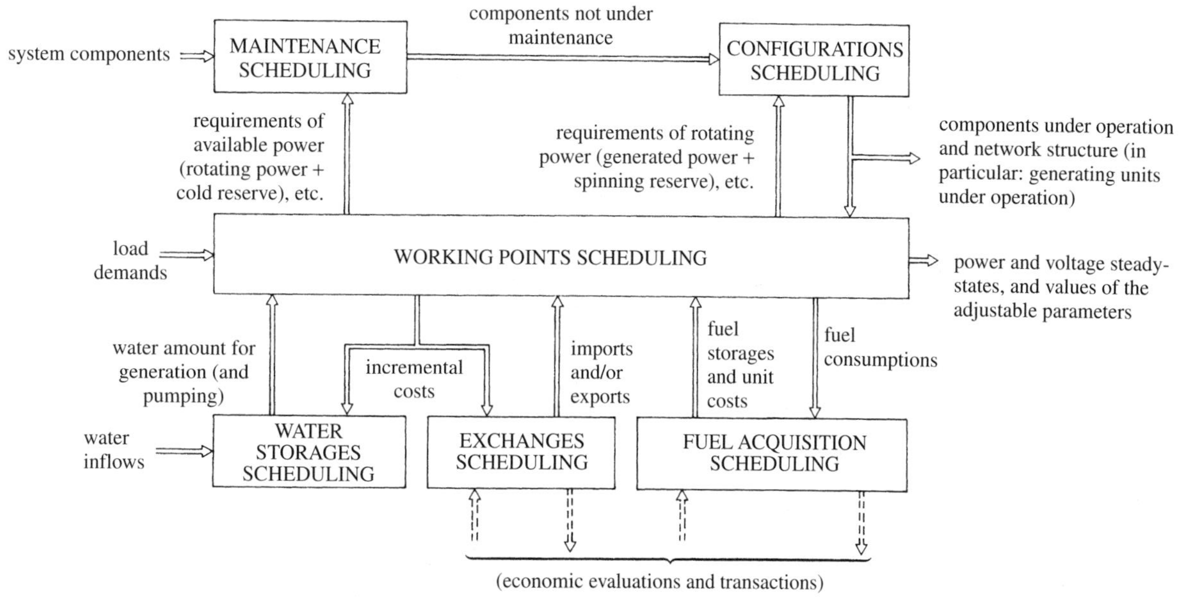

Figure 2.37. Basic problems of previsional scheduling.

As a general guideline, the following may be assumed:

- long-term scheduling: $T = 1$ year, $\tau = 1$ week;
- medium-term scheduling: $T = 1$ week, $\tau = 1$ day (or 4 hours, or 2 hours);
- short-term scheduling: $T = 1$ day, $\tau = 30$ minutes (or 15 minutes)[(13)].

The relevance of the problems and their subdivision into the three mentioned levels may depend on the characteristics of the system under examination. The hints reported in the following reference examples of formulations and procedures, with particular concern to active power generation (and exchange) schedules, and under the assumption that the hydroelectric generation has a significant, though not prevailing, importance.

(a) Long-Term Scheduling

With the long-term, the choice $T = 1$ *year* appears particularly reasonable, because it corresponds to the longest time interval by which load demands and hydraulic inflows vary more or less cyclically. The scheduling may be updated at each month (with reference to the following 12 months), so that the most updated results concerning the next month can be available for the medium-term scheduling, as time goes on.

Similarly, the choice $\tau = 1$ *week* appears to be convenient, as it corresponds to a significant duration with respect to load demands and maintenance requirements, without getting into too much detail for what concerns possible daily or weekly generating-pumping cycles etc., according to the following.

With such choices and with $R \triangleq T/\tau$ (i.e., $R = 52$ or $R = 53$, excluding truncations at the first and last weeks) we may assume as *known, for each week* ($r = 1, \ldots, R$):

- the set of generating units (thermal and hydroelectric ones);
- the total energy $E_c^{(r)}$ demanded by loads and, in a typical, properly simplified form, the total load diagram or the corresponding "duration" diagram (see footnote[(12)]);
- the total energy $E_a^{(r)}$ at disposal, due to the whole set of inflows (both natural and due to hydraulic coupling).

The *maintenance scheduling* (based on estimation of wear, as a function of past operating conditions) allows definition of—at least broadly and apart from

[(13)] Assuming $\tau = 1$ day for medium-term scheduling, any overlapping is avoided, with time detail, between two consecutive schedulings. In fact, when passing from long- to medium-term scheduling, the elementary interval becomes the total interval, as well as passing from medium- to short-term scheduling.

However, smaller values of τ for medium term may look preferable, e.g., to schedule in a better way (already on a weekly base) the operating hours of thermal units.

the risk of outages — the set of available (i.e., not under maintenance) units within each week[(14)].

In this regard, it is important to consider the following, within each time interval:

- the set of available units must guarantee not only a sufficient spinning reserve, but also an adequate *"cold" reserve* (i.e., available but not operating units) to be capable of handling critical situations caused by loads much larger than forecasted or lack of inflows, unit outages, loss of imported powers, etc.;
- the cold reserve must (see also Section 1.7.1) be geographically spread in a relatively uniform way (similar as per the spinning reserve);
- the cold reserve must furthermore include an adequate portion of "quick" reserve — typically constituted by hydroelectric or gas-turbine units — capable of intervening within minutes (the slower reserve, constituted by steam-turbine units, which can startup in hours, can be useful in case of critical conditions which are foreseen in advance, such as systematic load increases in the day, etc.).

The maintenance schedule also may be influenced by other requirements (e.g., to avoid overflows in a hydroelectric plant, the available power should not be reduced for that plant when inflows are large), and it must be coordinated with the availability of maintenance personnel, etc.

The maintenance schedule initially chosen may undergo corrections and refinements in a shorter-term context.

The *scheduling of water storage* has the aim of defining, for each week, the total energy $E_w^{(r)} (r = 1, \ldots, R)$ to be hydroelectrically generated without considering pumping. By indicating with $V^{(0)}$ the total (hydraulic) energy stored at the beginning of the year, and with $V^{(r)}$ the value assumed for energy at the end of the r-th week, with constraints:

$$V^{(r)} \in [V_{\min}, V_{\max}] \qquad [2.4.1]$$

that will be discussed later, the full sequence $V^{(1)}, \ldots, V^{(R)}$ is given by[(15)]:

---

(14) The time required for maintaining a generating unit is weeks per year (e.g., approximately 2–5 weeks/year for hydroelectric or gas-turbine units, and 5–10 weeks/year or more for steam-turbine thermal units).

The maintenance of other components (e.g., lines, transformers, etc.) requires instead smaller durations and can be carried out during low-load intervals (night or weekends), consistently with the medium- and short-term scheduling.

(15) Possible overflows and spillages, the effects of which would be equivalent to a decrement in $E_a^{(r)}$, are disregarded here for simplicity. Furthermore, it is assumed that generating-pumping cycles are daily or, at most, weekly. When pumping is present, by indicating with $E_g^{(r)} \triangleq E_w^{(r)} + \Delta E_g^{(r)}$ the total energy hydroelectrically generated within the r-th week (Section 2.3.3), the term $\Delta E_g^{(r)}$ corresponds to the use of pumped volumes of water (pumping inflows) within the r-th week itself, so that Equation [2.4.2] remains valid.

$$V^{(r)} = V^{(r-1)} + E_a^{(r)} - E_w^{(r)} = V^{(0)} + \sum_1^r {}_j \left(E_a^{(j)} - E_w^{(j)}\right) \quad (r = 1, \ldots, R)$$

[2.4.2]

It may be reasonable to impose that the stored hydraulic energy has the same value at the beginning and at the end of the year, i.e., $V^{(R)} = V^{(0)}$; thus it holds:

$$\sum_1^R {}_r E_w^{(r)} = \sum_1^R {}_r E_a^{(r)} \qquad [2.4.3]$$

which states that the total annual hydroelectrically generated energy is equal to that (known) available from inflows for the whole year.

The problem is then the choice of the subsequent values $E_w^{(1)}, \ldots, E_w^{(R)}$, with their sum known. If the storage capability is small, the solution cannot differ significantly from $E_w^{(1)} = E_a^{(1)}, \ldots, E_w^{(R)} = E_a^{(R)}$, corresponding to the simple use of inflows, week by week (see also the following Section (b), with reference to medium term). On the contrary, assume that the system also includes large-capacitance reservoirs, specifically "multiweeks" or "seasonal" reservoirs which, in the absence of inflows and for nominal conditions of initial storage and (available) generated power, are characterized by an "emptying time" larger than 1 week. To determine the sequence $E_w^{(1)}, \ldots, E_w^{(R)}$ it is then possible, for each week:

- to consider some possible values of the energy $E_w^{(r)}$;
- to carry out, for each value $E_w^{(r)}$ (based on the presumed load diagram or on the "duration" diagram), a preliminary scheduling of:
  - the configurations (specifically, operating intervals of thermal units);
  - the working points (specifically, generated powers of thermal units during the week, with possible pumpings);
  - the exchanged powers;

  by adopting, in a simplified form, criteria similar to those adopted in medium-term scheduling (specifically, a priority list based on $(C_i/P_{ti})^*$ values; see Section (b));
- to determine the resulting total cost $J^{(r)}$ (and fuel requirements) for the week, as a function of the energy $E_w^{(r)}$.

The values $E_w^{(1)}, \ldots, E_w^{(R)}$ can be then determined by imposing:

$$\min \sum_1^R {}_r J^{(r)}(E_w^{(r)})$$

with the equality constraint [2.4.3] and, for any given value $V^{(0)}$, with the inequality constraints [2.4.1], where each energy $V^{(r)}$ is defined by Equation [2.4.2].

With conditions [2.4.1], it should be preliminarily assumed that they account for storage limits of plants, with prudential margins caused by uncertainties on data (availability of units and values of energy $E_c^{(r)}$ and $E_a^{(r)}$). For each admissible sequence $V^{(0)}, V^{(1)}, \ldots, V^{(R)}$, with $V^{(R)} = V^{(0)}$, it is possible, at the beginning of each interval r (monitoring data uncertainties), to evaluate the risk of not meeting the energy demand $(E_c^{(r)} + \cdots + E_c^{(R)})$ for the rest of the year, with the actually available units and inflows.

Specifically, it is possible to define for any generic $V^{(0)}$ a "lower-limit" curve, corresponding to a sequence $V^{(0)}, V^{(1)*}, \ldots, V^{(R-1)*}, V^{(0)}$ at a constant risk (equal to the one at the beginning of the year). To guarantee an always adequate "*energy*" *reserve*, with risks not larger than the above-mentioned value, it must be imposed $V^{(r)} \geq V^{(r)*}$ (that is, $V^{(r)} \in [V^{(r)*}, V_{\max}]$), with $V^{(r)*}$ dependent on $r (r = 1, \ldots, R-1)$. The larger values of $V^{(r)*}$, i.e., of the strictly required hydraulic storages, correspond to weeks preceding the heavier load periods.

The scheduling of configurations, working points, and power exchanges, obtained for the whole year in correspondence with the chosen values of $E_w^{(1)}, \ldots, E_w^{(R)}$, may be considered as largely preliminary ones, in view of the most adequate corrections and refinements which can be achieved by medium- and short-term scheduling.

However, the *convenience of exchanges* (active power imports or exports) in given periods of the year may already suggest negotiations and preliminary long-term agreements with the involved utilities (also the possibility of temporal compensations, according to Section 2.3.4, should be considered).

Finally, the *scheduling of fuel acquisition* may already use valuable indications concerning fuel needs along the year. Within multiyear agreements it is possible, for example, to specify minimum and maximum amounts of fuel requested in different months (with the possibility of further details and possible new contracting, in a shorter-term scheduling), also accounting for possible cost fluctuations. Additionally, the need of adequate storages (see Section 2.3.2b) must be considered, also with reference to nonforecasted increments in energy demands, risks of interruptions in supply, delays in transportation, and so on.

(b) Medium-Term Scheduling

The fundamental problems to be solved in medium-term scheduling concern configuration and water storages; precisely, the typical goals are:

- determination of water storages at the end of subsequent days (for any value of the storage at the beginning of the week), for each hydroelectric plant;
- determination of the operating schedule of units (*unit commitment*), specifically thermal units;

in addition to the revision and refinement of exchange programs, etc.

The choice of a total interval duration $T = 1$ *week*, as assumed in the following, may be efficient and meaningful, as:

- it allows a good link with the long-term scheduling;

- the operating schedule of thermal units (with possible startups and shutdowns) must be chosen along a time interval including at least 1 week because of the alternating working days and holidays, with significantly different load demands;
- considering a period of 1 week, it is possible to use data reliable enough, with respect to the described goals.

With the elementary interval, the duration $\tau = 1$ *day* is significant for load demands and water storage schedules, and for deciding thermal unit shutdowns in the lower load days (e.g., holidays). However, when evaluating the efficiency of shutdowns within the same day (typically at night), it is necessary to adopt smaller values (e.g., 4 *hours* or 2 *hours*).

Data for medium-term scheduling are basically the following(16):

- *from updated previsions:*
  - the load diagram (adequately sketched) along the week;
  - the daily energy made available by natural inflows (having subtracted possible spillages) in the different days of the week, for each hydroelectric plant;
- *based on long-term scheduling and updated information*:
  - the set of (thermal and hydro) generating units available for the whole or partial week, considering maintenance schedules and forced outages;
  - the (hydraulic) total energy stored for all hydroplants at the beginning of the week (V_{in}), and the desired one at the end of the week (V_{fin});
  - a preliminary schedule of active power exchanges during the week.

It is possible to add *particular constraints*, caused by contingent reasons, for example:

- assignments concerning the *operating intervals* of some units and, possibly, their generated powers (*assigned generation*); for example, units belonging to flowing water or geothermal or cogeneration plants, units under testing, etc.;
- constraints on the *geographical distribution of generated powers* (accounting for that of loads); for instance to avoid excessively large currents in lines surviving the loss of important network link.

The *scheduling of water storages* or of energy stored in hydroelectric plants may be detailed as follows:

(16) Further data concerning fuels are not considered here for simplicity (forecasted incoming of fuel, quantities used in the week, etc., similarly to data concerning hydraulic resources; see Section 2.3.2b), by assuming that their availability is always guaranteed.

- variations in energy $(V_{\text{fin}} - V_{\text{in}})_{(j)} (j = 1, 2, \ldots)$ stored in different plants are selected, accounting for the constraint $\sum_j (V_{\text{fin}} - V_{\text{in}})_{(j)} = V_{\text{fin}} - V_{\text{in}}$;
- energy $E_{w(j)}$ generated in the week by each plant (not considering pumping) is deduced;
- each $E_{w(j)}$ is subdivided between the different days of the week;
- finally, the storage values at the end of the different days are determined for each plant.

The subdivision of $(V_{\text{fin}} - V_{\text{in}})$ between plants may be decided (at least as a first approximation) based on respective storage capacities. Specifically, for smaller-capacity plants it may be assumed $(V_{\text{fin}} - V_{\text{in}})_{(j)} = 0$, i.e., by imposing the same storage at the beginning and end of the week; such a choice typically concerns the so-called *weekly* basins for which the storage decrement during weekdays is compensated at weekends, and (with all the reasons) *daily* basins for which the storage decrements at the heavier load hours are compensated during the remaining hours, so that the initial storage can be reset at the end of each day.

Once the differences $(V_{\text{fin}} - V_{\text{in}})_{(j)}$ are assigned, it is possible to derive each energy $E_{w(j)}$ by using the relationship $E_{w(j)} = E_{a(j)} - (V_{\text{fin}} - V_{\text{in}})_{(j)}$, indicating by $E_{a(j)}$ the available weekly energy caused by water inflows (natural and due to hydraulic couplings) in the j-th plant(17). The total energy $\sum_j E_{w(j)}$ generally differs from that forecasted in the long-term scheduling, because it is evaluated from more realistic data.

If values $E_{w(j)}$ are inconsistent with available unit power and other constraints — concerning for instance, the geographical distribution of generation — the partition of $(V_{\text{fin}} - V_{\text{in}})$ should be rearranged.

Moreover, the partition of $E_{w(j)}$ between days of the week must be determined, considering the requirement of satisfying the load diagram and storage constraints, according to the following. However, this partition may be considered compulsory (or almost that) because of storage constraints (e.g., regarding daily basin plants for which the energy generated in each day should practically equal that corresponding to water inflows). Moreover, the need for adequately coordinating generation schedules of the plants in the same valley is evident.

Once the initial storage is known, *storages at the end of successive days* can be determined for each plant, based on daily generated energy and total inflow. If these storage values are inconsistent with the respective constraints, the partition of $E_{w(j)}$ between the days of the week should be adequately modified. For generating-pumping plants, the nondaily cycle pumping schedules must be considered, according to the following.

(17) The energies $E_{w(j)}$'s can be computed at once only for isolated plants and for those at the top of possible "valleys," for which $E_{a(j)}$ is a data, depending only on natural inflows. However, knowing the energies generated through the top plants in each valley, enables evaluation of the total water inflows to downstream plants supplied by them, and thus it is possible to know all corresponding $E_{a(j)}$'s.

Regarding the *matching of load demand*, the diagram of requested power P'_c during the week (having subtracted exchanges and assigned generations, and disregarding network losses) must first be derived from available data. Once this diagram is known, the hydroelectric generation can be scheduled in such a way to minimize the total generation cost. Remember what is expressed in Section 2.3.3b2 with reference to always insufficient hydroelectric generation, specifically assuming that storage constraints are not activated and disregarding pumping.

Under the hypothesis that the (until now unknown) operating thermal plants remain unchanged for the whole week, each energy $E_{w(j)}$ must be used to "level" at maximum effort, during the whole week, the diagram of the total nonassigned thermal power P_t, starting from the diagram of power P'_c (accounting for power limits of hydroelectric units and other constraints already considered).

Usually, it is efficient, if not necessary, to modify the set of operating thermal units, for reasons specified later in this section. Indicating by P'_t the total power generated by thermal units in continuous operation, and by P''_t the total power generated by *units in noncontinuous operation*, not considering assigned generation, it is possible to impose that the hydroelectric generation "levels" at the best the diagram of P'_t starting from the diagram of $(P'_c - P''_t)$. However, as the operating schedule of units and the time behavior of P''_t are unknown, it is possible to proceed as follows (see Fig. 2.38):

- an adequate preliminary time diagram for P''_t is assumed (e.g., $P''_t \equiv 0$);
- the corresponding diagram for P'_t is derived (by leveling it at the most, as already said) and then that of $P_t = P'_t + P''_t$;
- based on the diagram of P_t, we may derive (according to the below information) the operating schedule of thermal units and the subdivision of P_t between the units, so that new diagrams for both P'_t and P''_t may be found;

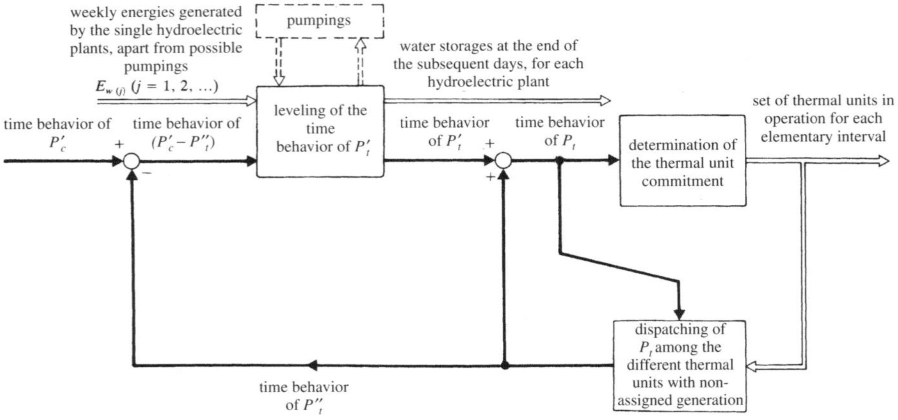

Figure 2.38. A computational block diagram in the presence of thermal units in noncontinuous operation.

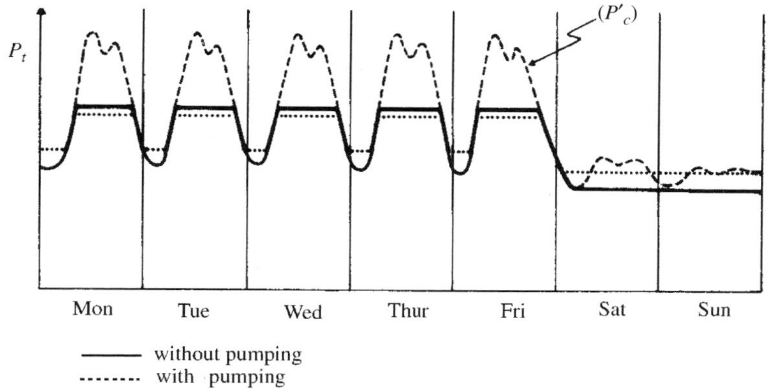

Figure 2.39. Example of weekly diagram of thermal generation. Power P'_c is the total load demand, having subtracted exchanges and assigned generations. Powers P_t and $(P'_c - P_t)$ are, respectively, the total thermally and hydroelectrically generated powers (apart from assigned generations).

- as a second attempt, the new P''_t diagram determined as above is adopted;
- new diagrams for P'_t, P''_t, etc. may be derived, by iterating until the final solution[(18)].

The resulting thermal generation diagram of working days is significantly different from that of holidays (see the example of Fig. 2.39), essentially because of:

- significant differences between working day and holiday loads;
- efficiency or necessity (according to what is already mentioned) of using inflows within short times (and thus of generating even in holidays) for hydroelectric plants having small storage capability.

On the other hand, pumping actions, which will be considered later, may not substantially reduce, in practice, the mentioned difference.

Once the diagram $P_t(t)$ of thermal generation is known, the determination of the unit's operating schedule still requires the evaluation of the *spinning reserve requirements* (see Section 2.3.1d) during the week, starting from the diagrams of load and of exchanged powers.

More simply, it is possible to determine for each day the maximum value of the strictly necessary spinning reserve or, in a better way (e.g., if $\tau = 2$ hours), the two maximum values respectively corresponding to daytime hours (the value

[(18)] If the operating intervals of each thermal unit remain unchanged in subsequent iterations (or if they are assigned), the procedure may be used to determine the time behavior of generated powers; however, this determination becomes more interesting in short-term scheduling.

to be used, for instance, between 6:00 a.m. and 10:00 p.m.) and night hours (the value to be used in the remaining 8 hours).

Once the spinning reserve supplied by hydroelectric plants (by assuming that these units are stopped during intervals at no power) is known, it is possible to derive the minimum spinning reserve required to thermal units (apart from further checks, concerning the geographical distribution of spinning reserve, its capability of coping with the largest plant outage, etc.), and thus the minimum value of rotating power during the week.

To determine the *operating intervals of thermal units* (with these intervals multiple of τ), it is then possible to apply the following procedure as a first approximation:

- by indicating $(C_i/P_{ti})^*$ as the minimum value of the "specific cost" of the generic i-th unit (see Fig. 2.31b), units are ordered so that $(C_1/P_{t1})^* < (C_2/P_{t2})^* < \ldots$;
- for each elementary interval τ, units are scheduled into operation adopting the above priority list 1,2,..., until the total rotating power (generation and spinning reserve) becomes sufficient for the whole interval considered[(19)].

The procedure illustrated is extremely simple and avoids consideration of time dependency $P_c'(t)$, so that it can be usefully applied (e.g., in the long term) even with only knowing the "duration" diagram of P_c' (which is much easier to be forecasted). As a partial justification of the procedure, it may be also noticed that:

- usually, the minimum specific cost $(C_i/P_{ti})^*$ corresponds (exactly, or almost exactly) to $P_{ti} = P_{ti\,\max}$;
- if the total power P_t to be supplied is equal to $P_{t1\,\max}$, $(P_{t1\,\max} + P_{t2\,\max})$, ..., the generation cost is minimum when respectively using (at full power), only unit 1, units 1 and 2, and so on (see also Section 2.3.3c).

On the other hand, one might easily object that, when using units only at full power, they cannot contribute to spinning reserve. Moreover, the minimization of the generation cost is not always guaranteed, for values of P_t different from those considered above (see the elementary example in Fig. 2.40). Furthermore, variations in the demanded rotating power generally lead to modifying the set of operating units during the week, through startups and shutdowns. Thus,

---

(19) If the operating intervals of some units are already assigned, this must be considered in advance in a straightforward way.

Moreover, it is efficient to place "older"—and therefore less reliable—units at the end of the list, possibly using them as cold reserve; the same applies, for different reasons, for gas-turbine units (which have rather high $(C_i/P_{ti})^*$ values and can serve as a possible "quick" cold reserve).

Generally, it must be considered that "equivalent" unit costs of fuels, and thus the above-mentioned specific costs, may undergo changes related to the actual availability of fuels themselves (see Section 2.3.2b).

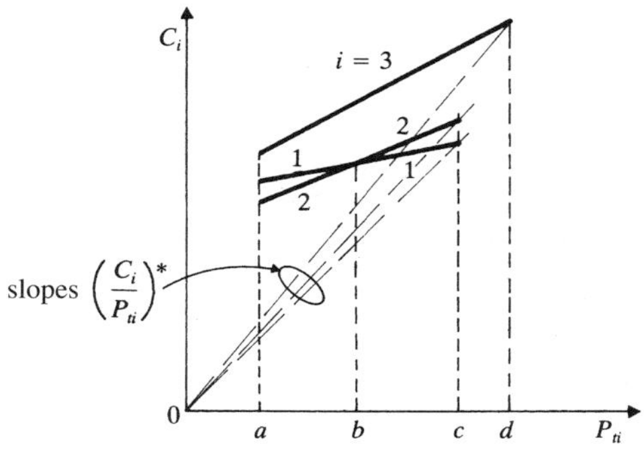

Figure 2.40. Example involving three units, with:

$$P_{t1\,\min} = P_{t2\,\min} = P_{t3\,\min} = a$$

$$C_1(b) = C_2(b)$$

$$P_{t1\,\max} = P_{t2\,\max} = c$$

$$P_{t3\,\max} = d$$

and furthermore:

$$(C_1/P_{t1})^* < (C_2/P_{t2})^* < (C_3/P_{t3})^*$$

Indicating by P_t the total demanded power, the minimization of $(C_1 + C_2 + C_3)$ is achieved by putting into operation:

- only unit 2, for $P_t \in [a, b)$;
- only unit 1, for $P_t \in (b, c]$;
- only unit 3, for $P_t \in (c, d]$;
- only units 1 and 2, for $P_t \in (d, 2c]$;
- all units, for $P_t \in (2c, 2c + d]$.

additional costs (and constraints) related to such operations should be considered, by accounting for the dependence on time.

More precisely, if a thermal plant is, after shutdown, left undergoing a natural cooldown, there is a loss of energy E — approximately increasing with time according to an exponential law $E = E^o(1 - \epsilon^{-t/T^o})$ — which must be supplied before the next restoration to operation. Such energy supply must be provided gradually to avoid too fast variations of temperatures. The startup then implies a consumption of fuel and a further time delay Δt, both of which are nearly proportional to E and then increase with the duration t of the cooldown. (The time constant T^o may be 30–40 hours; in the worst conditions, i.e., after a complete cooldown ($E = E^o$), the fuel consumption is not usually excessive, as it may be the same amount necessary to generate the rated power for 30 minutes of operation, but the time delay Δt can be significant, e.g., 5 hours.). As an alternative, the plant also could be kept in a "warm" state, with fuel consumption proportional to the duration of nonoperating condition. This can be efficient only if such a duration is relatively modest (e.g., a few hours)[(20)]. However, the cost to be considered is not only the cost related to the fuel consumption. The different operations required, in fact, imply a personnel cost and contribute to the deterioration process of the plant, which requires additional maintenance costs. Due to limitations in personnel, the maximum number of such operations that can be performed for a given plant, within each assigned elementary time interval τ, should be considered.

Usually, the above-mentioned procedure is acceptable to determine *the set of units* (thermal units, with nonassigned generation schedule) *to be operated at the most severe conditions*, i.e., within the interval or intervals τ for which the rotating power that is required to them is maximum. Then, it should be verified whether it is efficient to stop the operation of some units in other time intervals, according to the above procedure itself (by directly accounting for time constraints related to startups and shutdowns, further than for constraint on spinning power), thus translating the original problem into a *unit decommitment* problem starting from the above-mentioned set[(21)].

In such a view, when stopping the h-th unit from operation in a generic interval θ_h:

- startup and shutdown operations imply a cost (A_h);
- within the interval θ_h, the cost of generation for the h-th unit becomes zero, but that of the remaining units increases because they must generate a larger amount of power; as a consequence, denoting by the superscript "o" the case in which the h-th unit is kept into operation, there is a saving:

$$G_h^o = \int_{(\theta_h)} C_h(P_{th}^o)\,\mathrm{d}t$$

and simultaneously an extra cost:

$$\Delta G' = \int_{(\theta_h)} \sum_{i \neq h} (C_i(P_{ti}) - C_i(P_{ti}^o))\,\mathrm{d}t \qquad [2.4.4]$$

(whereas $\sum_i P_{ti}^o = \sum_{i \neq h} P_{ti} = P_t$, and thus $\sum_{i \neq h} (P_{ti} - P_{ti}^o) = P_{th}^o$).

---

(20) The fuel consumption required to restore the kinetic energy lost at shutdown may be disregarded. Note that for a hydroelectric unit the energy to be restored is, practically, only kinetic; this similarly implies a negligible consumption of water.

(21) Even not considering startups and shutdowns, the leveling of $P_t(t)$ can be positively judged not only for the reason considered up to now (minimization of generation cost, for any given set of units), but also for the possible reduction of the maximum value of rotating power requested and, consequently, of the set of units to be operated.

Consequently, it is efficient to remove the h-th unit from operation in the interval θ_h, only if it results[22]:

$$G_h^o > \Delta G' + A_h \tag{2.4.5}$$

When evaluating the generation costs, it is assumed that they are minimized, by imposing (for units in operation, with nonassigned generation) the *equality of the incremental costs* (see Section 2.3.1, neglecting specific constraints on currents etc., which also could suggest some rearrangement to the set of considered units).

If $P_{th}^o \ll P_t$, the argument within the integral of Equation [2.4.4], i.e., the overcost per unit of time resulting from the increment of power P_{th}^o required to the remaining units, may be approximated by $\lambda^o P_{th}^o$, assuming that λ^o is the incremental cost when the h-th unit is in operation[23]. It then results:

$$\Delta G' \cong \int_{(\theta_h)} \lambda^o P_{th}^o \, \mathrm{d}t$$

and condition [2.4.5] can be approximated by:

$$\int_{(\theta_h)} (C_h(P_{th}^o) - \lambda^o P_{th}^o) \, \mathrm{d}t > A_h$$

or even:

$$(C_h(P_{th}^o) - \lambda^o P_{th}^o) T_h > A_h \tag{2.4.6}$$

if P_{th}^o, λ^o are nearly constant in the whole interval θ_h, lasting T_h (i.e., with a total power P_t slightly varying in θ_h, and the same assumption for the number of units in operation). Therefore, to decide which units must be excluded in a given interval, with λ^o known, it may be useful to evaluate the respective values $(C_h(P_{th}^o) - \lambda^o P_{th}^o)$; see for instance (under the assumption that (C_h, P_{th}) are linear characteristics) Figure 2.41, by which the importance of the specific cost $(C_h/P_{th})^*$, as well as that of $C_h(P_{th\,\min})$, $P_{th\,\min}$, $P_{th\,\max}$, is evident.

Alternatively, it is possible to adopt a procedure based on dynamic programming, by applying a formula:

$$\mathcal{F}^o(r, X_k) = \min_{X_j}(\mathcal{F}^o(r-1, X_j) + A(r-1, X_j; r, X_k)) + G(r, X_k)$$

---

[22] The (connected) interval θ_h can, for instance, correspond to the two holidays in the weekend or to night hours of a given day. In some cases, the cost A_h, which depends on the duration of nonoperation, must be evaluated by considering the contiguous weeks. Furthermore, because of the technical minima $P_{ti\,\min}$'s, the shutdown of some units becomes a necessity, without further consideration, if $P_t < \sum_i P_{ti\,\min}$.

[23] Actually, λ^o should be replaced by a proper value, somewhere between the incremental costs with and without the h-th unit in operation.

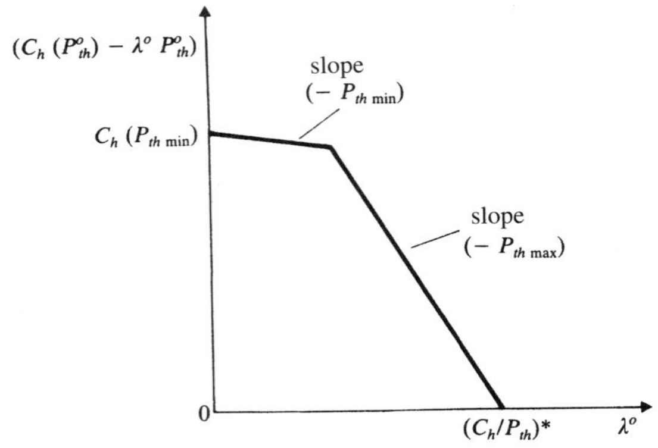

Figure 2.41. Overall saving on the generation cost per unit of time, if the h-th unit is excluded by operation in a given interval. The characteristic $C_h(P_{th})$ is assumed to be linear and it is assumed that:

$$P^o_{th} \begin{cases} = P_{th\,\min} \\ \in [P_{th\,\min}, P_{th\,\max}] \\ = P_{th\,\max} \end{cases} \text{depending on whether } \lambda^o \lesseqgtr \frac{\mathrm{d}C_h}{\mathrm{d}P_{th}}$$

where, assuming that the initial set of units is known:

- $\mathcal{F}^o(r, X_k)$ is the minimum total cost that can be obtained in the first r elementary intervals, when passing from the initial set $I = I^{(0)}$ at $t = 0^-$ to the set $I = X_k$ at $t \in ((r-1)\tau, r\tau)$ (for $r = 1$ it must be assumed $\mathcal{F}^o(1, X_k) = A(0, I^{(0)}; 1, X_k) + G(1, X_k)$);
- $A(r-1, X_j; r, X_k)$ is the cost of possible startups, required to pass from the set $I = X_j$ to the set $I = X_k$ at the instant $t = (r-1)\tau$;
- $G(r, X_k)$ is the generation cost (minimized by imposing the equality of incremental costs) that corresponds to the set $I = X_k$ at $t \in ((r-1)\tau, r\tau)$;

by assuming that, at each step, only pairs of sets X_j, X_k consistent with all constraints (rotating power constraints, possible assignments on operating intervals for some units, etc.) are considered.

For any given pair of initial and final sets $I^{(0)}$ and $I^{(R)}$, the optimal solution is defined by $\mathcal{F}^o(R, I^{(R)}))$ and the corresponding sets for $r = 1, \ldots, R-1$.

If only costs (and constraints) concerning generation are considered, thus assuming $A = 0$, the time dependencies actually become noninfluential, so that the problem is reduced to the choice of the set that minimizes the generation cost G, for each r. In such a situation, to improve the priority criterion with respect to that described above (based on the values $(C_i/P_{ti})^*$), it is again possible to use dynamic programming in analogy to what is illustrated in Section 2.2.5d, also considering the possibility of setting generated power to zero for those units that are not in operation.

In the more general case that also accounts for startup costs, the procedure may be possibly simplified:

- considering only preassigned sets, e.g., (1; 1,2; 1,2,3; ...), which satisfy the priority criterion based on values $(C_i/P_{ti})^*$;
- considering at each r only the sets X_j that correspond to the smallest values of $\mathcal{F}^o(r-1, X_j)$.

Once the operating schedule for units is chosen, knowing the behavior of the incremental cost λ during the week enables the evaluation of the efficiencies of:

- pumpings, with weekly and/or daily generating-pumping cycles,
- adjustments to the schedule concerning exchanged powers (which affect the diagram of $P_c'(t)$).

Since both actions modify the thermal generation diagram, it must be considered that the unit operating schedule might undergo some arrangements.

If *pumping* actions are performed in the interval (or set of intervals) $\theta_{(p)}$, they imply a thermal generation overcost $\Delta G_{(p)}$ that can be approximated by:

$$\Delta G_{(p)} = \int_{(\theta_{(p)})} \lambda P_{(p)}\, \mathrm{d}t = \lambda_p \int_{(\theta_{(p)})} P_{(p)}\, \mathrm{d}t$$

assuming that power $P_{(p)}$ requested for pumping is small with respect to the total generated power P_t, and indicating by λ_p a proper mean value of the incremental cost in $\theta_{(p)}$.

Then, indicating by $\theta_{(g)}$ the interval or set of intervals in which the amounts of water related to pumping are used for generation, at power $P_{(g)}$, the saving obtained in $\theta_{(g)}$ is similarly given by:

$$-\Delta G_{(g)} = \int_{(\theta_{(g)})} \lambda P_{(g)}\, \mathrm{d}t = \lambda_g \int_{(\theta_{(g)})} P_{(g)}\, \mathrm{d}t$$

with λ_g a proper mean value of the incremental cost in $\theta_{(g)}$.

(Typically $\theta_{(p)}$ includes night hours and/or holidays, whereas $\theta_{(g)}$ corresponds to workday load peaks; see also Figure 2.39.)

Therefore, pumping is convenient only if $\Delta G_{(p)} + \Delta G_{(g)} < 0$, or equivalently:

$$\lambda_p < \eta \lambda_g \qquad [2.4.7]$$

where:

$$\eta \triangleq \frac{\displaystyle\int_{(\theta_{(g)})} P_{(g)}\, \mathrm{d}t}{\displaystyle\int_{(\theta_{(p)})} P_{(p)}\, \mathrm{d}t}$$

represents the overall efficiency of generating-pumping cycles (η may be 0.6–0.7); such a result agrees with that obtained (under the more stringent hypothesis of constant configuration) in Section 2.3.3b3, according to which the optimal solution corresponds (if constraints allow it) to the condition $\lambda_p = \eta\lambda_g$; see Figures 2.29 and 2.30.

However, the convenience of pumping must be evaluated in relation to consequences on the operating program of thermal units. Specifically, the reduced thermal generation required in $\theta_{(g)}$ also may permit the permanent exclusion of one or more thermal units from operation, provided that the requirements of rotating power are still met, with the contribution of the generating-pumping units themselves. For instance, in the elementary case in Figure 2.42, the exclusion of the h-th unit implies, assuming that the considered powers are small with respect to P_t:

- a saving:

$$C_h(P_{th(p)})T_p + C_h(P_{th(o)})T_o + \left[C_h(P_{th(g)}) + \lambda_g P_{(g)}\right] T_g$$

- an overcost:

$$\lambda_p \left[P_{(p)} + P_{th(p)}\right] T_p + \lambda_o P_{th(o)} T_o + \lambda_g P_{th(g)} T_g$$

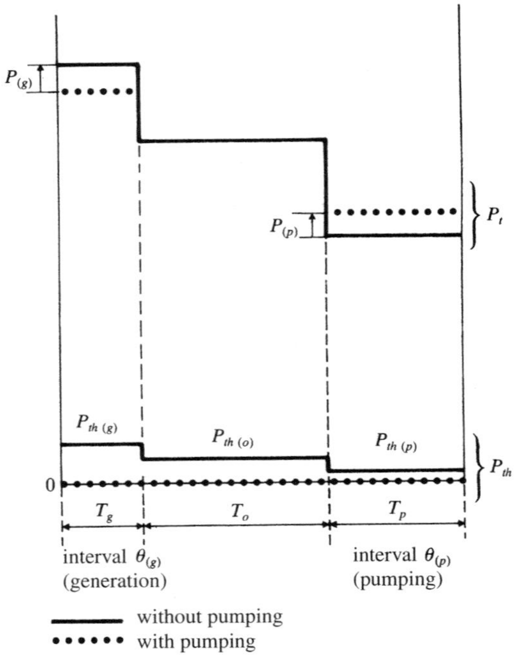

Figure 2.42. Diagrams of the total thermal power P_t and power P_{th} generated by the h-th unit.

By assuming that $P_{(p)}T_p = P_{(g)}T_g/\eta$ and, for simplicity, $P_{(g)} = P_{th(g)}$ — so that the generation shortage caused by the absence of the h-th unit is exactly compensated in $\theta_{(g)}$ — the exclusion of the h-th unit is effective only if:

$$\left[C_h(P_{th(p)}) - \lambda_p P_{th(p)}\right] T_p + \left[C_h(P_{th(o)}) - \lambda_o P_{th(o)}\right] T_o + \left[C_h(P_{th(g)}) - \frac{\lambda_p}{\eta} P_{th(g)}\right] T_g > 0$$

On the other hand, in $\theta_{(p)}$ the augmented thermal generation demand may keep some units in operation which, without pumping, might have been efficiently excluded (recall condition [2.4.5]).

Finally, with the schedule of the *exchanged powers*, it may be possibly corrected, with respect to what is arranged in the long-term scheduling, based on the incremental costs and by means of agreements with other interested utilities (see Section 2.3.4), by considering effects on the operating schedule of thermal units.

(c) Short-Term Scheduling

The fundamental aim of short-term scheduling is to accurately define the hydroelectrical and thermal generation schedules and, more generally, the working points of the system with reference, for instance, to the next day.

The choice $T = 1$ *day* allows a precise connection with medium-term scheduling, at least for water storages in hydroelectric plants; whereas errors in forecasting load behavior, etc. may be considered acceptable, without adopting a shorter duration, because their effects can be compensated by real-time scheduling.

Correspondingly, the duration of the elementary intervals may be chosen as $\tau = 30$ *minutes* (or 15 or 60), assuming that powers absorbed at load nodes, as well as water inflows and exchange powers, have a stepwise temporal behavior with constant values at each elementary interval. This avoids further detail that would be difficult to forecast.

Short-term scheduling may be based on the following data:

- *from updated previsions*:
  - the diagram of active and reactive powers absorbed during the day at each load node (reference is made to a relatively small number of "equivalent loads");
  - the diagram of natural inflows (considering possible spillages) during the day, for different hydroelectric plants;
- *based on medium-term scheduling and updated information*:
  - the set of available components (considering maintenance schedules and forced outages);
  - the set of thermal units scheduled for operation for the whole day or part of it;

- storage volumes at the beginning and at the end of the day, for each hydroelectric plant (further than preliminary schedules of possible pumping);
- a preliminary diagram of active exchange powers during the day for boundary nodes.

Furthermore, it is possible to add *particular constraints* to such data (e.g., assigned generation), as indicated for medium-term scheduling; whereas it is assumed, for simplicity, that the fuel availability is always guaranteed.

First, available data allow a coordinated choice of *hydro- and thermal generation schedules*, accounting for:

- minimum and maximum limits on powers P_{wj} and P_{ti}, respectively, generated by hydro- and thermal units (possibly $P_{wj} < 0$ in case of pumping);
- the characteristics (q_j, P_{wj}) and (C_i, P_{ti}) concerning water consumptions and generation costs, respectively, for hydro- and thermal units;
- minimum and maximum limits for storage volumes in hydroplants;
- network losses, adequately estimated (possibly as a function of the dispatching of powers P_{wj} and P_{ti}).

Recall Sections 2.3.3b1 and 2.3.3b3, assuming an always insufficient hydrogeneration.

The problem is then similar to that of the medium-term scheduling, but it is now solved using less-simplified approaches (particularly, storage limits are accounted for even within each day); on the contrary, the operating schedule of thermal units may be now considered assigned[(24)], apart from the accidental outage of some units and corrections that might appear efficient.

Actually, the solution must be adjusted if *limits on currents* and/or *security constraints* are violated (see Sections 2.3.1c,d); particularly, requirements of *spinning reserve* — also for what concerns its geographical distribution — may impose some rearrangements to the unit operating schedule; for instance by avoiding the shutdown of hydroelectric units or by increasing the set of thermal units at peak load.

Further refinements may be required when powers generated by thermal units undergo fast variations (see Section 2.3.2a).

Finally, corrections also may concern the *exchange power* schedule, under conditions similar to those in medium-term scheduling.

What is reported up to now actually implies a solution of only the "active dispatching" in the subsequent elementary intervals, without considering interactions with network voltage and reactive power steady-states.

[(24)] If some units are in operation only for a part of the day, the solution may be obtained by applying an iterative procedure similar to Figure 2.38.

For a detailed scheduling of working points, at least for the most critical situations it is optimal to use the *combined solution of active and reactive dispatchings*, according to Section 2.2.6. This solution correctly accounts for network losses and currents in the different branches and for their dependence on active power dispatching. It may happen, more generally, that variations to system configuration (and possibly operating schedule of units) become necessary because of specific requirements from the reactive power dispatching to improve the "voltage support" and increase the reactive power margins at network nodes.

2.5. REAL-TIME SCHEDULING

The goals to be achieved by means of real-time scheduling are:

- to check the actual working point with relation to quality, security, and economy requirements, based on measurements performed on the system during real operation;
- to determine necessary corrective actions (on control system set-points, parameters, and system configuration itself), to obtain the most satisfying working point.

First, system measurements must be adequately selected and processed, so that a reliable "*state estimation*," i.e., estimation of the actual operating state (configuration and working point), is achievable, according to the following.

The check and determination of corrections essentially imply the solution of the two mentioned problems—*active and reactive dispatching* (recall Fig. 2.15)—starting from:

- active and reactive powers absorbed by loads (apart from critical cases for which it would be effective to disconnect some loads, to meet security requirements);
- other possible assignments, e.g., concerning:
    - active powers generated by hydrounits, according to previsional scheduling stage decisions (however, because of the difference between actual and forecasted loads, powers actually generated may assume other values, as a consequence of the f/P control);
    - active powers exchanged through boundary nodes (these powers are kept at scheduled values by means of the f/P control; see Section 3.4), apart from rearrangements agreed in real-time.

Differently from the short-term previsional scheduling, the hydrogeneration may be now considered as an input data. Therefore, active dispatching results as purely instantaneous, apart from problems related to fast variations of thermally generated powers; see Section 2.3.2a.

Real-time scheduling assumes that the operating state is a steady-state that is kept unchanged for a sufficiently long time interval, considering the unavoidable delay between the measurement achievement and the corrective actions. This delay may be some minutes, because of state estimation and dispatching (with security checks, etc.), *teletransmission of corrective signals*, and their subsequent *actuation*.

Specifically:

- it must be assumed that, in the meantime, powers absorbed by loads do not vary significantly;
- measurements must be adequately filtered, to avoid the effect of transient components overlapping the searched steady-state values.

According to Figure 2.43, the actuation of corrections also may be interpreted in terms of a "*tertiary*" *control*, which acts on the set points of secondary (and primary) f/P and v/Q controls, and possibly (by means of "adaptive" type actions) on parameters and configuration. Such tertiary control is, of course, the slowest type of control, because it is based on filtered and sampled (i.e., taken only at given instants) measurements and its action is necessarily delayed, according to what is stated above.

Specifically, if the interactions between the active and reactive dispatchings are disregarded, the distinction between the tertiary f/P and v/Q controls, respectively involving active and reactive dispatchings, is evident. In Section 3.3.2, an approximated type of

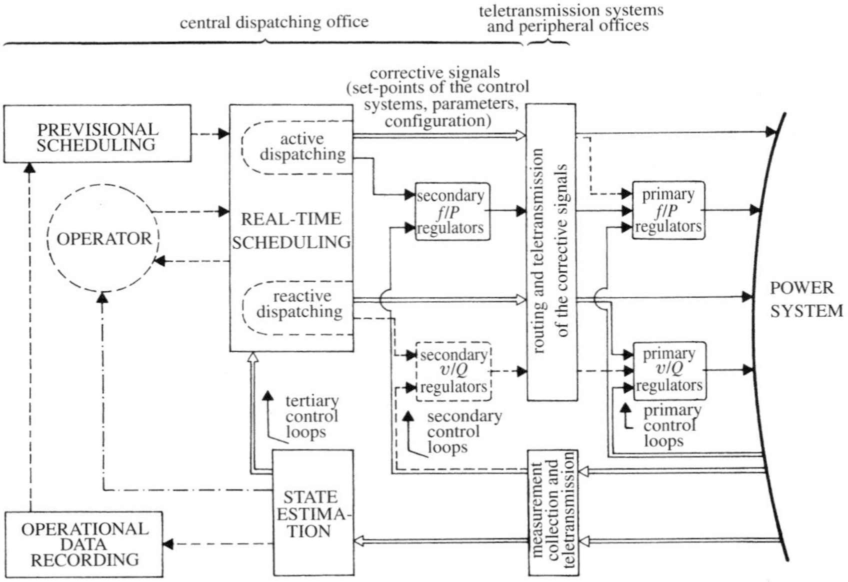

Figure 2.43. General diagram for scheduling and control. Secondary regulators are assumed to be located at the central dispatching office.

tertiary f/P control will be referenced, based on the value of "level" signal of secondary f/P regulator, without need for "state estimation." Corrective actions for the two mentioned types of tertiary control also may be actuated at different times; for instance, every 5–10 minutes for the f/P control and every 30 minutes or more for the v/Q control.

State estimation and real-time scheduling may be performed into a single location that we will generically denote by *central "dispatching" office*. The link with plants may be achieved by a digital data teletransmission system, having a "tree" structure. The central office is connected to several *peripheral offices* (e.g., at the regional level) which are connected to other offices at the zonal levels and so on, down to the single parts of the system. If measurements and corrections are "analog," they, respectively, require an analog-to-digital conversion and vice versa.

Peripheral offices, further than surveillance tasks on plants concerning respective areas (and interventions to handle incorrect operation or outages, etc.), must:

- collect different measurements from plants and transmit them to the central office (available measurements must be selected, based on preliminary checks of reliability; transmitted data also may be reduced by proper processing, with reference to simplified "equivalents" adopted at the central office location);
- receive corrective signals from the central office and route them toward plants.

As indicated in Figure 2.43, the central office can additionally perform, by means of one or more digital computers, the following tasks:

- *updated information to the operator* (with display and alarm in case of noncredible measurements and/or critical situations, as nonacceptable voltages or currents, nonadequate spinning reserve, etc.), for surveillance purposes and possible actions on scheduling;
- *operational data recording* for reconstruction and interpretation of specific events, for statistical analyses of different nature (also as an aid to load forecasting, etc.), and so on;
- long-medium-short term *previsional scheduling*;
- *f/P secondary control* (for instance, with a sampling period of seconds, for frequency and exchanged powers measurements) and, possibly, v/Q secondary control.

With the perturbed operation (see Section 1.7), other facilities (achieved in more or less simplified manners) may be added, such as:

- *preventive determination of possible intervention plans*;
- *real-time diagnosis*;
- *choice of corrective actions*.

Such functions can be, at least partially, delegated to peripheral offices; the actuation of corrective actions (e.g., breaker opening or closing) may be performed through the

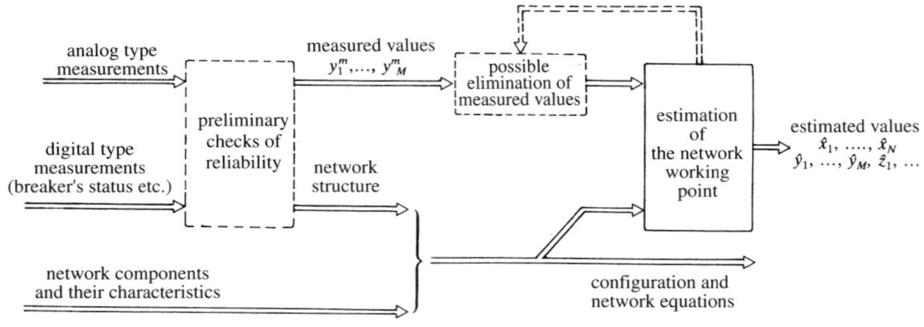

Figure 2.44. Outline of the "state estimation" procedure.

teletransmission system. However, with insufficient rotating power, the diagnosis and the actuation of corrective actions (typically load-shedding) also may be achieved by simple devices located at the proximity of loads (see Section 3.5).

The "state estimation" may be performed according to Figure 2.44. Analog-type *measurements* to be converted into digital may concern active and reactive generated powers, active exchanged powers, voltage magnitudes, active and reactive power flows and, rarely, active and reactive powers absorbed by loads. Digital measurements typically concern the status (open or closed) of circuit breakers and disconnectors, and the values of adjustable parameters (discontinuous ones, such as transformer tap-changer position).

The *preliminary checks* of reliability consider, for instance:

- consistency between each measurement and the range of the measured variable (similarly, a too fast varying measured value may not appear as credible, and so on);
- consistency between different measurements; for instance, power flows that are different from zero only if the relative breaker is closed, etc.

Apart from disregarding measured values, presumably affected by "abnormal" errors (as considered in the following), the *estimation of the working point of the network* is derived starting from:

- network equations;
- measured values.

Therefore, the estimation will be affected by errors, because of:

- approximations in network equations (recall Sections 1.2.2 and 2.1.2, e.g., physical dissymmetries, nonlinearities, inductive coupling between proximal lines, etc.);

- inaccuracy in measured values caused by overlapping of disturbances and/or errors (normal and abnormal) in measuring devices and in conversion and teletransmission systems.

Usually it is assumed that estimation errors are practically caused by inaccurately measured values, which may be defined (with the exception of abnormal errors) in terms of:

- systematic errors, that are assumed known;
- random errors, with zero mean and known probabilistic distributions.

Other uncertainties can be added because the operating condition is not exactly a steady-state, and the measured values are not all taken at the same instant[(25)]. It is then efficient to reduce, as much as possible, the time interval within which all measurements are achieved (e.g., 30 seconds or less). Generally, variables could be estimated with time, by attributing to them a probabilistic model and applying a procedure based on the use of the "Kalman filter."

If by $x_1, \ldots, x_N$ and $y_1 = f_1(x_1, \ldots, x_N), \ldots, y_M = f_M(x_1, \ldots, x_N)$, one generically indicates respectively the N independent variables defining the working point — for the given network equations — and the M variables which are under measurement, the estimation problem may be translated, by obvious notation, to the determination of proper "estimated values" $\hat{x}_1, \ldots, \hat{x}_N$, starting from the "measured values" $y_1^m, \ldots, y_M^m$.

Based on the solution, it will be possible to determine the estimated values of $y_1, \ldots, y_M$; i.e., $\hat{y}_1 \triangleq f_1(\hat{x}_1, \ldots, \hat{x}_N), \ldots, \hat{y}_M \triangleq f_M(\hat{x}_1, \ldots, \hat{x}_N)$, as well as those of any other dependent variable $z_j = g_j(x_1, \ldots, x_N)$; i.e., $\hat{z}_j \triangleq g_j(\hat{x}_1, \ldots, \hat{x}_N) (j = 1, 2, \ldots)$.

It is evident that:

- If $M < N$, measurements will be insufficient to solve the problem.
- If $M = N$, measurements can be used to derive $\hat{x}_1, \ldots, \hat{x}_N$ (apart from singular cases), by assuming:

$$\hat{y}_k = y_k^m - E\left(y_k^m - y_k\right) \qquad (k = 1, \ldots, M)$$

where $E(y_k^m - y_k)$ is the possible systematic error related to the measure of y_k, i.e., the "expected value" (or mean value in probabilistic terms) of the error $(y_k^m - y_k)$. In this case, it should be a question of solving network

(25) Furthermore, measured values do not reach the centralized computer simultaneously; it is possible to process them in sequential blocks, which prevents waiting until all are at disposal, by using (at each step) the last block for updating the estimation.

equations, simply based on measured data and the knowledge of systematic errors.

However, because of random measured values (and the risk of temporary unavailability of some of them), it is efficient to use a number of measurements M somewhat larger than N (*measurement "redundancy"*), to reduce by a statistical compensation the effects of errors affecting the measured values. Such redundancy should be adequately spread among the different parts of the network. Doing so reduces the effects of errors and helps to overcome the possible lack of some measurements. Anyway, it is important to detect the presence of abnormal errors and identify those measurements affected by them, so that they can be disregarded.

If values of adjustable parameters can be considered known, it holds $N = (2n - 1)$, with n the number of terminal nodes (see Section 2.1.5a). The n voltage magnitudes $v_1, \ldots, v_n$, and $(n - 1)$ phase shifts $\alpha_{21}, \ldots, \alpha_{n1}$ may be assumed as independent variables.
Possible minimal solutions (i.e., with $M = N$) to be assumed as reference basis for achieving measurement redundancy, are:

- measurement of the voltage (magnitude) or injected reactive power at each of the n nodes, and measurement of injected active powers at $(n - 1)$ nodes (if there are no other measurements, the "load-flow" problem illustrated in Section 2.2 can be solved);
- measurement of voltage at just one node and measurement of active and reactive power flows at only one side of $(n - 1)$ branches, chosen in a "tree" configuration to involve all nodes without any mesh.

However, the former solution implies measurements at all terminal nodes (including load nodes), whereas the latter can be accepted as a starting point only if several voltage measurements are added; in fact, to achieve reliable estimations, it is usually necessary to have a sufficient number of accurate voltage measurements.

At a proper redundancy ($M > N$), the estimation procedure may, for instance, be based on the "*weighted*" *least square method* (apart from possible variations, more or less approximated). In the following, for simplicity, systematic errors are assumed to be zero, i.e., $E(y_k^m - y_k) = 0$ $(k = 1, \ldots, M)$. If this is not true, each generic value y_k^m should be replaced by $(y_k^m - E(y_k^m - y_k))$. In matrix notation, if x denotes the column matrix constituted by $x_1, \ldots, x_N$, and so on, the estimated value $\hat{x}$ is then defined by the value of x which minimizes a function like:

$$J(x) \triangleq (y^m - f(x))^T W (y^m - f(x)) \qquad [2.5.1]$$

where W is a (M, M) symmetrical and positive definite matrix, to be properly chosen.

Usually it is assumed:

$$W = V^{-1}$$

where $V \triangleq E((y^m - y)(y^m - y)^T)$ is (because of the hypothesis $E(y^m - y) = 0$) the "covariance matrix" of errors related to measurements. This matrix can be assumed known, based on characteristics of measuring, conversion, and teletransmission systems.

It is often simply assumed that each error $(y_k^m - y_k)$ $(k = 1, \ldots, M)$ is "statistically independent" of other errors, further than having zero mean; it then follows $V_{kh} \triangleq E((y_k^m - y_k)(y_h^m - y_h)) = 0$ for each $h \neq k$, so that the matrix V is diagonal (its generic element $V_{kk} \triangleq E((y_k^m - y_k)^2)$ is the "variance" related to the k-th measurement, whereas $\sqrt{V_{kk}}$ is the so-called "standard deviation").

Actually, errors may be instead correlated with each other because of concurring causes. This is the case, for instance, of measuring devices partially shared, to evaluate active and reactive powers at a given point of the network.

Furthermore, it often may be assumed that each error has a Gaussian distribution, i.e., a probability density equal to:

$$\Pi(y_k^m - y_k) = \frac{\epsilon^{-(y_k^m - y_k)^2/2V_{kk}}}{\sqrt{2\pi V_{kk}}}$$

(it then results:

$$\int_{-3\sqrt{V_{kk}}}^{+3\sqrt{V_{kk}}} \Pi(y_k^m - y_k)\,\mathrm{d}(y_k^m - y_k) > 0.99$$

i.e., the standard deviation $\sqrt{V_{kk}}$ is approximately one-third of the maximum error which, in probabilistic terms, occurs in more than 99% of the cases).

The convenience of choosing $W = V^{-1}$ may have several justifications; for instance, by assuming that errors are statistically independent with a Gaussian distribution, and that $y_k = f_k(x)$, the use of the *"maximum likelihood" criterion* leads to assume, as estimation $\hat{x}$, the value of x which maximizes the following product:

$$\Pi(y_1^m - f_1(x)) \cdot \ldots \cdot \Pi(y_M^m - f_M(x))$$

(i.e., the probability density for the set of measured values $y_1^m, \ldots, y_M^m$), and thus minimizes the sum:

$$\frac{(y_1^m - f_1(x))^2}{V_{11}} + \cdots + \frac{(y_M^m - f_M(x))^2}{V_{MM}}$$

which is just the function defined by Equation [2.5.1] with $W = V^{-1}$.

Under the hypothesis of convexity for the function $J(x)$, it follows:

$$0 = \frac{\mathrm{d}J}{\mathrm{d}x}(\hat{x}) = -2(y^m - f(\hat{x}))^T W H(\hat{x})$$

with $H(x) \triangleq (\mathrm{d}f/\mathrm{d}x)(x)$ ((M, N) matrix), that is the matrix equation:

$$0 = H(\hat{x})^T W(y^m - f(\hat{x})) \qquad [2.5.2]$$

which is equivalent to N scalar equations, sufficient for deducing $\hat{x}_1, \ldots, \hat{x}_N$ (apart from singular cases, which are not considered here)[(26)].

Note that, if $y = f(x) = H^o x$, with H^o constant, the solution of Equation [2.5.2] is:

$$\hat{x} = (H^{oT} W H^o)^{-1} H^{oT} W y^m$$

from which $\hat{y}$ can be derived ($\hat{y} = H^o \hat{x}$), as well as the other dependent variables. In this case, moreover:

- the estimation errors concerning the independent variables are defined by:

$$\hat{x} - x = \left(H^{oT} W H^o\right)^{-1} H^{oT} W \left(y^m - y\right) \qquad [2.5.3]$$

  where $(y^m - y)$ is the column matrix of errors related to measurements (we can also derive $\hat{y} - y = H^o(\hat{x} - x)$, etc.);

- it holds:

$$\begin{cases} y^m - \hat{y} = y^m - H^o \hat{x} \triangleq A y^m \\ \hat{J} = J(\hat{x}) \triangleq y^{mT} B y^m \end{cases}$$

  or even:

$$\left.\begin{aligned} y^m - \hat{y} &= A(y^m - y) \\ \hat{J} &= (y^m - y)^T B (y^m - y) \end{aligned}\right\} \qquad [2.5.4]$$

---

(26) By means of numerical calculations, the solution $\hat{x}$ may be derived by the gradient method, considering that:

$$\text{grad } J(x) = \left(\frac{\mathrm{d}J}{\mathrm{d}x}\right)^T = -2H(x)^T W \left(y^m - f(x)\right)$$

Alternatively, the equation $0 = \text{grad } J(x)$ may be solved by the Newton-Raphson method, by imposing at the generic step:

$$x^{(i)} = x^{(i-1)} - \left[\frac{\mathrm{d}\ \text{grad } J}{\mathrm{d}x}\left(x^{(i-1)}\right)\right]^{-1} \text{grad } J\left(x^{(i-1)}\right)$$

or even more simply (disregarding the term with $\mathrm{d}H/\mathrm{d}x$, so that $(\mathrm{d}\ \text{grad } J)/\mathrm{d}x \cong 2H(x)^T W H(x)$):

$$x^{(i)} = x^{(i-1)} + \left([H(x)^T W H(x)]^{-1} H(x)^T W \left(y^m - f(x)\right)\right)_{x = x^{(i-1)}}$$

As initial value for x, the one adopted in the short-term scheduling can be assumed.

with:

$$\left.\begin{aligned} A &\triangleq I_{(M)} - H^o(H^{oT}WH^o)^{-1}H^{oT}W \\ B &\triangleq A^TWA = WA = W - WH^o(H^{oT}WH^o)^{-1}H^{oT}W \end{aligned}\right\} \quad [2.5.5]$$

(in fact, $AH^o = 0$ and thus $BH^o = 0$, $H^{oT}B = 0$, so that $Ay = By = 0$, $y^TB = 0$);

- finally, from the hypothesis $E(y^m - y) = 0$ (zero systematic errors), it follows:

$$E(\hat{x} - x) = 0, \quad E(\hat{y} - y) = 0, \ldots \quad [2.5.6]$$

(i.e., estimation errors have zero mean) and similarly:

$$E(y^m - \hat{y}) = 0 \quad [2.5.7]$$

In the general case, assuming that the estimated value $\hat{x}$ is close to the "true" value x and thus $f(\hat{x}) \cong f(x) + H(\hat{x})(\hat{x} - x)$ (where $f(\hat{x}) = \hat{y}$, $f(x) = y$), from Equation [2.5.2] it can be derived:

$$0 \cong H(\hat{x})^TW(y^m - y) - (H(\hat{x})^TWH(\hat{x}))(\hat{x} - x)$$

so that finally the quantities given by Equations [2.5.3] and [2.5.4] can be obtained, provided it is assumed that $H^o = H(\hat{x})$ (which is a known matrix, once $\hat{x}$ has been evaluated); whereas Equations [2.5.6] and [2.5.7] may be accepted if the dependence of H^o on $\hat{x}$ (as well as the randomness of the matrix H^o itself, which is related to the randomness of measured values) is disregarded.

By the adopted approximation, the estimation errors concerning the measured variables are expressed by:

$$\hat{y} - y = H^o(H^{oT}WH^o)^{-1}H^{oT}W(y^m - y) \quad [2.5.8]$$

similarly, those concerning other dependent variables (generically defined by the column matrix $z = g(x)$) are given by:

$$\hat{z} - z = K^o(H^{oT}WH^o)^{-1}H^{oT}W(y^m - y) \quad [2.5.9]$$

with $K^o \triangleq (\mathrm{d}g/\mathrm{d}x)(\hat{x})$.

The quality of estimation may be worsened by the presence of *"abnormal" errors*, related to one or more measurements. On the other hand, the possibility of detecting such error presence and identifying the affected measurements is based only on measured and estimated values (whereas "true" values still remain unknown), further than on the knowledge of probabilistic distributions of errors $(y_k^m - y_k)$ in normal conditions, i.e., in the absence of abnormal errors. In this concern, Equations [2.5.4] become useful, as they express the known quantities

$(y^m - \hat{y})$ and $\hat{J}$ as functions of $(y^m - y)$. Then, the probabilistic distributions of $(y_k^m - \hat{y}_k)$ and $\hat{J}$ in normal conditions can be derived (in a simple way, if the randomness of H^o is disregarded), and the probability of abnormal errors affecting one or more measurements can be consequently detected.

If, for instance, in normal conditions:

$$E(y^m - y) = 0, \quad E((y^m - y)(y^m - y)^T) = V$$

(with V known) and $W = V^{-1}$ is assumed, from the first part of Equations [2.5.4] it is possible to derive—further than Equation [2.5.7], disregarding the randomness of H^o—

$$E((y^m - \hat{y})(y^m - \hat{y})^T) = AVA^T = V - H^o(H^{oT}V^{-1}H^o)^{-1}H^{oT}$$

Letting, to shorten the notation, $C_y \triangleq H^o(H^{oT}V^{-1}H^o)^{-1}H^{oT}$, it results:

$$E(y_k^m - \hat{y}_k) = 0, E((y_k^m - \hat{y}_k)^2) = V_{kk} - (C_y)_{kk} \quad (k = 1, \ldots, M)$$

so that the generic measured value y_k^m can be considered less credible as the larger is the ratio $|y_k^m - \hat{y}_k|/\sqrt{V_{kk} - (C_y)_{kk}}$. By means of the second part of Equations [2.5.4], i.e.,

$$\hat{J} = \sum_1^M {}_k \left[B_{kk}(y_k^m - y_k)^2 + \sum_1^M {}_{h \neq k} B_{kh}(y_k^m - y_k)(y_h^m - y_h) \right]$$

the mean value can finally be derived:

$$E\left(\hat{J}\right) = \sum_1^M {}_k \left[B_{kk}V_{kk} + \sum_1^M {}_{h \neq k} B_{kh}V_{kh} \right]$$

to which $\hat{J}$ should be quite close, in normal conditions. In particular, if it is assumed that errors $(y_k^m - y_k)$ are statistically independent (and thus V is diagonal), it results:

$$\begin{aligned} E\left(\hat{J}\right) &= tr(BV) = tr\left(I_{(M)} - V^{-1}H^o(H^{oT}V^{-1}H^o)^{-1}H^{oT}\right) \\ &= M - tr\left(V^{-1}H^o(H^{oT}V^{-1}H^o)^{-1}H^{oT}\right) \end{aligned}$$

where:

$$tr\left(V^{-1}H^o(H^{oT}V^{-1}H^o)^{-1}H^{oT}\right) = tr\left(H^{oT}V^{-1}H^o(H^{oT}V^{-1}H^o)^{-1}\right) = N$$

from which, very simply:

$$E\left(\hat{J}\right) = M - N$$

With estimation errors, further to Equation [2.5.6] it can be similarly derived:

$$\begin{cases} E((\hat{x}-x)(\hat{x}-x)^T) = (H^{oT}V^{-1}H^o)^{-1} \triangleq C_x \\ E((\hat{y}-y)(\hat{y}-y)^T) = H^o(H^{oT}V^{-1}H^o)^{-1}H^{oT} \triangleq C_y \\ E((\hat{z}-z)(\hat{z}-z)^T) = K^o(H^{oT}V^{-1}H^o)^{-1}K^{oT} \triangleq C_z \end{cases}$$

and particularly:

$$\begin{cases} E(\hat{x}_i - x_i) = 0, \quad E((\hat{x}_i - x_i)^2) = (C_x)_{ii} \quad (i = 1, \ldots, N) \\ E(\hat{y}_k - y_k) = 0, \quad E((\hat{y}_k - y_k)^2) = (C_y)_{kk} \quad (k = 1, \ldots, M) \\ E(\hat{z}_j - z_j) = 0, \quad E((\hat{z}_j - z_j)^2) = (C_z)_{jj} \quad (j = 1, 2, \ldots) \end{cases}$$

Based on the value of $\hat{J}$, it is possible to detect (at the desired probability, evaluated from probability distributions) the presence of abnormal errors. In case of positive answer, it is also possible to identify as wrong measurements those that appear less reliable, in accordance with what was mentioned above. The estimation procedure can be applied again, without considering such measurements, and so on.

Before concluding, recall the risk of *some unavailable measured values*, because of the outage of measuring devices or of a peripheral system for measurement collection or of a teletransmission channel, etc. In extreme situations, it might happen that $M < N$. It may then be necessary to compensate for the missing measurements with information of lower quality (with the possible aid of the operator), such as:

- transmission from a power plant regarding active and reactive generated powers;
- deduction of load active powers based on the "conformity" assumption, evaluating the total active load as the difference between the total generated active power and the estimated losses;
- substitution of a recently missing value y_k^m, with its last estimation $\hat{y}_k$;

and so on, by obviously assuming that the "weight" of such data in $J(x)$ is reduced; for instance by attributing a sufficiently large variance[(27)].

With a part of the network that presents a limited redundancy or an uncertain configuration, all measurements related to it may be disregarded and estimations made only for the remaining part of the network. In this case, the power flows between the two parts of the network, if measured, may be considered injections for that under estimation; otherwise, the disregarded part must be accounted for by means of a proper equivalent circuit.

[(27)] On the contrary, if active and reactive powers injected at a given terminal node are definitely zero (for instance because neither load nor generator is connected to it), we may assume that exact (zero) injection measurements are available, to which a significant weight in $J(x)$ has to be assigned, by means of sufficiently small variance values.

ANNOTATED REFERENCES

Among the works of general interest, the following may be listed: 21, 23, 25, 37, 44, 47, 129, 152, 160, 166, 179, 200, 218, 220, 222, 234, 235, 236, 249, 260, 261, 285, 327.

More specifically:

- for Section 2.1 (network equations): 1, 20, 35;
- for Section 2.2 (choice of working point, "load-flow" techniques): 22, 70, 76, 87, 107, 120, 178, 240, 252, 253, 266, 268, 269, 281, 282, 324; with particular concern to security: 148, 168, 177, 267;
- for Section 2.3 (economical optimization): 2, 12, 13, 30, 41, 43, 49, 100, 137, 185; with particular concern to
  - loss formula: 75, 110;
  - hydroelectric generation: 86, 90, 91, 92, 94, further than some notes prepared by the author (in particular, for the case of hydropower supply always sufficient), in view of the writing of 53;
  - choice of exchanged powers: 143, 174;
- for Section 2.4 (previsional scheduling): 30, 49, 86, 90, 91, 92, 94, 141, 150, 163, 205, 271, 312; with particular concern to the "unit commitment" problem: 84, 121, 128, 149, 204, 237;
- for Section 2.5 (real-time scheduling): 49, 138, 187, 246, 276, 299, 303, 305, 307, 314, 335; with particular concern to state estimation: 131, 132, 165, 167, 169, 182, 189, 197, 257, 278, 321.

CHAPTER 3

FREQUENCY AND ACTIVE POWER CONTROL

3.1. SPEED REGULATION OF A SINGLE UNIT IN ISOLATED OPERATION

3.1.1. Preliminaries

The regulation of the network frequency constitutes one of the essential elements for the "quality" of operation; excessive frequency variations would not, in fact, be tolerated by many end-users, nor by auxiliary equipment of the generating power stations themselves.

As an introduction, let us consider a system with only one synchronous generator. Since the frequency generated in the network is proportional to the rotation speed of the generator, the problem of frequency regulation may be directly translated into a speed regulation problem of the turbine-generator unit. It will be also assumed, for simplicity, that the system does not include other synchronous machines, compensators, and/or motors; for the case with these devices considered, refer to the end of Section 3.1.3.

More precisely, if Ω_m is the mechanical angular speed (i.e., the effective rotation speed) of the rotor, and N_p is the number of pole pairs, the so-called "electric" angular speed $\Omega = \Omega_m N_p$ may be assumed as a measure of the generated frequency. If Ω is expressed in rad/sec, the frequency f in Hz is then given by $f = \Omega/2\pi$.

The hypothesis that the system includes only one generator cannot be considered realistic for practical cases. However, it makes it possible to dedicate greater attention to the characteristics of the single generating station in relation to the fundamental requirements of frequency regulation. This is also important when examining the role of each station in frequency regulation for multiple generator systems.

In addition, the problem of one unit speed regulation is of direct interest as far as start-up and "no-load" operation are concerned, before the machine is connected to the network, or in operation after disconnection from the network; in addition to temporary situations, in which the unit is called to operate alone on a relatively small network.

The constancy of the speed of a turbine-generator unit implies that the mechanical driving power provided by the turbine is exactly counterbalanced by the active electrical power generated, increased by the mechanical-type losses that occur in the unit itself.

Actually, this power balance may be altered for various reasons, such as variations in the driving power (caused by disturbances in the supply system), and above all, with the machine connected to the network, changes in the electric power generated resulting from changes in the load demanded by users. If the turbine valves were blocked, the speed of the unit, i.e., the frequency, would be at the mercy of the above-listed "disturbances," and it could easily reach intolerable values. From here stems the necessity of acting on the turbine valves, to reset the power balance at the desired frequency and to maintain the frequency, even during transient operation, within an acceptable range of values.

To study the frequency regulation of only one unit, it is possible to make a schematic reference to Figure 3.1, where the following notation is used:

- Ω = electrical angular speed, i.e., frequency (according to above);
- Ω_{rif} = "frequency reference";
- $\varepsilon_f \triangleq \Omega_{\text{rif}} - \Omega$ = frequency error;
- β = "output" of the speed governor (β can be for instance the position of a mechanical device actuated by the governor, such as the rotation angle of the so-called "regulation shaft," which by itself acts on the valve positioning system);

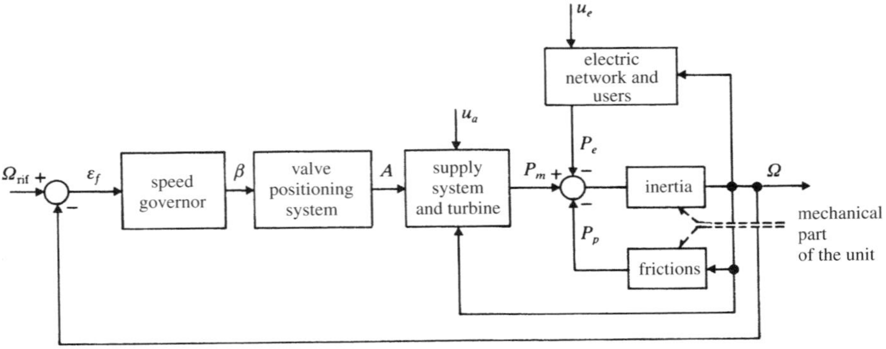

Figure 3.1. Speed regulation of an isolated unit: broad block diagram.

- A = opening of the turbine valves;
- P_m = driving mechanical power;
- P_e = active electric power generated (sum of the power absorbed by users and of electrical losses);
- P_p = mechanical power losses of the unit;

whereas u_a and u_c generically indicate disturbances capable of altering the power balance in accordance with what is indicated above.

In Figure 3.1, the dependent ties of P_m, P_e, and P_p with the speed (i.e., the frequency) are qualitatively emphasized.

With the block termed "inertia," it is to observe that the difference $P_m - (P_e + P_p)$ constitutes the accelerating power and therefore can be written as:

$$P_m - (P_e + P_p) = J\Omega_m \frac{\mathrm{d}\Omega_m}{\mathrm{d}t} \qquad [3.1.1]$$

J being the inertia moment of the unit and Ω_m the rotational speed of the rotor (the torsional-type phenomena discussed in Section 4.3.4 are neglected as, due to their quickness, they interact very little with frequency regulation). In practice, the speed variations with respect to the nominal value $\Omega_{m\,\text{nom}}$ are, due to the regulation itself, fairly modest to enable Equation [3.1.1] to be replaced by(1):

$$P_m - (P_e + P_p) = J\Omega_{m\,\text{nom}} \frac{\mathrm{d}\Omega_m}{\mathrm{d}t}$$

In terms of electrical angular speed, it can be then deduced that:

$$P_m - (P_e + P_p) = M \frac{\mathrm{d}\Omega}{\mathrm{d}t} \qquad [3.1.2]$$

where:

$$M \triangleq \frac{J\Omega^2_{m\,\text{nom}}}{\Omega_{\text{nom}}} \qquad [3.1.3]$$

is the "*inertia coefficient*" of the unit, which is the ratio between:

- twice the kinetic energy of the unit at the nominal equilibrium condition,
- the nominal electrical angular speed (the value of Ω_{nom} in rad/sec is 2π times that of the nominal frequency in Hz; for instance, at a 50-Hz nominal frequency, it corresponds $\Omega_{\text{nom}} \cong 314$ rad/sec).

---

(1) In fact, for small variations with respect to a steady-state equilibrium condition at nominal speed, it can be written that:

$$P_m - (P_e + P_p) = \Delta\left(J\Omega_m \frac{\mathrm{d}\Omega_m}{\mathrm{d}t}\right) \cong J\Delta\Omega_m \left(\frac{\mathrm{d}\Omega_m}{\mathrm{d}t}\right)_{\text{nom}} + J\Omega_{m\,\text{nom}}\Delta\frac{\mathrm{d}\Omega_m}{\mathrm{d}t}$$

where $(\mathrm{d}\Omega_m/\mathrm{d}t)_{\text{nom}} = 0$.

The coefficient M also can be posed in the form:

$$M = \frac{P_{\text{nom}} T_a}{\Omega_{\text{nom}}} \qquad [3.1.4]$$

where:

$$T_a \triangleq \frac{J \Omega_{m\,\text{nom}}^2}{P_{\text{nom}}} \qquad [3.1.5]$$

is the so-called *"start-up" time* (or "acceleration" time) of the unit. The usefulness of Equation [3.1.4] derives from the fact that the start-up time T_a is normally included in a relatively narrow range of values (e.g., $T_a = 6\text{–}10$ sec, and more often $T_a \cong 8$ sec), independent of the type of turbine and alternator.

The term "start-up" time can be justified by observing that, if the unit — initially at standstill — were subject to the nominal (accelerating) torque $P_{\text{nom}}/\Omega_{m\,\text{nom}}$, it would reach the nominal speed $\Omega_{m\,\text{nom}}$ after the time T_a. In such conditions, it in fact results $P_{\text{nom}}/\Omega_{m\,\text{nom}} = J\,\text{d}\Omega_m/\text{d}t$, and thus $\Delta\Omega_m = \Omega_{m\,\text{nom}}$ for $\Delta t = J\Omega_{m\,\text{nom}}^2/P_{\text{nom}} = T_a$. Therefore the term "start-up" must be interpreted here in a purely mechanical sense, with reference only to the inertia of the unit, disregarding problems associated with the supply system (startup of the thermal part etc., see Section 2.4.2b).

3.1.2. Basic Criteria for the Control Loop Synthesis

Significant characteristics of most operating conditions may be obtained by examining the behavior of the system in Figure 3.1 for small changes around an equilibrium condition (at nominal speed). Thus, it is possible to deduce the block diagram of Figure 3.2, by assuming the following:

$$\Delta P_r \triangleq \frac{\partial P_m}{\partial A}(s)\Delta A = \text{variation in the "regulating" (mechanical) power}$$

ΔP_L = variation in the "demanded power of the unit" due to the disturbances u_a and u_c indicated in Figure 3.1

$$G_c(s) \triangleq \frac{\partial P_e}{\partial \Omega}(s)$$

$$G_g(s) \triangleq \frac{\partial (P_p - P_m)}{\partial \Omega}(s)$$

and furthermore:

$$G_r(s) \triangleq \frac{\Delta\beta}{\Delta\varepsilon_f}(s) = \text{transfer function of the speed governor}$$

$$G_v(s) \triangleq \frac{\Delta A}{\Delta\beta}(s) = \text{transfer function of the valve-positioning system}$$

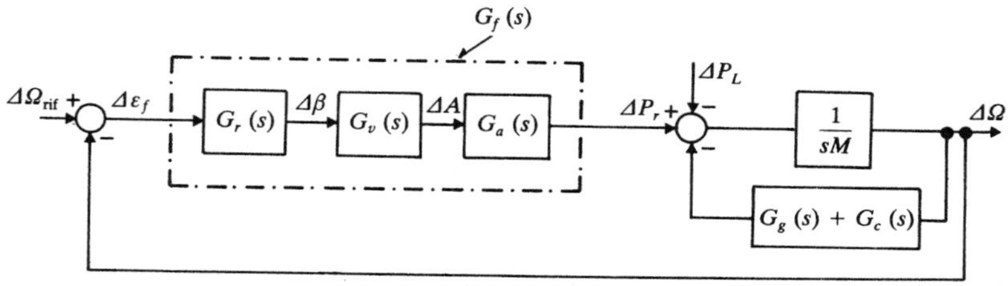

Figure 3.2. Block diagram of the system in Figure 3.1 for small variations.

$$G_a(s) \triangleq \frac{\partial P_m}{\partial A}(s) = \frac{\Delta P_r}{\Delta A}(s) = \text{transfer function of the supply system}$$

and of the turbine (with reference to the only input ΔA)

$$G_f(s) \triangleq G_r(s)G_v(s)G_a(s) = \frac{\partial P_m}{\partial \varepsilon_f}(s) = \frac{\Delta P_r}{\partial \varepsilon_f}(s)$$

If the unit is disconnected from the network ($P_e = 0$), ΔP_L is caused by only the disturbance u_a, and $G_c(s) = 0$.

If the unit is connected to the network, the effects of u_a instead usually become negligible with respect to those of u_c, and therefore, in such conditions, we may assume $\Delta P_e = \Delta P_L + G_c(s)\Delta\Omega$. The term ΔP_L can be interpreted as a consequence of the disturbances inside the network, caused by loads etc.

Furthermore, as will be shown, the transfer function G_c depends essentially on the power-frequency characteristics of loads, whereas G_g, G_v, G_a depend on the specific type of plant.

To identify the fundamental characteristics of the transfer function $G_f(s)$, and therefore of the transfer function $G_r = G_f/(G_v G_a)$ of the regulator, it is convenient to disregard the usually minor effects of the transfer functions $G_g(s)$ and $G_c(s)$, which will be considered in Section 3.1.3.

If a zero static error is desired, i.e., $\Delta\varepsilon_f = 0$ under every possible steady-state condition, it is necessary to include an "integral action" into G_f (i.e., in the regulator) by realizing a transfer function $G_f(s)$ with one pole at the origin. However, it is easy to verify that a purely integral $G_f(s)$—even independently of its feasibility, for the given G_v, G_a—would not be acceptable. In fact, if it were $G_f(s) = K_{If}/s$, it would be possible to derive the following characteristic equation for the system in Figure 3.2 (with $G_g + G_c = 0$):

$$Ms^2 + K_{If} = 0$$

which has, assuming $K_{If} > 0$, two complex conjugate imaginary roots ($s = \pm\tilde{\jmath}\nu_o$, with $\nu_o \triangleq \sqrt{K_{If}/M}$)(2), so that the system would be at the stability limit, with

(2) The symbol $\tilde{\jmath}$ represents the imaginary unit in the phasors' plane (see Appendix 1).

possible persistent oscillations at the "resonance" frequency ν_o, even in the absence of external signals(3).

However, a G_f of the "proportional-integral" type looks acceptable, i.e.,

$$G_f(s) = K_{If}\frac{1+sT_2}{s} \qquad [3.1.6]$$

where T_2 is a "time constant" having a proper (positive) value. With such a G_f, in fact, the following characteristic equation can be derived:

$$Ms^2 + K_{If}(1+sT_2) = 0$$

which has two roots with negative real part, in accordance with the stability requirement. More precisely, the two characteristic roots can be written in the form:

$$s = (-\zeta \pm \tilde{j}\sqrt{1-\zeta^2})\nu_o$$

where $\zeta \triangleq \nu_0 T_2/2$. By assuming for instance that $\zeta \in (0.5, 1)$, i.e., $\nu_0 T_2 \in (1, 2)$, it is possible to obtain a behavior that is oscillatory but sufficiently damped (with a "damping factor" equal to ζ). In such conditions, the "Bode diagrams" (for $s = \tilde{j}\nu$) of the loop transfer function:

$$\mathcal{G}(s) = G_f(s)\frac{1}{sM}$$

are like those indicated in Figure 3.3, with a "cutoff frequency" ν_t close to the "asymptotic" one, given by $\overline{\nu}_t = \nu_0^2 T_2$, whereas the "phase margin" is positive (as required by stability) and is:

$$\gamma = \text{artg}(\nu_t T_2)$$

for instance $\gamma \cong 52°$ for $\zeta = 0.5 (\nu_o T_2 = 1,\ \overline{\nu}_t = \nu_0,\ \nu_t \cong 1.27\overline{\nu}_t)$, and $\gamma \cong 76°$ for $\zeta = 1$ $(\nu_0 T_2 = 2,\ \overline{\nu}_t = 2\nu_o,\ \nu_t \cong 1.03\overline{\nu}_t)$.

For reasons clarified in Section 3.3.1 (in relation to the case of more than one regulating unit), one may forsake to have a zero static error in response to the disturbance ΔP_L, by tolerating small deviations $\Delta\varepsilon_f$ at steady-state conditions. Then, the transfer function G_f may be generalized as follows:

$$G_f(s) = K_{If}T_1\frac{1+sT_2}{1+sT_1} \qquad [3.1.7]$$

(3) Actually, it would be possible to expect some stabilizing effects from $(G_g + G_c)$, but it would not be practical to rely on this effect since it is minor (and unpredictable, because of uncertainties in the characteristics of the loads). In practice, the stability would be easily compromised by the presence of various response delays which have not been considered here.

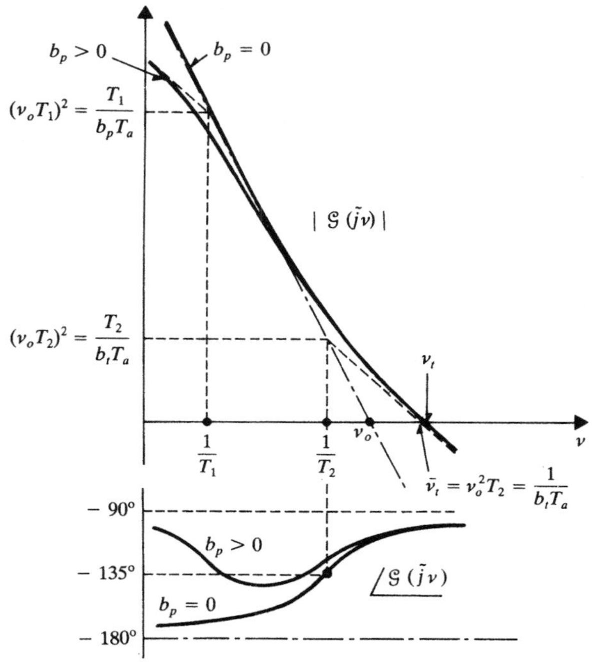

Figure 3.3. Frequency response of the transfer function of the loop.

with T_1 fairly large to have at steady-state conditions:

$$\Delta\varepsilon_f = \frac{\Delta P_r}{K_{If} T_1} = \frac{\Delta P_L}{K_{If} T_1}$$

sufficiently small, leaving the behavior of $\mathcal{G}$ around the "cutoff" (particularly the values of ν_t and γ) practically unchanged as indicated in Figure 3.3 (under such an assumption, Equation [3.1.6] can be viewed as a particular case, for $T_1 \to \infty$).

The static gain:

$$K_{fp} \triangleq G_f(0) = \frac{\Delta P_r}{\Delta\varepsilon_f}(0) = \left(-\frac{\Delta P_r}{\Delta\Omega}(0)\right)_{\Delta\Omega_{\text{rif}}=0} = K_{If} T_1$$

which is equal to the ratio, at steady-state conditions, between the variation of regulating power (i.e., of load power) and the corresponding variation of frequency, with the sign changed, is termed "(permanent) *regulating energy* due to the regulation" of the unit. (This is actually energy per radian, or even, by premultiplying it by 2π, energy per cycle in units of MW/Hz.)

On the other hand, the ratio between the relative variations of frequency and power, with the sign changed, is:

$$b_p \triangleq \left(\frac{-\Delta\Omega/\Omega_{\text{nom}}}{\Delta P_r/P_{\text{nom}}}(0)\right)_{\Delta\Omega_{\text{rif}}=0} = \frac{P_{\text{nom}}}{\Omega_{\text{nom}} K_{fp}}$$

which is termed “(permanent) speed *droop* due to the regulation” of the unit.

In an analogous way, by assuming $s = \infty$ (instead of $s = 0$) into Equation [3.1.7], the transient regulating energy may be defined:

$$K_{ft} \triangleq G_f(\infty) = \frac{\Delta P_r}{\Delta \varepsilon_f}(\infty) = \left(-\frac{\Delta P_r}{\Delta\Omega}(\infty)\right)_{\Delta\Omega_{\text{rif}}=0} = K_{If} T_2$$

as well as the transient (speed) droop:

$$b_t \triangleq \left(\frac{-\Delta\Omega/\Omega_{\text{nom}}}{\Delta P_r/P_{\text{nom}}}(\infty)\right)_{\Delta\Omega_{\text{rif}}=0} = \frac{P_{\text{nom}}}{\Omega_{\text{nom}} K_{ft}}$$

related to the previous definitions by $K_{ft} = K_{fp} T_2/T_1$, $b_t = b_p T_1/T_2$.

On the basis of these definitions, the transfer function in Equation [3.1.7] may thus be rewritten as:

$$G_f(s) = K_{fp}\frac{1 + sT_2}{1 + sT_1} = K_{ft}\frac{1 + sT_2}{\dfrac{T_2}{T_1} + sT_2}$$

or even:

$$G_f(s) = \frac{P_{\text{nom}}}{\Omega_{\text{nom}} b_p}\frac{1 + sT_2}{1 + sT_1} = \frac{P_{\text{nom}}}{\Omega_{\text{nom}}}\frac{1 + sT_2}{b_p + sT_2 b_t} \qquad [3.1.8]$$

whereas Equation [3.1.6] corresponds to the particular case $T_2/T_1 = b_p = 0$; furthermore, it can be derived:

$$\begin{cases} \nu_o = \sqrt{\dfrac{K_{ft}}{MT_2}} = \sqrt{\dfrac{1}{b_t T_a T_2}} \\ \zeta = \dfrac{1}{2}\sqrt{\dfrac{T_2 K_{ft}}{M}} = \dfrac{1}{2}\sqrt{\dfrac{T_2}{b_t T_a}} \\ \nu_t \cong \overline{\nu}_t = \dfrac{K_{ft}}{M} = \dfrac{1}{b_t T_a} \\ \gamma = \text{artg}(\nu_t T_2) \cong \text{artg}\left(\dfrac{T_2 K_{ft}}{M}\right) = \text{artg}\left(\dfrac{T_2}{b_t T_a}\right) \end{cases}$$

where T_a is the start-up time defined by Equation [3.1.5].

It is clear that, for given values of b_t and T_2, a permanent droop $b_p > 0$ has a stabilizing effect, since the pole $s = -1/T_1 = -b_p/(T_2 b_t)$ of $G_f(s)$ corresponds, in terms of frequency response, to a phase delay equal to 90° if $b_p = 0$, and smaller if $b_p > 0$ (refer also to Fig. 3.3). However, as previously noted, normally T_1 is assumed to be fairly large, and hence we have $T_1 \gg T_2$ (that is $b_p \ll b_t$: for example $b_p = 0$–5%, $b_t = 25\%$–40%), so that the cutoff frequency results much greater than $1/T_1$, and the above-mentioned stabilizing effect may be considered negligible.

In particular, the response to the disturbance ΔP_L is then defined by:

$$\frac{\partial \Omega}{\partial P_L} = -\frac{1}{G_f + sM}$$

that is, in relative terms:

$$\frac{\partial \Omega / \Omega_{\text{nom}}}{\partial P_L / P_{\text{nom}}} = -\frac{b_p + sT_2 b_t}{1 + s(b_p T_a + T_2) + s^2 b_t T_a T_2}$$

where $b_p T_a = T_2/(\overline{\nu}_t T_1) \ll T_2$, so that the most interesting parameters, from the speed of response point of view (besides stability), remain b_t and T_2. To emphasize the effect of such parameters, the trends of $\Delta\Omega/\Omega_{\text{nom}}$ following a step change in the load power (having an amplitude ΔP_L) are reported in Figure 3.4 as a function of time, for different values of $T_2/(b_t T_a)(= 4\zeta^2 = \overline{\nu}_t T_2 \cong tg\gamma)$ and $b_p = 0$. Nonzero, modest-sized values of the permanent droop may practically modify only the final part of the transient, through which $\Delta\Omega/\Omega_{\text{nom}}$ tends toward the steady-state value $(-b_p \Delta P_L / P_{\text{nom}})$.

Note that, based on the above discussion (see also Fig. 3.4), it would appear that the speed of response of the system could increase at will, with a satisfactory phase margin, by assuming suitably small values of b_t and T_2. In practice, however, it is necessary

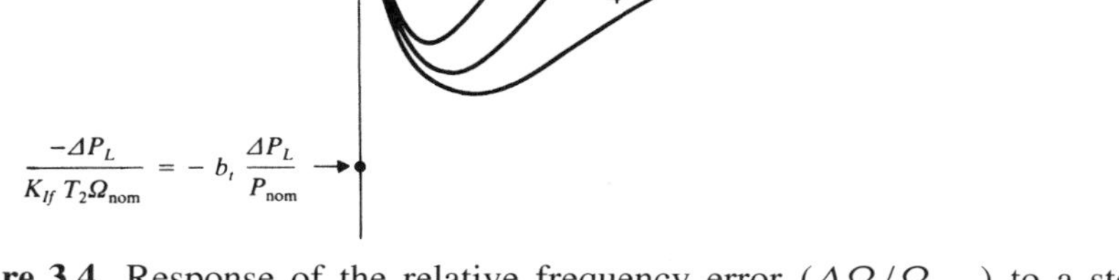

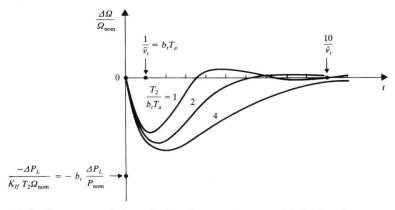

Figure 3.4. Response of the relative frequency error $(\Delta\Omega/\Omega_{\text{nom}})$ to a step of load variation (ΔP_L), under the assumption $b_p = 0$.

to bear in mind that, for values of ν_t that are not sufficiently small, the effect of other phenomena (up to now neglected) related to the response delays of the valves and supply system, other than those of the regulator itself, would lead to an unavoidable worsening of the phase margin, with a negative impact on the stability. In other words, a $G_f(s)$ of the type considered up to now can be realized for $s = \tilde{j}\nu$ only in a narrow band of frequencies ν, such as $\nu < \sim 0.5$ rad/sec.[4]

In some cases (e.g., hydraulic plants with Pelton turbines), the destabilizing effect of such phenomena increases so much for increasing values of the cut-off frequency, that one is forced to maintain the latter within well-defined limits. Having chosen a reasonable value of ν_t, or equivalently of $\overline{\nu}_t$, it is possible to derive b_t and T_2 in a very simple manner ($b_t = 1/(\overline{\nu}_t T_a)$, $T_2 = \tan\gamma/\nu_t$). Usually it may be assumed, for example, that $\nu_t \cong 0.3$ rad/sec, $T_a = 8$ sec, so that it is necessary that $b_t \cong 40\%$. Assuming $T_2 \cong 4$ sec, it is possible to have a phase margin $\gamma \cong 50°$, apart from negative contributions caused by the above-mentioned delays.

In addition to the above considerations, it also must be remembered that, in general, large and not sufficiently slow load variations cannot be tolerated by the plant. They would cause, for example, excessive overpressures or underpressures in hydraulic plants and heavy thermal stresses (and/or unacceptable operating conditions for the boiler) in steam plants. As a consequence, each plant is generally protected (intrinsically or through proper devices), so that the regulating power cannot increase or decrease too rapidly. It is evident that an excessive gain of the regulator, achieved to increase the speed of response, would cause the intervention of these protections for (even moderate) load disturbances, making unreliable the advantages foreseen based on the linearized analysis. Moreover, the mechanical stresses in the elements driven by the regulator could be increased intolerably. On the other hand, the analysis should also account for the effects of possible "insensitivities" in the regulator and in the elements controlled by it, which will be discussed in the following.

Finally, the static behavior of the regulation results defined by $\Delta P_r = G_f(0)\Delta\varepsilon_f$, that is by:

$$\Delta P_r = K_{fp}(\Delta\Omega_{\text{rif}} - \Delta\Omega) = \frac{P_{\text{nom}}}{\Omega_{\text{nom}} b_p}(\Delta\Omega_{\text{rif}} - \Delta\Omega)$$

where $\Delta P_r = \Delta P_L$ (recall Fig. 3.2, with $G_g + G_c = 0$).

This corresponds, for large variations, to a *static characteristic* as indicated in Figure 3.5, where P_{rif} is the value of P_r for $\Omega = \Omega_{\text{rif}}$. For varying K_{fp} (i.e., b_p), the characteristic rotates around the point (P_{rif}, Ω_{rif}), whereas for varying P_{rif} or Ω_{rif}, the characteristic is simply translated in the direction of the axis P_r or Ω, respectively. However, if $b_p = 0$, it simply follows $\Omega = \Omega_{\text{rif}}$, and the value of P_{rif} has no effect; similarly, if it were $b_p = \infty$, it would be $P_r = P_{\text{rif}}$ independent of Ω_{rif}.

[4] If the variable ν is regarded as the "inverse of a time," it may be clearly expressed in sec^{-1} (as in Figs. 3.15, 3.25, etc.). To keep the meaning of "frequency," it is sufficient to replace "sec^{-1}" with "rad/sec."

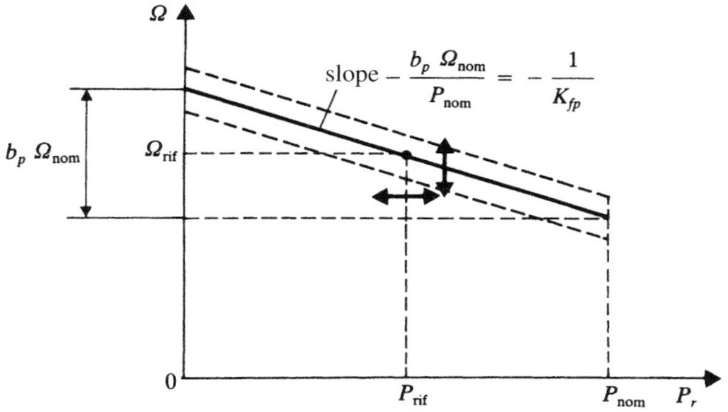

Figure 3.5. Static characteristic for a regulating unit.

Usually, it is convenient to keep the frequency reference "blocked" at the value $\Omega_{rif} = \Omega_{nom}$ and use P_{rif} (*power reference*) as a further input of the regulator, so that, for $b_p \neq 0$, it is possible to adequately set the static characteristic in the plane (P_r, Ω). This opportunity is evident in a system with more than one unit. By acting on the power references we can obviously realize the desired sharing of active powers in the operation at nominal frequency (active dispatching) and, as illustrated in Section 3.3.2, the frequency regulation itself, through the contribution of the different units. The regulation scheme must be able to accept, in a proper position, the signal P_{rif} for each unit (see Section 3.2), and the previous equation must be rewritten as:

$$\Delta P_r = \Delta P_{rif} + K_{fp}(\Delta\Omega_{rif} - \Delta\Omega) = \Delta P_{rif} + \frac{P_{nom}}{\Omega_{nom} b_p}(\Delta\Omega_{rif} - \Delta\Omega) \quad [3.1.9]$$

Instead of acting on P_{rif}, it would be possible to act on Ω_{rif}, since a variation $\Delta\Omega_{rif}$ is equivalent to a $(\Delta P_{rif})_{eq} = (P_{nom}/(\Omega_{nom} b_p))\Delta\Omega_{rif}$. However, by so doing, the value of b_p also should be considered, and variations of it (e.g., motivated by local operating requirements for the unit) would act as undesired power perturbations.

Actually, the definition of the (regulating) power P_r for large variations does not present difficulties if the driving power P_m depends only on the opening A of valves, so that it can be assumed $P_r = P_m$. On the contrary, it is necessary to intend that P_r is the driving power, without considering the further contributions (assumed as separable) independent of the opening of valves (see Fig. 3.1).

In addition, for simplicity of representation, it is assumed here as well as later that the static characteristic is linear, i.e., that K_{fp} (or b_p) is constant in the full variation range $[0, P_{nom}]$ of the regulating power. Generally, the static characteristic may be nonlinear, above all because of the nonlinearities present, and not adequately compensated, in the valve-positioning system and supply system. For small variations, this fact corresponds to different values of the permanent droop, depending on the operating point.

3.1.3. Influence of the Natural Characteristics of the Unit and Characteristics of Network and Loads

Before concluding, it may be useful to provide general information on the transfer functions:

$G_g(s) \triangleq \dfrac{\partial(P_p - P_m)}{\partial\Omega}(s)$ (which accounts for the "natural" characteristics of the unit)

$G_c(s) \triangleq \dfrac{\partial P_e}{\partial\Omega}(s)$ (which accounts for the characteristics of the electric network and of loads)

the effects of which have been neglected in the analysis presented above.

From Figure 3.2, it is evident that, in the response $\Delta\Omega$ to a load disturbance ΔP_L, the effects of these transfer functions are added to those of $G_f(s)$, since:

$$\frac{\partial\Omega}{\partial P_L} = -\frac{1}{G_f + G_g + G_c + sM} \qquad [3.1.10]$$

or alternatively, in relative terms:

$$\frac{\partial(\Omega/\Omega_{\text{nom}})}{\partial(P_L/P_{L\text{ nom}})} = -\frac{1}{\dfrac{\Omega_{\text{nom}}}{P_{\text{nom}}}(G_f + G_g + G_c) + sT_a}$$

For this reason, the term "permanent regulating energy" is applied to the static gains $G_g(0)$, $G_c(0)$, as already done for $G_f(0)$. More precisely, the gain $G_g(0)$ is termed "natural permanent regulating energy of the unit," and $G_c(0)$ — which essentially depends on the characteristics of the users — is called "permanent regulating energy of the load," whereas the sum:

$$E \triangleq (G_f + G_g + G_c)(0)$$

is the total permanent regulating energy.

Correspondingly, the dimensionless quantities $b_g \triangleq P_{\text{nom}}/(\Omega_{\text{nom}} G_g(0))$, and $b_c \triangleq P_{\text{nom}}/(\Omega_{\text{nom}} G_c(0))$ can respectively be called "natural permanent droop of the unit" and "permanent droop of the load," whereas the "resulting permanent droop" is given by:

$$b \triangleq \frac{P_{\text{nom}}}{\Omega_{\text{nom}} E} = \left(\frac{1}{b_p} + \frac{1}{b_g} + \frac{1}{b_c}\right)^{-1}$$

The transfer function $G_g(s)$ depends on the type of plant (particularly, of turbine) and has usually modest effects, similar to those of the static gain $G_g(0)$. On the other hand, $G_g(s)$ appears in the feedback path of $1/(sM)$ (see Fig. 3.2), and it originates a closed-loop transfer function $1/(G_g(s) + sM)$. This function may be approximated by $1/(G_g(0) + sM)$, as the cutoff frequency of this loop (the value of which is approximately $G_g(0)/M = 1/(b_g T_a)$) is sufficiently smaller than the critical frequencies of $G_g(s)$ (see for instance $(\partial P_m/\partial \Omega)(s)$ for an hydroelectric plant, in Section 3.2.2). The value of $G_g(0)$, which can be deduced by the "static" mechanical characteristics of the unit with turbine valves blocked, has a relatively modest impact, as it corresponds to a very large natural droop (with respect to the usual values of b_p); for instance, $b_g = 200\%–1000\%$, according to the type of turbine and the operating point. As a first approximation, the transfer function $G_g(s)$ can be neglected without an appreciable error.

For reasons similar to those expressed above, the transfer function $G_c(s)$ also may be approximated by accounting for the slower components that characterize the response of ΔP_e to $\Delta \Omega$. In particular, by neglecting the electrical-type transients, the electric part of the system can be represented by the same equations that hold at steady-state (i.e., in terms of phase variables: sinusoidal operation of the positive sequence), but assuming that the frequency Ω is (slowly) varying.

Despite these approximations, the determination of $G_c(s)$ proves to be generally very complex, not only because of the large number (and the variety) of loads, but also because the function $G_c(s)$ can be viewed as the result of strictly interacting phenomena.

In effect, a variation in the frequency Ω, with no changes in the system configuration, can be translated into:

- a variation of the reactances in the electrical part;
- a variation of the slip for the asynchronous motors;

and it causes a variation in voltages and currents, with consequences on active powers absorbed by loads and on losses, i.e., on the power P_e, which is their sum. All this is further complicated by the overlapping effects of the speed transients of electromechanical-type loads and the voltage regulations (more generally of the v/Q control; see also Section 5.6.1).

Fortunately, the effects of $G_c(s)$ usually may be considered modest with respect to those of $G_f(s)$, so that the uncertainties on $G_c(s)$ are not a critical problem.

If, for simplicity, the variation of reactances is not considered, the active power P_{cj} absorbed by the generic load can dynamically depend:

(1) on the voltage v_{cj} applied to it, and on the frequency Ω;

(2) only on the voltage v_{cj}.

The former case is typical for electromechanical-type loads that include asynchronous motors. For them, the power P_{cj} (as well as the generated mechanical power P_{mcj}) can be considered a function of v_{cj} and the relative slip $\sigma' \triangleq (\Omega - \Omega_{cj})/\Omega$, where the electrical speed of the rotor Ω_{cj} dynamically depends on the P_{mcj} itself, further than on the resistant mechanical power. (Refer to the end of this section for synchronous motors and compensators.)

The latter case is typical for loads of the "static" type (i.e., without rotating parts) that can be assimilated to purely electrical equivalent circuits with P_{cj} function of only v_{cj}. Actually, it is also possible to consider within this category some electromechanical-type loads, e.g., including dc motors, for which P_{cj} and P_{mcj} depend on the speed, which however is dynamically related to P_{mcj} itself and to the resistant mechanical power, and is not affected by the frequency Ω.

After these premises, a frequency variation generally causes voltage variations in the whole system, and consequently variations in the absorbed powers also of (2)-type loads, unless their voltages are not suitably regulated. This clearly holds even if reactances remain unchanged, provided that the system includes (1)-type loads as well.

In general, the transfer function $G_c(s)$ can be determined by using a block diagram such as the one in Figure 3.6. This representation assumes that the "electrical part" of the system (which has $\Omega, \ldots, \Omega_{cj}, \ldots$ as inputs, and $P_e, \ldots, P_{mcj}, \ldots$ as outputs) includes also the v/Q control. Considering the further possible inputs (generically indicated by dotted lines) constant, the following equation can be then written, for small variations:

$$\begin{bmatrix} \Delta P_e \\ \vdots \\ \Delta P_{mcj} \\ \vdots \end{bmatrix} = H \begin{bmatrix} \Delta \Omega \\ \vdots \\ \Delta \Omega_{cj} \\ \vdots \end{bmatrix}$$

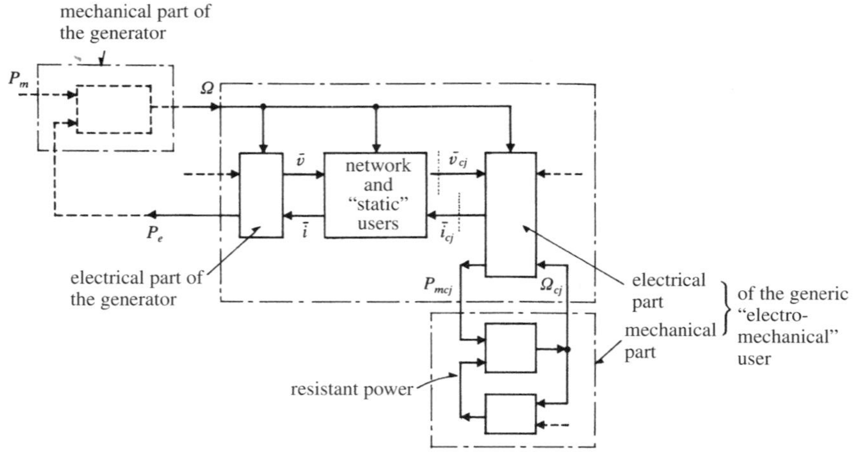

Figure 3.6. Effect of network and loads on the relationship between the generated power and frequency: broad block diagram.

where H is a proper transfer matrix (or even, more simply, a constant matrix if the v/Q control can be approximately accounted for by means of algebraic equations).

For the mechanical part of the generic load (of the electromechanical type), it may be assumed that[5]:

$$\Delta P_{mcj} = (D_{mj} + sM_{cj})\Delta\Omega_{cj} \qquad [3.1.11]$$

where D_{mj} accounts for the dependence of the mechanical resistant power on Ω_{cj}, and where M_{cj} is the "inertia coefficient" of the load (electric motor and mechanical load driven by it) similar to the coefficient M of the generating unit.

From the above equations, eliminating variations ΔP_{mcj}, $\Delta\Omega_{cj}$, it is possible to derive the function $G_c(s) \triangleq (\partial P_e/\partial\Omega)(s)$. Note that, in the absence of electromechanical users, the solution is simply $G_c(s) = H$, with H a scalar.

The problem is noticeably simplified if:

- the variations of the electrical losses (caused by variations of the frequency Ω) are neglected;
- it is assumed that the loads are supplied in "radial way" from nodes having a constant voltage (amplitude) v_j, as indicated in Figure 3.7.

Actually, the nodes at constant voltage may be suitable "equivalent" ones; the same applies for the parameters of the links between these nodes and the respective loads.

Under such conditions, it results (with the symbols of Figure 3.7) $\Delta P_e = \sum_j \Delta P_j = \sum_j \Delta P_{cj}$ and thus:

$$G_c(s) = \sum_j G_{cj}(s) \qquad [3.1.12]$$

where the single $G_{cj}(s) \triangleq (\partial P_{cj}/\partial\Omega)(s)$ can be evaluated separately in a trivial way.

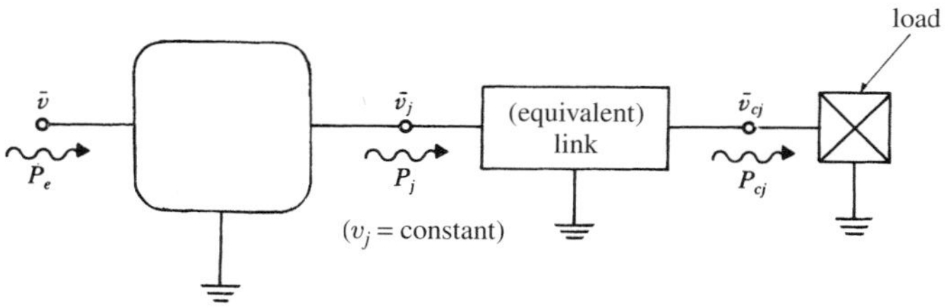

Figure 3.7. Basic example of an electrical system.

---

[5] It is assumed that the mechanical load is not defined by further parameters, e.g., concerning elastic couplings etc.

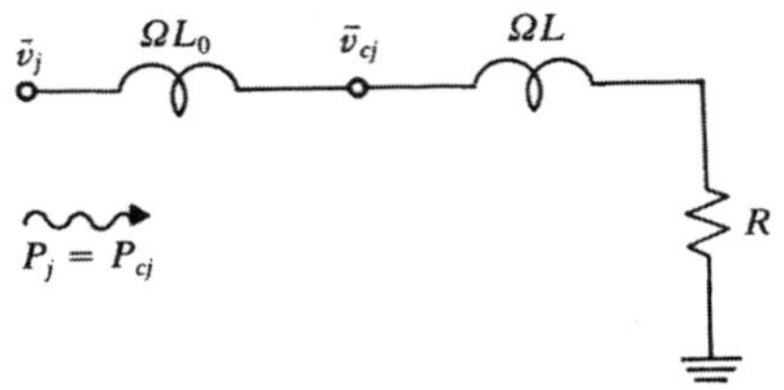

Figure 3.8. Example of an equivalent circuit in the case of "static" load.

For instance, in the case of a *"static" load* that may be viewed as an impedance $(R + j\Omega L)$, by representing the link as a simple series impedance $j\Omega L_0$, it can be obtained (see Fig. 3.8):

$$P_{cj} = \frac{v_j^2 R}{R^2 + \Omega^2(L_0 + L)^2}$$

from which, assuming v_j to be constant and indicating by the superscript "o" the values corresponding to the operating point:

$$G_{cj}(s) = -\frac{a P_{cj}^o}{\Omega^o} \qquad [3.1.13]$$

(independently of s) with:

$$a \triangleq \frac{2\mu^2}{1 + \mu^2} \qquad [3.1.14]$$

having set, for brevity, $\mu \triangleq \Omega^o(L_0 + L)/R$. The dependence of $G_{cj}\Omega^o/P_{cj}^o = -a$ on μ is represented by a dotted line in Figure 3.11.

By recalling Equations [3.1.10] and [3.1.12], it can be concluded that the load examined contributes, in a negative way, only to the total (permanent) regulating energy; nevertheless this contribution, which is exclusively related to the variations in the reactances, usually is negligible with respect to $G_f(0) = P_{\text{nom}}/(\Omega_{\text{nom}} b_p)$ (recall Equation [3.1.8]). For instance, if all the static loads were of the present type with the same value of μ, and it were $\Omega^o = \Omega_{\text{nom}}$ and $\sum_j P_{cj}^o = 0.4 P_{\text{nom}}$, it would follow $\sum_j G_{cj}(0)/G_f(0) = -0.4ab_p$, which is very small in absolute value (intending, of course, that these last sums are extended only to the static loads).

In the case of an *electromechanical load* with an asynchronous motor, by adopting the well-known equivalent circuit indicated in Figure 3.9a (see also Section 5.6.2), and representing the link by a simple series impedance $j\Omega L_0$, it

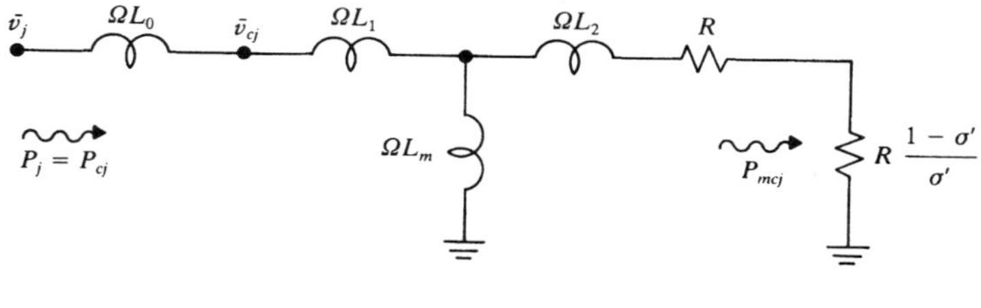

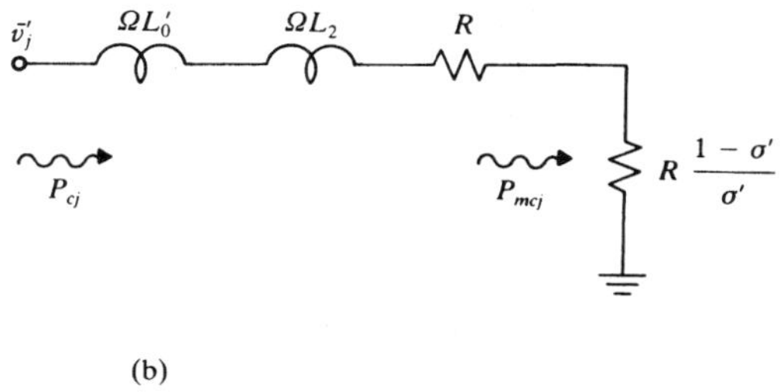

Figure 3.9. Case of an electromechanical load with an asynchronous motor: (a) example of equivalent circuit; (b) reduction of the equivalent circuit.

follows (see Fig. 3.9b) that:

$$P_{cj} = \frac{v'^2_j \dfrac{R}{\sigma'}}{\left(\dfrac{R}{\sigma'}\right)^2 + \Omega^2(L'_0 + L_2)^2}, \qquad P_{mcj} = (1-\sigma')P_{cj}$$

where:

$$v'_j \triangleq v_j \frac{L_m}{L_0 + L_1 + L_m}, \qquad L'_0 \triangleq \frac{(L_0 + L_1)L_m}{L_0 + L_1 + L_m}$$

whereas $\sigma' \triangleq (\Omega - \Omega_{cj})/\Omega$ is the relative slip.

It can be then derived, by assuming v_j to be constant:

$$\Delta P_{cj} = \left(-a\frac{\Delta\Omega}{\Omega^o} + (1-a)\frac{\Delta\sigma'}{\sigma'^o}\right) P^o_{cj}$$

where a is given by Equation [3.1.14], but now intending that $\mu \triangleq \Omega^o(L'_0 + L_2)/(R/\sigma'^o)$ (note that, for $\Delta\sigma' = 0$, there would be only the term $-a = \Delta\Omega\, P^o_{cj}/\Omega^o$, resulting from the only variation of reactances); and furthermore:

$$\Delta P_{mcj} = (1 - \sigma'^o)\Delta P_{cj} - P^o_{cj}\Delta\sigma'$$

By considering that $\Delta\sigma' = ((1-\sigma'^{o})\Delta\Omega - \Delta\Omega_{cj})/\Omega^{o}$, it can be then derived that:

$$\Delta P_{cj} = H_0\Delta\Omega - H_3\Delta\Omega_{cj}, \qquad \Delta P_{mcj} = H_1\Delta\Omega - H_2\Delta\Omega_{cj}$$

with:

$$\begin{cases} H_0 \triangleq \left(\dfrac{1-a}{\sigma'^{o}} - 1\right)\dfrac{P_{cj}^{o}}{\Omega^{o}} \\ H_1 \triangleq (1-\sigma'^{o})\left(\dfrac{1-a}{\sigma'^{o}} - 2\right)\dfrac{P_{cj}^{o}}{\Omega^{o}} \\ H_2 \triangleq \left(\dfrac{1-a}{\sigma'^{o}} - 2 + a\right)\dfrac{P_{cj}^{o}}{\Omega^{o}} \\ H_3 \triangleq \dfrac{1-a}{\sigma'^{o}}\dfrac{P_{cj}^{o}}{\Omega^{o}} \end{cases}$$

Furthermore, the relationship between ΔP_{mcj} and $\Delta\Omega_{cj}$ can be represented using Equation [3.1.11], where the inertia coefficient M_{cj} is (obviously) proportional to the electrical speed $\Omega_{cj}^{o} = (1-\sigma'^{o})\Omega^{o}$ at the operating point, as given by:

$$M_{cj} = (1-\sigma'^{o})M_{cjo}$$

in which M_{cjo} is the inertia coefficient evaluated for $\Omega_{cj}^{o} = \Omega^{o}$, i.e., for operation at null slip.

The previous equations are summarized in Figure 3.10, finally leading to the following transfer function:

$$G_{cj}(s) = H_0 - \frac{H_1 H_3}{H_2 + D_{mj} + sM_{cj}}$$

which can be written in the form:

$$G_{cj}(s) = G_{cj}(0) + \frac{sM_{cj}^{*}}{1 + sT_{cj}} \qquad [3.1.15]$$

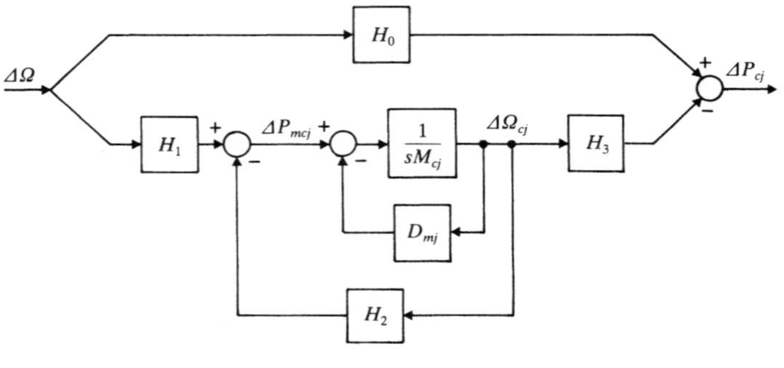

Figure 3.10. Case of an electromechanical load: block diagram.

in which:

$$G_{cj}(0) = H_0 - \frac{H_1 H_3}{H_2 + D_{mj}}$$

and furthermore:

$$T_{cj} \triangleq \frac{M_{cj}}{H_2 + D_{mj}}, \qquad M^*_{cj} \triangleq \frac{H_1 H_3}{(H_2 + D_{mj})^2} M_{cj}$$

If it is assumed that:

$$D_{mj} \triangleq \frac{P^o_{mcj}\alpha}{\Omega^o_{cj}} = \frac{P^o_{cj}\alpha}{\Omega^o}$$

then the dimensionless quantities $G_{cj}(0)\Omega^o/P^o_{cj}$, $T_{cj}P^o_{cj}/(M_{cjo}\Omega^o)$, M^*_{cj}/M_{cjo} can be determined starting from a (or from μ; see Equation [3.1.14]), α and σ'^o; see for example Figure 3.11, for the case $\alpha = 2$, $\sigma'^o = 1\%$. The value of α is independent of the operating point if it is assumed that the resistant mechanical power is, apart from a possible constant, proportional to $(\Omega_{cj})^\alpha$. Usually $\alpha \cong 1$–3, depending on the type of mechanical load.

Moreover, the inertia coefficient M_{cjo} can be put in the form:

$$M_{cjo} = \frac{P_{cj\,\text{nom}} T_{ac}}{\Omega_{\text{nom}}}$$

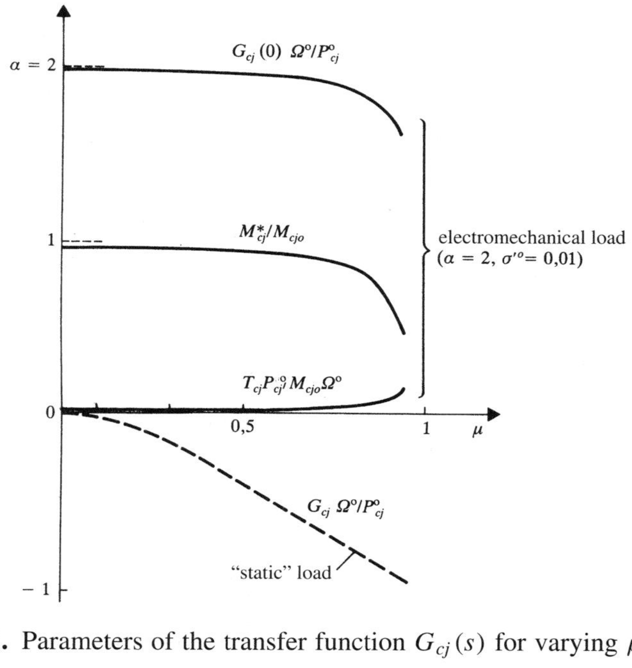

Figure 3.11. Parameters of the transfer function $G_{cj}(s)$ for varying μ (see text).

(in all similar to Equation [3.1.4]), where the "start-up time" T_{ac}, which is quite variable from case to case, is much smaller than that of generating units. It may be, for example, in the range of 0.5–6 seconds or smaller.

By accounting for the previous relationships, it is easy to ascertain that usually:

- the time constant T_{cj} is smaller than T_{ac} (e.g., $T_{ac}/100$) and equal to hundredths of a second or less; within the frequency regulation problems, Equation [3.1.15] may be approximated by:

$$G_{cj}(s) \cong G_{cj}(0) + sM^*_{cj} \qquad [3.1.16]$$

- the values of $G_{cj}(0)$, M^*_{cj} are slightly smaller than D_{mj}, M_{cjo}, respectively.

By recalling Equations [3.1.10] and [3.1.12], the effects of the considered load can be then summarized into as:

- an increase, slightly smaller than D_{mj}, of the total (permanent) regulating energy;
- an increase, slightly smaller than M_{cjo}, of the resulting inertia coefficient.

If it is assumed that all the electromechanical loads are of the type discussed above, with the same values $\alpha' \triangleq \alpha P^o_{cj}/P_{cj\,\text{nom}}$ and T_{ac}, and $\Omega^o = \Omega_{\text{nom}}$ and $\sum_j P_{cj\,\text{nom}} = 0.4\text{–}0.6\, P_{\text{nom}}$:

- the increment in the regulating energy is slightly smaller than:

$$\sum_j D_{mj} = (0.4\text{–}0.6)\,\alpha' \frac{P_{\text{nom}}}{\Omega_{\text{nom}}}$$

  and thus much smaller than $G_f(0) = P_{\text{nom}}/(\Omega_{\text{nom}} b_p)$;
- the increment in the inertia coefficient is slightly smaller than:

$$\sum_j M_{cjo} = (0.4\text{–}0.6)\, T_{ac} \frac{P_{\text{nom}}}{\Omega_{\text{nom}}}$$

  and thus small, even if not negligible, with respect to $M = T_a P_{\text{nom}}/\Omega_{\text{nom}}$ (e.g., an increase of approximately 15%–20% for $T_{ac} \cong 3$ sec, $T_a \cong 8$ sec).

On the contrary, the generic $G_{cj}(s)$ may significantly change in some (singular) cases, particularly for relatively large values of μ, for which (Fig. 3.11) $G_{cj}(0)$ and M^*_{cj} may be further reduced, and T_{cj} may no longer be negligible (in qualitative terms, the effect of T_{cj} is to make the increment of the inertia coefficient in the faster parts of the transients practically zero, because of Equation [3.1.15]).

Up to now, it was assumed that the system does not include other synchronous machines apart from the considered generator. If an electromechanical load includes a synchronous motor, it may be assumed, because of the synchronizing actions (Section 1.6):

$$\Delta\Omega_{cj} = \Delta\Omega$$

apart from relatively fast electromechanical oscillations, which, for practical purposes, do not influence the frequency regulation. Because of Equation [3.1.11] (neglecting the electrical losses variations), the effects of the load can be translated into an increase D_{mj} of the total regulating energy and an increase M_{cj} of the inertia coefficient.

Similar considerations hold for the case of a synchronous compensator, for which we may assume $D_{mj} = 0$.

3.2. TYPICAL SCHEMES FOR SPEED GOVERNORS

3.2.1. Preliminaries

The speed regulator (currently called "governor") of a given unit must be considered necessary equipment of the unit itself, not only for its participation in the frequency regulation, but also for requirements of a local nature, such as speed regulation during the phases preceding the parallel connection to the network (startup and no-load operation), or after a disconnection from the network, and, generally, for the control of the driving power supplied by the turbine, e.g., to follow the generation schedule.

The speed governor is thus also called "turbine (or machine) regulator" or "*primary* frequency regulator," since it constitutes a necessary equipment for the frequency regulation.

Based on the discussions of Section 3.1.2, the primary regulator can be generally required to follow not only the frequency set point, but also the power set point, in such a way to realize, through the control of the driving power, a static characteristic as indicated in Figure 3.5. Therefore, if it is not imposed that the permanent droop is zero ($b_p = 0$), the primary regulator can no longer be considered purely a frequency regulator (and by imposing $b_p = \infty$, it is even possible to use it as a purely power regulator).

In whichever manner the primary regulator is used, it is clear that its effects on the driving power are influenced strictly by the characteristics of the valve-positioning system, the supply system, and the turbine (recall Fig. 3.1), which must then be considered in regulator synthesis.

The objectives of this section are to:

- summarize the characteristics of typical hydroelectric and thermal plants with particular reference to the frequency range relevant to regulation;
- illustrate the most common schemes of primary regulators and their respective transfer functions.

In generic terms, by recalling Section 3.1, it is useful to consider that:

- the cutoff frequency ν_t of the primary regulation may be, for example, within the range of 0.3–0.5 rad/sec, whereas larger values of ν_t are practically unacceptable for the reasons above (e.g., risks of instability, excessive stress);
- at least within the frequency range $(0, \nu_t)$, it is convenient to realize a transfer function $G_f(s) \triangleq (\partial P_m/\partial \varepsilon_f)(s)$ of the type [3.1.8], i.e.,

$$G_f(s) = \frac{P_{\text{nom}}}{\Omega_{\text{nom}} b_p} \frac{1 + sT_2}{1 + sT_1} = \frac{P_{\text{nom}}}{\Omega_{\text{nom}}} \frac{1 + sT_2}{b_p + sT_2 b_t}$$

  with T_2 approximately 3–5 sec, and $T_1 = b_t T_2/b_p$ significantly larger (e.g., 15–25 sec if $b_t = 25\%$ and $b_p = 5\%$, or even $T_1 = \infty$ if $b_p = 0$).

Such conclusions have come about by assuming that the system includes only a single generating unit. However, they can be generically extended to more than one unit, as presented in Section 3.3.1.

3.2.2. Case of Hydrounits

(a) Dependence of the Driving Power on the "Output" of the Speed Governor

In the case of a hydroelectric unit, the relationship between β ("output" of the speed governor) and P_m (driving power) can be generically represented by Figure 3.12.

The rotation angle of the so-called "regulation shaft," that governs the turbine gate by means of an adequately powered positioning system, may be represented as variable β.

Particular reference will be made to the case of "high-head" plants (equipped with Pelton turbines), which is the most interesting case when considering the control of the driving

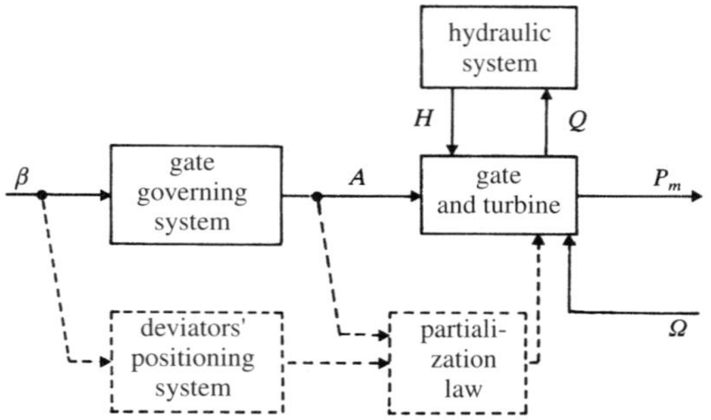

Figure 3.12. Dependence of the driving power on the output of the primary regulator, for a hydroelectric unit: broad block diagram.

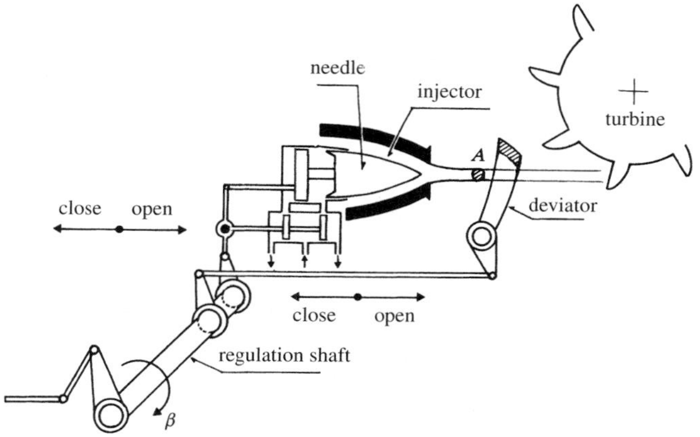

Figure 3.13. Example of the system that governs the gate and the deviators, for the case of a Pelton turbine.

power and the participation to the frequency regulation. In such a case, the gate is generally constituted by several injectors and the respective needles, actuated in parallel (through the same number of servomechanisms) by the regulation shaft. In the following, reference will be made to only a single "equivalent" pair injector-needle. For a more complete analysis, which is of interest for large variations, it is, however, necessary to consider the control of the deviators (directly linked to the regulation shaft, to avoid response delays) and "partialization" of the driving power when the deviators intercept the respective jets (refer to Figs. 3.12 and 3.13).

"Opening" A may be assumed to be the "useful cross section" of the water flowing into the turbine. The other variables indicated in Figure 3.12 are:

Q = water flow (volume per unit of time) at the gate output;

H = water energy per unit of weight within the distributor.

The operation of the plant also depends on the water level in the supply reservoir, basin, or tank, and for the case of reaction turbines on the level at the water discharge. In the following, it will be assumed that such levels are constant, thus neglecting their variations with Q and with (possible) inflows and spillages.

The transfer function $G_v(s) \triangleq (\Delta A/\Delta\beta)(s)$ (see Fig. 3.2), which specifically accounts for the above-mentioned servomechanism, may be considered to be:

$$G_v(s) = \frac{K_v}{1 + sT_v} \qquad [3.2.1]$$

where:

- the static gain $K_v \triangleq G_v(0)$ may depend on the operating point, if the static characteristic (β, A) is nonlinear;
- the time constant T_v may be, for example, 0.1–0.3 seconds.

For large variations, it also is necessary to consider that the speed of the servomechanism is actually limited, in the opening and the closing phase, to avoid excessive pressure stresses in the hydraulic plant.

The nonlinearity of the static characteristic (β, A) also can be achieved intentionally, through the use of proper "cams," e.g., to make $K_v = \mathrm{d}A/\mathrm{d}\beta$ smaller for small opening values so to reduce automatically $G_f(0)$ in the operation at "no-load"; see Figure 3.2 and the following footnote[(7)] (Section 3.3.1). The dependence of A on β may present an "insensitiveness" (which may not always be negligible), which in terms of β can be, for example, 0.2%–0.5% with respect to the value that corresponds to the full opening.

With reference to the block "gate and turbine," it may be assumed that P_m and Q depend on A, Ω, H according to algebraic-type relationships, i.e., without any dynamic delay. For small variations around a given operating point, characterized by the superscript "o," it can be then generally written:

$$\Delta P_m = P_m^o \left(K_{PA} \frac{\Delta A}{A^o} + K_{P\Omega} \frac{\Delta \Omega}{\Omega^o} + K_{PH} \frac{\Delta H}{H^o} \right) \quad [3.2.2]$$

$$\Delta Q = Q^o \left(K_{QA} \frac{\Delta A}{A^o} + K_{Q\Omega} \frac{\Delta \Omega}{\Omega^o} + K_{QH} \frac{\Delta H}{H^o} \right) \quad [3.2.3]$$

with proper values (generally variable with the operating point) of the coefficients K_{PA}, $K_{P\Omega}$, etc. Note that $\Delta H/H^o$ also represents the relative pressure variation in the gate if the kinetic energy per unit of weight is, in H, negligible with respect to the potential energy, as it occurs, for instance, in plants equipped with Pelton turbines.

If it is assumed that:

$$P_m = \eta \gamma Q H$$

where:

η = turbine efficiency
γ = water specific weight

and the variations of η and γ are considered as negligible, Equation [3.2.2] can be substituted by:

$$\frac{\Delta P_m}{P_m^o} = \frac{\Delta Q}{Q^o} + \frac{\Delta H}{H^o}$$

(whereas $K_{PA} = K_{QA}$, $K_{P\Omega} = K_{Q\Omega}$, $K_{PH} = K_{QH} + 1$).

The hypothesis that $\Delta\eta = 0$ is acceptable if the operating point is close to that at maximum efficiency. With Kaplan turbines, it is possible to reduce the efficiency variations by acting not only on the opening of the gate, i.e., on A, but also on position (B) of the blades. In such a case, it is necessary to simultaneously account for the effects of A, B on P_m, Q.

The blades control may be subjected to that of the gate, or the two controls may be realized in parallel, by two distinct, adequately interacting servomechanisms.

Furthermore, in the case of a Pelton turbine, it can be stated that, per weight unit, the energy H is converted into the kinetic energy $(Q/A)^2/(2g)$ of the outflowing jet, where g is the acceleration due to gravity (the potential energy of the jet is zero, assuming that the pressures are referred to the atmospheric pressure, and the heights to that of the gate). As a result:

$$Q = A\sqrt{2gH}$$

and Equation [3.2.3] becomes:

$$\Delta Q = Q^o\left(\frac{\Delta A}{A^o} + \frac{1}{2}\frac{\Delta H}{H^o}\right)$$

$(K_{QA} = 1,\ K_{Q\Omega} = 0,\ K_{QH} = 1/2)$.

Finally, regarding the block "hydraulic system," the dynamic dependence of H on Q can be generally expressed for small variations by the equation:

$$\Delta H = -Z_w(s)\Delta Q \qquad [3.2.4]$$

where $Z_w(s)$ is the "impedance" of the hydraulic system as viewed from the gate.

Once $Z_w(s)$ has been derived, Equations [3.2.2], [3.2.3], and [3.2.4] allow the derivation of the transfer function $G_a(s) \triangleq (\partial P_m/\partial A)(s)$, which can be usefully put into the form:

$$G_a(s) = K_{PA}\frac{P_m^o}{A^o}\frac{1 - a_1\dfrac{Q^o}{H^o}Z_w(s)}{1 + a_2\dfrac{Q^o}{2H^o}Z_w(s)} \qquad [3.2.5]$$

where:

$$\begin{cases} a_1 \triangleq \dfrac{K_{QA}K_{PH}}{K_{PA}} - K_{QH} \\ a_2 \triangleq 2K_{QH} \end{cases}$$

In the case of a Pelton turbine, it then holds that $a_2 = 1$, and $K_{PA} = a_1 = 1$ if the efficiency variations are negligible; in the other cases, the values of K_{PA}, a_1, a_2 are slightly different from unity. Moreover:

$$\frac{\partial P_m}{\partial \Omega}(s) = \frac{P_m^o}{\Omega^o}\left(K_{P\Omega} - \frac{K_{PH}K_{Q\Omega}\dfrac{Q^o}{H^o}Z_w(s)}{1 + a_2\dfrac{Q^o}{2H^o}Z_w(s)}\right)$$

which contributes, usually in a modest way, to the $G_g(s)$ (Fig. 3.2).

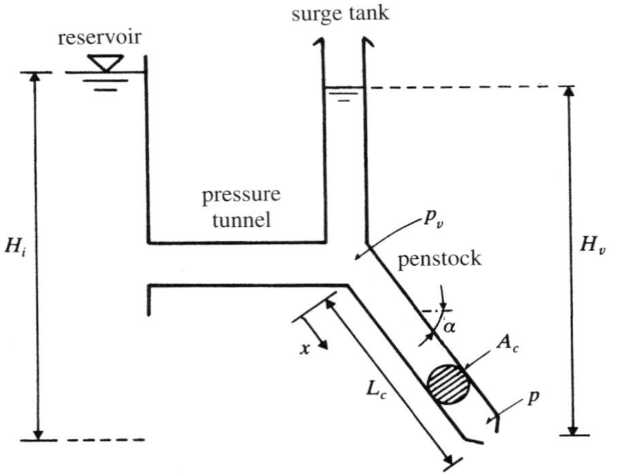

Figure 3.14. Example of a "high-head" hydraulic system. The scheme does not respect the geometrical proportions. Furthermore, the section variations in the single parts are not represented, particularly for the surge tank.

As a first approximation, the impedance $Z_w(s)$ can be derived:

- assuming that the gate is supplied, through only the penstock, by a reservoir with a constant level (with reference to the typical scheme of Fig. 3.14, it is assumed $H_v = H_i =$ constant, thus ignoring the presence of the pressure tunnel and of the surge tank);
- neglecting the head losses in the penstock and the compressibility of the water (moreover, supposing as rigid the walls of the penstock).

Using generally acceptable simplifications (and by symbols that are known or defined in the figure), it is possible to write:

$$H_v - H = L_c \sin\alpha + \frac{p_v - p}{\gamma}$$

Moreover, because of the assumptions made:

$$\gamma A_c L_c \sin\alpha + (p_v - p)A_c = \frac{\gamma A_c L_c}{g} \frac{\mathrm{d}(Q/A_c)}{\mathrm{d}t}$$

where the left side term, resulting from the weight of the water column in the penstock and the difference between the mean pressures at the terminal sections, constitutes the accelerating force that acts on the column itself, the mass of which is $(\gamma L_c A_c)/g$, and the speed of which is Q/A_c.

It then follows:

$$H_v - H = J_c \frac{\mathrm{d}Q}{\mathrm{d}t} \qquad [3.2.6]$$

thus $H^o = H_v$, and furthermore:

$$Z_w(s) = sJ_c \tag{3.2.7}$$

where $J_c \triangleq L_c/gA_c$ is the so-called "*inertance*" of the penstock. It has been assumed, for simplicity, that the penstock is cylindrical. Generally, if the section A_c depends on the abscissa x, the inertance is given by $J_c = 1/g \int_0^{L_c} \mathrm{d}x/A_c(x)$.

Substituting into Equation [3.2.5], it can be finally obtained:

$$G_a(s) = K_a \frac{1 - a_1 s T_w}{1 + a_2 s \dfrac{T_w}{2}} \tag{3.2.8}$$

where a_1, a_2 have the values already seen (equal to one or almost so), whereas:

$$G_a(0) = K_a \triangleq K_{PA} \frac{P_m^o}{A^o} \cong \frac{P_m^o}{A^o}, \qquad T_w \triangleq \frac{Q^o}{H^o} J_c$$

In terms of frequency response, the magnitude of $G_a(\tilde{j}\nu)$ then increases from K_a (for $\nu = 0$) to $2a_1K_a/a_2$ (for $\nu \to \infty$), whereas the phase delay increases from $0°$ to $180°$, passing through the value $90°$ at the frequency $\nu = \sqrt{2}/(T_w\sqrt{a_1a_2})$; see Figure 3.15.

For a Pelton turbine, if the efficiency variations are neglected:

- the power P_m is proportional to QH and thus to $AH^{3/2}$;
- because of the adopted assumptions, it then holds $H^o = H_v$ = constant, $K_a = P_m^o/A^o$.

Consequently, the static gain $K_a \triangleq G_a(0)$ is independent of the operating point; in actual cases, K_a may significantly vary because of the efficiency variations (it may be reduced for small A^o).

The quantity T_w, called "*water inertia time*" *of the penstock*, is proportional to Q^o for any given H^o, so that:

$$T_w = \frac{Q^o}{Q_{\text{nom}}} T_{wn}$$

where Q_{nom} is the nominal flow, and T_{wn} is the corresponding water inertia time; the usual values of T_{wn} (particularly in the case of Pelton turbines) are 0.5–1.5 seconds. Note that T_w also can be written in the form:

$$T_w = \frac{(\gamma/g)L_cA_c(Q^o/A_c)^2}{\gamma Q^o H^o}$$

thus it is equal to the ratio between twice the kinetic energy of the water in the penstock (at the speed Q^o/A_c) and the power ($\gamma Q^o H^o$) of the outflowing jet.

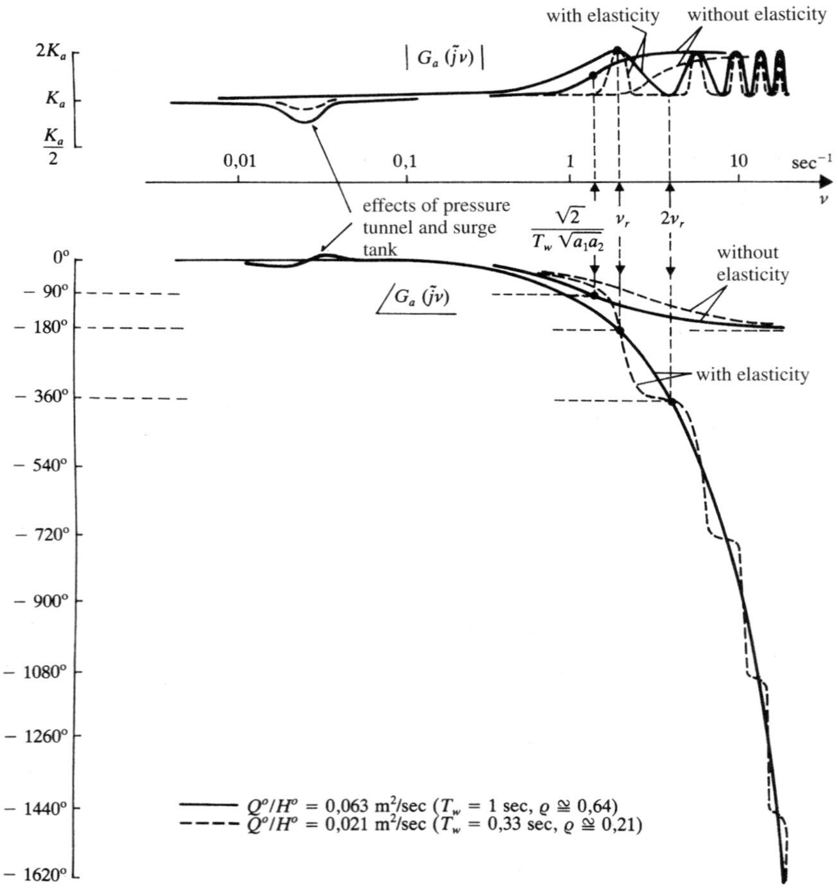

Figure 3.15. Frequency response of the transfer function $G_a(s)$ (supply system and turbine): case of a hydroelectric plant having $T_e = 1.57$ sec ($\nu_r = 2$ rad/sec), $K_c = 0$, $J_c = 16\ \text{sec}^2/\text{m}^2$, $a_1 = a_2 = 1$ (see also footnote[(4)]).

It may be observed that T_w also represents the time taken for a mass, placed on an inclined plane with a slope $\alpha' \triangleq \text{arsin}(H^o/L_c)$ (note that α' is different from α; see Fig. 3.14), to reach the speed Q^o/A_c in the absence of friction.

For a better approximation, the impedance $Z_w(s)$ can be derived considering:

(1) the head losses (caused by friction effects) in the penstock;
(2) the elasticity of the water and the penstock walls;
(3) the presence of other elements of the hydraulic plant, i.e., (for high-head plants, with Pelton turbine; see Fig. 3.14) the pressure tunnel and the surge tank.

To account for (1), a term $K_c|Q|Q$ (*head losses* of the quadratic type) may be added to the right-hand side of Equation [3.2.6] to obtain:

$$H_v - H = K_c|Q|Q + J_c\frac{\mathrm{d}Q}{\mathrm{d}t}$$

from which $H^o = H_v - K_c Q^{o2}$ and furthermore, by linearizing:

$$Z_w(s) = 2K_c Q^o + sJ_c \qquad [3.2.9]$$

Therefore, it is sufficient to formally replace sT_w by $((2K_c Q^{o2})/H^o + sT_w)$ into Equation [3.2.8]. However, the relative head loss $(K_c Q^{o2})/H^o$ can be a few percentages at the nominal flow, so that the effect of the head losses on $G_a(s)$ can be considered negligible (in particular, the static gain $G_a(0)$ is slightly reduced with respect to the value $K_{PA} P_m^o / A^o$ previously found).

In any case, if energies per unit of weight and flows are respectively correlated to electric voltages and currents, the linearized behavior of the penstock is similar to that of a "series" branch having a resistance $2K_c Q^o$ and an inductance J_c.

To account for the effects (2) of the *elasticity*, it must be generically assumed to be $h = h(x, t)$, $q = q(x, t)$, where h is the energy per unit of weight, q the flow, and x the abscissa defined in Figure 3.14. Assuming, for simplicity, that the penstock is cylindrical and that its characteristics are independent of x (and neglecting the head losses), the Equation [3.2.6] can be substituted for the generic element of the penstock by:

$$-\frac{\partial h}{\partial x} = \frac{1}{gA_c}\frac{\partial q}{\partial t} \qquad [3.2.10]$$

to which the continuity equation must be added, of the type:

$$-\frac{\partial q}{\partial x} = c_c\frac{\partial h}{\partial t} \qquad [3.2.11]$$

whereas $h(0, t) = H_v$, $q(L_c, t) = Q(t)$, $h(L_c, t) = H(t)$. The effects of the elasticity are of more practical interest for high-head plants with Pelton turbines. For such plants, the kinetic component in h is negligible, and thus h is almost equal to the piezometric head $(p/\gamma + z)$, being $p =$ pressure and $z =$ height. It may be further assumed that $c_c = \mathrm{d}(\gamma A_c)/\mathrm{d}p =$ constant.

By again correlating the variables h and q, respectively, to electric voltages and currents, the penstock behaves as an electric line, for which $1/(gA_c)$ and c_c are, respectively, the (series) inductance and the (shunt) capacitance per unit of length. Through developments similar to those in Section 5.4 with reference to electrical lines, from Equations [3.2.10] and [3.2.11], it is possible to deduce:

$$H(s) = \frac{H_v(s)}{\text{ch}\theta_c} - Z_{oc}\text{th}\theta_c Q(s) \qquad [3.2.12]$$

where $H^o = H_v$, and further:

$$Z_w(s) = Z_{oc}\text{th}\theta_c \qquad [3.2.13]$$

where:

$$Z_{oc} \triangleq \frac{1}{\sqrt{gA_c c_c}}, \qquad \theta_c \triangleq s\sqrt{\frac{c_c}{gA_c}}L_c$$

Such relationships specify the well-known phenomena of perturbation propagation along the penstock; more precisely, the speed of propagation is given by:

$$a_c \triangleq \sqrt{\frac{gA_c}{c_c}}$$

which is within 700–1200 m/sec (typical value: approximately 1000 m/sec), whereas it can be written $Z_{oc} = a_c/gA_c$, $\theta_c = sL_c/a_c$. Finally, substituting the above into Equation [3.2.5] it follows:

$$G_a(s) = K_a \frac{1 - a_1 \dfrac{Q^o}{H^o} Z_{oc}\text{th}\theta_c}{1 + a_2 \dfrac{Q^o}{2H^o} Z_{oc}\text{th}\theta_c} = K_a \frac{1 - a_1 \dfrac{Q^o}{H^o}\dfrac{a_c}{gA_c}\text{th}\dfrac{sL_c}{a_c}}{1 + a_2 \dfrac{Q^o}{2H^o}\dfrac{a_c}{gA_c}\text{th}\dfrac{sL_c}{a_c}} \qquad [3.2.14]$$

instead of Equation [3.2.8], whereas it still holds $G_a(0) = K_a \triangleq K_{PA}P_m^0/A^0$.

In terms of frequency response, by writing:

$$\nu_r \triangleq \frac{\pi}{2}\frac{a_c}{L_c}$$

(usually, 1–10 rad/sec), it can be deduced that the magnitude of $Z_w(\tilde{j}\nu)$ becomes infinite at the frequencies $\nu = \nu_r, 3\nu_r, \ldots$, called "*resonance frequencies*" of the penstock, and zero at the frequencies $\nu = 0, 2\nu_r, 4\nu_r, \ldots$, called "*antiresonance frequencies*" of the penstock. Consequently, the magnitude of $G_a(\tilde{j}\nu)$ varies alternatively between K_a at the antiresonance frequencies, and $2a_1K_a/a_2$ at the resonance frequencies, whereas the phase delay increases from 0° (at $\nu = 0$) to 180°, 360°, 540°, ..., respectively at $\nu = \nu_r, 2\nu_r, 3\nu_r, \ldots$, (see Fig. 3.15). Such a behavior, which is quite different from the one corresponding to Equation [3.2.8], is independent (as well as ν_r) of the value of Q^o.

These conclusions remain practically unchanged in the presence of head losses, the effects of which, although rather modest, particularly affect the static gain $G_a(0)$ as already pointed out. To account for $K_c \neq 0$, the term $K_c|q|q/L_c$ can be added at the right-hand side of Equation [3.2.10], and for small variations, this term corresponds to a series resistance per unit of length, equal to $2K_c Q^o/L_c$.

The hypothesis of zero elasticity corresponds to $c_c \to 0$, $a_c \to \infty$, $\nu_r \to \infty$, and furthermore $\theta_c \to \infty$, $Z_w(s) = Z_{oc}\text{th}\theta_c \to Z_{oc}\theta_c = sL_c/gA_c = sJ_c$, according to Equation [3.2.7]. The quantity $T_e \triangleq 2L_c/a_c = \pi/\nu_r$, equal to the time required by the generic perturbation to go through the whole penstock in the two senses, going back to the starting point, is called "*reflection time*" (usually, T_e = 0.3–3 sec). To avoid excessive pressure variations in the penstock it is necessary, based on the "water hammer" theory, that the times of complete opening or closing of the turbine gate are sufficiently longer than T_e.

Another typical parameter is the so-called "*Allievi parameter*" (or "water hammer number") as given by:

$$\rho \triangleq \frac{Q^o}{2H^o}\frac{a_c}{gA_c} = \frac{\nu_r T_w}{\pi} = \frac{T_w}{T_e}$$

the value of which determines the type of transient (oscillatory or aperiodic) of the variation ΔP_r of the regulating power, following a hypothetical opening step ΔA. More precisely, $G_a(s)$ can be put into the form:

$$G_a(s) = K_a \frac{1 - a_1\, 2\rho\, \text{th}\dfrac{sT_e}{2}}{1 + a_2\, \rho\, \text{th}\dfrac{sT_e}{2}}$$

Assuming, for simplicity, that $a_1 = a_2 = 1$, it can be recognized that the response is oscillatory if $\rho < 1$, and aperiodic if $\rho > 1$; see the examples of Figure 3.16. According to what is shown in the same figure, if instead (disregarding the elasticity) the transfer function of Equation [3.2.8] was assumed, with $a_1 = a_2 = 1$, a simple exponential response would result, which starts from the initial value $\Delta P_r = -2K_a\Delta A$ at $t = 0^+$.

Finally, the effects (3) of the other hydraulic plant elements, such as the *pressure tunnel* and the *surge tank*, are generally modest and limited to a range of low frequencies (e.g., $\nu < 0.1$ rad/sec), where the amplitude of the term $(Q^o/H^o)Z_w(\tilde{j}\nu)$ is fairly small, with little effect on $G_a(\tilde{j}\nu)$; refer to Figure 3.15 for a qualitative example. It is in fact intuitive that, with the normal values of the surge tank cross section (or, more generally, of the tank feeding the penstock), the piezometric head at the inlet of the penstock (or the water level in the tank) may be subject to relatively slow variations, whereas fast variations are necessarily of reduced magnitude.

The magnitude of the slow variations (e.g., oscillations with a 150–250 second period) may, however, not be negligible. It is necessary to check that the opening and/or closing of the turbine gate, do not cause excessive oscillations of the water level, such as to empty

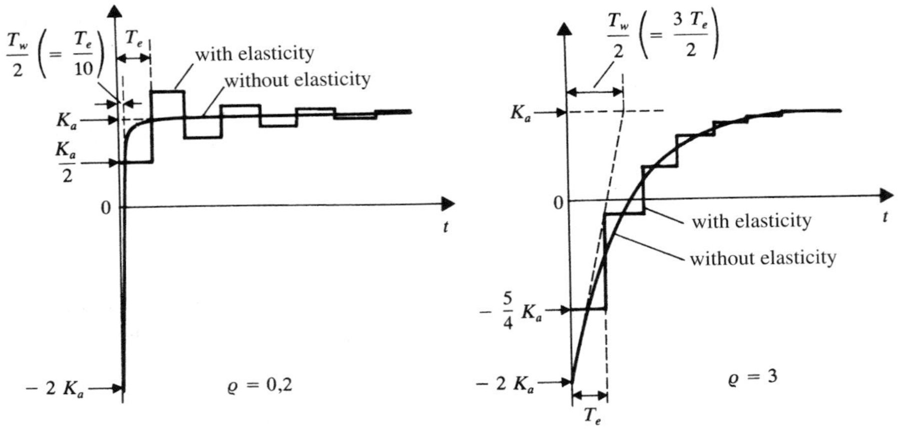

Figure 3.16. Response of the variation of regulating power to an unit step opening of the gate.

the tank, with serious damage to the plant because of introduction of air into penstock and tunnel, etc.

Regarding the cross section of the surge tank in high-head plants, too small a value might cause significant effects on the $G_a(s)$ and even prejudice the regulation stability. To determine the minimum admissible cross section, the elasticity in the penstock and the pressure tunnel can be reasonably neglected, because of the slowness of the phenomena in which the pressure tunnel and surge tank are mostly involved. Thus, we can assume that the penstock has an impedance $(2K_c Q^o + sJ_c)$ (see Equation [3.2.9]) and the tunnel similarly has an impedance $(2K_g Q^o + sJ_g)$. Representing the surge tank by the (capacitive) impedance $1/(sA_p)$, shunt connected between the above impedances (where A_p is the surge tank cross section, assumed to be constant within the range of variation for the water level), an approximate expression of $Z_w(s)$ useful for describing the considered phenomena can be derived. Usually, the penstock impedance can be disregarded, so it holds that:

$$Z_w(s) \cong \frac{2K_g Q^o + sJ_g}{1 + (2K_g Q^o + sJ_g)sA_p}$$

which, by Equation [3.2.5], corresponds to function $G_a(s)$ with two zeros and two poles. If, as it usually occurs, the total transfer function of the regulating loop (see Fig. 3.2) has, with $s = \tilde{j}\nu$, a large magnitude in the low-frequency range, then the two zeros of $G_a(s)$ are practically translated into two closed-loop poles. By imposing (for stability requirements) that these zeros have a negative real part, the following condition, called "*Thoma condition*," can be derived:

$$A_p > a_1 \frac{J_g}{2K_g H^o}$$

(besides $2a_1 K_g Q^{o2} \leq H^o$, which can be undoubtedly considered as verified). Such a condition can be improved, by also accounting for the penstock parameters.

(b) Typical Regulation Schemes

Based on the above information, it may be concluded that, for primary regulation:

- For sufficiently low frequencies, e.g., $\nu < \nu_t$ with $\nu_t = 0.3$–0.5 rad/sec, it is possible to assume $G_v(\tilde{j}\nu) \cong G_v(0)$, $G_a(\tilde{j}\nu) \cong G_a(0)$. In fact, the variations of $G_a(\tilde{j}\nu)$, caused by pressure tunnel and surge tank (see Fig. 3.15), are confined in a frequency range where the gain of the primary regulation loop is very high. Therefore, provided that the Thoma condition or a similar one holds, their consequences on the regulation characteristics are quite negligible.
- For higher frequencies, one should consider Equation [3.2.1] for the effects of the gate-positioning system, and Equation [3.2.14] (or even Equation [3.2.8], provided that $\nu \ll \nu_r$) for the effects of the supply system and the turbine.

Therefore, the desired form for the function $G_f(s)$—which is given, at least within the frequency range $(0, \nu_t)$, by Equation [3.1.8]—must be essentially achieved through the primary regulator, by imposing:

$$G_r(s) \triangleq \frac{\Delta\beta}{\Delta\varepsilon_f}(s) \cong \frac{G_f(s)}{G_v(0)G_a(0)} = \frac{P_{\text{nom}}}{G_v(0)G_a(0)\Omega_{\text{nom}}} \frac{1 + sT_2}{b_p + sT_2 b_t} \qquad [3.2.15]$$

(whereas, by known notation, it then holds $T_1 = b_t T_2 / b_p$).

It is evident that, as to the cutoff frequency ν_t and then the rapidity of regulation, the most restrictive effects are a result of $G_a(s)$, which is a "nonminimum-phase" function, responsible for phase delays that progressively increase with frequency, without a corresponding magnitude decrease. By considering, for simplicity, the case of an isolated unit, it is clear that ν_t must be:

- sufficiently lower than $1/T_w$ (e.g., $\nu_t T_w < 0.4$ to have a phase delay $<33°$ in $G_a(\tilde{j}\nu_t)$, having assumed Equation [3.2.8] with $a_1 = a_2 = 1$);
- in addition, sufficiently lower than the resonance frequency ν_r (Equation [3.2.14] and Fig. 3.15 should be remembered).

This latter condition may be determinant in the case of high-head plants (large L_c, small ν_r), at least for small values of the flow Q^o (small T_w, with $1/T_w$ of the same order as ν_r). In this connection, it may be noted that according to both Equations [3.2.8] and [3.2.14], the magnitude and the phase delay of $G_a(\tilde{j}\nu)$ decrease as Q^o decreases for every given value of $\nu < \nu_r$ (refer to Fig. 3.15), but Equation [3.2.8] corresponds to the assumption $\nu_r = \infty$ and thus it may be optimistic. For example, if Q^o is such that $T_w < \sqrt{2}/(\nu_r\sqrt{a_1 a_2})$, the phase delay of $G_a(\tilde{j}\nu_r)$ is lower than 90° by Equation [3.2.8], whereas it is already equal to 180° by Equation [3.2.14].

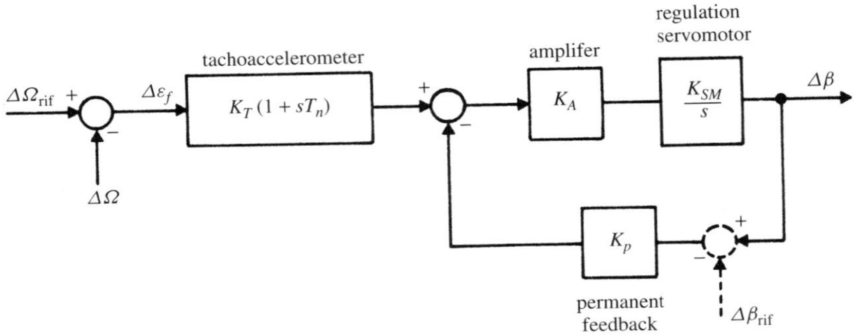

Figure 3.17. Block diagram of the regulator with tachoaccelerometer for small variations.

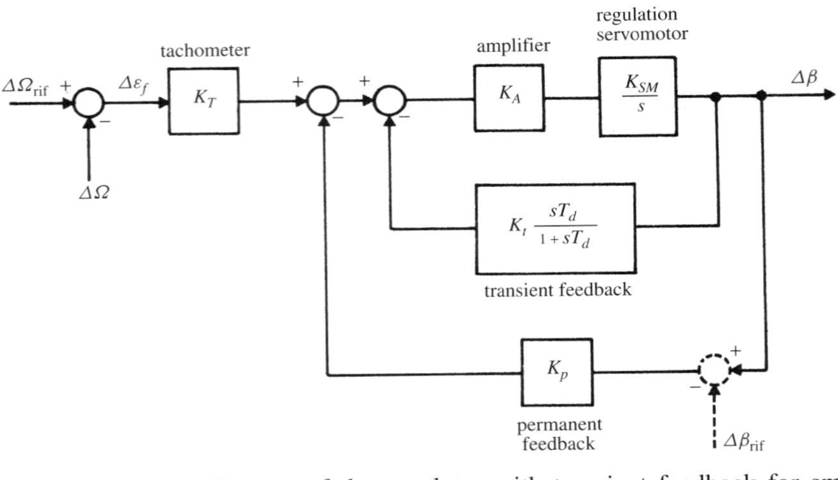

Figure 3.18. Block diagram of the regulator with transient feedback for small variations.

The implementation of a transfer function as per Equation [3.2.15] is traditionally obtained with comparable approximations through one of the following solutions:

- regulator with tachoaccelerometer: see Figure 3.17;
- regulator with transient feedback: see Figure 3.18.

The block diagrams refer to the linearized behavior for small variations, disregarding "insensitivities" in the tachometer and in the servomotor distributing valve. In frequency terms, at steady-state there can be insensitivities of approximately 0.05 Hz for mechanical regulators of more ancient production, and 0.005 Hz or lower for regulators with electric/electronic tachometer or tachoaccelerometer.

For large variations, it is necessary to consider the limits on the servomotor stroke, which correspond to the full opening and closing, and of the speed limits (in opening and closing) of the servomotor itself. The output of the amplifier and the input $(1/K_{SM})\,\mathrm{d}\beta/\mathrm{d}t$ of the servomotor are related by a nonlinear characteristic, having

- a possible deadband around the origin (responsible for the insensitivities, as already said), due to an overlapping in the servomotor distributing valve;
- saturations (responsible for the speed limits of the servomotor), corresponding to the extreme positions of the distributing valve in one or in the other sense.

Furthermore, in the case of Pelton turbines, such a characteristic is usually nonsymmetrical, in such a way to force, during the closing action, the intervention of deviators (whereas the speed of the needles remains limited through the respective servopositioners).

By the symbols in the figure, and having set for brevity:

$$K'_T \triangleq K_T G_v(0) G_a(0) \frac{\Omega_{\text{nom}}}{P_{\text{nom}}}$$

in the case of the *regulator with tachoaccelerometer* it follows:

$$G_r(s) = K_T K_A K_{SM} \frac{1 + sT_n}{K_A K_{SM} K_p + s}$$

and thus, by comparison with Equation [3.2.15]:

$$\begin{cases} T_2 = T_n \\ b_p = \dfrac{K_p}{K'_T} \\ b_t = \dfrac{1}{K_A K_{SM} K'_T T_n} \end{cases}$$

whereas $T_1 = 1/(K_A K_{SM} K_p)$. However, the transfer function of the tachoaccelerometer can be approximated by $K_T(1 + sT_n)$ as in Figure 3.17 (thus neglecting any delay) only for frequencies lower than, for example, 2–10 rad/sec, where the smaller and the larger values, respectively, correspond to the mechanical and the electrical tachoaccelerometers.

Instead, in the case of the *regulator with transient feedback* it holds that:

$$G_r(s) = K_T K_A K_{SM} \frac{1 + sT_d}{K_A K_{SM} K_p + s(1 + K_A K_{SM}(K_p + K_t)T_d) + s^2 T_d}$$

which is a transfer function with one zero and two poles, and thus not exactly referable to Equation [3.2.15]. Nevertheless, the cutoff frequency ($\sim K_A K_{SM} K_t$)

of the transient feedback loop (Fig. 3.18) — limited only by the presence of further small delays in the loop, which were neglected in the figure — can be assumed to be quite high, with $K_A K_{SM} K_t \gg 1/T_d$ (e.g., $K_A K_{SM} K_t = 10$ rad/sec, $T_d = 3$ sec), whereas it is $K_p \ll K_t$, so that it results in:

$$G_r(s) \cong K_T \frac{1 + sT_d}{K_p + sT_d(K_p + K_t)} \frac{1}{1 + sT_3}$$

with $T_3 \cong 1/(K_A K_{SM} K_t)$ negligible (e.g., $T_3 = 0.1$ sec). Then, the function $G_r(s)$ can be still considered as in Equation [3.2.15], with:

$$\begin{cases} T_2 = T_d \\ b_p = \dfrac{K_p}{K'_T} \\ b_t \cong \dfrac{K_p + K_t}{K'_T} \end{cases}$$

whereas $T_1 \cong ((K_p + K_t)/K_p)T_d$.

The possible dependence of $G_v(0)G_a(0)$ on the operating point, as a result of the (already mentioned) nonlinearities between the regulator output and the driving power, also affects (for both the schemes, and for given values of the different parameters) K'_T and thus b_p and b_t.

Furthermore, the operation with $b_p = 0$ (purely frequency regulation) can be obtained by excluding the permanent feedback ($K_p = 0$). Similarly, the operation with $b_p = \infty$ (purely power regulation, for example through the signal β_{rif} as specified below) can be obtained by excluding the tachometer or the tachoaccelerometer ($K_T = 0$). It is, however, convenient that such an exclusion is made by means of a dead-zone for a limited range of frequencies, so that it is possible to reconnect the regulator to the frequency error when this exceeds determined limits.

To act on the driving power independent of the frequency error, so as to realize (by the desired value of P_{rif}) a static characteristic as reported in Figure 3.5, it is possible to use the *opening reference* β_{rif} onward to the permanent feedback block (see Figs. 3.17 and 3.18). For small variations it then results, at steady-state:

$$\Delta\beta = \Delta\beta_{\text{rif}} + \frac{K_T}{K_p}(\Delta\Omega_{\text{rif}} - \Delta\Omega)$$

to which, accounting for the expressions of K'_T and b_p, the following variation of regulating power corresponds:

$$\Delta P_r = G_v(0)G_a(0)\Delta\beta = G_v(0)G_a(0)\Delta\beta_{\text{rif}} + \frac{P_{\text{nom}}}{\Omega_{\text{nom}} b_p}(\Delta\Omega_{\text{rif}} - \Delta\Omega)$$

as if it were (by obvious symbols recalling Equation [3.1.9]):

$$(\Delta P_{\text{rif}})_{\text{eq}} = G_v(0)G_a(0)\Delta\beta_{\text{rif}}$$

The schemes under examination can seem insufficient, as the setting of β_{rif} (in terms of P_{rif}) usually remains subject to some uncertainties, because of:

- the nonlinearities downward to the regulator, because of which the product $G_v(0)G_a(0)$ somewhat depends on the operating point;
- disturbances on the supply system etc., that lead to operating conditions different from the ones foreseen.

To counteract the effects of the nonlinearities, or at least to reduce them, it is possible to impose a proper nonlinear relationship between the values of β_{rif} and the desired ones of P_{rif}. Alternatively, it is possible to insert a nonlinearity (similar to the one that must be compensated) in the permanent feedback path, onward to the β_{rif} comparing node. By this latter solution, it is also possible to eliminate the dependence of the permanent droop b_p on the operating point.

To eliminate, at steady-state, all the mentioned inconveniences (thus making also b_p independent of the operating point), the permanent feedback with a proper gain K'_p may instead be realized starting from the difference between the active power delivered by the unit and the (exact) power reference P_{rif}, instead of the difference between β and β_{rif}, with a gain K_p, as discussed up to now.

Neglecting the variations of the electrical losses in the unit, the delivered active power variations can be confused with those of the generated power, i.e., (neglecting also the effects of the possible disturbance u_a, indicated in Figure 3.1):

$$\Delta P_e = \Delta P_r - (G_g(s) + sM)\Delta\Omega$$

Therefore, with the solution under examination, which is similar to that in Section 3.2.3b for thermal units, it is as if:

(1) the permanent feedback were connected to the difference between P_r and P_{rif}, so as to produce a permanent droop equal to $b_p = (K'_p P_{\text{nom}})/(K_T \Omega_{\text{nom}})$;

(2) the signal $K'_p(G_g(s) + sM)\Delta\Omega \cong K'_p(G_g(0) + sM)\Delta\Omega$ was added to the amplifier input (Section 3.1.3).

Both the circumstances (1) and (2) may actually imply a destabilizing effect, respectively on the permanent feedback loop and on the frequency regulation loop; in fact:

- because of (1), the permanent feedback loop now includes the transfer functions $G_v(s)$ and $G_a(s)$, which are responsible of further response delays;
- because of (2), the frequency regulation loop includes, between the frequency and the amplifier input, the transfer function $-K_T(1 + sT_n) + K'_p(G_g(0) + sM) = -K_T(1 - b_p/b_g + s(T_n - b_p T_a))$ instead of $-K_T(1 + sT_n)$ (with $T_n = 0$ in the case of regulator with transient feedback), where the ratio $(1 - b_p/b_g + s(T_n - b_p T_a))/(1 + sT_n)$ practically implies some further delays (and moreover an increase of the droop, according to the ratio $1/(1 - b_p/b_g)$).

However, such effects can be usually disregarded because of the modest value of b_p and thus the cutoff frequency of the permanent feedback loop.

For a comparison between the regulator with tachoaccelerometer and the one with transient feedback, the following can be noted:

- With the transfer function $G_r(s)$, the two solutions can be considered practically equivalent, as the time constant T_3 (which cannot be avoided in regulators with transient feedback) may represent a modest delay, comparable to the tachoaccelerometer delay.
- In response to the opening set-point, the transfer function $(\partial\beta/\partial\beta_{\text{rif}})(s)$ is equal to $1/(1+sT_1)$ in the case of the regulator with tachoaccelerometer, and to $(K_p/K_T)G_r(s) \cong (1+sT_2)/((1+sT_1)(1+sT_3))$ in the case of regulator with transient feedback (if β_{rif} were put into both (permanent and transient) feedback paths, the transfer function would become about equal to $1/(1+sT_3)$, which would correspond to a very fast response). Note that in the frequency range for which $G_v(s) \cong G_v(0)$ and $G_a(s) \cong G_a(0)$, such expressions also hold for the transfer function $(\partial P_r/\partial(P_{\text{rif}})_{\text{eq}})(s)$, by intending $(\Delta P_{\text{rif}})_{\text{eq}} = G_v(0)G_a(0)\Delta\beta_{\text{rif}}$.
- The closure of the transient feedback loop allows operation with relatively large values of K_A thus reducing the effects, in terms of frequency Ω, of possible disturbances and insensitivities in the servomotor and in its control. In this regard note, particularly, that the product $K_A K_{SM} K'_T$ is equal to $1/(b_t T_2)$ with the tachoaccelerometer and $\sim 1/(b_t T_3)$ with the transient feedback. At equal b_t, K_{SM} and K'_T, the amplifier gain K_A is then, with transient feedback, approximately T_2/T_3 times larger than that with tachoaccelerometer, where T_2/T_3 can be, for instance, 30–50.
- Following perturbations which are not small, the presence of the speed limits in the regulating servomotor causes different effects for the two types of regulators. For a purely qualitative comparison, refer to the diagrams of Figure 3.19, which refer to the response to a step-load variation ΔP_L in the case of a single unit, with $b_p = 0$, $T_2/(b_t T_a) = 3$, $T_3/(b_t T_a) = 0.1$, $T_v = 0$, $a_1 = a_2 = 1$, $T_w/(b_t T_a) = 0.4$ (see Equations [3.2.1], [3.2.8], and [3.2.15]), where T_a is the start-up time of the unit, and $T_{\min}$ is the minimum time for a complete opening or closing of the regulating servomotor.

The simultaneous use of a tachoaccelerometer and transient feedback might give some benefits, provided that it does not cause an increase of the cutoff frequency (ν_t) near to the resonance frequency (ν_r) of the penstock, with a consequent impact on the stability margins.

Before concluding, the progressive improvement of the regulator's functional characteristics, which has been achieved through the implementation of low-power electric or electronic components, should be considered. Significant advantages have been obtained in the past, by passing from traditional "mechanical-hydraulic" regulators to "electrohydraulic" ones. In the latter, both the tachoaccelerometer (or the tachometer) and the permanent and transient feedbacks, are electrically implemented, and the first stages of the amplifier are electronic, whereas the servomotor is still used as an integrator in the implementation of the function $G_r(s)$.

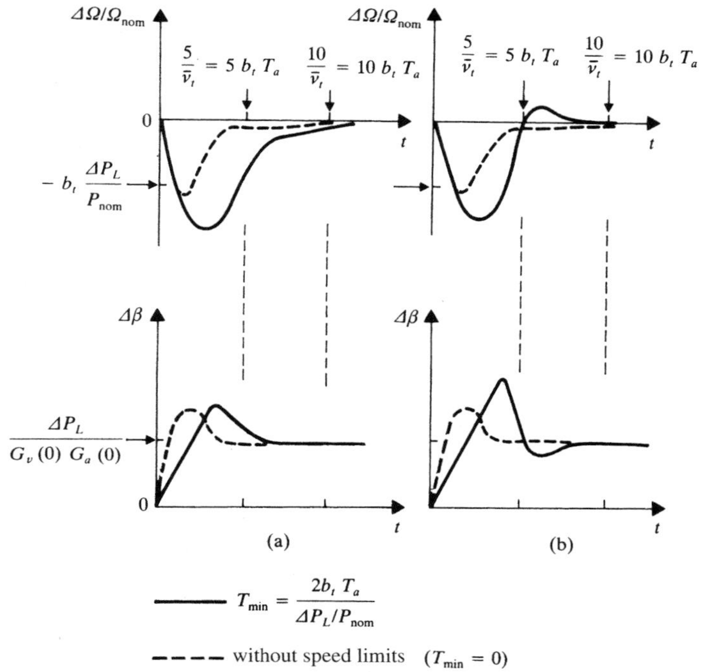

Figure 3.19. Response of the relative frequency error ($\Delta\Omega/\Omega_{\text{nom}}$) to a step of load variation (ΔP_L): effect of the speed limits of the regulating servomotor, with: (a) regulator with tachoaccelerometer; (b) regulator with transient feedback.

Further advantages may be obtained by electronically implementing the desired whole function $G_r(s)$ and by using the servomotor (with a suitable feedback) simply as a positioner, with a relatively wide bandwidth. In this way, the effect of disturbances, insensitivities, etc. can be greatly reduced.

The opportunity of sending other signals to the regulator should be considered, to achieve a more coordinated use of the plant. Signals sensitive to the penstock pressure or to the reservoir storage, are examples of these signals.

3.2.3. Case of Thermal Units

(a) Dependence of the Driving Power on the "Output" of the Speed Governor

Because of the relative complexity of thermal plants, we will limit discussion here to the basic characteristics relating to the problems of f/P control. More precisely, reference will be made to the typical plant in Figure 3.20a, with a reheater and three turbine sections (respectively, *HP*, *MP*, and *LP*, i.e., high, medium, and low pressure). For this plant type, it is possible to associate, in the sequence $1, \ldots, 10$ and apart from obvious details, the thermodynamic cycle

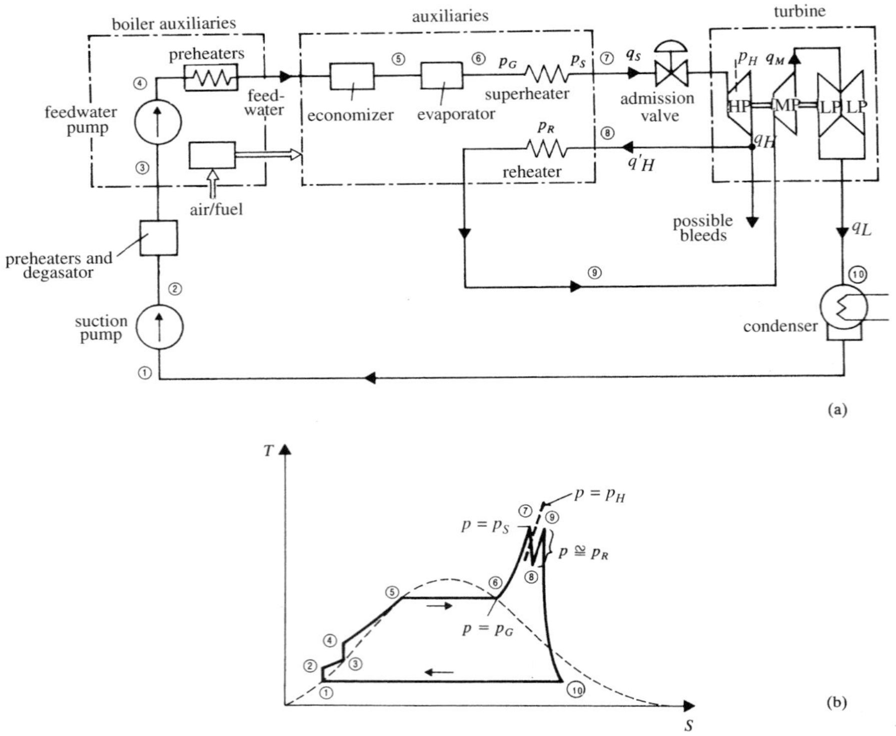

Figure 3.20. Typical example of a thermal plant: (a) broad scheme; (b) thermodynamic cycle.

reported in Figure 3.20b (where T and S are, respectively, fluid temperature and entropy). Based on the following considerations, the changes to be adopted for plants different from the one considered may be obvious (e.g., plants without reheater, more than one reheater, etc.).

The generic plant includes several valves, with different functions, and not only between the superheater and the *HP* turbine section, but also between the reheater and the *MP* section, etc. In particular, between the superheater and the *HP* section there may be a "throttling" valve, followed by more "partial admission" valves, each one of which supplies a circular sector of the first stage, of the "impulse" type, of the *HP* section. To obtain a better efficiency, the supply is usually done in a "partial arc" mode, by keeping the throttling valve fully open and by sequentially operating the partial admission valves. In some situations—for instance, at startup or at low load—the supply is instead done in a "full arc" mode, by simultaneously operating the partial admission valves, or by keeping them fully open and by operating the throttling valve. With the partial arc mode, the static characteristic that relates the steam flow to the opening signal exhibits a so-called "valve point," with an abrupt increase in the slope (e.g., three to four times) every time that the opening of a new valve is initiated. The operation of further valves, located downward to

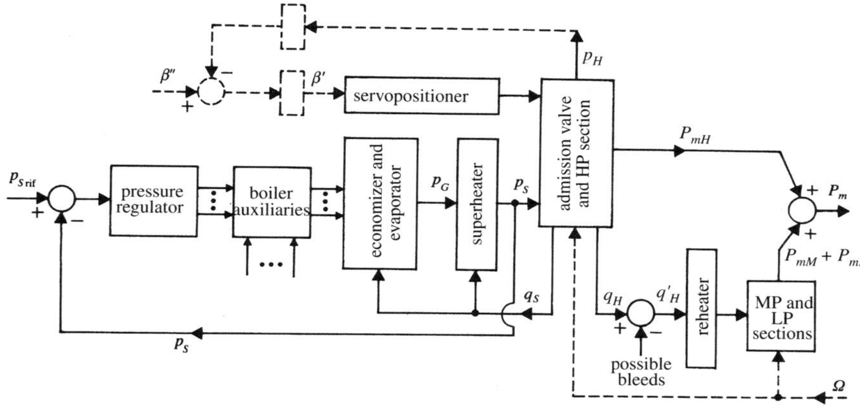

Figure 3.21. Dependence of the driving power on the output of the primary regulator (and on the pressure set point; see the text), in the case of a thermal plant: general block diagram.

the reheater, may be related to the opening signal, particularly under emergency conditions. The terminology concerning the valves varies. However, according to Figure 3.20a, we will simply consider a single equivalent valve, called "admission" valve, located between the superheater and the *HP* turbine section.

The relationship between β ("output" of the primary regulator) and P_m (driving power) may be generically drawn schematically, as in Figure 3.21.

The variable β may be assumed to represent the position β' of an actuator that governs the turbine valves through a servopositioner (the actuator itself may be constituted by a servopositioner, having a smaller power and a relatively negligible response delay).

As an alternative, the servopositioner may be inserted in a closed loop, with a feedback from the pressure p_H in the "wheel chamber" (i.e., in the chest downstream of the impulse stage of the *HP* section). In such a case, it can be seen that the output β of the regulator is constituted by the signal β'' (e.g., an electric voltage), indicated in Figure 3.21.

The "opening" A can be assumed to be the "useful cross section" of the fluid jet into the admission valve. The further variables used in the figures are:

- p_G = pressure at the outlet of the evaporator
- p_S = pressure at the outlet of the superheater
- p_R = pressure at the outlet of the reheater
- q_S = (mass) flow at the outlet of the superheater
- q_R = flow at the outlet of the reheater
- q_H, q_M, q_L = flows, respectively, at the outlet of the *HP*, *MP*, *LP* turbine sections

- P_{mH}, P_{mM}, P_{mL} = driving powers that respectively correspond to the *HP*, *MP*, *LP* turbine sections

whereas $P_m = P_{mH} + P_{mM} + P_{mL}$ is the resulting driving power, and Ω the electrical speed.

If $\beta = \beta'$, the transfer function $G_v(s) = (\Delta A/\Delta\beta)(s)$ (see Fig. 3.2), which particularly accounts for the above-mentioned servopositioner, can be considered to be:

$$G_v(s) = G'_v(s) = \frac{K_v}{1 + sT_v} \qquad [3.2.16]$$

where:

- the static gain $K_v \triangleq G'_v(0)$ may depend on the operating point, if the static characteristic (β', A) is nonlinear;
- the time constant T_v can be, for example, 0.1–0.3 seconds or even smaller.

For large variations, the limitations on the servomotor speed must be considered, both in opening and closing. They prevent excessive stresses that can result from abrupt variations of the thermal exchanges, in the turbine and in other elements of the plant.

Usually, the possible nonlinearities are adequately compensated (e.g., through complementary nonlinearities, realized by means of "cams") so as to obtain a static characteristic (β', A) (or alternatively $(\beta', q_H)_{\Delta p_s=0}$) that is sufficiently linear. The dependence of A on β' can furthermore present an "insensitivity," that is not always negligible, and which in terms of β', can be 0.5% with respect to the value corresponding to the full opening.

The possible closed loop indicated in Figure 3.21, with a feedback from the pressure p_H, usually includes a proportional-integral element (electronically implemented) in the "forward" path, onward to β', and a simple transducer in the feedback path. The cutoff frequency of such a loop may be, e.g., 3–7 rad/sec, so that (even if the dynamic relationship between ΔA and Δp_H, or between ΔA and Δq_H, is accounted for) the resulting response delays between $\Delta\beta''$ and Δq_H also may be considered as negligible, as a first approximation, because of the relative slowness of the primary regulation. More specifically, at steady-state, the dependence of the pressure p_H on β'' is only determined, because of the integral action of the forward element, by the characteristic of the transducer. If this is linear, the considered loop allows then to directly obtain a linear static characteristic (β'', p_H). The same can be said for the (β'', q_H), as at steady-state q_H can be considered proportional to p_H.

With the block "admission valve and *HP* section," as a first approximation, accounting for the relative slowness of the primary regulation, it can be assumed that:

- the temperature of the superheated steam (because of the rapidity of its regulation), and the temperature distribution along the turbine remain constant;
- the flows q_S and q_H are equal to each other, even during transients;
- the flow $(q_S = q_H)$ depends algebraically on A, p_S, p_H, according to the static characteristics of the valve and the impulse stage of the *HP* section;
- the pressure p_H and the flow are proportional to each other (in particular, it is assumed that in the reaction stages down to the wheel chamber, the flow is "critical," i.e., reaches the sound speed);
- the power P_{mH} is itself simply proportional to the flow (actually, the "work" done by the fluid per unit of mass, i.e., the "total" enthalpic head, may vary, but the effects of such variations can be disregarded with respect to those of flow variations; the variations of P_{mH} with Ω, already mentioned in Section 3.1.3, are furthermore disregarded for simplicity).

Consequently, the variables $q_S = q_H$, p_H, P_{mH} (all proportional to one another) can be considered algebraic functions of A, p_S, and for small variations around a given operating point it is possible to write equations of the form:

$$\Delta q_H = \Delta q_S = h_A \Delta A + \frac{\Delta p_S}{R_v} \quad [3.2.17]$$

$$\Delta p_H = R_H \Delta q_H$$

$$\Delta P_{mH} = \left(\frac{P_{mH}}{q_S}\right)_{\text{nom}} \Delta q_S \quad [3.2.18]$$

according to Figure 3.22.

With the thermal system onward, as a first approximation it also may be assumed, as specified in the following, that the pressure p_G at the output of the evaporator is constant. Also, in such a case it is necessary to account for the block "superheater," the effect of which (predominantly of the resistive type) is translated into a pressure drop according to an equation such as:

$$\Delta p_S = \Delta p_G - R_S \Delta q_S$$

As a result of Equation [3.2.17], it can be finally deduced that:

$$\Delta q_H = \Delta q_S = \frac{R_v h_A \Delta A + \Delta p_G}{R_v + R_S} \quad [3.2.19]$$

that is, a reduction of the gain $\Delta q_S / \Delta A$, as:

$$\left(\frac{\Delta q_S}{\Delta A}\right)_{\Delta p_G = 0} = \frac{R_v}{R_v + R_S} h_A < h_A$$

where $R_v/(R_v + R_S)$ can be, for example, in the range of 0.8–1.0.

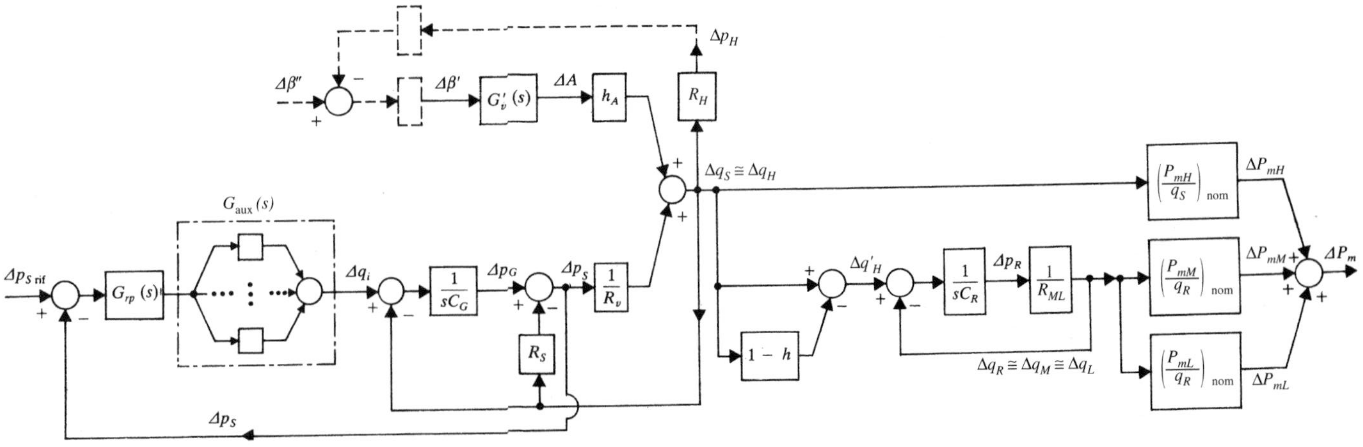

Figure 3.22. Block diagram of the system in Figure 3.21 for small variations.

The flow variations Δq_S reverberate down through the block "reheater," the effect of which (predominantly capacitive) can be defined by an equation such as:

$$\Delta q'_H - \Delta q_R = V_R \frac{\mathrm{d}\Delta\rho_R}{\mathrm{d}t} \triangleq C_R \frac{\mathrm{d}\Delta p_R}{\Delta t}$$

where q'_H is the input flow (and it will be assumed $q'_H = hq_H$, with h a proper constant, smaller than unity in the presence of bleeds), V_R is the reheater volume, ρ_R is the fluid density (evaluated at a proper temperature, such as a mean value along the reheater), and $C_R \triangleq V_R(\mathrm{d}\rho_R/\mathrm{d}p_R)^o$.

Finally, with the block "*MP* and *LP* sections," by approximations similar to those already adopted, and disregarding for simplicity the existence of further bleeds, it may be assumed that:

- the flows q_R, q_M, q_L equal one another;
- the pressure p_R is proportional to $q_R = q_M = q_L$ (hypothesis of critical flow);
- the powers P_{mM}, P_{mL} are themselves simply proportional to the considered flow (neglecting the effects of the variations of the "work" done by the fluid per unit mass, and of the variations of Ω).

It can be then derived:

$$\Delta q_M = \Delta q_L = \Delta q_R$$

$$\Delta p_R = R_{ML} \Delta q_R$$

$$\Delta P_{mM} = \left(\frac{P_{mM}}{q_R}\right)_{\text{nom}} \Delta q_M$$

$$\Delta P_{mL} = \left(\frac{P_{mL}}{q_R}\right)_{\text{nom}} \Delta q_L$$

where $q_{R\,\text{nom}} = q'_{H\,\text{nom}} = hq_{S\,\text{nom}}$, and thus, accounting for the equation of the reheater (being $\Delta q'_H = h\Delta q_S$):

$$\Delta q_R = h \frac{1}{1 + sT_R} \Delta q_S$$

$$\Delta P_{mM} + \Delta P_{mL} = \left(\frac{P_{mM} + P_{mL}}{q_R}\right)_{\text{nom}} \Delta q_R = \left(\frac{P_{mM} + P_{mL}}{q_S}\right)_{\text{nom}} \frac{1}{1 + sT_R} \Delta q_S \qquad [3.2.20]$$

having posed $T_R \triangleq R_{ML} C_R$ (T_R is usually termed the "*reheater*" *time constant*, and it results within the range of 5–15 sec).

Finally, by intending that $P_{\text{nom}} = (P_{mH} + P_{mM} + P_{mL})_{\text{nom}}$, $\alpha \triangleq P_{mH\,\text{nom}}/P_{\text{nom}}$ (usually α is within the range 0.25–0.30), from Equations [3.2.18] and [3.2.20] it follows:

$$\frac{\Delta P_m}{\Delta q_S}(s) = \frac{\Delta P_{mH} + \Delta P_{mM} + \Delta P_{mL}}{\Delta q_S}(s)$$
$$= \frac{P_{\text{nom}}}{q_{S\,\text{nom}}}\left(\alpha + \frac{1-\alpha}{1+sT_R}\right) = \frac{P_{\text{nom}}}{q_{S\,\text{nom}}}\frac{1+s\alpha T_R}{1+sT_R}$$

where Δq_S is given by Equation [3.2.19]. Therefore, if the pressure at the output of the evaporator was constant ($\Delta p_G = 0$)—as it also may be assumed, apart from the slower phases of the transients—the transfer function:

$$G_a(s) \triangleq \frac{\partial P_m}{\partial A}(s) = \frac{\Delta P_m}{\Delta q_S}(s)\frac{\Delta q_S}{\Delta A}(s)$$

would result to be:

$$G_a(s) = K_a'\frac{1+s\alpha T_R}{1+sT_R} \tag{3.2.21}$$

in which, for simplicity:

$$K_a' \triangleq \frac{h_A R_v}{R_v + R_S}\frac{P_{\text{nom}}}{q_{S\,\text{nom}}}$$

In terms of frequency response, Equation [3.2.21] leads to diagrams as indicated by the continuous line in Figure 3.23.

With reference to the faster parts of the regulation transients, further capacitive effects may be accounted for, particularly between the valve and the *HP* section (and in its "wheel chamber") and in the connections between the *MP* and *LP* sections (and within the *LP*

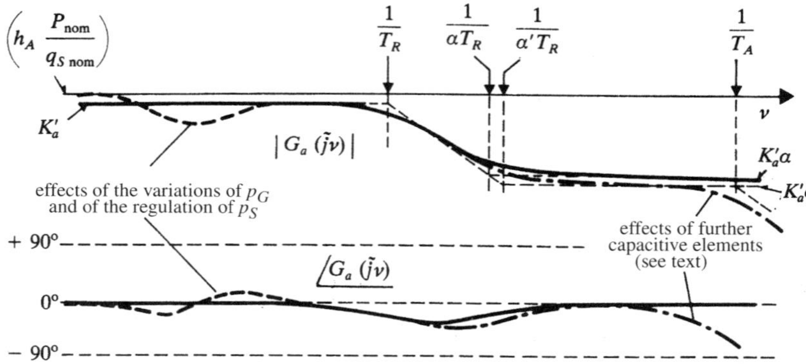

Figure 3.23. Frequency response of the transfer function $G_a(s)$ (supply system and turbine): case of a thermal unit.

section). These effects are, respectively, responsible for some dynamic delays between Δq_S and Δq_H, and between Δq_R ($= \Delta q_M$) and Δq_L. With powers it can be assumed, for instance, instead of Equations [3.2.18] and [3.2.20], that:

$$\Delta P_{mH} = \left(\frac{P_{mH}}{q_S}\right)_{\text{nom}} \frac{1}{1 + sT_A} \Delta q_S$$

$$\Delta P_{mM} + \Delta P_{mL} = \left[\left(\frac{P_{mM}}{q_S}\right)_{\text{nom}} + \left(\frac{P_{mL}}{q_S}\right)_{\text{nom}} \frac{1}{1 + sT_L}\right] \frac{1}{(1 + sT_R)(1 + sT_H)} \Delta q_S$$

(T_H may be, for example, 0.1–0.4 sec, and T_L 0.3–0.6 sec), from which, through simple derivations:

$$\frac{\Delta P_m}{\Delta q_S}(s) = \frac{\Delta P_{mH} + \Delta P_{mM} + \Delta P_{mL}}{\Delta q_S}(s) = \frac{P_{\text{nom}}}{q_{S\,\text{nom}}} \frac{(1 + s\alpha' T_R)(1 + sT_L')}{(1 + sT_R)(1 + sT_H)(1 + sT_L)}$$

$$\cong \frac{P_{\text{nom}}}{q_{S\,\text{nom}}} \frac{1 + s\alpha' T_R}{(1 + sT_R)(1 + sT_H)}$$

where α' is slightly smaller than α ($\alpha' \cong \alpha - (P_{L\,\text{nom}}/P_{\text{nom}})T_L/T_R$, for example $\alpha' \cong 0.28$ for $\alpha = P_{H\,\text{nom}}/P_{\text{nom}} = 0.30$, $P_{L\,\text{nom}}/P_{\text{nom}} = 0.40$ and $T_L/T_R = 0.05$), whereas T_L' is very close (slightly smaller) to T_L; in particular, for $\Delta p_G = 0$, Equation [3.2.21] can be then replaced by:

$$G_a(s) \cong K_a' \frac{1 + s\alpha' T_R}{(1 + sT_R)(1 + sT_H)} \quad [3.2.22]$$

to which corrections like those indicated by the dotted-dashed line in Figure 3.23 correspond. For higher frequencies, further delays should be considered.

However, during the slower transients, and particularly during the phase of approaching the steady-state conditions, the hypothesis $\Delta p_G = 0$ no longer is applicable. In fact, the block "economizer and evaporator" has a predominantly capacitive effect, by which a variation in the flow q_S tends to cause a progressively increasing variation, in the opposite sense, in the pressure p_G. Within an acceptable approximation for the present aims, this can be accounted for by equation:

$$\Delta q_i - \Delta q_S = C_G \frac{\mathrm{d}\Delta p_G}{\mathrm{d}t}$$

where Δq_i is an equivalent variation in the flow that inlets the boiler, as a consequence of the overall action of the auxiliary equipment (refer again to the block diagram of Fig. 3.22).

If it were $\Delta q_i = 0$, by reminding Equation [3.2.19], we would obtain:

$$\frac{\Delta q_S}{\Delta A}(s) = \frac{sh_A R_v C_G}{1 + sT_G}$$

and thus, instead of Equation [3.2.21]:

$$G_a(s) = K'_a \frac{sT_G(1 + s\alpha T_R)}{(1 + sT_G)(1 + sT_R)}$$

where $T_G \triangleq (R_v + R_S)C_G$ is several tens of seconds for once-through boilers and hundreds of seconds for drum boilers.

It then becomes essential to consider the effects of the "boiler auxiliaries" and the "pressure regulator," according to Figures 3.21 and 3.22.

The transfer function $G_{\text{aux}}(s)$, equivalent to the whole set of the auxiliary equipment, is variable with the type of plant and often not easily determined. It can be generically stated, in terms of frequency response, that it can imply significant delays at low frequencies, for example, 0.05 rad/sec for once-through boilers, and even lower frequencies for drum boilers (in particular, for the coal supplied plants a "transportation" delay in tens of seconds may result).

The transfer function $G_{rp}(s)$ of the pressure regulator is usually of the proportional-integral type, such that the regulation loop with $\Delta A = 0$ has a cutoff frequency, e.g., 0.01–0.03 rad/sec (comparable, as in Section 3.3.2, with that of the secondary regulation loop), or even smaller.

The pressure to be regulated is that (p_S) at the output of the superheater, i.e., the pressure at the admission onward to the valve. Because of the integral action of the regulator, at steady-state (for $\Delta p_{S\,\text{rif}} = 0$) $\Delta p_S = 0$, so that the static gain of $G_a(s)$ is given by:

$$G_a(0) = h_A \frac{P_{\text{nom}}}{q_{S\,\text{nom}}}$$

(recall Equations [3.2.17], [3.2.18], and [3.2.20]), somewhat larger than K'_a. More generally:

$$\frac{\Delta q_S}{\Delta A}(s) = \frac{h_A R_v(sC_G + G_{rp}G_{\text{aux}})}{1 + sT_G + R_v G_{rp} G_{\text{aux}}}$$

and thus:

$$G_a(s) = h_A \frac{P_{\text{nom}}}{q_{S\,\text{nom}}} \frac{R_v(sC_G + G_{rp}G_{\text{aux}})(1 + s\alpha T_R)}{(1 + sT_G + R_v G_{rp} G_{\text{aux}})(1 + sT_R)} \qquad [3.2.23]$$

instead of Equation [3.2.21]; refer to the corrections made by the dashed line in Figure 3.23.

The elements that constitute the pressure regulation loop have an effect on $G_a(s)$, similar to that of a pressure tunnel and surge tank for a hydraulic plant (Fig. 3.15). In this latter case, on the other hand, the pressure (as well as the piezometric head) at the turbine gate may be considered intrinsically regulated, as a result of the presence of the reservoir.

(b) Typical Regulation Schemes

Based on the discussions above and with reference to Figure 3.20, the following conclusions can be drawn.

- For sufficiently low frequencies (e.g., $\nu < \nu_t$, being $\nu_t = 0.3\text{–}0.5$ rad/sec), it also may be written that:

$$\frac{\partial P_m}{\partial \beta}(s) \cong \frac{\partial P_m}{\partial \beta}(0)\frac{1 + s\alpha T_R}{1 + sT_R} \qquad [3.2.24]$$

as:

  - in the absence of feedback from the pressure $p_H(\beta = \beta')$: the transfer function $G_v(s)$ can be approximated by its static gain $G_v(0) = K_v$, whereas for $G_a(s)$ the Equation [3.2.21] can be assumed to hold (in fact, the corrections related to the boiler dynamics — variations of p_G and regulation of p_S — lay in a frequency range (Fig. 3.23) for which the primary regulation loop gain is quite high, and thus they have modest influence on the regulation characteristics), so that finally Equation [3.2.24] with $(\partial P_m/\partial\beta)(0) = K_v K_a'$ can be deduced;
  - in the presence of feedback from the pressure p_H $(\beta = \beta'')$: the effects of the boiler dynamics are even more attenuated, and the transfer function $(\Delta q_S/\Delta\beta)(s) \cong (\Delta q_H/\Delta\beta)(s)$ can be approximated by its static gain $(\Delta q_H/\Delta\beta)(0)$, whereas it can be assumed that $(\Delta P_m/\Delta q_S)(s) = (P_{\text{nom}}/q_{S\,\text{nom}})(1 + s\alpha T_R)/(1 + sT_R)$, so that the Equation [3.2.24] is again obtained, with $(\partial P_m/\partial\beta)(0) = (\Delta q_H/\Delta\beta)(0)P_{\text{nom}}/q_{S\,\text{nom}}$.

  By accepting Equation [3.2.24], the difference between the two cases then consists into the static gain $(\partial P_m/\partial\beta)(0)$; it must be remembered that K_v and K_a' may vary significantly with the operating point, whereas this does not happen for $(\Delta q_H/\Delta\beta)(0)$. Therefore the feedback from the pressure p_H also prevents, or at least acts to reduce, the nonlinearities of the static characteristic (β, P_m).

- For higher frequencies, we must also consider the delays associated with the time constants T_v, T_H (recall Equations [3.2.16] and [3.2.22]), or alternatively those between β'' and q_H, and moreover the effects of T_L and of further delays of the thermal system.

Therefore, the desired form of the transfer function $G_f(s) \triangleq (\partial P_m/\partial\varepsilon_f)(s)$, defined, at least within the frequency range $(0, \nu_t)$, by Equation [3.1.8], is that of Equation [3.2.24] with $T_1 = T_R$ and $T_2 = \alpha T_R$. Because the values of T_R and αT_R actually are comparable to those required for T_1 and T_2, it is also possible to adopt a primary regulator of the purely proportional type, by writing:

$$G_r(s) \triangleq \frac{\Delta\beta}{\Delta\varepsilon_f}(s) \cong \frac{G_f(0)}{(\partial P/\partial\beta)(0)} = \frac{P_{\text{nom}}}{\Omega_{\text{nom}} b_p (\partial P_m/\partial\beta)(0)} \triangleq G_r'(0) \qquad [3.2.25]$$

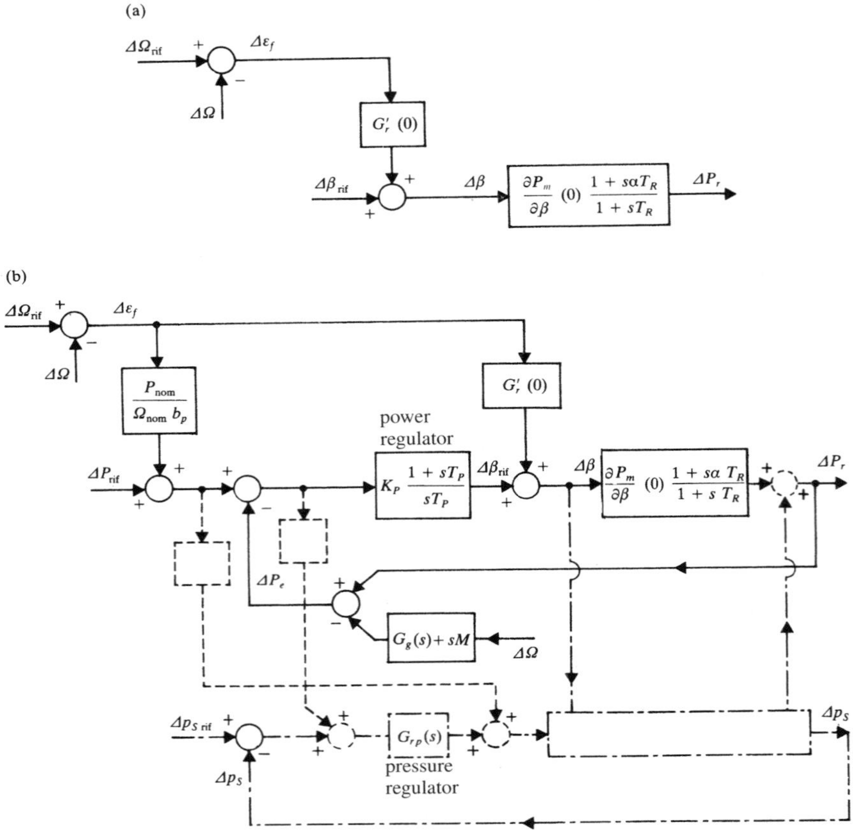

Figure 3.24. Block diagram of the regulation for small variations: (a) elementary solution, for a first level approximation; (b) solution with a power regulator and its interactions with the pressure regulation.

according to the block diagram of Figure 3.24a. The variations for plants different from the one considered may be obvious; for example, in the absence of the reheater it can be specified that $T_R = 0$, so that the situation becomes similar to that described in Section 3.2.2b with reference to hydroelectric units.

By this simple solution, the desired value of P_{rif} (recall the static characteristic of Fig. 3.5) can be realized through the *opening reference* β_{rif} as indicated in the block diagram, to obtain for small variations:

$$\Delta\beta = \Delta\beta_{rif} + G_r'(0)(\Delta\Omega_{rif} - \Delta\Omega)$$

At steady-state, the variation in the regulating power then results in:

$$\Delta P_r = \frac{\partial P_m}{\partial\beta}(0)\Delta\beta = \frac{\partial P_m}{\partial\beta}(0)\Delta\beta_{rif} + \frac{P_{nom}}{\Omega_{nom} b_p}(\Delta\Omega_{rif} - \Delta\Omega)$$

as if it were (recalling Equation [3.1.9]):

$$(\Delta P_{\text{rif}})_{\text{eq}} = \frac{\partial P_m}{\partial \beta}(0)\Delta\beta_{\text{rif}}$$

Generally, the dynamic effects of β_{rif} on the regulating power then are equal to those of β, so that (by Equation [3.2.24]) it also can be written that:

$$\frac{\partial P_r}{\partial (P_{\text{rif}})_{\text{eq}}}(s) = \frac{1 + s\alpha T_R}{1 + sT_R}$$

The previous scheme may seen unsatisfactory, as the setting of β_{rif} (in terms of P_{rif}) remains affected by noticeable uncertainties, caused by:

- nonlinearities down to the regulator, because of which the gain $(\partial P_m/\partial\beta)(0)$ somewhat depends on the operating point (recall the beneficial effect of the feedback from the pressure p_H);
- model approximations, for which the actual value of the above-mentioned gain may be different from the assumed one (recall in particular the variations of the enthalpic heads, of the efficiencies etc., up to now neglected, which cannot be efficiently counteracted even by the feedback from p_H);
- possible disturbances on the thermal system etc., which lead to operating conditions different from the assumed ones.

Furthermore, the dependence of $(\partial P_m/\partial\beta)(0)$ on the operating point affects (at equal $G'_r(0)$) also the resulting value of the permanent droop b_p (and thus also the value of the transient droop $b_t = b_p T_1/T_2$). Finally, the operation with $b_p = \infty$ (purely power regulation) can be obtained by imposing $G'_r(0) = 0$, whereas the operation with $b_p = 0$ (purely frequency regulation) cannot be realized.

It is then convenient to resort to a scheme as indicated in Figure 3.24b, which includes a *power regulator* of the proportional-integral type (with a transfer function $K_P(1 + sT_P)/(sT_P)$), sensitive to:

- the difference between the (exact) power reference P_{rif} and the active power delivered by the unit (the variations of the latter can be taken as the variations of the electric power generated P_e, apart from the variations in the electrical losses of the unit);
- the signal $(P_{\text{nom}}/(\Omega_{\text{nom}} b_p))(\Omega_{\text{rif}} - \Omega)$ (also named "*frequency bias*").

Because of the integral action of the regulator, it is possible to obtain, at steady-state, the Equation [3.1.9], i.e.,

$$\Delta P_r = \Delta P_{\text{rif}} + \frac{P_{\text{nom}}}{\Omega_{\text{nom}} b_p}(\Delta\Omega_{\text{rif}} - \Delta\Omega)$$

without any uncertainties on the values of P_{rif} and b_p (at steady-state it can be assumed $\Delta P_r = \Delta P_e$, also disregarding the variation of the mechanical losses and the dependence of the driving power on the speed, and thus assuming $G_g(s) = 0$); furthermore, it is possible not only to operate at $b_p = \infty$ (by excluding the "frequency bias"), but also at $b_p = 0$ (by excluding the signals P_{rif}, P_e).

With transfer functions, if the (even transient) differences between ΔP_r and ΔP_e are disregarded and Equation [3.2.24] is accepted, it then follows that:

$$\begin{cases} F_p(s) \triangleq \dfrac{\partial P_r}{\partial P_{\text{rif}}}(s) = \dfrac{(1+sT_P)(1+s\alpha T_R)}{1+s(T_P+\alpha T_R+T')+s^2T_R(\alpha T_P+T')} \\ G_f(s) \triangleq \dfrac{\partial P_r}{\partial \varepsilon_f}(s) = \dfrac{P_{\text{nom}}}{\Omega_{\text{nom}} b_p}\dfrac{1+sT''}{1+sT_P}F_p(s) \\ \qquad = \dfrac{P_{\text{nom}}}{\Omega_{\text{nom}} b_p}\dfrac{(1+sT'')(1+s\alpha T_R)}{1+s(T_P+\alpha T_R+T')+s^2T_R(\alpha T_P+T')} \end{cases}$$

where, for simplicity:

$$T' \triangleq \frac{T_P}{K_P(\partial P_m/\partial \beta)(0)}, \qquad T'' \triangleq \left(1+\frac{G_r'(0)\Omega_{\text{nom}} b_p}{K_P P_{\text{nom}}}\right)T_P$$

As a simplified example, if $T_P = T_R$ it follows that:

$$\begin{cases} F_p(s) = \dfrac{1+s\alpha T_R}{1+s(\alpha T_R+T')} \\ G_f(s) = \dfrac{P_{\text{nom}}}{\Omega_{\text{nom}} b_p}\dfrac{(1+sT'')(1+s\alpha T_R)}{(1+sT_R)(1+s(\alpha T_R+T'))} \end{cases}$$

as indicated in the frequency response diagrams of Figure 3.25. Note that the frequency error acts also through the (faster) path constituted by $G_r'(0)$; actually, as a result of the presence of $G_r'(0)$ it holds that $G_f(s)/F_p(s) = (P_{\text{nom}}/(\Omega_{\text{nom}} b_p))(1+sT'')/(1+sT_P)$, with $T'' > T_P$, so that the response to $\Delta\varepsilon_f$ is less delayed than that to ΔP_{rif}.

Because of the effect of the differences between ΔP_r and ΔP_e, by more generally assuming (see also Section 3.2.2b):

$$\Delta P_e = \Delta P_r - (G_g(s) + sM)\Delta\Omega$$

with $G_g(s) \cong G_g(0)$, it may be seen, as indicated in Figure 3.24, that the signal $(G_g(0) + sM)\Delta\Omega$ is added to the "frequency bias" $(P_{\text{nom}}/(\Omega_{\text{nom}} b_p))(\Delta\Omega_{\text{rif}} - \Delta\Omega)$; the resulting signal then depends on $\Delta\Omega$ according to the transfer function:

$$-\frac{P_{\text{nom}}}{\Omega_{\text{nom}} b_p} + G_g(0) + sM = -\frac{P_{\text{nom}}}{\Omega_{\text{nom}} b_p}\left(1 - \frac{b_p}{b_g} - sb_pT_a\right)$$

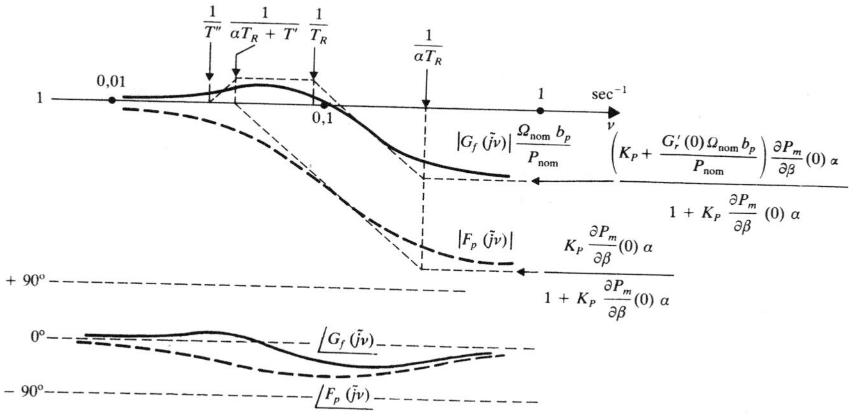

Figure 3.25. Frequency response of the transfer functions $G_f(s)$ and $F_p(s)$ (simplified example; see text).

responsible for some response delay and for an increase in the droop (in the ratio $1/(1 - b_p/b_g)$). Such effects usually are negligible, because of the small value of b_p.

Furthermore, the cutoff frequency of the power regulation loop is relatively low (on the order of $1/T'$: e.g., approximately 0.05 rad/sec). Actually, the functions $F_p(s)$ and $G_f(s)$ may be affected in a nonnegligible way by the effects of the boiler dynamics, at least when the feedback from the pressure p_H is not present (the Equation [3.2.24] may then be inadequate, as it ignores such dynamics).

Finally, based on the previous derivations, the intervention of the primary regulation causes, through $\Delta\beta$, perturbations in the pressure Δp_S (Figure 3.22) which only later are corrected by pressure regulation. For this reason, this operation is called "*boiler following*" mode. Better performance may be obtained by sending into the pressure regulation loop proper signals sensitive to P_{rif}, ε_f, P_e. As illustrated by a dashed line in Figure 3.24b, with this approach (named "*coordinated control*") it is possible to "force" the intervention of the boiler controls, thus accelerating the response of the driving power. Conversely, the power regulator and the valve-positioning system may be enslaved through threshold devices to the pressure p_S and the pressure error $(p_{S\,\text{rif}} - p_S)$, to avoid unacceptable values of such variables.

3.3. "f/P" CONTROL IN AN ISOLATED SYSTEM

3.3.1. Characteristics of the Primary Control

With a system of multiple units, the frequency regulation is called "*primary*" when it is the result of more local speed regulations, achieved by means of primary regulators.

According to the following sections, the "secondary" frequency regulation is instead assigned to a unique centralized regulator, called secondary regulator, which acts on the power (or opening) references of the primary regulators, to realize the desired sharing of the regulating powers.

Usually it is assumed that the electric speed is, even during transients, equal for all the units, as if they were, at the same number of pole pairs, mechanically coupled one another. This hypothesis, which is a simplifying one, appears generally acceptable since, if the network is not too large, the transient slips between the machines vanish quite rapidly because of the synchronizing actions. On the other hand, the slower components of the speed transients (on which the intervention of the regulations essentially depends) may be evaluated, with an approximation that is often satisfactory, by assuming that the units maintain synchronism (see also Section 8.5.1).

With this simplification, the variations of the electric speed Ω, common to all the units, become dependent on a mechanical balance as per Equation [3.1.2], i.e.,

$$P_m - (P_e + P_p) = M\frac{d\Omega}{dt}$$

applied to all the units as a whole, so that:

$$P_m \triangleq \sum_1^N {}_i P_{mi}, \quad P_e \triangleq \sum_1^N {}_i P_{ei}, \quad P_p \triangleq \sum_1^N {}_i P_{pi}, \quad M \triangleq \sum_1^N {}_i M_i$$

where N is the number of units. Furthermore, the resulting inertia coefficient M can be expressed as in Equation [3.1.4], i.e.,

$$M = \frac{P_{\text{nom}} T_a}{\Omega_{\text{nom}}}$$

where $P_{\text{nom}} \triangleq \sum_1^N {}_i P_{\text{nom}\,i}$ represents the total nominal power; consequently, the "start-up time" T_a relative to the whole set of units results defined by:

$$T_a \triangleq \frac{\sum_1^N {}_i P_{\text{nom}\,i} T_{ai}}{\sum_1^N {}_i P_{\text{nom}\,i}}$$

which is a weighted mean of the single start-up times T_{ai}. (Usually T_a is approximately 8 sec, since the single T_{ai} are distributed regularly around this value.)

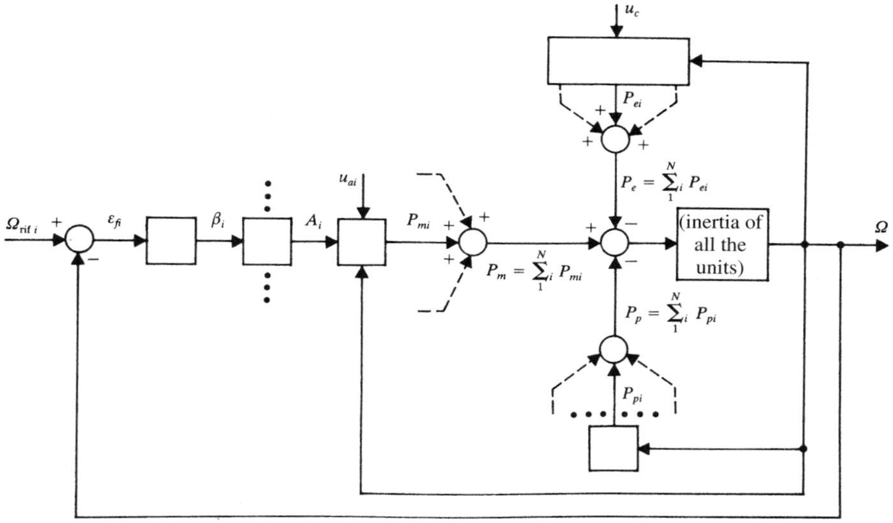

Figure 3.26. Primary frequency regulation for more than one unit: broad block diagram.

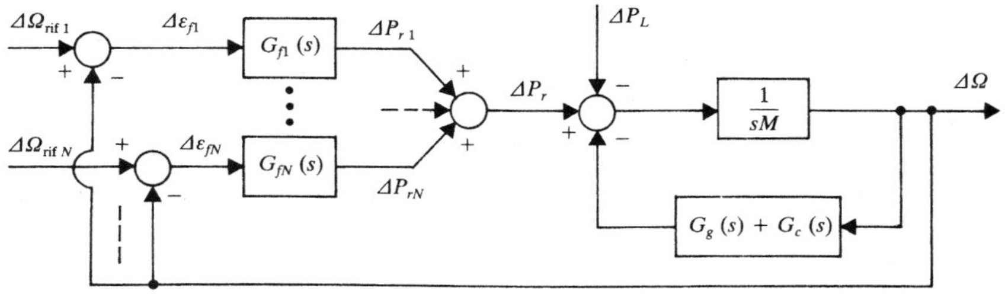

Figure 3.27. A block diagram of the system in Figure 3.26 for small variations.

To analyze the primary regulation, Figures 3.1 and 3.2 can be generalized to Figures 3.26 and 3.27, respectively (in the latter, it is $G_g(s) \triangleq \sum_i G_{gi}(s)$, $G_c(s) \triangleq \sum_j G_{cj}(s)$, and the disturbance ΔP_L may be caused by load perturbations, generator trip etc.).

However, the situation is different for units fed by the same supply system through separate valves, since the driving power P_{mi} of each of these units (and that of the ones for which the valves are kept blocked) generally may depend on the opening of all the unit valves. In this case, it is necessary to consider the dependence of the overall driving power on the frequency errors ε_{fi} acting on the corresponding regulators (or rather the dependence on the frequency Ω, for given values of the references $\Omega_{\text{rif}\, i}$).

In response to the disturbance ΔP_L, the set of transfer functions $G_{f1}(s), \ldots, G_{fN}(s)$ may be replaced by only one equivalent transfer function of the form:

$$G_f(s) \triangleq \sum_i G_{fi}(s) \qquad [3.3.1]$$

so that it can be deduced that:

$$-\frac{\partial \Omega}{\partial P_L} = \frac{1}{G_f + G_g + G_c + sM} \qquad [3.3.2]$$

as in the case of only one generator (see Equation [3.1.10]).

Hence it is evident that, because of the approaches discussed, the problem of the primary frequency regulation in a multimachine system can be managed with the same considerations made in Section 3.1 for a single generator, provided that one considers the problem of the distribution, among the different generators, of the overall regulating power required for the regulation itself ("primary" regulating power).

With respect to a single generator, the increase in P_{nom} as a result of the presence of the other units, even if accompanied by a larger total load power demanded, may already have, on its own account, a beneficial effect on the maintenance of the frequency value; in fact:

- by Equation [3.3.2] it is possible to determine that, in the absence of regulation — or even in the initial response to sudden disturbances, which is practically dominated only by the inertias of the units — the frequency variations are practically proportional to $\Delta P_L / P_{\text{nom}}$ (in fact M, because of the modest variability of T_a, can be considered proportional to P_{nom}, as well as G_g and G_c);
- therefore, as the value of P_{nom} increases, smaller values of $\Delta\Omega$ can be expected, not only following a single disturbance corresponding to a given value of ΔP_L (e.g., the trip of a generator), but also under the effect of normal load disturbances, which statistically tend to compensate one another among the users, hence the average value (in a probabilistic sense) of $\Delta P_L / P_{\text{nom}}$ decreases.

The above-mentioned beneficial effect, caused by the increase of P_{nom}, would then maintain itself practically unchanged even in the presence of regulation, if $G_f(s)$ were selected by the same criteria indicated for a single unit, i.e., (for given values of the cutoff frequency ν_t and of the phase margin γ) proportional to $P_{\text{nom}} T_a$, or rather approximately to P_{nom}.

It has been assumed up to now that all the units remain in synchronism even during transients; however, the f/P control and the electromechanical oscillations may somewhat interact with each other. Let us assume that the system is comprised of two subsystems A

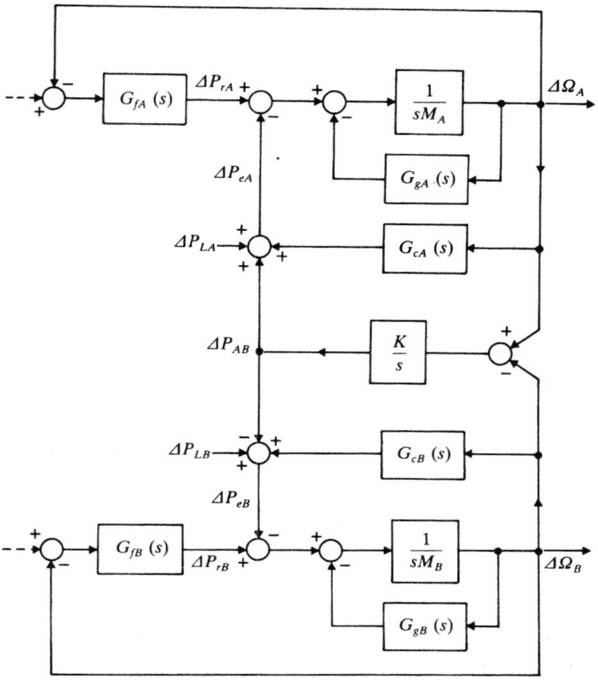

Figure 3.28. Interaction with the electromechanical oscillations: elementary example of block diagram.

and B oscillating against each other, because of the synchronizing actions (Section 1.6), and that the units in each subsystem can be considered synchronous, with electrical speeds Ω_A and Ω_B, respectively. For small variations, and disregarding the variations in the electrical losses, it is possible to refer to Figure 3.28 (in which P_{AB} is the active power exchanged from A to B, and ΔP_{LA} and ΔP_{LB} are assumed to depend only on network perturbations). Disregarding for simplicity the transfer functions G_{gA}, G_{cA} and G_{gB}, G_{cB}, the effects of which are respectively similar to those of G_{fA} and G_{fB}, it follows that:

$$\Delta\Omega_A = \frac{\left(s^2 + \dfrac{K}{M_B}\right)(\Delta P_{rA} - \Delta P_{LA}) + \dfrac{K}{M_B}(\Delta P_{rB} - \Delta P_{LB})}{M_A s\left(s^2 + \dfrac{K}{M_{AB}}\right)}$$

and similarly for $\Delta\Omega_B$ (it only requires to interchange each other the indices A and B), with:

$$M_{AB} \triangleq \frac{M_A M_B}{M_A + M_B}$$

Under the adopted hypotheses, such equations reveal the presence of an undamped resonance, at the frequency $\sqrt{K/M_{AB}}$, which is the frequency of the electromechanical oscillation between A and B (such a frequency usually results in 3–10 rad/sec, and thus much larger than the cutoff frequency of the primary regulation). By imposing $\Delta P_{rA} = -G_{fA}(s)\Delta\Omega_A$, $\Delta P_{rB} = -G_{fB}(s)\Delta\Omega_B$, it is possible to derive that:

$$\Delta\Omega_A = \frac{-\left(s^2 + s\dfrac{G_{fB}}{M_B} + \dfrac{K}{M_B}\right)\Delta P_{LA} - \dfrac{K}{M_B}\Delta P_{LB}}{M_A\left[s\left(s + \dfrac{G_{fA}}{M_A}\right)\left(s + \dfrac{G_{fB}}{M_B}\right) + \dfrac{K}{M_{AB}}\left(s + \dfrac{G_{fA} + G_{fB}}{M_A + M_B}\right)\right]}$$

(and analogously for $\Delta\Omega_B$), where the polynomial inside squared parenthesis at the denominator is the characteristic polynomial, the roots of which (in s) define the dynamic behavior of the linearized system.

Because of the relatively high value of the frequency $\nu = \sqrt{K/M_{AB}}$, at which the functions $G_{fA}(\tilde{\jmath}\nu)$, $G_{fB}(\tilde{\jmath}\nu)$ usually have a very small magnitude, the above-mentioned polynomial can, for all practical purposes, be approximated by:

$$\left(s + \frac{G_{fA} + G_{fB}}{M_A + M_B}\right)\left(s^2 + s\left(\frac{G_{fA}}{M_A^2} + \frac{G_{fB}}{M_B^2}\right)M_{AB} + \frac{K}{M_{AB}}\right)$$

so that the characteristic roots can be evaluated by:

$$0 = s + \frac{G_{fA} + G_{fB}}{M_A + M_B}$$

$$0 = s^2 + s\left(\frac{G_{fA}}{M_A^2} + \frac{G_{fB}}{M_B^2}\right)M_{AB} + \frac{K}{M_{AB}}$$

These results become exact in the case (not far from usual practice) in which $G_{fA}/M_A = G_{fB}/M_B$.

The former of these equations is the characteristic equation obtained under the hypothesis of synchronism between A and B; the latter equation accounts for the oscillatory phenomenon and the damping effect (positive or negative) that the f/P control has on it. Regarding this last aspect, the stability condition, i.e., a positive damping of the oscillation, can practically be translated into:

$$\mathcal{R}e\left(\left(\frac{G_{fA}}{M_A^2} + \frac{G_{fB}}{M_B^2}\right)(\tilde{\jmath}\sqrt{K/M_{AB}})\right) > 0$$

which must be properly accounted for in the synthesis of $G_{fA}(s)$ and $G_{fB}(s)$ if the system did not actually include other damping elements (see also Sections 7.2.2a and 8.5.1).

However, primary regulation alone is not sufficient enough to achieve a zero frequency error and an acceptable *load sharing between units*, at steady-state.

In fact, if the frequency regulation were achieved by the speed regulation of one unit, letting the other unit speeds conform to the regulating unit under the effect of the synchronizing actions, this unit would be requested to face, alone, the total load variation ΔP_L. Besides the consequences associated with the risk of a regulating unit outage, this situation would not be acceptable in practice for various reasons, including:

(1) the power of the regulating unit may be insufficient to cover the total load variations;
(2) the rapidity of the regulation may be poor, even for relatively modest load disturbances, since as already mentioned the power of the regulating unit cannot vary too rapidly[(6)];
(3) allowing a single unit to provide all variations in regulating power, with the generation of the other units fixed at the scheduled values, can lead to an unsatisfactory sharing of the generated powers, with overloads on some lines and possible detrimental impacts on stability itself.

The suitability of sharing the burden of the regulation among several units, which have sufficient power and are conveniently located, thus remains confirmed.

On the other hand, if the regulation is performed by means of more units of which only one has a zero permanent droop ($b_p = 0$), the slowest components of the load variations (as the corresponding transients exhaust) still tend to charge only the zero-droop unit, with the same inconveniences (1) and (3) mentioned above.

Purely for illustrative purposes, some examples of trends of ΔP_{r1}, ΔP_{r2} in response to a step ΔP_L are shown in Figure 3.29, assuming that the regulation is provided only by units 1 and 2, with:

$$G_{f1}(s) = \frac{P_{\text{nom}\,1}}{\Omega_{\text{nom}}} \frac{1 + sT_2}{sT_2 b_{t1}} \qquad \text{(i.e., } b_{p1} = 0\text{)}$$

$$G_{f2}(s) = \frac{P_{\text{nom}\,2}}{\Omega_{\text{nom}}} \frac{1 + sT_2}{b_{p2} + sT_2 b_{t2}}$$

and disregarding the effects of $G_g(s)$ and $G_c(s)$.

In the initial part of the transient—of a relatively short duration, approximately 5–10 $b_t T_a$, with:

$$b_t \triangleq \left[\frac{1}{P_{\text{nom}}}\left(\frac{P_{\text{nom}\,1}}{b_{t1}} + \frac{P_{\text{nom}\,2}}{b_{t2}}\right)\right]^{-1}$$

(see Equation [3.3.5])—the regulating powers vary in the ratio of $\Delta P_{r1}/\Delta P_{r2} \cong (P_{\text{nom}\,1}/b_{t1})/(P_{\text{nom}\,2}/b_{t2})$, with a peak in $(\Delta P_{r1} + \Delta P_{r2})$ that is slightly larger than ΔP_L, to bring back the frequency error (not indicated in the figure) to relatively negligible

(6) For given values of ν_t and γ, the gain of the regulator should be approximately proportional to P_{nom} and, therefore, should usually assume very large values. On the other hand, because of limits on the regulating power rate of change, there is no practical value in increasing the gain of the regulator beyond prefixed values, hence one should be forced to accept a relatively low value of the cutoff frequency ν_t.

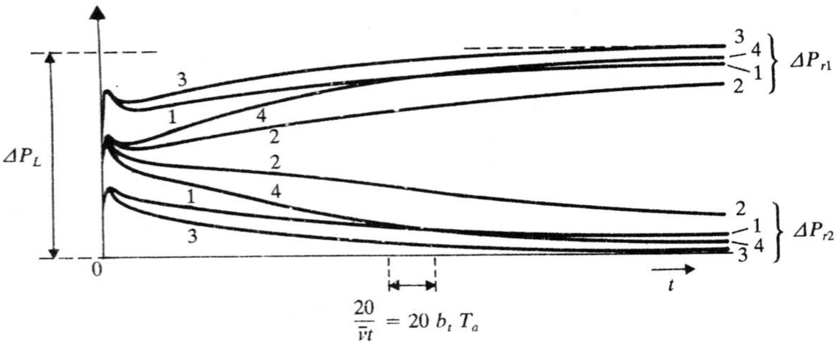

Figure 3.29. Response of the regulating power variations (ΔP_{r1}, ΔP_{r2}) to a step of load variation (ΔP_L), in the case of two units, with:

$$b_t = \left[\frac{1}{P_{\text{nom}}}\left(\frac{P_{\text{nom}1}}{b_{t1}} + \frac{P_{\text{nom}2}}{b_{t2}}\right)\right]^{-1}$$

$$\overline{\nu}_t = \frac{1}{b_t T_a}$$

and furthermore:

	$\dfrac{P_{\text{nom}1}/b_{t1}}{P_{\text{nom}}/b_t}$	$\overline{\nu}_t T_1 = \dfrac{b_{t2} T_2}{b_{p2} b_t T_a}$	$\overline{\nu}_t T_2$
1	0.7	150	3
2	0.5	150	3
3	0.7	60	3
4	0.5	60	3

values; after which the transient exhibits a long "tail," during which the second unit "discharges" itself at the expense of the zero-droop unit, which then ends up by taking all the load variation.

To share the load among several units at steady-state conditions as well, one might decide to assume $b_{pi} = 0$ for more than one unit. However, even such a solution would not be acceptable because the load sharing over the zero-droop units would, in practice, be indeterminate and extremely sensitive to the frequency references of such units. More precisely, because of unavoidable differences between the frequency references $\Omega_{\text{rif}\,i}$ on the various zero-droop units, the frequency Ω would conform to the frequency reference of only one unit, whereas the remaining units would go to the full open position or to the full closed position (according to whether $\Omega_{\text{rif}\,i} \gtrless \Omega$).

Therefore, with only primary regulation it is usually necessary to renounce attempts to $\Delta\varepsilon_f = 0$ at steady-state, and to assume *nonzero permanent droops for all the regulating units*. By generically posing that:

$$G_{fi}(s) = K_{Ifi} T_{1i} \frac{1 + sT_{2i}}{1 + sT_{1i}} = \frac{P_{\text{nom}\,i}}{\Omega_{\text{nom}}} \frac{1 + sT_{2i}}{b_{pi} + sT_{2i} b_{ti}} \qquad [3.3.3]$$

(see Equations [3.1.7] and [3.1.8]), the permanent regulating energy caused by the whole set of regulators is then given by the static gain:

$$G_f(0) \triangleq \sum_i G_{fi}(0) = \sum_i K_{Ifi} T_{1i} = \frac{1}{\Omega_{\text{nom}}} \sum_i \frac{P_{\text{nom}\,i}}{b_{pi}}$$

and corresponds to a permanent droop:

$$b_p \triangleq \left(\frac{1}{P_{\text{nom}}} \sum_i \frac{P_{\text{nom}\,i}}{b_{pi}} \right)^{-1} \qquad [3.3.4]$$

The transient regulating energy and the transient droop are similarly respectively given by the following:

$$G_f(\infty) \triangleq \sum_i G_{fi}(\infty) = \sum_i K_{Ifi} T_{2i} = \frac{1}{\Omega_{\text{nom}}} \sum_i \frac{P_{\text{nom}\,i}}{b_{ti}}$$

$$b_t \triangleq \left(\frac{1}{P_{\text{nom}}} \sum_i \frac{P_{\text{nom}\,i}}{b_{ti}} \right)^{-1} \qquad [3.3.5]$$

The summations above must be extended only to the regulating units; therefore, for example, if the regulating units have equal permanent droop values $b_{pi} = 3\%$, and an overall power $P_{\text{nom}}/3$, then it follows $b_p = (b_{pi} P_{\text{nom}})/(\sum_i P_{\text{nom}\,i}) = 3b_{pi} = 9\%$, and so on.

By selecting the regulators' parameters using criteria similar to those already illustrated with a single unit, the cutoff frequency of the primary regulation is then still expressed by:

$$\nu_t \cong \overline{\nu}_t = \frac{1}{b_t T_a}$$

and, therefore, it is approximately proportional to $1/b_t$ because of the slight variability of the resulting start-up time T_a[(7)].

Finally, with reference to Equation [3.3.3], it is possible to state that, in response to load disturbances:

[(7)] However, the regulating units usually constitute only a part of the units of the system, whereas the remaining units are kept at a fixed load, and it is thus opportune that the gain of their regulators (and in particular $1/b_{ti}$) is larger than the value selected for isolated operation, to avoid too small values of $1/b_t$ and therefore of ν_t.

- during the relatively rapid phases of the frequency transients (e.g, under the normal fluctuations of the load, or in the initial part of the transient caused by a step of load), the regulating units react roughly proportionate to the respective values of $P_{\text{nom}\,i}/b_{ti}$, i.e., to the respective transient regulating energies (see also Fig. 3.29);
- during the slowest phases of the frequency transients (e.g., in response to slow load variations, or in the final part of the transient caused by a step of load), the regulating units tend instead to react to the disturbance proportionate to the respective values of $P_{\text{nom}\,i}/b_{pi}$, i.e., to the respective permanent regulating energies (usual values of the permanent droop: e.g., $b_{pi} = 2\%–5\%$ for thermal units and for hydrounits with Pelton turbines, and much larger values, e.g., 15%–20% for other units in primary regulation).

At steady-state conditions, it then holds:

$$\Delta P_{ri} = -\frac{P_{\text{nom}\,i}}{\Omega_{\text{nom}} b_{pi}} \Delta\Omega, \quad \sum_i \Delta P_{ri} = \Delta P_L$$

and therefore:

$$\Delta\Omega = -\frac{\Omega_{\text{nom}} b_p}{P_{\text{nom}}} \Delta P_L, \quad \Delta P_{ri} = \frac{P_{\text{nom}\,i}}{b_{pi}} \frac{b_p}{P_{\text{nom}}} \Delta P_L$$

in accordance to Figure 3.30, by using the static characteristics (Ω, P_{ri}) of the different regulating units.

In particular, the static frequency error is not zero, and to overcome this drawback, other actions must be taken in the form of "secondary" regulation, as discussed in the following sections.

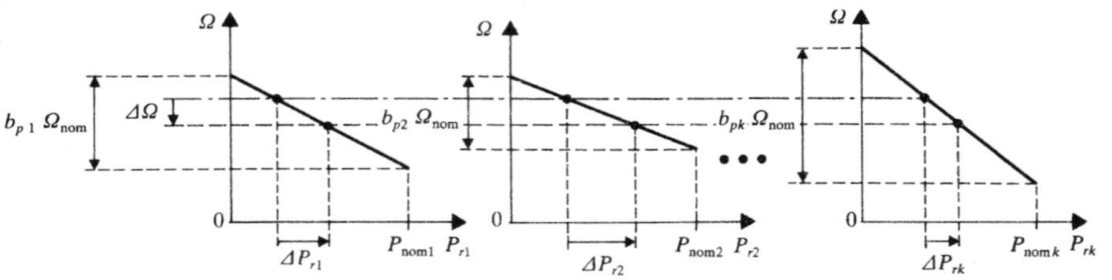

k = number of regulating units

$\sum_1^k{}_i \Delta P_{ri} = \Delta P_L$

Figure 3.30. Static characteristics of units under primary regulation and sharing of the regulating powers for varying frequency.

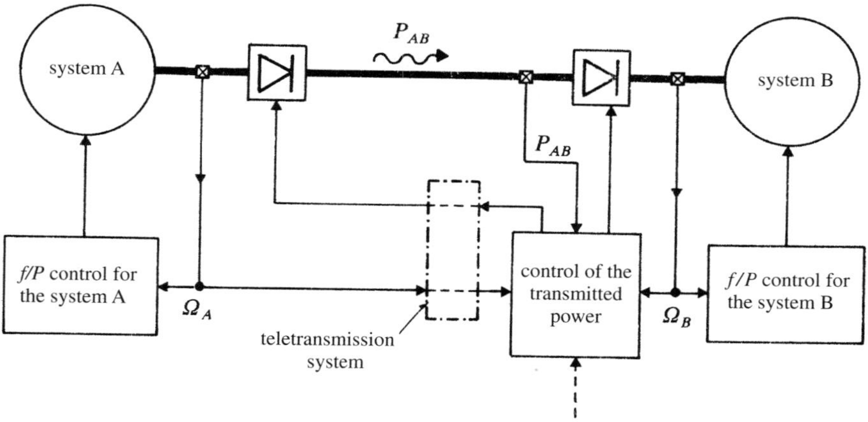

Figure 3.31. Control of the power transmitted through a direct current link, as a function of the frequencies at the two terminals: broad block diagram.

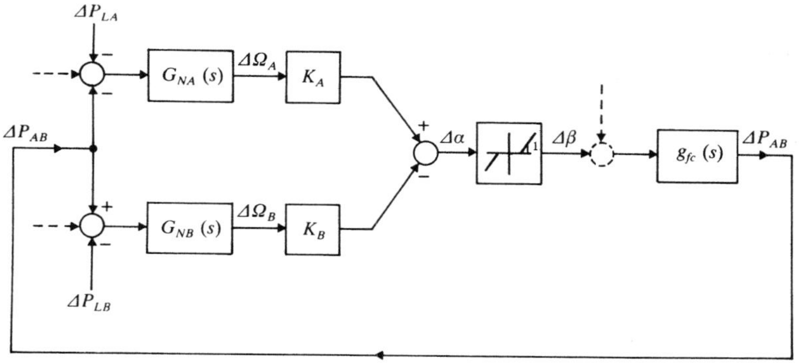

Figure 3.32. Example of block diagram for the system in Figure 3.31 for small variations. (The dashed inputs indicate possible signals from the secondary f/P controls.)

In the case of two systems A and B connected through a *dc link* (see Fig. 3.31 and Section 5.5), by acting on the transmitted power P_{AB} it is possible to realize a reciprocal support between the two systems, even at the primary regulations level. In this concern, it is clearly convenient that the power P_{AB} increases for increasing values of Ω_A and diminishes for increasing values of Ω_B. A possible control scheme is shown in Figure 3.32, for which the transfer functions $G_{NA} \triangleq (-\partial\Omega_A/\partial P_{LA})_{\Delta P_{AB}=0}$, $G_{NB} \triangleq (-\partial\Omega_B/\partial P_{LB})_{\Delta P_{AB}=0}$ are defined as in Equation [3.3.2], and therefore they account for the primary frequency regulations in the two systems. The loss variations in the link are disregarded for simplicity. The control of P_{AB} implies also a communication system between the two stations; in case of an outage in such system, the control must be properly "locked."

If the "deadband" between $\Delta\alpha$ and $\Delta\beta$ is omitted (thus assuming $\Delta\beta = \Delta\alpha$, see Fig. 3.32), assuming $G_A \triangleq K_A g_{fc}$, $G_B \triangleq K_B g_{fc}$ it follows that:

$$\begin{cases} \Delta P_{AB} = \dfrac{-G_A G_{NA} \Delta P_{LA} + G_B G_{NB} \Delta P_{LB}}{1 + G_A G_{NA} + G_B G_{NB}} \\ \\ \Delta \Omega_A = \dfrac{-G_{NA}\left[(1 + G_B G_{NB}) \Delta P_{LA} + G_B G_{NB} \Delta P_{LB}\right]}{1 + G_A G_{NA} + G_B G_{NB}} \\ \\ \Delta \Omega_B = \dfrac{-G_{NB}\left[G_A G_{NA} \Delta P_{LA} + (1 + G_A G_{NA}) \Delta P_{LB}\right]}{1 + G_A G_{NA} + G_B G_{NB}} \end{cases}$$

In the absence of control on P_{AB} it would be $\Delta P_{AB} = 0$, $\Delta\Omega_A = -G_{NA}\Delta P_{LA}$, $\Omega_B = -G_{NB}\Delta P_{LB}$, as in separate operation. If the link were in ac, disregarding the transient differences between Ω_A and Ω_B, it would instead hold:

$$\Delta P_{AB} = \frac{-G_{NA}\Delta P_{LA} + G_{NB}\Delta P_{LB}}{G_{NA} + G_{NB}}, \qquad \Delta\Omega_A = \Delta\Omega_B = \frac{-G_{NA} G_{NB}(\Delta P_{LA} + \Delta P_{LB})}{G_{NA} + G_{NB}}$$

without the possibility of adjusting the reciprocal support between the two systems.

By means of a proper choice of G_A and G_B, the effect of ΔP_{LA} on $\Delta\Omega_A$ can be conveniently reduced (with respect to the case $\Delta P_{AB} = 0$), but at the cost of a perturbation on the system B, accompanied by an "induced" variation $\Delta\Omega_B$; similar conclusions then hold in response to ΔP_{LB}. The frequency variations induced in the two cases relate to each other by:

$$\frac{\partial\Omega_A/\partial P_{LB}}{\partial\Omega_B/\partial P_{LA}} = \frac{G_B}{G_A} = \frac{K_B}{K_A}$$

and this must be considered in the choice of the ratio K_B/K_A. In this concern,

- the assumption $K_A = K_B$ may appear to penalize the system in which the load perturbation (ΔP_{LA} or ΔP_{LB}) is likely minor, i.e., in practice for the system with smaller power;
- by assuming that system B is the one with smaller power, it is then convenient to choose K_B adequately larger than K_A, thus giving more importance to the frequency variations that occur in the system B (remember that $\Delta\alpha = K_A\Delta\Omega_A - K_B\Delta\Omega_B$).

In the extreme case in which A is much larger than B, and it is assumed $K_A = 0$, it specifically follows that:

$$\Delta P_{AB} = -G_B\Delta\Omega_B = \frac{G_B G_{NB}}{1 + G_B G_{NB}}\Delta P_{LB}$$

so that the link behaves, for the system B, as an additional unit under primary regulation. A form similar to Equation [3.3.3] can be imposed to the function $g_{fc}(s)$, at least within the frequency range of interest for the primary regulation.

Generally, the interposition of the deadband between $\Delta\alpha$ and $\Delta\beta$ (see Fig. 3.32) avoids the control of P_{AB} when the perturbations are relatively modest. In any case, restoring the desired frequency values in the two systems is then left to the respective secondary regulations, with or without involving the P_{AB} control.

Furthermore, it may be useful that $\Delta\beta$ also depends on the speeds of variation of $\Delta\Omega_A$ and $\Delta\Omega_B$, particularly when facing relatively large perturbations, or perturbations that might even lead one of the two systems to emergency conditions. In such a way, in fact, the support of the other system may be conveniently anticipated.

However, it is important to avoid the situation in which, under large perturbations, one of the two systems drags the other to unacceptable frequency values (and more specifically to a frequency collapse, when the total available power results to be insufficient; see Section 3.5). It is then convenient to limit the variation range of the transmitted power and not allow further increases of P_{AB} when Ω_A is already too low (or Ω_B too high), and further decreases when Ω_B is already too low (or Ω_A too high).

3.3.2. Characteristics of the Secondary Control

In Section 3.3.1, it was seen that the primary regulation alone does not permit the load to be distributed over several units, under steady-state conditions with $\Delta\varepsilon_f = 0$. The zero-droop unit could benefit only temporarily from other regulating units, hence the available regulating power for the entire system, under steady-state conditions, would only be the regulating power of this unit. Such an inconvenience may be overcome by using a proper "secondary" regulation, in accordance with Figure 3.33.

The *"secondary regulator"* (also called "network regulator") is unique for the whole system and is generally located at the central dispatching office, mentioned in Section 2.5. The input of the secondary regulator is constituted by the frequency error $\varepsilon_R = \Omega_{\text{RIF}} - \Omega$, where Ω_{RIF} is the "secondary" frequency reference usually locked at the nominal value, whereas Ω is the actual frequency value (locally measured, or possibly deduced as a mean value from several measurements, performed at different network locations). The output y_R of the regulator (also called *"level"* of the secondary regulation) is then translated by the block called *"dispatcher"* into the power references $P_{\text{rif}\,i}$ of the primary regulators, to act on the driving powers of the respective units.

The scheme implies the existence of a communication system (not shown in the figure) that connects the central office to different generating units. The dispatcher functions may be partially delegated to the peripheral offices (Section 2.5), thus realizing the passage from y_R to each $P_{\text{rif}\,i}$ according to a scheme having a tree structure, with a more and more increasing detail. Usually, only some of the units may be on secondary regulation, whereas the remaining units might be under (only) primary regulation or maintained at locked generation.

To have a zero frequency error under steady-state conditions, it is no longer necessary to impose that the permanent droop b_p of the primary regulation be zero, since it is sufficient to have an "integral" action in the secondary regulator. In this way, under steady-state, it follows that $\varepsilon_R = 0$ (provided that the

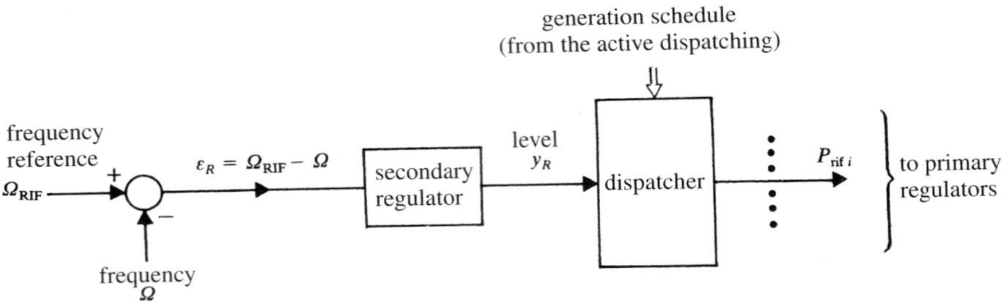

Figure 3.33. Secondary regulator and dispatcher: block diagram in the case of isolated system.

total available regulating power sufficiently covers load variations), and thus the frequency Ω is driven to the desired value Ω_{RIF}.

By assuming $b_{pi} \neq 0$ for all the units (and that the frequency references are constant and equal one another), the power variations caused by the primary regulation are zero because $\Delta\Omega = 0$, whereas the distribution of the load power over the various units depends (at steady-state) only on the signals $P_{rif\,i}$, and then it may be achieved as desired through the suitable selection of such signals.

For units at the same power plant or pertaining to neighboring plants, and operating in a more or less occasional way on a relatively small network, it would be possible to think of regulating the frequency by a single zero-droop unit, by "interlocking" the primary regulators of the other (nonzero droop) units with the output of the regulator (or with the generated power) of the zero-droop unit. In this way, such a regulator would behave as a primary regulator for its own unit, and as a secondary regulator for the others. However, this solution—conditioned to a precise choice of the roles of the different units—must be properly corrected to avoid unacceptable operation whenever the zero-droop unit is disconnected from the network.

For small variations, the dependence of the single $P_{rif\,i}$'s on y_R can be generally defined by equations such as $\partial P_{rif\,i}/\partial y_R = R_i$, where $R_1, R_2, \ldots$ are suitable constants that determine the sharing of the active power generated at steady-state[(8)].

Such a share must account for both the generation schedule determined at the active dispatching stage, and the differences between the actual load situation and the foreseen one; it then seems reasonable to assume the following equations:

$$\Delta P_{di} = \Delta P_{bi} + r_i(\Delta P_d - \Delta P_b) \qquad [3.3.6]$$

[(8)] Actually, transfer functions $R_i(s)$ should be considered to account for the response delays of the communication system. Nevertheless, such delays may be considered negligible in the present analysis, since their effects, in terms of frequency response ($s = \tilde{j}\nu$), fall into a range of relatively high frequencies (e.g., $\nu > 0.3$ rad/sec) with respect to the cutoff frequency of the secondary regulation.

with:

$$\sum_i r_i = 1 \tag{3.3.7}$$

where:

P_{di} = ("desired") power demanded to the i-th unit

P_{bi} = ("base") power scheduled for the i-th unit

$$P_d \triangleq \sum_i \Delta P_{di}$$

$$P_b \triangleq \sum_i \Delta P_{bi}$$

and where r_i, called "participation factor" of the i-th unit, may be chosen based on an economic criterion (recall the factors g_i defined in Section 2.3.1a).

By assuming $\Delta P_{\text{rif}\,i} = \Delta P_{di}$, the desired share can be then realized by:

$$\Delta P_{\text{rif}\,i} = \Delta P_{bi} + R_i \Delta y_R$$

and furthermore (where $R \triangleq \sum_i R_i$):

$$\Delta y_R = \frac{\Delta P_d - \Delta P_b}{R}, \qquad \frac{R_i}{R} = r_i$$

For a thermal generation system, the economic dispatching also can be achieved by assuming the level y_R as the "incremental cost" (see Section 2.3.1) equal for all the units under secondary regulation, and thus imposing that the single P_{di}'s vary with y_R according to their respective incremental cost characteristics, which are generally nonlinear. The values P_{di}'s then become directly dependent on y_R, avoiding any preventive scheduling of the base values P_{bi}; the result is a completely automatic updated dispatching, which, however, is based only on the economic criterion. The scheme may be modified, at the cost of evident complications, to include the network losses by means of a suitable "loss formula."

The variation of the regulating power ΔP_{ri} of the generic i-th unit also may be seen as the sum of two contributions $\Delta P'_{ri}$ and $\Delta P''_{ri}$, which respectively define the variations of the "primary" regulating power (which depends on the primary frequency error ε_{fi}) and of the "secondary" one (which depends on $\Delta P_{\text{rif}\,i} = \Delta P_{di}$). The block diagram of Figure 3.34 finally results, in which $G_F(s) \triangleq G_o(s)/R = (\Delta y_R/\Delta \varepsilon_R)(s)$ is the transfer function of the secondary regulator, with a pole at the origin, whereas $G_{fi}(s) \triangleq (\partial P_{ri}/\partial \varepsilon_{fi})(s)$ and $F_{pi}(s) \triangleq (\partial P_{ri}/\partial P_{\text{rif}\,i})(s)$ are the transfer functions (already considered in previous sections) realized by the primary regulator of the i-th unit, with:

$$F_{pi}(0) = 1 \tag{3.3.8}$$

(it then follows at steady-state: $\Delta P_R \triangleq \sum_i \Delta P''_{ri} = \sum_i \Delta P_{di} = \Delta P_d$).

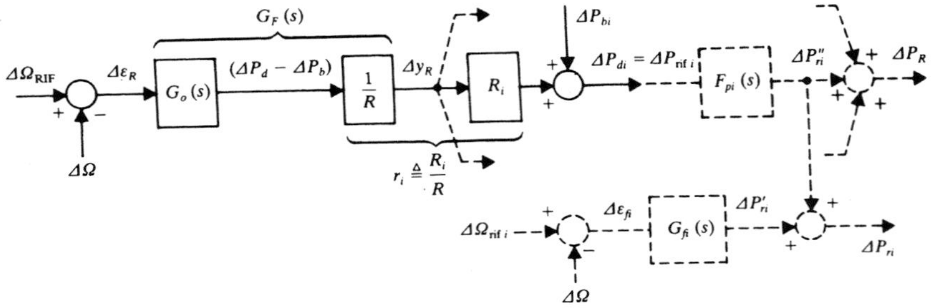

Figure 3.34. Block diagram of the system in Figure 3.33 for small variations (and interactions with the primary regulation).

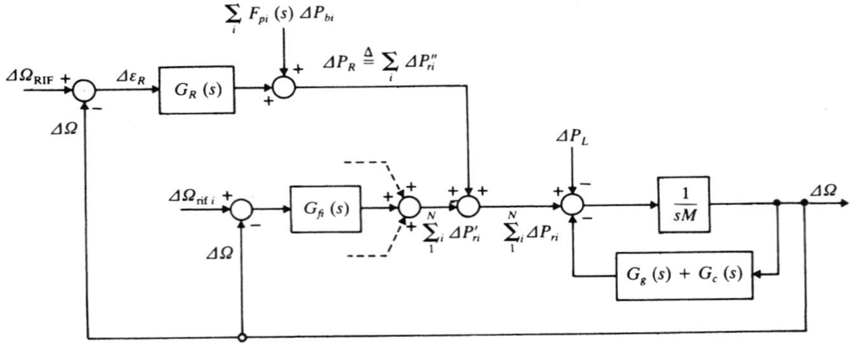

Figure 3.35. Overall block diagram for small variations, corresponding to the solution of Figure 3.34.

Finally, the overall control scheme is as shown in Figure 3.35, with:

$$G_R(s) \triangleq G_F(s) \sum_i R_i F_{pi}(s) = G_o(s) \sum_i r_i F_{pi}(s) \qquad [3.3.9]$$

where $R_i = r_i = 0$ for those units which do not take part into the secondary regulation, and where $\sum_1^N i \Delta P'_{ri}$ and $\Delta P_R \triangleq \sum_i \Delta P''_{ri}$ are the total variations of the primary and secondary regulating powers, respectively. The former summation must be extended to all the units under primary regulation, and the latter to those under secondary regulation. Thus, in general, only a portion of the units takes part to both the summations. A unit also might be active only under the secondary regulation, i.e., without taking part to the primary regulation. It is, in fact, sufficient that its regulator was used purely as a power regulator, with frequency thresholds, according to Section 3.2.

Actually, each single $\Delta P''_{ri}$ also contains a "tertiary" component that depends on possible variations ΔP_{bi} decided under the real-time scheduling (see Section 2.5). However, at steady-state, the tertiary components have influence on Δy_R and the different $\Delta P''_{ri}$'s,

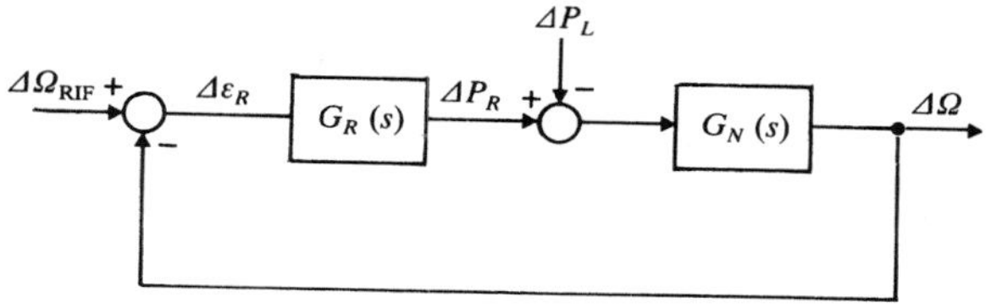

Figure 3.36. Block diagram equivalent to that of Figure 3.35 for $\Delta\Omega_{\mathrm{rif}\,i} = 0$ and $\Delta P_{bi} = 0\ \forall i$.

but not on the total variation ΔP_R (in fact, because of the only ΔP_{bi}, it follows $\Delta P_d = \Delta P_R = \Delta P_L = 0$, $\Delta y_R = -\Delta P_b/R$, $\Delta P''_{ri} = \Delta P_{bi} - r_i \Delta P_b$).
By the real-time scheduling, the base powers P_{bi} and the participation factors r_i themselves are actually updated, e.g., every 5 minutes or more. Such an update may be determined based on the "state estimation" of the system operation (Section 2.5) or, more simply, by accounting for the value P_d (total desired power) deducible from P_b and from the value of the "level" y_R at the output of the regulator.

From Figure 3.35, and by assuming all $\Delta\Omega_{\mathrm{rif}\,i}$ and ΔP_{bi} to be zero, it is possible to derive Figure 3.36, which is particularly suitable for examining the dependence of the system behavior on the characteristics of the secondary regulator. In such a block diagram, for simplicity, it has been assumed that:

$$G_N(s) \triangleq \frac{1}{G_f(s) + G_g(s) + G_c(s) + sM} \tag{3.3.10}$$

where $G_f(s) \triangleq \sum_1^N {}_i G_{fi}(s)$ (recall Equations [3.3.1] and [3.3.2]). The transfer function $G_R(s)$ is defined by Equation [3.3.9].

In general, because of the integral action of the secondary regulator, at steady-state it holds that $\Delta\varepsilon_R = 0$ and thus:

$$\Delta\Omega = \Delta\Omega_{\mathrm{RIF}}$$

whereas:

$$\sum_1^N {}_i \Delta P'_{ri} = \sum_1^N {}_i G_{fi}(0)(\Delta\Omega_{\mathrm{rif}\,i} - \Delta\Omega_{\mathrm{RIF}}) \tag{3.3.11}$$

$$\Delta P_d = \Delta P_R = \Delta P_L + E\,\Delta\Omega_{\mathrm{RIF}} - \sum_1^N {}_i G_{fi}(0)\Delta\Omega_{\mathrm{rif}\,i} \tag{3.3.12}$$

where:

$$E \triangleq G_f(0) + G_g(0) + G_c(0) = \frac{1}{G_N(0)}$$

is the total (permanent) regulating energy of the system (Section 3.1.3), from which simply $\Delta\Omega = 0$, $\sum_1^N {}_i\Delta P'_{ri} = 0$, $\Delta P_d = \Delta P_R = \Delta P_L$, if all the frequency references are maintained locked. In any case, at steady-state, it then holds that:

$$\Delta P_{di} = \Delta P''_{ri} = \Delta P_{bi} + r_i(\Delta P_R - \Delta P_b) \qquad [3.3.13]$$

(i.e., $\Delta P''_{ri} = r_i \Delta P_R$ if the base powers are kept constant), so that the total variation ΔP_R is shared among the units according to Equation [3.3.6].

As to the synthesis of the secondary regulator (with reference to Fig. 3.36), the function $G_N(s)$ has generally a very high dynamic order (since the number of units under primary regulation is very large) and can be actually known only with some approximation. Also for this reason, it is convenient to accept a cutoff frequency ν_t relatively low (e.g., 0.01–0.02 rad/sec) so that, for the synthesis of $G_o(s)$, the functions $G_N(s)$ and $\sum_i r_i F_{pi}(s)$ may be approximated by their respective static gains, which can be known in an easier way:

$$\begin{cases} G_N(0) = 1/E \\ \sum_i r_i F_{pi}(0) = 1 \end{cases}$$

As seen in Section 3.2, the functions $G_{fi}(s)$, which contribute to the $G_N(s)$, and $F_{pi}(s)$ may somewhat be affected by slow phenomena that occur at the same low-frequency range in which the secondary regulation operates. These slow phenomena may be related, for example, to boiler controls or to the behavior of pressure tunnel and surge tank.

By adopting a $G_o(s)$ of the purely integral type, i.e.,[(9)]:

$$G_o(s) = \frac{K_o}{s} \qquad [3.3.14]$$

with K_o sufficiently small, it then results that $\nu_t \cong \overline{\overline{\nu}}_t \triangleq K_o/E$, so that it must be simply assumed that:

$$K_o = E\overline{\overline{\nu}}_t \qquad [3.3.15]$$

with $\overline{\overline{\nu}}_t$ within the above-mentioned range. For the secondary regulator, the following transfer function (see Fig. 3.34) finally holds:

$$G_F(s) \triangleq \frac{G_o(s)}{R} = \frac{K_o}{sR} \qquad [3.3.16]$$

[(9)] The use of a proportional-integral regulator, i.e., a $G_o(s)$ of the type $G_o(s) = K_o(1 + sT_o)/s$, does not appear to be strictly justified; on the contrary, it may adversely affect the stability if it implies an excessive increase of the cutoff frequency.

With Equation [3.3.16], the gain of the secondary regulator must be inversely proportional to $R \triangleq \sum_i R_i$; this must be considered, e.g., with a disconnection from the network of one or more units under secondary regulation, with consequent reduction of R (at the same values R_i for the units which remain connected).

From the realization point of view, the most traditional solution is directly based on the scheme of Figure 3.34, with:

- communications (at least for the largest distances) only of the level signal y_R;
- dispatching (according to the agreed values $R_1, R_2, \ldots$) delegated to the peripheral offices or even to the power plants;
- timely communication, to the central office, of possible modifications in the participation to the secondary regulation, so that the gain of the regulator can be adjusted (see Equation [3.3.16]) based on the updated value of R.

More updated solutions require communication of several signals to realize a complete control by the central office of the active powers delivered by the units under secondary regulation, and thus of the total power dispatching (*"automatic generation control"*). The involved signals are useful for the real-time scheduling (Section 2.5).

Apart from possible variations with specific details, such solutions may be of the form shown in Figure 3.37, in which P_{ei} represents the active power delivered by the generic unit under secondary regulation (i.e., being the losses in the unit negligible, the generated power), and $P_e \triangleq \sum_i P_{ei}$ is the active power delivered (or even generated) as a whole by these units.

The scheme includes, for each unit under secondary regulation, an integral "power regulator" (k_i/s), which acts on the primary regulator. The functions $G_{pi}(s)$, $G_{fi}(s)$ are achieved by means of the primary regulator. The output of the power regulator also may be $P_{\text{rif}\,i}$, in which case it holds $G_{pi}(s) = F_{pi}(s)$ and thus, because of Equation [3.3.8], $G_{pi}(0) = 1$.

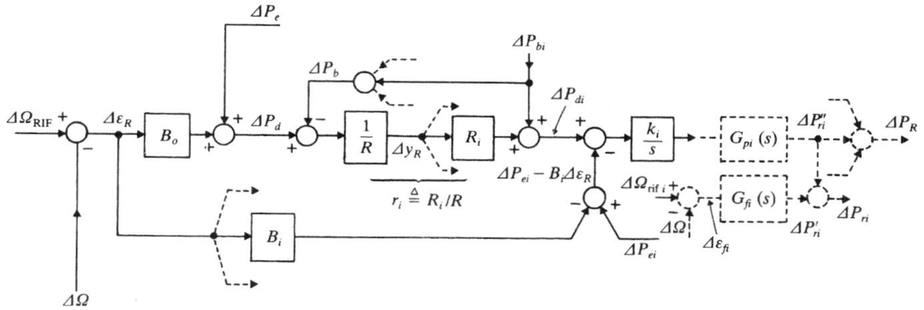

Figure 3.37. Variation of the block diagram of Figure 3.34 with centralized control of the active powers generated ("automatic generation control").

As a consequence of the integral action of the power regulators, at steady-state $\Delta P_{di} = \Delta P_{ei} - B_i \Delta \varepsilon_R$, and thus $\Delta P_d = \Delta P_e - \sum_i B_i \Delta \varepsilon_R$. As it holds that $\Delta P_d = \Delta P_e + B_o \Delta \varepsilon_R$, it follows, as desired, that $\Delta \varepsilon_R = 0$ (i.e., $\Delta \Omega = \Delta \Omega_{\text{RIF}}$), independently of the disturbance ΔP_L and variations ΔP_{bi}.

Furthermore it can be written:

$$\Delta P'_{ri} = G_{fi}(s)(\Delta \Omega_{\text{rif}\,i} - \Delta \Omega)$$

$$\Delta P_{ei} = \Delta P''_{ri} + G_{fi}(s)\Delta \Omega_{\text{rif}\,i} - \frac{1}{G_{Ni}(s)}\Delta \Omega$$

where:

$$\frac{1}{G_{Ni}(s)} \triangleq G_{fi}(s) + G_{gi}(s) + sM_i$$

At steady-state Equation [3.3.11] holds, and Equations [3.3.12] and [3.3.13] are replaced by:

$$\begin{cases} \Delta P_d = \Delta P_e = \Delta P_R + \sum_i G_{fi}(0)\Delta \Omega_{\text{rif}\,i} - \sum_i E_i \Delta \Omega_{\text{RIF}} = \Delta P_L + E_o \Delta \Omega_{\text{RIF}} \\ \Delta P_{di} = \Delta P_{ei} = \Delta P_{bi} + r_i(\Delta P_e - \Delta P_b) \end{cases}$$

where $E_i \triangleq 1/G_{Ni}(0)$, $E_o \triangleq E - \sum_i E_i$. Based on these relationships, it can be observed that the dispatching defined by Equation [3.3.6] is now achieved, more suitably, with reference to the total variation ΔP_e, rather than with reference to the total variation ΔP_R (which is relative only to the secondary regulating power). However, the results are coincident if all the frequency references remain locked.

For $\Delta \Omega_{\text{rif}\,i} = 0$:

$$\Delta P_{ei} = \Delta P''_{ri} - \frac{1}{G_{Ni}(s)}\Delta \Omega \qquad [3.3.17]$$

$$\Delta P_e = \Delta P_R - \sum_i \frac{1}{G_{Ni}(s)}\Delta \Omega$$

so that the variation $\Delta \Omega = G_N(s)(\Delta P_R - \Delta P_L)$ also can be written as:

$$\Delta \Omega = G_{No}(s)(\Delta P_e - \Delta P_L) \qquad [3.3.18]$$

where:

$$\frac{1}{G_{No}(s)} \triangleq \frac{1}{G_N(s)} - \sum_i \frac{1}{G_{Ni}(s)}$$

Recalling Equation [3.3.10], it is evident that $1/G_{No}(s)$ is constituted by the sum of $G_c(s)$ and of all the functions $(G_{fi}(s) + G_{gi}(s) + sM_i)$ which concern the units excluded by the secondary regulation.

If it were possible to assume:

$$\left.\begin{aligned} B_o &= \frac{1}{G_{No}(s)} \\ B_i &= \frac{1}{G_{Ni}(s)} \end{aligned}\right\} \qquad [3.3.19]$$

then, the signals ΔP_d, $(\Delta P_{ei} - B_i \Delta\varepsilon_R)$ shown in Figure 3.37 would, because of Equations [3.3.17] and [3.3.18], then be given by:

$$\Delta P_d = \Delta P_e + B_o \Delta\varepsilon_R = \Delta P_L + \frac{1}{G_{No}(s)} \Delta\Omega_{\text{RIF}}$$

$$\Delta P_{ei} - B_i \Delta\varepsilon_R = \Delta P''_{ri} - \frac{1}{G_{Ni}(s)} \Delta\Omega_{\text{RIF}}$$

so that the overall control scheme might turn into that shown in Figure 3.38, which is typical of a simple open-loop regulation with a "compensation" for the disturbance ΔP_L.

Assuming, for simplicity, all $\Delta\Omega_{\text{rif}\,i}$ and ΔP_{bi} to be zero, the block diagram of Figure 3.39a can be derived, in which:

$$\begin{cases} H_1(s) \triangleq \sum_i (r_i B_o + B_i) \dfrac{k_i G_{pi}(s)}{s + k_i G_{pi}(s)} \\[2ex] H_2(s) \triangleq \dfrac{1}{s \sum_i \dfrac{r_i}{s + k_i G_{pi}(s)}} \\[2ex] H_3(s) \triangleq \sum_i \left(r_i \sum_k \dfrac{1}{G_{Nk}(s)} - \dfrac{1}{G_{Ni}(s)} \right) \dfrac{k_i G_{pi}(s)}{s + k_i G_{pi}(s)} \end{cases}$$

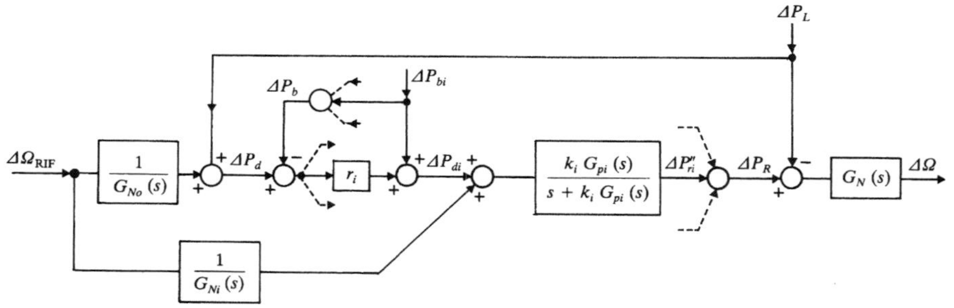

Figure 3.38. Overall block diagram for small variations, corresponding to the solution of Figure 3.37 under the assumptions indicated in text.

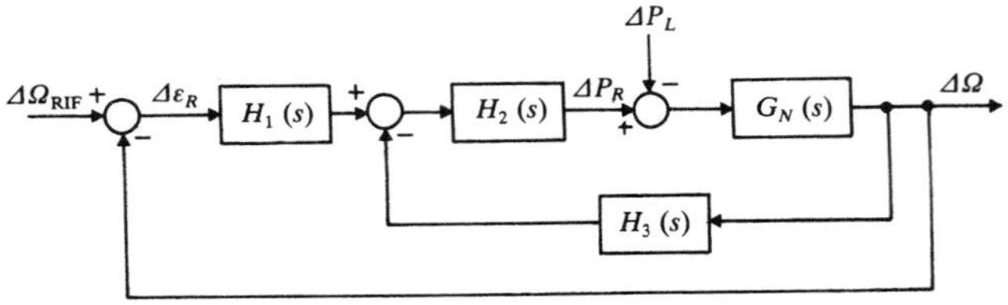

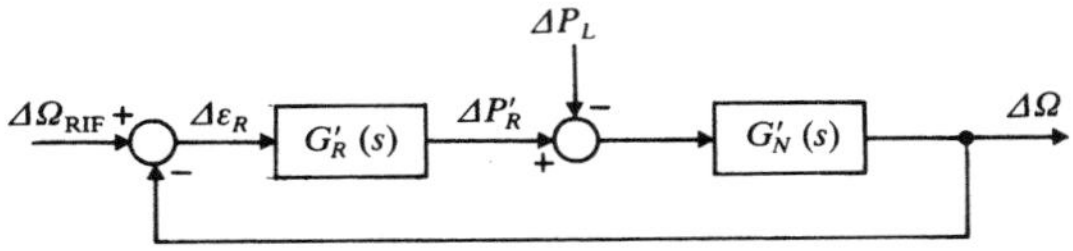

Figure 3.39. Overall block diagram for small variations, corresponding to the solution of Figure 3.37 for $\Delta\Omega_{\text{rif}\,i} = 0$ and $\Delta P_{bi} = 0\ \forall i$. From (a), one can derive the block diagram (b), which is more directly comparable to that of Figure 3.36.

from which the block diagram shown in Figure 3.39b, with:

$$\begin{cases} G'_{(R)}(s) \triangleq H_1(s)H_2(s) \\ G'_{(N)}(s) \triangleq \dfrac{G_N(s)}{1 + H_2(s)H_3(s)G_N(s)} \end{cases}$$

and $\Delta P'_R \triangleq \Delta P_R + H_2(s)H_3(s)\Delta\Omega$. In particular, in $H_2(s)$, there is a pole at the origin that is also present in $G'_R(s)$.

For relatively low frequencies, which may be of interest for the synthesis of the control system, if it is assumed that $G_{pi}(s) \cong G_{pi}(0)$ and imposing $k_i G_{pi}(0) = K$ for all the units, it can be simply written that[10]:

$$\begin{cases} G'_R(s) \cong (B_o + \sum_i B_i)\dfrac{K}{s} \\ G'_N(s) \cong \dfrac{1}{\dfrac{1}{G_{No}(s)} + \sum_i \dfrac{1}{G_{Ni}(s)}} = G_N(s) \end{cases}$$

[10] More generally, if $k_i G_{pi}(s) = K(s)$ for all the units, then $H_3(s) = 0$, $G'_R(s) = H_1(s)H_2(s) = \left(B_o + \sum_i B_i\right) K(s)/s$, $G'_N(s) = G_N(s)$, and furthermore:

$$\Delta P_R = \Delta P'_R = G'_R(s)\Delta\varepsilon_R = \left(B_o + \sum_i B_i\right)\frac{K(s)}{s}\Delta\varepsilon_R$$

which also can be directly derived from the scheme shown in Figure 3.37, observing that the signals ΔP_e, ΔP_{ei} would evidently have no overall effect on ΔP_R.

As $G'_N(s) \cong G_N(s)$, the function $G'_R(s)$ may be chosen as the function $G_R(s)$ previously considered, or even (by Equation [3.3.9], with $\sum_i r_i F_{pi}(s) \cong \sum_i r_i F_{pi}(0) = 1$) as the function $G_o(s)$. Therefore, by writing, according to Equations [3.3.14] and [3.3.15]:

$$G'_R(s) = \frac{K_o}{s} \cong \frac{E \nu_t}{s}$$

it can be derived that:

$$\left(B_o + \sum_i B_i\right) K \cong E \nu_t$$

where $\nu_t \cong \overline{\overline{\nu}}_t \triangleq K_o/E$ is the cutoff frequency (which is quite low) of the secondary frequency regulation.

The above condition may be satisfied, for example, by assuming that:

$$\begin{cases} B_o = \dfrac{1}{G_{No}(0)} = E_o \\ B_i = \dfrac{1}{G_{Ni}(0)} = E_i \end{cases}$$

to realize Equations [3.3.19] at least in the low-frequency range. It then follows that $B_o + \sum_i B_i = 1/G_N(0) = E$, and thus:

$$K \cong \nu_t$$

The schemes discussed above are primarily for illustrative purposes, and there are a variety of schemes in actual use. Furthermore, for simplicity, it has been assumed that the control is realized "in continuous time." The most updated solutions use digital computers as elements of the controlling system, and this may allow the realization of further particular functions, which can overlap the ones described above. The control by computer can, for example, enable the adaptive setting of limits or thresholds on some signals, and the possible interlocking of the power regulators for which the error $(\Delta P_{di} - \Delta P_{ei} + B_i \Delta \varepsilon_R)$ does not have the same sign as $\Delta \varepsilon_R$, and so on. Furthermore, the input signals must be adequately filtered, according to the sampling period adopted (e.g., 2–5 sec).

As a concluding remark, the following provides a description of the overall operation, concerning frequency and active power:

- By means of the previsional scheduling adequately corrected in real time, a dispatching of the generated active powers is determined. This is defined, for all the units, by the *"base" values* P_{bi}.
- However, the actual operating situation may differ from the foreseen one because of different reasons, which may be classified into:
  (1) random fluctuations of the load having zero mean and which may be relatively fast;

(2) structural perturbations (particularly, connection or disconnection of a group of loads, disconnection of one or more units), which determine, through an initial discontinuity, a variation in the mean load and/or in the generation;

(3) slow variations of the mean load or of the generation (load rampings, etc.).

- The consequent imbalance between the driving powers and active powers generated by the individual units cause speed variations which, because of the synchronizing actions, tend to have (in relatively short time frames and assuming that there is no loss of synchronism) the same time behavior, which can be assimilated to that of the variation $\Delta\Omega(t)$ of the network frequency.
- With a good approximation it is then possible to write, disregarding the fastest phenomena for which the synchronizing transients should be accounted for, and disregarding the mechanical lost powers P_{pi}, that:

$$P_{mi} - P_{ei} = M_i \frac{\mathrm{d}\Omega}{\mathrm{d}t} \qquad [3.3.20]$$

  for all the units ($i = 1, 2, \ldots$). The sharing of the generated powers P_{ei} is then determined by the values of the *inertia coefficients* M_i and of the driving powers P_{mi} (which vary because of the control actions), through a consequent readjustment of the phase-shifts between the units.
- Furthermore, regarding the driving powers, the intervention of the primary control:
  - limits the transient frequency variations, particularly those caused by the slower fluctuations (1) and by the perturbations (2), by means of a sharing of the regulating powers (primary variations $\Delta P'_{ri}$ of the driving powers) substantially determined by the values of the *transient droops* b_{ti};
  - tends to lead the system, in response to the perturbations (2) or (3), to a quasi–steady-state situation, with $\Delta\Omega$ constant but nonzero, and a sharing of primary regulating powers determined by the values of the *permanent droops* b_{pi}.
- The much slower intervention of the secondary regulation, in response to the perturbations (2) or (3), finally results in the conditions $\Delta\Omega = 0$, $\Delta P'_{ri} = 0$, with a sharing of the regulating powers (secondary variations $\Delta P''_{ri}$ of the driving powers) determined by the values of the *participation factors* r_i.

With practical effects, the result of the frequency regulations usually can be considered satisfactory. For example, the frequency in Europe usually remains within the range of 49.95–50.05 Hz, with a relative deviation smaller than 10^{-3}, approximately 99% of the time.

The satisfactory operation of the secondary regulation nevertheless requires caution in a few situations.

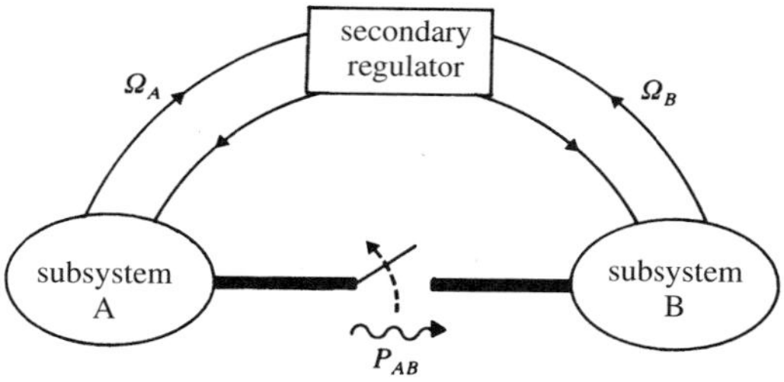

Figure 3.40. Example of conditions for the possible exclusion of secondary regulation in a part of the system.

If the disconnection of the system into two parts as indicated in Figure 3.40 might be possible, the secondary regulator should be sensitive to both the frequency values Ω_A and Ω_B (normally coincident, apart from small and fast slips). Let us assume, for example, that $P_{AB} > 0$ during the normal operation, so that the disconnection causes an increase in Ω_A and a decrease in Ω_B. If the secondary regulator were sensitive only to Ω_A, it would then command a reduction of the driving power for all the units, whereas the units in B should be commanded in the opposite sense. By measuring both frequencies Ω_A and Ω_B, the secondary regulation instead can be properly excluded, at least in one of the two subsystems, when such frequencies are very different from each other. In the above example, the intervention of the emergency control in the subsystem B also may occur, if the deficit of power were excessive (see Section 3.5).

Furthermore, it is necessary that the total regulating power P_R available for secondary regulation may vary within a sufficiently large range $[P_{R\,\min}, P_{R\,\max}]$, to efficiently face the variations of mean load (caused by forecasting errors), the unavailability of some units, etc. (otherwise a steady-state frequency error results, depending also on the primary regulation).

Therefore, also this range of variation (called *"secondary regulating power band"*) must be adequately scheduled (in the short-term and in real time, and in a measure possibly varying along the day). From the behavior of the "level" y_R, useful indications may be deduced for the updating of the values P_{bi}, r_i (as already seen) and of the above mentioned "band" and the set of operating units itself.

Usually, the various types of units contribute in differing measures to the secondary regulating power band. Typical relative values are as follows:

- for hydroelectric units:
  - with Pelton turbine: $P_{mi} = 0 - 100\%$;
  - with Francis turbine: $P_{mi} = 25 - 100\%$;
  - fluent water: $P_{mi} =$ constant (that is $\Delta P_{mi} = 0$);
- for thermal units:

- with traditional fuel:
  $P_{mi} = P^o_{mi} \pm 10\%$(that is $(\Delta P_{mi})_{\max} = 20\%$) for not large P^o_{mi}, e.g., $P^o_{mi} = 50\%$, $P_{mi} = P^o_{mi} \pm 5\%$(that is $(\Delta P_{mi})_{\max} = 10\%$) for large P^o_{mi}, e.g., $P^o_{mi} = 90\%$;
- nuclear units: $P_{mi} = 100\%$ (that is $\Delta P_{mi} = 0$), if the total nuclear generation is relatively modest.

3.3.3. Identification of the Power-Frequency Transfer Function

It has been noted that the function:

$$\frac{\Delta\Omega}{\Delta P_R - \Delta P_L}(s) \triangleq G_N(s) \triangleq \frac{1}{G_f(s) + G_g(s) + G_c(s) + sM} \qquad [3.3.21]$$

(see Equation [3.3.10]), named the *"power-frequency transfer function"* of the system, generally has a very high dynamic order, and can be known only approximately. In effect, the function $G_f(s) \triangleq \sum_i G_{fi}(s)$ references to the large number of units under primary regulation, and the sum $(G_g(s) + G_c(s) + sM)$ is made up of the contributions of all units (with their respective inertias and "natural" characteristics), the network, and all the loads.

Nevertheless, it also has been seen that it is usually possible to assume, within acceptable approximation, relationships like the following:

$$G_f(s) = \sum_i G_{fi}(0)\frac{1 + sT_{2i}}{1 + sT_{1i}}$$

with time constants T_{1i} of the same order of magnitude (e.g., 15–25 sec), and furthermore:

$$G_g(s) = G_g(0), \qquad G_c(s) = G_c(0) + sM_c^*$$

(where M_c^* represents the overall contribution of the electromechanical loads to the resulting inertia coefficient).

Assuming that $T_{1i} \cong T_1^m$ for all the units under primary regulation, it can be deduced, as if formally there were a single regulating unit:

$$G_f(s) \cong G_f(0)\frac{1 + sT_2^m}{1 + sT_1^m}$$

with $G_f(0) = \sum_i G_{fi}(0)$, $T_2^m \triangleq \sum_i G_{fi}(0)T_{2i}/G_f(0)$. Thus:

$$G_N(s) \cong \frac{1}{E}\frac{1 + sT_1^m}{1 + 2\zeta\dfrac{s}{\nu_o} + \dfrac{s^2}{\nu_o^2}} \qquad [3.3.22]$$

where:

$$E \triangleq G_f(0) + G_g(0) + G_c(0)$$

is the total (permanent) regulating energy of the system, whereas:

$$\begin{cases} \nu_o \triangleq \sqrt{\dfrac{E}{T_1^m (M + M_c^*)}} \\ \zeta \triangleq \dfrac{\nu_o}{2}\left(T_2^m + \dfrac{(G_g(0) \ + G_c(0))(T_1^m - T_2^m) + M + M_c^*}{G_f(0)}\right) \end{cases}$$

respectively define the resulting resonance frequency and damping factor, which are practically in the same order of magnitude as those defined in Section 3.1.2 when referring to a single unit (e.g., $\nu_o = 0.2–0.4$ rad/sec, $\zeta = 0.6–0.9$).

The adopted hypothesis might appear too simple; on the other hand, relatively precise estimations of $G_N(s)$ would introduce complexities that are unnecessary for practical purposes. It also must be considered that the linearized models of the individual equipment are susceptible to variations under operation, with a consequent dispersion in the parameter values, because of the different nonlinearities (particularly, the insensitivities in the valve positioning systems etc. may constitute an important cause of dispersion, just in the small variation range). Furthermore, it should be remembered that the definition itself, $G_N(s) \triangleq \Delta\Omega/(\Delta P_R - \Delta P_L)$, is based on the simplifying hypothesis of "coherency" between all the machines of the system.

In particular, it can then be presumed that the response of $\Delta\Omega$ to a step $(\Delta P_R - \Delta P_L)$ applied at the instant $t = 0$, exhibits an initial slope equal to:

$$\frac{\mathrm{d}\Delta\Omega}{\mathrm{d}t}(0^+) = \frac{\Delta P_R - \Delta P_L}{M + M_c^*} \qquad [3.3.23]$$

and then, if Equation [3.3.22] is acceptable, a single damped "dominant" oscillation, up to the final value:

$$\Delta\Omega(\infty) = \frac{\Delta P_R - \Delta P_L}{E} \qquad [3.3.24]$$

On the other hand, even a general (but sufficiently credible) knowledge of $G_N(s)$ appears to be appropriate, at least for:

- the synthesis of the secondary regulator, particularly for evaluating the possibility of increasing the gain K_o and, thus, the cutoff frequency (and the response speed) of the secondary regulation (recall Equations [3.3.14] and [3.3.15]);
- the diagnosis of emergency conditions and the choice of possible load shedding, according to Section 3.5.

The use of experimental tests may overcome the difficulties and uncertainties related to the updated knowledge of the large amount of data required; nevertheless, particular problems may arise, both in the organization of the tests and in the interpretation of their results.

Through *tests of the deterministic type* it is necessary to cause frequency variations that are clearly distinguishable from those caused by the casual perturbations under normal operation. Such tests may be realized, as specified in the following, by imposing a sudden disconnection of a power plant (or of the interconnection with other systems, assumed as the unique one), with the secondary regulator locked ($\Delta P_R = 0$). Sudden disconnections or connections of loads having sufficient power, would imply greater difficulties, whereas tests in response to ΔP_R (caused by a suitable variation in the level y_R) would require estimation of the actual behavior $\Delta P_R(t)$ itself (which might be quite slow, because of limits in the response speed of the plants involved).

Instead, the use of *statistical identification methods* does not imply any large programmed perturbation; such methods are based on the normal casual perturbations (which are assumed to be characterized in a statistical sense) and may even, theoretically, allow an on-line identification through continuous measuring and processing of the operating system data. However, the computations required are much more complex than those of the deterministic methods, and significant uncertainties may arise from inadequate knowledge of statistical properties of the load demands. Furthermore, the choice of long recording durations (e.g., hours) to reduce the truncation errors in statistical analysis, may be made useless by the variability of the system during the operation itself.

The effects of insensitivities may have a determinant weight in the $G_N(s)$ identified by statistical methods, usually based on very small fluctuations, whereas they may even scarcely affect the results obtained by deterministic methods. The two methods can, therefore, provide different information with an equally significant meaning in defining the system behavior when reacting to perturbations of different amplitude.

By tests based on a generating plant (or tie-line) disconnection, with the secondary regulator locked ($\Delta P_R = 0$), it is possible to verify with sufficient approximation that a $G_N(s)$ of the type like Equation [3.3.22] is usually acceptable, with reasonable values of parameters (particularly $M + M_c^*$, E, ν_0, ζ), notwithstanding the simplifications made for its deduction.

Usually it is convenient for the perturbation size to be some percent of the total nominal power P_{nom}, as in Figure 3.41 (e.g., if the resulting permanent droop ($P_{\text{nom}}/\Omega_{\text{nom}}E$) is 10% and the nominal frequency is 50 Hz, a perturbation equal to 2% of P_{nom} leads, at steady-state conditions, to a variation of 0.1 Hz).

Higher perturbation values might be undesirable, not only because of the disturbance to the system operation, but also for a significant identification of the linearized model. On the other hand, modest perturbations would be inadvisable, because of the difficulties in distinguishing their effects from those of accidental perturbations present in the system.

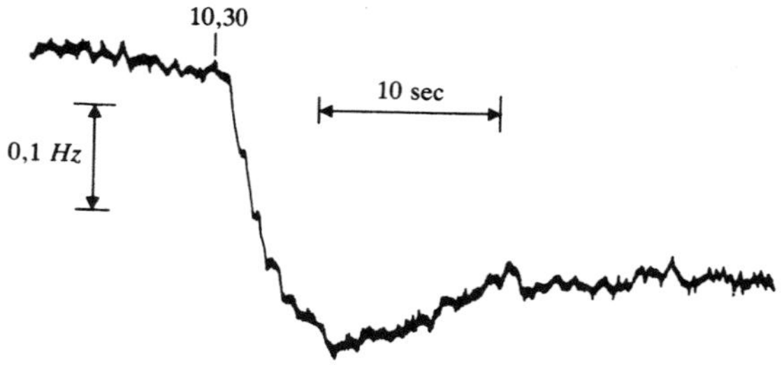

Figure 3.41. Italian electrical system under isolated operation and with secondary regulator locked; response of the network frequency to the trip of the S. Massenza power plant (at 10.30 a.m. of November, 9, 1964), with a predisturbance output of: $P = 300$ MW, $Q = 34.4$ MVAr, and with a total rotating power equal to approximately 13,000 MW (see reference 154).

However, some caution may be necessary to avoid large errors in the interpretation of the test results. In particular:

- During transient conditions, the measurable frequencies in the different network points have the same *average trend*, to which however, even ignoring the accidental perturbations, components different from point to point (caused by unavoidable slips between the machines) are superimposed. Therefore, the response $\Delta\Omega(t)$ must be evaluated as a significant estimate of only average trend (see also Section 8.2.5).
- The magnitude of the *actual perturbation* applied to the system must be suitably evaluated, considering the transients (of voltages and currents) produced in the network.

With reference to the latter consideration, a structural change in the network (corresponding, for example, to the disconnection here considered) causes a redistribution of voltages and currents, such that the active power perturbation ΔP_L actually produced may be quite different from the value of disconnected active power. Furthermore, the phenomena surely call into action the voltage regulations, so that $\Delta P_L(t)$ generally does not result exactly constant after the disconnection. In other words, the hypothesis that the actual perturbation has a step-wise behavior may be accepted only from a simplified point of view.

In effect, the results obtained assuming that ΔP_L is simply equal to the active power disconnected have often revealed themselves to be surprising ones, of doubtful interpretation: e.g., values of the resulting start-up time much larger—*even twice or more*—than the expected (which cannot reasonably be much larger than 8–10 sec, even with the contribution of the loads). Furthermore, the unreliability of the above hypothesis, and therefore of the consequent interpretations, has been confirmed by performing different subsequent tests at the same node and for the same system configuration, and by observing the large dispersion of the results.

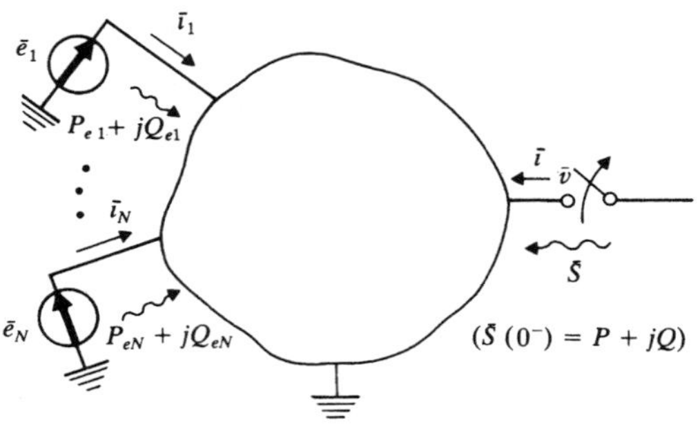

Figure 3.42. Representation of the electrical system for evaluating the total perturbation caused at the instant of disconnection.

To justify the above statements, it may be sufficient to examine the total perturbation $\Delta P_L(0^+)$ of active power on the machines, caused by the sudden disconnection (at $t = 0$) of power $P + jQ$ in accordance with Figure 3.42. To determine the response at $t = 0^+$, each synchronous machine can be reasonably represented by an emf. $\bar{e}_i$ of constant amplitude and phase, in series with a proper impedance, whereas it is possible to disregard the effects of the voltage regulation. By assuming that the above-mentioned emfs are connected to one another and to the node at which the disconnection is operated, by a linear passive network (including also loads), it is possible to write equations:

$$\begin{cases} \bar{\imath} = \overline{Y}_{oo}\bar{v} + \sum_1^N {}_k\overline{Y}_{ok}\bar{e}_k \\ \bar{\imath}_k = \overline{Y}_{ko}\bar{v} + \sum_1^N {}_i\overline{Y}_{ki}\bar{e}_i \quad (k = 1, \ldots, N) \end{cases}$$

with $\bar{\imath}(0^-) = (P - jQ)/\bar{v}^*(0^-)$, $\bar{\imath}(0^+) = 0$; it then follows:

$$\Delta\bar{v}(0^+) \triangleq \bar{v}(0^+) - \bar{v}(0^-) = \frac{P - jQ}{-\overline{Y}_{oo}\bar{v}^*(0^-)}$$

$$\Delta\bar{\imath}_k(0^+) \triangleq \bar{\imath}_k(0^+) - \bar{\imath}_k(0^-) = \overline{Y}_{ko}\Delta\bar{v}(0^+) = \frac{\overline{Y}_{ko}(P - jQ)}{-\overline{Y}_{oo}\bar{v}^*(0^-)} \quad (k = 1, \ldots, N)$$

and therefore:

$$\sum_1^N {}_k(\Delta P_{ek} + j\Delta Q_{ek})(0^+) = \sum_1^N {}_k\bar{e}_k\Delta\bar{\imath}_k^*(0^+) = \left(\sum_1^N {}_k\overline{Y}_{ko}^*\bar{e}_k\right)\frac{P + jQ}{-\overline{Y}_{oo}^*\bar{v}(0^-)}$$

By observing that:

$$\overline{v}(0^+) = \left(\sum_1^N {}_k\overline{Y}_{ok}\overline{e}_k\right)\frac{1}{-\overline{Y}_{oo}}$$

and posing, for brevity:

$$m\epsilon^{j\beta} \triangleq \frac{\sum_1^N {}_k\overline{Y}^*_{ko}\overline{e}_k}{\sum_1^N {}_k\overline{Y}_{ok}\overline{e}_k}\frac{\overline{Y}_{oo}}{\overline{Y}^*_{oo}} \quad [3.3.25]$$

it can be written:

$$\sum_1^N {}_k(\Delta P_{ek} + j\Delta Q_{ek})(0^+) = m\epsilon^{j\beta}(P' + jQ') \quad [3.3.26]$$

where $P' + jQ' \triangleq \overline{v}(0^+)\overline{\imath}^*(0^-)$, that is:

$$P' + jQ' = \frac{\overline{v}(0^+)}{\overline{v}(0^-)}(P + jQ) \quad [3.3.27]$$

or also, as $\overline{v}(0^+)/\overline{v}(0^-) = 1 + \Delta\overline{v}(0^+)/\overline{v}(0^-) = 1 + (P - jQ)/(-\overline{Y}_{oo}v(0^-)^2)$:

$$P' + jQ' = P + jQ + \frac{P^2 + Q^2}{-\overline{Y}_{oo}v(0^-)^2} \quad [3.3.28]$$

Finally, as $\Delta P_L(0^+) = \sum_1^N {}_k\Delta P_{ek}(0^+)$ (see also Figures 3.26 and 3.27), from Equation [3.3.26] it can be derived:

$$\Delta P_L(0^+) = m(P'\cos\beta - Q'\sin\beta) \quad [3.3.29]$$

or also, because of Equation [3.3.28]:

$$\Delta P_L(0^+) = m\left(P\cos\beta - Q\sin\beta + \sigma\frac{P^2 + Q^2}{v(0^-)^2}\right) \quad [3.3.30]$$

where $\sigma \triangleq \mathcal{R}e\left(\epsilon^{j\beta}/(-\overline{Y}_{oo})\right)$.

It is evident that $\Delta P_L(0^+)$ depends not only on P but also on Q (and on $v(0^-)$). The computation of the parameters m and β (and σ), from the given system configuration and operating point, and for the assigned disconnection node, may however not be trivial.

The difference $(\Delta P_L(0^+) - P)$ represents the variation of the resulting active power injected into the network (which includes also the loads). Therefore, if the variation in the losses is disregarded, such a difference constitutes the variation of the total active power absorbed by the loads, caused by the voltage variations on the different loads.

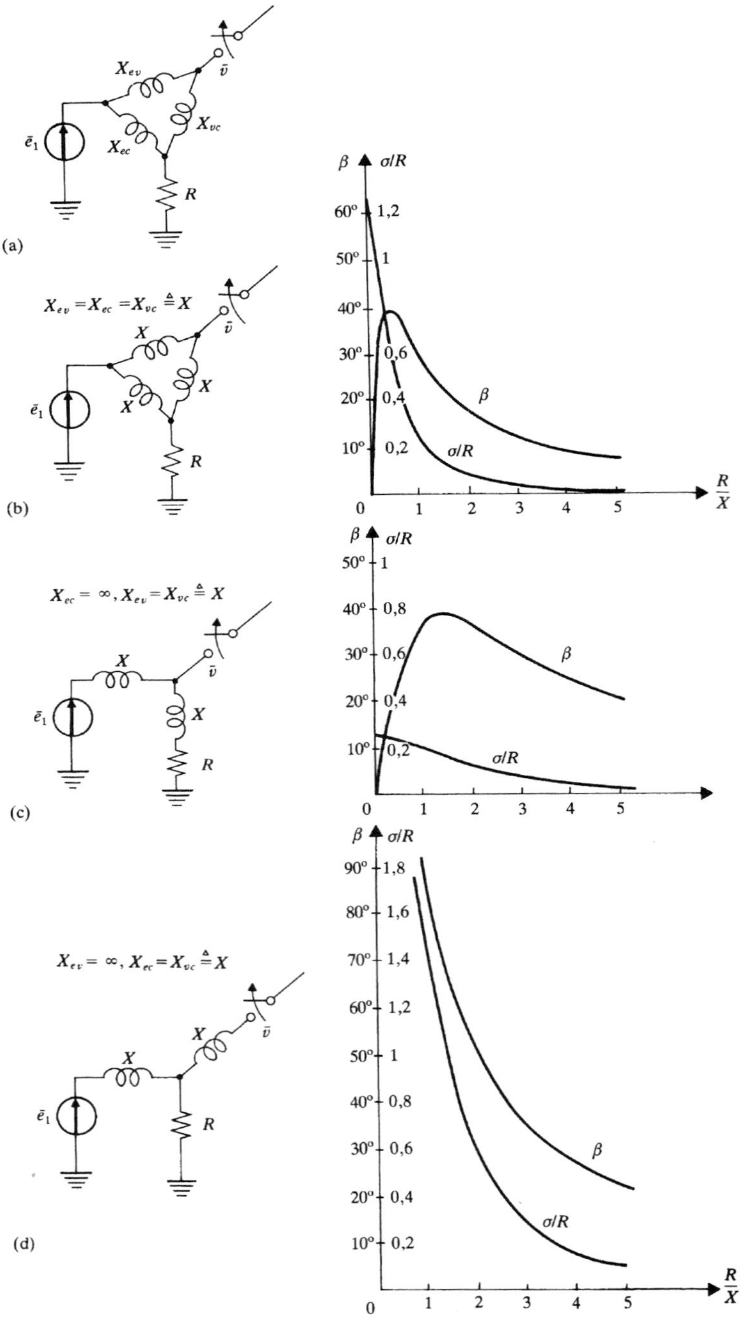

Figure 3.43. Deduction of the parameters β and σ in an elementary case (see text): (a) reference equivalent circuit; (b), (c), (d) particular examples.

In an illustrative way, some elementary numerical examples are reported in Figure 3.43, assuming that the system includes a single machine ($N = 1$) and a single load, with zero transmission losses. The diagrams of m are not reported as, for $N = 1$, it is simply $m = 1$ (as it can be deduced by Equation [3.3.25]). It is interesting to note that, by e.g., assuming:

$$\frac{Q}{P} = 0.75 \quad \text{(power factor equal to 0.8)}, \qquad \frac{P}{v(0^-)^2/R} = 0.05, \qquad \frac{R}{X} \cong 3$$

it follows $\Delta P_L(0^+)/P \cong 0.82$, $\Delta P_L(0^+)/P \cong 0.49$, and even $\Delta P_L(0^+)/P \cong 0.36$, respectively in the cases of Figures 3.43b,c,d.

On the other hand, from Equations [3.3.23] (with $\Delta P_R = 0$) and [3.3.29], it can be derived:

$$\frac{-\dfrac{\mathrm{d}\Delta\Omega}{\mathrm{d}t}(0^+)}{P'} = \frac{m}{M + M_c^*}\left(\cos\beta - \frac{Q'}{P'}\sin\beta\right) \qquad [3.3.31]$$

where P', Q' (defined by Equation [3.3.27]) and $(\mathrm{d}\Delta\Omega/\mathrm{d}t)(0^+)$ can be experimentally determined.

By performing two (or more) tests with different values of P, Q, it is possible to evaluate β and $(M + M_c^*)/m$, according to Figure 3.44 (assuming that m and β — or equivalently, because of Equation [3.3.25], the emfs $\overline{e}_1, \ldots, \overline{e}_N$ — do not vary from one test to another). The knowledge of β allows the determination of $\Delta P_L(0^+)/m$ for each test, based on Equation [3.3.29].

Then, for the identification of $(M + M_c^*)$ and $\Delta P_L(0^+)$, it is necessary to evaluate the parameter m. However, it is reasonable to expect that:

$$m \cong 1$$

as, in Equation [3.3.25],

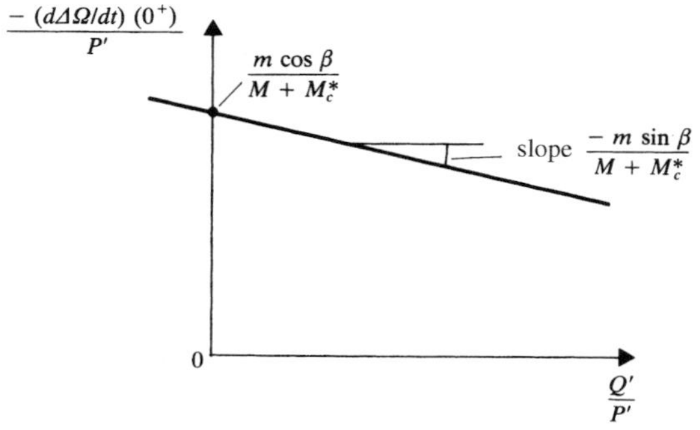

Figure 3.44. Characteristic to be identified (by at least two tests) for the experimental determination of β and $(M + M_c^*)/m$ (see text).

- the prevailing terms in the sums correspond to the $\overline{Y}_{ko}, \overline{Y}_{ok}$ having the largest magnitudes, and thus correspond to the emfs "closer" to the disconnection node (and relatively close to one another), the phases of which are normally not different;
- usually, the mutual admittances $\overline{Y}_{1o}, \ldots\ldots, \overline{Y}_{No}$ are predominantly reactive (i.e., with a dominant imaginary part), and furthermore, $\overline{Y}_{k0} = \overline{Y}_{0k}$.

In the limit case in which all emfs have the same phase, in the ratio $\left(\sum_1^N {}_k\overline{Y}^*_{ko}\overline{e}_k\right) / \left(\sum_1^N {}_k\overline{Y}_{ok}\overline{e}_k\right)$ the vectors $\overline{e}_k$ may be replaced by their respective magnitudes; therefore, with $\overline{Y}_{ko} = \overline{Y}_{ok}$, a ratio between complex conjugate numbers (as in the case $N = 1$) is obtained. In the limit case in which the admittances $\overline{Y}_{ko} = \overline{Y}_{ok}$ are purely imaginary, the ratio under examination is equal to -1. In both cases, then, it is $m = 1$.

As a consequence of the above simplification, it is then possible to derive, at a presumably acceptable approximation, the resulting inertia coefficient $(M + M_c^*)$ and the values assumed by the actual perturbation $\Delta P_L(0^+)$ in the different tests.

For the procedure described above, the following can be noted:

- The hypothesis that $m \cong 1$ proved acceptable (with an approximation even better than 1%) not only through the computation of m for test networks, based on Equation [3.3.25], but also through the application of the procedure itself to real systems, which led to reasonable values of $(M + M_c^*)$, i.e., of the resulting start-up time (e.g., 9–11 sec). Furthermore, the execution of more than two tests has confirmed the practical validity of the linear relationship of Equation [3.3.31], according to Figure 3.44.
- To evaluate P' and Q' it is necessary, by Equation [3.3.27], to measure $\overline{v}(0^+)/\overline{v}(0^-)$ in magnitude and phase. However, the powers P' and Q' may be usually substituted by the (disconnected) powers P and Q respectively, as in Equation [3.3.30] the last term is normally negligible. It then follows that $\Delta P_L(0^+) \cong P \cos\beta - Q \sin\beta = \mathcal{Re}\left(\epsilon^{j\beta}(P + jQ)\right)$, where $\epsilon^{j\beta}(P + jQ)$ is formally obtainable through a simple "rotation," of an angle β, of the complex power disconnected.
- Experimentally, values of β on the order of 10°–40° have been found. The ratio $\Delta P_L(0^+)/P$ may be then significantly smaller than unity if the disconnection is performed, as it often occurs, with $Q/P > 0$: e.g., $\Delta P_L(0^+)/P \cong$ 0.55–0.70, for $\beta = 30°$ and $Q/P = 0.4$–0.6. This may justify the excessive values of the resulting start-up time obtained by a priori assuming $\Delta P_L(0^+) = P$ (similarly, this provides an explanation for the consequent dispersion among the test results).
- The assumptions made at $t = 0^+$ may be considered acceptable for the first part of the transient (e.g., some seconds), whereas the subsequent part is influenced by voltage regulation. Disregarding the variation of the losses, it can be assumed that $\Delta P_L(t)$ tends, for increasing t, to a value closer to P, because the load voltages are regulated toward their initial values. This

must be considered, in particular, for the identification of the regulating energy E, based on Equation [3.3.24].

- In addition to providing a better identification of $G_N(s)$, the procedure also allows (through the knowledge of β, for the generic disconnection node and for significant operating points) more correct evaluation of the actual perturbation at the preventive stage, as when choosing the load shedding to be performed under emergency conditions (Section 3.5).

3.4. "f/P" CONTROL IN THE PRESENCE OF INTERCONNECTIONS

3.4.1. Preliminaries

In comparison with several systems (or "areas") in isolated operation, the realization of a set of interconnected areas may lead to considerable benefits in holding the frequency within narrow limits. This is valid not only for normal operating conditions, for the reduced incidence of load perturbations related to the total power of the system and for the tendency to compensate each other statistically (see Section 3.3.1), but also for a power "deficit" in an area, possibly caused by the disconnection of several generators, when aid from other areas via interconnections becomes essential. (Refer to Sections 1.3.3. and 2.3.4. for information on advantages offered by interconnections.)

In addition to regulating frequency, it is necessary to regulate, according to schedules, the power exchanged between the different areas, i.e., the powers P_{ij}, with $i, j = 1, \ldots, n, i \neq j$, assuming that:

- $n =$ number of areas;
- $P_{ij} =$ total active power supplied by area i to area j, corresponding to the boundary nodes between the two areas (it then results $P_{ji} = -P_{ij}$).

The number of exchanged powers to be regulated depends on the topology of the overall system, and is within $(n-1)$ and $n(n-1)/2$; this latter value corresponds to the case in which each area is connected to all the others. However, for the reasons in Section 2.3.4 (recall for instance Figure 2.34), it may be assumed that the steady-state values of the exchanged powers essentially depend only on $(n-1)$ degrees of freedom.

On the other hand, out of the n powers $P_{Ei} = \sum_{j \neq i} P_{ij} (i = 1, \ldots, n)$ in total exported by the individual areas, only $(n-1)$ can be independently assigned as:

$$\sum_{1}^{n}{}_{i} P_{Ei} = 0 \qquad [3.4.1]$$

and thus any one of the P_{Ei} can be derived from the others.

Therefore, it may be sufficient to impose the regulation of n variables (as many as the areas), i.e., *the frequency and* $(n-1)$ *exported powers*. The possible correction of the individual P_{ij}'s, at equal sum P_{Ei}, may be left to local actions in each area ($i = 1, \ldots, n$) through suitable adjustments to the operating point (particularly to the v/Q regime).

The frequencies $\Omega_1, \ldots, \Omega_n$ in different areas also may differ from one another in transient conditions, and become equal at steady-state because of the synchronizing actions. The $(n-1)$ transient differences between the frequencies are in such a case dynamically related to the exported powers (see also Figure 3.47 and Equation [3.4.16]), and the number of variables to be regulated still remains, in all, equal to n.

Section 3.3.2 described how the frequency regulation in an isolated area may be entrusted to the secondary regulator, characterized by an integral action on the frequency error. To regulate the frequency and exported powers with n interconnected areas, it is sufficient for the secondary regulator of each area to be sensitive with an integral action, to a linear combination of the errors of frequency and of power exported from the area itself. By doing so, under steady-state conditions (and for given frequency and exported power references), n conditions are achieved, as:

$$K_i \Delta\Omega + K_i^* \Delta P_{Ei} = 0 \qquad (i = 1, \ldots, n) \qquad [3.4.2]$$

from which (recall Equation [3.4.1]) $\left(\sum_1^n {}_i K_i/K_i^*\right) \Delta\Omega = 0$, i.e.,:

$$\Delta\Omega = 0$$

and thus, also:

$$\Delta P_{Ei} = 0 \quad \forall i = 1, \ldots, n$$

as desired.

The same result is obtained if a number of secondary regulators, but not all, are sensitive only to the exported power error (i.e., if for them it is $K_i = 0$), which is, for example, adopted for small areas connected to one or more relatively powerful areas.
Moreover, the secondary regulator of only one area may be sensitive only to the frequency error ($K_i^* = 0$), to realize $\Delta\Omega = 0$.
Generally, the generic secondary regulator might be sensitive to the power exported errors of other areas, but such a solution does not appear to be convenient (nor strictly desirable) because of the required complexity of the communications system.

Under general dynamic operating conditions, considering that the frequency may transiently vary between areas, the dependence of the secondary regulating powers (P_{Ri}) on the respective local frequency errors ($\varepsilon_{Fi} \triangleq \Omega_{\text{RIF}\,i} - \Omega_i$) and exported power errors ($\varepsilon_{Pi} \triangleq P_{E\,\text{RIF}\,i} - P_{Ei}$) may be expressed, for small variations, by equations (see Fig. 3.45a):

$$\Delta P_{Ri} = G_{Ri}(s)\Delta\varepsilon_{Fi} + G_{Ri}^*(s)\Delta\varepsilon_{Pi} \qquad (i = 1, \ldots, n) \qquad [3.4.3]$$

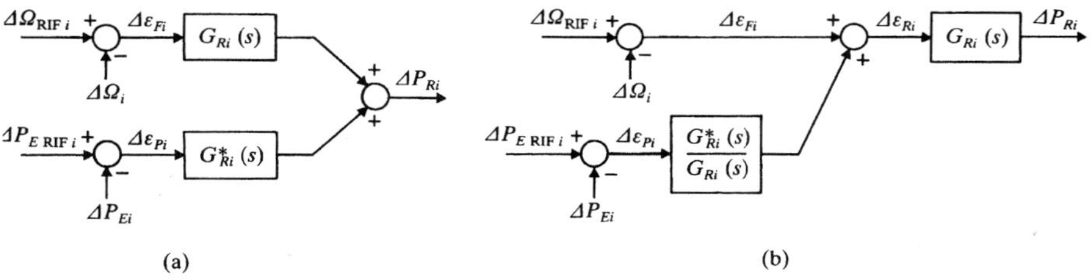

Figure 3.45. Dependence of the secondary regulating powers (P_{Ri}) on the errors of frequency (ε_{Fi}) and exported power (ε_{Pi}) in more interconnected areas: (a) block diagram for small variations; (b) equivalent diagram for $G_{Ri}(s) \neq 0$.

where $\Omega_{\mathrm{RIF}\,i}$, $P_{E\mathrm{RIF}\,i}$ are the references of frequency and exported power for the generic area i, and $G_{Ri}(s)$, $G^*_{Ri}(s)$ are suitable transfer functions, having in common a pole at the origin (in Equation [3.4.2] it may be then written that $K_i \triangleq \lim_{s\to 0}(sG_{Ri}(s))$, $K_i^* \triangleq \lim_{s\to 0}(sG^*_{Ri}(s))$).

For $G_{Ri}(s) \neq 0$, Equation [3.4.3] also may be written (see Fig. 3.45b):

$$\Delta P_{Ri} = G_{Ri}(s)\Delta\varepsilon_{Ri} \tag{3.4.4}$$

where:

$$\Delta\varepsilon_{Ri} \triangleq \Delta\varepsilon_{Fi} + \frac{G^*_{Ri}(s)}{G_{Ri}(s)}\Delta\varepsilon_{Pi} \tag{3.4.5}$$

constitutes the so-called *"network error,"* with the dimension of a frequency. Alternatively, it is possible to derive the error $(G_{Ri}(s)/G^*_{Ri}(s))\Delta\varepsilon_{Fi} + \Delta\varepsilon_{Pi}$, which has the dimension of a power and is usually named, by a term similar to the previous one, *"area control error,"* i.e., ACE.

The simplicity of the solution described is evident; the problem of the regulation of n variables is solved through n *"local" regulators*, i.e., each sensitive to quantities measurable in the respective area. In this regard, it must be noted that—under suitable assumptions which are generally satisfied—the use of local regulators is the only situation that permits complete "autonomy" and/or "noninteraction," according to the following sections. The measurements of the powers exchanged with other networks must be communicated, and algebraically added to obtain the signal of exported power P_{Ei} to be sent to the regulator.

The solution under examination may imply a redundant number of references (generically: n frequency references $\Omega_{\mathrm{RIF}\,i}$, and n power references $P_{E\,\mathrm{RIF}\,i}$), which should ideally satisfy, at each instant, the obvious conditions:

$$\left.\begin{aligned} &\Omega_{\mathrm{RIF}\,1} = \ldots = \Omega_{\mathrm{RIF}\,n} \\ &\sum_1^n {}_i P_{E\,\mathrm{RIF}\,i} = 0 \end{aligned}\right\} \tag{3.4.6}$$

If conditions [3.4.6] are not verified because of variations in some references, one may consider, instead of the set of real references, a set of *"ideal" references* (satisfying conditions [3.4.6]):

$$\Omega^o_{\text{RIF}},\ P^o_{E\,\text{RIF}\,1}, \ldots,\ P^o_{E\,\text{RIF}\,n}$$

equivalent to the set of real references (at least for small variations) in the effects on the P_{Ri}'s.

In fact, based on Equation [3.4.3], it can be imposed (omitting the indication of variable s):

$$\begin{aligned}\Delta P_{Ri} &= G_{Ri}(\Delta\Omega_{\text{RIF}\,i} - \Delta\Omega_i) + G^*_{Ri}(\Delta P_{E\,\text{RIF}\,i} - \Delta P_{Ei}) \\ &= G_{Ri}(\Delta\Omega^o_{\text{RIF}} - \Delta\Omega_i) + G^*_{Ri}(\Delta P^o_{E\,\text{RIF}\,i} - \Delta P_{Ei})\end{aligned} \qquad [3.4.7]$$

i.e.,

$$\frac{G_{Ri}}{G^*_{Ri}}\Delta\Omega_{\text{RIF}\,i} + \Delta P_{E\,\text{RIF}\,i} = \frac{G_{Ri}}{G^*_{Ri}}\Delta\Omega^o_{\text{RIF}} + \Delta P^o_{E\,\text{RIF}\,i}$$

for $i = 1, \ldots, n$; so that, adding up (and imposing $\sum_1^n {}_i P^o_{E\,\text{RIF}\,i} = 0$):

$$\Delta\Omega^o_{\text{RIF}} = \frac{\displaystyle\sum_1^n {}_k \left(\frac{G_{Rk}}{G^*_{Rk}}\Delta\Omega_{\text{RIF}\,k} + \Delta P_{E\,\text{RIF}\,k}\right)}{\displaystyle\sum_1^n {}_k \frac{G_{Rk}}{G^*_{Rk}}} \qquad [3.4.8]$$

whereas:

$$\Delta P^o_{E\,\text{RIF}\,i} = \Delta P_{E\,\text{RIF}\,i} + \frac{G_{Ri}}{G^*_{Ri}}\left(\Delta\Omega_{\text{RIF}\,i} - \Delta\Omega^o_{\text{RIF}}\right) \qquad [3.4.9]$$

The effect (on ΔP_{Ri}) of small variations $\Delta\Omega_{\text{RIF}\,i}$, $\Delta P_{E\,\text{RIF}\,i}$ of the (real) references of the regulators may be evaluated starting from the variations $\Delta\Omega^o_{\text{RIF}\,i}$, $\Delta P^o_{E\,\text{RIF}\,i}$ defined above, according to the block diagram in Figure 3.46.

(The quantities $\Delta\varepsilon^o_{Fi} \triangleq \Delta\Omega^o_{\text{RIF}} - \Delta\Omega_i$, $\Delta\varepsilon^o_{Pi} \triangleq \Delta P^o_{E\,\text{RIF}\,i} - \Delta P_{Ei}$ may be different from $\Delta\varepsilon_{Fi}$, $\Delta\varepsilon_{Pi}$, even if the effects on ΔP_{Ri} remain unchanged.)

If the secondary regulator of one area (e.g., area 1) is sensitive to the frequency error only, i.e., if $G^*_{R1} = 0$, we simply have:

$$\Delta\Omega^o_{\text{RIF}} = \Delta\Omega_{\text{RIF}1}$$

whereas:

$$\Delta P^o_{E\,\text{RIF}\,j} = \Delta P_{E\,\text{RIF}\,j} + \frac{G_{Rj}}{G^*_{Rj}}(\Delta\Omega_{\text{RIF}\,j} - \Delta\Omega^o_{\text{RIF}}) \quad (j = 2, \ldots, n)$$

$$\Delta P^o_{E\,\text{RIF}1} = -\sum_2^n {}_j \Delta P^o_{E\,\text{RIF}\,j}$$

Specifically, if $G_{R2} = \ldots = G_{Rn} = 0$ (i.e., if the areas $2, \ldots, n$ are only under power regulation, whereas area 1 is only under frequency regulation), we have:

$$\Delta P^o_{E\,\text{RIF}\,j} = \Delta P_{E\,\text{RIF}\,j} \quad (j = 2, \ldots, n) \qquad [3.4.10]$$

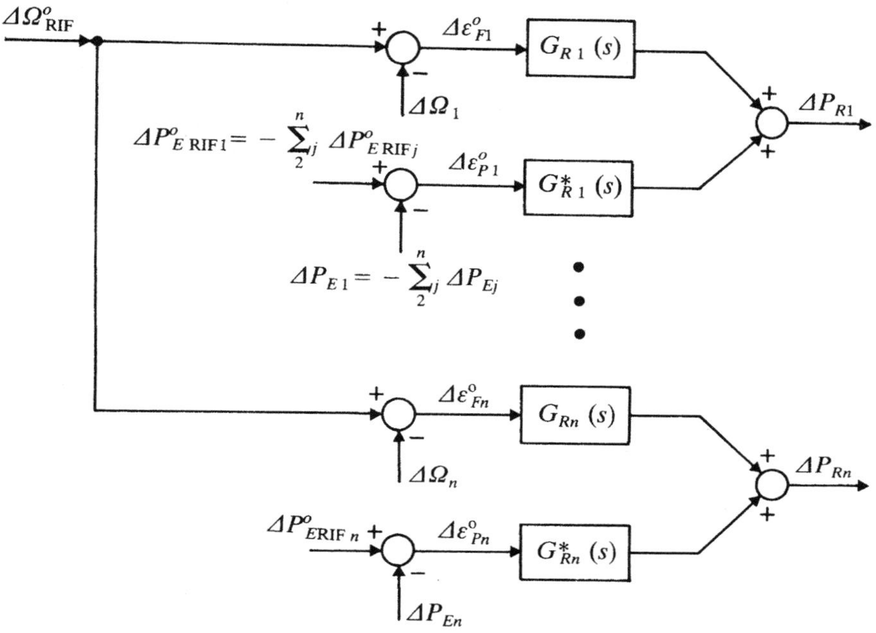

Figure 3.46. Definition of "ideal" references.

i.e., any possible variations $\Delta P_{E\,\mathrm{RIF}\,j}$ of the (exported power) references in the regulators $2, \ldots, n$ assume a clear and unequivocal meaning, since they result equal, respectively, to the equivalent "ideal" variations $\Delta P^o_{E\,\mathrm{RIF}\,j}$ to be considered.

To derive the block diagram of the secondary regulation, note that, for the generic area i, the variations of the exported power ΔP_{Ei} have an effect similar to the load variations ΔP_{Li}. Therefore it is possible to write (disregarding for simplicity the variations of network losses, related to $\Delta P_{E1}, \ldots, \Delta P_{En}$):

$$\Delta P'_{Ri} \triangleq \Delta P_{Ri} - \Delta P_{Li} = \frac{1}{G_{Ni}(s)} \Delta \Omega_i + \Delta P_{Ei} \qquad [3.4.11]$$

where $G_{Ni}(s)$ is the power-frequency transfer function of the considered area (recall Equation [3.3.21]).

Moreover, within a sufficient approximation, it is possible to assume (as already noted) that the exported power variations depend on the phase-shift variations between the voltages at the n "area nodes" (see Section 2.3.4 and Fig. 2.34). Therefore, the following equations may be written:

$$\Delta P_{Ei} = \sum_{1}^{n}{}_h \frac{k_{ih} \Delta \Omega_h}{s} \qquad (i = 1, \ldots, n) \qquad [3.4.12]$$

with:

$$\sum_{1}^{n}{}_{h} k_{ih} = 0 \quad \forall i = 1, \ldots, n \tag{3.4.13}$$

and in addition, because of Equation [3.4.1]:

$$\sum_{1}^{n}{}_{i} k_{ih} = 0 \quad \forall h = 1, \ldots, n \tag{3.4.14}$$

(in the $\{k_{ih}\}$ matrix, the sum of the elements of each row is zero, as is that of each column).

Equation [3.4.12] may be replaced by:

$$\Delta P_{Ej} = \sum_{2}^{n}{}_{h} \frac{k_{jh}(\Delta\Omega_h - \Delta\Omega_1)}{s} \quad (j = 2, \ldots, n) \tag{3.4.15}$$

or in matrix form:

$$\begin{bmatrix} \Delta P_{E2} \\ \vdots \\ \Delta P_{En} \end{bmatrix} = \frac{1}{s}[\overline{k}] \begin{bmatrix} \Delta\Omega_2 - \Delta\Omega_1 \\ \vdots \\ \Delta\Omega_n - \Delta\Omega_1 \end{bmatrix} \tag{3.4.16}$$

with:

$$[\overline{k}] \triangleq \begin{bmatrix} k_{22} & \cdots & k_{2n} \\ \vdots & & \vdots \\ k_{n2} & \cdots & k_{nn} \end{bmatrix}$$

and moreover:

$$\Delta P_{E1} = -\sum_{2}^{n}{}_{j} \Delta P_{Ej} \tag{3.4.17}$$

From the present equations and from Equation [3.4.3], it is possible to derive the block diagram of Figure 3.47. Such a diagram remains valid, based on the information presented above, also when replacing all the $\Delta\Omega_{\text{RIF}i}$ by $\Delta\Omega^o_{\text{RIF}}$, and $\Delta P_{E\,\text{RIF}i}$ by $\Delta P^o_{E\,\text{RIF}i}$, and thus also $\Delta\varepsilon_{Fi}$, $\Delta\varepsilon_{Pi}$, respectively, by $\Delta\varepsilon^o_{Fi}$, $\Delta\varepsilon^o_{Pi}$ $(i = 1, \ldots, n)$.

Generally, it should be considered that the variations $\Delta P_{E1}, \ldots, \Delta P_{En}$ are accompanied, in different areas, by variations in losses. The loss variations, with the model defined in Figure 2.34, depend on the phase-shift variations between the voltages at area nodes, and may be expressed as a function of $\Delta P_{E2}, \ldots, \Delta P_{En}$. Therefore, if loss variations are not negligible, a linear combination of all the $\Delta P_{E2}, \ldots, \Delta P_{En}$ should be added (for each $i = 1, \ldots, n$) to the right-hand side of Equation [3.4.11].

Usually, the transient differences between the frequencies $\Omega_1, \ldots, \Omega_n$ measurable in the different areas, attenuate so rapidly with respect to the phenomena

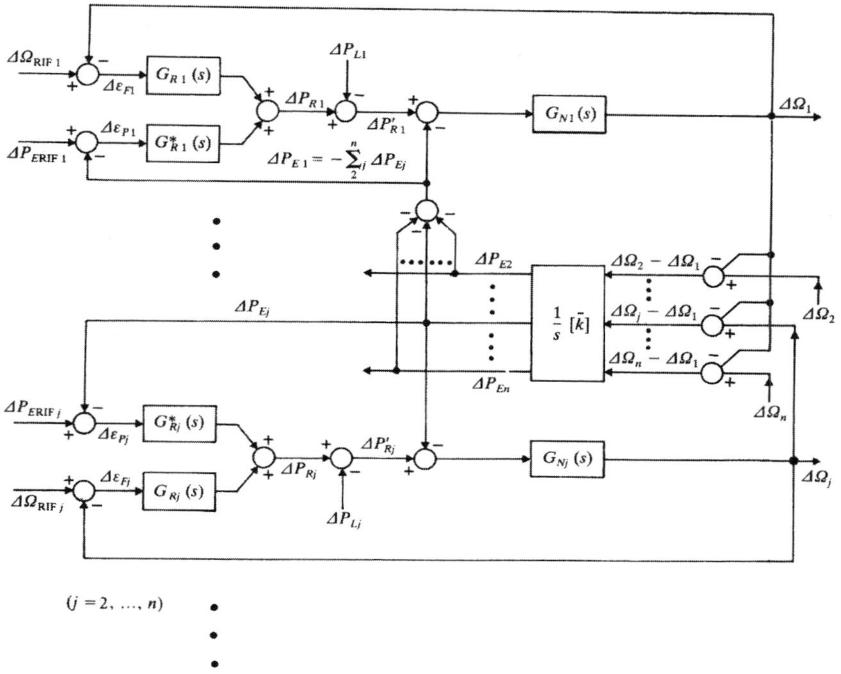

Figure 3.47. Regulation of frequency and exported powers: overall block diagram for small variations.

involving the secondary regulation, that they can be disregarded. Hence, by assuming $\Omega_1 = \ldots = \Omega_n \triangleq \Omega$, Equations [3.4.11] and [3.4.12] are replaced by the following much simpler equations:

$$\Delta P'_{Ri} \triangleq \Delta P_{Ri} - \Delta P_{Li} = \frac{1}{G_{Ni}(s)}\Delta\Omega + \Delta P_{Ei} \qquad (i = 1, \ldots, n) \qquad [3.4.18]$$

By recalling Equation [3.4.1] it can be derived:

$$\sum_{1}^{n} {}_i \Delta P'_{Ri} \triangleq \sum_{1}^{n} {}_i (\Delta P_{Ri} - \Delta P_{Li}) = \left(\sum_{1}^{n} {}_i \frac{1}{G_{Ni}(s)}\right)\Delta\Omega$$

and thus:

$$\Delta\Omega = G_N(s)\sum_{1}^{n} {}_k \Delta P'_{Rk} = G_N(s)\sum_{1}^{n} {}_k (\Delta P_{Rk} - \Delta P_{Lk}) \qquad [3.4.19]$$

$$\Delta P_{Ei} = \Delta P'_{Ri} - \frac{1}{G_{Ni}(s)}\Delta\Omega = \left(1 - \frac{G_N}{G_{Ni}}\right)\Delta P'_{Ri} - \frac{G_N}{G_{Ni}}\sum_{1}^{n} {}_{j\neq i} \Delta P'_{Rj}$$

[3.4.20]

where:

$$G_N(s) \triangleq \left(\sum_1^n {}_i \frac{1}{G_{Ni}(s)}\right)^{-1} \qquad [3.4.21]$$

constitutes the power-frequency transfer function of the whole system, seen as a single area.

In a matrix form, Equations [3.4.19] and [3.4.20] lead to:

$$\begin{bmatrix} \Delta\Omega \\ \Delta P_{E2} \\ \vdots \\ \Delta P_{En} \end{bmatrix} = \begin{bmatrix} G_N & G_N & \cdots & G_N \\ \dfrac{-G_N}{G_{N2}} & \left(1 - \dfrac{G_N}{G_{N2}}\right) & \cdots & \dfrac{-G_N}{G_{N2}} \\ \vdots & \vdots & & \vdots \\ \dfrac{-G_N}{G_{Nn}} & \dfrac{-G_N}{G_{Nn}} & \cdots & \left(1 - \dfrac{G_N}{G_{Nn}}\right) \end{bmatrix} \begin{bmatrix} \Delta P'_{R1} \\ \Delta P'_{R2} \\ \vdots \\ \Delta P'_{Rn} \end{bmatrix} \qquad [3.4.22]$$

where $\Delta P'_{Ri} \triangleq \Delta P_{Ri} - \Delta P_{Li}$ $(i = 1, \ldots, n)$, whereas $\Delta P_{E1} = -\sum_2^n {}_j \Delta P_{Ej}$.

Similarly, from Equation [3.4.7] it is possible to derive, in terms of ideal references:

$$\begin{bmatrix} \Delta P_{R1} \\ \Delta P_{R2} \\ \vdots \\ \Delta P_{Rn} \end{bmatrix} = \begin{bmatrix} G_{R1} & -G^*_{R1} & \cdots & -G^*_{R1} \\ G_{R2} & G^*_{R2} & \cdots & 0 \\ \vdots & \vdots & & \vdots \\ G_{Rn} & 0 & \cdots & G^*_{Rn} \end{bmatrix} \begin{bmatrix} \Delta\varepsilon^o_F \\ \Delta\varepsilon^o_{P2} \\ \vdots \\ \Delta\varepsilon^o_{Pn} \end{bmatrix} \qquad [3.4.23]$$

where:

$$\Delta\varepsilon^o_F \triangleq \Delta\Omega^o_{\text{RIF}} - \Delta\Omega,\ \Delta\varepsilon^o_{Pj} \triangleq \Delta P^o_{E\,\text{RIF}\,j} - \Delta P_{Ej}\,(j = 2, \ldots, n)$$

and

$$\Delta\varepsilon^o_{P1} = -\sum_2^n {}_j \Delta\varepsilon^o_{Pj}.$$

It is possible to associate the block diagram of Figure 3.48 to these matrix equations, assuming that the matrices S (corresponding to the system to be regulated) and R (corresponding to the regulators) are respectively those specified in Equations [3.4.22] and [3.4.23]. Through trivial developments, it follows that:

$$\begin{bmatrix} \Delta\Omega \\ \Delta P_{E2} \\ \vdots \\ \Delta P_{En} \end{bmatrix} = [I_{(n)} + SR]^{-1} S \left(R \begin{bmatrix} \Delta\Omega^o_{\text{RIF}} \\ \Delta P^o_{E\,\text{RIF2}} \\ \vdots \\ \Delta P^o_{E\,\text{RIF}\,n} \end{bmatrix} - \begin{bmatrix} \Delta P_{L1} \\ \Delta P_{L2} \\ \vdots \\ \Delta P_{Ln} \end{bmatrix} \right) \qquad [3.4.24]$$

It also results:

$$\left[I_{(n)} + SR\right]^{-1} S = \left[S^{-1} + R\right]^{-1} \qquad [3.4.25]$$

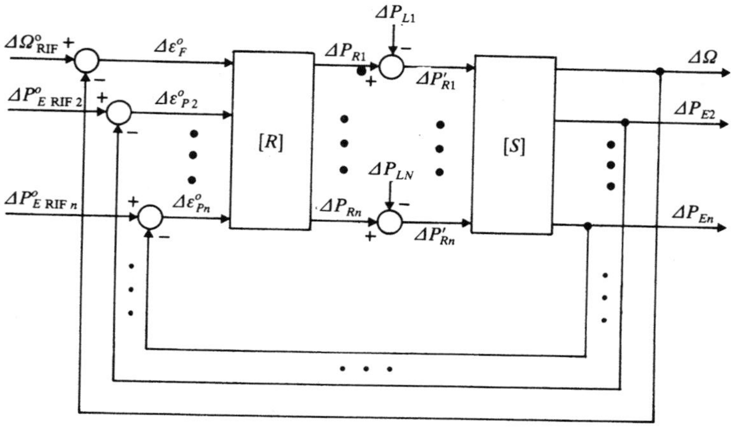

Figure 3.48. Block diagram (with "ideal" references) equivalent to that in Figure 3.47, under the hypothesis of equal frequency in all the areas.

where the matrix S^{-1}, which is directly obtained from Equation [3.4.18], is given by:

$$S^{-1} = \begin{bmatrix} \dfrac{1}{G_{N1}} & -1 & \cdots & -1 \\ \dfrac{1}{G_{N2}} & 1 & \cdots & 0 \\ \vdots & \vdots & & \vdots \\ \dfrac{1}{G_{Nn}} & 0 & \cdots & 1 \end{bmatrix} \qquad [3.4.26]$$

and thus has the same zero elements of the matrix R obtained by using "local" regulators (see Equation [3.4.23]).

However, this last property (the interest in which will be underlined later) no longer holds if the variation of the losses in different areas, consequent to the exported power variations, must be considered. In such a case, it is necessary to add, in the right-hand side of Equation [3.4.18] (as already pointed out), a linear combination of all the $\Delta P_{E2}, \ldots, \Delta P_{En}$, so that the matrix S^{-1} may become a "full" matrix.

3.4.2. Response to Disturbances and Autonomy Criterion

With the response to the disturbances $\Delta P_{L1}, \ldots, \Delta P_{Ln}$, from the block diagram of Figure 3.47 (or alternatively from that of Fig. 3.48, under the hypothesis $\Delta\Omega_1 = \ldots = \Delta\Omega_n \triangleq \Delta\Omega$), it appears evident that each ΔP_{Ri} may depend, transiently, not only on the respective ΔP_{Li}, but also on the $(\Delta P_{Lj})_{j\neq i}$ relative to other areas. By the examination of Figure 3.47, it is easy to determine that such an interdependency is realized through the variation ΔP_{Ei} caused by the interconnected operation. (At steady-state, the regulation leads instead to $\Delta P_{Ei} = 0$, so that P_{Ri} is no longer influenced by the perturbations in other areas.)

A rather usual criterion adopted in the synthesis of the secondary regulation avoids such interdependency, even if transient, simply by imposing that the secondary regulating power P_{Ri} be unaffected by the variations of the respective exported power P_{Ei}. Since from Equations [3.4.3] and [3.4.11], or (more directly) from the figure itself, it is possible to derive:

$$\Delta P_{Ri} = \frac{G_{Ri}\Delta\Omega_{\text{RIF}\,i} + G^*_{Ri}\Delta P_{E\,\text{RIF}\,i} + G_{Ri}G_{Ni}\Delta P_{Li} + (G_{Ri}G_{Ni} - G^*_{Ri})\Delta P_{Ei}}{1 + G_{Ri}G_{Ni}}$$

thus the above criterion leads to the condition $G_{Ri}G_{Ni} - G^*_{Ri} = 0$, i.e.,

$$\frac{G^*_{Ri}}{G_{Ri}} = G_{Ni} \qquad [3.4.27]$$

For the generic area into which such a condition is adopted, the (single loop) block diagram of Figure 3.49 can be then derived, with:

$$\frac{\partial P_{Ri}}{\partial P_{Li}} = \frac{G_{Ni}G_{Ri}}{1 + G_{Ni}G_{Ri}} = \frac{G^*_{Ri}}{1 + G^*_{Ri}} \qquad [3.4.28]$$

whereas, as desired:

$$\frac{\partial P_{Ri}}{\partial P_{Lj}} = 0 \quad \forall j \neq i \qquad [3.4.29]$$

Note that, due to Equation [3.4.9], in the block diagram $\Delta\Omega_{\text{RIF}\,i} + G_{Ni}\Delta P_{E\,\text{RIF}\,i} = \Delta\Omega^o_{\text{RIF}} + G_{Ni}\Delta P^o_{E\,\text{RIF}\,i}$, $\Delta\varepsilon_{Ri} = \Delta\varepsilon^o_{Fi} + G_{Ni}\Delta\varepsilon^o_{Pi}$.
Furthermore, if condition [3.4.27] is realized in all the areas, the overall block diagram is constituted by n single loops of the type indicated, not interacting one another. The single $\Delta\Omega_i$ may be derived starting from $\Delta\Omega_i + G_{Ni}\Delta P_{Ei} = G_{Ni}\Delta P'_{Ri}$, based on the equation (deducible from Equations [3.4.11] and [3.4.12]):

$$\begin{bmatrix}\Delta\Omega_1\\ \vdots \\ \Delta\Omega_n\end{bmatrix} = \left(I_{(n)} + \begin{bmatrix} G_{N1} & \cdots & 0 \\ \vdots & & \vdots \\ 0 & \cdots & G_{Nn}\end{bmatrix}\begin{bmatrix} k_{11} & \cdots & k_{1n} \\ \vdots & & \vdots \\ k_{n1} & \cdots & k_{nn}\end{bmatrix}\frac{1}{s}\right)^{-1}\begin{bmatrix} G_{N1}\Delta P'_{R1} \\ \vdots \\ G_{Nn}\Delta P'_{Rn}\end{bmatrix}$$

whereas the ΔP_{ei}'s are given by Equation [3.4.12]. If it is assumed that $\Omega_1 = \ldots = \Omega_n \triangleq \Omega$, these last equations have instead to be replaced by Equations [3.4.19] and [3.4.20].

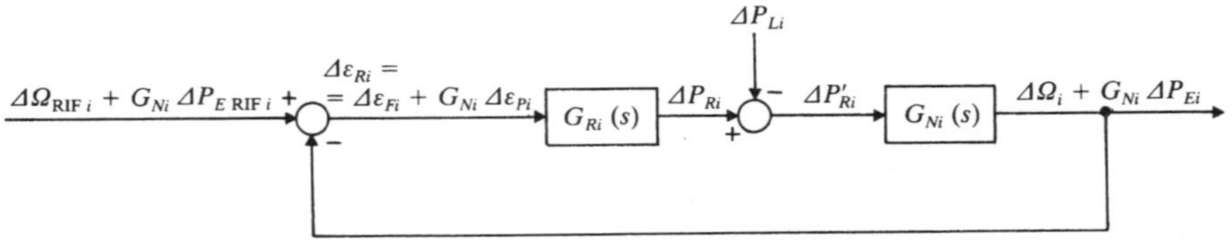

Figure 3.49. Equivalent regulation loop under the "autonomy" condition.

For each area, the condition [3.4.27] may be regarded as an *"autonomy"* condition of the respective secondary regulation, in the sense that the area regulator produces variations in the secondary regulating power only if a load variation in its own area (or a variation in its own references) has occurred, ignoring variations in other areas. However, the primary regulation intervenes, because of frequency variations, in all the areas, regardless of where the load variation has occurred.

It is important to underline what follows:

- The adoption (or not) of the autonomy criterion in a given area may be decided independently of what is done in the other areas.
- The adoption of the autonomy criterion in a given area leads to a "local" regulator, i.e., a regulator sensitive only to the errors of frequency and exported power relative to the area itself (this holds if the effects of the loss variations related to the exported power variations are disregarded).
- *The loop in Figure* 3.49 *is similar to that in Figure* 3.36, concerning the secondary frequency regulation in an isolated area. As a consequence, all considerations detailed in Section 3.3.2, relative to the synthesis of the secondary regulator may be fully applied to the present case, as far as the transfer function $G_{Ri}(s)$ is concerned, whereas condition [3.4.27] then allows the derivation of function $G^*_{Ri}(s)$ in correspondence to the exported power error.
- To the advantage of having a "local" regulator, it is thus added that of having to solve a synthesis problem independent of the rest of the system, since the required information regards only the respective power-frequency transfer function $G_{Ni}(s)$.

In this last regard, normally one is satisfied with approximating $G_{Ni}(s)$ with its static gain $G_{Ni}(0) = 1/E_i$ (where E_i is the total regulating energy of the i-th considered area), not only in the synthesis of $G_{Ri}(s)$ but also in that of $G^*_{Ri}(s)$, by simply assuming:

$$G^*_{Ri}(s) = \frac{G_{Ri}(s)}{E_i} \qquad [3.4.30]$$

Based on Figure 3.34, where it may be written that (because of Equations [3.3.14] and [3.3.15]) $G_o(s) = E\overline{\overline{v}}_t/s$, the block diagram of the secondary regulator then becomes the one indicated in Figure 3.50, where (for simplicity) the index i has been omitted. The modifications required to any possible scheme, such as that of Figure 3.37, are obvious.

The response of the regulated variables, i.e., of the frequency (or frequencies) and the exported powers, to the load disturbances $\Delta P_{L1}, \ldots, \Delta P_{Ln}$ may be deduced, *in general*, from the equations reported in Section 3.4.1 (see also Fig. 3.47).

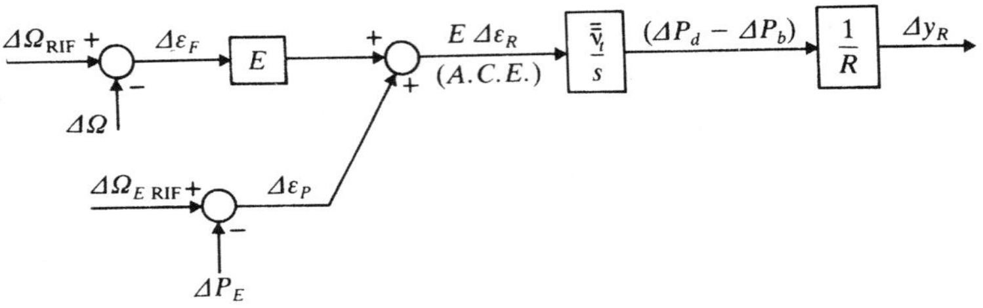

Figure 3.50. Block diagram of the secondary regulator under the "autonomy" condition (approximated solution; see text).

Assuming for simplicity that $\Omega_1 = \ldots = \Omega_n \triangleq \Omega$, it can be deduced, through trivial developments, that:

$$\frac{\partial \Omega}{\partial P_{Li}} = \frac{-g}{1 + G_{Ri}^*} \quad (i = 1, \ldots, n) \tag{3.4.31}$$

$$\frac{\partial P_{Ei}}{\partial P_{Li}} = -\left(1 - \frac{g}{G_{Ni}} \frac{1 + G_{Ni} G_{Ri}}{1 + G_{Ri}^*}\right) \frac{1}{1 + G_{Ri}^*} \quad (i = 1, \ldots, n) \tag{3.4.32}$$

$$\frac{\partial P_{Ej}}{\partial P_{Li}} = \frac{g}{G_{Nj}} \frac{1 + G_{Nj} G_{Rj}}{1 + G_{Rj}^*} \frac{1}{1 + G_{Ri}^*} \quad (i, j = 1, \ldots, n; j \neq i) \tag{3.4.33}$$

having omitted, for brevity, the indication of the variable s, and having written:

$$g \triangleq \left(\sum_1^n {}_k \frac{1 + G_{Nk} G_{Rk}}{G_{Nk}(1 + G_{Rk}^*)} \right)^{-1} \tag{3.4.34}$$

In particular, if the autonomy condition [3.4.27] is realized in all the areas, it results in the simplified expressions (further than Equations [3.4.28] and [3.4.29]):

$$\frac{\partial \Omega}{\partial P_{Li}} = \frac{-G_N}{1 + G_{Ri}^*} \quad (i = 1, \ldots, n) \tag{3.4.35}$$

$$\frac{\partial P_{Ei}}{\partial P_{Li}} = -\left(1 - \frac{G_N}{G_{Ni}}\right) \frac{1}{1 + G_{Ri}^*} \quad (i = 1, \ldots, n) \tag{3.4.36}$$

$$\frac{\partial P_{Ej}}{\partial P_{Li}} = \frac{G_N}{G_{Nj}} \frac{1}{1 + G_{Ri}^*} \quad (i, j = 1, \ldots, n; j \neq i) \tag{3.4.37}$$

and $g = G_N$ (recall Equation [3.4.21]).

From the operating point of view, it is interesting to define the *"phase" error* $\Delta\alpha \triangleq \int_0^t \Delta\Omega \, dt$ and the *"exported energy" errors* $\Delta E_{Ej} \triangleq \int_0^t \Delta P_{Ej} \, dt$, caused by generic load perturbations. With Laplace transforms, the effect of ΔP_{Li} on such errors is defined by:

$$\Delta\alpha(s) = \frac{1}{s}\frac{\partial\Omega}{\partial P_{Li}}\Delta P_{Li}(s)$$

$$\Delta E_{Ej}(s) = \frac{1}{s}\frac{\partial P_{Ej}}{\partial P_{Li}}\Delta P_{Li}(s)$$

and it is then deducible from the previous equations.

For a constant value of ΔP_{Li}, at steady-state it holds that:

$$\Delta\alpha = \left(\lim_{s\to 0}\frac{1}{s}\frac{\partial\Omega}{\partial P_{Li}}\right)\Delta P_{Li} = -g(0)\frac{\Delta P_{Li}}{K_i^*}$$

$$\Delta E_{Ei} = \left(\lim_{s\to 0}\frac{1}{s}\frac{\partial P_{Ei}}{\partial P_{Li}}\right)\Delta P_{Li} = -\left(1 - g(0)\frac{K_i}{K_i^*}\right)\frac{\Delta P_{Li}}{K_i^*}$$

$$\Delta E_{Ej} = \left(\lim_{s\to 0}\frac{1}{s}\frac{\partial P_{Ej}}{\partial P_{Li}}\right)\Delta P_{Li} = g(0)\frac{K_j}{K_j^*}\frac{\Delta P_{Li}}{K_i^*} \qquad (j \neq i)$$

where $K_i \triangleq \lim_{s\to 0}(sG_{Ri}(s))$, $K_i^* \triangleq \lim_{s\to 0}(sG_{Ri}^*(s))$, $g(0) = \left(\sum_1^n{}_k \frac{K_k}{K_k^*}\right)^{-1}$. It is also possible to obtain these results directly, by observing that at steady-state:

$$-K_j\Delta\alpha - K_j^*\Delta E_{Ej} = \Delta P_{Rj} = \begin{cases}\Delta P_{Li} & \text{if} j = i\\ 0 & \text{if} j \neq i\end{cases}$$

and recalling that $\sum_1^n{}_j \Delta E_{Ej} = 0$.

Some caution is necessary when, for example, area 1 is only under frequency regulation (that is $G_{R1}^* = 0$) so that $g(0) = 0$, $K_1^* = 0$, and the ratio $g(0)/K_1^*$ takes an indeterminate form. Through elementary considerations, it can be verified that the effects at steady-state are as follows:

- for a perturbation ΔP_{L1}, i.e., within the area under frequency regulation only:

$$\Delta\alpha = -\frac{\Delta P_{L1}}{K_1}$$

$$\Delta E_{E1} = -\left(\sum_2^n{}_j \frac{K_j}{K_j^*}\right)\frac{\Delta P_{L1}}{K_1}$$

$$\Delta E_{Ej} = \frac{K_j}{K_j^*}\frac{\Delta P_{L1}}{K_1} \qquad (j = 2, \ldots, n)$$

  thus the errors of exported energies are all zero if $K_2 = \ldots = K_n = 0$, i.e., if areas $2, \ldots, n$ are only under power regulation;

- for a perturbation ΔP_{Li} with $i \neq 1$:

$$\Delta\alpha = 0$$

$$\Delta E_{E1} = \frac{\Delta P_{Li}}{K_i^*}$$

$$\Delta E_{Ei} = -\frac{\Delta P_{Li}}{K_i^*}$$

$$\Delta E_{Ej} = 0 \quad \forall j \neq 1, i$$

i.e., there are only errors of exported energies, opposite each other, in area 1 and in the perturbed area.

In general, for transient responses, assuming for simplicity that autonomy conditions hold true, and assuming as a first approximation $G_{Ni}(s) \cong G_{Ni}(0) = 1/E_i$, $G_{Ri}(s) \cong E_i\overline{\overline{\nu}}_{ti}/s$, $G^*_{Ri}(s) \cong \overline{\overline{\nu}}_{ti}/s$, and thus $G_N(s) \cong G_N(0) = 1/E_{\text{tot}}$ (where $E_{\text{tot}} \triangleq \sum_1^n {}_i E_i$ is the total regulating energy of the set of the n areas), it follows that:

$$\Delta\alpha \cong -\frac{1}{E_{\text{tot}}}\frac{\Delta P_{Li}(s)}{s + \overline{\overline{\nu}}_{ti}}$$

$$\Delta E_{Ei} \cong -\left(1 - \frac{E_i}{E_{\text{tot}}}\right)\frac{\Delta P_{Li}(s)}{s + \overline{\overline{\nu}}_{ti}}$$

$$\Delta E_{Ej} \cong \frac{E_j}{E_{\text{tot}}}\frac{\Delta P_{Li}(s)}{s + \overline{\overline{\nu}}_{ti}} \qquad (j \neq i)$$

from which the time behavior of such errors is easily obtained, once the behavior of ΔP_{Li} is known (in particular, note the effect of the parameter $\overline{\overline{\nu}}_{ti}$, which constitutes, under the adopted approximations, the cutoff frequency of the secondary regulation in the area i).

3.4.3. Response to Reference Settings and Noninteraction Criterion

Assuming for simplicity that $\Omega_1 = \ldots = \Omega_n \triangleq \Omega$ and considering the ideal references (recall Equations [3.4.8] and [3.4.9]), the block diagram of Figure 3.51 is particularly useful to give individual emphasis to the frequency regulation loop (which is sensitive, because of Equation [3.4.19], only to the summation $\sum_1^n {}_k \Delta P'_{Rk}$) and to the regulation loops of exported powers[(11)].

Generally, these regulation loops are interacting with one another, as each reference setting influences not only the respective regulated variable, but also the other ones. Then, it may be helpful to establish methods to avoid this situation completely or partially. In this regard, the following cases appear particularly interesting.

(1) If it is desired that $\Delta\Omega$ *not be influenced by the regulation of the exported powers*, and therefore by the references $P^o_{E\,\text{RIF}\,i}$, it is necessary and

(11) The present analysis may be extended to the general case with frequencies Ω_i's transiently different, by intending, for example, that the frequency regulation is that of a given Ω_i, or that of a suitable mean frequency of the system, e.g., defined by $\Delta\Omega = G_N(s)\sum_1^n {}_i \Delta\Omega_i/G_{Ni}(s)$ for which Equation [3.4.19] is still valid.

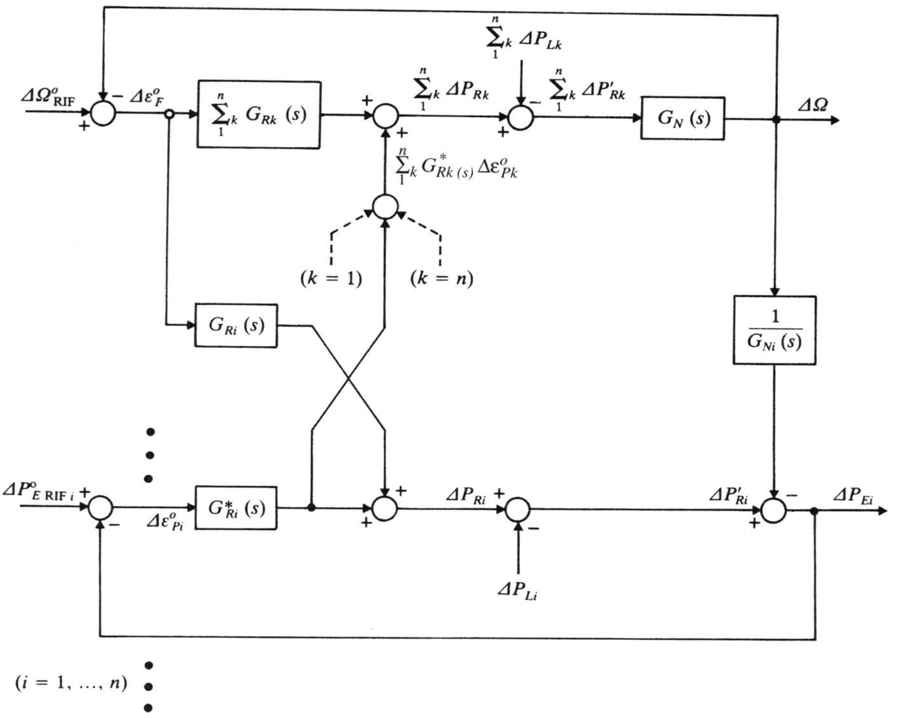

Figure 3.51. Detailed block diagram equivalent to that of Figure 3.48.

sufficient that $\sum_1^n {}_k \Delta P_{Rk}$ does not depend on the $\Delta\varepsilon^o_{Pi}$ errors, i.e., (see Fig. 3.51):

$$\sum_1^n {}_i G^*_{Ri}(s)\Delta\varepsilon^o_{Pi} = 0$$

whatever the $\Delta\varepsilon^o_{pi}$ are (or rather $(n-1)$ of them are, since $\sum_1^n {}_i \Delta\varepsilon^o_{pi} = 0$). Then it is possible to derive $(n-1)$ conditions as:

$$G^*_{R1} = \ldots = G^*_{Rn} \qquad [3.4.38]$$

In this case $\Delta\Omega$ and $\sum_1^n {}_k \Delta P_{Rk}$ depend on, in addition to $\Delta\Omega^o_{\mathrm{RIF}}$, the *sum* $\left(\sum_1^n {}_k \Delta P_{Lk}\right)$ of the load variations, and therefore, in particular, the response of $\Delta\Omega$ to a given load variation does not depend on the area in which the variation occurred (*"uniformity"* of the response of $\Delta\Omega$). This can be deduced also from Equation [3.4.31], with the functions G^*_{Ri}'s equal to one another. In this case, Equation [3.4.31] is then translated into the following Equation [3.4.39].

For the frequency regulation loop (see the upper side of Fig. 3.51) it then holds that:

$$\frac{\partial \Omega}{\partial P_{Li}} = \frac{-G_N}{1 + G_N \sum_1^n {}_k G_{Rk}} \qquad [3.4.39]$$

$$\frac{\partial \sum_1^n {}_k P_{Rk}}{\partial P_{Li}} = \frac{G_N \sum_1^n {}_k G_{Rk}}{1 + G_N \sum_1^n {}_k G_{Rk}} \qquad [3.4.40]$$

regardless of $i = 1, \ldots, n$, where the synthesis of $\sum_1^n {}_k G_{Rk}$ may be performed in a manner similar to that used for G_R in Figure 3.36, by considering the whole n-areas system as an isolated area for which it is required to regulate only frequency.

The choice of the G_{Ri}^*'s (equal to one another) and of the individual G_{Ri} must be made in relation to the regulation requirements of the respective exported powers P_{Ei}. In this regard, the conditions [3.4.38] also achieve the *noninteraction between the power regulations* (however, each one of these regulations is, in general, influenced by the frequency regulation).

(2) Reciprocally, if it is desired that $\Delta P_{E1}, \ldots, \Delta P_{En}$ *not be influenced by the frequency regulation*, and therefore by the reference Ω_{RIF}^o, it is necessary and sufficient that the overall effect of $\Delta\varepsilon_F^o$ on each ΔP_{Ei} is zero, and therefore that (see Fig. 3.51):

$$G_{Ri}(s) = \left(\sum_1^n {}_k G_{Rk}(s)\right) \frac{G_N(s)}{G_{Ni}(s)}$$

from which the $(n - 1)$ following conditions are obtained:

$$G_{R1} G_{N1} = \ldots = G_{Rn} G_{Nn} \qquad [3.4.41]$$

In this case, however, the power regulations, in addition to influencing the frequency regulation, generally interact with one another, since (see also Fig. 3.51):

$$\Delta P_{Ei} = G_{Ri}^* \Delta\varepsilon_{Pi}^o - \Delta P_{Li} - \frac{G_N}{G_{Ni}} \sum_1^n {}_k (G_{Rk}^* \Delta\varepsilon_{Pk}^o - \Delta P_{Lk})$$

i.e., ΔP_{Ei}, generally, also depends on power errors in other areas.

(3) As a consequence of the two previous cases, if the *complete noninteraction* between the n regulations (of frequency and of exported powers) is

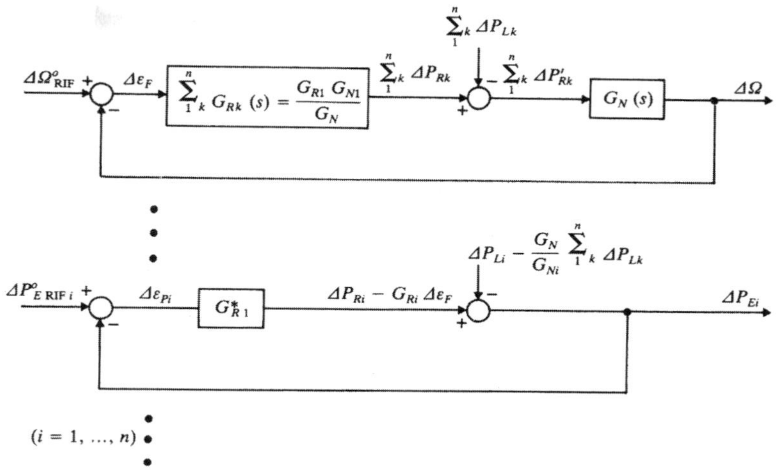

Figure 3.52. Equivalent regulation loops under the conditions of complete "noninteraction."

desired, it is necessary and sufficient to satisfy both conditions [3.4.38] and [3.4.41].

The block diagram of the whole system may then be reduced to n noninteracting single loops, as shown in Figure 3.52. Actually, $(n + 1)$ single loops are indicated in the figure, but the one relative to any one of the P_{Ei}'s is a consequence of the others.

In particular (recall also Equations [3.4.31], [3.4.32], and [3.4.33], with $g = G_N(1 + G^*_{R1})/(1 + G_{R1}G_{N1})$), it holds:

$$\frac{\partial \Omega}{\partial P_{Li}} = \frac{-G_N}{1 + G_{R1}G_{N1}} \quad (i = 1, \ldots, n) \qquad [3.4.42]$$

$$\frac{\partial P_{Ei}}{\partial P_{Li}} = -\left(1 - \frac{G_N}{G_{Ni}}\right)\frac{1}{1 + G^*_{R1}} \quad (i = 1, \ldots, n) \qquad [3.4.43]$$

$$\frac{\partial P_{Ej}}{\partial P_{Li}} = \frac{G_N}{G_{Nj}}\frac{1}{1 + G^*_{R1}} \quad (i, j = 1, \ldots, n; j \neq i) \qquad [3.4.44]$$

where Equation [3.4.42] is equivalent to Equation [3.4.39]. Moreover, Equation [3.4.40] holds.

Figure 3.53 schematically summarizes the cases of interaction or noninteraction, in accordance with what is detailed above. As a concluding remark, it is interesting to observe the following:

- The conditions [3.4.38] and [3.4.41] are not incompatible with each other, nor with the "autonomy" condition [3.4.27]; on the contrary, they are mutually implied if the autonomy conditions are satisfied in all the areas.

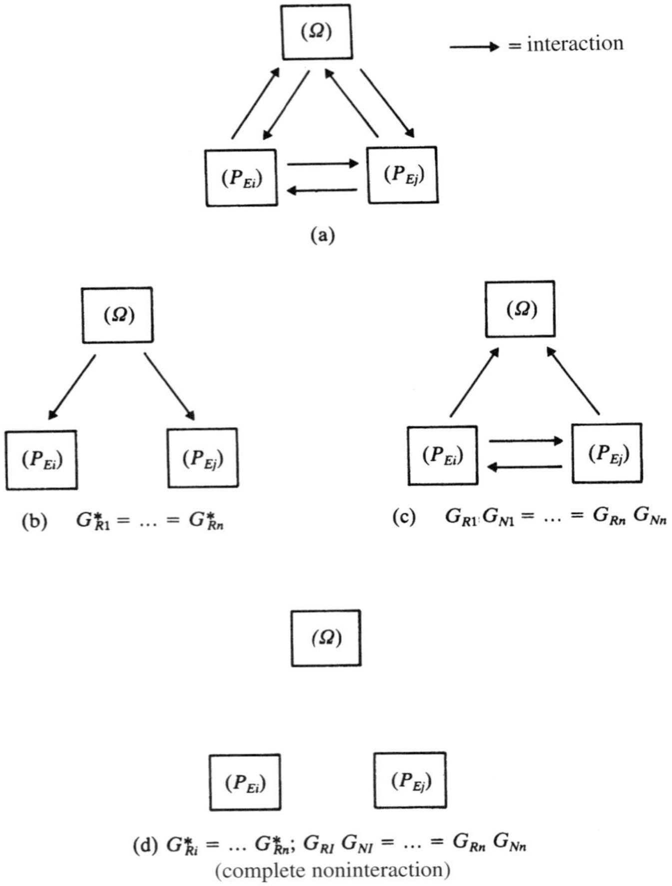

Figure 3.53. Interaction or noninteraction between the regulations of the frequency [block (Ω)] and of two generic exported powers [blocks (P_{Ei}), (P_{Ej}), with $j \neq i$]: (a) general case; (b), (c), (d) particular cases.

In particular, it is then possible to achieve *both the complete autonomy and the complete noninteraction*, by assuming:

$$G^*_{R1} = \ldots = G^*_{Rn} = G_{R1}G_{N1} = \ldots = G_{Rn}G_{Nn} \qquad [3.4.45]$$

(hence, Equations [3.4.28] and [3.4.29], as well as Equations [3.4.42], [3.4.43] and [3.4.44], hold for all $i = 1, \ldots, n$). On the other hand, the choice of conditions [3.4.45] appears to be reasonable, as it leads to the same loop transfer function (and then to similar synthesis criteria) for all control loops indicated in Figures 3.49 and 3.52. With reference to the scheme of

Figure 3.50, all this is then translated, using the stated approximations, into the equality of the parameter $\overline{\overline{\nu}}_t$ for the regulators of the different areas.

- In comparison with the alternatives detailed here, which are all achievable by means of "local" regulators (one for each area), it does not appear to be practical to consider more sophisticated solutions for which the matrix R in Figure 3.48 has not an assigned structure (as the matrix indicated in Equation [3.4.23]) but in general might be "full." In fact, this could imply difficulties in implementation. In this regard, generally the autonomy criterion in all the areas ($\partial P_{Ri}/\partial P_{Lj}$ $\forall j \neq i; i, j = 1, \ldots, n$) implies that the RS matrix is diagonal, as may be derived directly from Figure 3.48. It is immediately possible to conclude that R must have the same zero elements as the S^{-1} matrix defined by the Equation [3.4.26], and more precisely that it *must* be achieved through local regulators, in accordance to what is described.
Similarly, the complete noninteraction among the regulations first of all implies (see Fig. 3.48) that SR is diagonal. Hence, R must again have the same zero elements as S^{-1}. However, it is necessary to have:

$$(SR)_{22} = \ldots = (SR)_{nn}$$

(so that $\Delta P_{E1} = -\sum_{2\ j}^{n} \Delta P_{Ej} = -\sum_{2\ j}^{n} (SR)_{jj} \Delta\varepsilon_{Pj}^{o}$ depends only on $\Delta\varepsilon_{P1}^{o} = -\sum_{2j}^{n} \Delta\varepsilon_{Pj}^{o}$ (apart from load variations), and by imposing such conditions the solution already described is obtained. In the case of partial noninteractions, R *may*, however, be realized through nonlocal regulators. The above implies that the variations in losses may be disregarded, as already stated.

- The conditions of complete noninteraction between the regulations guarantee that the variation of a given reference only acts on the corresponding regulated variable (similar conclusions hold for cases of partial noninteraction). However, the present considerations hold with regard to the variations of the "ideal" references, i.e., they presume that the real references satisfy the conditions [3.4.6]. Consequently, a variation of a given (real) reference that does not respect the conditions [3.4.6]—and then is equivalent, in general, to variations both of $\Omega_{\mathrm{RIF}}^{o}$ and of all the $P_{E\,\mathrm{RIF}\,i}^{o}$—may end up influencing all the regulated variables, even if the noninteraction conditions defined above are met.

3.5. EMERGENCY CONTROL

3.5.1. Preliminaries

In Section 1.7, the possibility that a system may be operated under emergency conditions was presented. The characteristics of an emergency condition depend on a number of factors, including:

- the originating perturbation (type, entity, point of application, etc.);
- particular characteristics of the power system (configuration, operating point, admissibility limits for single equipment, etc.);
- characteristics of protection and control systems.

The preventive analysis of emergency situations may allow intervention plans to be implemented following a suitable "diagnosis" of the operating system. But the cases to be examined and the consequent operating solutions, involve such a large number of specific problems that it is generally impractical to derive a systematic treatment of general usefulness.

As a consequence, the information here is limited to qualitative, although basic, considerations regarding the typical emergency that involves the operation of the f/P control, i.e., the case of a system experiencing a *severe power "deficit"* caused by a possible opening of interconnections, with a loss in imported power, or tripping of a set of generators.

The frequency decrement caused by the initial perturbation may be unacceptable, and moreover, may cause the automatic tripping of some units, with a further reduction of the available power and collapse of the frequency itself. (In particular, the auxiliary system protections of a power plant may allow their operation only within a given frequency range, e.g., 47.5–52.5 Hz for a nominal frequency of 50 Hz.)

It is therefore necessary to avoid, even during transients, frequency excursions below a predetermined value $\Omega_{\lim}$ (e.g., 48.5 Hz, to maintain a prudent margin over the value of 47.5 Hz indicated above), also considering the different nonlinearities that characterize the response to large perturbations, such as the limits in opening (and in the opening speed) of the unit valves.

During the operation, an indication of the severity of the situation may be obtained by proper measurement of the frequency Ω and the "deceleration" $(-\mathrm{d}\Omega/\mathrm{d}t)$. In effect, the deceleration measurement allows the evaluation of the tendency of Ω to decline before the frequency has reached problematic values; thus, it offers the possibility of early diagnoses and timely interventions.

Nevertheless, it is necessary that the frequency measurement and the deceleration measurement are properly filtered to reduce the effect of load fluctuations and machine oscillations. In fact, similarly to what was indicated in Section 3.3.3, $\Omega(t)$ should be the mean behavior of the frequency and $(\mathrm{d}\Omega/\mathrm{d}t)(t)$ should be the time derivative of such mean behavior. For frequency values close to that at normal operation, the emergency implies relatively large decelerations, and uncertainties in the deceleration measurement have a more modest weight.

Finally, interventions on the system typically involve the disconnection of some loads, having a suitable power and geographical distribution (*"load-shedding"*), and/or of units under pumping operation. Some further advantages may be obtained, for example, by directly activating the full opening of hydroelectric units to the maximum opening speed (*"power stimulation"*), etc. Such interventions are normally made using threshold devices, sensitive to Ω

and $d\Omega/dt$. If such actions do not stop the frequency decline, the "isolation" of the thermal units from the network (by means of suitable switching, set for example at 47.5–48 Hz) may be a useful option, as mentioned in Section 1.7.2.

It is here intended that the initial perturbation (e.g., the opening of interconnections or the tripping of generators) is persistent, i.e., not followed by the restoration of the original configuration (see Fig. 1.8d). Moreover, it is assumed that the perturbation causes a power deficit, as the opposite case (power "surplus") normally implies noncritical situations, because of intervention of the primary regulation (fast closing of valves, etc.); recall Section 2.3.1d.

A different problem arises when the initial configuration is restored, within relatively small times (e.g., tenths of a second), because of automatic breaker reclosing. Referencing to Figure 1.8d, we must consider only a temporary disconnection, during which the two subsystems indicated in the figure are, respectively, under power deficit and power surplus, after which they should recover their synchronism. Under such conditions, the frequency variations (in opposite senses) in the two subsystems may be of concern, not because of their magnitude (by itself perhaps modest, because of the shortness of the time disconnection), but rather because of the risk of loss of synchronism after the reconnection, as a result of phenomena similar to that illustrated in Section 1.6. To suitably limit the frequency variations, it is necessary to intervene within very short times (considering the status, open or closed, of breakers), by initiating, for example:

- the temporary disconnection of some loads (*"load-skipping"*) in the subsystem under power deficit;
- the temporary interruption of the steam flow to the turbines of thermal units (*"fast valving"*) or the temporary insertion of resistive loads (*"braking resistors"*), in the subsystem under power surplus.

Further remedies to avoid the loss of synchronism may involve, for example, the control of the generator excitation, as discussed in Section 7.3.

3.5.2. Example of Simplified Procedure

In simplified terms, let us consider an isolated system for which the block diagram of Figure 3.54a, or Figure 3.54b apply. In both cases:

- it is simply assumed (with known symbols) that:

$$G_f(s) = E\frac{1 + sT_2}{1 + sT_1}$$

$$G_g(s) = G_c(s) = 0$$

so that the power-frequency transfer function is of the second order, and more precisely of the type:

$$G_N(s) = \frac{1}{G_f(s) + sM} = \frac{1}{E}\frac{1 + sT_1}{1 + 2\zeta\dfrac{s}{\nu_o} + \dfrac{s^2}{\nu_o^2}}$$

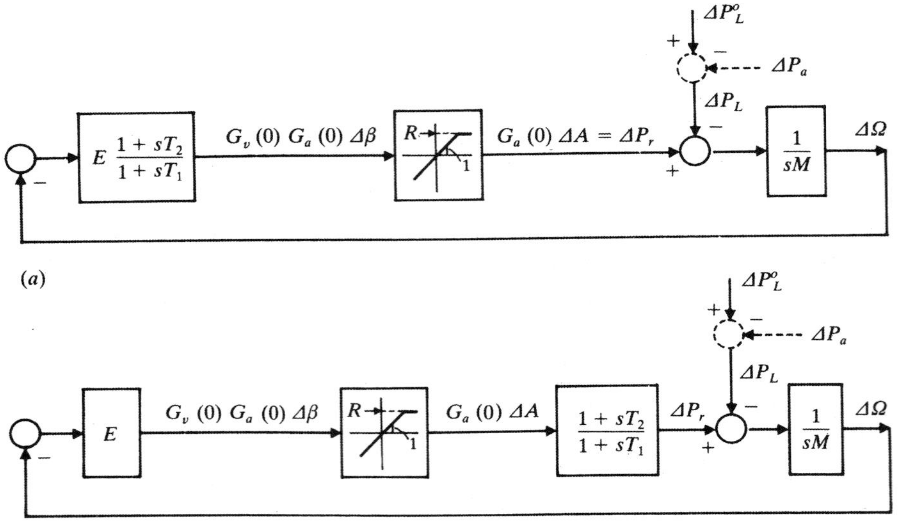

(b)

Figure 3.54. Example of simplified block diagram for a system in isolated operation and under only primary regulation: (a) case of hydroregulating units; (b) case of thermal regulating units.

where $\nu_o \triangleq \sqrt{E/(T_1 M)}$, $\zeta \triangleq (\nu_o/2)(T_2 + M/E)$ (recall also Equation [3.3.22]);

- it is assumed that the intervention of the secondary regulation can be disregarded (because of its relative slowness) and that the primary frequency references are constant;
- only the nonlinearities caused by opening limits of the unit valves are considered, moreover assuming that reaching such limits occurs simultaneously for all units (the symbol R indicates the spinning reserve available to the primary regulation; e.g., $R \cong 0.05 P_{\text{nom}}$).

The two schemes differ, however, for the different positions of the nonlinear element. More precisely, as seen in Section 3.2, the scheme in Figure 3.54a may correspond, under the above simplifications, to the case in which the primary regulation is performed by hydro units. Instead, with Figure 3.54b, it can be assumed that the units under primary regulation are thermal, disregarding "power loops" with a feedback from delivered powers.

The resulting behavior is the same in both cases if, even during transient conditions, it is:

$$G_v(0)G_a(0)\Delta\beta < R$$

i.e., if the opening limits are not activated.

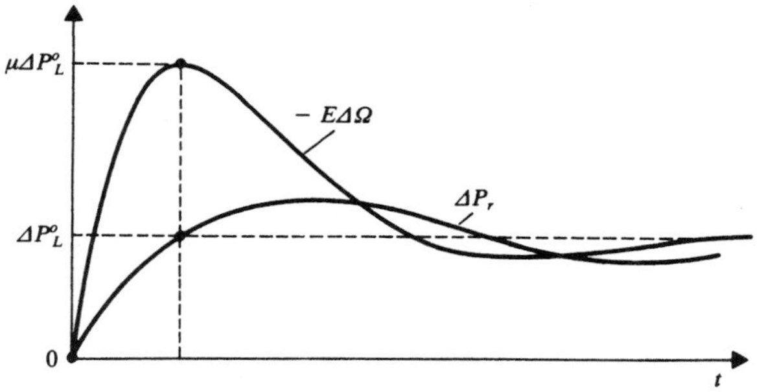

Figure 3.55. Responses to a step ΔP_L^o under the hypothesis that the opening limits remain unactivated. Note that ΔP_r and $-E\Delta\Omega$ represent $G_v(0)G_a(0)\Delta\beta$, respectively, in (a) and (b) of Figure 3.54.

Under such conditions, and assuming, for example, that:

- the system is initially at steady-state (with $\Omega = \Omega^o$ constant, and $\Delta P_L = 0$);
- the perturbation ΔP_L is constituted by a step having an amplitude ΔP_L^o, at $t = 0$ (whereas $\Delta P_a = 0$: see Fig. 3.54);

responses like those indicated in Figure 3.55 can be then determined. In particular, the minimum value reached by the frequency Ω results in:

$$\Omega_{\min} = \Omega^o - \mu \frac{\Delta P_L^o}{E}$$

where E is the permanent regulating energy, and the coefficient $\mu > 1$ depends on $\nu_o T_1$ and ζ.

Furthermore, as $G_N(s)$ is of the second order, the knowledge of Ω and $\mathrm{d}\Omega/\mathrm{d}t$ at a generic instant allows (once the system parameters are known) the value ΔP_L^o of the perturbation to be determined. Thus, the value $\Omega_{\min}$ may be estimated even before it is reached. The "diagnosis" becomes an emergency if this value $\Omega_{\min}$ is smaller than $\Omega_{\lim}$. In such a case, it is necessary to intervene in time, by means of a "load-shedding" of proper size (power ΔP_a). Other actions, such as the power stimulation, etc., are not considered here.

More particularly, the following procedure may be adopted.

(1) Preventive analysis of the effects of ΔP_L^o (assumed to be a step), with:
   - deduction, starting from the characteristic $(\Delta P_L^o, \Omega_{\min})$, of the limit value $\Delta P_L^o = \Delta P_{L\lim}^o$ which leads to $\Omega_{\min} = \Omega_{\lim}$;
   - deduction, for each generic value $\Omega = \Omega_a > \Omega_{\lim}$ reached at $t = t_a$ during the initial frequency decrease, of the characteristic $(\Delta P_L^o, M\,(-\mathrm{d}\Omega/\mathrm{d}t))_{\Omega=\Omega_a}$, and thus the deceleration limit value $(-\mathrm{d}\Omega/\mathrm{d}t)_{\lim}$ corresponding to $\Delta P_L^o = \Delta P_{L\lim}^o$, over which it is necessary to activate the shedding (see qualitatively Fig. 3.56).

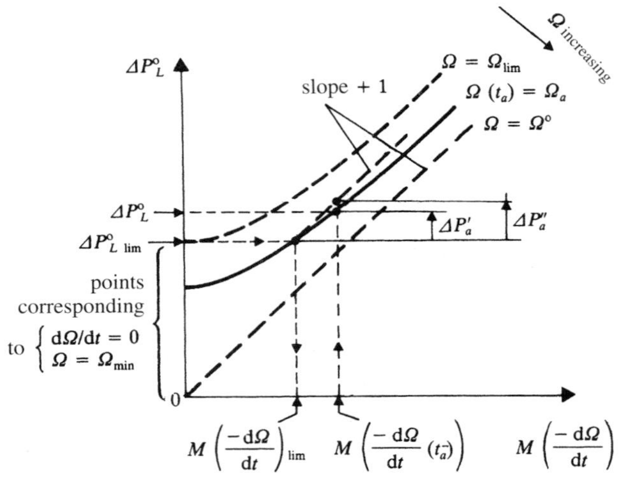

Figure 3.56. Characteristics to be used for the deduction of $(-\mathrm{d}\Omega/\mathrm{d}t)_{\lim}$, $\Delta P_a'$ and $\Delta P_a''$ (see text).

(2) Preventive analysis of the effects of ΔP_a (which too is assumed to be a step), with deduction of the value $\Delta P_{a\,\min}$ which, starting from the generic Ω_a with $(-\mathrm{d}\Omega/\mathrm{d}t)(t_a^-) > (-\mathrm{d}\Omega/\mathrm{d}t)_{\lim}$, leads to $\Omega_{\mathrm{MIN}} = \Omega_{\lim}$ (Ω_{MIN} indicates the minimum value reached by the frequency after the shedding).

(3) Measurement, during the operation, of the deceleration that occurs in correspondence to a preassigned value Ω_a of the frequency (obviously larger than $\Omega_{\lim}$) and consequent activation of the shedding if the deceleration is over the limit value (function of Ω_a). All this may be translated, in terms of command logic, into the use of "threshold" devices, with regard to the measured values of frequency and deceleration.

Regarding (2), for a given Ω_a, the order of magnitude of the minimum required value $\Delta P_{a\,\min}$ as a function of $(-\mathrm{d}\Omega/\mathrm{d}t)(t_a^-) > (-\mathrm{d}\Omega/\mathrm{d}t)_{\lim}$, is easily determined from the characteristic $\left(\Delta P_L^o, M(-\mathrm{d}\Omega/\mathrm{d}t)\right)_{\Omega=\Omega_a}$ used at the point (1). In fact, with the notations of Figure 3.56, it is possible to state that generically, for $\Omega_a \in (\Omega_{\lim}, \Omega^o)$:

- a load-shedding $\Delta P_a = \Delta P_L^o - \Delta P_{L\,\lim}^o \triangleq \Delta P_a'$, corresponding to the excess of load, is insufficient, i.e., smaller than $\Delta P_{a\,\min}$, as it leads to $\Omega_{\mathrm{MIN}} < \Omega_{\lim}$;
- a load-shedding $\Delta P_a = M(-\mathrm{d}\Omega/\mathrm{d}t)(t_a^-) - M(-\mathrm{d}\Omega/\mathrm{d}t)_{\lim} \triangleq \Delta P_a''$, corresponding to the excess of deceleration, is larger than $\Delta P_{a\,\min}$, as it leads to $\Omega_{\mathrm{MIN}} > \Omega_{\lim}$;

so that we have:

$$\Delta P_{a\,\min} \in (\Delta P_a', \Delta P_a'')$$

The previous procedure, described for a linear response, may be generally applied when, under the assumptions made, the *limits in valve opening* are activated.

However, it is important to note the difference of behavior for the two schemes of Figure 3.54. In effect, under the linear operation, the quantity $G_v(0)G_a(0)\Delta\beta$ is equal to ΔP_r in case (1) and to $(-E\Delta\Omega)$ in case (2). By examining Figure 3.55, it is evident that, in case (2), the opening limits become active during the initial decrease of the frequency ($\Delta P_r \leqslant \Delta P_L^0$) already for $\Delta P_L^0 \geqslant R/\mu$. Instead, in case (1) this occurs for $\Delta P_L^0 \geqslant R$, i.e., for a larger value of ΔP_L^0 (recall that $\mu > 1$). As an example, review the behaviors of $\Omega(t)$ reported in Figure 3.57, assuming $\nu_0 T_1 = 5$, $\zeta = 1/\sqrt{2}$, to which it corresponds $\nu_0 T_2 = 2\zeta - 1/(\nu_0 T_1) \cong 1.21$, $1/\mu \cong 0.37$.

In effect, the results obtained may be very different in the two cases, as it appears evident, with reference to the same numerical example given above, by the examination of the characteristics:

$$(\Delta P_L^o, \Omega_{\min}), \quad \left(\Delta P_L^o, M\left(\frac{-\mathrm{d}\Omega}{\mathrm{d}t}\right)\right)_{\Omega=\Omega_a}, \quad \left(\Delta P_{a\min}, M\left(\frac{-\mathrm{d}\Omega}{\mathrm{d}t}\right)\right)_{\Omega=\Omega_a}$$

(for different values of Ω_a) reported in Figures 3.58, 3.59, and 3.60. For actual applications, the importance of the nonlinear model adopted is then evident.

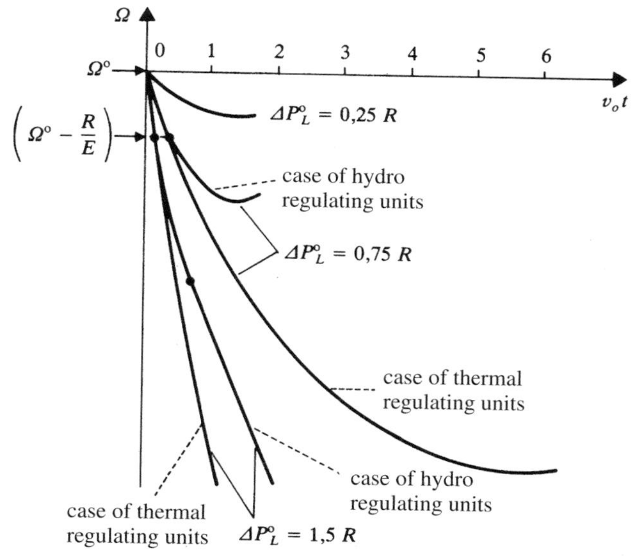

Figure 3.57. Examples of response of the frequency Ω to a step ΔP_L^0, with $\nu_0 T_1 = 5$, $\zeta = 1/\sqrt{2}$.

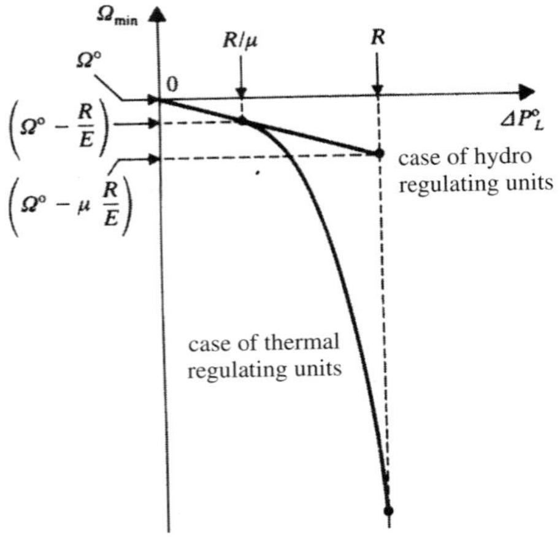

Figure 3.58. Characteristic to be used for the deduction of $\Delta P^0_{L\,\lim}$ (value of ΔP^0_L which corresponds to $\Omega_{\min} = \Omega_{\lim}$): numerical example of Figure 3.57.

In particular, it is easy to determine that, independently of the model:

- for $\Omega_a = \Omega^o$, i.e., for $t_a = 0$, it simply holds that $M(-\mathrm{d}\Omega/\mathrm{d}t)(0^-) = \Delta P^o_L$ and thus $(-\mathrm{d}\Omega/\mathrm{d}t)_{\lim} = \Delta P^o_{L\lim}/M$, so that, if $(-\mathrm{d}\Omega/\mathrm{d}t)(0^-) > \Delta P^o_{L\lim}/M$,:

$$\Delta P_{a\,\min} = \Delta P^o_L - \Delta P^o_{L\,\lim} = M\left(\frac{-\mathrm{d}\Omega}{\mathrm{d}t}(0^-)\right) - M\left(\frac{-\mathrm{d}\Omega}{\mathrm{d}t}\right)_{\lim}$$

  (in this case, it then is $\Delta P_{a\,\min} = \Delta P'_a = \Delta P''_a$);

- for $\Omega_a = \Omega_{\lim}$, it is instead necessary to avoid a further decrease of the frequency, so that $(-\mathrm{d}\Omega/\mathrm{d}t)_{\lim} = 0$ and moreover, if $(-\mathrm{d}\Omega/\mathrm{d}t)(t_a^-) > 0$:

$$\Delta P_{a\,\min} = M\left(\frac{-\mathrm{d}\Omega}{\mathrm{d}t}(t_a^-)\right)$$

  (that is $\Delta P_{a\,\min} = \Delta P''_a$).

Therefore, the corresponding characteristics $(P_{a\,\min}, M(-\mathrm{d}\Omega/\mathrm{d}t))_{\Omega=\Omega_a}$ are simply constituted by the half-straight lines, having unitary slope, indicated in Figure 3.60, whereas the characteristics relative to the intermediate values of Ω_a, which may correspond to frequency thresholds of practical interest, are included between these two half-straight lines.

The procedure proposed above is an illustrative example and, for practical applications, the following also must be considered.

- The hypothesis that ΔP^0_L and ΔP_a have a step-like behavior implies some approximations, in addition to those concerning the model of the

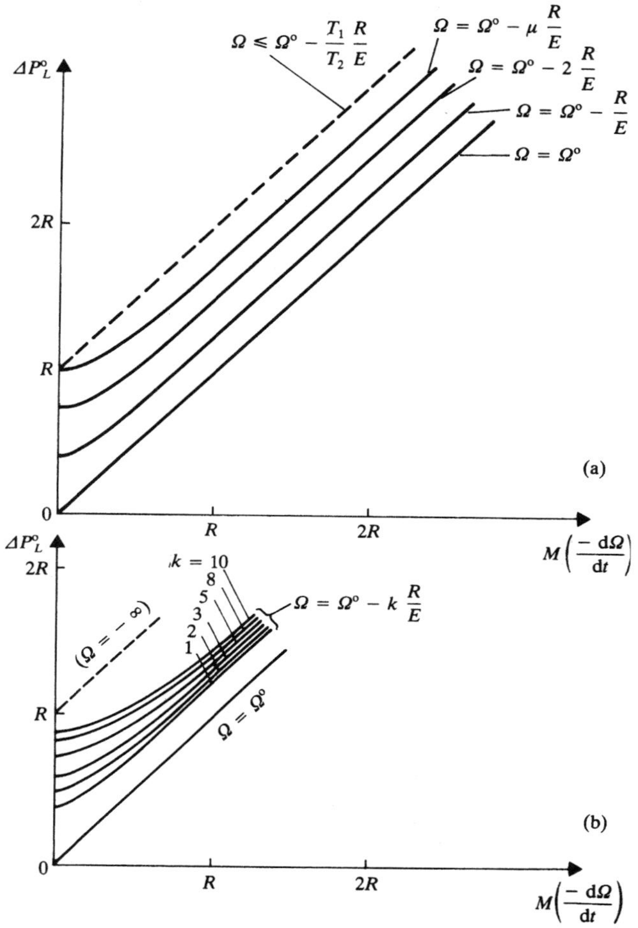

Figure 3.59. Characteristics as in Figure 3.56 for the numerical example of the previous figures: (a) case of hydroregulating units; (b) case of thermal regulating units.

system. Moreover (particularly for the more immediate effects and thus corresponding to the thresholds Ω_a closer to $\Omega_{\lim}$), it is necessary to properly evaluate the actual size of ΔP_a, which corresponds to the sheddings; recall Section 3.3.3.

- For reasons related to users, it is convenient to limit the load-sheddings (and to plan them in accordance to the load type). On the other hand, intervening too quickly might imply unnecessary sheddings (perhaps a consequence of false alarms), whereas delayed interventions might imply sheddings of larger size. Therefore, it may be convenient to arrange more frequency thresholds Ω_{ai} $(i = 1, 2, \ldots)$, and furthermore:

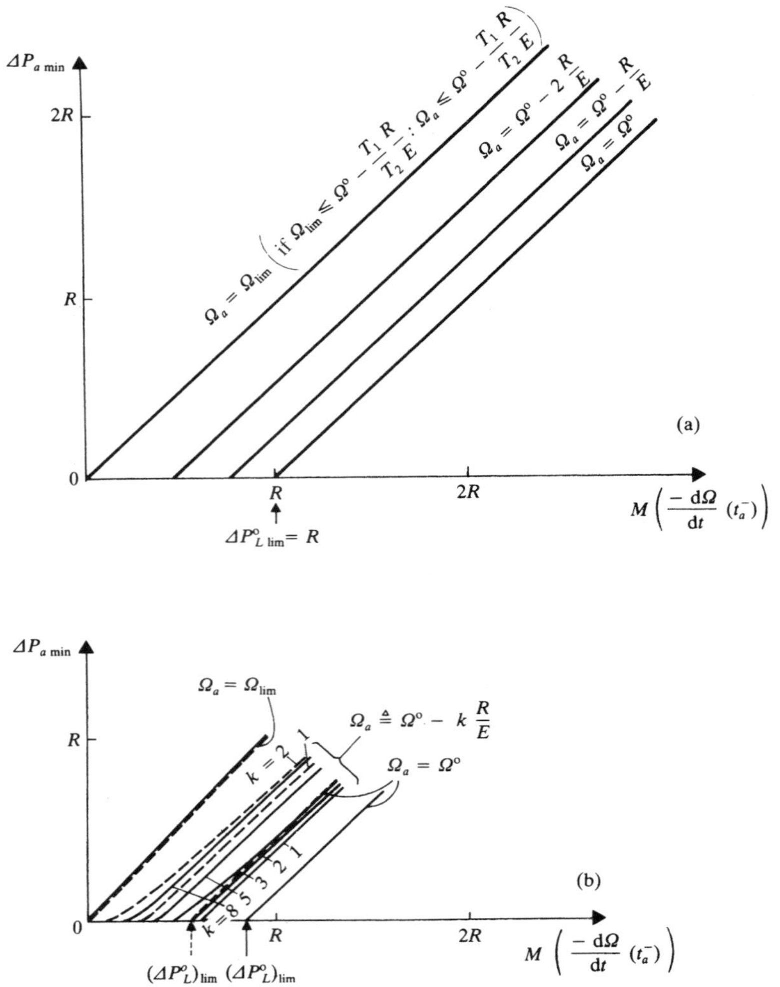

Figure 3.60. Dependence of the minimum shedding size ($\Delta P_{a\,\min}$) on the values of frequency ($\Omega_a = \Omega(t_a)$) and deceleration ($(-\mathrm{d}\Omega/\mathrm{d}t)(t_a^-)$) at the generic instant t_a, for the numerical example of the previous figures: (a) case of hydroregulating units, with $\Omega_{\lim} \leq \Omega^o - \mu R/E$; (b) case of thermal regulating units, with $\Omega_{\lim} = \Omega^o - 10R/E$ (continuous line) or with $\Omega_{\lim} = \Omega^o - 3R/E$ (dashed line).

- to increase the values of the deceleration thresholds, for the values Ω_{ai} closer to Ω^o (this also contributes to reduce the effect of the measuring errors on $\mathrm{d}\Omega/\mathrm{d}t$),
- to plan values ΔP_a significantly larger than what may seem strictly necessary (in particular, $\Delta P_{a\,\min} = \Delta P_a''$) for the more worrisome Ω_{ai}, close to $\Omega_{\lim}$.

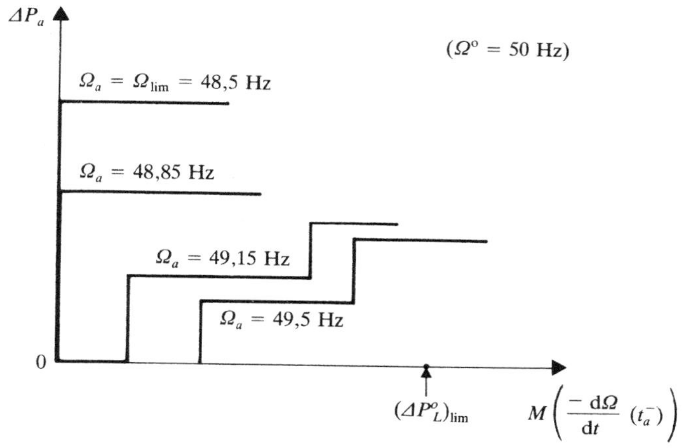

Figure 3.61. An example of load-shedding characteristics.

- The characteristics $(\Delta P_a, M(-\mathrm{d}\Omega/\mathrm{d}t))_{\Omega=\Omega_a}$ to be planned are discontinuous, as qualitatively indicated in Figure 3.61; they correspond to the disconnection of given load sets, in a limited number. For the Ω_{ai}'s closer to $\Omega_{\lim}$, it may be convenient to assume a preassigned value ΔP_a, independent of the measured value of deceleration (larger than the threshold). In doing so, a smaller weight is attributed not only to the uncertainties on the deceleration measurement, but also to uncertainties about the system parameters.
- The geographical distribution of different sheddings is important for reasons related to the subsequent behavior of network voltages and currents (without excluding oscillatory phenomena between the machines), and for guaranteeing a uniform presence of generation and load in the different parts of the system with relation to the risk of their possible (subsequent) disconnection.
- The shedding actuation occurs with some delay (e.g., $\tau \cong 0.2$ sec or more) with respect to the generic instant t_a (this may be the result of delays in measurements, breakers opening, etc.), so that it may be convenient to anticipate reaching frequency thresholds. For example, supposing that the deceleration slightly varies in the time τ, the frequency measurement may be corrected, by generically replacing $\Omega(t)$ with:

$$\Omega'(t) \triangleq \Omega(t) - \left(\frac{-\mathrm{d}\Omega}{\mathrm{d}t}(t)\right)\tau \cong \Omega(t+\tau)$$

ANNOTATED REFERENCES

Among the works of more various or general interest, the following are mentioned: 13, 25, 28, 31, 37, 51, 58, 67, 136 (terminology aspects), 157, 159, 211.

Furthermore, regarding

- the primary control: 162, 195, 289, and more particularly:
  - for the hydroplants: 3, 85, further than 62 (about the Thoma condition) and 101 (about possible generalizations of the condition itself);
  - for the thermal plants: 175, 247, 265, 298, 333 (with a wide bibliography), 313;
- the secondary control schemes: 219, 300, 329, 338 (page 13);
- the use of direct current links: 139;
- the identification of the power-frequency transfer function: 78, 82, 83, 89, 102, 122, 154;
- the control in the presence of interconnections: 14, 66, 73, 96, 97, 133, 213 (moreover, the text takes into particular account what exposed in 37);
- the emergency control: 186, 190, 199, 207, 226, 250, 284, 320, 332, 339, further than some notes prepared by the author, in view of the writing of 53.

CHAPTER 4

DYNAMIC BEHAVIOR OF THE SYNCHRONOUS MACHINE

4.1. BASIC FORMULATIONS

4.1.1. Preliminaries

This chapter summarizes the basic dynamic characteristics of the synchronous machine, by recalling the mathematical models usually employed with various degrees of approximation in the analysis of the electric and electromechanical transients.

Roughly speaking, the synchronous machine may simply be regarded as a three-phase generator of sinusoidal voltages (of positive sequence and with a frequency equal to the electric angular speed Ω of the rotor) with suitable output linear impedances (evaluated at the frequency Ω), symmetrically related to the three phases. By accepting such simplifications, the machine may be represented, in steady-state operation at the frequency Ω and with the convention of the generators, by an equation

$$\overline{v} = \overline{e} - \overline{Z}\,\overline{\imath} \tag{4.1.1}$$

where $\overline{v}$, $\overline{\imath}$, $\overline{e}$ are the vectors obtained by applying the Park's transformation (see Appendix 2) with a reference rotating at the equilibrium speed Ω^o and (respectively) representing the voltage and the current at the stator terminals, and the internal emf (equal to the open-circuit voltage), whereas $\overline{Z}$ is the corresponding output impedance (see Fig. 4.1).

For vector $\overline{e}$, assume, for simplicity, that its magnitude is a function of the excitation current alone, and its phase may vary following variations of the rotor speed in respect of normal operating speed.

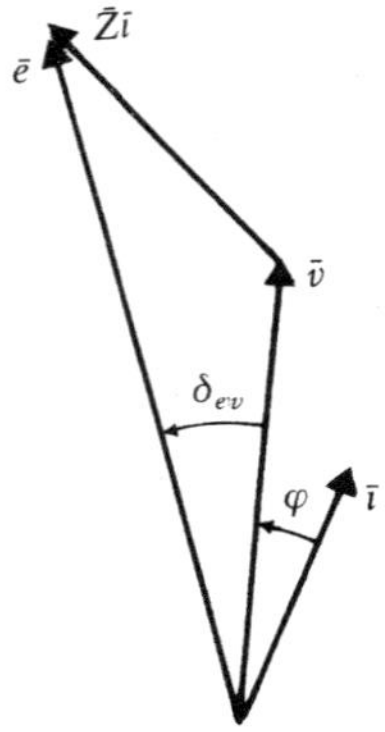

Figure 4.1. A vector diagram of first approximation.

In their turn, the speed transients of the rotor may be represented, again with the convention of the electric generators, by the equation:

$$C_m = J\frac{d\Omega_m}{dt} + C_p(\Omega_m) + C_e \quad [4.1.2]$$

(equivalent to Equation [3.1.1]), in which:

- J = total moment of inertia;
- $\Omega_m = \Omega/N_p$ = mechanical angular speed of the rotor (N_p = number of pole pairs in the rotor);
- C_m = torque supplied from outside (positive if driving);
- C_p = torque corresponding to mechanical losses;
- C_e = torque generated by the machine (positive if opposite to the motion; negative if driving, that is if generated by the machine acting as a motor).

We can assume, in addition, that the power $C_e\Omega_m$ is equal to the (active) electric power generated:

$$P_e = \langle \bar{e}, \bar{\imath} \rangle = ei\cos(\delta_{ev} + \varphi) \quad [4.1.3]$$

in which the symbol $\langle \ldots, \ldots \rangle$ means "scalar product," whereas e and i denote the magnitudes of $\bar{e}$ and $\bar{\imath}$, respectively, and δ_{ev}, φ, $(\delta_{ev} + \varphi)$ are the phase differences between $\bar{e}$ and $\bar{v}$, $\bar{v}$ and $\bar{\imath}$, $\bar{e}$ and $\bar{\imath}$, respectively (see Fig. 4.1). By assuming $\overline{Z} \triangleq \overline{R} + j\overline{X}$, it also follows:

$$P_e = \left\langle \bar{e}, \frac{\bar{e} - \bar{v}}{\overline{Z}} \right\rangle = \frac{1}{R^2 + X^2}(Xev\sin\delta_{ev} + R(e^2 - ev\cos\delta_{ev})) \quad [4.1.3']$$

$$P_e = \langle \bar{v} + \overline{Z}\bar{\imath}, \bar{\imath} \rangle = vi\cos\varphi + Ri^2 \quad [4.1.3'']$$

A model of the type described here is often used for studies of first approximation, but it is easy to realize how inadequate it is for representing the real behavior of a synchronous machine, except for special operating conditions.

First, the definition of the output impedance $\overline{Z}$ itself may appear doubtful, even during normal operation. If the machine is anisotropic, it is not possible to define a single output reactance value, and the internal voltage drop cannot simply be expressed by a term $\overline{Z}\bar{\imath}$, as in Equation [4.1.1], since it also depends on the phase of the vector $\bar{\imath}$ in respect of the polar axis (i.e., in respect of the vector $\bar{e}$).

In addition, the concept of frequency itself (and therefore of reactance and impedance) may seem anything but clear, during disturbed operating conditions, in which the stator voltages and currents are not positive-sequence sinusoidal.

Finally, Equation [4.1.1] alone does not make it possible to consider the real electromagnetic phenomena associated with delayed variation of the magnetic fluxes in the machine during transients.

The following sections will show that it is possible, through application of the Park's transformation (and then of the vector representations) even in transient conditions, to define mathematical models of different approximation, that evaluate the dynamic phenomena of more interest.

The following considerations may apply to normal types of synchronous machines; any extension to special types may require supplementary considerations that are fairly obvious, at least from the formal point of view. For the set up of the equations, it may be advisable, without prejudicing the degree of generality required here, to refer to Figure 4.2a, related to a *bipolar machine* with salient poles, with three "phase" stator circuits (marked a, b, and c), and at least one rotor circuit, consisting of the "field" circuit (marked f).

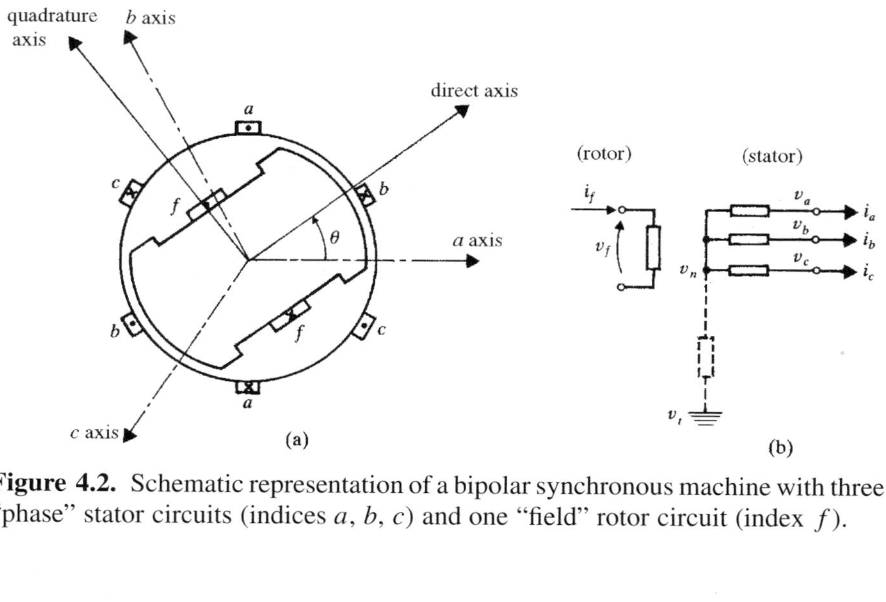

Figure 4.2. Schematic representation of a bipolar synchronous machine with three "phase" stator circuits (indices a, b, c) and one "field" rotor circuit (index f).

In Figure 4.2a, the aforementioned circuits are represented fairly schematically. If it is assumed that the mmfs generated by them are positive when the current circulation takes place as in the figure, it is possible to associate each circuit with an oriented axis along which it acts magnetically. The phase circuits act along the axes a, b, and c respectively, whereas the field circuit acts along the polar axis, and more precisely (bearing in mind the orientation) along the so-called *"direct" axis*. In addition, the (interpolar) axis oriented 90° in advance in respect of the direct axis is called the *"quadrature" axis*.

Furthermore, the "electrical" angle of the direct axis in advance in respect of the axis a is indicated by the symbol θ. In the case of the bipolar machine, the electrical angles coincide with the actual mechanical angles (see for example θ in Fig. 4.2a). For a machine with N_p pairs of poles, the electrical angles are, on the other hand, N_p times the mechanical ones.

Lastly, it is intended, as in Figure 4.2b, that the three "phase" stator circuits are "star" connected, with a suitable impedance between the "neutral" and the "earth" (such impedance is infinite if the neutral is isolated).

4.1.2. Equations of the Electrical, Magnetic, and Mechanical Parts

In this section, the mathematical model of the synchronous machine is derived with the following *basic assumptions*:

- there are no rotor circuits other than the field circuit (i.e., additional rotor circuits are not considered, which may correspond to the so-called "damper bars" or to induced eddy currents present in the body of the rotor when the latter is solid);
- there is no magnetic saturation;
- the unit shaft is a rigid connection, without any torsional phenomena.

Other simplifying assumptions, which we assume of minor significance, may relate to, for example, the conductor distribution and the rotor profile (as specified further on), as well as effects relatively negligible, such as the effects of the slots, winding radial sizes, magnetic hysteresis, and so on.

The machine equations may be derived by separately analyzing:

- the electrical part
- the magnetic part
- the mechanical part

(a) Equations of the Electrical Part

For the electrical part of the machine, the following equations may be assumed:

- For the *field circuit*, with the convention of the loads:

$$v_f = R_f i_f + \frac{\mathrm{d}\psi_f}{\mathrm{d}t} \qquad [4.1.4]$$

in which:

- v_f, i_f, ψ_f = field voltage, current, flux,
- R_f = field resistance.

- For the *phase circuits*, with the convention of the generators:

$$\left.\begin{aligned} v_a - v_n &= -Ri_a + \frac{d\psi_a}{dt} \\ v_b - v_n &= -Ri_b + \frac{d\psi_b}{dt} \\ v_c - v_n &= -Ri_c + \frac{d\psi_c}{dt} \end{aligned}\right\} \qquad [4.1.5]$$

in which:

- $(v_a - v_n), i_a, \psi_a$ = voltage, current, flux of phase a (and similarly for phases b, c), v_n being the neutral voltage (the voltages v_a, v_b, v_c, v_n are intended to be defined with respect to an arbitrary reference);
- R = armature resistance, per phase.

By applying the Park's transformation (see Appendix 2) with angular reference $\theta_r = \theta$[1], such that:

$$\frac{d\theta_r}{dt} = \frac{d\theta}{dt} = \Omega = \text{"electrical" angular speed of the rotor}$$

from Equations [4.1.5] we may derive:

$$\left.\begin{aligned} v_d &= -Ri_d + \frac{d\psi_d}{dt} - \Omega\psi_q \\ v_q &= -Ri_q + \frac{d\psi_q}{dt} + \Omega\psi_d \\ v_o - \sqrt{3}v_n &= -Ri_o + \frac{d\psi_o}{dt} \end{aligned}\right\} \qquad [4.1.5']$$

in which the first two equations may be rewritten in vector terms:

$$\overline{v} = -R\overline{\imath} + (p + j\Omega)\overline{\psi} \qquad [4.1.5'']$$

by generically assuming $\overline{w} \triangleq w_d + jw_q (\overline{w} = \overline{v}, \overline{\imath}, \overline{\psi})$, $p \triangleq d/dt$. (According to the usual denominations, we have:

[1] For this assumption, the index "r" will be omitted in the subsequent notations, relating to the variables d, q, and o. Furthermore, the multiplying constants of the Park's transformation (see Appendix 2) are assumed $K_{dq} = \sqrt{2/3}$, $K_o = 1/\sqrt{3}$.

- v_d = "direct-axis" voltage,
- v_q = "quadrature-axis" voltage,

and similarly for the currents i_d, i_q and for the fluxes ψ_d, ψ_q.)

(b) Equations of the Magnetic Part

With regard to the magnetic part of the machine, the following equations may be written (here, it is assumed that the magnetic part is linear and nondissipative, ruling out any saturation, hysteresis etc.), with $L_{ik} = L_{ki}$ for $i, k = a, b, c, f$:

$$\psi_f = L_{ff} i_f - (L_{fa} i_a + L_{fb} i_b + L_{fc} i_c) \qquad [4.1.6]$$

$$\left.\begin{aligned} \psi_a &= L_{af} i_f - (L_{aa} i_a + L_{ab} i_b + L_{ac} i_c) \\ \psi_b &= L_{bf} i_f - (L_{ba} i_a + L_{bb} i_b + L_{bc} i_c) \\ \psi_c &= L_{cf} i_f - (L_{ca} i_a + L_{cb} i_b + L_{cc} i_c) \end{aligned}\right\} \qquad [4.1.7]$$

Since we have adopted the convention of the generators for the phase circuits, it must be assumed that positive phase currents i_a, i_b, i_c generate negative mmfs ("armature reaction"): hence the negative signs in Equations [4.1.6] and [4.1.7].

Recalling Figure 4.2, it is easy to realize that the inductances L_{af}, L_{bf}, L_{cf} (i.e., the mutual inductances between the phase circuits and the field circuit) are periodic functions of the position θ, with a period $\Delta\theta = 360°$ (more precisely, L_{af} is maximum for $\theta = 0$, minimum for $\theta = 180°$, and zero for $\theta = 90°$ and $\theta = 270°$; likewise L_{bf}, L_{cf}, for values of θ greater than the foregoing, by the quantities 120° and 240°, respectively).

The self-inductance L_{ff} of the field circuit is constant, since the profile of the stator can be considered smooth.

Lastly, for the inductances L_{ik} with $i, k = a, b, c$, (i.e., the self and mutual inductances relating to the phase circuits), these may be considered constant if the rotor is cylindrical (round). If, on the other hand, the rotor has salient poles, such inductances also vary periodically with θ, but with a period $\Delta\theta = 180°$ (L_{aa} is maximum for $\theta = 0$ and minimum for $\theta = 90°$; L_{ab} is negative, with maximum absolute value for $\theta = -30°$ and minimum for $\theta = 60°$; etc.).

In practice, the above dependencies on θ may, with fair approximation, be considered sinusoidal, as a consequence of constructional characteristics (in particular, distribution of the conductors and profile of the poles)[(2)], so that the following equations may be assumed:

$$\begin{cases} L_{fa} = L_{af} = M_f \cos\theta \\ L_{fb} = L_{bf} = M_f \cos\theta' \\ L_{fc} = L_{cf} = M_f \cos\theta'' \end{cases}$$

---

(2) It is therefore possible to achieve the steady-state operating conditions described in Section 4.4.1, in which both the phase voltages and the phase currents are positive-sequence sinusoidal, without any harmonic content.

$$\begin{cases} L_{aa} = L_s + L_m \cos 2\theta \\ L_{bb} = L_s + L_m \cos 2\theta' \\ L_{cc} = L_s + L_m \cos 2\theta'' \end{cases}$$

$$\begin{cases} L_{ab} = L_{ba} = -M_s - L_m \cos(2\theta + 60°) \\ L_{bc} = L_{cb} = -M_s - L_m \cos(2\theta' + 60°) \\ L_{ca} = L_{ac} = -M_s - L_m \cos(2\theta'' + 60°) \end{cases}$$

In these equations, L_s, M_s, L_m, M_f are constant (L_m smaller than both L_s and M_s, and negligible if the rotor is round), $\theta' \triangleq \theta - 120°$, $\theta'' \triangleq \theta - 240°$.

By applying the Park's transformation (see Appendix 2) with $\theta_r = \theta$, the following equations can then be obtained from Equations [4.1.6] and [4.1.7]:

$$\psi_f = L_f i_f - L_{md} i_d \qquad [4.1.6']$$

$$\left.\begin{aligned} \psi_d &= L_{md} i_f - L_d i_d \\ \psi_q &= -L_q i_q \\ \psi_o &= -L_o i_o \end{aligned}\right\} \qquad [4.1.7']$$

in which the inductances:

$$L_f \triangleq L_{ff}$$

$$L_{md} \triangleq \sqrt{\tfrac{3}{2}} M_f$$

$$L_d \triangleq L_s + M_s + \tfrac{3}{2} L_m$$

$$L_q \triangleq L_s + M_s - \tfrac{3}{2} L_m$$

$$L_o \triangleq L_s - 2M_s$$

are all independent of θ (the difference $L_d - L_q = 3L_m$ considers possible anisotropy of the rotor, and is disregarded if the rotor is round).

In vector terms, from the first two of the Equations [4.1.7′] we also can derive:

$$\overline{\psi} = L_{md} i_f - (L_d i_d + L_q j i_q) = L_{md} i_f - L_q \overline{\imath} - (L_d - L_q) i_d \qquad [4.1.7'']$$

and in particular:

$$\overline{\psi} = L_{md} i_f - L_q \overline{\imath}$$

if $L_d = L_q$ ($L_m = 0$, i.e., round rotor).

As an alternative, Equation [4.1.6′] also may be replaced by:

$$i_f = \frac{\psi_f + L_{md} i_d}{L_f} \qquad [4.1.6'']$$

so that the first part of Equations [4.1.7′] leads to:

$$\psi_d = \frac{L_{md}}{L_f} \psi_f - \hat{L}'_d i_d$$

with:

$$\hat{L}'_d \triangleq L_d - \frac{L_{md}^2}{L_f} \qquad [4.1.8]$$

and [4.1.7″] may be rewritten:

$$\overline{\psi} = \frac{L_{md}}{L_f}\psi_f - (\hat{L}'_d i_d + L_q j i_q) = \frac{L_{md}}{L_f}\psi_f - L_q \overline{\imath} + (L_q - \hat{L}'_d) i_d \qquad [4.1.7''']$$

(c) Equations of the Mechanical Part

For the mechanical part of the machine, by disregarding possible torsional phenomena, we can assume Equation [4.1.2], i.e.,

$$C_m = J\frac{\mathrm{d}\Omega_m}{\mathrm{d}t} + C_p(\Omega_m) + C_e \qquad [4.1.9]$$

where the symbols have the already defined meaning.

The ("electromagnetic") torque C_e may be expressed as a function of electric and magnetic variables of the machine. As indicated in the following, we may write:

$$C_e = N_p(\psi_d i_q - \psi_q i_d) = N_p\langle j\overline{\psi}, \overline{\imath}\rangle \qquad [4.1.10]$$

N_p being the number of pole pairs. Power corresponding to the torque C_e is:

$$P_e = C_e\Omega_m = \Omega(\psi_d i_q - \psi_q i_d) = \langle j\Omega\overline{\psi}, \overline{\imath}\rangle \qquad [4.1.10']$$

(it must be remembered that $\Omega = N_p\Omega_m$).

Lastly, remembering Equation [4.1.7″] we also can write:

$$\frac{C_e}{N_p} = \frac{P_e}{\Omega} = (L_{md} i_f - (L_d - L_q) i_d) i_q$$

from which it is evident that, if $L_d \neq L_q$, the machine is capable of generating torque even in the absence of excitation current. The component of torque:

$$(C_e)_{i_f=0} = -N_p(L_d - L_q) i_d i_q$$

caused by the anisotropy of the rotor, is said to be the "reluctance torque."

Similarly, Equation [4.1.7‴] makes it possible to obtain:

$$\frac{C_e}{N_p} = \frac{P_e}{\Omega} = \left(\frac{L_{md}}{L_f}\psi_f + (L_q - \hat{L}'_d) i_d\right) i_q$$

To demonstrate Equation [4.1.10], assume that the magnetic part of the machine is nondissipative, with characteristics depending only on its instantaneous configuration (in particular, hysteresis phenomena are ignored). In addition, reference will be made, for the sake of simplicity, to the case in which the rotor includes only the field circuit. On the other hand, the extension to the case of several rotor circuits is obvious. Finally, the treatment is still valid in presence of magnetic saturation.

Under these conditions, the total magnetic energy W stored in the machine is defined by the integral:

$$W = \int_{(1)}^{(2)} (i_f \, \mathrm{d}\psi_f - (i_a \, \mathrm{d}\psi_a + i_b \, \mathrm{d}\psi_b + i_c \, \mathrm{d}\psi_c))$$

(in which the currents i_f, i_a, i_b, i_c, as well as W, are functions of the fluxes ψ_f, ψ_a, ψ_b, ψ_c, and of the angular position θ), calculated by assuming as integration extremes:

(1): zero flux linkages;

(2): flux linkages at their present value;

and by assuming θ to be constant, equal to its present value. On the other hand, by using the Park's variables with θ as angular reference, we obtain:

$$i_a \, \mathrm{d}\psi_a + i_b \, \mathrm{d}\psi_b + i_c \, \mathrm{d}\psi_c = i_d \, \mathrm{d}\psi_d + i_q \, \mathrm{d}\psi_q + i_o \, \mathrm{d}\psi_o + (\psi_d i_q - \psi_q i_d) \, \mathrm{d}\theta$$

in which, because of Equations [4.1.6′] and [4.1.7′], the currents i_f, i_d, i_q, i_o can be expressed as functions only of the flux linkages ψ_f, ψ_d, ψ_q, ψ_o, irrespective of θ. We can then derive:

$$\mathrm{d}W = i_f \, \mathrm{d}\psi_f - (i_d \, \mathrm{d}\psi_d + i_q \, \mathrm{d}\psi_q + i_o \, \mathrm{d}\psi_o)$$

(and thus W too can be expressed as a function of only ψ_f, ψ_d, ψ_q, ψ_o, irrespective of θ). Alternatively, it holds:

$$\mathrm{d}W = \mathrm{d}W_e + (\psi_d i_q - \psi_q i_d) \, \mathrm{d}\theta \qquad [4.1.11]$$

where:

$$\mathrm{d}W_e \triangleq i_f \, \mathrm{d}\psi_f - (i_a \, \mathrm{d}\psi_a + i_b \, \mathrm{d}\psi_b + i_c \, \mathrm{d}\psi_c)$$

is the variation in total electric energy absorbed by the magnetic part of the machine.

For conservation of the energy, the last term in Equation [4.1.11] must therefore be equal to the variation $C_e \, \mathrm{d}\theta_m$ of absorbed mechanical energy (with $\mathrm{d}\theta_m = \Omega_m \, \mathrm{d}t = \Omega \, \mathrm{d}t / N_p = \mathrm{d}\theta / N_p$, N_p being the number of pole pairs), from which the "electromagnetic" torque equals:

$$C_e = (\psi_d i_q - \psi_q i_d) \frac{\mathrm{d}\theta}{\mathrm{d}\theta_m} = N_p (\psi_d i_q - \psi_q i_d) \qquad [4.1.12]$$

according to Equation [4.1.10].

The same result may be achieved, albeit somewhat laboriously, by applying the property:

$$C_e = + \left(\frac{\partial W}{\partial \theta_m} \right)_{\psi_f, \psi_a, \psi_b, \psi_c \text{ constant}} \qquad [4.1.13]$$

after having expressed W as a function of ψ_f, ψ_a, ψ_b, ψ_c, θ, and remembering that $\mathrm{d}\theta_m = \mathrm{d}\theta / N_p$. Often, instead of the Equation [4.1.13], it is assumed:

$$C_e = -\left(\frac{\partial W}{\partial \theta_m}\right)_{i_f, i_a, i_b, i_c \text{ constant}} \qquad [4.1.14]$$

which, however, is equivalent to Equation [4.1.13] only if the magnetic part of the machine is, apart from being conservative, linear. In other words, in the presence of saturation, Equation [4.1.13] remains valid, but Equation [4.1.14] does not.

Additionally, in Equation [4.1.10], the scalar product $\langle j\overline{\psi}, \overline{\imath}\rangle$ does not depend on the angular reference of the Park's transformation; therefore, in general, we also can write:

$$C_e = N_p \langle j\overline{\psi}_r, \overline{\imath}_r\rangle$$

with arbitrary reference θ_r.

Remark 1: Equivalent Circuits

From Equations [4.1.4], [4.1.5′], [4.1.6′], and [4.1.7′], relating to the electrical and magnetic parts, it is possible to derive the set of equivalent circuits shown in Figure 4.3.

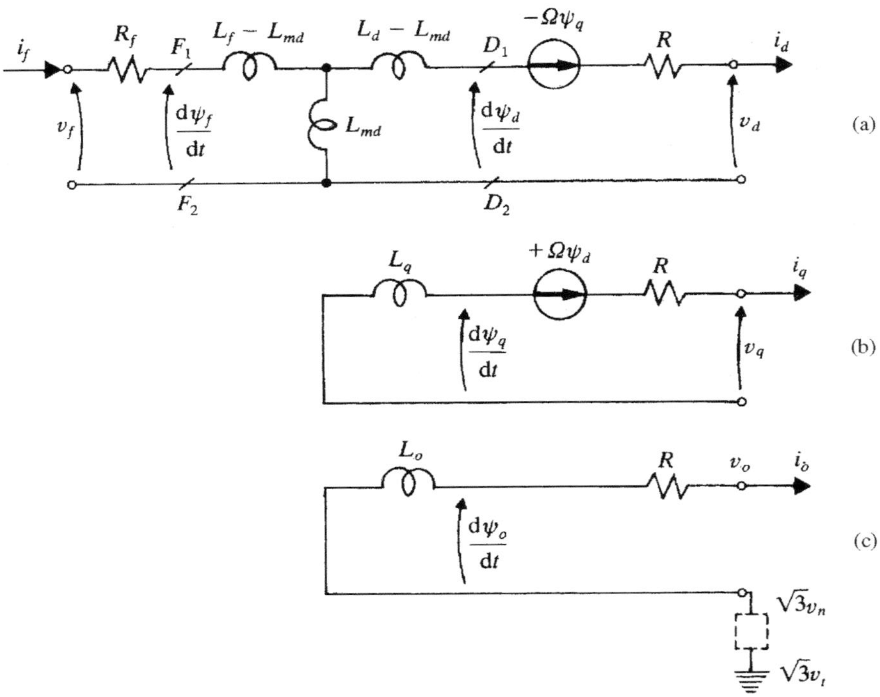

Figure 4.3. Equivalent circuits of the synchronous machine (electrical and magnetic parts), without additional rotor circuits and without magnetic saturation.

(More generally, the circuit in Fig. 4.3a remains valid even if we replace:

$$\begin{cases} v_f & \text{with} \quad \tau_f v_f \\ i_f & \text{with} \quad i_f/\tau_f \\ \psi_f & \text{with} \quad \tau_f \psi_f \end{cases} \qquad \begin{cases} R_f & \text{with} \quad \tau_f^2 R_f \\ L_f & \text{with} \quad \tau_f^2 L_f \\ L_{md} & \text{with} \quad \tau_f L_{md} \end{cases}$$

where τ_f is an arbitrary transformation ratio.)

Note that, because of Equation [4.1.10′], the power P_e is equal to the total power supplied by the two voltage generators indicated in Figure 4.3a,b. The active power P delivered through the stator terminals is, on the other hand, expressed by:

$$\begin{aligned} P &= (v_a - v_n)i_a + (v_b - v_n)i_b + (v_c - v_n)i_c = v_d i_d + v_q i_q + (v_o - \sqrt{3}v_n)i_o \\ &= P_e - R(i_d^2 + i_q^2 + i_o^2) + \left(\frac{d\psi_d}{dt}i_d + \frac{d\psi_q}{dt}i_q + \frac{d\psi_o}{dt}i_o\right) \end{aligned}$$

In addition, the inductances L_d and $\hat{L}'_d$ (the latter is defined by Equation [4.1.8]) are equal to the inductance seen from the terminals (D_1, D_2), when the pair of terminals (F_1, F_2) is open or short-circuited, respectively.

Similarly, L_f is the inductance seen from (F_1, F_2) when (D_1, D_2) is open, whereas the inductance seen with (D_1, D_2) short-circuited is expressed by $L_f - L_{md}^2/L_d = (\hat{L}'_d/L_d)L_f$.

Finally, regarding the circuit in Figure 4.3c, it results $i_o = 0$ if the neutral is isolated from earth (see also Section 5.2.1).

Remark 2: Per Unit (pu) Reduction

The best choice for the base values may not be obvious, at least for the rotor variables. In the following, it will be assumed to refer:

$$\begin{cases} v_d, v_q, v_o & \text{to} \quad \sqrt{3}V_{(F)\text{nom}} \\ i_d, i_q, i_o & \text{to} \quad \sqrt{3}I_{(F)\text{nom}} \\ \psi_d, \psi_q, \psi_o & \text{to} \quad \sqrt{3}V_{(F)\text{nom}}/\omega_{\text{nom}} \end{cases} \qquad \begin{cases} v_f & \text{to} \quad V_{\text{f}}^* \\ i_f & \text{to} \quad I_{\text{f}}^* \\ \psi_f & \text{to} \quad \Psi_f^* \end{cases}$$

and, in addition, the torques:

$$C_m, C_p, C_e \quad \text{to} \quad \frac{A_{\text{nom}}}{\Omega_{m\text{nom}}}$$

(also referring the powers to A_{nom}), by denoting:

$V_{(F)\text{nom}}$ = nominal effective value of phase voltages
$I_{(F)\text{nom}}$ = nominal effective value of phase currents
$A_{\text{nom}} = 3V_{(F)\text{nom}}I_{(F)\text{nom}}$ = nominal apparent power
$\omega_{\text{nom}} = \Omega_{\text{nom}}$ = nominal frequency

$\Omega_{m\text{nom}} = \dfrac{\Omega_{\text{nom}}}{N_p}$ = nominal mechanical speed

V_f^*, I_f^*, Ψ_f^* = values of v_f, i_f, ψ_f corresponding, in open circuit operation (with $\Omega = \omega_{\text{nom}}$ and without saturation), to nominal stator voltage.

(As indicated in Section 4.4.1, Equation [4.4.1], it results:

$$I_f^* = \frac{\sqrt{3}V_{(F)\text{nom}}}{\omega_{\text{nom}} L_{md}}$$

while $V_f^* = R_f I_f^*$, $\Psi_f^* = L_f I_f^*$.)

By generically setting:

$$\boldsymbol{v}_\alpha \triangleq \frac{v_\alpha}{\sqrt{3}V_{(F)\text{nom}}}, \quad \boldsymbol{i}_\alpha \triangleq \frac{i_\alpha}{\sqrt{3}I_{(F)\text{nom}}}, \quad \boldsymbol{\psi}_\alpha \triangleq \frac{\omega_{\text{nom}}\psi_\alpha}{\sqrt{3}V_{(F)\text{nom}}} \qquad (\alpha = d, q, o)$$

$$\boldsymbol{v}_f \triangleq \frac{v_f}{V_f^*}, \quad \boldsymbol{i}_f \triangleq \frac{i_f}{I_f^*}, \quad \boldsymbol{\psi}_f \triangleq \frac{\psi_f}{\Psi_f^*}, \quad \boldsymbol{C}_\alpha \triangleq \frac{\Omega_{m\text{nom}} C_\alpha}{A_{\text{nom}}} \qquad (\alpha = m, p, e)$$

and also:

$$\left.\begin{aligned} x_d &\triangleq \frac{\omega_{\text{nom}} L_d}{Z_{(F)\text{nom}}}, & \hat{x}'_d &\triangleq \frac{\omega_{\text{nom}} \hat{L}'_d}{Z_{(F)\text{nom}}} \\ x_q &\triangleq \frac{\omega_{\text{nom}} L_q}{Z_{(F)\text{nom}}}, & r &\triangleq \frac{R}{Z_{(F)\text{nom}}} \\ x_o &\triangleq \frac{\omega_{\text{nom}} L_o}{Z_{(F)\text{nom}}} \end{aligned}\right\} \qquad [4.1.15]$$

where $Z_{(F)\text{nom}} \triangleq V_{(F)\text{nom}}/I_{(F)\text{nom}}$ is the nominal impedance per phase, the Equations [4.1.4], [4.1.5′], [4.1.6′], [4.1.7′], [4.1.9], and [4.1.10] therefore give us, respectively:

$$\boldsymbol{v}_f = \boldsymbol{i}_f + \hat{T}'_{do}\frac{\mathrm{d}\boldsymbol{\psi}_f}{\mathrm{d}t} \qquad [4.1.16]$$

$$\left.\begin{aligned} &\left.\begin{aligned} \boldsymbol{v}_d &= -r\boldsymbol{i}_d + \frac{1}{\omega_{\text{nom}}}\left(\frac{\mathrm{d}\boldsymbol{\psi}_d}{\mathrm{d}t} - \Omega\boldsymbol{\psi}_q\right) \\ \boldsymbol{v}_q &= -r\boldsymbol{i}_q + \frac{1}{\omega_{\text{nom}}}\left(\frac{\mathrm{d}\boldsymbol{\psi}_q}{\mathrm{d}t} + \Omega\boldsymbol{\psi}_d\right) \end{aligned}\right\} \text{that is (similarly to Equation [4.1.5″]) :} \\ &\qquad \bar{\boldsymbol{v}} = -r\bar{\boldsymbol{\imath}} + \frac{p + j\Omega}{\omega_{\text{nom}}}\bar{\boldsymbol{\psi}} \qquad [4.1.17'] \\ &\boldsymbol{v}_o - \frac{v_n}{V_{(F)\text{nom}}} = -r\boldsymbol{i}_o + \frac{1}{\omega_{\text{nom}}}\frac{\mathrm{d}\boldsymbol{\psi}_o}{\mathrm{d}t} \end{aligned}\right\} \qquad [4.1.17]$$

$$\boldsymbol{\psi}_f = \boldsymbol{i}_f - (x_d - \hat{x}'_d)\boldsymbol{i}_d \qquad [4.1.18]$$

$$\left.\begin{array}{l} \left.\begin{array}{l}\boldsymbol{\psi}_d = \boldsymbol{i}_f - x_d \boldsymbol{i}_d \\ \boldsymbol{\psi}_q = -x_q \boldsymbol{i}_q\end{array}\right\} \text{ that is (similarly to Equations [4.1.7''] and [4.1.7''']) :} \\ \overline{\boldsymbol{\psi}} = \boldsymbol{i}_f - (x_d \boldsymbol{i}_d + x_q j \boldsymbol{i}_q) = \boldsymbol{\psi}_f - (\hat{x}'_d \boldsymbol{i}_d + x_q j \boldsymbol{i}_q) \quad [4.1.19'] \\ \boldsymbol{\psi}_o = -x_o \boldsymbol{i}_o \end{array}\right\} \quad [4.1.19]$$

$$\boldsymbol{C}_m = \frac{T'_a}{\omega_{\text{nom}}} \frac{d\Omega}{dt} + \boldsymbol{C}_p(\Omega) + \boldsymbol{C}_e \quad [4.1.20]$$

$$\boldsymbol{C}_e = \boldsymbol{\psi}_d \boldsymbol{i}_q - \boldsymbol{\psi}_q \boldsymbol{i}_d = (\boldsymbol{i}_f - (x_d - x_q)\boldsymbol{i}_d)\boldsymbol{i}_q = (\boldsymbol{\psi}_f + (x_q - \hat{x}'_d)\boldsymbol{i}_d)\boldsymbol{i}_q \quad [4.1.21]$$

where:

- $\hat{T}'_{do} \triangleq L_f/R_f$ is the "field time constant," i.e., the time constant seen from the terminals of the field circuit when the phase circuits are open or, more particularly, when the pair (D_1, D_2) in Figure 4.3a is open;
- $T'_a \triangleq J\Omega^2_{m\text{ nom}}/A_{\text{nom}}$ is the so-called "start-up (or acceleration) time" of the unit, referred to the nominal apparent power (by remembering Equation [3.1.5], it then results $T'_a = T_a P_{\text{nom}}/A_{\text{nom}} = T_a \cos\varphi_{\text{nom}}$).

4.1.3. Transfer Functions

In terms of transfer functions, from Equations [4.1.4], [4.1.6′], and [4.1.7′] it is possible to deduce the following equations:

$$\left.\begin{array}{l} i_f = B(s)v_f + C(s)i_d \\ \psi_f = B'(s)v_f - C'(s)i_d \\ \psi_d = A(s)v_f - \mathcal{L}_d(s)i_d \end{array}\right\} \text{ for the } d \text{ axis} \quad [4.1.22]$$

$$\psi_q = -\mathcal{L}_q(s)i_q \quad \text{for the } q \text{ axis} \quad [4.1.23]$$

$$\psi_o = -\mathcal{L}_o(s)i_o \quad \text{for the } o \text{ axis} \quad [4.1.24]$$

with:

$$\left.\begin{array}{l} C(s) = sA(s) \\ B'(s) = \dfrac{1 - R_f B(s)}{s} \\ C'(s) = \dfrac{R_f C(s)}{s} = R_f A(s) \end{array}\right\} \quad [4.1.25]$$

$$\left.\begin{array}{l} A(s) = \dfrac{L_{md}}{R_f(1 + s\hat{T}'_{do})} \\ B(s) = \dfrac{1}{R_f(1 + s\hat{T}'_{do})} \\ \mathcal{L}_d(s) = \dfrac{L_d + s\hat{T}'_{do}\hat{L}'_d}{(1 + s\hat{T}'_{do})} = L_d \dfrac{1 + s\hat{T}'_d}{1 + s\hat{T}'_{do}} \\ \mathcal{L}_q(s) = \text{constant} = L_q \\ \mathcal{L}_o(s) = \text{constant} = L_o \end{array}\right\} \quad [4.1.26]$$

where the time constant $\hat{T}'_{do}$ has the known meaning, whereas:

$$\hat{T}'_d \triangleq \frac{\hat{L}'_d}{L_d}\hat{T}'_{do} = \frac{\hat{x}'_d}{x_d}\hat{T}'_{do} = \hat{T}'_{do} - \frac{L^2_{md}}{L_d R_f}$$

is the time constant seen from the terminals of the field circuit when the pair (D_1, D_2) in Figure 4.3a is short-circuited.

Turning to "per unit" values, the previous equations may be translated into the following equations (directly deducible from Equations [4.1.16], [4.1.18], and [4.1.19]):

$$\left.\begin{aligned} \boldsymbol{i}_f &= b(s)\boldsymbol{v}_f + c(s)\,\boldsymbol{i}_d \\ \boldsymbol{\psi}_f &= b'(s)\boldsymbol{v}_f - c'(s)\boldsymbol{i}_d \\ \boldsymbol{\psi}_d &= a(s)\boldsymbol{v}_f - l_d(s)\boldsymbol{i}_d \end{aligned}\right\} \quad \text{for the } d \text{ axis} \qquad [4.1.22']$$

$$\boldsymbol{\psi}_q = -l_q(s)\boldsymbol{i}_q \quad \text{for the } q \text{ axis} \qquad [4.1.23']$$

$$\boldsymbol{\psi}_o = -l_o(s)\boldsymbol{i}_o \quad \text{for the } o \text{ axis} \qquad [4.1.24']$$

with:

$$\left.\begin{aligned} c(s) &= \frac{s}{\omega_{\text{nom}}}\frac{A_{\text{nom}}}{P^*_f}a(s) = s(x_d - \hat{x}'_d)\hat{T}'_{do}a(s) \\ b'(s) &= \frac{1 - b(s)}{s\hat{T}'_{do}} \\ c'(s) &= \frac{c(s)}{s\hat{T}'_{do}} = (x_d - \hat{x}'_d)\,a(s) \end{aligned}\right\} \qquad [4.1.25']$$

($P^*_f \triangleq V^*_f I^*_f = R_f I^{*2}_f$ is the power absorbed by the field circuit in open-circuit operation with nominal stator voltage and $\Omega = \omega_{\text{nom}}$, without saturation), and in addition:

$$\left.\begin{aligned} a(s) &= b(s) = \frac{1}{1 + s\hat{T}'_{do}} \\ l_d(s) &= \frac{x_d + s\hat{T}'_{do}\hat{x}'_d}{1 + s\hat{T}'_{do}} = x_d\frac{1 + s\hat{T}'_d}{1 + s\hat{T}'_{do}} \\ l_q(s) &= \text{constant} = x_q \\ l_o(s) &= \text{constant} = x_o \end{aligned}\right\} \qquad [4.1.26']$$

As it will be noted further on, Equations [4.1.22], [4.1.23], [4.1.24], and [4.1.25], as well as the corresponding ones in pu, i.e., Equations [4.1.22′], etc., remain valid even for models of a more general type, whereas Equations [4.1.26] (and Equations [4.1.26′]) must be suitably adapted (in particular, it may be that $a(s) \neq b(s)$).

Also note that $P^*_f = R_f(3V^2_{(F)\text{nom}}/(\omega_{\text{nom}}L_{md})^2)$, and therefore, in the first part of Equations [4.1.25′]:

$$\frac{A_{\text{nom}}}{P_f^*} = \frac{(\omega_{\text{nom}} L_{md})^2}{Z_{(F)\text{nom}} R_f} = \omega_{\text{nom}}(x_d - \hat{x}'_d)\hat{T}'_{do} = \omega_{\text{nom}} x_d (\hat{T}'_{do} - \hat{T}'_d)$$

Equations [4.1.22] give i_f, ψ_f, ψ_d starting from v_f and i_d; alternatively, it is possible to derive i_f, ψ_f, i_d starting from v_f and ψ_d. In pu (see Equations [4.1.22′]), this leads to:

$$\left.\begin{aligned} \boldsymbol{i}_f &= \left(b(s) + \frac{a(s)c(s)}{l_d(s)}\right)\boldsymbol{v}_f - \frac{c(s)}{l_d(s)}\boldsymbol{\psi}_d \\ \boldsymbol{\psi}_f &= \left(b'(s) - \frac{a(s)c'(s)}{l_d(s)}\right)\boldsymbol{v}_f + \frac{c'(s)}{l_d(s)}\boldsymbol{\psi}_d \\ \boldsymbol{i}_d &= \frac{a(s)}{l_d(s)}\boldsymbol{v}_f - \frac{1}{l_d(s)}\boldsymbol{\psi}_d \end{aligned}\right\} \qquad [4.1.27]$$

where, because of Equations [4.1.26′]:

$$\frac{a(s)}{l_d(s)} = \frac{1}{x_d + s\hat{T}'_{do}\hat{x}'_d} = \frac{1}{x_d(1 + s\hat{T}'_d)}$$

$$\frac{1}{l_d(s)} = \frac{1 + s\hat{T}'_{do}}{x_d + s\hat{T}'_{do}\hat{x}'_d} = \frac{1 + s\hat{T}'_{do}}{x_d(1 + s\hat{T}'_d)} = \frac{1}{x_d} + \left(\frac{1}{\hat{x}'_d} - \frac{1}{x_d}\right)\frac{s\hat{T}'_d}{1 + s\hat{T}'_d}$$

Furthermore, Equations [4.1.5′] account for the behavior of the voltages v_d, v_q, v_o, as well as of the speed Ω; in this connection, the first two expressions in Equations [4.1.5′] are however nonlinear, given the presence of the products $\Omega\psi_d$, $\Omega\psi_q$.

In per-unit values (see Equations [4.1.17]), for small variations around a general condition $\boldsymbol{\psi}_d = \boldsymbol{\psi}_d^o$, $\boldsymbol{\psi}_q = \boldsymbol{\psi}_q^o$, $\Omega = \Omega^o = \omega_{\text{nom}}$, we can derive:

$$\left.\begin{aligned} \Delta\boldsymbol{\psi}_d &= \frac{\dfrac{s}{\omega_{\text{nom}}}(\Delta\boldsymbol{v}_d + r\Delta\boldsymbol{i}_d) + (\Delta\boldsymbol{v}_q + r\Delta\boldsymbol{i}_q) + \left(-\boldsymbol{\psi}_d^o + \dfrac{s}{\omega_{\text{nom}}}\boldsymbol{\psi}_q^o\right)\dfrac{\Delta\Omega}{\omega_{\text{nom}}}}{1 + \dfrac{s^2}{\omega_{\text{nom}}^2}} \\ \Delta\boldsymbol{\psi}_q &= \frac{-(\Delta\boldsymbol{v}_d + r\Delta\boldsymbol{i}_d) + \dfrac{s}{\omega_{\text{nom}}}(\Delta\boldsymbol{v}_q + r\Delta\boldsymbol{i}_q) - \left(\boldsymbol{\psi}_q^o + \dfrac{s}{\omega_{\text{nom}}}\boldsymbol{\psi}_d^o\right)\dfrac{\Delta\Omega}{\omega_{\text{nom}}}}{1 + \dfrac{s^2}{\omega_{\text{nom}}^2}} \end{aligned}\right\} \qquad [4.1.28]$$

while:

$$\boldsymbol{\psi}_o = \frac{\omega_{\text{nom}}}{s}\left(\boldsymbol{v}_o - \frac{v_n}{V_{(F)\text{nom}}} + r\boldsymbol{i}_o\right) \qquad [4.1.29]$$

and recalling Equations [4.1.27], [4.1.23′], and [4.1.24′]:

$$\left.\begin{aligned}
\Delta\boldsymbol{\psi}_d &= \frac{\left(\dfrac{s}{\omega_{\text{nom}}}+\dfrac{r}{l_q(s)}\right)\left(\dfrac{r}{l_d(s)}a(s)\Delta\boldsymbol{v}_f+\Delta\boldsymbol{v}_d\right)+\Delta\boldsymbol{v}_q+\left(-\boldsymbol{\psi}_d^o+\left(\dfrac{s}{\omega_{\text{nom}}}+\dfrac{r}{l_q(s)}\right)\boldsymbol{\psi}_q^o\right)\dfrac{\Delta\Omega}{\omega_{\text{nom}}}}{1+\left(\dfrac{s}{\omega_{\text{nom}}}+\dfrac{r}{l_d(s)}\right)\left(\dfrac{s}{\omega_{\text{nom}}}+\dfrac{r}{l_q(s)}\right)}\\
\Delta\boldsymbol{\psi}_q &= \frac{-\left(\dfrac{r}{l_d(s)}a(s)\Delta\boldsymbol{v}_f+\Delta\boldsymbol{v}_d\right)+\left(\dfrac{s}{\omega_{\text{nom}}}+\dfrac{r}{l_d(s)}\right)\Delta\boldsymbol{v}_q+\left(-\boldsymbol{\psi}_q^o-\left(\dfrac{s}{\omega_{\text{nom}}}+\dfrac{r}{l_d(s)}\right)\boldsymbol{\psi}_d^o\right)\dfrac{\Delta\Omega}{\omega_{\text{nom}}}}{1+\left(\dfrac{s}{\omega_{\text{nom}}}+\dfrac{r}{l_d(s)}\right)\left(\dfrac{s}{\omega_{\text{nom}}}+\dfrac{r}{l_q(s)}\right)}
\end{aligned}\right\} \qquad [4.1.30]$$

$$\boldsymbol{\psi}_o = \frac{\boldsymbol{v}_o - v_n/V_{F(\text{nom})}}{\dfrac{s}{\omega_{\text{nom}}}+\dfrac{r}{l_o(s)}} \qquad [4.1.31]$$

(from which we also have $\Delta\boldsymbol{i}_f$, $\Delta\boldsymbol{\psi}_f$, $\Delta\boldsymbol{i}_d$, $\Delta\boldsymbol{i}_q$, $\boldsymbol{i}_o$, by using Equations [4.1.27], [4.1.23′], and [4.1.24′]).

Among other things, we may note that the dependence of $\Delta\boldsymbol{\psi}_d$, $\Delta\boldsymbol{\psi}_q$ on $\Delta\boldsymbol{v}_f$ (for given $\Delta\boldsymbol{v}_d$, $\Delta\boldsymbol{v}_q$, $\Delta\Omega$) is the result of the term $(r\Delta\boldsymbol{i}_d)$ in Equations [4.1.28], and would therefore be lacking if $r=0$. Furthermore, if $\Omega = \text{constant} = \omega_{\text{nom}}$, Equations [4.1.28] and [4.1.30] could be used without the Δ variation symbol (and $\Delta\Omega = 0$).

Lastly, from Equations [4.1.20] and [4.1.21] we derive, for small variations (and in pu):

$$\Delta\boldsymbol{C}_m = \left(\frac{T_a'}{\omega_{\text{nom}}}s + \left(\frac{\mathrm{d}\boldsymbol{C}_p}{\mathrm{d}\Omega}\right)^o\right)\Delta\Omega + \Delta\boldsymbol{C}_e \qquad [4.1.32]$$

with:

$$\Delta\boldsymbol{C}_e = \boldsymbol{i}_q^o\Delta\boldsymbol{\psi}_d - \boldsymbol{i}_d^o\Delta\boldsymbol{\psi}_q + \boldsymbol{\psi}_d^o\Delta\boldsymbol{i}_q - \boldsymbol{\psi}_q^o\Delta\boldsymbol{i}_d \qquad [4.1.33]$$

or even (recalling Equations [4.1.23′], and [4.1.27]):

$$\Delta\boldsymbol{C}_e = -\boldsymbol{\psi}_q^o\frac{a(s)}{l_d(s)}\Delta\boldsymbol{v}_f + \left(\boldsymbol{i}_q^o + \frac{\boldsymbol{\psi}_q^o}{l_d(s)}\right)\Delta\boldsymbol{\psi}_d + \left(-\boldsymbol{i}_d^o - \frac{\boldsymbol{\psi}_d^o}{l_q(s)}\right)\Delta\boldsymbol{\psi}_q \qquad [4.1.33']$$

which, in turn, can be expressed as a function of $\Delta\boldsymbol{v}_f$, $\Delta\boldsymbol{v}_d$, $\Delta\boldsymbol{v}_q$, $\Delta\Omega$ using the expressions in Equations [4.1.30].

4.1.4. General Comments

Comment 1

In the dynamic analysis of a machine, the application of the Park's transformation proves particularly useful for various reasons, among which are the following:

(1) it makes it possible to extend, based on precise definitions, the use of vector representations even to the case of transient operation;
(2) the Equations [4.1.6] and [4.1.7] (relating to the magnetic part of the machine), containing inductances that vary with the position θ, are replaced by Equations [4.1.6′] and [4.1.7′], with constant inductances;
(3) for the normal "alternating current" steady-state, there is, in the Park's variables (see also the equivalent circuits in Fig. 4.3), a corresponding "dc" static condition; see Section 4.4.1.

These considerations also may be extended to the case of more sophisticated models, such as those outlined in Section 4.3.

With (1), emphasis must be placed on the strictly spatial meaning of the vector representations in general operating conditions, as will be pointed out in the following Comment 3.

In fact, it is possible for example to ascertain that the vector:

$$\psi_a + \psi_b \epsilon^{j120^\circ} + \psi_c \epsilon^{j240^\circ} = \sqrt{\tfrac{3}{2}}\,\overline{\psi}\epsilon^{j\theta}$$

(see Appendix 2) represents, in magnitude and position, the resulting flux linked with the stator circuits, with the assumption that the angular position of this flux is referred to axis a in Figure 4.2.

The same conclusion therefore applies, on an appropriate scale, to the vector $\overline{\psi}$, but with angular position referred to the direct axis instead of axis a.

The vector $\overline{\imath}$ may similarly be considered representative of the resulting mmf caused by the stator circuits, still assuming that its angular position refers to the direct axis.

These considerations can be used to justify — even intuitively — the transition (2) from Equations [4.1.6] and [4.1.7], to [4.1.6′] and [4.1.7′], with regard to the variables d, q. If the vectors $\overline{\psi}$ and $\overline{\imath}$ are decomposed according to the direct axis (components ψ_d, i_d) and the quadrature axis (components ψ_q, i_q), it is spontaneously expected that the variables ψ_f, i_f, ψ_d, i_d, ψ_q, i_q could be tied with each other irrespective of θ, as expressed by Equations [4.1.6′] and [4.1.7′].

Moreover, it follows from (3) that the analysis can be made in the traditionally simplest manner, since the "equilibrium" operating condition of the system is represented by a static state for which it is sufficient to assume formally $p \triangleq \mathrm{d}/\mathrm{d}t = 0$, whereas through the operator $p \triangleq \mathrm{d}/\mathrm{d}t$ account can be taken, in accordance with the usual criteria, of the variations in respect of that operating condition. In particular, any equilibrium state may be analyzed by assuming $s = 0$ in transfer functions, whereas the assumption $s = \infty$ may allow the study of the initial part of the transient following sudden disturbances. Moreover, a further advantage concerns the treatment of nonlinearities, for small variations around the equilibrium condition. For example, in the case of expressions $z = z(x, y)$ with x, y, z any variables of the system, we may derive, in "linearized" terms:

$$\Delta z = \frac{\partial z}{\partial x}(x^o, y^o)\Delta x + \frac{\partial z}{\partial y}(x^o, y^o)\Delta y$$

where x^o, y^o are constants equal to the equilibrium values of x, y (see Equations [4.1.28], [4.1.30], [4.1.32], [4.1.33], and [4.1.33′]).

Comment 2

The parameters that appear in Equations [4.1.26′] fall traditionally, with appropriate denominations, within the characteristic machine parameters, whose values are supplied by the manufacturer or can be determined through experimental tests. Among the denominations in use, we may recall the following:

- x_d = direct-axis synchronous reactance (in pu);
- $\hat{x}'_d$ = direct-axis transient reactance (in pu);
- x_q = quadrature-axis synchronous reactance (in pu);
- x_o = homopolar reactance (in pu);
- $\hat{T}'_{do}$ = open-circuit transient time constant, referring to the direct axis;
- $\hat{T}'_d = (\hat{x}'_d/x_d)\hat{T}'_{do}$ = short-circuit transient time constant, referring to the direct axis(3).

Regarding x_d (and similarly x_q), the adjective "*synchronous*" is justified by the fact that:

$$x_d = l_d(0)$$

is the reactance to be considered, for the direct axis, in the synchronous (equilibrium) state of the machine; on the other hand, immediately after the occurrence of a sudden disturbance the reactance to be considered for the direct axis is:

$$\hat{x}'_d = l_d(\infty)$$

(3) For round rotor machines, it usually holds $x_d \cong x_q \cong 2$–2.5 (x_q also may be smaller than x_d by the 5–10 %), $\hat{x}'_d \cong 0.2$–0.35, $\hat{T}'_{do} \cong 5$–9 seconds.

Usual values for salient pole machines are instead:

- for generators: $x_d \cong 1.0$–1.2, $x_q \cong 0.6$–0.9, $\hat{x}'_d \cong 0.25$–0.35;
- for compensators: $x_d \cong 2$–2.5, $x_q \cong 1.2$–1.6, $\hat{x}'_d \cong 0.35$–0.45;

whereas $\hat{T}'_{do} \cong 5$–10 seconds, with the higher values for the compensators.

which is therefore said to be "*transient*." Generally, $l_d(s)$ may be considered, for the third of the expressions in Equations [4.1.22′], as the "operational" reactance of the direct axis, equal to the operational inductance (in pu) seen from the terminals (D_1, D_2) in Figure 4.3a during any perturbed operation with v_f constant.

Moreover, the time constants $\hat{T}'_{do}$, $\hat{T}'_d$ are called "*open-circuit*" and "*short-circuit*," respectively, inasmuch as they, respectively, characterize the transient response for $i_d = 0$ (see Equations [4.1.22′]), i.e., (D_1, D_2) open, and for $\psi_d = 0$ (see Equations [4.1.27]), i.e., (D_1, D_2) short-circuited.

However, the parameters $\hat{x}'_d$, $\hat{T}'_d$, $\hat{T}'_{do}$ refer to the model so far described, i.e., are defined under the assumption that the rotor contains only the field circuit (and without magnetic saturation). In practical cases, because of the presence of additional rotor circuits, the denominations here assigned to $\hat{x}'_d$, $\hat{T}'_d$, $\hat{T}'_{do}$ become valid (more correctly and still ignoring possible effects of saturation) for the parameters x'_d, T'_d, T'_{do} defined in Section 4.3.2. In practice, however, the values of these last parameters are slightly different from previous ones.

Comment 3

In Sections 4.4.2 and 4.4.3, typical perturbed operating conditions of the machine will be outlined for which, in particular, the Park's variables are not constant but are sinusoidal functions of the time.

The analysis of the system (after possible linearization) in sinusoidal operation with pulsation ν, may be performed in accordance with the usual techniques, by associating (see Appendix 1) with each sinusoidal variable:

$$w = \sqrt{2} W \cos(\nu t + \varphi_w)$$

a *phasor* of the type:

$$\tilde{w} \triangleq W \epsilon^{\tilde{j}\varphi_w}$$

so that we have, with obvious symbols:

$$w = \sqrt{2}\mathcal{R}e(\tilde{w}\,\epsilon^{\tilde{j}\nu t}), \quad \widetilde{(w_1 + w_2)} = \tilde{w}_1 + \tilde{w}_2, \quad \left(\widetilde{\frac{dw}{dt}}\right) = \tilde{j}\nu\tilde{w}$$

By virtue of this last property, the relationships that hold between the different phasors may be directly obtained from the differential equations of the system, by replacing the variables (or their Laplace transforms) by the respective phasors, and the operator $p \triangleq d/dt$ (or the Laplace variable s in transfer functions) by the multiplying factor $\tilde{j}\nu$.

In present notations, the use of the circumflex symbol avoids confusion between the "phasors" and the (Park's) "vectors" previously considered, as well as the operators acting on them, respectively. The definition of the phasors is closely linked with the temporal assumption that the (scalar) variables in question are sinusoidal at a given frequency, whereas Park's vectors (e.g. $\overline{v} = v_d + j v_q$, etc.) are defined for any operating condition, based on pairs of scalar variables (d, q) related to direct and quadrature axes, and therefore have a strictly spatial meaning. In particular, $\tilde{\jmath}$ (not to be confused with j) represents the imaginary unit in the "plane of phasors," whereas the symbols $\mathcal{R}e$ and Im, applied to a given phasor, denote the real and imaginary parts of the phasor in the aforementioned plane, respectively.

Comment 4

According to above, the mathematical model of the synchronous machine consists of Equations [4.1.4], [4.1.5′], [4.1.6′], [4.1.7′], [4.1.9], and [4.1.10] (or, in pu, Equations [4.1.16],. . ., [4.1.21]), in which the variables d and q depend, by definition, on the reference θ, equal to the angular electric position of the rotor. Instead of θ, we may consider, more generally, the difference:

$$\delta \triangleq \theta - \theta_s$$

by denoting:

$$\frac{\mathrm{d}\theta_s}{\mathrm{d}t} = \Omega_s = \text{constant}$$

(θ_s may be interpreted as the angular electric position of a fictitious rotor, rotating at the constant speed Ω_s; by assuming $\Omega_s = \Omega^o$, equal to the electric speed at equilibrium state, it follows δ^o = constant). If we also consider the equation:

$$\frac{\mathrm{d}\delta}{\mathrm{d}t} = \Omega - \Omega_s$$

it is easy to ascertain that the model in question is dynamically of the *fifth order* (state variables: ψ_f, ψ_d, ψ_q, Ω, δ) for the axes d and q, and of the first order (state variable: ψ_o) for the axis o.

Moreover, vectors $\overline{v} = v_d + j v_q$, $\overline{\imath} = i_d + j i_q$ are defined by assuming θ as the reference of the Park's transformation. By denoting $\overline{v}_s = v_{ds} + j v_{qs}$, $\overline{\imath}_s = i_{ds} + j i_{qs}$, the vectors obtained by assuming as reference the position θ_s defined above (independent of the mechanical transients of the machine), it follows:

$$\overline{v} = \overline{v}_s \epsilon^{-j\delta}, \quad \overline{\imath} = \overline{\imath}_s \epsilon^{-j\delta}$$

where the effect of δ (that is of the mechanical transients) is evident.

For practical applications, the model so far described may be considered too complicated in some cases and too simple in others. Sections 4.2 and 4.3 describe usual criteria for obtaining, respectively, simpler models and better approximate models, according to the indications given in Figure 4.4.

However, note that — independent of the adopted model — the relationship between variables (possibly in pu) may be generally represented by a block diagram as shown in Figure 4.5.

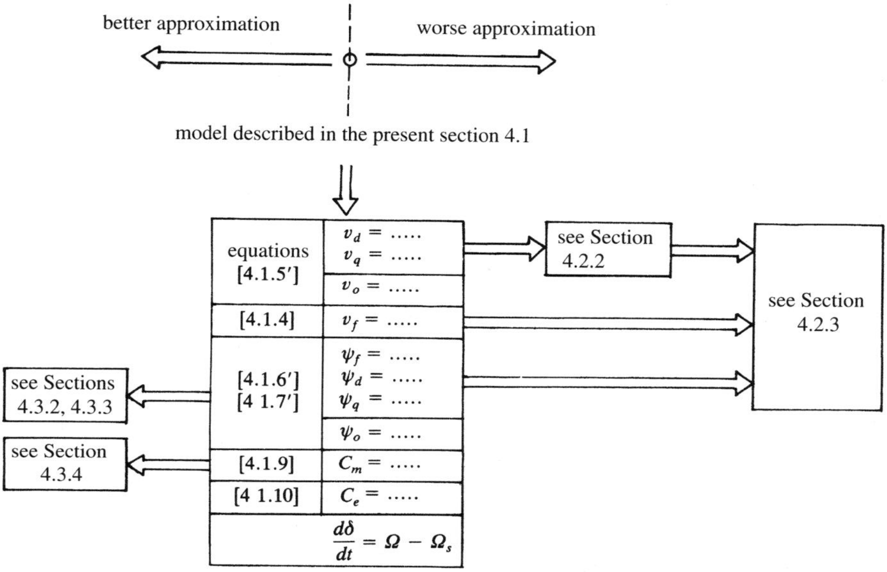

Figure 4.4. Summarizing scheme of the possible different mathematical models of the synchronous machine.

4.2. USUAL SIMPLIFICATIONS

4.2.1. Preliminaries

Despite the simplifying assumptions on which it is based, the model described in Section 4.1 may still seem too complicated for many applications (especially for the study of electromechanical phenomena in a multimachine system, even though relatively powerful computers are available for simulating the system), particularly with the axes d, q, and certainly a long way from the extremely simplified model in Section 4.1.1.

This section describes usual simplifications, with reference to the model defined above. Some of these simplifications do not appear to be particularly prejudicial with respect to the model's degree of approximation. Others, however, are mainly dictated by convenience and are useful only for analyses of first approximation or for cases in which the machine behavior can be considered of minor importance in relation to the overall problem.

4.2.2. Third-Order "dq" Model

In the analysis of electromechanical phenomena, a generally accepted simplification is obtained by dropping, in the first two expressions in Equations [4.1.17], the terms $(1/\omega_{\text{nom}})\,\mathrm{d}\psi_d/\mathrm{d}t$ and $(1/\omega_{\text{nom}})\,\mathrm{d}\psi_q/\mathrm{d}t$ (known as *"transformer" emfs*,

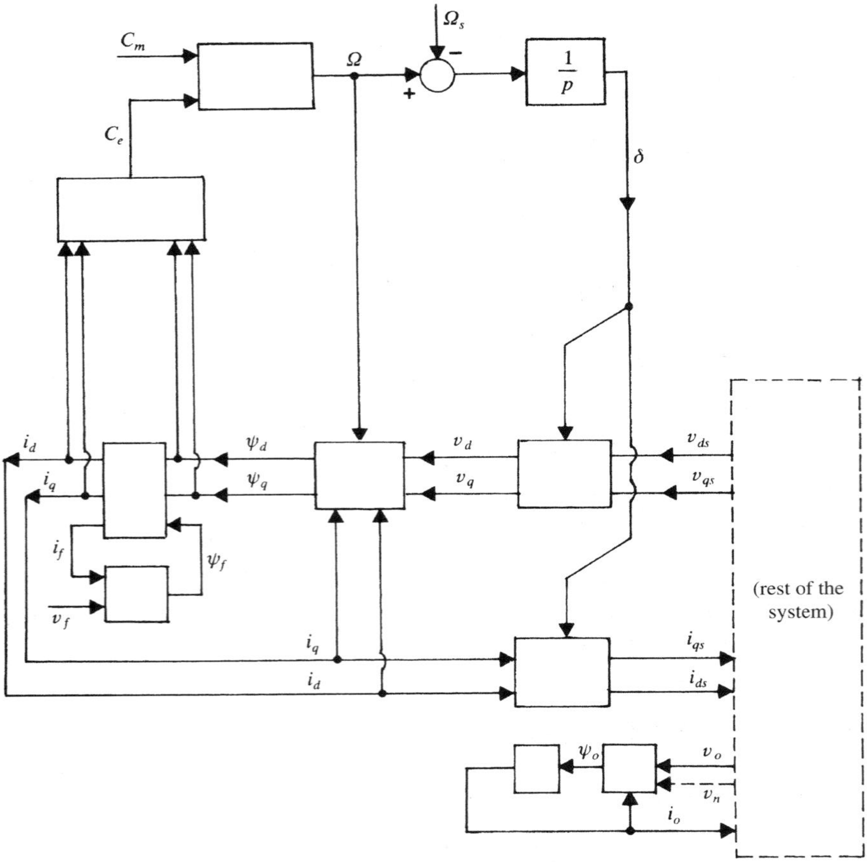

Figure 4.5. Block diagram of the synchronous machine: general structure.

in pu). If we do this, the model for the axes d and q becomes of the third order (state variables: ψ_f, Ω, δ).

This simplification is usually accompanied by the substitution, in the equations in question, of the terms $(-\Omega/\omega_{\text{nom}})\boldsymbol{\psi}_q$, $(\Omega/\omega_{\text{nom}})\boldsymbol{\psi}_d$ (*"motional" emfs*, in pu) by $-\boldsymbol{\psi}_q$ and $\boldsymbol{\psi}_d$ respectively. This is, therefore, a further simplification that leads to the linear equations (in pu):

$$\left.\begin{aligned} \boldsymbol{v}_d &= -r\boldsymbol{i}_d - \boldsymbol{\psi}_q \\ \boldsymbol{v}_q &= -r\boldsymbol{i}_q + \boldsymbol{\psi}_d \end{aligned}\right\} \overline{\boldsymbol{v}} = -r\overline{\boldsymbol{\imath}} + j\overline{\boldsymbol{\psi}} \qquad [4.2.1]$$

instead of the first two parts of Equations [4.1.17]. Consequently, the equivalent circuits (not in pu) in Figure 4.3a,b are replaced by those in Figure 4.6a.

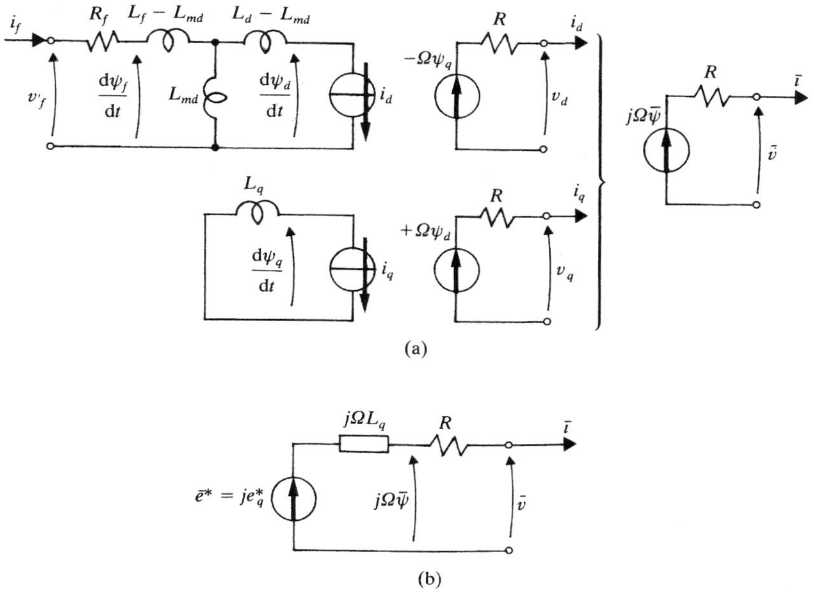

Figure 4.6. Equivalent circuits for d and q axes with simplifications of the transformer and motional emfs.

The following remarks must be made:

(1) The simplifications in question influence phenomena that are often of minor importance (e.g., the presence of unidirectional components in the phase fluxes and currents, during the short-circuit transients; see Section 4.4.4), leaving important operating characteristics unchanged. In this connection, it is important to note that if the transformer emfs are neglected, the simplification regarding the motional emfs may be convenient not only for obtaining simpler (and linear) equations, but also for avoiding a less accurate approximation; see for example, for $r = 0$, the characteristics described in Section 4.4.2b, the transfer functions that relate ΔC_e to $\Delta\delta$ and Δv_f in Section 7.2, and so on.

(2) The above simplifications follow normal practice, which represents the elements of the network (lines, transformers, etc.) with impedances or admittances at nominal frequency, disregarding their actual dynamic behavior (see Chapter 5), without generally prejudicing the resulting approximation.

(3) If such simplifications were not made, the network element dynamics should be consistently considered. This would, in practice, be prohibitive (even for the simulations on computer), except for extremely simple cases.

If we accept Equation [4.2.1], the model of the machine becomes, in pu:

$$\boldsymbol{v}_f = \boldsymbol{i}_f + \hat{T}'_{do}\frac{\mathrm{d}\boldsymbol{\psi}_f}{\mathrm{d}t} \qquad \text{[4.1.16 rep.]}$$

$$\left.\begin{array}{l} \left.\begin{array}{l} \boldsymbol{v}_d = -r\boldsymbol{i}_d - \boldsymbol{\psi}_q \\ \boldsymbol{v}_q = -r\boldsymbol{i}_q + \boldsymbol{\psi}_d \end{array}\right\} \overline{\boldsymbol{v}} = -r\overline{\boldsymbol{\imath}} + j\overline{\boldsymbol{\psi}} \\ \boldsymbol{v}_o - \dfrac{v_n}{V_{(F)\mathrm{nom}}} = -r\boldsymbol{i}_o + \dfrac{1}{\omega_{\mathrm{nom}}}\dfrac{\mathrm{d}\boldsymbol{\psi}_o}{\mathrm{d}t} \end{array}\right\} \qquad \text{[4.2.2]}$$

$$\boldsymbol{\psi}_f = \boldsymbol{i}_f - (x_d - \hat{x}'_d)\boldsymbol{i}_d \qquad \text{[4.1.18 rep.]}$$

$$\left.\begin{array}{l} \left.\begin{array}{l} \boldsymbol{\psi}_d = \boldsymbol{i}_f - x_d\boldsymbol{i}_d \\ \boldsymbol{\psi}_q = -x_q\boldsymbol{i}_q \end{array}\right\} \overline{\boldsymbol{\psi}} = \boldsymbol{i}_f - (x_d\boldsymbol{i}_d + x_q j\boldsymbol{i}_q) = \boldsymbol{\psi}_f - (\hat{x}'_d\boldsymbol{i}_d + x_q j\boldsymbol{i}_q) \\ \boldsymbol{\psi}_o = -x_o\boldsymbol{i}_o \end{array}\right\}$$

[4.1.19 rep.]

$$\boldsymbol{C}_m = \frac{T'_a}{\omega_{\mathrm{nom}}}\frac{\mathrm{d}\Omega}{\mathrm{d}t} + \boldsymbol{C}_p(\Omega) + \boldsymbol{C}_e \qquad \text{[4.1.20 rep.]}$$

$$\boldsymbol{C}_e = \boldsymbol{\psi}_d\boldsymbol{i}_q - \boldsymbol{\psi}_q\boldsymbol{i}_d = (\boldsymbol{i}_f - (x_d - x_q)\boldsymbol{i}_d)\boldsymbol{i}_q = (\boldsymbol{\psi}_f + (x_q - \hat{x}'_d)\boldsymbol{i}_d)\boldsymbol{i}_q$$

[4.1.21 rep.]

With the axes d, q (whereas for the axis o, the last parts of Equations [4.2.2] and [4.1.19] apply), the model may be expressed in more compact form, by putting:

$$\overline{\boldsymbol{e}}^* = j\boldsymbol{e}_q{}^* \triangleq j(\boldsymbol{i}_f - (x_d - x_q)\boldsymbol{i}_d) = j(\boldsymbol{\psi}_f + (x_q - \hat{x}'_d)\boldsymbol{i}_d)$$

and by eliminating:

$$\boldsymbol{i}_f = \boldsymbol{e}_q^* + (x_d - x_q)\boldsymbol{i}_d$$

$$\boldsymbol{\psi}_f = \boldsymbol{e}_q^* - (x_q - \hat{x}'_d)\boldsymbol{i}_d$$

$$\left.\begin{array}{l} \boldsymbol{\psi}_d = \boldsymbol{e}_q^* - x_q\boldsymbol{i}_d \\ \boldsymbol{\psi}_q = -x_q\boldsymbol{i}_q \end{array}\right\} \overline{\boldsymbol{\psi}} = \boldsymbol{e}_q^* - x_q\overline{\boldsymbol{\imath}}$$

In fact, by doing this, we obtain (see Fig. 4.6b):

$$\left.\begin{array}{l} \overline{\boldsymbol{v}} = \overline{\boldsymbol{e}}^* - (r + jx_q)\,\overline{\boldsymbol{\imath}} \\ \boldsymbol{C}_m = \dfrac{T'_a}{\omega_{\mathrm{nom}}}\dfrac{\mathrm{d}\Omega}{\mathrm{d}t} + \boldsymbol{C}_p(\Omega) + \boldsymbol{C}_e \\ \boldsymbol{C}_e = \boldsymbol{e}_q^*\boldsymbol{i}_q = \langle\overline{\boldsymbol{e}}^*, \overline{\boldsymbol{\imath}}\rangle \end{array}\right\} \qquad \text{[4.2.3]}$$

in which $\boldsymbol{e}_q^*$ is dynamically influenced by $\boldsymbol{v}_f$ and $\boldsymbol{i}_d$, according to the equation:

$$\boldsymbol{e}_q^* = \frac{\boldsymbol{v}_f - ((x_d - x_q) - s\hat{T}'_{do}(x_q - \hat{x}'_d))\boldsymbol{i}_d}{1 + s\hat{T}'_{do}} \quad [4.2.4]$$

It is important to emphasize the close formal analogy between the Equations [4.2.3] and the first approximation model in Equations [4.1.1], [4.1.2], and [4.1.3] considered in Section 4.1.1, by taking $\overline{\boldsymbol{e}}^* = j\boldsymbol{e}_q^*$ and $(r + jx_q)$ as the "equivalent" internal emf and the "equivalent" output impedance of the machine (in pu), respectively. In addition, we may ascertain that the angular position of the emf is fixed with respect to the polar axis (to be more precise, the emf lies on the quadrature axis).

However, Equation [4.2.4] is now added to Equations [4.2.3]. For Equation [4.2.4], the emf $\overline{\boldsymbol{e}}^*$ cannot be simply defined as the open-circuit voltage ($\boldsymbol{e}_q^*$ also depends on $\boldsymbol{i}_d$), even in steady-state operation (except for $x_d = x_q$).

Very often, the effects of the armature resistance can be ignored, so that it may be assumed $r = 0$. In the present case, the Equations [4.2.1] then become more simple:

$$\left.\begin{aligned} \boldsymbol{v}_d &= -\boldsymbol{\psi}_q \\ \boldsymbol{v}_q &= +\boldsymbol{\psi}_d \end{aligned}\right\} \overline{\boldsymbol{v}} = j\overline{\boldsymbol{\psi}} \quad [4.2.5]$$

and the output impedance is purely reactive.

From the equation:

$$\boldsymbol{\psi}_d = \boldsymbol{\psi}_f - \hat{x}'_d \boldsymbol{i}_d$$

(see Equation [4.1.19′]), it follows:

$$\frac{\mathrm{d}\boldsymbol{\psi}_d}{\mathrm{d}t} = \frac{\mathrm{d}\boldsymbol{\psi}_f}{\mathrm{d}t} - \hat{x}'_d \frac{\mathrm{d}\boldsymbol{i}_d}{\mathrm{d}t}$$

in which, for Equation [4.1.16]:

$$\frac{\mathrm{d}\boldsymbol{\psi}_f}{\mathrm{d}t} = \frac{\boldsymbol{v}_f - \boldsymbol{i}_f}{\hat{T}'_{do}}$$

To obtain the desired reduction in the dynamic order of the model (from 5 to 3, for the axes d, q), we also may approximate, in the first expression in Equations [4.1.17], the transformer emf $(1/\omega_{\text{nom}})\mathrm{d}\boldsymbol{\psi}_d/\mathrm{d}t$ with $(1/\omega_{\text{nom}})\,\mathrm{d}\boldsymbol{\psi}_f/\mathrm{d}t = (\boldsymbol{v}_f - \boldsymbol{i}_f)/(\omega_{\text{nom}}\hat{T}'_{do})$, by dropping only the term $-(\hat{x}'_d/\omega_{\text{nom}})\,\mathrm{d}\boldsymbol{i}_d/\mathrm{d}t$.

The criterion in question may be extended to more sophisticated models (see Section 4.3), with reference to both the transformer emfs $(1/\omega_{\text{nom}})\,\mathrm{d}\boldsymbol{\psi}_d/\mathrm{d}t$, $(1/\omega_{\text{nom}})\,\mathrm{d}\boldsymbol{\psi}_q/\mathrm{d}t$. Expressing $\boldsymbol{\psi}_d$, $\boldsymbol{\psi}_q$ as functions of $\boldsymbol{i}_d$, $\boldsymbol{i}_q$ and of state variables of the machine ($\boldsymbol{\psi}_f, \ldots$), the criterion then leads to the elimination of the only resulting terms in $\mathrm{d}\boldsymbol{i}_d/\mathrm{d}t$, $\mathrm{d}\boldsymbol{i}_q/\mathrm{d}t$. However, the model complications may be far from negligible, especially if $\boldsymbol{\psi}_d$ and $\boldsymbol{\psi}_q$ are nonlinear functions of the aforementioned variables, e.g., in the presence of magnetic saturation.

4.2.3. Second-Order "dq" Models

The model described in Section 4.2.2 is liable to some acceptable simplifications for particular operating conditions. If, for example, the transients in question are fairly slow, from Equation [4.1.16] it follows, ignoring the term in $\mathrm{d}\boldsymbol{\psi}_f/\mathrm{d}t$:

$$\boldsymbol{i}_f = \boldsymbol{v}_f$$

and by putting:

$$\overline{\boldsymbol{e}} = j\boldsymbol{e}_q \triangleq j\boldsymbol{i}_f \qquad [4.2.6]$$

(the so-called "*synchronous*" *emf*, in pu), the Equations [4.2.3] and [4.2.4] can be replaced, respectively, by:

$$\left.\begin{aligned} \overline{\boldsymbol{v}} &= \overline{\boldsymbol{e}} - r\overline{\boldsymbol{\imath}} - jx_d\boldsymbol{i}_d + x_q\boldsymbol{i}_q \\ \boldsymbol{C}_m &= \frac{T'_a}{\omega_{\text{nom}}}\frac{\mathrm{d}\Omega}{\mathrm{d}t} + \boldsymbol{C}_p(\Omega) + \boldsymbol{C}_e \\ \boldsymbol{C}_e &= \boldsymbol{e}_q\boldsymbol{i}_q - (x_d - x_q)\boldsymbol{i}_d\boldsymbol{i}_q \end{aligned}\right\} \qquad [4.2.7]$$

and

$$\boldsymbol{e}_q = \boldsymbol{i}_f = \boldsymbol{v}_f \qquad [4.2.8]$$

Note that the assumption $\boldsymbol{i}_f = \boldsymbol{v}_f$ does not mean to suppose $\mathrm{d}\boldsymbol{\psi}_f/\mathrm{d}t = 0$ (i.e., $\boldsymbol{\psi}_f =$ constant as in the next examined case), but it corresponds to the assumption $s = 0$ in Equations [4.1.22′], as if the value of $\boldsymbol{\psi}_f$ would instantaneously obey to the equations (of the electrical and magnetic parts) which hold in the equilibrium steady-state. Since $\hat{T}'_{do}$ may be of 5–10 seconds, it appears evident that, because of Equation [4.1.16], the model under question may be accepted only when managing particularly slow phenomena.

Similarly, if the transients in question are fast enough to assume:

$$\boldsymbol{\psi}_f = \text{constant}$$

by putting:

$$\overline{\boldsymbol{e}}' = j\boldsymbol{e}'_q \triangleq j\boldsymbol{\psi}_f \qquad [4.2.9]$$

(the so-called "*transient*" *emf*, in pu), we obtain the equations:

$$\left.\begin{aligned} \overline{\boldsymbol{v}} &= \overline{\boldsymbol{e}}' - r\overline{\boldsymbol{\imath}} - j\hat{x}'_d\boldsymbol{i}_d + x_q\boldsymbol{i}_q \\ \boldsymbol{C}_m &= \frac{T'_a}{\omega_{\text{nom}}}\frac{\mathrm{d}\Omega}{\mathrm{d}t} + \boldsymbol{C}_p(\Omega) + \boldsymbol{C}_e \\ \boldsymbol{C}_e &= \boldsymbol{e}'_q\boldsymbol{i}_q + (x_q - \hat{x}'_d)\boldsymbol{i}_d\boldsymbol{i}_q \end{aligned}\right\} \qquad [4.2.10]$$

and

$$\boldsymbol{e}'_q = \boldsymbol{\psi}_f = \text{constant} \qquad [4.2.11]$$

Note that the assumption $\boldsymbol{\psi}_f =$ constant does not imply $\boldsymbol{i}_f = \boldsymbol{v}_f - \hat{T}'_{do}\mathrm{d}\boldsymbol{\psi}_f/\mathrm{d}t = \boldsymbol{v}_f$ as in the previous case, but it corresponds to the assumption $s = \infty$ in Equations [4.1.22′], applied to the different quantity variations. Generally, it also may be intended $\hat{T}'_{do}\mathrm{d}\boldsymbol{\psi}_f/\mathrm{d}t = \boldsymbol{v}_f - \boldsymbol{v}_f^o$, and thus $\boldsymbol{\psi}_f$ possibly variable. However, by doing this, the model remains of the third order (see Section 7.2.2b.)

Through both the approximations described above, the model for the axes d and q thus becomes of the second order (state variables: Ω, δ). Furthermore, in Equations [4.2.7], the expressions of $\boldsymbol{v}$ and $\boldsymbol{C}_e$ become simpler if $x_d = x_q$ (round rotor). The result is:

$$\overline{\boldsymbol{v}} = \overline{\boldsymbol{e}} - (r + jx_d)\overline{\boldsymbol{\imath}}$$

$$\boldsymbol{C}_e = \boldsymbol{e}_q\boldsymbol{i}_q = \langle\overline{\boldsymbol{e}}, \overline{\boldsymbol{\imath}}\rangle$$

and similarly for Equations [4.2.10], if it is assumed $\hat{x}'_d = x_q$.

Often the following model (see also Section 4.1.1) is assumed:

$$\left.\begin{aligned} \overline{\boldsymbol{v}} &= \overline{\boldsymbol{e}}_{\mathrm{eq}} - (r_{\mathrm{eq}} + jx_{\mathrm{eq}})\overline{\boldsymbol{\imath}} \\ \boldsymbol{C}_m &= \frac{T'_a}{\omega_{\mathrm{nom}}}\frac{\mathrm{d}\Omega}{\mathrm{d}t} + \boldsymbol{C}_p(\Omega) + \boldsymbol{C}_e \\ \boldsymbol{C}_e &= \langle\overline{\boldsymbol{e}}_{\mathrm{eq}}, \overline{\boldsymbol{\imath}}\rangle \end{aligned}\right\} \qquad [4.2.12]$$

where $\overline{\boldsymbol{e}}_{\mathrm{eq}}$, $(r_{\mathrm{eq}} + jx_{\mathrm{eq}})$ are the "equivalent" internal emf and the "equivalent" output impedance, suitably defined in relation to the particular problem. In addition, it is assumed that the emf has a fixed angular position with respect to the polar axis, and its magnitude $\boldsymbol{e}_{\mathrm{eq}}$ is independent of the armature current $\overline{\boldsymbol{\imath}}$ and dynamically influenced (according to a suitable law) by the field voltage $\boldsymbol{v}_f$. At the cost of further schematizations — often very rough — this type of model also is used to represent, jointly, the machine and its voltage regulation. In such a case, $\boldsymbol{e}_{\mathrm{eq}}$ may be considered depending on the reference of the voltage regulator.

By assuming $\boldsymbol{e}_{\mathrm{eq}}$ constant, the machine is represented, for the axes d and q, by an extremely simple second-order model. In particular, the definition of the output impedance prevents consideration of any anisotropy of the machine and makes it possible to treat this impedance like the network impedances, in accordance with the usual computer programs for calculating the impedance or admittance matrices between assigned nodes, etc. However, such a model can find justification only in special cases; for example, when it is possible to assume $\boldsymbol{i}_f$ constant and $x_d = x_q$ (we then find the Equations [4.2.7], assuming $\overline{\boldsymbol{e}}_{\mathrm{eq}} = \overline{\boldsymbol{e}}$, $r_{\mathrm{eq}} = r$, $x_{\mathrm{eq}} = x_d = x_q$), or $\boldsymbol{\psi}_f$ constant and $\hat{x}'_d = x_q$ (we then find the Equations [4.2.10], assuming $\overline{\boldsymbol{e}}_{\mathrm{eq}} = \overline{\boldsymbol{e}}'$, $r_{\mathrm{eq}} = r$, $x_{\mathrm{eq}} = \hat{x}'_d = x_q$)[4].

---

[4] The assumption $\boldsymbol{\psi}_f =$ constant, very frequently used in studies on "transient stability," is generally accompanied by the assumptions $r_{\mathrm{eq}} = r$, $x_{\mathrm{eq}} = \hat{x}'_d$, so that $\overline{\boldsymbol{e}}_{\mathrm{eq}} = \overline{\boldsymbol{v}} + (r + j\hat{x}'_d)\overline{\boldsymbol{\imath}}$ is the so-called *emf "behind the transient reactance."* By comparison with the first part of the Equations [4.2.10], this emf equals $\overline{\boldsymbol{e}}' + (x_q - \hat{x}'_d)\boldsymbol{i}_q$ (in which $\overline{\boldsymbol{e}}' \triangleq j\boldsymbol{\psi}_f =$ constant), and the assumption that it remains constant is equivalent to assuming that, for $\hat{x}'_d \neq x_q$, the variations of $\boldsymbol{i}_q$ are slight.

The emf $\overline{e}_{eq}$ and the impedance $(r_{eq} + jx_{eq})$ are called "*pendular emf*" and "*pendular impedance*" when they have been determined for reproducing, with sufficient approximation, the electromechanical oscillation (or "*pendulation*") of the machine with respect to the rest of the system, or at least the "dominant" oscillation (see also the end of Section 7.2.2b).

4.3. MORE SOPHISTICATED MODELS

4.3.1. Preliminaries

The model described in Section 4.1 is based on the following fundamental assumptions:

(a) there are no "additional" rotor circuits other than the field circuit;

(b) there is no magnetic saturation;

(c) there are no torsional phenomena in the unit shaft.

For many practical applications, this model is still too burdensome, and Section 4.2 describes the simplified models that are more often used. Nevertheless, this does not alter the fact that in other cases, when the machine is particularly involved in the dynamic phenomena in question, it is necessary to have recourse to models of closer approximation, by removing the simplifying assumptions referred to above, or at least some of them. The Equations (in pu) [4.1.18], [4.1.19], and [4.1.20] must then be replaced by more suitable ones, as indicated later in this section, whereas the Equations [4.1.16], [4.1.17], and [4.1.21] remain valid, since they are insensitive to the aforementioned assumptions.

Hereafter, special reference will be made to the axes d and q, for which the model defined in Section 4.1 is affected in a nonnegligible measure by all the assumptions under question.

In most applications, it is still advisable, in the Equations [4.1.17], to make the simplifications on the transformer emfs and on the motional emfs, as in Section 4.2.2 (Equations [4.2.1]). Indeed, recourse to a "complete" model giving up such simplifications may in practice (for the already seen reasons) only be justified in the study of special problems, most relating to systems with a few machines (e.g., problems relating to synchronous or asynchronous start-up, short-circuit near the terminals of the machine, etc.).

4.3.2. Effect of the Additional Rotor Circuits

To consider the effect of additional rotor circuits, let us first assume that the rotor includes (in addition to the field circuit, acting magnetically along the direct axis) a second circuit, acting along the quadrature axis.

On this assumption, the equations of the electrical part comprise, regarding the rotor circuits, no longer Equation [4.1.4] alone, i.e.,

$$v_f = R_f i_f + \frac{d\psi_f}{dt} \qquad [4.1.4 \text{ rep.}]$$

but also an equation:

$$v_B = R_B i_B + \frac{d\psi_B}{dt} \qquad [4.3.1]$$

with:

v_B, i_B, ψ_B, R_B = voltage, current, flux, resistance of the additional circuit in question.

(The reduction in pu is omitted here, although it is obvious how to apply it.)

Additionally, the second part of the Equations [4.1.7′] relating to the magnetic part, axis q, must replaced by equations:

$$\psi_B = L_B i_B - L_{mq} i_q \qquad [4.3.2]$$

$$\psi_q = L_{mq} i_B - L_q i_q \qquad [4.3.3]$$

with L_B and L_{mq} constant.

In fact, it is assumed that the magnetic part is linear, excluding saturation and hysteresis. In addition, it is assumed that, due to the constructional characteristics of the machine, the mutual inductances between the additional circuit considered and the phase circuits are sinusoidal functions of the position θ, i.e.,

$$\begin{cases} L_{aB} = L_{Ba} = -M_B \sin\theta \\ L_{bB} = L_{Bb} = -M_B \sin\theta' \\ L_{cB} = L_{Bc} = -M_B \sin\theta'' \end{cases}$$

so that it can be obtained, applying the Park's transformation, $L_{mq} = \sqrt{3/2}M_B =$ constant.

As a first consequence, Equation [4.1.7″] must be replaced by:

$$\overline{\psi} = L_{md} i_f + L_{mq} j i_B - (L_d i_d + L_q j i_q) \qquad [4.3.4]$$

and Equation [4.1.7‴] (expressing i_f as a function of ψ_f, i_d, and i_B as a function of ψ_B, i_q for Equation [4.3.2]) by:

$$\overline{\psi} = \frac{L_{md}}{L_f}\psi_f + \frac{L_{mq}}{L_B} j\psi_B - \left(L_d - \frac{L_{md}^2}{L_f}\right) i_d - \left(L_q - \frac{L_{mq}^2}{L_B}\right) j i_q \qquad [4.3.4']$$

If the voltage v_B is actually applied from outside, the additional circuit may be a *transverse excitation* circuit, present in some types (albeit rare) of machines. We

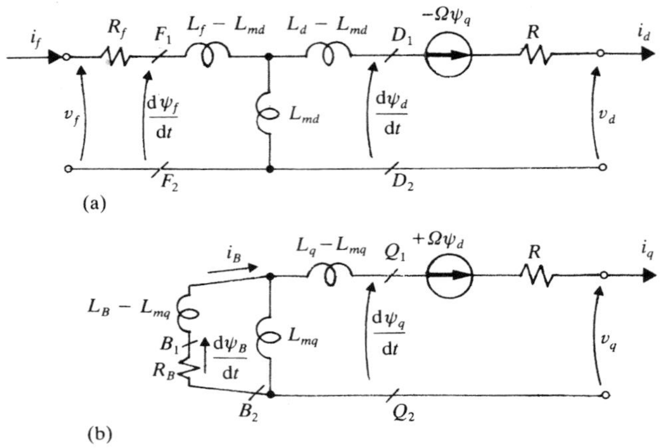

Figure 4.7. Equivalent circuits for d and q axes in the presence of an additional rotor circuit along the q axis.

will not consider such a case (to which, however, it is obvious how the analysis might be extended) and will therefore assume $v_B = 0$, i.e., that the additional circuit is short-circuited.

With the equivalent circuits in Figure 4.3a,b relating to the axes d and q, the second of these must then be modified as indicated in Figure 4.7b. The inductances L_q and $(L_q - L_{mq}^2/L_B)$ are equal to the inductance seen from the terminals (Q_1, Q_2), when the pair of terminals (B_1, B_2) is open or short-circuited, respectively, (similarly to what is already pointed out for the direct axis inductances L_d and $L_d - L_{md}^2/L_f = \hat{L}'_d$).

In the steady-state operation of the machine (see Section 4.4.1), corresponding to the static state (with constant voltages, currents, and fluxes) of the equivalent circuits, $i_B = 0$. Therefore, the effects of the additional rotor circuit only manifest themselves in the transient states. More precisely, the transfer functions defined in Equations [4.1.22], [4.1.23], and [4.1.24] (with the constraints [4.1.25]) are still expressed by Equations [4.1.26], except for $\mathcal{L}_q(s)$, which now becomes (see Fig. 4.7b);

$$\mathcal{L}_q(s) = \frac{L_q + s\left(L_q - \dfrac{L_{mq}^2}{L_B}\right)\dfrac{L_B}{R_B}}{1 + s\dfrac{L_B}{R_B}} \qquad [4.3.5]$$

that is, in pu:

$$l_q(s) \triangleq \frac{-\psi_q}{i_q} = \frac{\omega_{\text{nom}}}{Z_{(F)\text{nom}}}\frac{-\psi_q}{i_q} = \frac{\omega_{\text{nom}}}{Z_{(F)\text{nom}}}\frac{L_q + s\left(L_q - \dfrac{L_{mq}^2}{L_B}\right)\dfrac{L_B}{R_B}}{1 + s\dfrac{L_B}{R_B}} \qquad [4.3.5']$$

and then of the first order, with:

$$l_q(0) = \frac{\omega_{\mathrm{nom}}}{Z_{(F)\mathrm{nom}}} L_q \triangleq x_q, \quad l_q(\infty) = \frac{\omega_{\mathrm{nom}}}{Z_{(F)\mathrm{nom}}}\left(L_q - \frac{L_{mq}^2}{L_B}\right)$$

so that the model's dynamic order of the machine increases by one.

Similarly, it can be recognized that an additional rotor circuit acting along the direct axis leads, in pu — and on the usual assumptions on the magnetic part — to transfer functions $a(s)$, $b(s)$, $b'(s)$, $c(s)$, $c'(s)$, $l_d(s)$ of the second order, instead of simply of the first order as expressed by Equations [4.1.25′] and [4.1.26′]. Transfer functions of a higher order may then be obtained in the presence of further additional circuits.

In practice, the *"damping" windings* of the machine can generally be assimilated with two "equivalent" (short-circuited) additional rotor circuits, one of which acts magnetically along the direct axis, and the other along the quadrature axis. In this case we then obtain, in particular, an $l_d(s)$ of the second order and an $l_q(s)$ of the first order, respectively:

$$\left.\begin{aligned} l_d(s) &= x_d \frac{(1+sT_d')(1+sT_d'')}{(1+sT_{do}')(1+sT_{do}'')} \\ l_q(s) &= x_q \frac{1+sT_q''}{1+sT_{qo}''} \end{aligned}\right\} \qquad [4.3.6]$$

or even:

$$\left.\begin{aligned} \frac{1}{l_d(s)} &= \frac{1}{x_d} + \left(\frac{1}{x_d'} - \frac{1}{x_d}\right)\frac{sT_d'}{1+sT_d'} + \left(\frac{1}{x_d''} - \frac{1}{x_d'}\right)\frac{sT_d''}{1+sT_d''} \\ \frac{1}{l_q(s)} &= \frac{1}{x_q} + \left(\frac{1}{x_q''} - \frac{1}{x_q}\right)\frac{sT_q''}{1+sT_q''} \end{aligned}\right\} \qquad [4.3.6']$$

in which, with the usual denominations:

- x_d, x_d', x_d'' = "direct" reactances (in pu): synchronous, transient and subtransient, respectively;
- x_q, x_q'' = "quadrature" reactances (in pu): synchronous and subtransient, respectively;
- T_d', T_d'' = "short-circuit, direct" time constants: transient and subtransient, respectively;
- T_{do}', T_{do}'' = same as above, but at "open-circuit" instead of "short-circuit;"
- T_q'', T_{qo}'' = subtransient "quadrature" time constants: "short-circuit" and "open-circuit," respectively.

The "transient" parameters x_d', T_d', T_{do}' are implicitly defined by the previous expressions of $l_d(s)$ and $l/l_d(s)$, and do not coincide with the similar parameters $\hat{x}_d', \hat{T}_d', \hat{T}_{do}'$ which appear in $l_d(s)$ in the absence of additional circuits, even if practically (as already pointed out) the numerical differences are small (remember the values reported in the footnote[(3)]). The subtransient reactances x''_d and x''_q slightly differ from each other and are usually within the range of 0.15–0.3, whereas the subtransient time constants may vary from hundredths to tenths of a second. Note that it holds $T'_{do} > T'_d > T''_{do} > T''_d$, $T''_{qo} > T''_q$. Furthermore, through comparison between the Equations [4.3.6] and [4.3.6′], we have:

$$x''_d = l_d(\infty) = x_d \frac{T'_d T''_d}{T'_{do} T''_{do}}, \quad x''_q = l_q(\infty) = x_q \frac{T''_q}{T''_{qo}}$$

If the machine has a *solid rotor*, the occurrence of eddy currents distributed during the transients in the body of the rotor, may then be expressed schematically (through suitable and generally acceptable assumptions) as the effect of infinite additional rotor circuits. This leads to irrational transfer functions of an infinite order (not reducible to ratios between polynomials in s). For practical applications, it is necessary to consider only suitable approximations, with transfer functions of a sufficiently high order.

Figure 4.8 shows qualitatively, at the cost of some reasonable approximations, the effects of additional rotor circuits on the equivalent circuits relating to the axes d and q.

If the rotor is not solid, the more usual assumptions are the following:

(1) an additional circuit acting only along the q axis (see Fig. 4.7 and Equation [4.3.5]); in this case, we have Equations like [4.3.6] or [4.3.6′], by assuming that $T''_d = T''_{do}$, $x''_d = x'_d = \hat{x}'_d$, $T'_d = \hat{T}'_d$, $T'_{do} = \hat{T}'_{do}$;

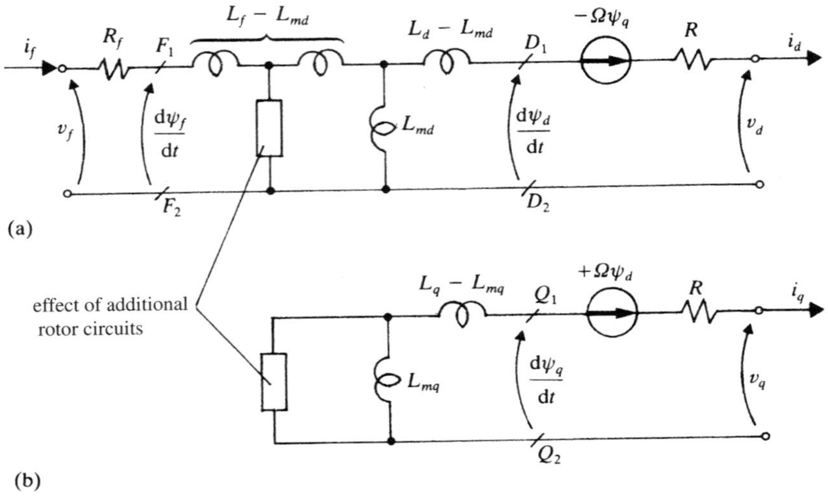

Figure 4.8. Equivalent circuits for d and q axes in the presence of additional rotor circuits along both axes.

(2) two additional circuits, one of which acts along the d axis, and the other along the q axis; in this case, Equations [4.3.6] or [4.3.6′] hold.

If the rotor is solid, it may sometimes be advisable to have recourse to the following assumption:

(3) three additional circuits, one of which acts along the d axis, and two along the q axis: transfer functions of the second order are then obtained on both the axes d and q;

or to other approximations of a higher order.

To assumptions (1), (2), and (3) there correspond, for the whole machine (apart from axis o), models of *dynamic order* 6, 7, and 8, respectively, or (more frequently, as already pointed out) of order 4, 5, and 6 if we accept the simplifications on the transformer emfs described in Section 4.2.2.

In practice, (2) is used far more often than (1); however, in favor of the latter it may be said that, for many applications, transfer functions of the first order for the direct axis may still be accepted, whereas the equation $l_q(s) = \text{constant} = x_q$ (see Equations [4.1.26′]) is too much of a simplification and one that cannot consider any electromagnetic transient on the quadrature axis. For this reason, assumption (1) may enable a considerable improvement in approximation to be achieved, and recourse to (2) may, all in all, prove to be inconvenient. Similar considerations apply for (3), in relation to possible approximations of a higher order; see the examples in Figure 4.9.

4.3.3. Effect of the Magnetic Saturation

The problem becomes substantially different when considering the magnetic saturation, because in this case the nonlinearities of the magnetic part make it very difficult (or improper) to have a rigorous treatment on the basis of Park's transformation. The problem in question may, however, be solved with practical interest, if we consider the physical meaning of the equivalent circuits and of the variables (voltages, currents, fluxes) that appear in them. By so doing, in fact, we can easily guess the type of corrections to make in the circuits themselves, and therefore in the model of the machine, to take the magnetic saturation into account.

In this connection, the use of d and q variables would appear helpful, assuming that the effect of the saturation is to render nonlinear (without increasing the dynamic order of the model) the links between fluxes and currents, e.g., between the fluxes ψ_f, ψ_d, ψ_q and the currents i_f, i_d, i_q in the absence of additional rotor circuits, or between the fluxes ψ_f, ψ_B, ψ_d, ψ_q and the currents i_f, i_B, i_d, i_q in the case of Figure 4.7.

To more clearly state the aforementioned nonlinear links, it is useful to subdivide the saturation (at the cost of a few schematic assumptions that are, in practice, acceptable) as follows:

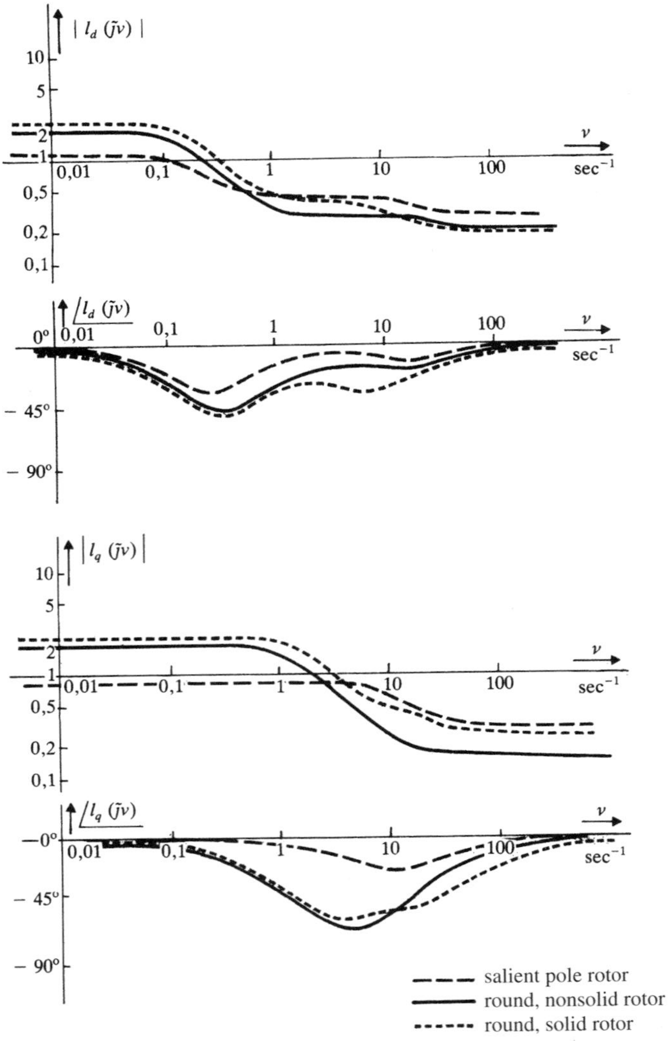

Figure 4.9. Numerical examples of frequency response of the synchronous machine (functions $l_d(s)$ and $l_q(s)$, with $s = \tilde{j}\nu$).

- saturation in the stator and more particularly just in the stator teeth (where magnetic induction may actually reach higher values);
- saturation in the rotor.

Usually, the *saturation only in the stator teeth* are considered and related to the so-called "air-gap flux" (in *dq* terms: $\overline{\psi}_S = \psi_{Sd} + j\psi_{Sq}$). This flux involves both the stator and the rotor, crossing the stator teeth and the air-gap.

The air-gap flux $\overline{\psi}_S$ differs from the total flux $\overline{\psi}$ linked with the stator circuits. The difference is the result of the flux $\overline{\psi}_l$ linked with the stator windings only, through magnetic circuits that practically involve only the stator slots (armature leakage). For this last reason, flux $\overline{\psi}_l$ may be considered practically exempt from saturation phenomena, so that (bearing in mind the sign convention used for stator current $\overline{\imath}$) we may write:

$$\begin{aligned}\overline{\psi}_l &= -L_l\overline{\imath}\\ \overline{\psi}_S &= \overline{\psi} - \overline{\psi}_l = \overline{\psi} + L_l\overline{\imath}\end{aligned} \qquad [4.3.7]$$

L_l being a suitable inductance, known as "(armature) leakage inductance". Generally, it also would be possible to assume two distinct values L_{ld} and L_{lq} for the axes d, q respectively, with $\overline{\psi}_l = -L_{ld}i_d - L_{lq}ji_q$ (see also Fig. 4.12, further on.)

Equation [4.3.7] makes it possible to identify the air-gap flux $\overline{\psi}_S$, now assumed to be the only flux prone to saturation.

The corrections in the equivalent circuits to consider the saturation thus appear fairly obvious. If the rotor includes only the field circuit, these corrections are shown schematically in Figure 4.10, in which the currents i_{Sd} and i_{Sq} correspond to the mmfs required on the axes d and q to compensate for the magnetic voltage drops due to saturation, whereas (see also Section 4.1.2, Remark 1)

$$\tau_f = \frac{L_d - L_l}{L_{md}} \qquad [4.3.8]$$

is the transformation ratio according to which the currents i_f and i_d influence the air-gap flux (with the assumption of having saturation only in the stator, this ratio is the so-called "armature reaction coefficient").

The vector $\overline{\imath}_S = i_{Sd} + ji_{Sq}$ is, by definition, a function of the air-gap flux $\overline{\psi}_S = \psi_{Sd} + j\psi_{Sq}$, and moreover it appears reasonable, especially with round rotor, to assume that $\overline{\imath}_S$ has the same phase as $\overline{\psi}_S$ and a magnitude suitably depending on that of $\overline{\psi}_S$, i.e.,

$$i_S = G_S(\psi_S)$$

in accordance with the magnetic characteristic of the stator; the following equations are then obtained:

$$\left.\begin{aligned}i_{Sd} &= \frac{G_S(\psi_S)}{\psi_S}\psi_{Sd}\\ i_{Sq} &= \frac{G_S(\psi_S)}{\psi_S}\psi_{Sq}\end{aligned}\right\} \qquad [4.3.9]$$

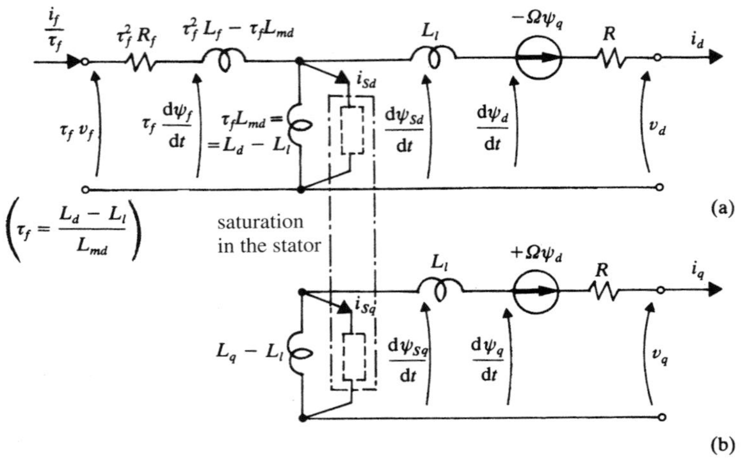

Figure 4.10. Equivalent circuits for d and q axes in the presence of magnetic saturation only in the stator.

As a final consequence, should the rotor not comprise any additional circuits, it is derived from the equivalent circuits in Figure 4.10 that the equations:

$$\begin{cases} \psi_f = L_f i_f - L_{md} i_d \\ \psi_d = L_{md} i_f - L_d i_d \\ \psi_q = -L_q i_q \end{cases}$$

(see Equations [4.1.6′] and [4.1.7′]) must be replaced to consider the saturation in the stator teeth, by the following ones:

$$\left.\begin{aligned} &\psi_f = \frac{1}{\tau_f}\left((\tau_f^2 L_f - \tau_f L_{md})\frac{i_f}{\tau_f} + \psi_{Sd}\right) = \left(L_f - \frac{L_{md}^2}{L_d - L_l}\right) i_f + \frac{L_{md}}{L_d - L_l}\psi_{Sd} \\ &\psi_d = \psi_{Sd} - L_l i_d \\ &\psi_q = \psi_{Sq} - L_l i_q \\ &\psi_{Sd} = (L_d - L_l)\left(\frac{i_f}{\tau_f} - (i_d + i_{Sd})\right) = L_{md} i_f - (L_d - L_l)(i_d + i_{Sd}) \\ &\psi_{Sq} = -(L_q - L_l)(i_q + i_{Sq}) \end{aligned}\right\} \quad [4.3.10]$$

in which i_{Sd} and i_{Sq} depend on ψ_{Sd} and ψ_{Sq} as expressed by Equations [4.3.9].

If we use the pu reduction defined in Section 4.1.2, Remark 2, relating;

$$\begin{cases} i_{Sd}, i_{Sq}, i_S = G_S & \text{to} \quad \sqrt{3} I_{(F)\text{nom}} \\ \psi_{Sd}, \psi_{Sq}, \psi_S & \text{to} \quad \sqrt{3}\dfrac{V_{(F)\text{nom}}}{\omega_{\text{nom}}} \end{cases}$$

Equations [4.3.9] and [4.3.10] then become, with obvious symbols:

$$\left.\begin{aligned} \boldsymbol{i}_{Sd} &= \frac{g_S(\boldsymbol{\psi}_S)}{\boldsymbol{\psi}_S}\boldsymbol{\psi}_{Sd} \\ \boldsymbol{i}_{Sq} &= \frac{g_S(\boldsymbol{\psi}_S)}{\boldsymbol{\psi}_S}\boldsymbol{\psi}_{Sq} \end{aligned}\right\} \qquad [4.3.9']$$

and

$$\left.\begin{aligned} \boldsymbol{\psi}_f &= \frac{\hat{x}'_d - x_l}{x_d - x_l}\boldsymbol{i}_f + \frac{x_d - \hat{x}'_d}{x_d - x_l}\boldsymbol{\psi}_{Sd} \\ \boldsymbol{\psi}_d &= \boldsymbol{\psi}_{Sd} - x_l\boldsymbol{i}_d \\ \boldsymbol{\psi}_q &= \boldsymbol{\psi}_{Sq} - x_l\boldsymbol{i}_q \\ \boldsymbol{\psi}_{Sd} &= \boldsymbol{i}_f - (x_d - x_l)(\boldsymbol{i}_d + \boldsymbol{i}_{Sd}) \\ \boldsymbol{\psi}_{Sq} &= -(x_q - x_l)(\boldsymbol{i}_q + \boldsymbol{i}_{Sq}) \end{aligned}\right\} \qquad [4.3.10']$$

where:

$$x_l \triangleq \omega_{\text{nom}} L_l / Z_{(F)\text{nom}} = \text{(armature) leakage reactance, in pu}$$

(such equations take the place of Equation [4.1.18] and of the first two parts in Equations [4.1.19]).

As a consequence of the above, let us now consider the following steady-state operating conditions of the machine (which are not influenced by the possible presence of additional rotor circuits, as already remarked):

(1) open-circuit (see also Section 4.4.1);

(2) reactive load only ($\cos\varphi = 0$), with current $\overline{\imath}$ lagging versus voltage $\overline{v}$;

by also assuming $\Omega = \omega_{\text{nom}}$.

In the first case, it may be derived in pu (recalling also the Equations [4.1.17]):

$$\boldsymbol{i}_f = \boldsymbol{v} + (x_d - x_l)g_S(\boldsymbol{v}) \triangleq f(\boldsymbol{v}) \qquad [4.3.11]$$

so that function g_s may be derived from the open-circuit characteristic $\boldsymbol{i}_f = f(\boldsymbol{v})$ as indicated in Figure 4.11a.

In the second case, ignoring the armature resistance r, we find, on the other hand:

$$\boldsymbol{i}_f = \boldsymbol{v} + x_d\boldsymbol{i} + (x_d - x_l)g_S(\boldsymbol{v} + x_l\boldsymbol{i})$$

i.e.,

$$\boldsymbol{i}_f - (x_d - x_l)\boldsymbol{i} = f(\boldsymbol{v} + x_l\boldsymbol{i}) \qquad [4.3.12]$$

so that the characteristic $(\boldsymbol{i}_f, \boldsymbol{v})$ may, for each given value of the current $\boldsymbol{i}$, be derived graphically (*"Potier construction"*) from the open-circuit characteristic, by a simple translation of the triangle shown in dashed lines in Figure 4.11b.

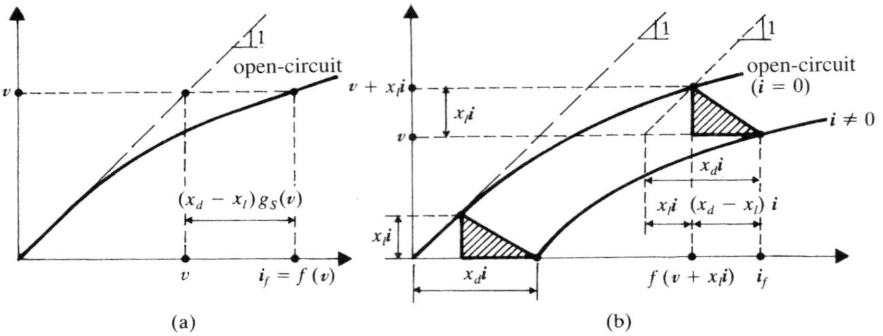

Figure 4.11. Open-circuit characteristic (with magnetic saturation) and Potier construction under the hypothesis of saturation only in the stator.

This triangle is known when we know, e.g., $x_d \boldsymbol{i}$ and $(x_d - x_l)\boldsymbol{i}$. In this connection, it may be pointed out that:

- for $\boldsymbol{v} = 0$ we have $\boldsymbol{i}_f - (x_d - x_l)\boldsymbol{i} \cong x_l \boldsymbol{i}$, that is $\boldsymbol{i}_f \cong x_d \boldsymbol{i}$ (the saturation has a slight effect; see the short-circuit characteristic (b) in Fig. 4.16): $x_d \boldsymbol{i}$ is therefore equal to the excitation current (in pu) $\boldsymbol{i}_f$ required to have the given current $\boldsymbol{i}$ in short-circuit operation;
- because of Equation [4.3.8], we have:

$$x_d - x_l = \frac{\omega_{\text{nom}}(L_d - L_l)}{Z_{(F)\text{nom}}} = \frac{\omega_{\text{nom}} L_{md} \tau_f}{Z_{(F)\text{nom}}} = \frac{\sqrt{3} I_{(F)\text{nom}}}{I_f^*} \tau_f$$

  so that $(x_d - x_l)$ is the "armature reaction coefficient" with the pu adopted here.

The model defined above justifies the Potier construction for representing saturation only in the stator; the acceptability of this latter assumption may be evaluated by observing the actual trend of the characteristics in question, measured experimentally.

In this connection, on the opposite assumption of *saturation only in the rotor*, the Potier construction (Fig. 4.11b) may still be accepted, provided that x_l is replaced by $\hat{x}'_d$. For clarification, we observe that in a steady-state operation at $\cos\varphi = 0$ (ignoring the armature resistance r), the rotor saturation can be attributed to the field flux ψ_f only. We then have equations, in pu:

$$\boldsymbol{\psi}_f = \boldsymbol{i}_f - g_R(\boldsymbol{\psi}_f) - (x_d - \hat{x}'_d)\boldsymbol{i}$$
$$\boldsymbol{v} = \boldsymbol{\psi}_d = \boldsymbol{i}_f - g_R(\boldsymbol{\psi}_f) - x_d \boldsymbol{i}$$

(whereas $\boldsymbol{v}_d = \boldsymbol{\psi}_q = \boldsymbol{i}_q = 0$), in which $g_R(\boldsymbol{\psi}_f)$ corresponds, in terms of field current, to the mmf required for compensating the magnetic voltage drop caused by the saturation; it follows then:

$$\boldsymbol{\psi}_f = \boldsymbol{v} + \hat{x}'_d \boldsymbol{i}$$
$$\boldsymbol{i}_f = \boldsymbol{v} + x_d \boldsymbol{i} + g_R(\boldsymbol{v} + \hat{x}'_d \boldsymbol{i})$$

(from which, in open-circuit operation, $\boldsymbol{i}_f = \boldsymbol{v} + g_R(\boldsymbol{v}) \triangleq f(\boldsymbol{v})$), so that Equation [4.3.12] still holds, with $\hat{x}'_d$ in place of x_l.

In reality, the Potier construction applies (although not exactly), provided that x_l is replaced by a suitable value x_p, known as the "*Potier reactance*," which is intermediate between x_l and $\hat{x}'_d$. From the comparison between x_p and x_l, $\hat{x}'_d$, we can obtain an indication of the relative "weight" of saturation in the stator and in the rotor, at least regarding the direct axis (in fact, in the operating conditions considered here, the quadrature components of fluxes and currents equal zero)[5].

In general, and at the cost of some schematic representation, we may decide to assume equivalent circuits as indicated in Figure 4.12, considering additional rotor circuits (see Fig. 4.8) as well as saturation both in the stator and in the rotor, and not considering further complications caused by the simultaneous presence of additional circuits and saturation. The corresponding dynamic model of the

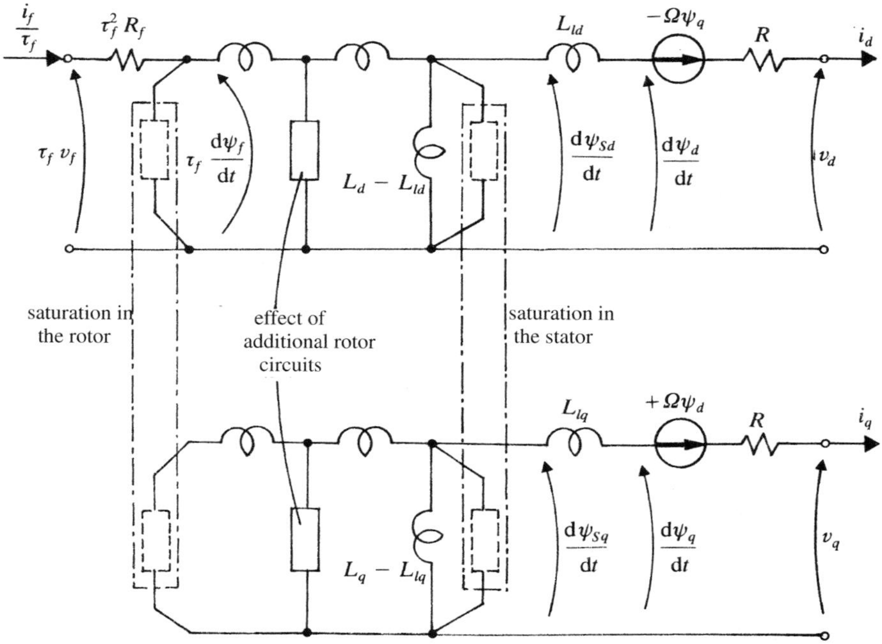

Figure 4.12. Equivalent circuits for d and q axes in the presence of additional rotor circuits along both axes, and of magnetic saturation in the stator and in the rotor.

[5] It usually results $x_l \cong 0.1 - 0.2$, $\hat{x}'_d \cong 0.2 - 0.45$ (see footnote[3]), $x_p \cong 0.15 - 0.4$.

machine would, however, require hard experimental evaluations for its identification, and its complexity would (generally speaking) be out of proportion to practical purposes.

For the greater part of the applications, it is preferred to avoid such model complications, still assuming a model formally analogous to the one described above for the case of saturation only in the stator, but replacing the leakage reactance x_l by the Potier reactance x_p.

Due to the nonlinearities of the magnetic part, the linearized model of the machine is *structurally different* from the model without saturation, because it includes some parameters not found in the latter. To show this difference it is sufficient to observe, for example, that (neglecting for simplicity the additional rotor circuits) Equations [4.1.22′] (or Equations [4.1.27]) and [4.1.23′] are no longer valid even in terms of small variations (with transfer functions dependent on the operating point), since $\Delta \boldsymbol{i}_f$, $\Delta \boldsymbol{\psi}_f$, $\Delta \boldsymbol{\psi}_d$ do not depend only on $\Delta \boldsymbol{v}_f$, $\Delta \boldsymbol{i}_d$ (as in the absence of saturation), but also on $\Delta \boldsymbol{i}_q$; similarly, $\Delta \boldsymbol{\psi}_q$ depends on $\Delta \boldsymbol{v}_f$ and $\Delta \boldsymbol{i}_d$, further than on $\Delta \boldsymbol{i}_q$.

4.3.4. Effect of the Torsional Phenomena

The shaft of a generating unit does not actually constitute a perfectly rigid connection between the different components, i.e., the turbine (in one or more "sections"), the alternator, the possible rotating exciter, etc. As a good approximation, the rotors of the single components may be assimilated with rigid lumped masses, torsionally connected through elastic elements of negligible mass. The mechanical part of the unit may be seen as a set of N masses connected by $(N - 1)$ springs, where N is usually within 2 and 6. The case $N = 2$ corresponds, for example, to the hydroelectric units (a single turbine section, and the generator) with static exciter, whereas the case $N = 6$ may correspond to the case of thermal units, with four turbine sections (one high and one medium pressure section, and two low pressure sections), the generator and a rotating exciter. Assume:

- C_i = torque applied from the external to the i-th mass (positive if driving, negative if resistant);
- $C_{(i-1)i}$, $C_{i(i+1)}$ = torques transmitted through the connections, respectively, from the $(i - 1)$-th mass to the i-th mass, and from the i-th mass to the $(i + 1)$-th mass;
- Ω_{mi}, θ_{mi} = mechanical angular speed and position of the i-th mass, respectively;

disregarding the mechanical losses and any dependence of the C_i's on the speeds, it is possible to write the following equations:

$$\begin{cases} C_i + C_{(i-1)i} - C_{i(i+1)} = J_i \dfrac{\mathrm{d}\Omega_{mi}}{\mathrm{d}t} \\ \Omega_{mi} = \dfrac{\mathrm{d}\theta_{mi}}{\mathrm{d}t} \end{cases} \qquad (i = 1, \ldots, N) \qquad [4.3.13]$$

and moreover:

$$C_{01} = 0$$
$$C_{(i-1)i} = K_{(i-1)i}(\theta_{m(i-1)} - \theta_{mi}) \quad (i = 2, \ldots, N)$$

where $J_1, \ldots, J_N$ represent the single inertia moments, and $K_{12}, \ldots, K_{(N-1)N}$ are the torsional stiffness coefficients of the connections.

The block diagram of Figure 4.13a can be associated with the above-mentioned equations; in particular, for each turbine section, it must be assumed that $C_i = C_{mi}$ (driving torque supplied by the section under consideration), whereas for the generator it results $C_i = -C_e$ (where C_e is the resistant electromagnetic torque).

For the simple case $N = 2$ (Fig. 4.13b), we have $C_1 = C_m$ and $C_2 = -C_e$, and furthermore:

- J_1, J_2 = turbine and generator inertia moments, respectively;
- K_{12} = torsional stiffness coefficient of the connection between turbine and generator;

whereas:

$$\Omega_{m2} = \Omega_m = \text{mechanical angular speed of the generator}$$

(The electrical angular speed instead is $N_p\Omega_{m2} = \Omega$, where N_p is the number of pole pairs.) From the previous equations, it then follows, in terms of transfer functions:

$$\Omega_m = \frac{1}{s(J_1 + J_2)} \frac{C_m - \left(1 + s^2\dfrac{J_1}{K_{12}}\right) C_e}{1 + s^2\dfrac{J_1 J_2}{K_{12}(J_1 + J_2)}} \qquad [4.3.14]$$

instead of:

$$\Omega_m = \frac{1}{s(J_1 + J_2)}(C_m - C_e)$$

which corresponds (with $J \triangleq J_1 + J_2$, and $C_p = 0$) to Equation [4.1.9].

Due to Equation [4.3.14], the response of the speed to the driving torque of the turbine therefore implies also an undamped *resonance*, at the frequency:

$$\nu_r = \sqrt{\frac{K_{12}(J_1 + J_2)}{J_1 J_2}}$$

whereas the response to the electromagnetic torque exhibits, in addition, an undamped *antiresonance* at the frequency $\sqrt{K_{12}/J_1} < \nu_r$.

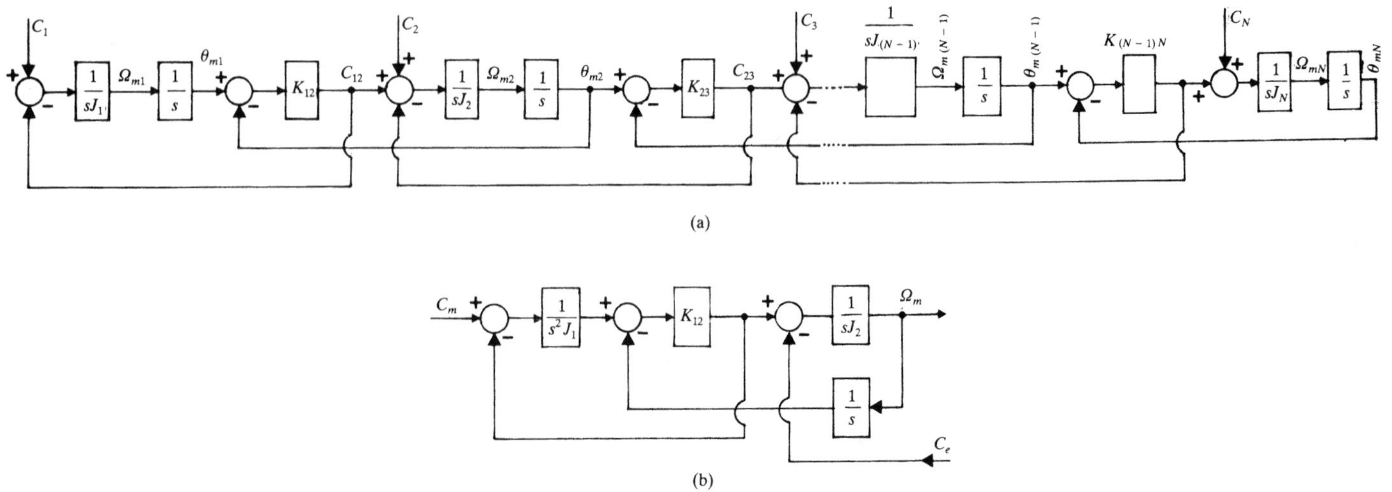

Figure 4.13. Block diagram of the mechanical part in the presence of torsional phenomena: (a) case with N lumped masses; (b) case with two lumped masses (hydraulic turbine and electric generator).

In practice, for the case $N = 2$ (which may occur, as pointed out, for a hydroelectric unit) the resonance frequency ν_r is normally lower than 10 Hz. Generally, for a given N, there are $(N - 1)$ resonances (and the same number of antiresonances, in response to C_e), at frequencies normally from 5 Hz to several tens of hertz. Therefore, the torsional phenomena constitute an example of relatively fast mechanical phenomena, which obviously do not lay among those (mechanical, and slow) up to now considered, according to Section 1.8.2.

Note that, in any case, it results (see for example Equation [4.3.14]):

- for $s \to \infty$: $\Omega_m \to -\dfrac{1}{sJ_2}C_e$
- for $s \to 0$: $\Omega_m \to +\dfrac{1}{s(J_1 + J_2)}(C_m - C_e)$

in accordance to the fact that, following abrupt perturbations, Ω_m initially responds only to C_e (and with only the moment of inertia of the generator), whereas the long-term response (also accounting for actual dampings) tends to coincide with that defined by Equation [4.1.9].

Actually, in the first of Equations [4.3.13], the following should be considered:

- resistant torques of the type:

$$C_{p1}(\Omega_{m1}, \Omega_{m2}), C_{p2}(\Omega_{m1}, \Omega_{m2}, \Omega_{m3}), \ldots, C_{pN}(\Omega_{m(N-1)}, \Omega_{mN})$$

  corresponding to the mechanical losses, where the dependence of the generic C_{pi} on the speeds $\Omega_{m(i-1)}$ and $\Omega_{m(i+1)}$ (of the adjacent masses) also may be related to mechanical hysteresis phenomena in the connections;
- the dependence of the driving torques on the speeds of the respective turbine sections.

In linearized terms, this can be translated into some damping (difficult to be determined, and modest) of the resonances and antiresonances defined above.

In any case, the torsional phenomena may imply significant stresses on the connections of the unit, particularly for sudden variations of C_e, repeated at time intervals which are critical for the above-mentioned resonances. This must be considered, in relation to erroneous paralleling operations, short-circuits followed by fast breaker reclosing, etc.

Moreover, the phenomena under examination may be particularly amplified — or even cause instability, up to the shaft break — in the case of units connected to the network by lines compensated with series condensers (see Section 7.2.4).

4.4. SOME OPERATIONAL EXAMPLES

4.4.1. Steady-State Operation

By "equilibrium" (steady-state) operation of the synchronous machine, we refer to (see also Section 1.2) the operation with:

- constant field voltage;
- positive-sequence sinusoidal stator currents;
- electrical speed Ω equal to the frequency ω of the stator currents.

This is characterized by:

$$v_f = \text{constant}$$

$$\begin{cases} i_a = \sqrt{2} I_{(F)} \cos(\omega t + \alpha_I) \\ i_b = \sqrt{2} I_{(F)} \cos(\omega t + \alpha_I - 120°) \\ i_c = \sqrt{2} I_{(F)} \cos(\omega t + \alpha_I - 240°) \end{cases}$$

$$\Omega = \text{constant} = \omega$$

with $I_{(F)}$, α_I constant (in particular, $I_{(F)}$ is the rms value of the phase currents).

From the expressions of i_a, i_b, i_c, we can derive, by applying the Park's transformation with reference $\theta = \int \Omega \, \mathrm{d}t = \omega t + \theta_o$:

$$\begin{cases} \left.\begin{aligned} i_d &= \sqrt{3} I_{(F)} \cos(\alpha_I - \theta_o) \\ i_q &= \sqrt{3} I_{(F)} \sin(\alpha_I - \theta_o) \end{aligned}\right\} \bar{\imath} = \sqrt{3} I_{(F)} \epsilon^{j(\alpha_I - \theta_o)} = \text{constant} \\ i_o = 0 \end{cases}$$

This means that, since v_f and Ω are also constant, the equivalent circuits in Figure 4.3 are in static operation (i.e., with constant voltages, currents, and fluxes). We can easily ascertain that also the phase voltages and fluxes — as well as the currents — are positive-sequence sinusoidal, at the frequency ω, of the type: $v_a = \sqrt{2} V_{(F)} \cos(\omega t + \alpha_v)$, etc. Moreover, under present conditions, any additional rotor circuits have no effect (see Section 4.3.2 and, in particular, Fig. 4.7), since they do not absorb any current.

In particular, we have:

$$i_f = \frac{v_f}{R_f}$$

$$\begin{cases} \left.\begin{aligned} v_d &= -R i_d - \omega \psi_q \\ v_q &= -R i_q + \omega \psi_d \end{aligned}\right\} \bar{v} = -R \bar{\imath} + j \omega \bar{\psi} \\ v_o - \sqrt{3} v_n = 0 \end{cases}$$

$$C_m = C_p \left(\frac{\omega}{N_p} \right) + C_e$$

in which C_e is given by Equation [4.1.10], whereas ψ_f, ψ_d, ψ_q, $\bar{\psi}$ are deducible (in the absence of saturation) from Equations [4.1.6′], [4.1.7′], and [4.1.7″] (or Equation [4.1.7‴]), and $\psi_o = 0$. The magnitude of the vector $\bar{\imath}$ is given by $i = \sqrt{3} I_{(F)}$; similarly, we have $v = \sqrt{3} V_{(F)}$, $\psi = \sqrt{3} \Psi_{(F)}$, with $V_{(F)}$ and $\Psi_{(F)}$ being the rms values of the phase voltages and flux linkages.

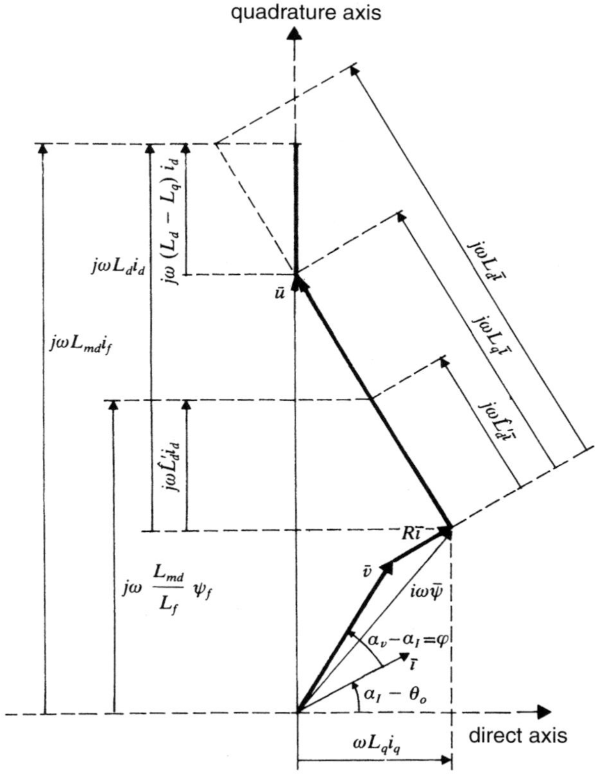

Figure 4.14. Vector diagram at steady-state without magnetic saturation.

In the absence of saturation, we may derive the *vector diagram* in Figure 4.14, in which the real and imaginary axes are denominated "direct" axis and "quadrature" axis, respectively. These axes may easily be identified from the vectors $\bar{v}$ and $\bar{\imath}$, since it results:

$$\bar{u} \triangleq \bar{v} + (R + j\omega L_q)\bar{\imath} = j\omega(L_{md} i_f - (L_d - L_q)i_d)$$
$$= j\omega\left(\frac{L_{md}}{L_f}\psi_f + (L_q - \hat{L}'_d)i_d\right)$$

and therefore this vector (see Fig. 4.14) lies, by definition, on the quadrature axis, with:

$$u_d = 0, \quad u_q = v_q + Ri_q + \omega L_q i_d = \omega(L_{md} i_f - (L_d - L_q)i_d)$$
$$= \omega\left(\frac{L_{md}}{L_f}\psi_f + (L_q - \hat{L}'_d)i_d\right)$$

Additionally, we have, in general:

$$P_e = \langle j\omega\overline{\psi}, \overline{\imath}\rangle = \omega(\psi_d i_q - \psi_q i_d)$$

and the active and reactive powers delivered are, respectively, given by:

$$P = 3V_{(F)}I_{(F)}\cos\varphi = \langle \overline{v}, \overline{\imath}\rangle = v_d i_d + v_q i_q = \omega(\psi_d i_q - \psi_q i_d) - Ri^2 = P_e - Ri^2$$

$$Q = 3V_{(F)}I_{(F)}\sin\varphi = \langle \overline{v}, j\overline{\imath}\rangle = v_q i_d - v_d i_q = \omega(\psi_d i_d + \psi_q i_q)$$

Under the simplifying hypothesis $R = 0$ (generally acceptable), the previous equations allow the determination of a relationship between (P/v^2), (Q/v^2), and $i_f/v = v_f/(R_f v)$, according to diagrams like those in Figure 4.15. Such diagrams are constituted by circles in the case of round rotor (with $L_d = L_q$), and by "Pascal spirals" in the case of salient pole rotor (for $i_f = 0$, that is $v_f = 0$, the spiral degenerates again into a circle, the internal points of which define the operation in the *counter-excitation* mode)[6].

As a particular case, in *open-circuit* operation ($i_a = i_b = i_c = 0$) we have $\overline{\imath} = 0$ and in the absence of saturation:

$$\overline{v} = j\omega\overline{\psi} = j\omega L_{md} i_f$$

---

[6] More precisely, if $u_q = v_q + \omega L_q i_d$ is the imaginary part of the vector $(\overline{v} + j\omega L_q\overline{\imath})$ — which for $R = 0$ certainly lies, as seen, on the quadrature axis — it results:

$$P = \frac{v_d u_q}{\omega L_q}, \quad Q = \frac{v_q u_q - v^2}{\omega L_q}$$

with:

$$\frac{u_q}{\omega L_q} = \frac{L_{md}}{L_d} i_f + \left(\frac{1}{\omega L_q} - \frac{1}{\omega L_d}\right) v_q, \quad v_d^2 + v_q^2 = v^2$$

For what concerns the sign of i_f (and v_f), it is here assumed that $u_q > 0$, so that v_d and v_q have, respectively, the same sign as (P/v^2) and $(Q/v^2 + 1/\omega L_q)$, whereas (as it can be derived by the previous equations) i_f and v_f will have the same sign as:

$$\left(\frac{P}{v^2}\right)^2 + \left(\frac{Q}{v^2} + \frac{1}{\omega L_q}\right)^2 - \left(\frac{1}{\omega L_q} - \frac{1}{\omega L_d}\right)\left(\frac{Q}{v^2} + \frac{1}{\omega L_q}\right)$$

In particular, for $P = 0$, we then have counterexcitation (i_f and v_f negative, and generally of the opposite sign with respect to u_q) for:

$$\frac{Q}{v^2} \in \left(\frac{-1}{\omega L_q}, \frac{-1}{\omega L_d}\right)$$

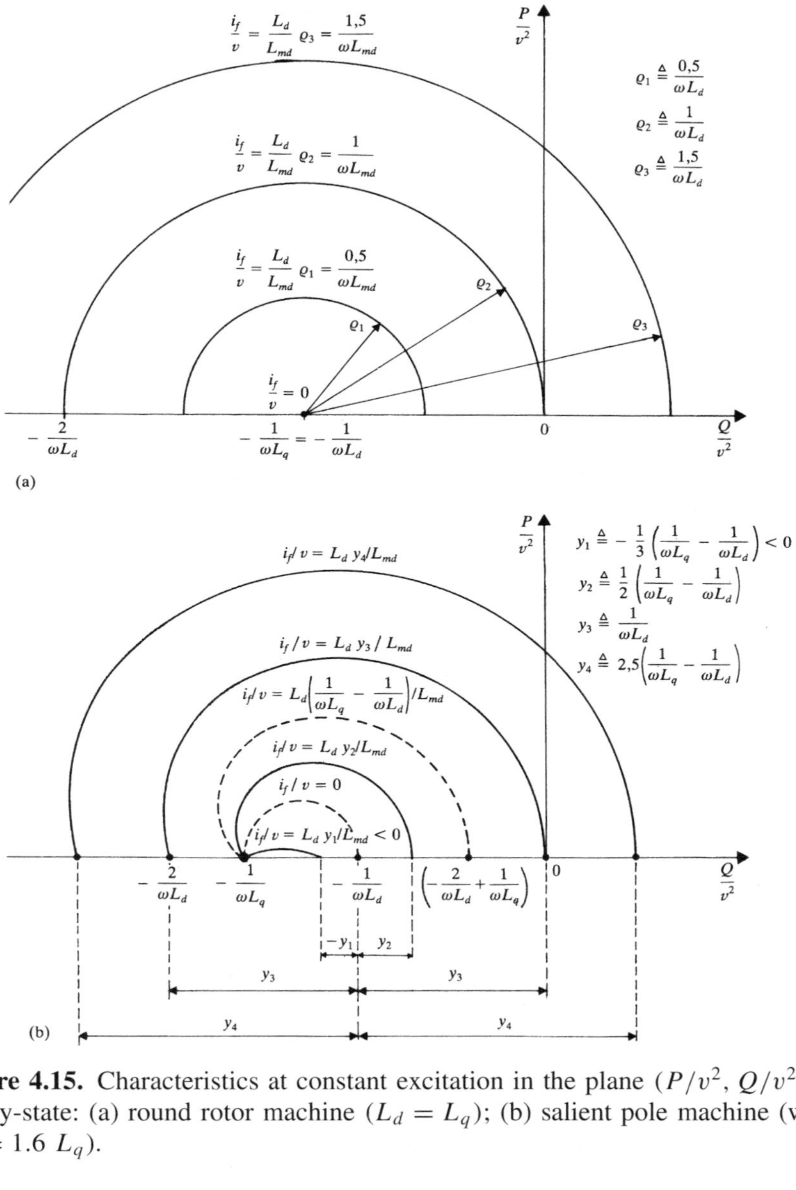

Figure 4.15. Characteristics at constant excitation in the plane $(P/v^2, Q/v^2)$ at steady-state: (a) round rotor machine ($L_d = L_q$); (b) salient pole machine (with $L_d = 1.6\ L_q$).

(besides $\psi_f = L_f i_f$, $i_f = v_f/R_f$), i.e., $v_d = 0$, $v_q = \omega L_{md} i_f$; the rms value of the phase voltages is therefore:

$$V_{(F)} = \frac{v}{\sqrt{3}} = \frac{\omega L_{md} i_f}{\sqrt{3}}$$

so that the field current I_f^* corresponding to nominal stator voltage and frequency ($V_{(F)} = V_{(F)\text{nom}}$, $\omega = \omega_{\text{nom}}$) is given by:

$$I_f^* = \frac{\sqrt{3}V_{(F)\text{nom}}}{\omega_{\text{nom}} L_{md}} \qquad [4.4.1]$$

On the other hand, in *short-circuit* operation ($v_a = v_b = v_c = 0$) we have $\overline{v} = 0$, and:

$$\overline{\imath} = (\omega L_q + jR)\frac{\omega L_{md}}{\omega^2 L_d L_q + R^2} i_f$$

whereas the rms value of the phase currents is given by $I_{(F)} = i/\sqrt{3}$; usually, the effect of R may however be ignored, so that:

$$\overline{\imath} \cong i_d \cong \frac{L_{md}}{L_d} i_f$$

while i_q is negligible, and the field current I_{fc} corresponding to the nominal stator current ($I_{(F)} = I_{(F)\text{nom}}$) is given by:

$$I_{fc} \cong \frac{\sqrt{3}L_d I_{(F)\text{nom}}}{L_{md}} \qquad [4.4.2]$$

Here too, we assume that the magnetic saturation is ignored, but this simplifying assumption is in practice quite acceptable under present short-circuit conditions, because of the low value of the fluxes (for current values of practical interest).
The ratio:

$$K_c \triangleq \frac{I_f^*}{I_{fc}} \cong \frac{V_{(F)\text{nom}}}{\omega_{\text{nom}} L_d I_{(F)\text{nom}}} \qquad [4.4.3]$$

is known as "short-circuit ratio."

If we use the *per-unit* reduction defined in Section 4.1.2, Remark 2 and assume $\omega = \omega_{\text{nom}}$, we have:

$$\left.\begin{cases} \boldsymbol{i}_d = \boldsymbol{i}\cos(\alpha_I - \theta_o) \\ \boldsymbol{i}_q = \boldsymbol{i}\sin(\alpha_I - \theta_o) \end{cases}\right\} \overline{\boldsymbol{\imath}} = \boldsymbol{i}\epsilon^{j(\alpha_I - \theta_o)}$$
$$\boldsymbol{i}_o = 0$$

$$\boldsymbol{i}_f = \boldsymbol{v}_f$$

$$\left.\begin{cases} \boldsymbol{v}_d = -r\boldsymbol{i}_d - \boldsymbol{\psi}_q \\ \boldsymbol{v}_q = -r\boldsymbol{i}_q + \boldsymbol{\psi}_d \end{cases}\right\} \overline{\boldsymbol{v}} = -r\overline{\boldsymbol{\imath}} + j\boldsymbol{\psi}$$
$$\boldsymbol{v}_o - \frac{v_n}{V_{(F)\text{nom}}} = 0$$

$$C_m = C_p(\omega) + C_e$$

in which C_e is given by Equation [4.1.21], whereas $\boldsymbol{\psi}_f$, $\boldsymbol{\psi}_d$, $\boldsymbol{\psi}_q$, $\overline{\boldsymbol{\psi}}$ are deducible (in the absence of saturation) from Equations [4.1.18], [4.1.19], and [4.1.19′], and $\boldsymbol{\psi}_o = 0$.

The magnitudes of the vectors $\overline{\boldsymbol{\imath}}$, $\overline{\boldsymbol{v}}$, $\overline{\boldsymbol{\psi}}$ are, respectively, given by:

$$\boldsymbol{i} = \frac{I_{(F)}}{I_{(F)\text{nom}}}, \qquad \boldsymbol{v} = \frac{V_{(F)}}{V_{(F)\text{nom}}}, \qquad \boldsymbol{\psi} = \frac{\omega_{\text{nom}}\Psi_{(F)}}{V_{(F)\text{nom}}}$$

and the powers P_e, P, Q related to the nominal apparent power A_{nom} are worth:

$$\begin{cases} \dfrac{P_e}{A_{\text{nom}}} = \langle j\overline{\boldsymbol{\psi}}, \overline{\boldsymbol{\imath}}\rangle = \boldsymbol{\psi}_d\boldsymbol{i}_q - \boldsymbol{\psi}_q\boldsymbol{i}_d = C_e \\ \dfrac{P}{A_{\text{nom}}} = \boldsymbol{vi}\cos\varphi = \langle\overline{\boldsymbol{v}}, \overline{\boldsymbol{\imath}}\rangle = \boldsymbol{v}_d\boldsymbol{i}_d + \boldsymbol{v}_q\boldsymbol{i}_q = \boldsymbol{\psi}_d\boldsymbol{i}_q - \boldsymbol{\psi}_q\boldsymbol{i}_d - r\boldsymbol{i}^2 = C_e - r\boldsymbol{i}^2 \\ \dfrac{Q}{A_{\text{nom}}} = \boldsymbol{vi}\sin\varphi = \langle\overline{\boldsymbol{v}}, j\overline{\boldsymbol{\imath}}\rangle = \boldsymbol{v}_q\boldsymbol{i}_d - \boldsymbol{v}_d\boldsymbol{i}_q = \boldsymbol{\psi}_d\boldsymbol{i}_d + \boldsymbol{\psi}_q\boldsymbol{i}_q \end{cases}$$

In the vector diagram in Figure 4.14, we then must replace, respectively:

$\overline{v}, \overline{\imath}, \omega\overline{\psi}$	with $\overline{\boldsymbol{v}}, \overline{\boldsymbol{\imath}}, \overline{\boldsymbol{\psi}}$
i_d, i_q	with $\boldsymbol{i}_d, \boldsymbol{i}_q$
$R, \omega L_d, \omega\hat{L}'_d, \omega L_q$	with $r, x_d, \hat{x}'_d, x_q$
$j\omega L_{md}i_f$	with $j\boldsymbol{i}_f$
$j\omega\dfrac{L_{md}}{L_f}\psi_f$	with $j\boldsymbol{\psi}_f$

In open-circuit operation, we have in the absence of saturation:

$$\overline{\boldsymbol{v}} = j\overline{\boldsymbol{\psi}} = j\boldsymbol{i}_f$$

(besides $\boldsymbol{\psi}_f = \boldsymbol{i}_f = \boldsymbol{v}_f$) and in short-circuit:

$$\overline{\boldsymbol{\imath}} = (x_q + jr)\frac{1}{x_d x_q + r^2}\boldsymbol{i}_f \cong \frac{1}{x_d}\boldsymbol{i}_f$$

while the short-circuit ratio may be written:

$$K_c \cong \frac{1}{x_d} \qquad [4.4.3']$$

Figure 4.16 summarizes the open-circuit (a) and short-circuit (b) characteristics. In practice, magnetic saturation modifies appreciably only the open-circuit characteristic, which is of the (a′) type. The characteristic (a) is also known as the "air-gap characteristic," since the excitation mmf is practically (ignoring the saturation) caused only by the presence of the air-gap. For a given value of voltage, the difference between the values of $\boldsymbol{i}_f$ with the characteristics (a′) and (a) is the result of the further mmf required by the saturation.

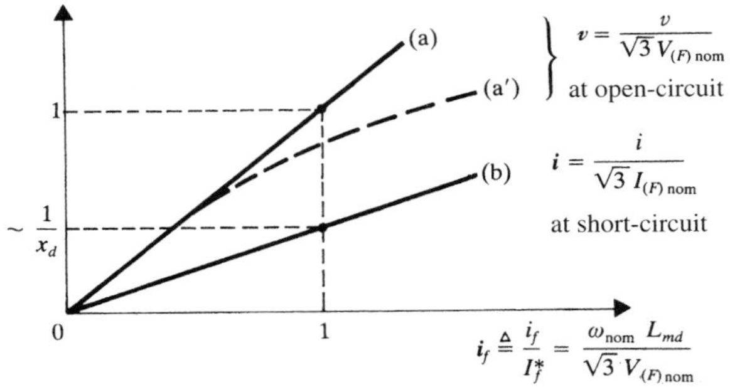

Figure 4.16. Open-circuit characteristic (with and without magnetic saturation) and short-circuit characteristic.

The vector diagram obviously must be modified in the presence of *magnetic saturation*. Assuming the model described in Section 4.3.3, based on the use of the "Potier reactance" x_p, the diagram of Figure 4.17 (in pu, and assuming that $\omega = \omega_{nom}$) can be derived, where the magnitude $\boldsymbol{i}_S = g_S(\boldsymbol{\psi}_S)$ is intended to be derived based on the open-circuit characteristic, according to what is indicated. The direct and quadrature axes may be determined from the vectors $\overline{v}$, $\overline{\imath}$, as it results:

$$\begin{aligned}\overline{\boldsymbol{u}} &\triangleq \overline{v} + (r + jx_q)\overline{\imath} + j(x_q - x_p)\overline{\imath}_S = j(\boldsymbol{i}_f - (x_d - x_q)(\boldsymbol{i}_d + \boldsymbol{i}_{Sd}))\\ &= j(\boldsymbol{\psi}_f + (x_q - \hat{x}'_d)(\boldsymbol{i}_d + \boldsymbol{i}_{Sd}))\end{aligned}$$

so that such a vector (see Fig. 4.17) lies, by definition, on the quadrature axis.

4.4.2. Effects of a Constant Slip

(a) Slip with Respect to the Currents

Compared to the operating conditions considered in Section 4.4.1, let us assume now that the electric speed Ω is constant, but different from the frequency ω of the stator currents (positive-sequence sinusoidal), with a constant "slip" $\sigma \triangleq \Omega - \omega$.

With known symbols, applying the Park's transformation with reference $\theta = \int \Omega\, dt = \omega t + \theta_o + \sigma t$, we may derive:

$$\left.\begin{aligned}&\left.\begin{aligned} i_d &= \sqrt{3} I_{(F)} \cos(\sigma t + \theta_o - \alpha_I)\\ i_q &= -\sqrt{3} I_{(F)} \sin(\sigma t + \theta_o - \alpha_I)\end{aligned}\right\} \overline{\imath} = \sqrt{3} I_{(F)} \epsilon^{-j(\sigma t + \theta_o - \alpha_I)}\\ &i_o = 0\end{aligned}\right\} \qquad [4.4.4]$$

In these conditions, the equivalent circuits in Figure 4.3 (or other more general ones, as indicated in Sections 4.3.2 and 4.3.3) are subjected to a constant voltage v_f and to sinusoidal currents i_d, i_q at "slip frequency" (i.e., frequency σ), whereas Ω is constant and $i_o = 0$.

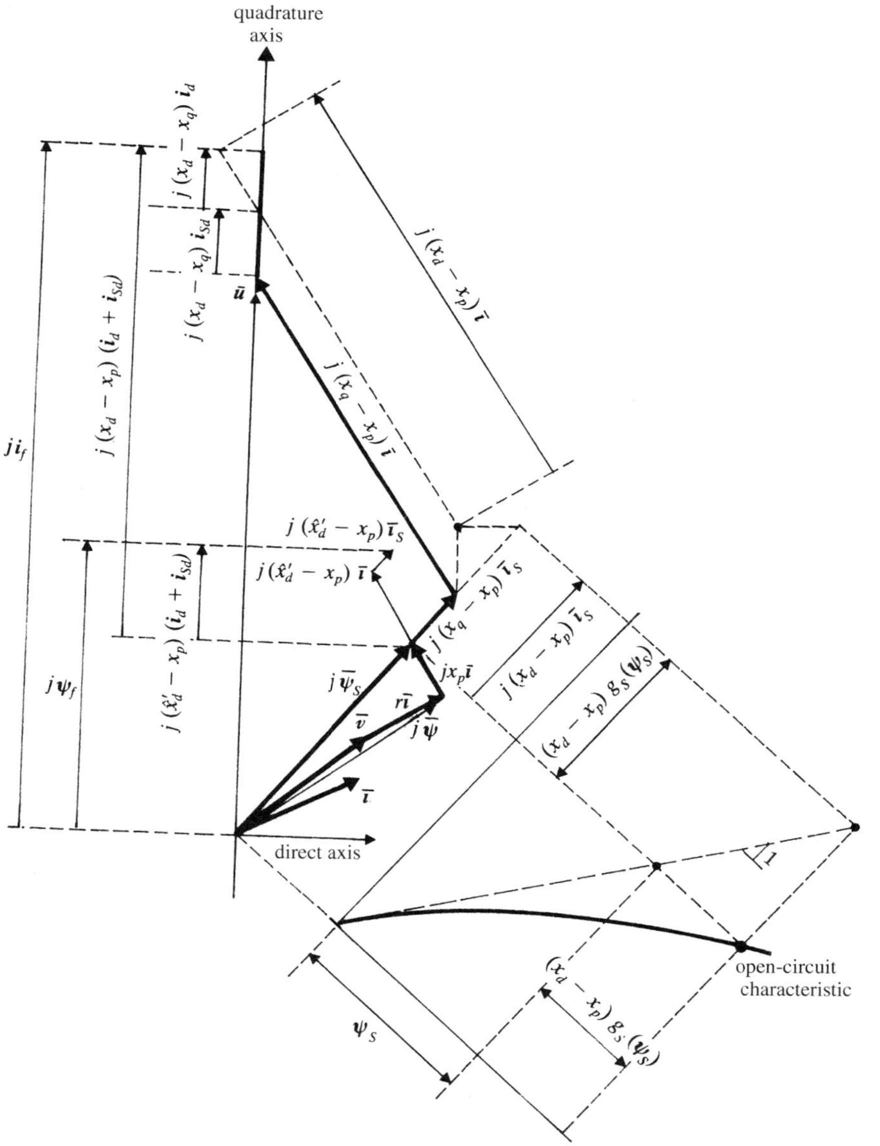

Figure 4.17. Vector diagram at steady-state in the presence of magnetic saturation (quantities in "per unit").

Since Ω — which appears in the voltage generators acting on the axes d and q (see Fig. 4.3) — is constant, we may apply the superposition of the effects in evaluating voltages, currents, and fluxes on the axes d and q, if magnetic saturation is ignored.

We may then deduce that:

- ψ_q, v_d are sinusoidal at slip frequency, due to the effect of i_d and i_q (without any contribution on the part of v_f);
- i_f, ψ_d, v_q include:
  - a slip-frequency sinusoidal component, due to i_d and i_q;
  - a constant component, due to v_f;

whereas $\psi_o = 0$, $v_o - \sqrt{3}v_n = 0$.

Due to the Equations [4.4.4], the sinusoids i_d and i_q have equal magnitude, and i_q is leading by 90° (in time) on i_d; in phasor terms, we therefore have:

$$\tilde{\imath}_q = \tilde{\jmath}\tilde{\imath}_d$$

The effects of these sinusoids may be evaluated in phasor terms by replacing, in the transfer functions, the variable s by $\tilde{\jmath}\sigma$; for example, remembering Equation [4.1.23′] we may derive (with obvious symbols and in pu):

$$\tilde{\boldsymbol{\psi}}_q = -l_q(\tilde{\jmath}\sigma)\tilde{\boldsymbol{\imath}}_q = -\tilde{\jmath}l_q(\tilde{\jmath}\sigma)\tilde{\boldsymbol{\imath}}_d$$

and similarly for the other variables.

The phase voltages and fluxes, obtainable by inverting the Park's transformation (see Appendix 2), are not positive-sequence sinusoidal like the currents.

Regarding the electromagnetic torque C_e, by using the indices (m) and (σ) for the constant components and the slip-frequency sinusoidal components, respectively, we get:

$$C_e = N_p((\psi_{d(m)} + \psi_{d(\sigma)})i_{q(\sigma)} - \psi_{q(\sigma)}i_{d(\sigma)})$$

in which the term $\psi_{d(m)}i_{q(\sigma)}$ is slip-frequency sinusoidal, whereas $(\psi_{d(\sigma)}i_{q(\sigma)} - \psi_{q(\sigma)}i_{d(\sigma)})$ consists of a sinusoidal component at double frequency, and of a constant component.

Because of the effect of the latter component (independent of v_f), the torque C_e therefore has an average value which is not zero and is given, in pu, by:

$$C_{e(m)} = -\tfrac{1}{2}\boldsymbol{i}^2\, Im(l_d(\tilde{\jmath}\sigma) + l_q(\tilde{\jmath}\sigma)) \qquad [4.4.5]$$

where $l_d(s)$ and $l_q(s)$ are defined by Equations [4.1.22′] and [4.1.23′], whereas $\boldsymbol{i} \triangleq I_{(F)}/I_{(F)\text{nom}}$. In particular, if we accept Equations [4.1.26′] we obtain:

$$C_{e(m)} = +\frac{1}{2}\boldsymbol{i}^2\frac{\sigma\hat{T}'_{do}(x_d - \hat{x}'_d)}{1 + (\sigma\hat{T}'_{do})^2}$$

In addition, sinusoidal components at the frequencies σ and 2σ are superposed on this average value, and the same happens for the power P_e, since the rotor speed is assumed to be constant.

The assumption $\Omega =$ constant implies, with a finite moment of inertia J, that at every instant the accelerating torque $(C_m - C_p - C_e)$ is zero, and therefore that the driving torque C_m is a suitable function of time. This situation is not realistic, but the assumption of constant speed may still seem acceptable for high values of the moment of inertia J, and an accelerating torque rapidly varying around an average value $(C_{m(m)} - C_{p(m)} - C_{e(m)})$ equal to zero.

(b) Slip with Respect to the Voltages

For practical applications, it may be of greater interest to consider the case in which the phase voltages and not the currents are positive-sequence sinusoidal, at the frequency ω, with constant slip $\sigma \triangleq \Omega - \omega$.

Through similar considerations to those advanced for the previous case, it may be seen that the equivalent circuits are now subject to a constant voltage v_f and to slip-frequency sinusoidal voltages v_d and v_q (with $\tilde{v}_q = \tilde{\jmath}\tilde{v}_d$), while Ω is constant and $v_o - \sqrt{3}v_n = 0$.

By again applying the superposition of the effects (without considering magnetic saturation), it may be derived that i_f, i_d, i_q, ψ_d, and ψ_q include:

- a slip-frequency sinusoidal component caused by v_d and v_q;
- a constant component caused by v_f;

whereas $i_o = 0$, $\psi_o = 0$ (however, the constant components of ψ_d, ψ_q, i_q are zero if $R = 0$).

Moreover, it may be seen that, in the present conditions, the phase currents i_a, i_b, i_c (as well as the fluxes ψ_a, ψ_b, ψ_c, unless $R = 0$) are not positive-sequence sinusoidal.

With the electromagnetic torque C_e, it includes slip-frequency and double-frequency sinusoidal components, and a constant component (see Section (a), as to the assumption $\Omega =$ constant); the latter is given by:

$$C_{e(m)} = N_p(\psi_{d(m)}i_{q(m)} - \psi_{q(m)}i_{d(m)}) + N_p(\psi_{d(\sigma)}i_{q(\sigma)} - \psi_{q(\sigma)}i_{d(\sigma)})_{(m)} \quad [4.4.6]$$

in which the first term is caused only by v_f (and is zero for $R = 0$), and is not influenced by the presence of additional rotor circuits, referred to in Section 4.3.2. The expression of $C_{e(m)}$ may be considerably simplified when R is negligible. In this case, in Equation [4.4.6], it is sufficient to consider only the last term (which, on the other hand, becomes much simpler if $R = 0$), and in the end we find, in pu (and with $\boldsymbol{v} \triangleq V_{(F)}/V_{(F)\text{nom}}$):

$$C_{e(m)} = \frac{1}{2}\boldsymbol{v}^2 \, Im\left(\frac{1}{l_d(\tilde{\jmath}\sigma)} + \frac{1}{l_q(\tilde{\jmath}\sigma)}\right)$$

from which, if we accept the Equations [4.1.26′]:

$$C_{e(m)} = \frac{1}{2}\boldsymbol{v}^2 \frac{\sigma \hat{T}'_{do}(x_d - \hat{x}'_d)}{x_d^2(1 + (\sigma \hat{T}'_d)^2)}$$

Before concluding, it is important to note that, if the armature resistance is disregarded, ψ_d and ψ_q are sinusoidal at slip-frequency, and in phasor terms we simply have:

$$\begin{cases} \tilde{\psi}_d = \dfrac{\tilde{j}\tilde{v}_d}{\Omega - \sigma} = \dfrac{\tilde{j}\tilde{v}_d}{\omega} \\ \tilde{\psi}_q = \dfrac{-\tilde{v}_d}{\Omega - \sigma} = \dfrac{-\tilde{v}_d}{\omega} \end{cases}$$

as can be derived directly from Equations [4.1.5′] for $R = 0$. This result also could have been achieved by assuming, instead of the first two parts of Equations [4.1.5′], the following equations:

$$\begin{cases} v_d = -\omega\psi_q \\ v_q = +\omega\psi_d \end{cases}$$

which correspond to a widely used approximation (as in Section 4.2.2; see, for $\omega = \omega_{\text{nom}}$, Equations [4.2.1]); therefore, in the present case, the above substitution may be fully accepted, inasmuch as it leaves unchanged not only the expressions of ψ_d and ψ_q, but also those of i_f, i_d, and i_q, as well as the expressions of C_e and $C_{e(m)}$.

4.4.3. Effects of Negative or Zero-Sequence Components

More generally than in the cases described in Sections 4.4.1 and 4.4.2, let us now assume:

- constant field voltage;
- sinusoidal stator currents at the frequency ω, with positive, negative, and zero-sequence components;
- constant electric speed Ω (in general, different from ω).

Similar considerations to those advanced hereafter can be extended to phase voltages or fluxes (not currents) sinusoidal at the three sequences.

From the phase current expressions:

$$\begin{cases} i_a = \sqrt{2}I_{(F0)}\cos(\omega t + \alpha_{I0}) + \sqrt{2}I_{(F1)}\cos(\omega t + \alpha_{I1}) + \sqrt{2}I_{(F2)}\cos(\omega t + \alpha_{I2}) \\ i_b = \sqrt{2}I_{(F0)}\cos(\omega t + \alpha_{I0}) + \sqrt{2}I_{(F1)}\cos(\omega t + \alpha_{I1} - 120^\circ) \\ \quad\quad + \sqrt{2}I_{(F2)}\cos(\omega t + \alpha_{I2} - 240^\circ) \\ i_c = \sqrt{2}I_{(F0)}\cos(\omega t + \alpha_{I0}) + \sqrt{2}I_{(F1)}\cos(\omega t + \alpha_{I1} - 240^\circ) \\ \quad\quad + \sqrt{2}I_{(F2)}\cos(\omega t + \alpha_{I2} - 120^\circ) \end{cases}$$

with $I_{(F0)}$, α_{I0}, $I_{(F1)}$, α_{I1}, $I_{(F2)}$, α_{I2} constant ($I_{(F0)}$, $I_{(F1)}$ and $I_{(F2)}$ represent the rms values of the phase currents at zero, positive, and negative sequence, respectively), by applying the Park's transformation with reference $\theta = \int \Omega \, dt = \Omega t + \theta_o$ we derive:

$$\left.\begin{aligned} i_d &= \sqrt{3} I_{(F1)} \cos(\sigma_1 t + \theta_o - \alpha_{I1}) + \sqrt{3} I_{(F2)} \cos(\sigma_2 t + \theta_o + \alpha_{I2}) \\ i_q &= -\sqrt{3} I_{(F1)} \sin(\sigma_1 t + \theta_o - \alpha_{I1}) - \sqrt{3} I_{(F2)} \sin(\sigma_2 t + \theta_o + \alpha_{I2}) \\ i_o &= \sqrt{6} I_{(F0)} \cos(\omega t + \alpha_{I0}) \end{aligned}\right\} \quad [4.4.7]$$

where we assume:

$\sigma_1 \triangleq \Omega - \omega =$ rotor slip with respect to positive sequence;

$\sigma_2 \triangleq \Omega + \omega = \Omega - (-\omega) =$ rotor slip with respect to negative sequence.

In the absence of magnetic saturation, we can see that the field current i_f, the fluxes ψ_d and ψ_q, and (because of the assumption $\Omega =$ constant) the voltages v_d and v_q include:

- a sinusoidal component at the frequency σ_1, caused by the presence of the positive sequence;
- a sinusoidal component at the frequency σ_2, caused by the presence of the negative sequence;

and, in addition, i_f, ψ_d, and v_q include a constant component because of the field voltage v_f, whereas ψ_o and $(v_o - \sqrt{3} v_n)$ are sinusoidal at the frequency ω, because of the effect of the zero sequence.

The phase voltages and fluxes (obtainable by inverting the Park's transformation) are not simply sinusoidal at the frequency ω like the currents (see also Equation [4.4.9]).

With the electromagnetic torque C_e, by respectively using the indices (m), $(\sigma 1)$, and $(\sigma 2)$ for the constant components, for the sinusoidal components at frequency σ_1, and for those at frequency σ_2, we derive:

$$\begin{aligned} C_e = N_p((&\psi_{d(m)} + \psi_{d(\sigma 1)} + \psi_{d(\sigma 2)})(i_{q(\sigma 1)} + i_{q(\sigma 2)}) \\ &- (\psi_{q(\sigma 1)} + \psi_{q(\sigma 2)})(i_{d(\sigma 1)} + i_{d(\sigma 2)})) \end{aligned}$$

so that C_e can be expressed as the sum of the following three terms (zero sequence makes no contribution):

(1) The term:

$$N_p((\psi_{d(m)} + \psi_{d(\sigma 1)}) i_{q(\sigma 1)} - \psi_{q(\sigma 1)} i_{d(\sigma 1)})$$

equal to the torque that would exist in the absence of negative sequence, and *caused only by the positive-sequence currents and the field voltage* v_f (see Section 4.4.2a, with slip $\sigma = \sigma_1 = \Omega - \omega$). This term includes a constant component and sinusoidal components at the frequencies $\sigma_1 = \Omega - \omega$, $2\sigma_1 = 2(\Omega - \omega)$.

(2) The term:

$$N_p(\psi_{d(\sigma 2)} i_{q(\sigma 2)} - \psi_{q(\sigma 2)} i_{d(\sigma 2)})$$

equal to the torque that would exist in the absence of positive sequence, and $v_f = 0$, and *caused only by the negative-sequence currents*. This term includes a constant component and a sinusoidal component at the frequency $2\sigma_2 = 2(\Omega + \omega)$.

(3) A further term, depending on the simultaneous presence of the field voltage and of the positive and negative sequences, and including sinusoids at the frequencies $\sigma_2 = \Omega + \omega$, $\sigma_1 + \sigma_2 = 2\Omega$ and $\sigma_2 - \sigma_1 = 2\omega$.

In most practical cases, with Ω close or equal to ω, only the slower components of C_e are, however, of interest. For these, it is sufficient to consider term (1) above, as well as the constant component included in term (2). This circumstance simplifies greatly, inasmuch as it makes it possible to separately consider the effects of the negative sequence (as if the positive sequence and the field voltage were zero), then adding them to the effects of the only positive sequence and of the field voltage.

In this connection, the average value of C_e caused by negative-sequence currents may be calculated, in pu, by applying Equation [4.4.5] with $\boldsymbol{i} = I_{(F2)}/I_{(F)\text{nom}}$ and $\sigma = \sigma_2 = \Omega + \omega$. However, this value can often be disregarded, since $l_d(\tilde{j}\sigma_2)$ and $l_q(\tilde{j}\sigma_2)$ are fairly close—for the relatively high value of the frequency σ_2—to $l_d(\infty)$ and $l_q(\infty)$, respectively, which are real constants.

In fact, if we accept these approximations, we simply have (in pu):

$$\left.\begin{aligned} \boldsymbol{\psi}_{d(\sigma 2)} &= -l_d(\infty)\boldsymbol{i}_{d(\sigma 2)} \\ \boldsymbol{\psi}_{q(\sigma 2)} &= -l_q(\infty)\boldsymbol{i}_{q(\sigma 2)} \end{aligned}\right\} \qquad [4.4.8]$$

where, because of Equations [4.4.7]:

$$\begin{cases} \boldsymbol{i}_{d(\sigma 2)} = \dfrac{I_{(F2)}}{I_{(F)\text{nom}}} \cos(\sigma_2 t + \theta_o + \alpha_{I2}) \\[2ex] \boldsymbol{i}_{q(\sigma 2)} = -\dfrac{I_{(F2)}}{I_{(F)\text{nom}}} \sin(\sigma_2 t + \theta_o + \alpha_{I2}) \end{cases}$$

and the torque in pu resulting from the negative-sequence currents is given by:

$$\boldsymbol{\psi}_{d(\sigma 2)}\boldsymbol{i}_{q(\sigma 2)} - \boldsymbol{\psi}_{q(\sigma 2)}\boldsymbol{i}_{d(\sigma 2)} = (l_d(\infty) - l_q(\infty))\frac{1}{2}\left(\frac{I_{(F2)}}{I_{(F)\text{nom}}}\right)^2 \sin^2(\sigma_2 t + \theta_o + \alpha_{I2})$$

with zero average value (this result can immediately be extended to the cases in which the phase voltages or fluxes (not the currents) are sinusoidal at the three sequences).

The behavior of the phase fluxes in pu $\boldsymbol{\psi}_a$, $\boldsymbol{\psi}_b$, $\boldsymbol{\psi}_c$ can be derived from those of $\boldsymbol{\psi}_d$, $\boldsymbol{\psi}_q$, $\boldsymbol{\psi}_o$ by inverting the Park's transformation. In the presence of negative-sequence currents only (and therefore $\boldsymbol{\psi}_o = 0$), if we accept the Equations [4.4.8], we then find:

$$\boldsymbol{\psi}_a \triangleq \frac{\omega_{\text{nom}}\psi_a}{V_{(F)\text{nom}}} = -\sqrt{2}\frac{I_{(F2)}}{I_{(F)\text{nom}}}\left(\frac{l_d(\infty)+l_q(\infty)}{2}\cos(\omega t+\alpha_{I2}) + \frac{l_d(\infty)-l_q(\infty)}{2}\cos((2\Omega+\omega)t+2\theta_o+\alpha_{I2})\right) \quad [4.4.9]$$

as well as similar expressions for $\boldsymbol{\psi}_b$ and $\boldsymbol{\psi}_c$. This means that, unlike the phase currents that are sinusoidal at frequency ω and negative sequence, the phase fluxes (and thus the voltages; see Equations [4.1.5]) consist of a set of three sinusoids at frequency ω and negative sequence, and a set of three sinusoids at frequency $(2\Omega + \omega)$.

If we ignore these last components, the ratios $-\boldsymbol{\psi}_a/\boldsymbol{i}_a$, $-\boldsymbol{\psi}_b/\boldsymbol{i}_b$, $-\boldsymbol{\psi}_c/\boldsymbol{i}_c$ are all equal to the reactance (in pu):

$$x_{2(I)} \triangleq \frac{-\boldsymbol{\psi}_a}{\boldsymbol{i}_a} = \frac{-\boldsymbol{\psi}_b}{\boldsymbol{i}_b} = \frac{-\boldsymbol{\psi}_c}{\boldsymbol{i}_c} = \frac{l_d(\infty)+l_q(\infty)}{2} \quad [4.4.10]$$

which is known as "negative-sequence reactance."

This definition of negative-sequence reactance, in addition to being based on the aforementioned simplifications, is also closely associated with the assumption of sinusoidal phase currents. If, on the other hand, we were to apply negative-sequence sinusoidal fluxes (instead of currents) at frequency ω, each of the phase currents would include two sinusoidal components, at the frequencies ω and $(2\Omega + \omega)$, respectively, and ignoring the latter we should obtain (accepting Equations [4.4.8]):

$$x_{2(\Psi)} \triangleq \frac{-\boldsymbol{\psi}_a}{\boldsymbol{i}_a} = \frac{-\boldsymbol{\psi}_b}{\boldsymbol{i}_b} = \frac{-\boldsymbol{\psi}_c}{\boldsymbol{i}_c} = 2\frac{l_d(\infty)l_q(\infty)}{l_d(\infty)+l_q(\infty)} \quad [4.4.10']$$

This last value might be assumed as the negative-sequence reactance, under the present conditions. It is smaller (but generally not much smaller) than the previous one.

In more general conditions, the negative-sequence reactance might be defined in a similar way, with a value depending on the harmonic content of phase currents and fluxes. However, its range of variation would generally be small, so that—also taking into account the simplifications made—we may reasonably agree to assume, for example, Equation [4.4.10] for all practical cases.

4.4.4. Short-Circuit Transient from Open-Circuit Operation

Let us assume that the machine, initially operating in open-circuit steady-state, is subject at $t = 0$ to a sudden three-phase armature short-circuit. In addition, let us assume, for the sake of simplicity, that the field voltage is kept constant, and that the speed is equal to its nominal value ($\Omega = \text{constant} = \omega_{\text{nom}}$). The extension to the case in which Ω is constant but different from ω_{nom}, is obvious. At $t = 0^-$, we have the following initial conditions, in pu:

$$\begin{aligned}
\boldsymbol{i}_d &= \boldsymbol{i}_q = \boldsymbol{i}_o = 0 \\
\boldsymbol{v}_d &= \boldsymbol{\psi}_q = 0 \\
\boldsymbol{v}_q &= \boldsymbol{\psi}_f = \boldsymbol{\psi}_d = \boldsymbol{i}_f = \boldsymbol{v}_f \\
\boldsymbol{v}_o - \frac{v_n}{V_{(F)\text{nom}}} &= \boldsymbol{\psi}_o = 0 \\
C_e &= 0
\end{aligned}$$

The sudden short-circuit causes, for $t > 0$, $\boldsymbol{v}_d = \boldsymbol{v}_q = \boldsymbol{v}_o - v_n/V_{(F)\text{nom}} = 0$, and is therefore equivalent to a step variation of $\boldsymbol{v}_q$ equal to $-\boldsymbol{v}_f$.

In terms of Laplace transforms, by applying Equations [4.1.30] for $\Omega =$ constant, we then deduce in general (ignoring magnetic saturation):

$$\left.\begin{aligned}
\Delta\boldsymbol{\psi}_d(s) &= \frac{\Delta\boldsymbol{v}_q(s)}{1 + \left(\dfrac{s}{\omega_{\text{nom}}} + \dfrac{r}{l_d(s)}\right)\left(\dfrac{s}{\omega_{\text{nom}}} + \dfrac{r}{l_q(s)}\right)} \\
\Delta\boldsymbol{\psi}_q(s) &= \frac{\left(\dfrac{s}{\omega_{\text{nom}}} + \dfrac{r}{l_d(s)}\right)\Delta\boldsymbol{v}_q(s)}{1 + \left(\dfrac{s}{\omega_{\text{nom}}} + \dfrac{r}{l_d(s)}\right)\left(\dfrac{s}{\omega_{\text{nom}}} + \dfrac{r}{l_q(s)}\right)}
\end{aligned}\right\} \qquad [4.4.11]$$

where:

$$\begin{aligned}
\Delta\boldsymbol{\psi}_d(s) &\triangleq \mathcal{L}\{\boldsymbol{\psi}_d(t) - \boldsymbol{\psi}_d(0^-)\} = \mathcal{L}\{\boldsymbol{\psi}_d(t)\} - \frac{\boldsymbol{v}_f}{s} \\
\Delta\boldsymbol{\psi}_q(s) &\triangleq \mathcal{L}\{\boldsymbol{\psi}_q(t) - \boldsymbol{\psi}_q(0^-)\} = \mathcal{L}\{\boldsymbol{\psi}_q(t)\} \\
\Delta\boldsymbol{v}_q(s) &\triangleq \mathcal{L}\{\boldsymbol{v}_q(t) - \boldsymbol{v}_q(0^-)\} = -\frac{\boldsymbol{v}_f}{s}
\end{aligned}$$

The previous equations make it possible to obtain the time behavior of $\boldsymbol{\psi}_d$ and $\boldsymbol{\psi}_q$, whereas the behavior of $\boldsymbol{i}_f$, $\boldsymbol{\psi}_f$, $\boldsymbol{i}_d$, $\boldsymbol{i}_q$ can be derived by applying Equations [4.1.27] and [4.1.23′] (from Equations [4.1.31] and [4.1.24′], we then have $\boldsymbol{\psi}_o = \boldsymbol{i}_o = 0$).

The analysis may be developed fairly simply if we ignore the armature resistance.

For $r = 0$, from Equations [4.4.11] we derive that the transfer functions $(\Delta\boldsymbol{\psi}_d/\Delta\boldsymbol{v}_q)(s)$ and $(\Delta\boldsymbol{\psi}_q/\Delta\boldsymbol{v}_q)(s)$ have poles $\pm\tilde{\jmath}\omega_{\text{nom}}$, and:

$$\left.\begin{aligned}
\boldsymbol{\psi}_d(t) &= \boldsymbol{v}_f\cos(\omega_{\text{nom}}t) \\
\boldsymbol{\psi}_q(t) &= -\boldsymbol{v}_f\sin(\omega_{\text{nom}}t)
\end{aligned}\right\} \qquad [4.4.12]$$

i.e., $\boldsymbol{\psi}_d$ and $\boldsymbol{\psi}_q$ are simply sinusoidal, at the frequency ω_{nom}. If we then accept Equations [4.1.26′], from Equations [4.1.27] we finally derive that $\boldsymbol{i}_f$, $\boldsymbol{\psi}_f$, and $\boldsymbol{i}_d$ consist of a constant component (resulting from the field voltage), of a sinusoidal component at frequency ω_{nom}, and of an exponential component corresponding to the time constant $\hat{T}'_d$; whereas for Equation [4.1.23′], we simply have:

$$\boldsymbol{i}_q(t) = \frac{-\boldsymbol{\psi}_q(t)}{x_q} = \frac{\boldsymbol{v}_f}{x_q}\sin(\omega_{\text{nom}}t) \qquad [4.4.13]$$

Developing the equations, we find that, since $\omega_{\text{nom}}\hat{T}'_d \gg 1$, the expressions of $\boldsymbol{i}_f(t)$, $\boldsymbol{\psi}_f(t)$, and $\boldsymbol{i}_d(t)$ may be approximated as follows:

$$\left.\begin{aligned}
\boldsymbol{i}_f(t) &\cong \boldsymbol{v}_f\left(1 + \frac{x_d - \hat{x}'_d}{\hat{x}'_d}\epsilon^{-t/\hat{T}'_d} - \cos(\omega_{\text{nom}}t)\right)\\
\boldsymbol{\psi}_f(t) &\cong \boldsymbol{v}_f\left(\frac{\hat{x}'_d}{x_d} + \frac{x_d - \hat{x}'_d}{x_d}\epsilon^{-t/\hat{T}'_d}\right)\\
\boldsymbol{i}_d(t) &\cong \boldsymbol{v}_f\left(\frac{1}{x_d} + \left(\frac{1}{\hat{x}'_d} - \frac{1}{x_d}\right)\epsilon^{-t/\hat{T}'_d} - \frac{\cos(\omega_{\text{nom}}t)}{\hat{x}'_d}\right)
\end{aligned}\right\} \qquad [4.4.14]$$

(in reality, the sinusoidal component in $\boldsymbol{\psi}_f(t)$ is very small).

With the phase fluxes and currents, by inverting the Park's transformation (with reference $\theta_r = \theta = \omega_{\text{nom}}t + \theta_o$), we derive:

- from Equations [4.4.12]:

$$\boldsymbol{\psi}_a \triangleq \frac{\omega_{\text{nom}}\psi_a}{V_{(F)\text{nom}}} = \sqrt{2}\boldsymbol{v}_f\cos\theta_o = \text{constant} \qquad [4.4.15]$$

  and similarly $\boldsymbol{\psi}_b$, $\boldsymbol{\psi}_c$, by replacing θ_o by $\theta_o - 120°$, $\theta_o - 240°$, respectively;
- from Equations [4.4.13] and [4.4.14]:

$$\begin{aligned}
\boldsymbol{i}_a \triangleq \frac{i_a}{I_{(F)\text{nom}}} &\cong \sqrt{2}\boldsymbol{v}_f\left[\left(\frac{1}{x_d} + \left(\frac{1}{\hat{x}'_d} - \frac{1}{x_d}\right)\epsilon^{-t/\hat{T}'_d}\right)\cos(\omega_{\text{nom}}t + \theta_o)\right.\\
&\left.-\frac{1}{2}\left(\frac{1}{\hat{x}'_d} + \frac{1}{x_q}\right)\cos\theta_o - \frac{1}{2}\left(\frac{1}{\hat{x}'_d} - \frac{1}{x_q}\right)\cos(2\omega_{\text{nom}}t + \theta_o)\right]
\end{aligned} \qquad [4.4.16]$$

  and similarly $\boldsymbol{i}_b$, $\boldsymbol{i}_c$, by replacing θ_o by $\theta_o - 120°$, $\theta_o - 240°$, respectively.

The fluxes $\boldsymbol{\psi}_a$, $\boldsymbol{\psi}_b$, $\boldsymbol{\psi}_c$ (for $r = 0$) are constant, according to the fact that $\boldsymbol{\psi}_d$ and $\boldsymbol{\psi}_q$ are sinusoidal with frequency $\omega_{\text{nom}} = \Omega$.

The currents $\boldsymbol{i}_a$, $\boldsymbol{i}_b$, $\boldsymbol{i}_c$, on the other hand, contain:

- an oscillatory term at frequency ω_{nom} with magnitude decreasing according to the time constant $\hat{T}'_d$;

- a "unidirectional" constant term;
- a persistent sinusoidal term at frequency $2\omega_{\text{nom}}$;

where the two last terms are the result of the sinusoidal components, at frequency ω_{nom}, present in $\boldsymbol{i}_d$ and $\boldsymbol{i}_q$ (such terms would be missing if we accepted the simplified Equations [4.2.1]).

The electromagnetic torque (in pu) $C_e = \boldsymbol{\psi}_d \boldsymbol{i}_q - \boldsymbol{\psi}_q \boldsymbol{i}_d$ has an average value other than zero—on which sinusoidal components at frequencies ω_{nom} and $2\omega_{\text{nom}}$ are superposed—because of the presence of sinusoidal components in the fluxes $\boldsymbol{\psi}_d$ and $\boldsymbol{\psi}_q$ and in the currents $\boldsymbol{i}_d$ and $\boldsymbol{i}_q$.

More precisely, the average value of C_e is entirely the result of the presence of the unidirectional components. In general (ignoring the magnetic saturation and the armature resistance), the unidirectional component in $\boldsymbol{i}_a$, $\boldsymbol{i}_b$, $\boldsymbol{i}_c$ is exactly equal to:

$$\frac{-\boldsymbol{v}_f}{\sqrt{2}}\left[\cos\theta_o \mathcal{Re}\left(\frac{1}{l_d(\tilde{j}\omega_{\text{nom}})}+\frac{1}{l_q(\tilde{j}\omega_{\text{nom}})}\right)\right.$$
$$\left.+\sin\theta_o\, Im\left(\frac{1}{l_d(\tilde{j}\omega_{\text{nom}})}+\frac{1}{l_q(\tilde{j}\omega_{\text{nom}})}\right)\right]$$

and the average value of C_e is equal to:

$$C_{e(m)} = \frac{\boldsymbol{v}_f^2}{2}\, Im\left(\frac{1}{l_d(\tilde{j}\omega_{\text{nom}})}+\frac{1}{l_q(\tilde{j}\omega_{\text{nom}})}\right)$$

By accepting Equations [4.1.26′], this average value is expressed by:

$$C_{e(m)} = \frac{\boldsymbol{v}_f^2}{2}\left(\frac{1}{\hat{x}'_d}-\frac{1}{x_d}\right)\frac{\omega_{\text{nom}}\hat{T}'_d}{1+(\omega_{\text{nom}}\hat{T}'_d)^2} \qquad [4.4.17]$$

and is therefore positive (although small in value, since $\omega_{\text{nom}}\hat{T}'_d \gg 1$), i.e., such as to cause, in practice (contrarily to the assumption of constant speed), a slowing down in the rotor, known as "*backswing*"(7).

If we consider that $r \neq 0$, the Equations [4.4.11] indicate that $\boldsymbol{\psi}_d$ and $\boldsymbol{\psi}_q$ are no longer simply sinusoidal.

(7) In general, in the case of a generator initially operating on load (instead of no-load, as here assumed), the "backswing" phenomenon may lead to a temporary initial slowing-down (after the short-circuit), followed by a progressive acceleration of the rotor caused by the dominance of the driving torque over the resistant torque. The effects of the latter are in fact decreasing with time, as remarked hereafter, because of the armature resistance.

The analysis may, however, be considerably simplified if r is fairly small, as normally happens. In such a case, the pair of poles $\pm \tilde{j}\omega_{nom}$ (for $r = 0$) in the Equations [4.4.11] is approximately replaced by the pair of poles:

$$\left[-\frac{r}{2} \mathcal{R}e \left(\frac{1}{l_d(\tilde{j}\omega_{nom})} + \frac{1}{l_q(\tilde{j}\omega_{nom})} \right) \pm \tilde{j} \right] \omega_{nom} \qquad [4.4.18]$$

so that $\boldsymbol{\psi}_d$ and $\boldsymbol{\psi}_q$ (as well as $\boldsymbol{\psi}_f$, and the currents $\boldsymbol{i}_f$, $\boldsymbol{i}_d$, and $\boldsymbol{i}_q$) include oscillatory components at frequency $\sim \omega_{nom}$, that are damped according to a damping factor:

$$\zeta \cong \frac{r}{2} \mathcal{R}e \left(\frac{1}{l_d(\tilde{j}\omega_{nom})} + \frac{1}{l_q(\tilde{j}\omega_{nom})} \right)$$

If, in addition, zeros and poles of $l_d(s)$ and $l_q(s)$ are, in absolute value, sufficiently smaller than ω_{nom}, the denominator in Equations [4.4.11] may be approximated to:

$$1 + \left(\frac{s}{\omega_{nom}} + \frac{r}{l_d(s)} \right) \left(\frac{s}{\omega_{nom}} + \frac{r}{l_q(s)} \right) \cong 1 + \left(\frac{s}{\omega_{nom}} + \frac{r}{l_d(\infty)} \right) \left(\frac{s}{\omega_{nom}} + \frac{r}{l_q(\infty)} \right)$$

and the pair of poles [4.4.18] may be approximated to:

$$\left(-\frac{r}{2} \left(\frac{1}{l_d(\infty)} + \frac{1}{l_q(\infty)} \right) \pm \tilde{j} \right) \omega_{nom} = \left(-\frac{r}{x_{2(\Psi)}} \pm \tilde{j} \right) \omega_{nom} \qquad [4.4.19]$$

corresponding to a damping factor $\zeta \cong r/x_{2(\Psi)}$, where $x_{2(\Psi)}$ is the negative-sequence reactance in Equation [4.4.10′].

All considered, the most important effect of the armature resistance is that of damping the oscillatory components in the fluxes $\boldsymbol{\psi}_d$, $\boldsymbol{\psi}_q$, $\boldsymbol{\psi}_f$, and in the currents $\boldsymbol{i}_d$, $\boldsymbol{i}_q$, $\boldsymbol{i}_f$. The magnitudes of the oscillatory components decrease according to an exponential law, with a time constant:

$$T_{ac} \cong \frac{1}{\zeta \omega_{nom}}$$

that is the so-called "(short-circuit) *armature time constant*" (if we accept the approximation [4.4.19], it results $T_{ac} \cong x_{2(\Psi)}/(\omega_{nom} r)$), and the same happens for the magnitude of the unidirectional components in the phase fluxes $\boldsymbol{\psi}_a$, $\boldsymbol{\psi}_b$, $\boldsymbol{\psi}_c$, and in the phase currents $\boldsymbol{i}_a$, $\boldsymbol{i}_b$, $\boldsymbol{i}_c$.

As a further effect of the armature resistance, the behaviors of $\boldsymbol{\psi}_d$, $\boldsymbol{\psi}_q$, and $\boldsymbol{i}_q$ are no longer simply oscillatory, but also — like $\boldsymbol{\psi}_f$, $\boldsymbol{i}_f$, $\boldsymbol{i}_d$ — include a constant component, and exponential components corresponding approximately, with the assumptions made, to the zeros of $l_d(s)$ and $l_q(s)$, i.e., $-1/\hat{T}'_d$ if we accept the Equations [4.1.26′].

By just assuming:

$$l_d(s) = x_d \frac{1 + s\hat{T}'_d}{1 + s\hat{T}'_{do}}$$

$$l_q(s) = x_q$$

the Equations [4.4.12], [4.4.13], and [4.4.14] are, as a first approximation (within the assumptions made), still applicable, provided that $\cos(\omega_{\text{nom}}t)$ and $\sin(\omega_{\text{nom}}t)$ are, respectively, replaced by $\epsilon^{-t/T_{ac}}\cos(\omega_{\text{nom}}t)$ and by $\epsilon^{-t/T_{ac}}\sin(\omega_{\text{nom}}t)$. The Equations [4.4.15] and [4.4.16] are correspondingly replaced by:

$$\boldsymbol{\psi}_a \cong \sqrt{2}\boldsymbol{v}_f \cos\theta_o \epsilon^{-t/T_{ac}}$$

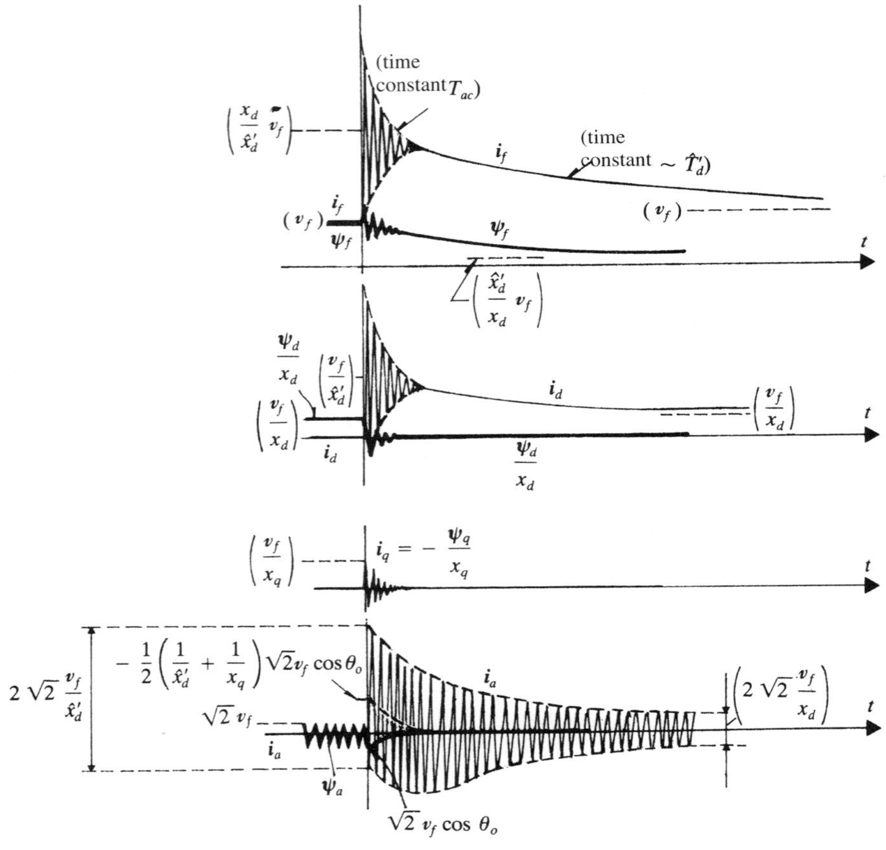

Figure 4.18. Short-circuit transient from open-circuit operation: (qualitative) time behavior of currents and fluxes in "per unit" in the absence of additional rotor circuits.

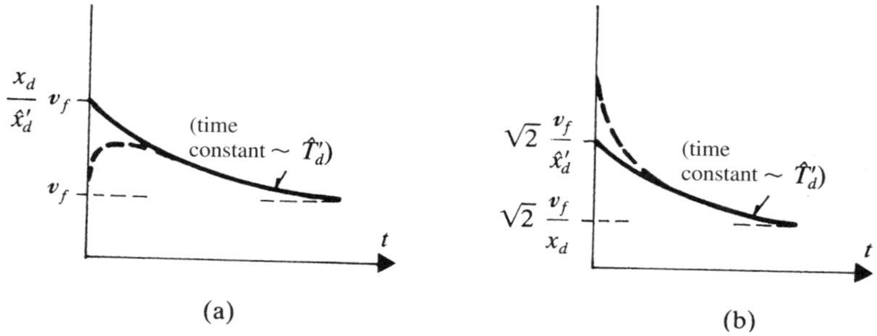

Figure 4.19. Short-circuit transient from open-circuit operation: time behavior of the "per-unit" magnitudes of: (a) the unidirectional component of the field current; (b) the oscillatory component (at nominal frequency) of the phase currents.

$$i_a \cong \sqrt{2} v_f \left[\left(\frac{1}{x_d} + \left(\frac{1}{\hat{x}'_d} - \frac{1}{x_d} \right) \epsilon^{-t/\hat{T}'_d} \right) \cos(\omega_{\text{nom}} t + \theta_o) \right.$$
$$\left. - \frac{1}{2} \left(\frac{1}{\hat{x}'_d} + \frac{1}{x_q} \right) \cos\theta_o \epsilon^{-t/T_{ac}} - \frac{1}{2} \left(\frac{1}{\hat{x}'_d} - \frac{1}{x_q} \right) \epsilon^{-t/T_{ac}} \cos(2\omega_{\text{nom}} t + \theta_o) \right]$$

and similarly for the indices b and c, by replacing θ_o by $\theta_o - 120°$ and $\theta_o - 240°$, respectively.

The behavior of the variables considered is indicated qualitatively in Figure 4.18, assuming that we may ignore, in the phase currents, the oscillatory component at frequency $2\omega_{\text{nom}}$. In particular, the unidirectional (not oscillatory) component of i_f and the magnitude of the oscillatory component (at frequency ω_{nom}) of the phase currents, have trends of the type indicated in Figure 4.19. If we assume the Equations [4.3.6] instead of Equations [4.1.26′], the latter diagrams are modified as indicated qualitatively by the dashed line.

ANNOTATED REFERENCES

Among the works of more general interest, the following ones are evidenced: 7, 10, 15, 16, 17, 34, 37, 65, 123, 151, 248.

More particularly, for what refers to

- the solid rotor: 116;
- the magnetic saturation: 151, 243, 311;
- the torsional phenomena (and the subsynchronous resonances): 36, 208, 230.

CHAPTER 5

DYNAMIC BEHAVIOR OF NETWORK ELEMENTS AND LOADS

5.1. PRELIMINARIES

One of the fundamental problems of the dynamic analysis of an electric system is the choice of the most suitable mathematical models for different components. This choice generally depends on the specific goals of the analysis. In fact, the use of complicated models can increase the complexity of a problem, often with no practical benefits (for some components, even the deduction of the mathematical model or the evaluation of the parameters may be uncertain).

When studying electromechanical phenomena, the most complicated models are generally related only to synchronous machines and their associated equipment (e.g., voltage and frequency regulators, excitation systems, turbine supplying systems, etc.). With network elements (lines, transformers, etc.) and loads, it is common to avoid representation of the actual transient characteristics, because their effects on the electromechanical phenomenon can be negligible.

Network elements are usually considered by simple equivalent impedances (or admittances), the value of which is the one assumed at the equilibrium steady-state at nominal frequency. With such a representation, the problem may be greatly simplified from the analytical and computational point of view, usually without significant errors in the analysis of electromechanical phenomenon.

However, before this approach is adopted, the acceptability of the simplifications should be determined. In some cases, the approximation obtained might be insufficient. Generally, the dynamic contribution of network elements is fundamental with respect to electrical phenomena, and may specifically affect the faster parts of electromechanical phenomena.

In Sections 5.2, 5.3, and 5.4, which refer to typical elements of ac three-phase networks, the dynamic characteristics are derived by using Park's transformation (see Appendix 2) and assuming that the elements are *linear and physically symmetrical*. Later, general criteria are described to directly determine operational impedances (or admittances) and transfer functions, which define the behavior in any dynamic operating condition.

Section 5.5 illustrates the most important dynamic characteristics of dc links and related control equipment.

In Section 5.6 load characteristics are considered, with more details for the dynamic model of the asynchronous machine.

Finally, in Section 5.7, typical electrical phenomena are described, with particular reference to the dynamic models of inductive and capacitive elements. The importance of such models will be confirmed in Sections 6.2.1 and 7.2.4, referring to the phenomena of "self-excitation" and "subsynchronous oscillations," respectively.

5.2. GENERALITIES ON NETWORK ELEMENTS

5.2.1. Elementary Equivalent Circuits

Resistors

The simplest three-phase element is constituted by a set of three equal resistors, not directly connected (see Fig. 5.1a), for which the "phase" voltages and currents $v_{(F)a} = v_{1a} - v_{2a}$, $i_{(F)a} = i_a$, etc. are related to one another by:

$$\begin{cases} v_{(F)a} = Ri_{(F)a} \\ v_{(F)b} = Ri_{(F)b} \\ v_{(F)c} = Ri_{(F)c} \end{cases}$$

Voltages v_{1a}, v_{2a}, etc. are defined with respect to an arbitrary voltage reference.

By applying Park's transformation with a generic angular reference θ_r, it can be derived (see Appendix 2):

$$\left.\begin{aligned} \overline{v}_{(F)r} &= \overline{v}_{1r} - \overline{v}_{2r} \\ v_{(F)o} &= v_{1o} - v_{2o} \end{aligned}\right\} \qquad [5.2.1]$$

$$\left.\begin{aligned} \overline{\imath}_{(F)r} &= \overline{\imath}_r \\ i_{(F)o} &= i_o \end{aligned}\right\} \qquad [5.2.2]$$

and furthermore:

$$\left.\begin{aligned} \overline{v}_{(F)r} &= R\overline{\imath}_{(F)r} \\ v_{(F)o} &= Ri_{(F)o} \end{aligned}\right\} \qquad [5.2.3]$$

according to the equivalent circuits of Figure 5.1b.

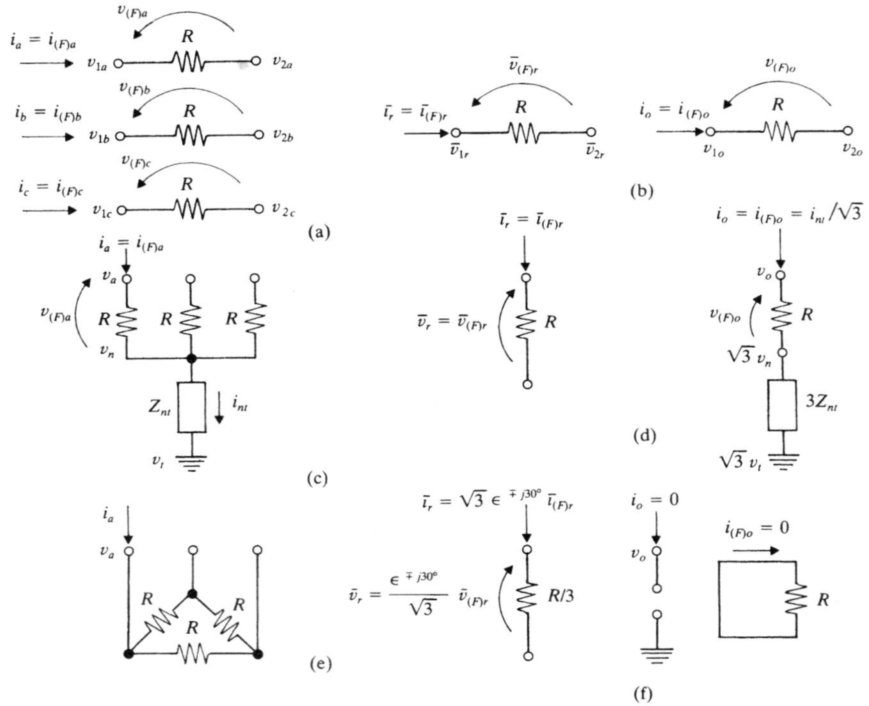

Figure 5.1. Resistive three-phase elements: set of three equal resistors not connected to one another (a), and equivalent circuits (b); wye-connected resistors (c), and equivalent circuits (d); delta-connected resistors (e), and equivalent circuits (f).

It should be noted that:

(1) Park's vectors $\overline{v}_{1r}$ and $\overline{v}_{2r}$ depend only on the differences between v_{1a}, v_{1b}, v_{1c} and v_{2a}, v_{2b}, v_{2c}, respectively, and thus they are independent of the voltage reference;

(2) the homopolar components v_{1o} and v_{2o}, instead, depend on the sums $(v_{1a} + v_{1b} + v_{1c})$ and $(v_{2a} + v_{2b} + v_{2c})$, respectively, and thus they remain defined with respect to the voltage reference adopted.

The term "homopolar" is more general than "of zero sequence," as it is not restricted to sinusoidal phase variables (see Appendices 1 and 2).

As a result of property (1), it appears unnecessary or even misleading to attribute a physical meaning to the common node to which $\overline{v}_{1r}$ and $\overline{v}_{2r}$ are referred, as well as to the voltage of this node. This observation can be extended to all equivalent circuits related to Park's vectors, such as the circuits reported in chapters 2 and 4, and those defined in the following. Usually, however, the common node is assumed to be a (fictitious) "ground" node for the mentioned equivalent circuits. This assumption also facilitates graphical representation.

If the three resistors are *wye-connected* (Fig. 5.1c) by assuming that $v_a = v_{1a}$, $v_b = v_{1b}$, $v_c = v_{1c}$, and that:

$$v_{2a} = v_{2b} = v_{2c} = v_n \qquad \text{(neutral voltage)}$$

$$i_a + i_b + i_c = i_{nt} \qquad \text{(earth current, flowing from neutral to earth)}$$

it can be derived:

$$\begin{cases} \overline{v}_{2r} = 0 \\ v_{2o} = \sqrt{3}v_n \end{cases}$$

and thus as a result of Equations [5.2.1]:

$$\begin{cases} \overline{v}_{(F)r} = \overline{v}_{1r} = \overline{v}_r \\ v_{(F)o} = v_{1o} - \sqrt{3}v_n = v_o - \sqrt{3}v_n \end{cases}$$

and further that:

$$i_o = \frac{i_{nt}}{\sqrt{3}}$$

whereas Equations [5.2.2] and [5.2.3] are still valid; see the equivalent circuits of Figure 5.1d, where:

$$\frac{\overline{v}_r}{\overline{\imath}_r} = \frac{\overline{v}_{(F)r}}{\overline{\imath}_{(F)r}} = R, \qquad \frac{v_o - \sqrt{3}v_n}{i_o} = \frac{v_{(F)o}}{i_{(F)o}} = R$$

As specified in Appendix 2 (Equations [A2.4]), the multiplicative constants of Park's transformation are assumed $K_{dq} = \sqrt{2/3}$, $K_o = 1/\sqrt{3}$; a generic choice of K_o leads to $v_{2o} = 3K_o v_n$, $i_o = K_o i_{nt}$.

As far as the *"earth" current* i_{nt} is concerned, it is zero if the neutral is isolated, whereas if it is connected to earth this current depends on the difference between the neutral voltage v_n and the earth voltage v_t (Fig. 5.1c), according to:

$$v_n - v_t = Z_{nt}(p)i_{nt}$$

where $Z_{nt}(p)$ is the operational impedance of the connection (with $p \triangleq \mathrm{d}/\mathrm{d}t$). It can be derived:

$$v_{2o} - \sqrt{3}v_t = 3Z_{nt}(p)i_o$$

(for a generic value K_o it would follow that $v_{2o} - 3K_o v_t = 3Z_{nt}(p)i_o$), so that the equivalent circuit related to homopolar variables must be completed, by considering an impedance equal to $3Z_{nt}(p)$ to the earth node. Note that the earth node has a precise physical meaning, unlike the "ground" used for other equivalent circuits, related to Park's vectors.

The impedance $Z_{nt}(p)$ can be, for instance, $Z_{nt}(p) = R_{nt}$ if the connection is made by a resistance R_{nt} (in particular: $R_{nt} = 0$ if the neutral is directly grounded, whereas we can assume $R_{nt} = \infty$ if the neutral is ungrounded), or $Z_{nt}(p) = pL_{nt}$ if the connection is made by an inductance L_{nt}.

If the three resistors are instead *delta-connected* as per Figure 5.1e, by assuming that $v_a = v_{1a} = v_{2c}$, $v_b = v_{1b} = v_{2a}$, $v_c = v_{1c} = v_{2b}$, and furthermore:

$$i_a = i_{(F)a} - i_{(F)c}, \quad i_b = i_{(F)b} - i_{(F)a}, \quad i_c = i_{(F)c} - i_{(F)b}$$

it follows that:

$$\begin{cases} \overline{v}_r = \overline{v}_{1r} = \overline{v}_{2r}\epsilon^{+j120^\circ} \\ v_o = v_{1o} = v_{2o} \end{cases}$$

and then as a result of Equations [5.2.1]:

$$\begin{cases} \overline{v}_{(F)r} = (1 - \epsilon^{-j120^\circ})\overline{v}_r = \sqrt{3}\epsilon^{+j30^\circ}\overline{v}_r \\ v_{(F)o} = 0 \end{cases}$$

and furthermore, instead of Equations [5.2.2]:

$$\begin{cases} \overline{\imath}_r = (1 - \epsilon^{+j120^\circ})\overline{\imath}_{(F)r} = \sqrt{3}\epsilon^{-j30^\circ}\overline{\imath}_{(F)r} \\ i_o = 0 \end{cases}$$

whereas Equations [5.2.3] are still valid; see the equivalent circuits of Figure 5.1f, where $\overline{v}_r/\overline{\imath}_r = \overline{v}_{(F)r}/(3\overline{\imath}_{(F)r}) = R/3$, $i_o = 0$, $v_{(F)o} = Ri_{(F)o} = 0$.

There is no change, as far as the external implications are considered, if the three phases are connected to have $v_a = v_{1a} = v_{2b}$, $v_b = v_{1b} = v_{2c}$, $v_c = v_{1c} = v_{2a}$. In this case, it is only necessary to substitute the terms $\epsilon^{\pm j120^\circ}$, $\epsilon^{\pm j30^\circ}$ by $\epsilon^{\mp j120^\circ}$, $\epsilon^{\mp j30^\circ}$, respectively, so that the phase-shifts between $\overline{v}_r$ and $\overline{v}_{(F)r}$, as well as those between $\overline{\imath}_r$ and $\overline{\imath}_{(F)r}$, change their signs, whereas it still holds $\overline{v}_r/\overline{\imath}_r = R/3$, $i_o = 0$, $v_{(F)o} = Ri_{(F)o} = 0$.

Inductors

For a set of three inductors, the phenomenon of mutual induction must be generally considered, by equations:

$$\begin{cases} v_{(F)a} = \dfrac{\mathrm{d}\psi_{(F)a}}{\mathrm{d}t} \\ v_{(F)b} = \dfrac{\mathrm{d}\psi_{(F)b}}{\mathrm{d}t} \\ v_{(F)c} = \dfrac{\mathrm{d}\psi_{(F)c}}{\mathrm{d}t} \end{cases} \qquad \begin{cases} \psi_{(F)a} = Li_{(F)a} - M(i_{(F)a} + i_{(F)b} + i_{(F)c}) \\ \psi_{(F)b} = Li_{(F)b} - M(i_{(F)a} + i_{(F)b} + i_{(F)c}) \\ \psi_{(F)c} = Li_{(F)c} - M(i_{(F)a} + i_{(F)b} + i_{(F)c}) \end{cases}$$

(with the hypotheses of linearity and physical symmetry), where self-inductances and mutual inductances are $(L - M)$ and $(-M)$, respectively.

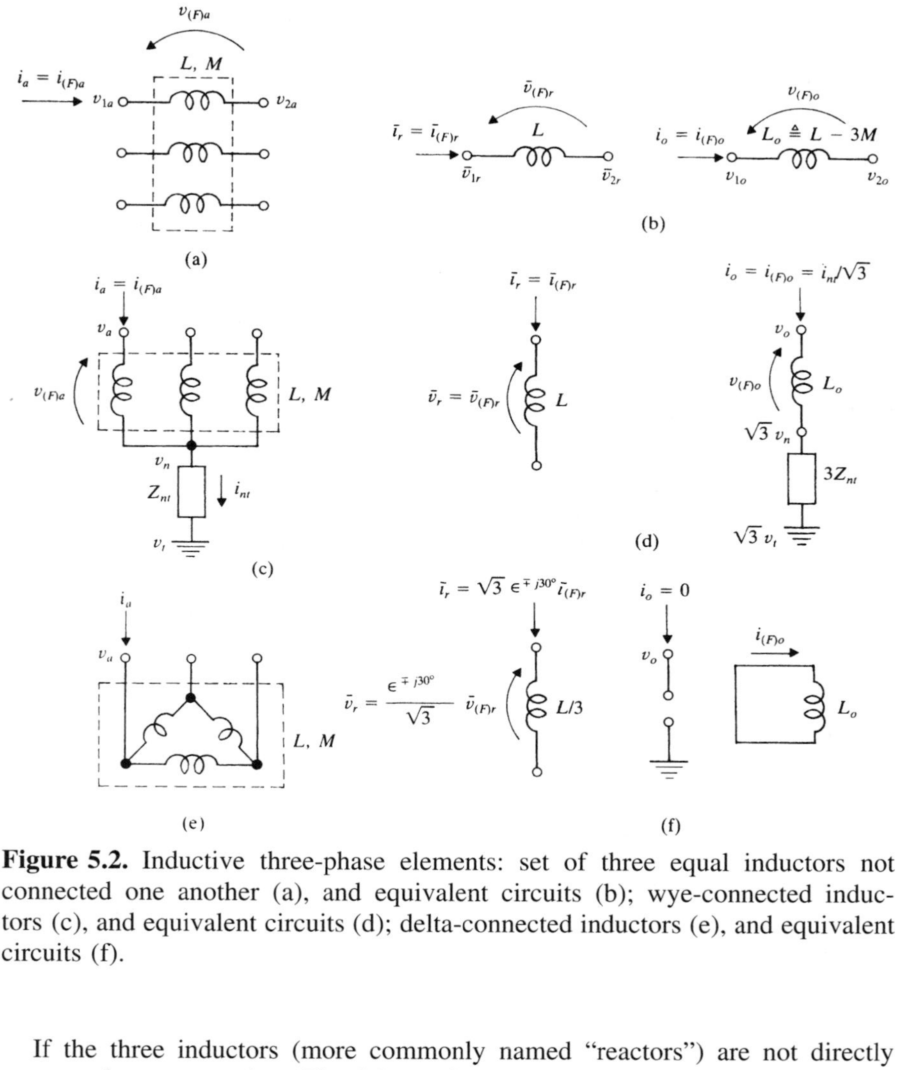

Figure 5.2. Inductive three-phase elements: set of three equal inductors not connected one another (a), and equivalent circuits (b); wye-connected inductors (c), and equivalent circuits (d); delta-connected inductors (e), and equivalent circuits (f).

If the three inductors (more commonly named "reactors") are not directly connected to one another (Fig. 5.2a), using the previously defined symbols one can again derive Equations [5.2.1] and [5.2.2], whereas Equations [5.2.3] are substituted by (see Equations [A2.5] in Appendix 2):

$$\left.\begin{aligned} \overline{v}_{(F)r} &= (p + j\Omega_r)\overline{\psi}_{(F)r} \\ v_{(F)o} &= p\psi_{(F)o} \end{aligned}\right\} \qquad \left.\begin{aligned} \overline{\psi}_{(F)r} &= L\overline{i}_{(F)r} \\ \psi_{(F)o} &= L_o i_{(F)o} \end{aligned}\right\} \qquad [5.2.4]$$

where $p \triangleq \mathrm{d}/\mathrm{d}t$, $\Omega_r \triangleq \mathrm{d}\theta_r/\mathrm{d}t$ and where the "homopolar" inductance L_o is given by:

$$L_o = L - 3M \qquad [5.2.5]$$

The value of M and consequently that of L_o depend on the magnetic circuit structure. In particular, $M = 0$ for single-phase inductors that are magnetically decoupled.

From the above equations, it is possible to determine the equivalent circuits of Figure 5.2b, where:

$$\frac{\overline{v}_{1r} - \overline{v}_{2r}}{\overline{\imath}_r} = \frac{\overline{v}_{(F)r}}{\overline{\imath}_{(F)r}} = (p + j\Omega_r)L, \qquad \frac{v_{1o} - v_{2o}}{i_o} = \frac{v_{(F)o}}{i_{(F)o}} = pL_o$$

Note that:

- in the equivalent circuit that refers to the homopolar components, the graphic symbol of the inductance L_o corresponds to the equation $v_{(F)o} = pL_o i_{(F)o}$;
- on the contrary, in the equivalent circuit referring to Park's vectors, the graphic symbol of the inductance L corresponds to the equation $\overline{v}_{(F)r} = (p + j\Omega_r)L\overline{\imath}_{(F)r}$ (instead of $\overline{v}_{(F)r} = pL\overline{\imath}_{(F)r}$).

Such inconsistency, concerning the latter equivalent circuit (and similar ones defined in the following) may be removed by adding the impedance $j\Omega_r L$ in series to the inductance L, or avoiding the use of the inductance symbol and considering the whole impedance $(p + j\Omega_r)L$. This, however, is not in common use. Usually, as indicated in the following, the transient behavior of the network element is disregarded ($p = 0$), thus adopting the equation $\overline{v}_{(F)r} = j\Omega_r L\overline{\imath}_{(F)r}$, a different equation from the two previously reported. To avoid misunderstandings, it is convenient to intend that, in the equivalent circuits concerning the Park's vectors, the graphic symbol for the inductance represents an inductive element, by separately specifying the corresponding equation (if it is not clear from the context).

If the three inductors are *wye-connected* (Fig. 5.2c) or *delta-connected* (Fig. 5.2e), the equivalent circuits shown in Figures 5.2d and 5.2f, respectively, can be derived with a similar procedure, adopting known notation. (In the case of delta-connection, it holds that $v_{(F)o} = pLi_{(F)o} = 0$, with the theoretical possibility of a persisting circulating current corresponding to $i_{(F)o} =$ constant.)

Condensers

For a set of three equal condensers not directly connected to one another (Fig. 5.3a), by assuming:

$$\begin{cases} i_{(F)a} = C\dfrac{dv_{(F)a}}{dt} \\ i_{(F)b} = C\dfrac{dv_{(F)b}}{dt} \\ i_{(F)c} = C\dfrac{dv_{(F)c}}{dt} \end{cases}$$

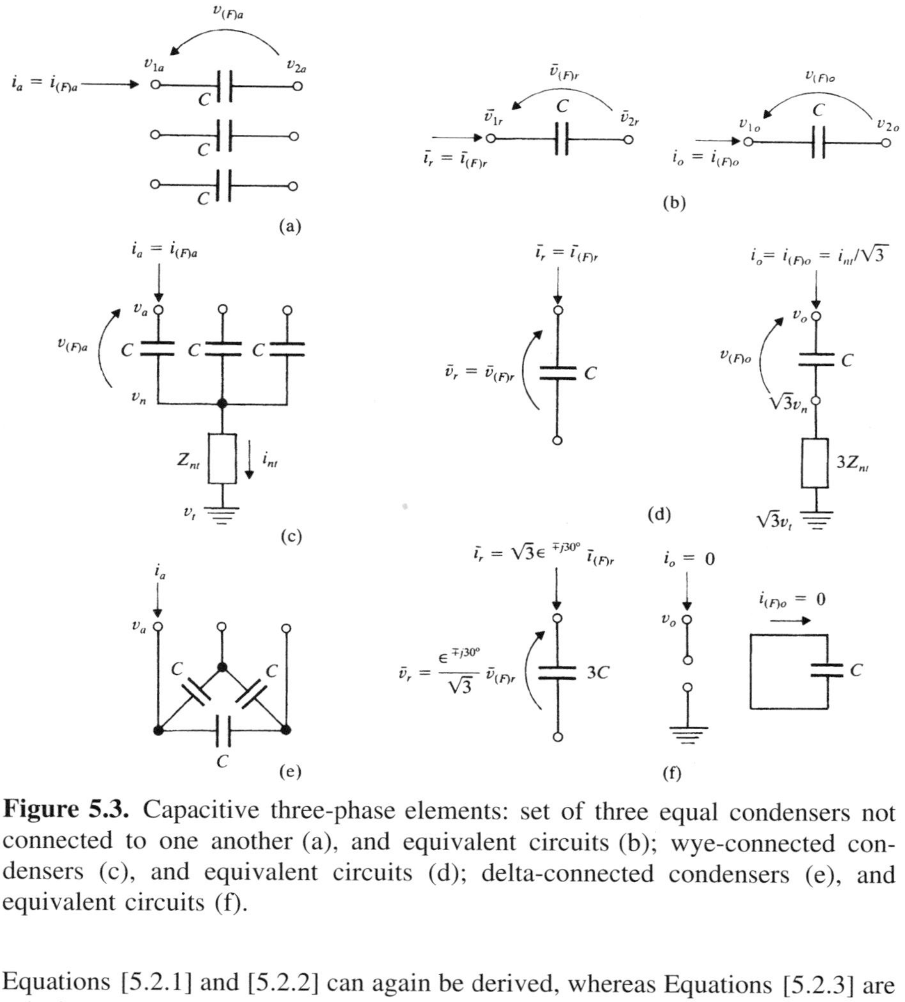

Figure 5.3. Capacitive three-phase elements: set of three equal condensers not connected to one another (a), and equivalent circuits (b); wye-connected condensers (c), and equivalent circuits (d); delta-connected condensers (e), and equivalent circuits (f).

Equations [5.2.1] and [5.2.2] can again be derived, whereas Equations [5.2.3] are substituted by (see Equations [A2.5] in Appendix 2):

$$\left.\begin{aligned} \overline{\imath}_{(F)r} &= (p + j\Omega_r)C\overline{v}_{(F)r} \\ i_{(F)o} &= pCv_{(F)o} \end{aligned}\right\} \qquad [5.2.6]$$

so that the equivalent circuits of Figure 5.3b can be derived, with:

$$\frac{\overline{\imath}_r}{\overline{v}_{1r} - \overline{v}_{2r}} = \frac{\overline{\imath}_{(F)r}}{\overline{v}_{(F)r}} = (p + j\Omega_r)C, \qquad \frac{i_o}{v_{1o} - v_{2o}} = \frac{i_{(F)o}}{v_{(F)o}} = pC$$

If the three condensers are *wye-connected* (Fig. 5.3c) or *delta-connected* (Fig. 5.3e), the equivalent circuits of Figure 5.3d and 5.3f can be, respectively, derived.

Similar to the above discussion regarding inductors:

- in the equivalent circuits that refer to the homopolar components, the symbol of the generic capacitance C stands for the equation $i_{(F)o} = pCv_{(F)o}$;
- on the contrary, in the equivalent circuits referring to Park's vectors, such a symbol represents the equation $\overline{\imath}_{(F)r} = (p + j\Omega_r)C\overline{v}_{(F)r}$ (instead of $\overline{\imath}_{(F)r} = pC\overline{v}_{(F)r}$), or even the equation $\overline{\imath}_{(F)r} = j\Omega_r C\overline{v}_{(F)r}$, when the transient behavior of the element considered is (clearly) disregarded.

5.2.2. Transfer Functions and Frequency Response for the Whole Network

By adopting Park's transformation and the corresponding equivalent circuits, each network element considered in Section 5.2.1 can be viewed as a simple series or shunt "branch" (see Section 2.1.2), defined by a proper operational impedance (or admittance).

In the *steady-state equilibrium condition*, with sinusoidal and positive sequence (phase) voltages and currents, the homopolar components are zero. Moreover, the Park's vectors are constant in magnitude and phase (see Appendix 2), provided the angular speed Ω_r of the Park's reference is equal to frequency ω of voltages and currents.

Therefore, under such conditions it can be assumed that $p = 0$ in the different operational impedances (and admittances). Specifically, it is:

- in the circuits of Figure 5.2b,d,f, respectively:

$$\overline{v}_{1r} - \overline{v}_{2r} = j\omega L\overline{\imath}_r, \qquad \overline{v}_r = j\omega L\overline{\imath}_r, \qquad \overline{v}_r = j\omega\frac{L}{3}\overline{\imath}_r$$

- in the circuits of Figure 5.3b,d,f, respectively:

$$\overline{\imath}_r = j\omega C(\overline{v}_{1r} - \overline{v}_{2r}), \qquad \overline{\imath}_r = j\omega C\overline{v}_r, \qquad \overline{\imath}_r = j\omega(3C)\overline{v}_r$$

in which ωL and $\omega L/3$ assume the meaning of reactances (evaluated at the frequency ω), and so on, according to the equations already considered in Chapter 2.

If instead $\Omega_r \neq \omega$, the Park's vectors are constant in magnitude but their phase vary with time (see Appendix 2). In fact, the generic Park's vector (with θ_r as reference) can be expressed in the form:

$$\overline{y}_r = \overline{y}_s \epsilon^{j(\theta_s - \theta_r)}$$

where $\overline{y}_s$ is the Park's vector, constant in magnitude and phase, that one would obtain assuming a "synchronous" angular reference θ_s (thus with $d\theta_s/dt = \omega$). However, based on this equation the factor $\epsilon^{j(\theta_s - \theta_r)}$ is common to all Park's vectors so that, specifically, the ratio between voltage and current vectors for a given branch is that obtained with a

synchronous reference, and has a constant value equal to the branch impedance evaluated (at steady-state condition) at the frequency ω.
Furthermore, by assuming $d\theta_r/dt = \Omega_r =$ constant, it can be written that:

$$\overline{y}_r = y\epsilon^{j(\alpha+(\omega-\Omega_r)t)}$$

where the magnitude y and the initial phase α (obtained by adding up, to the phase of $\overline{y}_s$, the possible initial difference $(\theta_s - \theta_r)_{t=0}$) are constant. Under such conditions, the components of the generic Park's vector $\overline{y}_r$ are consequently sinusoidal, both of magnitude y and at frequency $|\omega - \Omega_r|$. Furthermore, one leads the other by 90° (in time), and it is the sign of $(\omega - \Omega_r)$ that determines which of the two components leads the other. In fact:

$$\begin{cases} y_{dr} = y\cos(\alpha + (\omega - \Omega_r)t) \\ y_{qr} = y\sin(\alpha + (\omega - \Omega_r)t) \end{cases}$$

and thus the sinusoid y_{dr} leads or lags y_{qr} by 90° according to whether $\omega > \Omega_r$ or $\omega < \Omega_r$. In terms of phasors (Appendix 1), with reference to the sinusoids at frequency $\nu \triangleq |\omega - \Omega_r|$, this can be translated into:

$$\tilde{y}_{qr} = \mp \tilde{j}\tilde{y}_{dr} \qquad [5.2.7]$$

according to $\omega \gtrless \Omega_r$, i.e., $\omega = \Omega_r \pm \nu$.

Based on the preceding, for the elementary cases in Section 5.2.1, it is possible to determine that the relationship between voltage and current vectors for a generic branch, under *any operating condition*, is the same which holds at the steady-state equilibrium condition at frequency ω, provided that $j\omega$ is replaced by $(p + j\Omega_r)$.

In other words, if $Z(j\omega)$ is the branch impedance evaluated (at steady-state) at frequency ω, then the operational impedance of the branch (at any possible operating condition and with Park's reference equal to θ_r) is simply given by:

$$\overline{Z}_r(p) = Z(p + j\Omega_r) \qquad [5.2.8]$$

where $\Omega_r \triangleq d\theta_r/dt$. An analogous conclusion holds for the generic admittance.

The validity of Equation [5.2.8] can be extended to any generic linear, passive equivalent circuit (concerning Park's vectors), with lumped or distributed parameters, provided the angular speed Ω_r of the Park reference is assumed constant. Usually it is just assumed, for network elements, $\Omega_r =$ constant $= \omega_{nom}$, that is Ω_r equal to the nominal frequency.

Such important extension may be justified by considering Equations [A2.6′] in Appendix 2, which are a generalization (for derivatives of any k-th order) of Equations [A2.5]. Indeed, when passing from phase variables to the corresponding Park's vector, *the application of the p operator (to the phase variables) may be simply translated—for any order of derivation if Ω_r is constant—into the application of the operator $(p + j\Omega_r)$ (to the Park's vector),* *similar to the multiplication by $j\omega$ under the steady-state equilibrium condition at frequency ω.*

Furthermore, if $\Omega_r =$ constant, this applies not only to impedances and admittances of each branch of a generic N-bipole, but also to "nodal" admittances and impedances (see Section 2.1.4) that define the relationships between node voltages and currents.

The deduction of *transfer functions* between the d, q components of different voltage and/or current vectors is then immediate.

More precisely, by assuming $\Omega_r =$ constant, if under a steady-state at a (generic) frequency ω the relationship between two generic Park's vectors $\overline{w}_r$ and $\overline{y}_r$ is given by:

$$\overline{w}_r = A(j\omega)\overline{y}_r$$

then, in terms of transfer functions, it can be derived under any operating condition, that:

$$\overline{w}_r = \overline{A}_r(s)\overline{y}_r \qquad [5.2.9]$$

where:

$$\overline{A}_r(s) = A(s + j\Omega_r)$$

i.e., using the (scalar) d, q components:

$$\begin{cases} w_{dr} = G_r(s)y_{dr} - H_r(s)y_{qr} \\ w_{qr} = H_r(s)y_{dr} + G_r(s)y_{qr} \end{cases}$$

where:

$$\begin{cases} G_r(s) \triangleq \mathcal{Re}(\overline{A}_r(s)) = \mathcal{Re}(A(s + j\Omega_r)) \\ H_r(s) \triangleq Im(\overline{A}_r(s)) = Im(A(s + j\Omega_r)) \end{cases}$$

Because of the hypothesis of linearity, the extension to the case of a generic vector $\overline{w}_r$ depending on more vectors $\overline{y}_{r1}, \overline{y}_{r2}, \ldots$ is obvious.

However, such criterion for the deduction of transfer functions implies knowledge of the analytic expression of the generic function $A(j\omega)$, and may lead to significant difficulties, particularly in the determination of the real and imaginary parts of $\overline{A}_r(s)$. (The deduction of the transfer functions for homopolar components is not considered here, because it implies no particular problems.)

A substantial simplification can be obtained by considering the *frequency response* characteristics of the system, which can be determined by assuming that, in terms of d, q variables, the operating condition is sinusoidal (and, thus, not an equilibrium condition), at a generic frequency ν.

By putting $s = \tilde{j}\nu$ in the transfer functions ($\tilde{j}$ is the imaginary unit in the phasor plane; see Appendix 1), we obtain $\overline{A}_r(\tilde{j}\nu) = A(\tilde{j}\nu + j\Omega_r)$ and consequently $G_r(\tilde{j}\nu)$ and $H_r(\tilde{j}\nu)$, which are real and imaginary parts in the plane of Park's vectors (where the imaginary unit is j).

It can be demonstrated that:

$$\overline{A}_r(\tilde{j}\nu) = \frac{1 - j\tilde{j}}{2} A(j(\Omega_r + \nu)) + \frac{1 + j\tilde{j}}{2} A(j(\Omega_r - \nu)) \qquad [5.2.10]$$

so that, setting:

$$A(j\omega) \triangleq R(\omega) + jX(\omega)$$

(where $R(\omega)$ is an even function, whereas $X(\omega)$ is an odd function) and more concisely:

$$\begin{cases} R' \triangleq R(\Omega_r + \nu) \\ R'' \triangleq R(\Omega_r - \nu) \\ X' \triangleq X(\Omega_r + \nu) \\ X'' \triangleq X(\Omega_r - \nu) \end{cases}$$

the following equations can be derived:

$$\left. \begin{aligned} G_r(\tilde{j}\nu) &= \frac{R' + R''}{2} + \tilde{j}\frac{X' - X''}{2} \\ H_r(\tilde{j}\nu) &= \frac{X' + X''}{2} - \tilde{j}\frac{R' - R''}{2} \end{aligned} \right\} \qquad [5.2.10']$$

which can be directly used for the "polar" representation (with $s = \tilde{j}\nu$) of such transfer functions, according to Figure 5.4.

For instance, in the simple case of a branch including L, R, C series elements, the impedance under the equilibrium condition at frequency ω is:

$$Z(j\omega) = j\omega L + R + \frac{1}{j\omega C}$$

to which, at any possible operating condition (and with reference θ_r), the following operational impedance corresponds:

$$\overline{Z}_r(p) = (p + j\Omega_r)L + R + \frac{1}{(p + j\Omega_r)C}$$

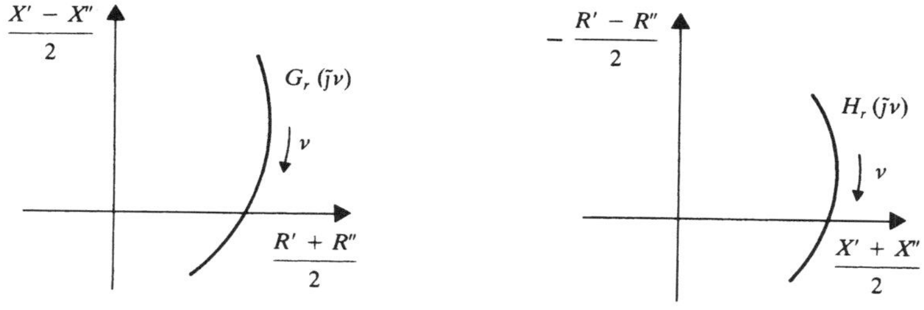

Figure 5.4. Frequency response polar diagrams for a generic function $G_r(s) + jH_r(s)$ (see text).

Substituting p by s, it can be derived:

$$\begin{cases} \mathcal{R}e(\overline{Z}_r(s)) = sL + R + \dfrac{s}{(s^2 + \Omega_r^2)C} \\ Im(\overline{Z}_r(s)) = \Omega_r L - \dfrac{\Omega_r}{(s^2 + \Omega_r^2)C} \end{cases}$$

from which, at $s = \tilde{j}\nu$:

$$\begin{cases} \mathcal{R}e(\overline{Z}_r(\tilde{j}\nu)) = R + \tilde{j}\nu \left(L + \dfrac{1}{(\Omega_r^2 - \nu^2)C} \right) \\ Im(\overline{Z}_r(\tilde{j}\nu)) = \Omega_r \left(L - \dfrac{1}{(\Omega_r^2 - \nu^2)C} \right) \end{cases}$$

These last expressions can be directly obtained by Equations [5.2.10′], by putting $Z(j\omega) = A(j\omega) = R(\omega) + jX(\omega)$ and thus:

$$\begin{cases} R(\omega) = R \\ X(\omega) = \omega L - \dfrac{1}{\omega C} \end{cases}$$

from which:

$$R' = R'' = R, \qquad X', X'' = (\Omega_r \pm \nu)L - \frac{1}{(\Omega_r \pm \nu)C}$$

Equation [5.2.10] can be demonstrated by assuming that:

$$y_{dr} = \sqrt{2}Y\cos(\nu t + \phi), \qquad y_{qr} = 0$$

which, in phasor terms, is equivalent to:

$$\left.\begin{aligned} \tilde{y}_{dr} &= Y\epsilon^{\tilde{j}\phi} \\ \tilde{y}_{qr} &= 0 \end{aligned}\right\} \tilde{\overline{y}}_r = \tilde{y}_{dr} \qquad [5.2.11]$$

whereas, because of Equation [5.2.9]:

$$\tilde{\overline{w}}_r = \overline{A}_r(\tilde{j}\nu)\tilde{y}_{dr} \qquad [5.2.12]$$

The situation expressed by Equations [5.2.11] can be interpreted as the overlapping of:

$$\left.\begin{aligned} \tilde{y}'_{dr} &= \frac{\tilde{y}_{dr}}{2} \\ \tilde{y}'_{qr} &= -\tilde{j}\frac{\tilde{y}_{dr}}{2} \end{aligned}\right\} \tilde{\overline{y}}'_r = (1 - j\tilde{j})\frac{\tilde{y}_{dr}}{2}$$

which correspond, because of Equation [5.2.7], to a steady-state at frequency $\omega = \Omega_r + \nu$, and of:

$$\left\{\begin{aligned} \tilde{y}''_{dr} &= \frac{\tilde{y}_{dr}}{2} \\ \tilde{y}''_{qr} &= +\tilde{j}\frac{\tilde{y}_{dr}}{2} \end{aligned}\right\} \tilde{\overline{y}}''_r = (1 + j\tilde{j})\frac{\tilde{y}_{dr}}{2}$$

which, instead, correspond to a steady-state at frequency $\omega = \Omega_r - \nu$. For $\nu > \Omega_r > 0$, the value $\omega < 0$ corresponds to a three-phase sinusoidal operating condition of the negative sequence, at frequency $\omega_{(2)} = -\omega = \nu - \Omega_r > 0$.

Using the previously defined notation, it is possible to derive from the superposition of effects:

$$\tilde{\overline{w}}_r = A(j(\Omega_r + \nu))\tilde{\overline{y}}'_r + A(j(\Omega_r - \nu))\tilde{\overline{y}}''_r = [A(j(\Omega_r + \nu))(1 - j\tilde{j}) + A(j(\Omega_r - \nu))(1 + j\tilde{j})]\frac{\tilde{y}_{dr}}{2}$$

from which, recalling Equation [5.2.12], Equation [5.2.10] is derived.

Note that, by using the adopted notation, Equation [5.2.10] can be rewritten as:

$$\overline{A}_r(\tilde{j}\nu) = \frac{1 - j\tilde{j}}{2}(R' + \tilde{j}X') + \frac{1 + j\tilde{j}}{2}(R'' - \tilde{j}X'') \qquad [5.2.13]$$

but this does not mean that $A(j(\Omega_r + \nu)) \triangleq R' + jX'$ and $A(j(\Omega_r - \nu)) \triangleq R'' + jX''$ are respectively equal to $(R' + \tilde{j}X')$ and to $(R'' - \tilde{j}X'')$.

A considerable advantage resulting from Equation [5.2.10] is constituted by the possibility of deducing the generic function $\overline{A}_r(\tilde{j}\nu)$ starting from the knowledge of R', R'', X', X'', i.e., of the function $A(j\omega)$ *corresponding to the steady-state* for different values ($\omega = \Omega_r \pm \nu$) of the network frequency. This knowledge can be easily achieved, e.g., by experimental tests on a network model, thus *avoiding any analytical complication* even for very complex networks.

By inspecting the behavior of transfer functions (see Equations [5.2.10′]) for varying ν, it is possible to quickly evaluate, for the most general cases, the actual importance of the transient characteristics of network elements.

In this context, a preliminary evaluation may suggest useful analytical approximations for the generic transfer functions $G_r(s)$ and $H_r(s)$, and possibly allow, with particular reference to electromechanical phenomena, a practical justification of the assumptions usually adopted ($\Omega_r = \omega_{nom}$, $p = 0$), which in reality are exact only at steady-state at nominal frequency. (In the study of particularly slow phenomena, it is also assumed $\Omega_r = \omega$, $p = 0$, with a varying ω; see Section 3.1.3.)

If phase variables are sinusoidal and of positive sequence (at a frequency $\omega > 0$), then $\tilde{y}_{qr} = \mp\tilde{j}\tilde{y}_{dr}$ (see Equation [5.2.7]), i.e., $\tilde{\overline{y}}_r = (1 \mp j\tilde{j})\tilde{y}_{dr}$, according to whether $\omega = \Omega_r \pm \nu \gtrless \Omega_r$. Under such conditions:

- from Equations [5.2.9] and [5.2.10] it can be derived:

$$\tilde{\bar{w}}_r = \overline{A}_r(\tilde{j}\nu)\tilde{\bar{y}}_r = A(j(\Omega_r \pm \nu))\tilde{\bar{y}}_r = \begin{cases} (R' + jX')\tilde{\bar{y}}_r & \text{if } \omega = \Omega_r + \nu > \Omega_r \\ (R'' + jX'')\tilde{\bar{y}}_r & \text{if } \omega = \Omega_r - \nu < \Omega_r \end{cases} \quad [5.2.14]$$

(in fact, it results $(1 \mp j\tilde{j})^2 = 2(1 \mp j\tilde{j})$, $(1 - j\tilde{j})(1 + j\tilde{j}) = 0$), but this does not mean that $\overline{A}_r(\tilde{j}\nu)$ is equal to $(R' + jX')$ or to $(R'' + jX'')$;

- similarly, from Equations [5.2.9] and [5.2.13]:

$$\tilde{\bar{w}}_r = \overline{A}_r(\tilde{j}\nu)\tilde{\bar{y}}_r = \begin{cases} (R' + \tilde{j}X')\tilde{\bar{y}}_r & \text{if } \omega = \Omega_r + \nu > \Omega_r \\ (R'' - \tilde{j}X'')\tilde{\bar{y}}_r & \text{if } \omega = \Omega_r - \nu < \Omega_r \end{cases} \quad [5.2.15]$$

without this implying the equality between $\overline{A}_r(\tilde{j}\nu)$ and $(R' + \tilde{j}X')$, or $\overline{A}_r(\tilde{j}\nu)$ and $(R'' - \tilde{j}X'')$.

Actually, Equations [5.2.14] and [5.2.15] are equivalent, as they both lead to the following final equations:

$$\left\{\begin{aligned} \tilde{w}_{dr} &= R(\omega)\tilde{y}_{dr} - X(\omega)\tilde{y}_{qr} = (R(\omega) + \tilde{j}X(\omega))\tilde{y}_{dr} \\ \tilde{w}_{qr} &= X(\omega)\tilde{y}_{dr} + R(\omega)\tilde{y}_{qr} = (R(\omega) + \tilde{j}X(\omega))\tilde{y}_{qr} \end{aligned}\right\} \quad \text{if } \omega = \Omega_r + \nu > \Omega_r$$

$$\left\{\begin{aligned} \tilde{w}_{dr} &= R(\omega)\tilde{y}_{dr} - X(\omega)\tilde{y}_{qr} = (R(\omega) - \tilde{j}X(\omega))\tilde{y}_{dr} \\ \tilde{w}_{qr} &= X(\omega)\tilde{y}_{dr} + R(\omega)\tilde{y}_{qr} = (R(\omega) - \tilde{j}X(\omega))\tilde{y}_{qr} \end{aligned}\right\} \quad \text{if } \omega = \Omega_r - \nu < \Omega_r$$

Such expressions can be extended to the case in which phase variables are sinusoidal of negative sequence (at frequency $\omega_{(2)} > 0$), by formally assuming $\omega = -\omega_{(2)} < 0$, $\nu = \Omega_r + \omega_{(2)}$. Because of Equation [5.2.7] $\tilde{y}_{qr} = +\tilde{j}\tilde{y}_{dr}$, $\tilde{\bar{y}}_r = (1 + j\tilde{j})\tilde{y}_{dr}$ and thus:

- because of the second part of Equations [5.2.14]:

$$\tilde{\bar{w}}_r = (R'' + jX'')\tilde{\bar{y}}_r = (R(\omega_{(2)}) - jX(\omega_{(2)}))\tilde{\bar{y}}_r$$

- because of the second part of Equations [5.2.15]:

$$\tilde{\bar{w}}_r = (R'' - \tilde{j}X'')\tilde{\bar{y}}_r = (R(\omega) - \tilde{j}X(\omega))\tilde{\bar{y}}_r = (R(\omega_{(2)}) + \tilde{j}X(\omega_{(2)}))\tilde{\bar{y}}_r$$

and both these expressions lead to:

$$\begin{cases} \tilde{w}_{dr} = R(\omega_{(2)})\tilde{y}_{dr} + X(\omega_{(2)})\tilde{y}_{qr} = (R(\omega_{(2)}) + \tilde{j}X(\omega_{(2)}))\tilde{y}_{dr} \\ \tilde{w}_{qr} = -X(\omega_{(2)})\tilde{y}_{dr} + R(\omega_{(2)})\tilde{y}_{qr} = (R(\omega_{(2)}) + \tilde{j}X(\omega_{(2)}))\tilde{y}_{qr} \end{cases}$$

5.3. TRANSFORMERS

5.3.1. Equivalent Circuits of the Two-Winding Transformer

Under the hypotheses of linearity and physical symmetry, a two-winding, three-phase transformer may be defined, in terms of phase voltages and currents ($v'_{(F)a}$, $i'_{(F)a}$ etc. at the primary side, and $v''_{(F)a}$, $i''_{(F)a}$ etc. at the secondary side; see Fig. 5.5a), by the following equations:

- for the primary winding, and adopting the load convention (currents are positive when entering):

$$\begin{cases} v'_{(F)a} = R' i'_{(F)a} + \dfrac{\mathrm{d}\psi'_{(F)a}}{\mathrm{d}t} \\ v'_{(F)b} = \dots \\ v'_{(F)c} = \dots \end{cases}$$

$$\begin{cases} \psi'_{(F)a} = L' i'_{(F)a} - M'(i'_{(F)a} + i'_{(F)b} + i'_{(F)c}) \\ \qquad\qquad + L_m(-i''_{(F)a}) - M_m(-i''_{(F)a} - i''_{(F)b} - i''_{(F)c}) \\ \psi'_{(F)b} = \dots \\ \psi'_{(F)c} = \dots \end{cases}$$

  where the expressions for $v'_{(F)b}$, $v'_{(F)c}$ and $\psi'_{(F)b}$, $\psi'_{(F)c}$ are, respectively, similar to $v'_{(F)a}$ and $\psi'_{(F)a}$ (it is sufficient to "rotate" indices a, b, c);
- for the secondary winding, and adopting the generator convention (currents are positive when outgoing):

$$\begin{cases} v''_{(F)a} = R''(-i''_{(F)a}) + \dfrac{\mathrm{d}\psi''_{(F)a}}{\mathrm{d}t} \\ v''_{(F)b} = \dots \\ v''_{(F)c} = \dots \end{cases}$$

$$\begin{cases} \psi''_{(F)a} = L_m i'_{(F)a} - M_m(i'_{(F)a} + i'_{(F)b} + i'_{(F)c}) \\ \qquad\qquad + L''(-i''_{(F)a}) - M''(-i''_{(F)a} - i''_{(F)b} - i''_{(F)c}) \\ \psi''_{(F)b} = \dots \\ \psi''_{(F)c} = \dots \end{cases}$$

  where the missing expressions are obtainable by a "rotation" of indices.

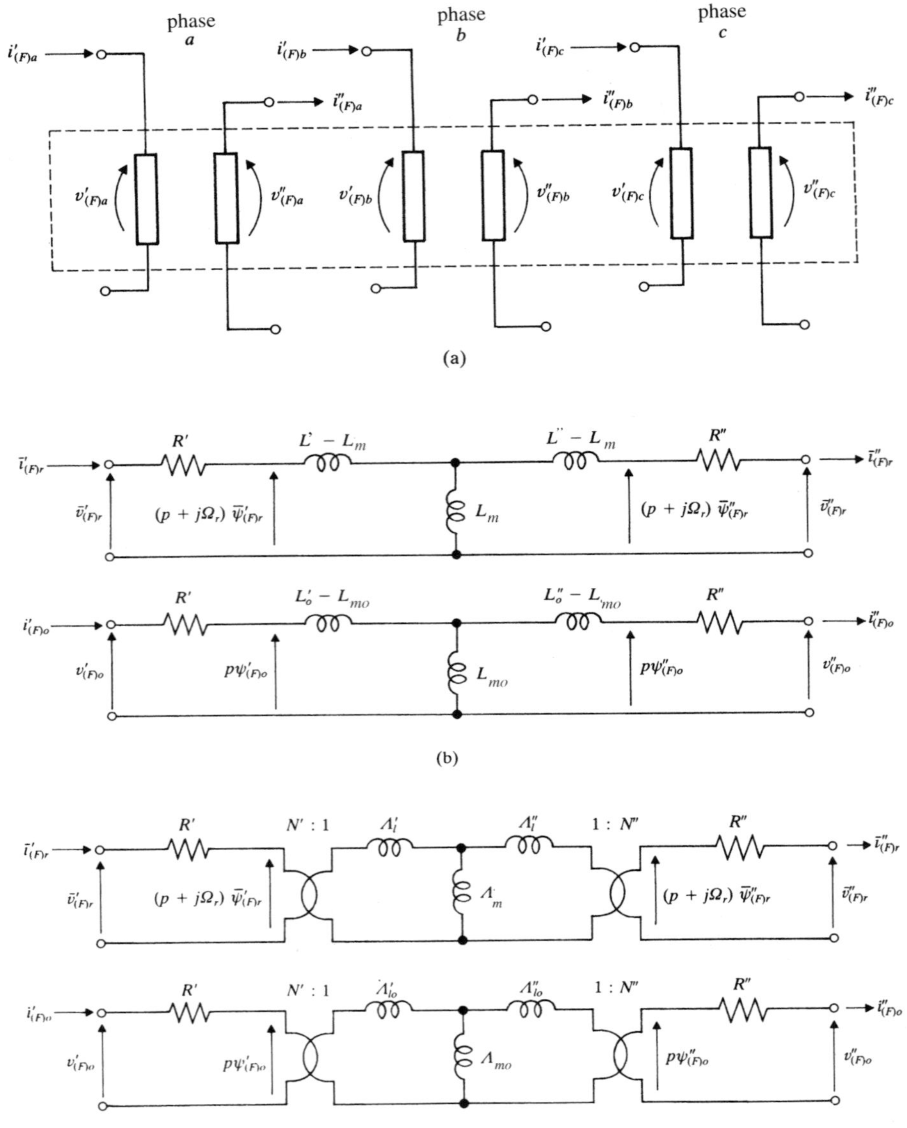

Figure 5.5. Two-winding three-phase transformer having phases (of each winding) not connected to one another: (a) schematic representation; (b), (c) equivalent circuits.

Such equations are similar, for the magnetic part, to those already written for the inductors (see Section 5.2.1b). Inductances L', M' refer to the magnetic couplings at the primary side, and L'', M'' refer to the secondary side, whereas L_m, M_m define the mutual couplings between primary and secondary sides. In the case of three single-phase transformers, it may be written that $M' = M_m = M'' = 0$.

By applying the Park's transformation with a generic angular reference θ_r, it can be derived (see Equations [A2.5] in Appendix 2):

$$\left.\begin{aligned} \overline{v}'_{(F)r} &= R'\overline{\imath}'_{(F)r} + (p + j\Omega_r)\overline{\psi}'_{(F)r} \\ v'_{(F)o} &= R'i'_{(F)o} + p\psi'_{(F)o} \end{aligned}\right\} \quad \left.\begin{aligned} \overline{\psi}'_{(F)r} &= L'\overline{\imath}'_{(F)r} - L_m\overline{\imath}''_{(F)r} \\ \psi'_{(F)o} &= L'_o i'_{(F)o} - L_{mo} i''_{(F)o} \end{aligned}\right\} \tag{5.3.1}$$

$$\left.\begin{aligned} \overline{v}''_{(F)r} &= -R''\overline{\imath}''_{(F)r} + (p + j\Omega_r)\overline{\psi}''_{(F)r} \\ v''_{(F)o} &= -R''i''_{(F)o} + p\psi''_{(F)o} \end{aligned}\right\} \quad \left.\begin{aligned} \overline{\psi}''_{(F)r} &= L_m\overline{\imath}'_{(F)r} - L''\overline{\imath}''_{(F)r} \\ \psi''_{(F)o} &= L_{mo} i'_{(F)o} - L''_o i''_{(F)o} \end{aligned}\right\} \tag{5.3.2}$$

where $p \triangleq \mathrm{d}/\mathrm{d}t$, $\Omega_r \triangleq \mathrm{d}\theta_r/\mathrm{d}t$, whereas the "homopolar" inductances are given by:

$$\left.\begin{aligned} L'_o &\triangleq L' - 3M' \\ L_{mo} &\triangleq L_m - 3M_m \\ L''_o &\triangleq L'' - 3M'' \end{aligned}\right\} \tag{5.3.3}$$

From these equations it is possible to determine the equivalent circuits in Figure 5.5b, meaning (as already specified for inductors) that the symbol of the generic inductance L represents, for any given branch:

- the operational impedance pL in the equivalent circuit for homopolar components;
- the operational impedance $(p + j\Omega_r)L$ (which is reduced to $j\Omega_r L$ at equilibrium conditions) in the equivalent circuit concerning the Park's vectors.

Resistances and inductances appearing in equations and equivalent circuits depend on the numbers of turns per phase N' (at primary side) and N'' (at secondary side). For a better account of the magnetic phenomena, reference circuits in Figure 5.5c, where the inductances:

$$\left\{\begin{aligned} \Lambda_m &\triangleq \frac{L_m}{N'N''} \\ \Lambda'_l &\triangleq \frac{L'}{N'^2} - \frac{L_m}{N'N''} \\ \Lambda''_l &\triangleq \frac{L''}{N''^2} - \frac{L_m}{N'N''} \end{aligned}\right.$$

$$\begin{cases} \Lambda_{mo} \triangleq \dfrac{L_{mo}}{N'N''} = \Lambda_m - 3\dfrac{M_m}{N'N''} \\ \Lambda'_{lo} \triangleq \dfrac{L'_o}{N'^2} - \dfrac{L_{mo}}{N'N''} \\ \Lambda''_{lo} \triangleq \dfrac{L''_o}{N''^2} - \dfrac{L_{mo}}{N'N''} \end{cases}$$

can be interpreted in terms of "permeances," substantially independent of N' and N'' and dependent on the magnetic circuits of the transformer (whereas it holds $L_m = N'N''\Lambda_m$, $L' = N'^2(\Lambda'_l + \Lambda_m)$, $L'' = N''^2(\Lambda''_l + \Lambda_m)$, etc.). Specifically, the current $(N'\overline{\imath}'_{(F)r} - N''\overline{\imath}''_{(F)r})$, flowing through the inductance Λ_m, represents the "magnetizing" magnetomotive force along the d and q axes, and so on.

The equivalent circuits may be used to reasonably take into account the *magnetic saturation* (replacing Λ_m by a proper nonlinear inductive element) and the *iron losses* (adding a proper resistive branch, in parallel to Λ_m). Furthermore, the magnetizing mmf can be usually considered negligible, at least for the d and q axes. In this case, in the circuit concerning the Park's vectors, it is possible to assume $\Lambda_m = \infty$, which is equivalent to neglect the shunt branch. Actually, the ratio $\Lambda_m/(\Lambda'_l + \Lambda''_l)$ can be some hundreds in value. The similar ratio $\Lambda_{mo}/(\Lambda'_{lo} + \Lambda''_{lo})$, which refers to the homopolar circuit, can instead vary over a wide range, such as from some units to some tens or hundreds, according to the structure of the magnetic circuit.

Adopting the *per-unit* reduction, voltages, currents, and fluxes appearing in the equivalent circuits are, respectively, referred as follows:

$$\begin{cases} \overline{v}'_{(F)r}, \quad v'_{(F)o} \quad \text{to } \sqrt{3}V'_{(F)\text{nom}} \\ \overline{\imath}'_{(F)r}, \quad i'_{(F)o} \quad \text{to } \sqrt{3}I'_{(F)\text{nom}} \\ \overline{\psi}'_{(F)r}, \quad \psi'_{(F)o} \quad \text{to } \dfrac{\sqrt{3}V'_{(F)\text{nom}}}{\omega_{\text{nom}}} \end{cases}$$

$$\begin{cases} \overline{v}''_{(F)r}, \quad v''_{(F)o} \quad \text{to } \sqrt{3}V''_{(F)\text{nom}} \\ \overline{\imath}''_{(F)r}, \quad i''_{(F)o} \quad \text{to } \sqrt{3}I''_{(F)\text{nom}} \\ \overline{\psi}''_{(F)r}, \quad \psi''_{(F)o} \quad \text{to } \dfrac{\sqrt{3}V''_{(F)\text{nom}}}{\omega_{\text{nom}}} \end{cases}$$

where:

- $V'_{(F)\text{nom}}$, $V''_{(F)\text{nom}}$ = nominal rms values of phase voltages at primary and secondary sides;

- $I'_{(F)\mathrm{nom}}$, $I''_{(F)\mathrm{nom}}$ = nominal rms values of phase currents at primary and secondary sides;
- ω_{nom} = nominal frequency.

As a consequence, the nominal values of the per-phase impedance are equal to:

$$Z'_{(F)\mathrm{nom}} \triangleq \frac{V'_{(F)\mathrm{nom}}}{I'_{(F)\mathrm{nom}}}, \qquad Z''_{(F)\mathrm{nom}} \triangleq \frac{V''_{(F)\mathrm{nom}}}{I''_{(F)\mathrm{nom}}}$$

for primary and secondary side, respectively.

Given that A_{nom} is the nominal apparent power of the transformer:

$$I'_{(F)\mathrm{nom}} = \frac{A_{\mathrm{nom}}}{3V'_{(F)\mathrm{nom}}}, \qquad I''_{(F)\mathrm{nom}} = \frac{A_{\mathrm{nom}}}{3V''_{(F)\mathrm{nom}}}$$

$$Z'_{(F)\mathrm{nom}} = \frac{3V'^2_{(F)\mathrm{nom}}}{A_{\mathrm{nom}}}, \qquad Z''_{(F)\mathrm{nom}} = \frac{3V''^2_{(F)\mathrm{nom}}}{A_{\mathrm{nom}}}$$

If, for purposes of uniformity with the other elements of the system, a power different from A_{nom} is assumed as reference, it is necessary to modify the reference currents and impedances.

In the general case for which the transformation ratio N'/N'' could be different from the nominal ratio $V'_{(F)\mathrm{nom}}/V''_{(F)\mathrm{nom}}$ (and N' and/or N'' could be varied), it is convenient to refer to a clearly stated situation, for which the turn numbers have given values ($N' = N'_{\mathrm{nom}}$, $N'' = N''_{\mathrm{nom}}$) with a ratio equal to the nominal one, so that:

$$\frac{Z'_{(F)\mathrm{nom}}}{N'^2_{\mathrm{nom}}} = \frac{Z''_{(F)\mathrm{nom}}}{N''^2_{\mathrm{nom}}}$$

Based on the above, the circuits of Figure 5.5c can be translated, in per-unit values, into those of Figure 5.6a, where the different variables in pu are indicated by bold letters, and $r' \triangleq R'/Z'_{(F)\mathrm{nom}}$, $r'' \triangleq R''/Z''_{(F)\mathrm{nom}}$, whereas the pu reactances (or inductances) x_m, x'_l, x''_l, x_{mo}, x'_{lo}, x''_{lo} are, respectively, obtained dividing the permeances Λ_m, Λ'_l, Λ''_l, Λ_{mo}, Λ'_{lo}, Λ''_{lo} by the reference permeance:

$$\Lambda_{\mathrm{nom}} \triangleq \frac{Z'_{(F)\mathrm{nom}}}{\omega_{\mathrm{nom}} N'^2_{\mathrm{nom}}} = \frac{Z''_{(F)\mathrm{nom}}}{\omega_{\mathrm{nom}} N''^2_{\mathrm{nom}}}$$

By recalling what has been already stated, the symbol of the generic pu reactance x represents, for each given branch:

- the pu operational impedance px/ω_{nom}, in the equivalent circuit concerning the homopolar components;

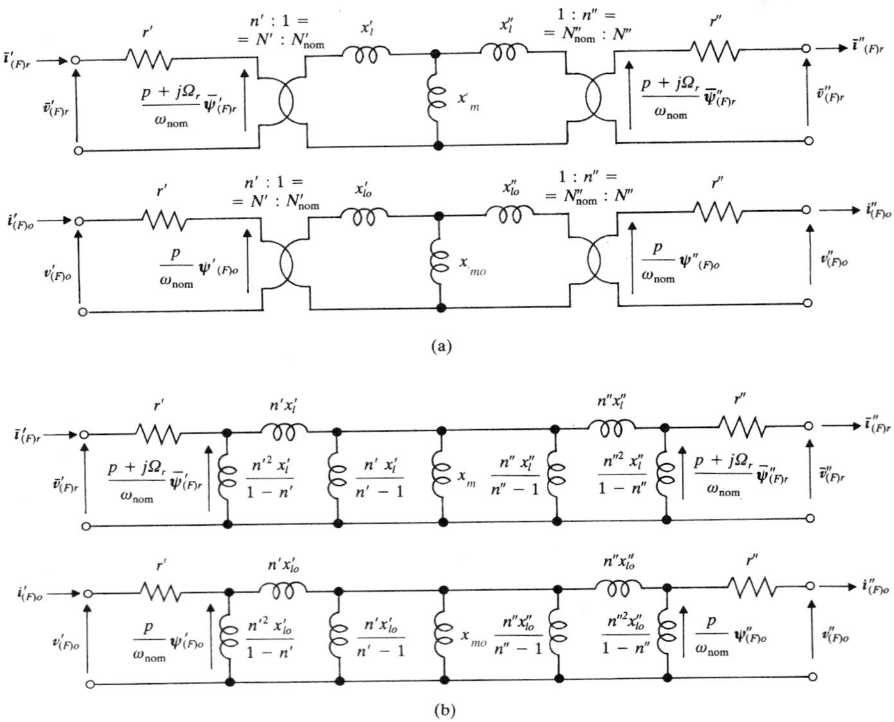

Figure 5.6. "Per-unit" equivalent circuits: (a) with ideal transformers (see Fig. 5.5c); (b) without ideal transformers.

- the pu operational impedance $(p + j\Omega_r)x/\omega_{\rm nom}$ (which is reduced to $j\Omega_r x/\omega_{\rm nom}$ at equilibrium conditions), in the equivalent circuit concerning the Park's vectors.

These new circuits include "ideal transformers" which can be avoided if their respective ratios ($N'/N'_{\rm nom}$ and/or $N''_{\rm nom}/N''$) are equal to unity. Otherwise, it is still possible to resort to equivalent circuits without ideal transformers, according to Figure 5.6b, by writing for brevity:

$$n' \triangleq \frac{N'}{N'_{\rm nom}}, \qquad n'' \triangleq \frac{N''}{N''_{\rm nom}}$$

These latter circuits can be significantly simplified if n' and/or n'' have unit values. In the opposite case, each of the new reactances appearing as shunt connections in the circuits of Figure 5.6b can be positive or negative, according to the sign of $(n' - 1)$ or of $(n'' - 1)$ (e.g., if $n' > 1$, the reactance $n'^2 x'_l/(1 - n')$ is of the type $x < 0$). At steady-state (where $p = 0$) the considered branch can be viewed as a capacitive element. However, this conclusion *does not* hold for all operating conditions, as the corresponding operational impedance remains of the type stated above, i.e.,

$$\frac{(p + j\Omega_r)x}{\omega_{\text{nom}}} = \frac{(-p - j\Omega_r)|x|}{\omega_{\text{nom}}}$$

Moreover, the resistance values somewhat depend on N', N'' or equivalently on n', n'', respectively.

In practice, it may be assumed that both ratios n' and n'' are equal to unity for transformer with constant ratio, and that one is unitary (whereas the other ratio varies, for instance, within the range 0.9–1.1) for transformers with variable ratios. If it may be assumed, as already stated, that $\Lambda_m = \infty$, i.e., $x_m = \infty$ and resistances may be disregarded, from the preceding Park's vector circuits (see Figs. 5.5c and 5.6a,b) it is possible to respectively determine the simplified circuits of Figures 5.7a,b, for the cases reported.

Up to now it has been assumed (Fig. 5.5a) that the phases of each winding are not directly connected to one another.

If the three primary and/or secondary phases are *wye- or delta-connected*, the equivalent circuits of Figure 5.5 must to be completed, similarly to Section 5.2.1, as indicated in Figure 5.8. In the case of a wye connection, the meaning of the operational impedance $Z'_{nt}(p)$ or $Z''_{nt}(p)$ is as specified for $Z_{nt}(p)$ in Section 5.2.1a.

Similarly, the pu circuits in Figure 5.6 (or simplified ones in Fig. 5.7) must be completed as in Figure 5.9.

For ideal transformers with unity transformation ratio as shown in the homopolar circuit with wye connections:

- if there is a wye connection at the primary only or at the secondary side only, the corresponding ideal transformer can be avoided;
- if there is a wye connection at both windings, it is necessary to retain at least one of the two ideal transformers, as neutral voltages (v'_n, v''_n) may be different from each other (both transformers under consideration may, however, be eliminated if their neutral nodes are directly grounded, with $v'_n = v'_t = v''_t = v''_n$).

The (two-winding) *autotransformer* may be derived from Figure 5.5a by connecting its primary and secondary windings, phase by phase, as per Figure 5.10a. Consequently, it must be assumed that voltages and currents of phase a are:

$$\begin{cases} (v'_{(F)a} + v''_{(F)a}) \text{ and } i'_{(F)a} & \text{at primary side (with } (N' + N'') \text{ turns per phase)} \\ v''_{(F)a} \text{ and } (i'_{(F)a} + i''_{(F)a}) & \text{at secondary side (with } N'' \text{ turns per phase)} \end{cases}$$

and similarly for phases b and c.

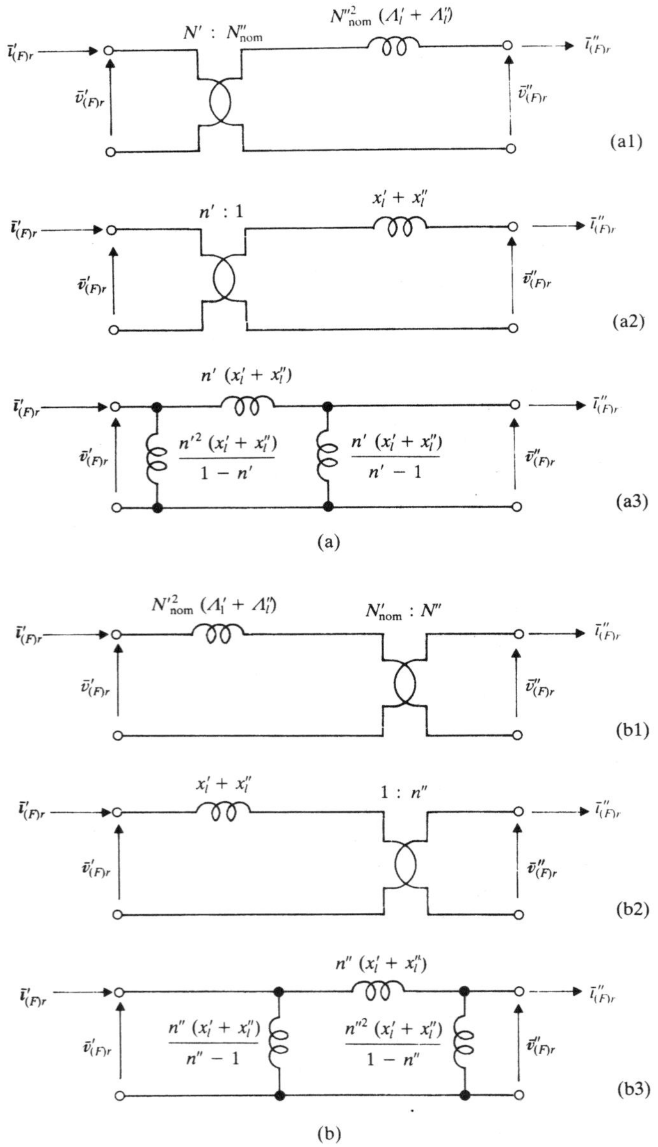

Figure 5.7. Simplified equivalent circuits relative to Park's vectors: (a1), (a2), (a3) case with nominal number of turns at the secondary side ($N'' = N''_{nom}$, $n'' = 1$); (b1), (b2), (b3) case with nominal number of turns at the primary side ($N' = N'_{nom}$, $n' = 1$). The circuits (a1), (b1) derive from that reported (for Park's vectors) in Figure 5.5c. The "per-unit" circuits (a2), (b2) and (a3), (b3), respectively, derive from those in Figures 5.6a and 5.6b.

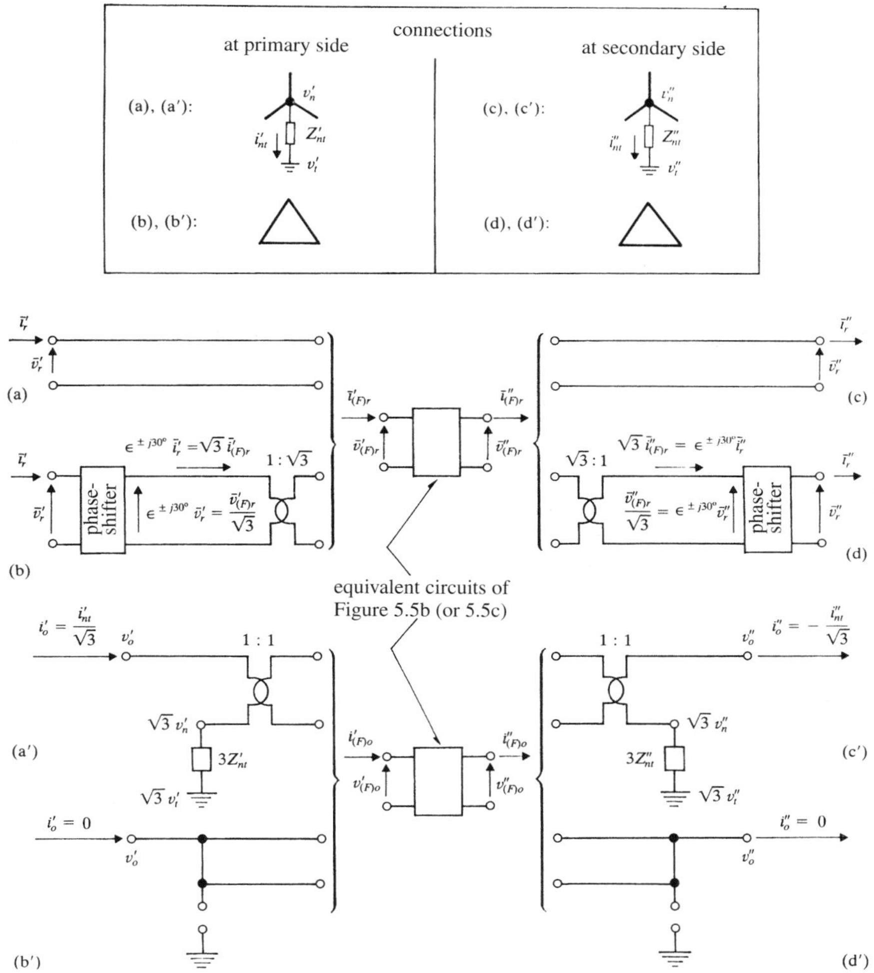

Figure 5.8. Effects of connections at primary and secondary sides: (a),(a') primary phases wye-connected; (b),(b') primary phases delta-connected; (c),(c') secondary phases wye-connected; (d),(d') secondary phases delta-connected.

By again using Equations [5.3.1] and [5.3.2], it is possible to determine the equivalent circuits in Figure 5.10b,c, instead of those in Figure 5.5b,c, respectively. If it is possible to assume $\Lambda_m = \infty$, the first circuit in Figure 5.10c can be simplified and changed into that in Figure 5.11.

It is then possible to determine the "per-unit" equivalent circuits, instead of those in Figure 5.6 or 5.7.

Finally, the type of phase connections for each winding can be considered as in Figure 5.8 or 5.9. In the homopolar circuit, the ideal transformers with

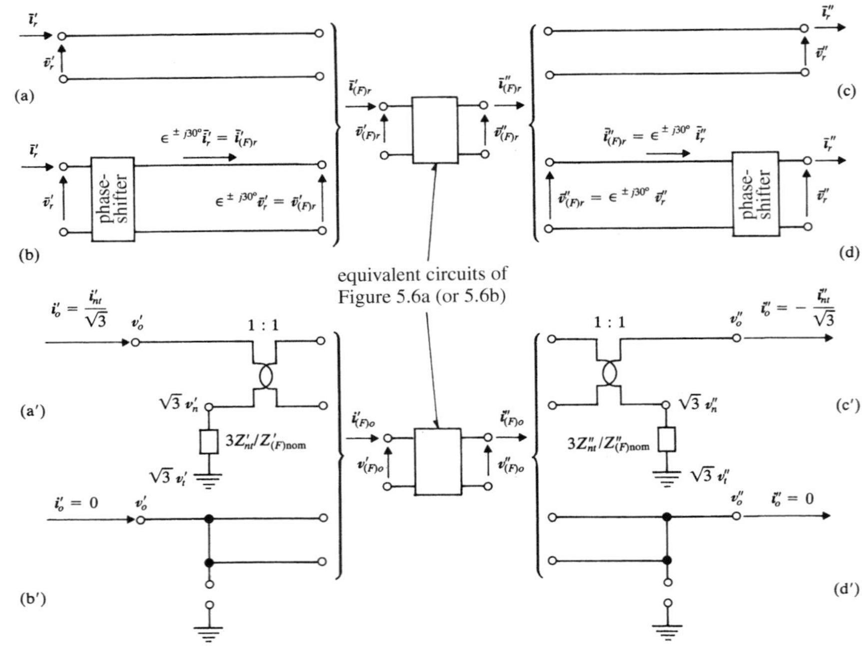

Figure 5.9. Effects of connections at primary and secondary sides in "per unit" (different cases correspond to those in Fig. 5.8).

unit ratio can be disregarded (wye-wye connections), because the neutral is in common and thus $v_n' = v_n''$.

5.3.2. Outline of Other Types of Transformers

The treatment becomes more complicated when addressing other transformer types, such as the three-winding transformer or the so-called "regulating" transformer.

Therefore, only qualitative information will be given by assuming, at least for the d and q axes, that the magnetizing mmf is negligible (recall the assumption $\Lambda_m = \infty$). In all the equivalent circuits, the symbol for the generic inductance retains the previously stated meaning, in terms of operational impedance.

In the above, the equivalent circuit of the *three-winding transformer* (or of the autotransformer, with the further "compensating" winding) is, for what concerns the Park's vectors (i.e., the d and q axes), a generalization of that corresponding to the two-winding case (see, from a qualitative point of view, the circuit reported in Fig. 5.12).

Using pu values, ideal transformers can be eliminated if transformation ratios are at nominal values, i.e., if, for the different windings, the numbers of turns per phase are

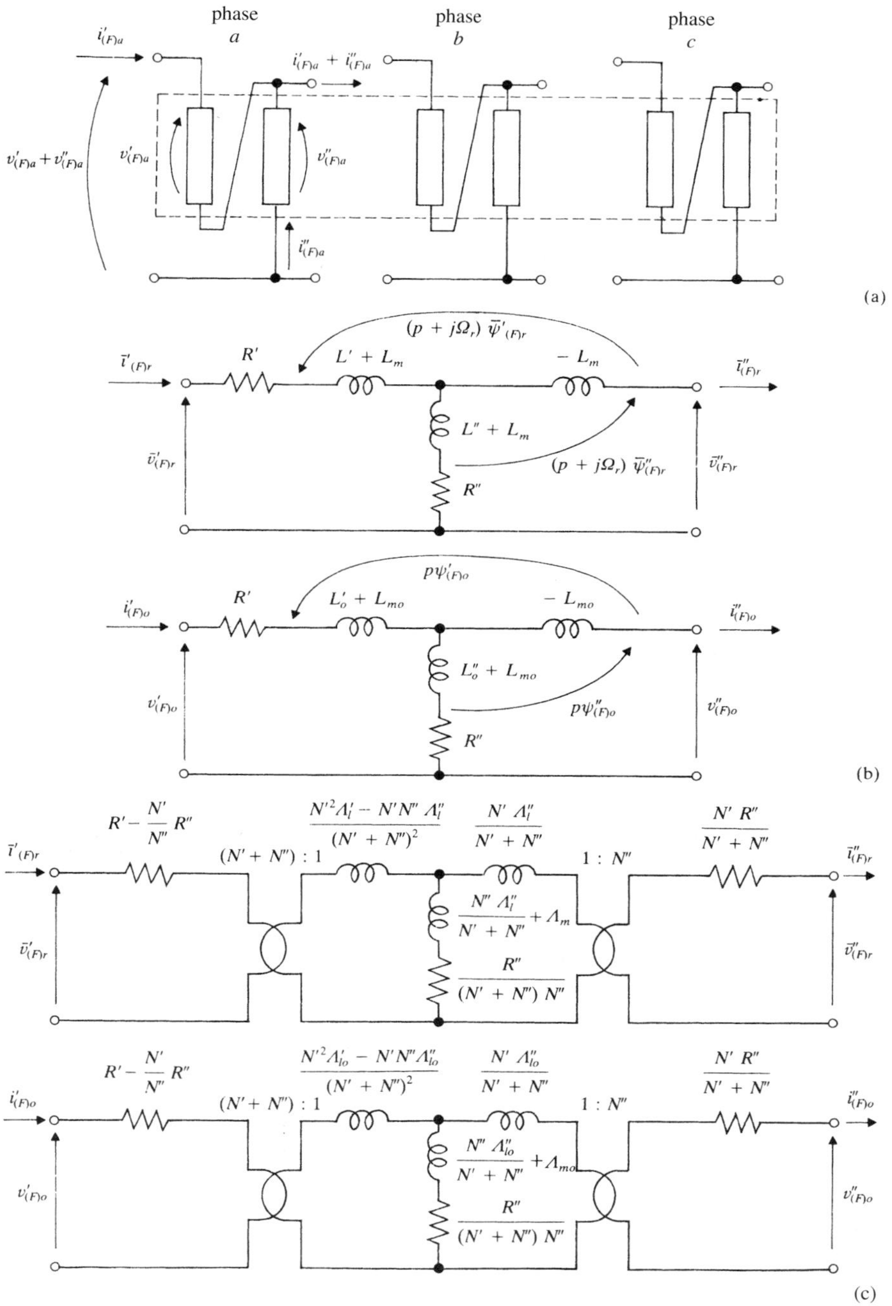

Figure 5.10. Three-phase autotransformer, having phases (of each winding) not connected to one another: (a) schematic representation; (b), (c) equivalent circuits.

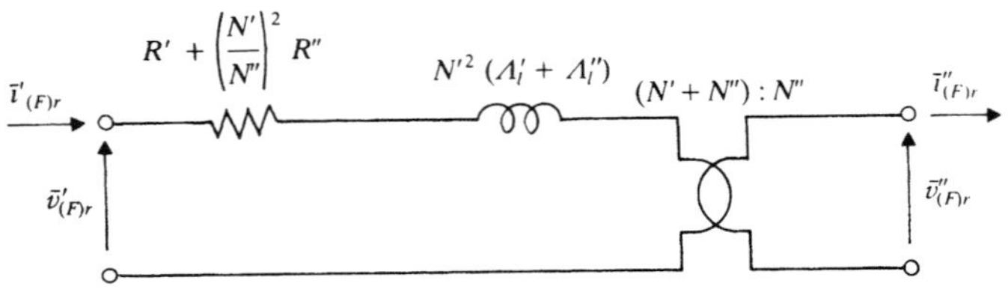

Figure 5.11. Simplified equivalent circuit, relative to Park's vectors, obtainable from that in Figure 5.10c.

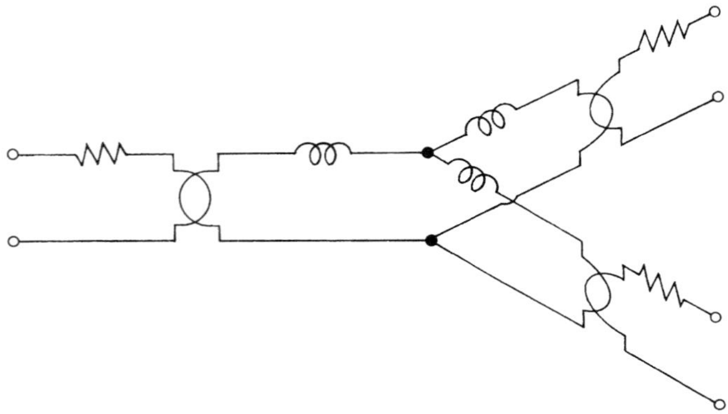

Figure 5.12. Three-winding three-phase transformer: structure of the simplified equivalent circuit relative to Park's vectors.

proportional to respective nominal voltages. Furthermore, the connections between phases can be considered based on Figure 5.8.

As a general guideline, the *"in-phase" regulating transformer* and the *"quadrature"-regulating transformer* (apart from actual realizations, possibly including combined types) may be represented as in Figures 5.13a and 5.14a.

If such representations are accepted, the generic regulating transformer can be seen as a variable ratio transformer, having:

- the primary side phases wye-connected in the case of in-phase regulation, and delta-connected in the case of quadrature regulation;
- the secondary side phases not directly connected to one another.

By considering further connections between primary and secondary sides (see Figs. 5.13a and 5.14a), it is possible to write:

- for the "in-phase" regulating transformer:

$$v'_{(F)a} = v_{1a} - v_n, \quad i'_{(F)a} = i_{1a} - i_{2a}, \quad v_{2a} = v_{1a} + v''_{(F)a}, \quad i_{2a} = i''_{(F)a}$$

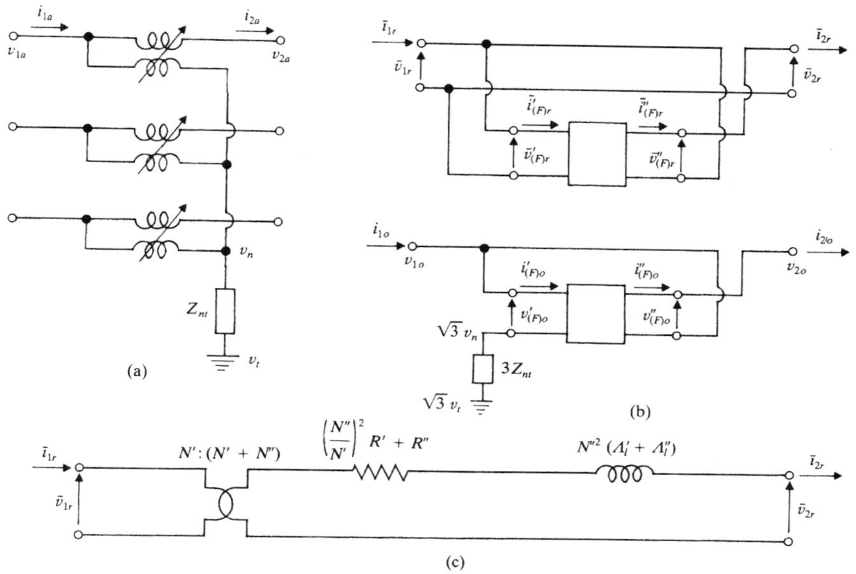

Figure 5.13. "In-phase" regulating transformer: (a) schematic representation; (b) equivalent circuits; (c) simplified equivalent circuit, relative to Park's vectors.

and similar equations (obtainable by "rotating" indices a, b, c), from which:

$$\begin{cases} \overline{v}'_{(F)r} = \overline{v}_{1r} \\ v'_{(F)o} = v_{1o} - \sqrt{3}v_n \end{cases} \quad \begin{cases} \overline{\imath}'_{(F)r} = \overline{\imath}_{1r} - \overline{\imath}_{2r} \\ i'_{(F)o} = i_{1o} - i_{2o} \end{cases} \quad \begin{cases} \overline{v}_{2r} = \overline{v}_{1r} + \overline{v}''_{(F)r} \\ v_{2o} = v_{1o} + v''_{(F)o} \end{cases} \quad \begin{cases} \overline{\imath}_{2r} = \overline{\imath}''_{(F)r} \\ i_{2o} = i''_{(F)o} \end{cases}$$

according to the equivalent circuits of Figure 5.13b;

- for the "quadrature"-regulating transformer:

$$v'_{(F)a} = \pm(v_{1c} - v_{1b}), \quad \pm(i'_{(F)a} - i'_{(F)b}) = i_{1c} - i_{2c},$$
$$v_{2a} = v_{1a} + v''_{(F)a}, \quad i_{2a} = i''_{(F)a}$$

and similar equations (obtainable by "rotating" indices a, b, c), from which:

$$\begin{cases} \overline{v}'_{(F)r} = \pm j\sqrt{3}\overline{v}_{1r} \\ v'_{(F)o} = 0 \end{cases} \quad \begin{cases} \overline{\imath}'_{(F)r} = \pm\dfrac{j}{\sqrt{3}}(\overline{\imath}_{1r} - \overline{\imath}_{2r}) \\ 0 = i_{1o} - i_{2o} \end{cases}$$

$$\begin{cases} \overline{v}_{2r} = \overline{v}_{1r} + \overline{v}''_{(F)r} \\ v_{2o} = v_{1o} + v''_{(F)o} \end{cases} \quad \begin{cases} \overline{\imath}_{2r} = \overline{\imath}''_{(F)r} \\ i_{2o} = i''_{(F)o} \end{cases}$$

according to the equivalent circuits of Figure 5.14b.

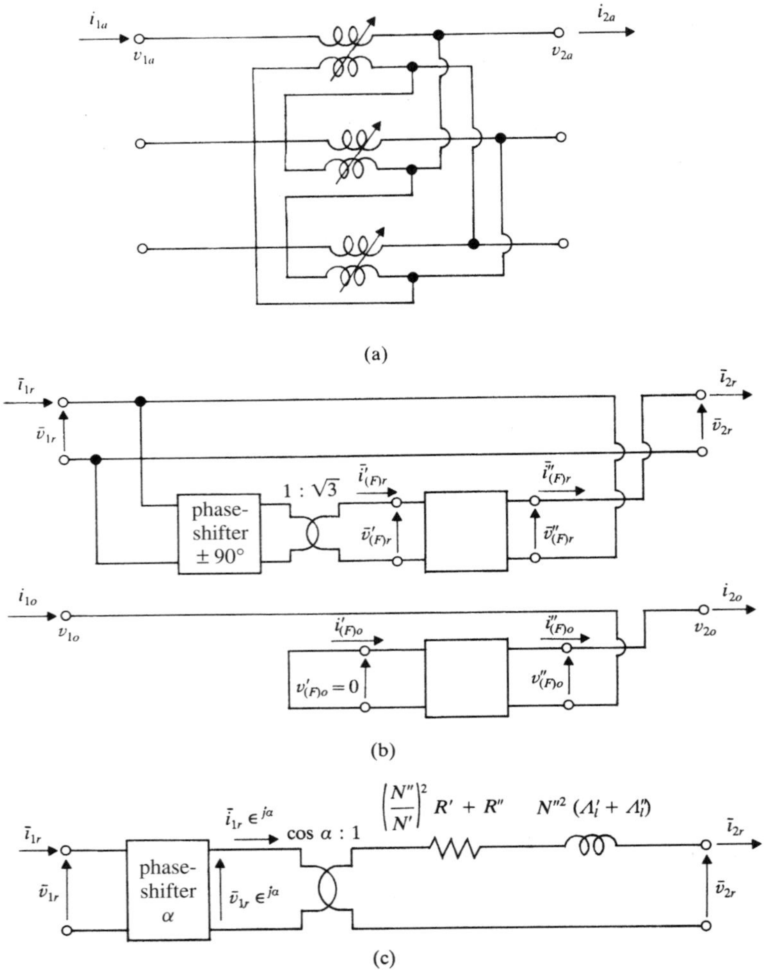

Figure 5.14. "Quadrature"-regulating transformer: (a) schematic representation; (b) equivalent circuits; (c) simplified equivalent circuit, relative to Park's vectors.

If the simplification $\Lambda_m = \infty$ is accepted, it follows that:

$$N'\overline{\imath}'_{(F)r} = N''\overline{\imath}''_{(F)r} = \frac{1}{(p + j\Omega_r)(\Lambda'_l + \Lambda''_l)} \left(\frac{\overline{v}'_{(F)r} - R'\overline{\imath}'_{(F)r}}{N'} - \frac{\overline{v}''_{(F)r} + R''\overline{\imath}''_{(F)r}}{N''} \right)$$

and by developing the equations it is possible to derive, for Park's vectors, the equivalent circuits in Figures 5.13c and 5.14c, respectively, for "in-phase" and "quadrature"-regulating transformers. In the latter case, $\alpha \triangleq \pm \arctan(\sqrt{3}N''/N') \gtrless 0$ has been written for brevity. In practical cases, the values of $|\alpha|$ may be a few degrees, so that $\alpha \cong \pm\sqrt{3}N''/N'$ (radians), $\cos\alpha \cong 1$.

It is evident, based on Figures 5.13a and 5.14a, that the "in-phase" regulating transformer essentially varies the amplitude of voltage and current vectors (in a similar way to tap-changing transformers), whereas the "quadrature"-regulating transformer varies the phase of the above-mentioned vectors. In both cases, the impedance indicated in the equivalent circuit may be considered, together with the impedances of the line to which the transformer is connected.

5.4. ALTERNATING CURRENT (AC) LINES

5.4.1. Basic Formulations

Because of its extension in length, the generic line is a typical "distributed" parameter element, the variables of which (voltages, currents, etc.) depend not only on time t but also on the distance x defined along the line itself.

Because of requirements of simplicity, it is assumed that the (three-phase) line is not only *linear* and *physically symmetrical* but also *uniform*, i.e., it has parameters that do not depend on distance x. Otherwise, the treatment can remain valid for any section of the line within which the uniformity hypothesis can be accepted.

We indicate by a the line length and label by indices A, B its variables at the two terminals, respectively corresponding to distances $x = 0$ and $x = a$.

Under the hypotheses of linearity and physical symmetry, the generic elementary section within the distances x and $(x + \Delta x)$ may be represented as the circuit in Figure 5.15a. More precisely, it may be intended that the different (constant) parameters shown in the figure represent, *per unit of length*:

- $l - m$ = phase self-inductance
- $-m$ = mutual inductance between phases
- r = phase resistance
- c' = capacitance between phases

and moreover, because of the presence of *"earth"* (and again per unit of length):

- g = phase-to-earth conductance
- c'' = phase-to-earth capacitance
- $z_t(p)$ = "earth$''$ operational impedance

The parameter g, which accounts for possible leakages between conductors and earth, is usually negligible. However, the situation is different in the presence of the *corona effect*, which is caused by particular environmental conditions. In such a case even the adoption of a value of g does not result in a good approximation, and it is convenient to make use of a nonlinear model. Moreover, in the case of overhead lines, *earth wires* should be considered as additional branches connected to earth in several locations.

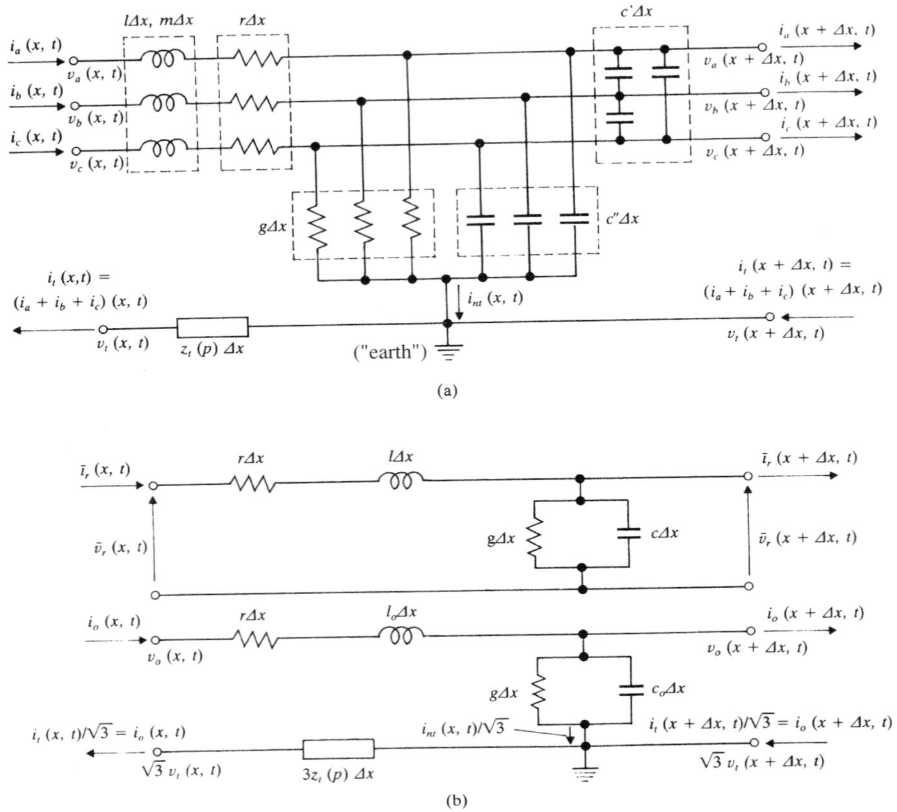

Figure 5.15. Elementary section of a (three-phase) line: (a) schematic representation; (b) equivalent circuits.

By writing the equations for each phase and applying Park's transformation (Appendix 2), it is possible to derive (similarly to Section 5.2.1, and by usual notation), for $\Delta x \to 0$:

$$\left.\begin{aligned}\frac{\partial \overline{v}_r(x,t)}{\partial x} &= -(r + (p + j\Omega_r)l)\overline{\imath}_r(x,t)\\ \frac{\partial \overline{\imath}_r(x,t)}{\partial x} &= -(g + (p + j\Omega_r)c)\overline{v}_r(x,t)\end{aligned}\right\} \quad [5.4.1]$$

$$\left.\begin{aligned}\frac{\partial (v_o - \sqrt{3}v_t)(x,t)}{\partial x} &= -(r + pl_o + 3z_t(p))i_o(x,t)\\ \frac{\partial i_o(x,t)}{\partial x} &= -(g + pc_o)(v_o - \sqrt{3}v_t)(x,t)\end{aligned}\right\} \quad [5.4.2]$$

having put:

$$\begin{cases} c \triangleq 3c' + c'' \\ l_o \triangleq l - 3m \\ c_o \triangleq c'' \end{cases}$$

The above equations, referring to the considered elementary section, correspond to the equivalent circuits of Figure 5.15b, where the symbols of inductances and capacitances assume, in terms of operational impedances, the meanings already specified in Section 5.2.1.

The parameters appearing in Equations [5.4.1] and [5.4.2] are the so-called "primary constants" of the line. At an indicative level, and assuming g negligible in normal conditions:

- for a high-voltage *overhead line*:
  - $l \cong 0.75-1.4$ mH/km (corresponding to approximately 0.23–0.45 ohm/km at 50 Hz);
  - $c \cong 8-14$ nF/km (approximately 2.5–4.5 μS/km at 50 Hz);
  - r, for instance, 0.25 ohm/km for 132 kV lines, and almost negligible (0.05 ohm/km, or less) for higher voltages;

  and furthermore, for homopolar components, by assuming $r + pl_o + 3z_t(p) \triangleq r_{o\,\text{tot}} + pl_{o\,\text{tot}}$ (where $r_{o\,\text{tot}}$ can be significantly affected by earth resistance);
  - $l_{o\,\text{tot}} \cong 2.5-5$ mH / km (corresponding to approximately 0.8–1.6 ohm/km at 50 Hz);
  - $c_o \cong 5-10$ nF/km (approximately 1.6–3.2 μS/km at 50 Hz);
  - $r_{o\,\text{tot}} \cong 0.2-0.4$ ohm/km;
- for a *cable line*:
  - $l \cong 0.2-0.4$ mH/km (corresponding to about 0.06–0.12 ohm/km at 50 Hz);
  - $c \cong 150-400$ nF/km (about 45–125 μS/km at 50 Hz);
  - r highly variable from case to case: for instance, about 1 ohm/km for 15 kV lines and even 0.05 ohm/km at high voltage;

  and furthermore, for homopolar components:
  - $l_{o\,\text{tot}}$ in the order of $2l$,
  - c_o somewhat smaller than c,
  - $r_{o\,\text{tot}}$ even up to $10r$.

For transfer functions, it is possible to derive equations similar to [5.4.1] and [5.4.2] (by assuming that Ω_r = constant) provided the operator $p \triangleq \mathrm{d}/\mathrm{d}t$ is formally replaced by the complex variable s, and the independent variables (x, t) by (x, s) (however, purely for graphical ease, we will retain the same symbols of functions $\overline{v}_r$, $\overline{\imath}_r$, etc.). The resulting equations are differential equations, with respect to the only independent variable x.

To define the behavior of the line for assigned conditions at its terminals (e.g., when voltages and currents are assigned at the sending terminal), it is necessary to integrate the mentioned equations. Under the hypothesis of uniformity it can be derived, for Park's vectors:

$$\left.\begin{aligned} \overline{v}_r(x,s) &= \frac{\overline{v}_{rA}+\overline{z}_r\overline{\imath}_{rA}}{2}\epsilon^{-\overline{n}_r x} + \frac{\overline{v}_{rA}-\overline{z}_r\overline{\imath}_{rA}}{2}\epsilon^{+\overline{n}_r x} \\ &= \overline{v}_{rA}\cosh(\overline{n}_r x) - \overline{z}_r\overline{\imath}_{rA}\sinh(\overline{n}_r x) \\ \overline{\imath}_r(x,s) &= \frac{\overline{v}_{rA}+\overline{z}_r\overline{\imath}_{rA}}{2\overline{z}_r}\epsilon^{-\overline{n}_r x} - \frac{\overline{v}_{rA}-\overline{z}_r\overline{\imath}_{rA}}{2\overline{z}_r}\epsilon^{+\overline{n}_r x} \\ &= -\frac{\overline{v}_{rA}}{\overline{z}_r}\sinh(\overline{n}_r x) + \overline{\imath}_{rA}\cosh(\overline{n}_r x) \end{aligned}\right\} \quad [5.4.3]$$

where $\overline{v}_{rA} = \overline{v}_{rA}(s)$ and $\overline{\imath}_{rA} = \overline{\imath}_{rA}(s)$ are the Laplace transforms of vectors $\overline{v}_r$, $\overline{\imath}_r$ at the sending terminal ($x = 0$) and having posed that:

$$\left.\begin{aligned} \overline{n}_r &= \overline{n}_r(s) \triangleq [(r+(s+j\Omega_r)l)(g+(s+j\Omega_r)c)]^{1/2} \\ \overline{z}_r &= \overline{z}_r(s) \triangleq \frac{\overline{n}_r(s)}{g+(s+j\Omega_r)c} = \frac{r+(s+j\Omega_r)l}{\overline{n}_r(s)} \end{aligned}\right\} \quad [5.4.4]$$

For the homopolar components, it is possible to derive similar equations, which for brevity are not reported and to which the following considerations can be extended in a trivial way.

In particular, for $x = a$, from Equations [5.4.3] the following result:

$$\left.\begin{aligned} \overline{v}_{rB} &= \overline{v}_{rB}(s) = \overline{v}_{rA}\cosh(\overline{n}_r a) - \overline{z}_r\overline{\imath}_{rA}\sinh(\overline{n}_r a) \\ \overline{\imath}_{rB} &= \overline{\imath}_{rB}(s) = -\frac{\overline{v}_{rA}}{\overline{z}_r}\sinh(\overline{n}_r a) + \overline{\imath}_{rA}\cosh(\overline{n}_r a) \end{aligned}\right\} \quad [5.4.5]$$

which, making reference to the whole line as seen by its terminals, correspond to the equivalent circuit of Figure 5.16, with the indicated operational impedances.

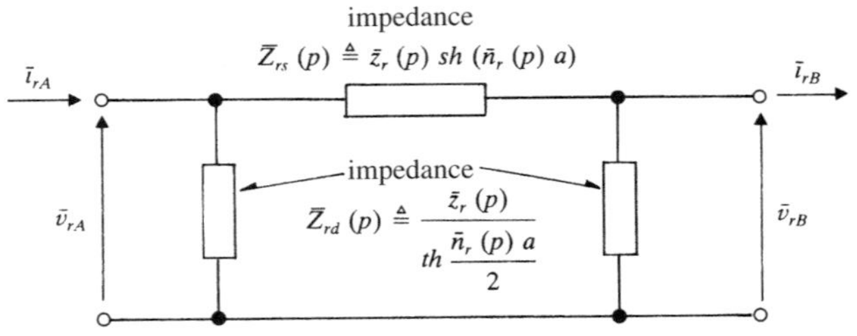

Figure 5.16. Equivalent circuit of a line (as seen by its terminals), relative to Park's vectors.

By considering [5.4.5], Equations [5.4.3] can be substituted by other equivalent equations, which relate the generic $\overline{v}_r(x, s)$ and $\overline{\imath}_r(x, s)$ to any two of the four functions $\overline{v}_{rA}$, $\overline{\imath}_{rA}$, $\overline{v}_{rB}$, $\overline{\imath}_{rB}$.

Within this concern, by generically assuming:

$$\left.\begin{aligned} \overline{v}'_r &= \overline{z}_r\overline{\imath}'_r \triangleq \frac{\overline{v}_r + \overline{z}_r\overline{\imath}_r}{2} \\ \overline{v}''_r &= -\overline{z}_r\overline{\imath}''_r \triangleq \frac{\overline{v}_r - \overline{z}_r\overline{\imath}_r}{2} \end{aligned}\right\} \qquad [5.4.6]$$

the Equations [5.4.3] can be substituted by:

$$\left.\begin{aligned} \overline{v}'_r(x, s) &= \overline{v}'_{rA}(s)\epsilon^{-\overline{n}_r(s)x} \\ \overline{v}''_r(x, s) &= \overline{v}''_{rB}(s)\epsilon^{-\overline{n}_r(s)(a-x)} \end{aligned}\right\} \qquad [5.4.7]$$

from which the block diagram of Figure 5.17 can be derived (note that it holds $\overline{v}_r = \overline{v}'_r + \overline{v}''_r$, $\overline{\imath}_r = \overline{\imath}'_r + \overline{\imath}''_r = (\overline{v}'_r - \overline{v}''_r)/\overline{z}_r$, and that, for given $\overline{v}_{rA}$ and $\overline{v}_{rB}$, it follows that $\overline{v}'_{rA} = \overline{v}_{rA} - \overline{v}''_{rA}$, $\overline{v}''_{rB} = \overline{v}_{rB} - \overline{v}'_{rB}$).

According to the following, such a block diagram is particularly suitable to derive some interesting interpretations for propagation phenomena (the different line sections, involved in such phenomena, are indicated aside each single block).

For reasons which will follow, $\overline{n}_r(s)$ is called the *"propagation function"* of the line, whereas $\overline{z}_r(s)$ is called *"characteristic" impedance* (or "natural" impedance) of the line. Furthermore, the line is called *"nondistorting"* if $r/l = g/c$ (and "distorting" for the contrary case). The simplest case of a nondistorting line is constituted by the *"nondissipative"* line, defined by $r = g = 0$.

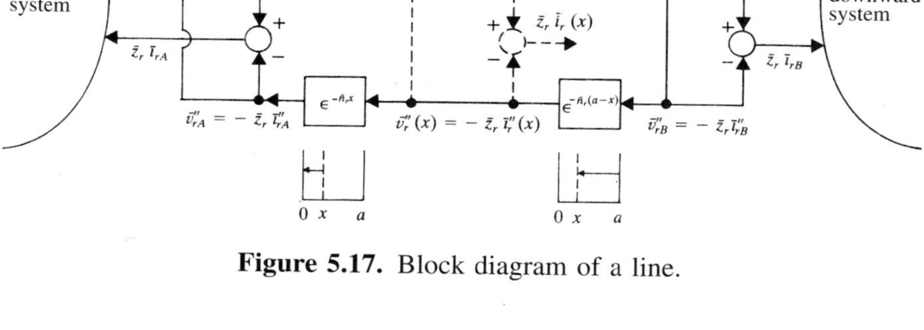

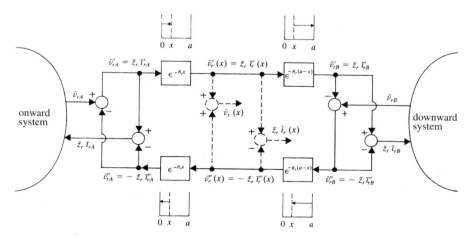

Figure 5.17. Block diagram of a line.

The first of Equations [5.4.4] implies two possible solutions, opposite each other, in $\overline{n}_r(s)$, and the same can be said for $\overline{z}_r(s)$, based on the second of Equations [5.4.4]. However, when passing from a solution $\overline{n}_r(s)$, $\overline{z}_r(s)$ to the other solution, we simply obtain a change between the definitions of $\overline{v}'_r$, $\overline{v}''_r$ as well as $\overline{\imath}'_r$, $\overline{\imath}''_r$.

For the *internal behavior of the line*, first assume, for simplicity, that $r = g = 0$, i.e., the line is (nondistorting, and) nondissipative.

As to the first of Equations [5.4.4], which becomes:

$$\overline{n}_r(s) = \left[(s + j\Omega_r)^2 lc\right]^{1/2}$$

assume the solution:

$$\overline{n}_r(s) = (s + j\Omega_r)\sqrt{lc} = \frac{s + j\Omega_r}{u}$$

where the parameter:

$$u \triangleq \frac{1}{\sqrt{lc}} \qquad [5.4.8]$$

is called the "*propagation speed.*" In a corresponding way, from the second of Equations [5.4.4], it can be derived:

$$\overline{z}_r(s) = z^{(o)}$$

where:

$$z^{(o)} \triangleq \sqrt{\frac{l}{c}} \qquad [5.4.9]$$

is a real positive constant, also called the "*wave impedance.*"

With such assumptions, in the block diagram of Figure 5.17:

$$\epsilon^{-\overline{n}_r x} = \epsilon^{-j\Omega_r x/u}\epsilon^{-sx/u}$$

where:

$\epsilon^{-j\Omega_r x/u}$ means a *rotation* equal to $-\Omega_r x/u$ (i.e., a phase-lag) in the plane of Park's vectors,

$\epsilon^{-sx/u}$ means a *pure delay* equal to x/u, in the time domain;

and similar conclusions hold for $\epsilon^{-\overline{n}_r(a-x)}$.

As a consequence, the transfer from $\overline{v}'_{rA}(s)$ to the generic $\overline{v}'_r(x, s)$, up to $\overline{v}'_{rB}(s)$ at $x = a$, corresponds to a pure propagation phenomenon (i.e., with no distortion or attenuation), at the speed u and in the direction of increasing x, accompanied by a progressive phase-lag; see the qualitative example of Figure 5.18, where $\tau \triangleq x/u = \sqrt{lc}x$ is the time necessary to cover the distance x.

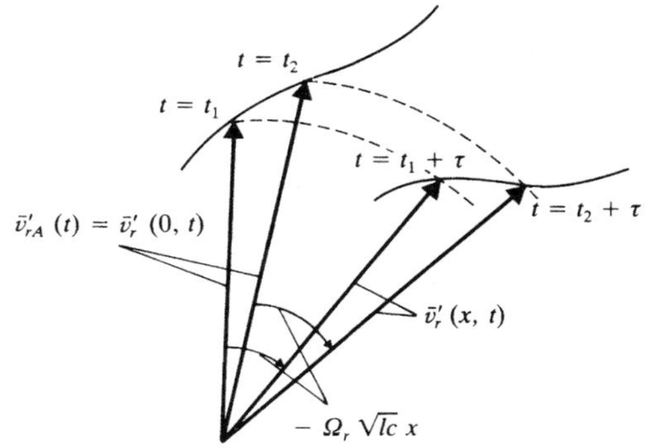

Figure 5.18. Nondissipative line: example of a "direct" traveling voltage wave (Park's vector). It is assumed $\tau = x/u = \sqrt{lc}x =$ time necessary to cover distance x.

Similarly, the transfer from $\overline{v}''_{rB}(s)$ to the generic $\overline{v}''_r(x, s)$, up to $\overline{v}''_{rA}(s)$ at $x = 0$, corresponds to a similar phenomenon but in the direction of decreasing x.

From the first of Equations [5.4.7] it can be derived, by inverse transformation:

$$\overline{v}'_r(x, t) = \overline{v}'_{rA}\left(t - \frac{x}{u}\right)\epsilon^{-j\Omega_r x/u}$$

In terms of phase variables (Fig. 5.15a), having set:

$$v_\alpha + jv_\beta \triangleq \overline{v}_r\epsilon^{j\theta_r} = \sqrt{\frac{2}{3}}(v_a + v_b\epsilon^{j120^\circ} + v_c\epsilon^{j240^\circ})$$

(Appendix 2) and thus:

$$\begin{cases} v_\alpha = \sqrt{\dfrac{2}{3}}\left(v_a - \dfrac{v_b + v_c}{2}\right) \\ v_\beta = \dfrac{v_b - v_c}{\sqrt{2}} \end{cases}$$

it is then possible to derive, with obvious symbols:

$$\begin{cases} v'_\alpha(x, t) = v'_{\alpha A}\left(t - \dfrac{x}{u}\right) \\ v'_\beta(x, t) = v'_{\beta A}\left(t - \dfrac{x}{u}\right) \end{cases}$$

i.e., the scalar signals v'_α and v'_β propagate at a speed u. Similar conclusions can be derived from the latter of Equations [5.4.7], when addressing the scalar signals v''_α and v''_β.

Consequently, the (vector) functions $\overline{v}'_r(x,t)$ and $\overline{\imath}'_r(x,t)$ can be interpreted in terms of *"direct"* (or "progressive") *traveling waves*, of voltage and current, respectively, whereas $\overline{v}''_r(x,t)$ and $\overline{\imath}''_t(x,t)$ can be similarly interpreted in terms of *"inverse"* (or "regressive") traveling waves. The name of "propagation speed" attributed to u appears to be justified.

Furthermore, the distance λ_r after which the phase-lag is 360° (if a is sufficiently large), so that the generic wave again assumes the same time behavior (with the same phase), is equal to:

$$\lambda_r \triangleq \frac{2\pi u}{\Omega_r} = \frac{2\pi}{\Omega_r \sqrt{lc}} \qquad [5.4.10]$$

and is called the *"wavelength"* of the line.

If the line is nondistorting but dissipative ($r/l = g/c \neq 0$) it can be set similarly:

$$\overline{n}_r(s) = \left(\frac{r}{l} + s + j\Omega_r\right)\sqrt{lc} = \left(\frac{r}{l} + s + j\Omega_r\right)\frac{1}{u}$$

(with the "propagation speed" $u \triangleq 1/\sqrt{lc}$), whereas it still results in:

$$\overline{z}_r(s) = \sqrt{\frac{l}{c}}$$

It then follows that:

$$\epsilon^{-\overline{n}_r x} = \epsilon^{-(r/l)x/u}\epsilon^{-j\Omega_r x/u}\epsilon^{-sx/u}$$

i.e., compared to the previously treated nondissipative case, we must consider the factor $\epsilon^{-(r/l)x/u}$, which implies an *attenuation* of the waves along the line (still with no distortion); a similar conclusion holds for $\epsilon^{-\overline{n}_r(a-x)}$.

Generally, if $r, g \neq 0$ and $r/l \neq g/c$ (distorting and dissipative line), the real and imaginary parts, in the plane of Park's vectors, of $\overline{n}_r(s)$ are no longer rational functions of s. Compared to the previous case, a *distortion* of the waves along the line is also added.

Based on the previous values for the "primary constants," the hypothesis of the nondissipative line is usually a good approximation for higher-voltage overhead lines (e.g., 220 kV, 380 kV, etc.), for which:

$$z^{(o)} \triangleq \sqrt{\frac{l}{c}} \cong 250 - 400 \text{ ohm}$$

$$u \triangleq \frac{1}{\sqrt{lc}} \cong 3 \cdot 10^5 \text{ km/sec}$$

and furthermore, at 50 Hz ($\Omega_r = 2\pi \cdot 50$ rad/sec):

$$\lambda_r \triangleq \frac{2\pi u}{\Omega_r} \cong 6000 \text{ km}$$

whereas at 60 Hz it is $\lambda_r \cong 5000$ km.

For cable lines, the propagation speed and the wavelength are smaller, e.g., $u \cong (1-1.4)10^5$ km/sec, and $\lambda_r \cong 2000-2800$ km at 50 Hz, 1700–2300 km at 60 Hz.

For the *behavior of the line at its terminals*, first assume that the downstream system is defined by the equation $\overline{v}_{rB} = \overline{Z}_{rB}(p)\overline{\imath}_{rB}$, i.e., the line is terminated, at $x = a$, with the operational impedance $\overline{Z}_{rB}(p)$ (which is itself defined by applying Park's transformation with reference θ_r).

For transfer functions, by remembering that as a result of Equations [5.4.6], $\overline{v}_{rB} = \overline{v}'_{rB} + \overline{v}''_{rB}$, $\overline{\imath}_{rB} = \overline{\imath}'_{rB} + \overline{\imath}''_{rB}$, with $\overline{v}'_{rB} = \overline{z}_r(s)\overline{\imath}'_{rB}$, $\overline{v}''_{rB} = -\overline{z}_r(s)\overline{\imath}''_{rB}$, it follows that:

$$\left.\begin{aligned} \overline{\imath}''_{rB} &= \overline{\rho}_B(s)\overline{\imath}'_{rB} \\ \overline{v}''_{rB} &= -\overline{\rho}_B(s)\overline{v}'_{rB} \end{aligned}\right\} \qquad [5.4.11]$$

having set:

$$\overline{\rho}_B(s) \triangleq \frac{\overline{z}_r(s) - \overline{Z}_{rB}(s)}{\overline{z}_r(s) + \overline{Z}_{rB}(s)} \qquad [5.4.12]$$

By interpreting $\overline{\imath}'_{rB}(t)$ and $\overline{v}'_{rB}(t)$ as *"incident" waves* (of current and voltage, respectively) at the receiving line terminal, the functions $\overline{\imath}''_{rB}(t)$ and $\overline{v}''_{rB}(t)$ constitute (see also Fig. 5.17) the waves *"reflected"* on the line itself. For this reason, $\overline{\rho}_B(s)$ is called the "reflection function" of the current at the receiving terminal, whereas $-\overline{\rho}_B(s)$ is the similar reflection function for the voltage. Furthermore, the functions $\overline{\imath}_{rB}(t)$ and $\overline{v}_{rB}(t)$, which are the sum of the respective incident and reflection waves, represent the waves *"refracted"* into the downstream system.

Note that $\overline{\rho}_B$ is generally a function of s. This means that current and voltage reflections are accompanied by distortion. This, however, does not happen in the following cases.

(1) If $\overline{Z}_{rB}(s) = \overline{z}_r(s)$, i.e., if the line is terminated with its characteristic impedance, $\overline{\rho}_B = 0$, and thus no (current or voltage) reflection occurs. At every $x \in [0, a]$ it can be derived $\overline{v}''_r(x, s) = 0$, $\overline{\imath}''_r(x, s) = 0$:

$$\overline{v}_r(x, s) = \overline{v}_{rA}(s)\epsilon^{-\overline{n}_r(s)x}$$

$$\overline{\imath}_r(x, s) = \overline{\imath}_{rA}(s)\epsilon^{-\overline{n}_r(s)x} = \frac{\overline{v}_{rA}(s)}{\overline{z}_r(s)}\epsilon^{-\overline{n}_r(s)x}$$

from which also:

$$\frac{\overline{v}_r(x, s)}{\overline{\imath}_r(x, s)} = \overline{z}_r(s) \quad \forall x \in [0, a]$$

and it is easy to determine that the line behaves like a section $[0, a]$ belonging to a line of infinite length. It should be noted that the present assumption $\overline{Z}_{rB}(s) = \overline{z}_r(s)$ is much more stringent than what might appear; in fact, *it must hold at any* s, and not for only $s = 0$ as at steady-state (see Section 5.4.2).

(2) If $\overline{Z}_{rB}(s) = \infty$, i.e., if the line is interfaced with an open circuit, $\overline{\rho}_B = -1$, so that the reflected current wave is opposite to the incident one (and $\overline{\imath}_{rB} = \overline{\imath}'_{rB} + \overline{\imath}''_{rB} = 0$, as obvious), whereas the voltage wave is reflected unaltered.

(3) If $\overline{Z}_{rB}(s) = 0$, i.e., if the line is short-circuited, $\overline{\rho}_B = +1$, so that the current wave is reflected unaltered, whereas the voltage wave simply changes its sign (and obviously, $\overline{v}_{rB} = \overline{v}'_{rB} + \overline{v}''_{rB} = 0$ holds).

(4) If the line is nondistorting (and then $\overline{z}_r(s) = z^{(o)} = \sqrt{l/c}$, which is real) and the downstream impedance is purely resistive (i.e., $\overline{Z}_{rB}(s) = R_B$, which is real, positive and s-independent), then $\overline{\rho}_B$ is real and independent of s, with an amplitude smaller than unity. In such a case, the reflections are accompanied by attenuation and a possible sign variation according to the sign of $\overline{\rho}_B$ (see Equations [5.4.11]). (If $R_B = z^{(o)}$ it then follows that $\overline{\rho}_B = 0$, and reflections are absent; see case (1).)

Similar considerations can be applied to the sending terminal ($x = 0$) of the line, by interpreting (Fig. 5.17) $\overline{\imath}''_{rA}(t)$ and $\overline{v}''_{rA}(t)$ as incident waves, and assuming that the upstream system is defined by an equation $\overline{v}_{rA} = \overline{Z}_{rA}(p)\overline{\imath}_{rA}$. The reflecting waves are then described by:

$$\begin{cases} \overline{\imath}'_{rA} = \overline{\rho}_A(s)\overline{\imath}''_{rA} \\ \overline{v}'_{rA} = -\overline{\rho}_A(s)\overline{v}''_{rA} \end{cases}$$

where:

$$\overline{\rho}_A(s) \triangleq \frac{\overline{z}_r(s) - \overline{Z}_{rA}(s)}{\overline{z}_r(s) + \overline{Z}_{rA}(s)}$$

is the reflection function for the current at the sending terminal of the line, whereas $-\overline{\rho}_A(s)$ is the analogous reflection function for the voltage, and so on.

Generally, if the up and downstream systems are defined by equations (Fig. 5.19):

$$\left.\begin{aligned} \overline{v}_{rA} &= \overline{e}_{rA} - \overline{Z}_{rA}(p)\overline{\imath}_{rA} \\ \overline{v}_{rB} &= \overline{e}_{rB} + \overline{Z}_{rB}(p)\overline{\imath}_{rB} \end{aligned}\right\} \qquad [5.4.13]$$

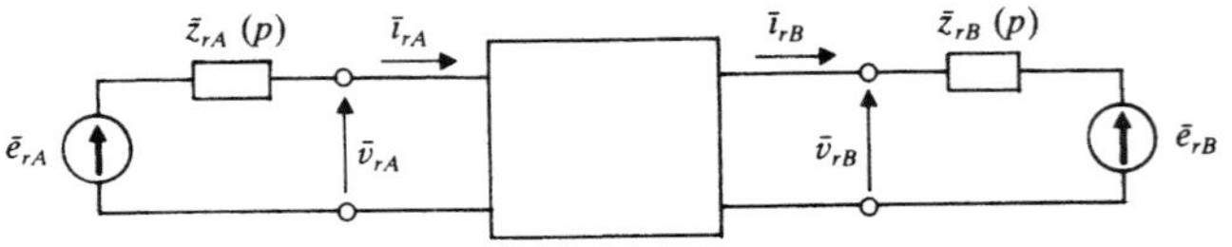

Figure 5.19. Example of a line closed at its terminals (the equivalent circuits refer to Park's vectors).

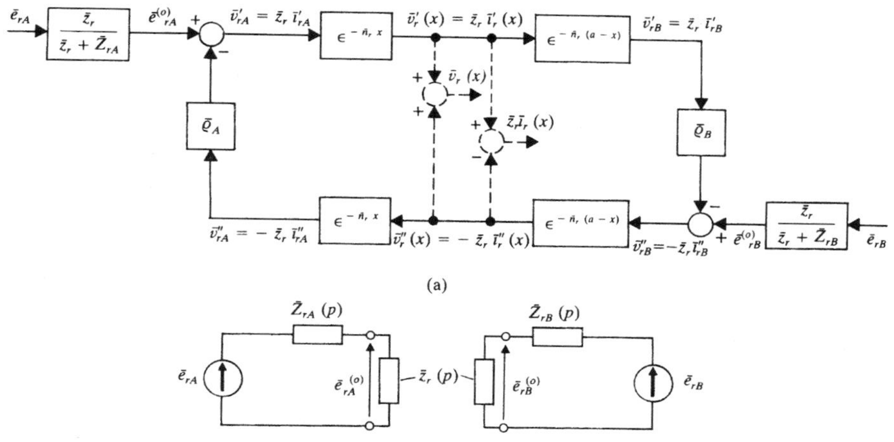

Figure 5.20. System in Figure 5.19: (a) block diagram; (b) deduction of voltages $\overline{e}_{rA}^{(o)}$, $\overline{e}_{rB}^{(o)}$ defined in the block diagram.

from Figure 5.17 we can derive Figure 5.20, which shows both the phenomena of propagation along the line and those of reflection at its terminals, as well as the dependence of the different line variables on "inputs" $\overline{e}_{rA}, \overline{e}_{rB}$.

According to the block diagram in Figure 5.20a, it is possible to consider as inputs the voltages:

$$\overline{e}_{rA}^{(o)} \triangleq \frac{\overline{z}_r}{\overline{z}_r + \overline{Z}_{rA}} \overline{e}_{rA}, \quad \overline{e}_{rBA}^{(o)} \triangleq \frac{\overline{z}_r}{\overline{z}_r + \overline{Z}_{rB}} \overline{e}_{rB}$$

defined by the circuits in Figure 5.20b, in which the line, as seen from each of its terminals, is simply represented by its characteristic impedance $\overline{z}_r$. Nevertheless, $\overline{Z}_{rA}, \overline{Z}_{rB}, \overline{z}_r$ play a role also in the expressions of $\overline{\rho}_A, \overline{\rho}_B$.

As an example, assume that:

- the line is nondissipative (and thus $\overline{n}_r(s) = (s + j\Omega_r)(1/u)$, $\overline{z}_r(s) = z^{(o)}$, with u and $z^{(o)}$ defined by Equations [5.4.8] and [5.4.9]), and the impedances $\overline{Z}_{rA}, \overline{Z}_{rB}$ are purely resistive (i.e., $\overline{Z}_{rA} = R_A$ and $\overline{Z}_{rB} = R_B$, are real, positive, and s-independent);
- the function $\overline{e}_{rA}(t)$ is constituted by a step of value $\overline{E}$, applied at the instant $t = 0$, whereas $\overline{e}_{rB}(t) = 0$.

The application of the step results, at the beginning, into direct waves of voltage (equal to $z^{(o)}\overline{E}/(z^{(o)} + R_A)$) and current (equal to $\overline{E}/(z^{(o)} + R_A)$), which travel in the direction of increasing x, with a progressive lag shift. At the receiving terminal $x = a$ of the line, such waves reflect according to Equations [5.4.11], and (if $R_B \neq z^{(o)}$) originate the inverse waves. Note that $\overline{\rho}_B = (z^{(o)} - R_B)/(z^{(o)} +$

R_B) is real and with an amplitude smaller than unity. In their turn, the inverse waves reflect when they reach the sending terminal $x = 0$, where $\overline{\rho}_A$ is real and with an amplitude smaller than unity, thus modifying the direct waves, and so on.

Therefore, the voltage and current transients are, if $R_B \neq z^{(o)}$:

- of the stepwise type, caused by the absence of distortion (both in the propagation along the line and in reflection at its terminals);
- damped, because of attenuations in the subsequent reflections (whereas the propagation along the line occurs without attenuation).

In particular, it is possible to recognize that the vector $\overline{v}_{rB} = \overline{v}'_{rB} + \overline{v}''_{rB} = (1 - \overline{\rho}_B)\overline{v}'_{rB}$ assumes in sequence the following values:

$$\overline{v}_{rB} = \begin{cases} 0 & \text{at } t \in \left[0, \dfrac{a}{u}\right) \\ \overline{V}_{rB(1)} \triangleq (1 - \overline{\rho}_B)\dfrac{z^{(o)}\overline{E}}{z^{(o)} + R_A}\epsilon^{-j\Omega_r \frac{a}{u}} & \text{at } t \in \left(\dfrac{a}{u}, \dfrac{3a}{u}\right) \\ \overline{V}_{rB(1)}\left(1 + \overline{\rho}_A\overline{\rho}_B\epsilon^{-j\Omega_r \frac{2a}{u}}\right) & \text{at } t \in \left(\dfrac{3a}{u}, \dfrac{5a}{u}\right) \\ \overline{V}_{rB(1)}\left(1 + \overline{\rho}_A\overline{\rho}_B\epsilon^{-j\Omega_r \frac{2a}{u}} + \left(\overline{\rho}_A\overline{\rho}_B\epsilon^{-j\Omega_r \frac{2a}{u}}\right)^2\right) & \text{at } t \in \left(\dfrac{5a}{u}, \dfrac{7a}{u}\right) \\ \ldots \end{cases}$$

(where $\overline{\rho}_A$ and $\overline{\rho}_B$ are real, and in particular $1 - \overline{\rho}_B = 2R_B/(z^{(o)} + R_B)$), with:

$$\overline{v}_{rB} \to \overline{V}_{rB(1)} \frac{1}{1 - \overline{\rho}_A\overline{\rho}_B\epsilon^{-j\Omega_r \frac{2a}{u}}} \qquad \text{at } t \to \infty$$

($|\overline{\rho}_A\overline{\rho}_B| < 1$). The vector $\overline{\imath}_{rB}$ is given by $\overline{v}_{rB}/R_B$.

More particularly, the transient of $\overline{v}_{rB}$ and $\overline{\imath}_{rB}$ is either aperiodic or oscillatory, according to $\overline{\rho}_A\overline{\rho}_B \gtrless 0$.

If instead, $R_B = z^{(o)} = \sqrt{l/c}$, $\overline{\rho}_B = 0$ holds, then more simply:

$$\overline{v}_{rB} = z^{(o)}\overline{\imath}_{rB} = \begin{cases} 0 & \text{at } t \in \left[0, \dfrac{a}{u}\right) \\ \dfrac{z^{(o)}\overline{E}}{z^{(o)} + R_A}\epsilon^{-j\Omega_r \frac{a}{u}} & \text{at } t > \dfrac{a}{u} \end{cases}$$

5.4.2. Steady-State Operation

If phase voltages and currents are sinusoidal and of the positive sequence, at frequency ω, then the homopolar components are zero, and furthermore (by assuming $\Omega_r = \omega$):

- vectors $\overline{v}_{rA}$, $\overline{\imath}_{rA}$, $\overline{v}_{rB}$, $\overline{\imath}_{rB}$ are constant;
- inside the line, voltage and current vectors are functions only of the distance x, so that it can be written that $\overline{v}_r = \overline{v}_r(x)$, $\overline{\imath}_r = \overline{\imath}_r(x)$.

As a consequence, it must be assumed that $p = 0$ in Equations [5.4.1] (in addition to $\Omega_r = \omega$) and similarly $s = 0$ in the different transfer functions defined in Section 5.4.1. In particular, from the first of Equations [5.4.4], it can be derived:

$$\overline{n}_r(0) = [(r + j\omega l)(g + j\omega c)]^{1/2} \triangleq n(j\omega)$$

and, since the phase of $(r + j\omega l)(g + j\omega c)$ is within $[0°, 180°]$, the two solutions in $\overline{n}_r(0) = n(j\omega)$ have phases within $[0°, 90°]$ and $[180°, 270°]$, respectively. By considering the former one, the constant $n(j\omega)$, called the "propagation constant" at the frequency ω, may be written:

$$n(j\omega) \triangleq \alpha + j\beta \tag{5.4.14}$$

where α (the *"attenuation constant"* at the frequency ω) and β (the *"phase constant"* at the frequency ω) are both nonnegative. From the second of Equations [5.4.4] it follows, as a unique solution, the value $\overline{z}_r(0) \triangleq z(j\omega)$ (the characteristic impedance, at the frequency ω). More particularly, if r and/or g are nonzero, it holds that:

$$\left.\begin{aligned} \alpha &= \sqrt{\frac{rg - \omega^2 lc + \sqrt{(r^2 + \omega^2 l^2)(g^2 + \omega^2 c^2)}}{2}} > 0 \\ \beta &= \frac{\omega(rc + gl)}{2\alpha} > 0 \end{aligned}\right\} \tag{5.4.15}$$

$$\left.\begin{aligned} \mathcal{Re}(z(j\omega)) &= \frac{r + g\sqrt{\dfrac{r^2 + \omega^2 l^2}{g^2 + \omega^2 c^2}}}{2\alpha} > 0 \\ Im(z(j\omega)) &= \frac{\omega\left(l - c\sqrt{\dfrac{r^2 + \omega^2 l^2}{g^2 + \omega^2 c^2}}\right)}{2\alpha} \end{aligned}\right\} \tag{5.4.16}$$

where $Im(z(j\omega)) \gtrless 0$ according to $gl \gtrless rc$ (usually $gl < rc$, and thus $Im(z(j\omega)) < 0$). If $r = g = 0$ (nondissipative line) it holds that:

$$\left.\begin{aligned} \alpha &= 0 \\ \beta &= \omega\sqrt{lc} = \frac{\omega}{u} \end{aligned}\right\} n(j\omega) = j\frac{\omega}{u} \tag{5.4.15'}$$

(e.g., $\beta \cong 1$ mrad/km for overhead lines at higher voltages and at 50 Hz), and furthermore, independently of ω:

$$\left.\begin{aligned} \mathcal{Re}(z(j\omega)) &= \sqrt{\frac{l}{c}} = z^{(o)} \\ Im(z(j\omega)) &= 0 \end{aligned}\right\} z(j\omega) = z^{(o)} \qquad [5.4.16']$$

By recalling Equations [5.4.6] and [5.4.7], we may derive $\overline{v}_r(x) = \overline{v}'_r(x) + \overline{v}''_r(x)$, with:

$$\left.\begin{aligned} \overline{v}'_r(x) &= \overline{v}'_{rA}\epsilon^{-n(j\omega)x} = \overline{v}'_{rA}\epsilon^{-\alpha x}\epsilon^{-j\beta x} \\ \overline{v}''_r(x) &= \overline{v}''_{rB}\epsilon^{-n(j\omega)(a-x)} = \overline{v}''_{rB}\epsilon^{-\alpha(a-x)}\epsilon^{-j\beta(a-x)} \end{aligned}\right\} \qquad [5.4.17]$$

whereas it then holds, at any given x, $\overline{\imath}_r(x) = \overline{\imath}'_r(x) + \overline{\imath}''_r(x)$, with $\overline{\imath}'_r(x) = \overline{v}'_r(x)/z(j\omega)$, $\overline{\imath}''_r(x) = -\overline{v}''_r(x)/z(j\omega)$.

If $\overline{v}_{rB} = z(j\omega)\overline{\imath}_{rB}$, i.e., if the line is terminated with an impedance which, at steady-state and at the assigned frequency ω, is equal to its characteristic impedance, then ($\forall x \in [0, a]$) $\overline{v}''_r(x) = 0$, $\overline{\imath}''_r(x) = 0$, and thus, similar to a section $[0, a]$ belonging to a line of infinite length:

$$\left.\begin{aligned} \overline{v}_r(x) &= \overline{v}_{rA}\epsilon^{-n(j\omega)x} = \overline{v}_{rA}\epsilon^{-\alpha x}\epsilon^{-j\beta x} \\ \overline{\imath}_r(x) &= \overline{\imath}_{rA}\epsilon^{-n(j\omega)x} = \overline{\imath}_{rA}\epsilon^{-\alpha x}\epsilon^{-j\beta x} \end{aligned}\right\} \qquad [5.4.18]$$

where $\overline{\imath}_{rA} = \overline{v}_{rA}/z(j\omega)$, from which it also follows that:

$$\frac{\overline{v}_r(x)}{\overline{\imath}_r(x)} = z(j\omega) \qquad [5.4.19]$$

In [5.4.18], the factor $\epsilon^{-n(j\omega)x} = \epsilon^{-\alpha x}\epsilon^{-j\beta x}$ defines (for positive α and β) an attenuation and a lag rotation, respectively, as a result of the terms $\epsilon^{-\alpha x}$, $\epsilon^{-j\beta x}$. Consequently, the vector $\overline{v}_r(x)$ describes, for varying x and for any given value of $\overline{v}_{rA}$, a logarithmic spiral (of the type reported in Fig. 5.21) which degenerates into a circle in the case of nondissipative line ($\alpha = 0$). The same can be said for $\overline{\imath}_r(x)$. Specifically, the magnitudes $v_r(x)$ and $i_r(x)$ decrease with x, being proportional to $\epsilon^{-\alpha x}$. Furthermore, the distance after which the phase-lag is 360° (if the line is sufficiently long) is in any case equal to $2\pi/\beta$, and it is called the "wavelength" of the line, at steady-state at the given frequency ω (instead, the Equation [5.4.10] refers to a nondissipative line under any operating condition).

By the first of Equations [5.4.18], it can be derived, with $\Omega_r = \omega$, $\theta_r = \theta_r(0) + \omega t$:

$$(v_\alpha + jv_\beta)(x, t) \triangleq \overline{v}_r(x)\epsilon^{j\theta_r} = \overline{v}_{rA}\epsilon^{-\alpha x}\epsilon^{j(\theta_r(0)+\omega t-\beta x)} = (v_\alpha + jv_\beta)\left(0, t - \frac{\beta x}{\omega}\right) \cdot \epsilon^{-\alpha x}$$

from which, for $v_a(x, t) + v_b(x, t) + v_c(x, t) = 0$ (see Fig. 5.15a and Appendix 2):

$$v_a(x, t) = \sqrt{\frac{2}{3}}v_\alpha(x, t) = v_a\left(0, t - \frac{\beta x}{\omega}\right) \cdot \epsilon^{-\alpha x}$$

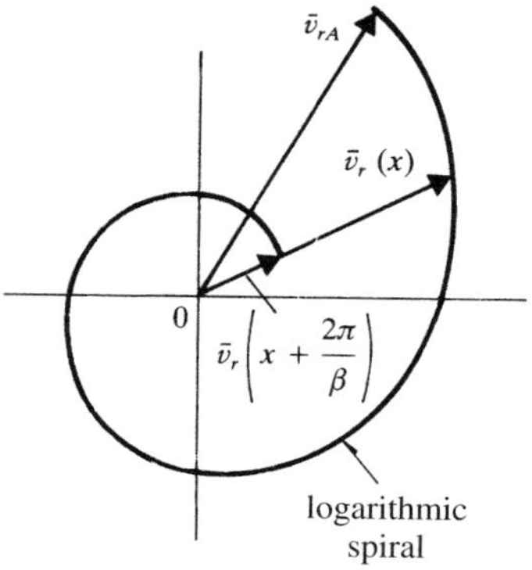

Figure 5.21. Line at steady-state terminated with its characteristic impedance: behavior of the voltage vector along the line.

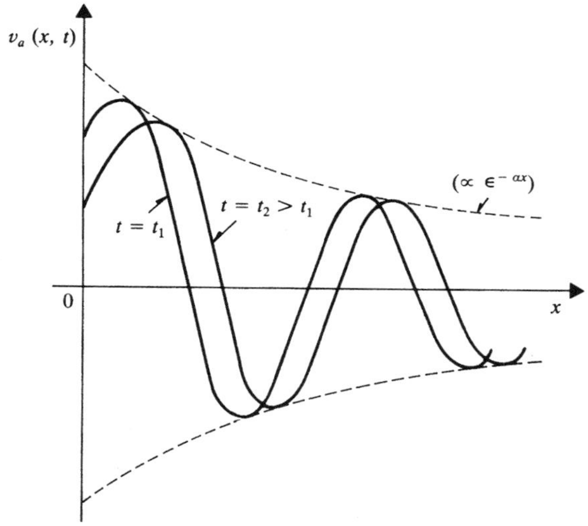

Figure 5.22. Line at steady-state terminated with its characteristic impedance: behavior of the generic phase voltage along the line at different instants.

as if the signal $v_a(0, t)$ would travel along the line at a speed ω/β, being attenuated according to the law $\epsilon^{-\alpha x}$ (and covering, in the period $2\pi/\omega$, a distance equal to the wavelength $2\pi/\beta$); see Figure 5.22. The same conclusion holds for $v_b(0, t)$ and $v_c(0, t)$ and for currents $i_a(0, t)$, $i_b(0, t)$, $i_c(0, t)$. In the case of nondissipative line $\omega/\beta = u$, and there is no attenuation.

Because of Equation [5.4.19] it is true that the phase-shift between voltage and current (at the same x), and thus the "power factor," is constant along the line. Usually $r/l > g/c$ and thus (as already seen) $Im(z(j\omega)) < 0$, whereas $\mathcal{R}e(z(j\omega)) > 0$. Therefore, at any x, current leads voltage.

Concerning the active (P) and reactive (Q) powers flowing into the line, which are defined by:

$$P(x) + jQ(x) = \overline{v}_r(x)\overline{\imath}_r^*(x) = \left|\frac{\overline{v}_{rA}}{z(j\omega)}\right|^2 z(j\omega)\epsilon^{-2\alpha x}$$

it can be derived, under the above-mentioned conditions, that $P(x) > 0$ and $Q(x) < 0$ (note that $P(x)$ and $Q(x)$ decrease in amplitude with x, in a way proportional to $\epsilon^{-2\alpha x}$). If $r = g = 0$ (nondissipative line), vectors $\overline{v}_r(x)$ and $\overline{\imath}_r(x)$ would be in phase with each other, and their amplitudes would remain constant along the line. Moreover, it would hold that $P(x) = \text{constant} > 0$, $Q(x) = 0$ for all the values of x. The equation $Q(x) = 0$ means, with reference to the first circuit in Figure 5.15b, that the reactive power absorbed by the element $l\Delta x$ is exactly balanced by the reactive power generated by the element $c\Delta x$.

Concerning the whole line as seen by its terminals, it is possible to refer to the *equivalent circuit* in Figure 5.16, with $p = 0$ and $\Omega_r = \omega$ in the operational impedances, and thus assuming $\overline{n}_r = \overline{n}_r(0) \triangleq n(j\omega)$ and $\overline{z}_r = \overline{z}_r(0) \triangleq z(j\omega)$, with $n(j\omega)$ and $z(j\omega)$ defined by Equations [5.4.14], [5.4.15], and [5.4.16]. The equivalent circuit in Figure 5.23a is obtained, in which the impedance of the series branch is equal to:

$$\overline{Z}_s \triangleq z(j\omega)\sinh(n(j\omega)a) = z(j\omega)(\sinh(\alpha a)\cos(\beta a) + j\cosh(\alpha a)\sin(\beta a)) \quad [5.4.20]$$

whereas that of the shunt branches is:

$$\overline{Z}_d \triangleq \frac{z(j\omega)}{\tanh\dfrac{n(j\omega)a}{2}} = z(j\omega)\frac{\sinh(\alpha a) - j\sin(\beta a)}{\cosh(\alpha a) - \cos(\beta a)} \quad [5.4.21]$$

($z(j\omega)$ is in general complex: recall Equation [5.4.16]).

The behavior of $\overline{v}_r(x)$ and $\overline{\imath}_r(x)$ *inside the line* is defined, for given $\overline{v}_{rA}$ and $\overline{\imath}_{rA}$, by Equations [5.4.3], with $\overline{n}_r = n(j\omega)$ and $\overline{z}_r = z(j\omega)$. In practical applications,

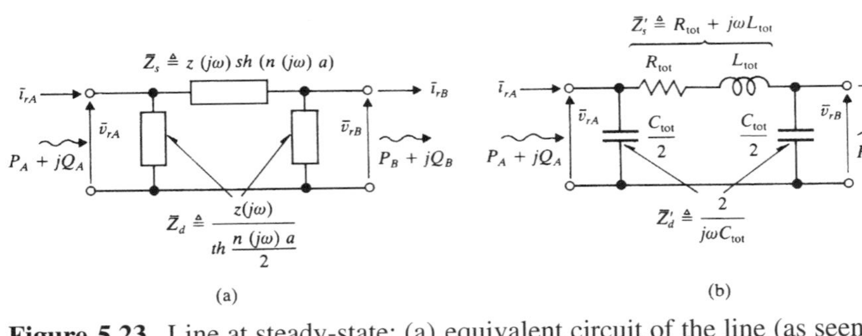

Figure 5.23. Line at steady-state: (a) equivalent circuit of the line (as seen by its terminals), relative to Park's vectors; (b) simplified circuit, deducible from the previous one in the case of a relatively short line.

however, it can appear more significant to refer to other assignments (instead of $\overline{v}_{rA}$, $\overline{\imath}_{rA}$). In particular:

(1) for a line interconnecting two networks of relatively large power, it can be assumed, at least as a limiting condition, that the terminal voltages (vectors) $\overline{v}_{rA}$ and $\overline{v}_{rB}$ are assigned; therefore, from the previous equations it follows, in terms of $\overline{v}_{rA}$ and $\overline{v}_{rB}$, that:

$$\left.\begin{aligned} \overline{v}_r(x) &= \frac{\overline{v}_{rA}\sinh(n(j\omega)(a-x)) + \overline{v}_{rB}\sinh(n(j\omega)x)}{\sinh(n(j\omega)a)} \\ \overline{\imath}_r(x) &= \frac{\overline{v}_{rA}\cosh(n(j\omega)(a-x)) - \overline{v}_{rB}\cosh(n(j\omega)x)}{z(j\omega)\sinh(n(j\omega)a)} \end{aligned}\right\} \quad [5.4.22]$$

(2) for a line which, through its receiving terminal, supplies a load or is possibly open-circuited ($\overline{\imath}_{rB} = 0$), it can be assumed that one of the voltages is assigned (i.e., $\overline{v}_{rA}$ or $\overline{v}_{rB}$), as well as the impedance $\overline{v}_{rB}/\overline{\imath}_{rB} = \overline{Z}_B$ with which the line is terminated; by assigning $\overline{v}_{rB}$ and $\overline{Z}_B$, it follows that:

$$\left.\begin{aligned} \overline{v}_r(x) &= \overline{v}_{rB}\left(\frac{z(j\omega)}{\overline{Z}_B}\sinh(n(j\omega)(a-x)) + \cosh(n(j\omega)(a-x))\right) \\ \overline{\imath}_r(x) &= \overline{v}_{rB}\left(\frac{1}{\overline{Z}_B}\cosh(n(j\omega)(a-x)) + \frac{1}{z(j\omega)}\sinh(n(j\omega)(a-x))\right) \end{aligned}\right\} \quad [5.4.23]$$

where $\overline{Z}_B$ is evaluated at steady-state at the given frequency ω, with $\overline{Z}_B = \infty$ for the case of open-circuited line; furthermore, in these conditions, the current $\overline{\imath}_{rB}$, and the active and reactive powers flowing out of the line are assigned.

Specifically, it is possible to derive Equations [5.4.18] and [5.4.19], if $\overline{v}_{rB} = \overline{v}_{rA}\epsilon^{-n(j\omega)a}$ in case (1), or $\overline{Z}_B = z(j\omega)$ in case (2).

By considering the values of the "primary constants" (see Section 5.4.1), it is immediate to determine that, for most of the lines, the magnitude of $n(j\omega)a$ is very small. Actually:

$$|n(j\omega)| = [(r^2 + \omega^2 l^2)(g^2 + \omega^2 c^2)]^{1/4}$$

can be, for instance, $(1000\ \text{km})^{-1}$ for an overhead line and $(400\ \text{km})^{-1}$for a cable line. Correspondingly, $|n(j\omega)a|$ can be approximately 0.05 for a 50-km length overhead line or for a 20-km cable line. Therefore, for *relatively short lines*, the hyperbolic functions that appear in the previous equations can be adequately approximated. More precisely, by neglecting the per unit of length conductance (g), Equations [5.4.20] and [5.4.21] can be substituted by:

$$\overline{Z}_s \cong z(j\omega)n(j\omega)a = (r + j\omega l)a = R_{\text{tot}} + j\omega L_{\text{tot}} \triangleq \overline{Z}'_s \quad [5.4.20']$$

$$\overline{Z}_d \cong \frac{z(j\omega)}{n(j\omega)a/2} = \frac{2}{j\omega ca} = \frac{2}{j\omega C_{\text{tot}}} \triangleq \overline{Z}'_d \quad [5.4.21']$$

according to Figure 5.23b. Note that $R_{tot} \triangleq ra$, $L_{tot} \triangleq la$ and $C_{tot} \triangleq ca$ represent the total resistance, inductance, and capacitance of the line, respectively. Often, for high-voltage overhead lines it may be assumed that $R_{tot} = 0$ and possibly (especially if the line length a is modest) that $C_{tot} = 0$, i.e., $\overline{Z}'_d = \infty$. In this case, the line is represented only by the parameter L_{tot}. The situation may be different for cable lines; in some cases, the most significant parameters may be R_{tot} and C_{tot}, whereas L_{tot} may, at a first approximation, be disregarded.

Similarly, Equations [5.4.22] and [5.4.23] may be approximated by:

$$\left.\begin{aligned} \overline{v}_r(x) &= \frac{\overline{v}_{rA}(a-x)+\overline{v}_{rB}x}{a} \\ \overline{\imath}_r(x) &= \frac{\overline{v}_{rA}\left(1+\dfrac{(n(j\omega)(a-x))^2}{2}\right)-\overline{v}_{rB}\left(1+\dfrac{(n(j\omega)x)^2}{2}\right)}{z(j\omega)n(j\omega)a} \\ &= \frac{\overline{v}_{rA}-\overline{v}_{rB}}{\overline{Z}'_s}+\frac{\overline{v}_{rA}(a-x)^2-\overline{v}_{rB}x^2}{a^2\overline{Z}'_d} \end{aligned}\right\} \qquad [5.4.22']$$

$$\left.\begin{aligned} \overline{v}_r(x) &= \overline{v}_{rB}\left(\frac{z(j\omega)}{\overline{Z}_B}n(j\omega)(a-x)+1+\frac{(n(j\omega)(a-x))^2}{2}\right) \\ &= \overline{v}_{rB}\left(1+\frac{\overline{Z}'_s}{\overline{Z}_B}\frac{a-x}{a}+\frac{\overline{Z}'_s}{\overline{Z}'_d}\left(\frac{a-x}{a}\right)^2\right) \\ \overline{\imath}_r(x) &= \overline{v}_{rB}\left(\frac{1}{\overline{Z}_B}\left(1+\frac{(n(j\omega)(a-x))^2}{2}\right)+\frac{1}{z(j\omega)}n(j\omega)(a-x)\right) \\ &= \overline{v}_{rB}\left(\frac{1}{\overline{Z}_B}\left(1+\frac{\overline{Z}'_s}{\overline{Z}'_d}\left(\frac{a-x}{a}\right)^2\right)+\frac{2}{\overline{Z}'_d}\frac{a-x}{a}\right) \end{aligned}\right\} \qquad [5.4.23']$$

However, by setting $\overline{v}_{rB}/\overline{\imath}_{rB} = \overline{Z}_B$, from Equations [5.4.22′] it is possible to deduce, for $x \in (0, a)$, equations that are somewhat different from [5.4.23′]. On the other hand, Equations [5.4.22′] are based on an approximation of only the line model, whereas Equations [5.4.23′] correspond to an approximation of the overall model (line, and impedance $\overline{Z}_B$).

In case (1) for which *the voltages* $\overline{v}_{rA}$ *and* $\overline{v}_{rB}$ *are assigned*, it can be derived:

$$\left.\begin{aligned} P_A &= \frac{v_{rA}}{(1+\rho^2)\omega L_{tot}}(v_{rB}\sin\delta+\rho(v_{rA}-v_{rB}\cos\delta)) \\ P_B &= \frac{v_{rB}}{(1+\rho^2)\omega L_{tot}}(v_{rA}\sin\delta-\rho(v_{rB}-v_{rA}\cos\delta)) \end{aligned}\right\} \qquad [5.4.24]$$

having set $\rho \triangleq R_{tot}/(\omega L_{tot}) = r/(\omega l)$, $\delta \triangleq \angle\overline{v}_{rA} - \angle\overline{v}_{rB}$ and by intending that P_A and P_B are, as indicated in Figure 5.23, the active power entering into the line

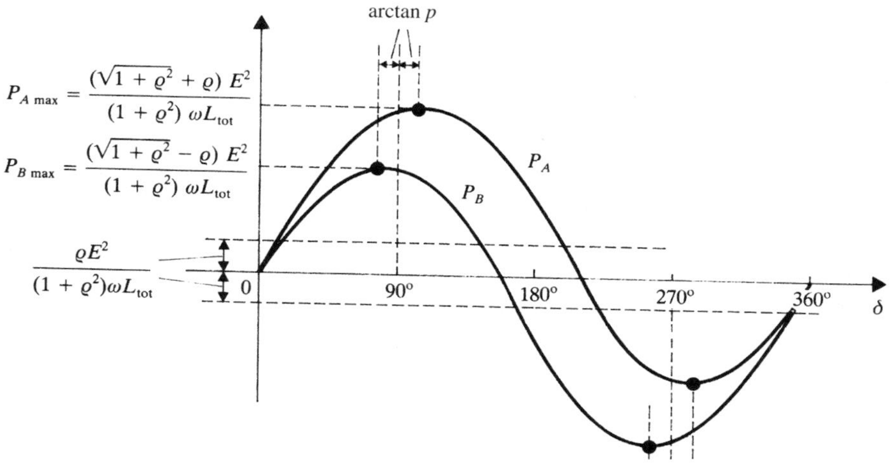

Figure 5.24. Case of a relatively short line, at steady-state: dependence of active powers (P_A) entering into the line and (P_B) flowing out of the line, on the phase-shift (δ) between voltage vectors at line terminals, under the assumption $v_{rA} = v_{rB} \triangleq E$.

at $x = 0$, and the active power flowing out at $x = a$ (if $R_{tot} = 0$, it simply holds $P_A = P_B = (v_{rA} v_{rB}/(\omega L_{tot})) \sin\delta$).

By assuming for simplicity $v_{rA} = v_{rB} \triangleq E$, the diagrams in Figure 5.24 can be deduced.

Furthermore, the reactive power entering into the line at $x = 0$ and the one flowing out at $x = a$ (Fig. 5.23) can be, respectively, put in the form:

$$\begin{cases} Q_A = Q_{As} - Q_{Ad} \\ Q_B = Q_{Bs} + Q_{Bd} \end{cases}$$

where Q_{As} and Q_{Bs} (related to the series branch) are equal to:

$$\left.\begin{aligned} Q_{As} &= \frac{v_{rA}}{(1+\rho^2)\omega L_{tot}}(v_{rA} - v_{rB}\cos\delta - \rho v_{rB}\sin\delta) \\ Q_{Bs} &= \frac{v_{rB}}{(1+\rho^2)\omega L_{tot}}(-v_{rB} + v_{rA}\cos\delta - \rho v_{rA}\sin\delta) \end{aligned}\right\} \qquad [5.4.25]$$

whereas Q_{Ad} and Q_{Bd} are the reactive powers generated by the (capacitive) shunt branches and are given by:

$$Q_{Ad} = \frac{\omega C_{tot} v_{rA}^2}{2}, \qquad Q_{Bd} = \frac{\omega C_{tot} v_{rB}^2}{2}$$

Assuming $v_{rA} = v_{rB} \triangleq E$ and $\rho = 0$, then $Q_{As} + Q_{Bs} = 0$, with $Q_{As} > 0$ and $Q_{Bs} < 0$ for all the values $\delta \neq 0$. If $\rho \neq 0$, then $Q_{As} < 0$ for $\delta \in (0, 2\arctan\rho)$, and $Q_{Bs} > 0$ for $\delta \in (2\pi - 2\arctan\rho, 2\pi)$.

The behaviors of $\overline{v}_r(x)$ and $\overline{\imath}_r(x)$ (from which those of $P(x)$ and $Q(x)$), are, within the adopted approximations, given by Equations [5.4.22′].

The analysis of case (2), when *the voltage* $\overline{v}_{rB}$ *and the impedance* $\overline{Z}_B$ *are assigned*, is similarly obvious (for $\overline{v}_r(x)$ and $\overline{\imath}_r(x)$, Equations [5.4.23′] should be recalled).

Having set:

$$\frac{\overline{Z}'_d\overline{Z}_B}{\overline{Z}'_d + \overline{Z}_B} = \frac{\overline{Z}_B}{1 + \dfrac{j\omega C_{\text{tot}}\overline{Z}_B}{2}} \triangleq \overline{Z}_{dB} \triangleq R_{dB} + jX_{dB}$$

it follows that:

$$\frac{\overline{v}_{rA}}{\overline{Z}'_s + \overline{Z}_{dB}} = \frac{\overline{v}_{rB}}{\overline{Z}_{dB}} = \frac{P_B - jQ_{Bs}}{\overline{v}^*_{rB}}$$

and thus, specifically:

- if $\angle\overline{Z}_{dB} = \angle\overline{Z}'_s$ (i.e., $X_{dB} = R_{dB}/\rho > 0$), $\overline{v}_{rA}$ and $\overline{v}_{rB}$ are in phase (i.e., $\delta = 0$), and the voltage drop is:

$$v_{rA} - v_{rB} = \omega L_{\text{tot}}\frac{Q_{Bs} + \rho P_B}{v_{rB}} = (1 + \rho^2)\omega L_{\text{tot}}\frac{Q_{Bs}}{v_{rB}}$$

- to obtain $v_{rA} = v_{rB}$ (zero voltage drop) it is necessary that $X_{dB} < 0$ and thus $Q_{Bs} < 0$ (more precisely, it is necessary that $X_{dB} = -((1 + \rho^2)(\omega L_{\text{tot}}/2) + \rho R_{dB}))$.

In the case of a line under *"open-circuit" operation* ($\overline{\imath}_{rB} = 0$, $\overline{Z}_B = \infty$), it can be derived:

$$\frac{v_{rB}}{v_{rA}} = \frac{1}{\sqrt{1 - \omega^2 L_{\text{tot}}C_{\text{tot}} + (1 + \rho^2)\dfrac{\omega^4 L^2_{\text{tot}}C^2_{\text{tot}}}{4}}}$$

and for a nondissipative line ($\rho = 0$), with $\omega^2 L_{\text{tot}}C_{\text{tot}}/2 < 1$ (i.e., $|n(j\omega)a| < \sqrt{2}$):

$$\frac{v_{rB}}{v_{rA}} = \frac{\overline{v}_{rB}}{\overline{v}_{rA}} = \frac{1}{1 - \dfrac{\omega^2 L_{\text{tot}}C_{\text{tot}}}{2}} \qquad [5.4.26]$$

and thus $v_{rB} > v_{rA}$ ("Ferranti effect"; see Equation [5.4.29], of which Equation [5.4.26] is an approximation when βa is sufficiently small).

If the line is not short enough, the previous approximations cannot be accepted. Regarding the behavior of the line as seen from its terminals, treatment is similar to that presented above, except that $\overline{Z}'_s = R_{\text{tot}} + j\omega L_{\text{tot}}$ and $\overline{Z}'_d = 2/(j\omega C_{\text{tot}})$ are substituted by the impedances $\overline{Z}_s$ and $\overline{Z}_d$ defined by Equations [5.4.20] and [5.4.21], respectively. Therefore, in the case for which $\overline{v}_{rA}$ and $\overline{v}_{rB}$ are assigned, the powers relative to the series branch remain expressed by equations similar to [5.4.24] and [5.4.25], and so on. The only significant difference can be caused by the presence of a nonzero real part in $\overline{Z}_d$. Moreover, for the internal behavior of the line, it is necessary to refer to Equations [5.4.22] or [5.4.23], as already specified.

In practice, however, the analysis can be reasonably simplified especially in the case of *relatively long lines* (e.g., having a length of hundreds of kilometers or more), typically constituted by overhead lines at high or extra-high voltage. By considering the values of the primary constants (see Section 5.4.1), the simplification $r = g = 0$ (i.e., the assumption of nondissipative line) indeed appears to be acceptable. It follows that:

$$\begin{cases} n(j\omega) = j\beta \\ z(j\omega) = z^{(o)} \end{cases}$$

where:

$$\begin{cases} \beta = \omega\sqrt{lc} = \dfrac{\omega}{u} = \dfrac{2\pi}{\lambda_r} \\ z^{(o)} = \sqrt{\dfrac{l}{c}} \end{cases}$$

Specifically, u is the propagation speed, approximately 3.10^5 km/sec, whereas λ_r is the wavelength at the frequency ω, equal to approximately 6000 km at 50 Hz.

Concerning the behaviors of voltages, currents, and powers along the line, it is interesting to note that because of Equations [5.4.1] it results that:

$$\frac{1}{v_r}\frac{\mathrm{d}v_r}{\mathrm{d}x} + j\frac{\mathrm{d}\angle\overline{v}_r}{\mathrm{d}x} = \frac{1}{\overline{v}_r}\frac{\mathrm{d}\overline{v}_r}{\mathrm{d}x} = -(r + j\omega l)\frac{\overline{\imath}_r}{\overline{v}_r} \qquad (\text{if } v_r \neq 0)$$

$$\frac{1}{i_r}\frac{\mathrm{d}i_r}{\mathrm{d}x} + j\frac{\mathrm{d}\angle\overline{\imath}_r}{\mathrm{d}x} = \frac{1}{\overline{\imath}_r}\frac{\mathrm{d}\overline{\imath}_r}{\mathrm{d}x} = -(g + j\omega c)\frac{\overline{v}_r}{\overline{\imath}_r} \qquad (\text{if } i_r \neq 0)$$

where $\overline{\imath}_r/\overline{v}_r = (P - jQ)/v_r^2$, $\overline{v}_r/\overline{\imath}_r = (P + jQ)/i_r^2$. Thus, if $v_r \neq 0$ it holds:

$$\begin{cases} \dfrac{\mathrm{d}v_r}{\mathrm{d}x} = -\dfrac{rP + \omega l Q}{v_r} \\ \dfrac{\mathrm{d}\angle\overline{v}_r}{\mathrm{d}x} = \dfrac{rQ - \omega l P}{v_r^2} \end{cases}$$

and analogously, if $i_r \neq 0$:

$$\begin{cases} \dfrac{\mathrm{d}i_r}{\mathrm{d}x} = \dfrac{-gP + \omega c Q}{i_r} \\ \dfrac{\mathrm{d}\angle \bar{\imath}_r}{\mathrm{d}x} = -\dfrac{gQ + \omega c P}{i_r^2} \end{cases}$$

Specifically, for a nondissipative line it can be derived:

$$\frac{\mathrm{d}v_r}{\mathrm{d}x} = -\frac{\omega l Q}{v_r}, \quad \frac{\mathrm{d}\angle \overline{v}_r}{\mathrm{d}x} = -\frac{\omega l P}{v_r^2}, \quad \frac{\mathrm{d}i_r}{\mathrm{d}x} = \frac{\omega c Q}{i_r}, \quad \frac{\mathrm{d}\angle \bar{\imath}_r}{\mathrm{d}x} = -\frac{\omega c P}{i_r^2}$$

so that:

- the voltage magnitude decreases, and the current magnitude increases, in the direction in which the reactive power Q flows (at the given x);
- both phases $\angle \overline{v}_r$ and $\angle \bar{\imath}_r$ decrease in the direction in which the active power P (constant along the line) flows.

In the case (1) for which *the voltages* $\overline{v}_{rA}$ *and* $\overline{v}_{rB}$ *are assigned*, Equations [5.4.22] become:

$$\begin{cases} \overline{v}_r(x) = \dfrac{\overline{v}_{rA} \sin(\beta(a-x)) + \overline{v}_{rB} \sin(\beta x)}{\sin(\beta a)} \\ \bar{\imath}_r(x) = \dfrac{\overline{v}_{rA} \cos(\beta(a-x)) - \overline{v}_{rB} \cos(\beta x)}{j z^{(o)} \sin(\beta a)} \end{cases}$$

from which, recalling that $P + jQ = \overline{v}_r \bar{\imath}_r^*$ and by assuming for simplicity that $v_{rA} = v_{rE} \triangleq E$, it is possible to derive:

$$P = P^{(o)} \frac{\sin \delta}{\sin(\beta a)} \tag{5.4.27}$$

$$Q(x) = P^{(o)} \frac{\sin(\beta(a - 2x))(\cos(\beta a) - \cos \delta)}{\sin^2(\beta a)}$$

where $\delta \triangleq \angle \overline{v}_{rA} - \angle \overline{v}_{rB}$ defines the phase-shift between terminal voltages, whereas:

$$P^{(o)} \triangleq \frac{v_{rB}^2}{z^{(o)}} = \frac{E^2}{z^{(o)}}$$

(the "*characteristic*" or "*natural*" *active power* of the line) is the active power that would flow along the line if it were terminated, at the given $\overline{v}_{rB}$, with its characteristic impedance.

Specifically, $Q_A = -Q_B$ and, in general, $Q(x) = -Q(a - x)$, whereas $v_r(x) = v_r(a - x)$. In the midpoint of the line, it holds that $Q(a/2) = 0$ and:

$$v_r\left(\frac{a}{2}\right) = E\left|\frac{\cos\dfrac{\delta}{2}}{\cos\dfrac{\beta a}{2}}\right| \qquad [5.4.28]$$

At $\delta = \beta a$ and $\delta = 2\pi - \beta a$ it, respectively, holds that $P = \pm P^{(o)}$, whereas $Q(x) = 0$, $v_r(x) = E\ \forall x \in [0, a]$.

If the line length is smaller than half of the wavelength (i.e., if $a < \lambda_r/2$, that is $a < 3000$ km for $\lambda_r = 6000$ km) and thus $\beta a < \pi$, the dependence of P on δ is of the type indicated in Figure 5.25a, with $P = P^{(o)}$ at $\delta = \beta a$, and with P at its maximum for $\delta = 90°$. This maximum is given by $P^{(o)}/\sin(\beta a)$ and can be approximated by $E^2/(z^{(o)}\beta a) = E^2/(\omega L_{\text{tot}})$ if βa is sufficiently small: see Figure 5.24, for $\rho = 0$.

It can be directly verified that $Q(x)$ becomes zero only in the midpoint of the line, where the value of v_r (expressed by Equation [5.4.28]) is the minimum or the maximum of $v_r(x)$ according to whether $\cos\delta \lesseqgtr \cos(\beta a)$: see the qualitative outlines of $v_r(x)$ and $Q(x)$ reported in Figure 5.25b for various values of δ.

Specifically, if $a < \lambda_r/4$ or, equivalently, if $\beta a < \pi/2$ ($a < 1500$ km for $\lambda_r = 6000$ km), for $\delta \in (0, 90°)$ it follows that:

$$P \gtrless P^{(o)}, \quad Q_A = -Q_B \gtrless 0, \quad v_r\left(\frac{a}{2}\right) \lesseqgtr E \quad \text{according to whether } \delta \gtrless \beta a$$

In practical cases, especially for lines which are not very long, $\delta \in (\beta a, 90°)$. In this regard, a typical value of $\beta a = |n(j\omega)a|$ for a 200-km length overhead line can be $200/1000 = 0.2$ (rad), which corresponds to an angle of approximately 12°.

In the infrequent case for which $a \in (\lambda_r/2, \lambda_r)$ and $\beta a \in (\pi, 2\pi)$, the dependence of P on δ is indicated in Figure 5.25c. Furthermore, $Q(x)$ becomes zero not only in the middle of the line ($x = a/2$), but at $x = a/2 \mp \pi/2\beta = a/2 \mp \lambda_r/4$, where v_r assumes the value:

$$v_r\left(\frac{a}{2} \mp \frac{\pi}{2\beta}\right) = E\left|\frac{\sin\dfrac{\delta}{2}}{\sin\dfrac{\beta a}{2}}\right|$$

smaller or greater than E according to whether $v_r(a/2) \gtrless E$: see, qualitatively, the diagrams of Figure 5.25d.

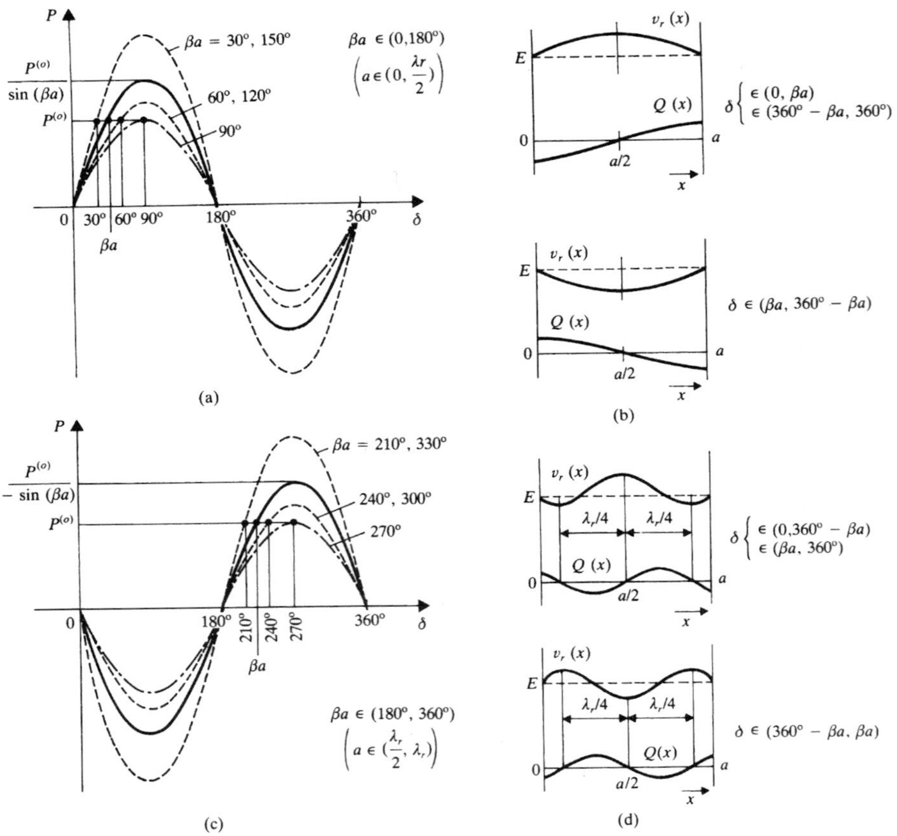

Figure 5.25. Case of a relatively long (and nondissipative) line, at steady-state and having voltage vectors assigned at its terminals: (a), (c) dependence of transmitted active power P on the phase-shift δ between voltage vectors at line terminals, respectively, for the cases $a < \lambda_r/2$ and $a \in (\lambda_r/2, \lambda_r)$; (b), (d) outlines of voltage magnitude v_r and reactive power Q along the line, for the same cases as above. It is intended a = line length, λ_r = wavelength and $v_{rA} = v_{rB} \triangleq E$.

Instead, in case (2) for which *the voltage* $\overline{v}_{rB}$ *and the impedance* $\overline{Z}_B$ *are assigned*, because of Equations [5.4.23]:

$$\begin{cases} \overline{v}_r(x) = \overline{v}_{rB} \left(\dfrac{jz^{(o)}}{\overline{Z}_B} \sin(\beta(a-x)) + \cos(\beta(a-x)) \right) \\ \overline{\imath}_r(x) = \overline{v}_{rB} \left(\dfrac{1}{\overline{Z}_B} \cos(\beta(a-x)) + \dfrac{j}{z^{(o)}} \sin(\beta(a-x)) \right) \end{cases}$$

By assuming for simplicity that $\overline{Z}_B = R_B$, i.e., $Q_B = 0$, it results that $P = v_{rB}^2/R_B$ and (with $P^{(o)} = v_{rB}^2/z^{(o)}$, $P/P^{(o)} = z^{(o)}/R_B$):

$$Q(x) = v_{rB}^2 \left(\frac{z^{(o)}}{R_B^2} - \frac{1}{z^{(o)}} \right) \frac{\sin(2\beta(a-x))}{2}$$

$$= P^{(o)} \left(\left(\frac{P}{P^{(o)}} \right)^2 - 1 \right) \frac{\sin(2\beta(a-x))}{2}$$

$$v_r^2(x) = v_{rB}^2 \left(\cos^2(\beta(a-x)) + \left(\frac{P}{P^{(o)}} \right)^2 \sin^2(\beta(a-x)) \right)$$

Moreover, the phase-shift $\delta \triangleq \angle\overline{v}_{rA} - \angle\overline{v}_{rB}$ is defined by:

$$\tan\delta = \frac{P}{P^{(o)}} \tan(\beta a)$$

and (for $P \neq 0$) is in the same quadrant as βa. Consequently, it is possible to write:

$$v_{rA} = v_{rB} \frac{\cos(\beta a)}{\cos\delta}$$

Specifically, the condition $P = P^{(o)}$ (i.e., $R_B = z^{(o)}$) implies $\delta = \beta a$, $v_{rA} = v_{rB}$, and more generally $Q(x) = 0$, $v_r(x) = v_{rB}$ along the whole line, according to information already presented.

From the previous equations, it follows that:

$$v_{rA} \gtrless v_{rB} \text{ according to whether } P \gtrless P^{(o)} \text{(i.e., } R_B \lessgtr z^{(o)})$$

and furthermore:

- if $a < \lambda_r/4$ (i.e., if $\beta a < \pi/2$):

  $Q \gtrless 0$, $\mathrm{d}v_r/\mathrm{d}x \lessgtr 0$ according to whether $P \gtrless P^{(o)}$, along the whole line,

  whereas $\delta \in (0, \pi/2)$;
- if $a \in (\lambda_r/4, \lambda_r/2)$ (i.e., if $\beta a \in (\pi/2, \pi)$):

$$\begin{cases} \text{as above,} & \text{for } x > a - \dfrac{\pi}{2\beta} = a - \dfrac{\lambda_r}{4} \\ \text{the opposite,} & \text{for } x < a - \dfrac{\pi}{2\beta} = a - \dfrac{\lambda_r}{4} \end{cases}$$

  whereas $\delta \in (\pi/2, \pi)$;

and so on (see Fig. 5.26a).

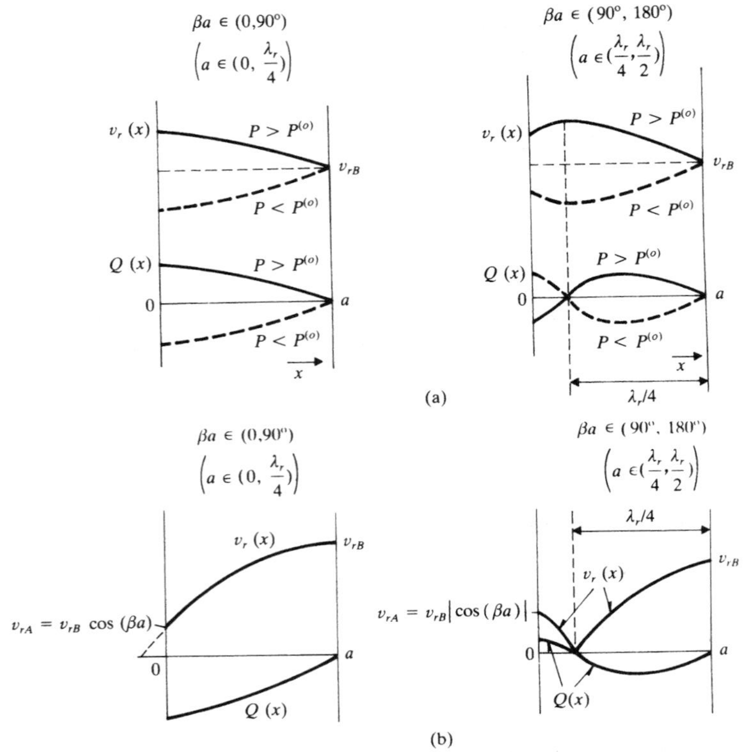

Figure 5.26. Case of a relatively long (and nondissipative) line, at steady-state, terminated with a load impedance: (a) outlines of voltage magnitude v_r and reactive power Q along the line, for resistive load impedance; (b) idem, with infinite load impedance (line under "open-circuit" operation). It is assumed $a < \lambda_r/2$.

Under *"open-circuit" operation* ($\bar{\imath}_{rB} = 0$, $\bar{Z}_B = \infty$):

$$\bar{v}_r(x) = \bar{v}_{rB} \cos(\beta(a - x))$$

$$\bar{\imath}_r(x) = \frac{j\bar{v}_{rB}}{z^{(o)}} \sin(\beta(a - x))$$

$$P = 0$$

$$Q(x) = -\frac{v_{rB}^2}{2z^{(o)}} \sin(2\beta(a - x))$$

and therefore:

- at the receiving terminal the voltage magnitude assumes its maximum value (*"Ferranti effect"*), with:

$$\frac{v_{rB}}{v_{rA}} = \frac{1}{|\cos(\beta a)|} \quad [5.4.29]$$

- the impedance seen at the generic distance x is purely reactive, and it is given by $\overline{v}_r(x)/\overline{\imath}_r(x) = -jz^{(o)}/\tan(\beta(a-x)) \triangleq jX_{\text{eq}}(x)$, with:

$$X_{\text{eq}}(x) = \frac{-z^{(o)}}{\tan(\beta(a-x))}$$

(at $x = 0$: $X_{\text{eq}}(0) = -z^{(o)}/\tan(\beta a)$);
- the generic $\overline{v}_r(x)$ is in phase or in opposition of phase with respect to $\overline{v}_{rB}$.

Specifically:

- if $a < \lambda_r/4$:

$$Q < 0, \quad \frac{\mathrm{d}v_r}{\mathrm{d}x} > 0, \quad X_{\text{eq}} < 0, \quad \angle\overline{v}_r - \angle\overline{v}_{rB} = 0, \quad \text{along the whole line}$$

(at $x = 0$: purely capacitive impedance, and $\delta = 0$);
- if $a \in (\lambda_r/4, \lambda_r/2)$:

$$\begin{cases} \text{as above,} & \text{for } x > a - \dfrac{\lambda_r}{4} \\ Q > 0, \dfrac{\mathrm{d}v_r}{\mathrm{d}x} < 0, X_{eq} > 0, \angle\overline{v}_r - \angle\overline{v}_{rB} = 180^\circ, & \text{for } x < a - \dfrac{\lambda_r}{4} \end{cases}$$

(at $x = 0$: purely inductive impedance, and $\delta = 180^\circ$);

and so on (see Fig. 5.26b).

5.4.3. Dynamic Models of Different Approximation

Whichever the operating condition, the line can be seen as a system having six scalar "inputs" (voltages and/or currents at its terminals) and several "outputs." To set concepts, in the following we will assume:

- as inputs: the voltages, i.e., the vectors $\overline{v}_{rA}$, $\overline{v}_{rB}$ (each one having two scalar components) and the zero-sequence components v_{oA}, v_{oB};
- as outputs: the currents at line terminals, i.e., the vectors $\overline{\imath}_{rA}$, $\overline{\imath}_{rB}$ and the zero-sequence components i_{oA}, i_{oB}.

Under the adopted assumptions (linearity, physical symmetry, uniformity) the dependence of $\overline{\imath}_{rA}$, $\overline{\imath}_{rB}$, on $\overline{v}_{rA}$, $\overline{v}_{rB}$ is defined by:

$$\left.\begin{aligned} \overline{\imath}_{rA} &= \frac{\overline{v}_{rA}\cosh(\overline{n}_r(p)a) - \overline{v}_{rB}}{\overline{z}_r(p)\sinh(\overline{n}_r(p)a)} = \frac{\overline{v}_{rA} - \overline{v}_{rB}}{\overline{\overline{Z}}_{rs}(p)} + \frac{\overline{v}_{rA}}{\overline{\overline{Z}}_{rd}(p)} \\ \overline{\imath}_{rB} &= \frac{\overline{v}_{rA} - \overline{v}_{rB}\cosh(\overline{n}_r(p)a)}{\overline{z}_r(p)\sinh(\overline{n}_r(p)a)} = \frac{\overline{v}_{rA} - \overline{v}_{rB}}{\overline{\overline{Z}}_{rs}(p)} - \frac{\overline{v}_{rB}}{\overline{\overline{Z}}_{rd}(p)} \end{aligned}\right\} \qquad [5.4.30]$$

(recall Equations [5.4.5] and Fig. 5.16), where $\overline{n}_r$ and $\overline{z}_r$ are expressed by Equations [5.4.4]. The operational impedances:

$$\begin{cases} \overline{Z}_{rs}(p) \triangleq \overline{z}_r(p)\sinh(\overline{n}_r(p)a) \\ \overline{Z}_{rd}(p) \triangleq \dfrac{\overline{z}_r(p)}{\tanh(\overline{n}_r(p)a/2)} \end{cases}$$

can be also put in the form:

$$\left.\begin{aligned} \overline{Z}_{rs}(p) &= \overline{Z}'_{rs}(p)\frac{\sinh(\overline{n}_r(p)a)}{\overline{n}_r(p)a} \\ \overline{Z}_{rd}(p) &= \overline{Z}'_{rd}(p)\frac{\overline{n}_r(p)a/2}{\tanh(\overline{n}_r(p)a/2)} \end{aligned}\right\} \quad [5.4.31]$$

by assuming:

$$\left.\begin{aligned} \overline{Z}'_{rs}(p) &\triangleq \overline{n}_r(p)\overline{z}_r(p)a = R_{\text{tot}} + (p + j\Omega_r)L_{\text{tot}} \\ \overline{Z}'_{rd}(p) &\triangleq \frac{2\overline{z}_r(p)}{\overline{n}_r(p)a} = \frac{2}{G_{\text{tot}} + (p + j\Omega_r)C_{\text{tot}}} \end{aligned}\right\} \quad [5.4.32]$$

The dependence of i_{oA}, i_{oB} on v_{oA}, v_{oB} is defined by similar equations, to which the following considerations can be extended.

With reference to Equations [5.4.20], [5.4.21], [5.4.20′], and [5.4.21′], which refer to the steady-state at frequency ω, it is also possible to write:

$$\overline{Z}_{rs}(p) = Z_s(p + j\Omega_r), \quad \overline{Z}_s = \overline{Z}_{rs}(j(\omega - \Omega_r)) = Z_s(j\omega)$$
$$\overline{Z}_{rd}(p) = Z_d(p + j\Omega_r), \quad \overline{Z}_d = \overline{Z}_{rd}(j(\omega - \Omega_r)) = Z_d(j\omega)$$

(recall Equation [5.2.8]), and similarly for $\overline{Z}'_{rs}$, $\overline{Z}'_{rd}$, $\overline{Z}'_s$, $\overline{Z}'_d$.

By applying the Laplace transformation, from Equations [5.4.30] it is possible to derive:

$$\left.\begin{aligned} \frac{\partial\overline{\imath}_{rA}}{\partial\overline{v}_{rA}} &= -\frac{\partial\overline{\imath}_{rB}}{\partial\overline{v}_{rB}} = \frac{1}{\overline{z}_r(s)\tanh(\overline{n}_r(s)a)} = \frac{1}{\overline{Z}_{rs}(s)} + \frac{1}{\overline{Z}_{rd}(s)} \\ -\frac{\partial\overline{\imath}_{rA}}{\partial\overline{v}_{rB}} &= \frac{\partial\overline{\imath}_{rB}}{\partial\overline{v}_{rA}} = \frac{1}{\overline{z}_r(s)\sinh(\overline{n}_r(s)a)} = \frac{1}{\overline{Z}_{rs}(s)} \end{aligned}\right\} \quad [5.4.33]$$

from which the *characteristic equation* is obtained:

$$0 = |\overline{z}_r(s)\sinh(\overline{n}_r(s)a)|^2 = |\overline{Z}_{rs}(s)|^2$$

or by accounting for the first of Equations [5.4.31] and recalling that $\sinh x/x = \prod_1^{\infty}{}_h(1 + (x/h\pi)^2)$:

$$0 = \left| \overline{Z}'_{rs}(s) \prod_{1}^{\infty}{}_h \left(1 + \left(\frac{\overline{n}_r(s)a}{h\pi} \right)^2 \right) \right|^2 \qquad [5.4.34]$$

from which:

$$0 = |\overline{Z}'_{rs}(s)|^2 = (R_{\text{tot}} + sL_{\text{tot}})^2 + (\Omega_r L_{\text{tot}})^2 \qquad [5.4.34']$$

$$0 = \left| 1 + \left(\frac{\overline{n}_r(s)a}{h\pi} \right)^2 \right|^2 \quad (h = 1, \ldots, \infty) \qquad [5.4.34'']$$

From Equation [5.4.34′] it is possible to derive two of the characteristic roots:

$$s = -\frac{R_{\text{tot}}}{L_{\text{tot}}} \pm \tilde{j}\Omega_r \qquad [5.4.35]$$

(in the s-plane, the imaginary unit is $\tilde{j}$), which correspond, for $R_{\text{tot}} \neq 0$, to a damped resonance. The other (infinite) characteristic roots can be derived from Equation [5.4.34″].

By assuming for simplicity that the line is nondissipative ($r = g = 0$), Equation [5.4.35] is reduced to:

$$s = \pm\tilde{j}\Omega_r$$

whereas Equation [5.4.34″] becomes:

$$0 = \left| 1 + \left(\frac{(s + j\Omega_r)\sqrt{lc}a}{h\pi} \right)^2 \right|^2$$

from which:

$$s = \pm\tilde{j}\left(\Omega_r - k\frac{\pi}{a\sqrt{lc}} \right) = \pm\tilde{j}\Omega_r \left(1 - k\frac{\lambda_r}{2a} \right)$$

where $k = \pm h$, that is $k = \pm 1, \pm 2, \ldots$, and where λ_r is the "wavelength" of the line, defined by Equation [5.4.10]. The above corresponds to *infinite* undamped *resonances*, at frequencies;

$$\nu_k = \Omega_r \left| 1 - k\frac{\lambda_r}{2a} \right|$$

with $k = 0, \pm 1, \pm 2, \ldots$, according to Figure 5.27.

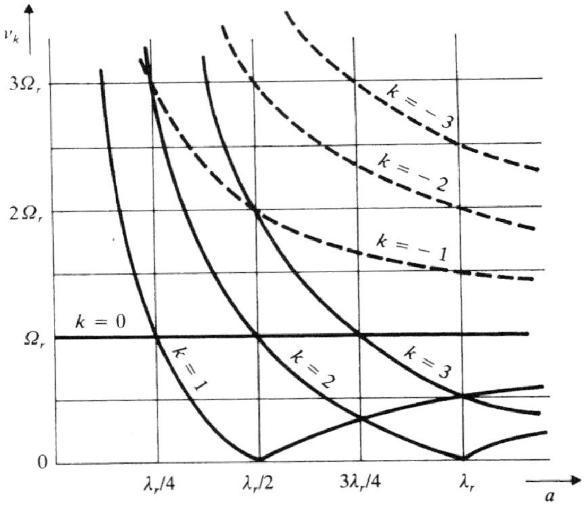

Figure 5.27. Nondissipative line: resonance frequencies as functions of line length (under the conditions specified in text).

Specifically, ordered at increasing values, the resonance frequencies are:

- if $a < \dfrac{\lambda_r}{4}$: $\quad \Omega_r, \left(\dfrac{\lambda_r}{2a} - 1\right)\Omega_r, \left(\dfrac{\lambda_r}{2a} + 1\right)\Omega_r, \ldots$
- if $a \in \left(\dfrac{\lambda_r}{4}, \dfrac{\lambda_r}{2}\right)$: $\quad \left(\dfrac{\lambda_r}{2a} - 1\right)\Omega_r, \Omega_r, \left(\dfrac{\lambda_r}{a} - 1\right)\Omega_r, \ldots$
- if $a \in \left(\dfrac{\lambda_r}{2}, \dfrac{3\lambda_r}{4}\right)$: $\quad \left(1 - \dfrac{\lambda_r}{2a}\right)\Omega_r, \left(\dfrac{\lambda_r}{a} - 1\right)\Omega_r, \Omega_r, \ldots$
- if $a \in \left(\dfrac{3\lambda_r}{4}, \lambda_r\right)$: $\quad \left(\dfrac{\lambda_r}{a} - 1\right)\Omega_r, \left(1 - \dfrac{\lambda_r}{2a}\right)\Omega_r, \left(\dfrac{3\lambda_r}{2a} - 1\right)\Omega_r, \Omega_r, \ldots$

and so on. Therefore, the smallest resonance frequency can be smaller than Ω_r (it is sufficient that $a > \lambda_r/4$) and it is very small if the line length a is close to $\lambda_r/2, \lambda_r, \ldots$ (at these values there are two zero characteristic roots). Therefore, for the same primary constants, the dynamic behavior of a line with a length $a = a_o < \lambda_r/4$ can be very different from that of a line with a length $a = a_o + \lambda_r/2$, etc.

In practice, the above-mentioned resonances are somewhat damped, because of the line resistances (see Equation [5.4.35]).

In the case of a nondissipative line, from the first of Equations [5.4.33] it is possible to derive:

$$\frac{\partial \overline{\imath}_{rA}}{\partial \overline{v}_{rA}} = -\frac{\partial \overline{\imath}_{rB}}{\partial \overline{v}_{rB}} = \frac{1}{z^{(o)} \tanh\left((s + j\Omega_r)\dfrac{a}{u}\right)} = G_r(s) + jH_r(s)$$

having set:

$$\begin{cases} G_r(s) \triangleq \dfrac{\sinh \dfrac{2sa}{u}}{2z^{(o)}\left[\left(\sinh \dfrac{sa}{u}\right)^2 + \left(\sin \dfrac{\Omega_r a}{u}\right)^2\right]} \\ H_r(s) \triangleq \dfrac{-\sin \dfrac{2\Omega_r a}{u}}{2z^{(o)}\left[\left(\sinh \dfrac{sa}{u}\right)^2 + \left(\sin \dfrac{\Omega_r a}{u}\right)^2\right]} \end{cases}$$

where $z^{(o)} = \sqrt{l/c}$, $u = 1/\sqrt{lc} = \Omega_r \lambda_r/(2\pi)$.

By decomposing vectors along the d and q axes, it follows:

$$\begin{cases} \dfrac{\partial i_{drA}}{\partial v_{drA}} = \dfrac{\partial i_{qrA}}{\partial v_{qrA}} = -\dfrac{\partial i_{drB}}{\partial v_{drB}} = -\dfrac{\partial i_{qrB}}{\partial v_{qrB}} = G_r(s) \\ -\dfrac{\partial i_{drA}}{\partial v_{qrA}} = \dfrac{\partial i_{qrA}}{\partial v_{drA}} = \dfrac{\partial i_{drB}}{\partial v_{qrB}} = -\dfrac{\partial i_{qrB}}{\partial v_{drB}} = H_r(s) \end{cases}$$

from which the meaning of $G_r(s)$ and $H_r(s)$ in terms of *transfer functions* is evident.

It is possible to reach similar conclusions starting from the second part of Equations [5.4.33], by deducing the real and imaginary parts of $1/(z^{(o)} \sinh((s + j\Omega_r)a/u))$. Furthermore, all the transfer functions under examination have, as their poles, the above-defined characteristic roots.

The difference in the dynamic behavior between lines of length $a = a_o < \lambda_r/4$ and $a = a_o + \lambda_r/2$, can be clearly shown by the frequency response characteristics. See for instance, the diagrams of:

$$\begin{cases} z^{(o)} G_r(\tilde{j}\nu)/\tilde{j} = \dfrac{\sin \dfrac{2\nu a}{u}}{2\left[-\left(\sin \dfrac{\nu a}{u}\right)^2 + \left(\sin \dfrac{\Omega_r a}{u}\right)^2\right]} \\ z^{(o)} H_r(\tilde{j}\nu) = \dfrac{-\sin \dfrac{2\Omega_r a}{u}}{2\left[-\left(\sin \dfrac{\nu a}{u}\right)^2 + \left(\sin \dfrac{\Omega_r a}{u}\right)^2\right]} \end{cases}$$

reported in Figure 5.28 for the two lines, by assuming $a_o = \lambda_r/20$ (e.g., $a_o = 300$ km if $\lambda_r = 6000$ km) and thus $\Omega_r a_o/u = \pi/10$.

The model described up to now is of an infinite dynamic order, with critical frequencies relatively high in the usual case of $a \ll \lambda_r/2$, for which the convenience of adopting simplified models appears evident. This is particularly

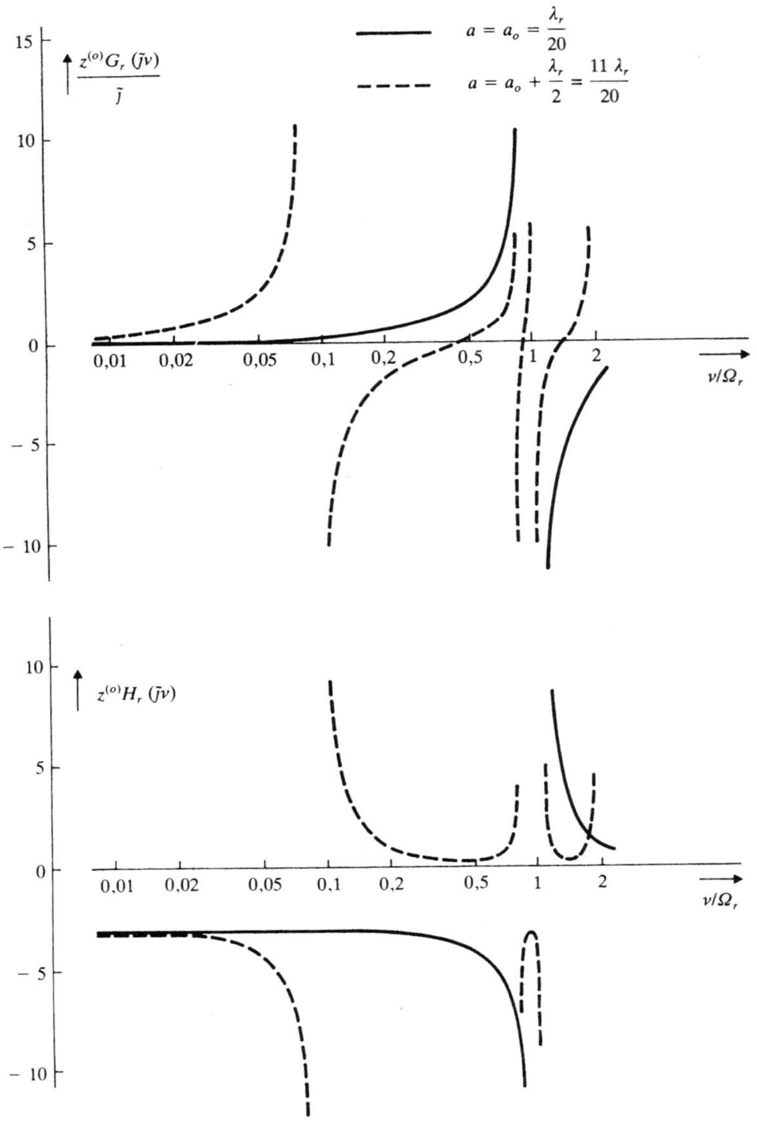

Figure 5.28. Frequency response of a line (under the conditions specified in text): comparison between two lines of different length, with a difference in length equal to half wavelength.

true for the analysis of mechanical or electromechanical phenomena in which the machine inertias, etc. can make the effects of lines' dynamics negligible. If $a \ll \lambda_r/2$, the generic line often can be represented by its *static model* ($s = 0$), i.e., by the model relative to the steady-state (see Section 5.4.2, at $\Omega_r = \omega$).

If we intend consider, in some measure, the dynamics of the lines (e.g., when addressing long lines or predominantly electrical phenomena), we may adopt different approaches that can be classified as follows:

- simplifications of the starting equations (i.e., of Equations [5.4.1] and [5.4.2]), through discretization of the independent variable x;
- simplification of the resulting equations, and specifically, of those defining the transfer functions (i.e., Equations [5.4.33]).

This latter approach allows adaptation of the simplifications to the specific problem, with direct estimation of the resulting degree of approximation. However, it generally requires that linearity holds.

The considerations developed below refer to the model for Park's vectors, with the extension to the model for the homopolar components being trivial.

To discretize the variable x, the most traditional way is to consider the line as a cascade of (one or) more *lumped parameter "cells."* In the case of a single cell, the equivalent circuit relative to Park's vectors is similar to the exact one (see Fig. 5.16), but with $\overline{Z}_{rs}(p)$, $\overline{Z}_{rd}(p)$ respectively substituted by:

$$\begin{cases} \overline{Z}'_{rs}(p) = R_{\text{tot}} + (p + j\Omega_r)L_{\text{tot}} \\ \overline{Z}'_{rd}(p) = \dfrac{2}{G_{\text{tot}} + (p + j\Omega_r)C_{\text{tot}}} \end{cases}$$

so that the series branch considers the total resistance and reactance of the line, whereas the total conductance (however negligible) and the total capacitance are shared, into equal parts, between the two shunt branches. Note that, for $p = 0$ and $\Omega_r = \omega$, the equivalent circuit in Figure 5.23b applies, whereas that in Figure 5.23a does not.

In the more general case of m cells, assuming for the sake of simplicity that each cell corresponds to a line section of a/m length, the equivalent circuit becomes the one reported in Figure 5.29. Instead of Equations [5.4.33],

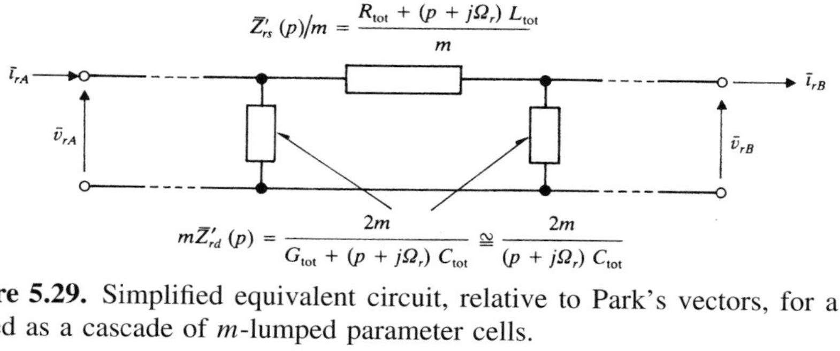

Figure 5.29. Simplified equivalent circuit, relative to Park's vectors, for a line treated as a cascade of m-lumped parameter cells.

it is possible to derive, by recalling that $\overline{Z}'_{rd}(s) = 2\overline{Z}'_{rs}(s)/(\overline{n}_r(s)a)^2$ (because of Equations [5.4.32]) and intending for brevity $\overline{n}_r = \overline{n}_r(s)$, $\overline{Z}'_{rs} = \overline{Z}'_{rs}(s)$, the following simplified expressions:

- for $m = 1$:

$$\left.\begin{aligned} \frac{\partial \overline{\imath}_{rA}}{\partial \overline{v}_{rA}} = -\frac{\partial \overline{\imath}_{rB}}{\partial \overline{v}_{rB}} &\cong \frac{1 + (\overline{n}_r a)^2/2}{\overline{Z}'_{rs}} \\ -\frac{\partial \overline{\imath}_{rA}}{\partial \overline{v}_{rB}} = \frac{\partial \overline{\imath}_{rB}}{\partial \overline{v}_{rA}} &\cong \frac{1}{\overline{Z}'_{rs}} \end{aligned}\right\} \qquad [5.4.36]$$

- for $m = 2$:

$$\left.\begin{aligned} \frac{\partial \overline{\imath}_{rA}}{\partial \overline{v}_{rA}} = -\frac{\partial \overline{\imath}_{rB}}{\partial \overline{v}_{rB}} &\cong \frac{1 + \dfrac{(\overline{n}_r a)^2}{2} + \dfrac{(\overline{n}_r a)^4}{32}}{\overline{Z}'_{rs}\left(1 + \dfrac{(\overline{n}_r a)^2}{8}\right)} \\ -\frac{\partial \overline{\imath}_{rA}}{\partial \overline{v}_{rB}} = \frac{\partial \overline{\imath}_{rB}}{\partial \overline{v}_{rA}} &\cong \frac{1}{\overline{Z}'_{rs}\left(1 + \dfrac{(\overline{n}_r a)^2}{8}\right)} \end{aligned}\right\} \qquad [5.4.37]$$

and so on.

The characteristic equation that results from Equations [5.4.36] exactly corresponds to Equation [5.4.34′] (with roots $s = -R_{\text{tot}}/L_{\text{tot}} \pm \tilde{\jmath}\Omega_r$), whereas Equation [5.4.34″] is ignored.

Instead, for $m = 2$, Equations [5.4.37] lead again to Equation [5.4.34′] and to:

$$0 = \left|1 + \frac{(\overline{n}_r(s)a)^2}{8}\right|^2$$

which can be considered an approximation of Equation [5.4.34″] for $h = 1$ (instead of π^2 there is 8). By assuming, for simplicity, that the line is nondissipative, the solutions of this last equation correspond to two resonance frequencies, respectively, equal to $\Omega_r|1 \mp \sqrt{8}\lambda_r/(2\pi a)|$.

Therefore, for $m = 2$, we obtain three resonance frequencies that correspond (one of them exactly, the other with some approximation) to the lowest ones for $a < \lambda_r/4$ (see Fig. 5.27). However, for larger values of a, for which the mentioned simplifications are more interesting, the results obtained may be unsatisfactory, and therefore the number of cells may have to be increased.

Other simplified models can be obtained, by adequately approximating Equations [5.4.33]. As a starting idea, the hyperbolic functions appearing in the operational impedances [5.4.31] or, more directly, in Equations [5.4.33], can be approximated by truncating the corresponding Taylor series. By recalling that $\sinh x/x = 1 + x^2/6 + x^4/120 + \ldots$, $\cosh x = 1 + x^2/2 + x^4/24 + \ldots$, it is possible to derive, by truncating the series at their second term, and by intending for brevity $\overline{n}_r = \overline{n}_r(s)$, $\overline{Z}'_{rs} = \overline{Z}'_{rs}(s)$:

$$\begin{cases} \dfrac{\partial \overline{\imath}_{rA}}{\partial \overline{v}_{rA}} = -\dfrac{\partial \overline{\imath}_{rB}}{\partial \overline{v}_{rB}} = \dfrac{\overline{n}_r a}{\overline{Z}'_{rs} \tanh(\overline{n}_r a)} \cong \dfrac{1 + \dfrac{(\overline{n}_r a)^2}{2}}{\overline{Z}'_{rs}\left(1 + \dfrac{(\overline{n}_r a)^2}{6}\right)} \\ -\dfrac{\partial \overline{\imath}_{rA}}{\partial \overline{v}_{rB}} = \dfrac{\partial \overline{\imath}_{rB}}{\partial \overline{v}_{rA}} = \dfrac{\overline{n}_r a}{\overline{Z}'_{rs} \sinh(\overline{n}_r a)} \cong \dfrac{1}{\overline{Z}'_{rs}\left(1 + \dfrac{(\overline{n}_r a)^2}{6}\right)} \end{cases}$$

From the corresponding characteristic equation, one can again derive Equation [5.4.34′], and furthermore:

$$0 = \left|1 + \frac{(\overline{n}_r(s)a)^2}{6}\right|^2$$

i.e., a new approximation of Equation [5.4.34″] for $h = 1$ (instead of π^2 there is 6). If the line is nondissipative, there are resonance frequencies $\Omega_r, \Omega_r|1 \mp \sqrt{6}\lambda_r/(2\pi a)|$, similarly, although at a slightly lower precision, to what can be obtained by the subdivision into two cells. The analysis can be extended, with truncation of the series done at higher-order terms.

A much more useful approach is that based on the "*modal decomposition*" of the transfer functions, in which the contributions corresponding to their "dominant" poles (the poles that can be considered the most important ones, for the problem under examination) are *exactly* retained and the contributions of the remaining poles are approximated. This approach will be specifically considered in Section 8.4.3.

The modal decomposition can be obtained by developing the functions $1/(\overline{z}_r(s)\tanh(\overline{n}_r(s)a))$, $1/(\overline{z}_r(s)\sinh(\overline{n}_r(s)a))$, which constitute the Equations [5.4.33], into a series of elementary fractions (the Mittag-Leffler development, which is similar to the usual Heaviside development related to functions with a finite number of poles).

By assuming for simplicity that:

- the line is nondissipative,
- its length a is not multiple of $\lambda_r/4$ (so that the characteristic roots are all simple ones; recall Fig. 5.27),

and thus $\overline{z}_r(s) = z^{(o)}, \overline{n}_r(s) = (s + j\Omega_r)\sqrt{lc}$, it is possible to derive the following equations (equivalent to Equations [5.4.33]):

$$\begin{cases} \dfrac{\partial \overline{\imath}_{rA}}{\partial \overline{v}_{rA}} = -\dfrac{\partial \overline{\imath}_{rB}}{\partial \overline{v}_{rB}} = \dfrac{1}{z^{(o)}} + \dfrac{1}{L_{\text{tot}}} \displaystyle\sum_{-\infty}^{+\infty}{}_k \dfrac{1}{s + j\Omega_r(1 - k\lambda_r/(2a))} \\ -\dfrac{\partial \overline{\imath}_{rA}}{\partial \overline{v}_{rB}} = \dfrac{\partial \overline{\imath}_{rB}}{\partial \overline{v}_{rA}} = \dfrac{1}{L_{\text{tot}}} \displaystyle\sum_{-\infty}^{+\infty}{}_k \dfrac{(-1)^k}{s + j\Omega_r(1 - k\lambda_r/(2a))} \end{cases}$$

(recall that $L_{\text{tot}} \triangleq la = z^{(o)}\sqrt{lca}$, $\lambda_r \triangleq 2\pi/(\Omega_r\sqrt{lc})$). The generic term in the sums corresponds to the poles $s = \pm\tilde{j}\Omega_r(1 - k\lambda_r/2a)$, defined by the equation $|s + j\Omega_r(1 - k\lambda_r/(2a))| = 0$.

The usefulness of such expressions is clear; if, for instance, the only two poles corresponding to $k = 0$ and $k = 1$ are considered as "dominant," as in the case of a line with $a \in (\lambda_r/4, \lambda_r/2)$ (see Fig. 5.27), such expressions can be respectively approximated by:

$$\begin{cases} \dfrac{1}{z^{(o)}} + \dfrac{1}{L_{\text{tot}}}\left(\dfrac{1}{s + j\Omega_r} + \dfrac{1}{s + j\Omega_r(1 - \lambda_r/(2a))}\right) + c_1 \\ \dfrac{1}{L_{\text{tot}}}\left(\dfrac{1}{s + j\Omega_r} - \dfrac{1}{s + j\Omega_r(1 - \lambda_r/(2a))}\right) + c_2 \end{cases}$$

where, to keep the behaviors at $s = 0$ unchanged, it must be assumed that:

$$\begin{cases} c_1 = \dfrac{1}{\Omega_r L_{\text{tot}}}\left[\dfrac{j(4a - \lambda_r)}{2a - \lambda_r} - \dfrac{2\pi a}{\lambda_r}\left(1 + j\dfrac{1}{\tan(2\pi a/\lambda_r)}\right)\right] \\ c_2 = \dfrac{-j}{\Omega_r L_{\text{tot}}}\left[\dfrac{\lambda_r}{2a - \lambda_r} + \dfrac{2\pi a}{\lambda_r}\dfrac{1}{\sin(2\pi a/\lambda_r)}\right] \end{cases}$$

By doing so, the simplified model leads to only two resonance frequencies, which are exactly equal to the "dominant" ones. Furthermore, it exactly considers the transient contributions corresponding to such resonances, as well as the static behavior ($s = 0$). On the contrary, the subdivision into two cells, for instance, leads to three resonance frequencies, only one of which is exact. Therefore, comparing the two approaches, we may state that the modal simplification can allow *a better approximation even by using a lower-order dynamic model*. Such advantages can become even more apparent when the line length is longer than $\lambda_r/2$.

As a concluding remark, it must be considered that the data given up to now refers to the dependence of currents $\overline{\imath}_{rA}$ and $\overline{\imath}_{rB}$ on voltages $\overline{v}_{rA}$ and $\overline{v}_{rB}$, by assuming that the time behavior of these voltages is assigned. In the case of other assignments (e.g., $\overline{v}_{rA}$ and the impedance $\overline{v}_{rB}/\overline{\imath}_{rB}$), the terms of comparison between the exact model and the possible simplified models change with respect to what is described above. All this considered, it appears convenient, for general cases, to refer to the assignments concerning the whole network under examination. To correctly simplify the whole network model, valuable information can be obtained by inspecting the steady-state behavior at different values of network frequency, according to Section 5.2.2 (recall Equations [5.2.10′] and subsequent comments).

5.5. DIRECT CURRENT (DC) LINKS

5.5.1. Generalities

The use of a dc link can be suitable in the following typical cases:

- the points to be connected are very distant, so that an ac link would imply difficulties in limiting voltage drops, in increasing transmittable active power (e.g., by inserting series capacitors, compensators at intermediate nodes, etc.) and in preventing loss of synchronism between the two sets of generators connected by the link;
- the link must be realized, at least for long sections, by cable (e.g., because of the necessity of crossing a sea or crowded metropolitan areas), and by adopting the ac solution difficulties would arise in compensating (by shunt reactors) the reactive power generated by the line;
- the networks to be connected operate at different frequencies.

In the following, we will refer to the typical "two-terminal" basic diagram in Figure 5.30, which includes:

- a dc line;
- two converters;
- between each converter and the line: a (series) "smoothing" inductor and a (shunt) dc filter;

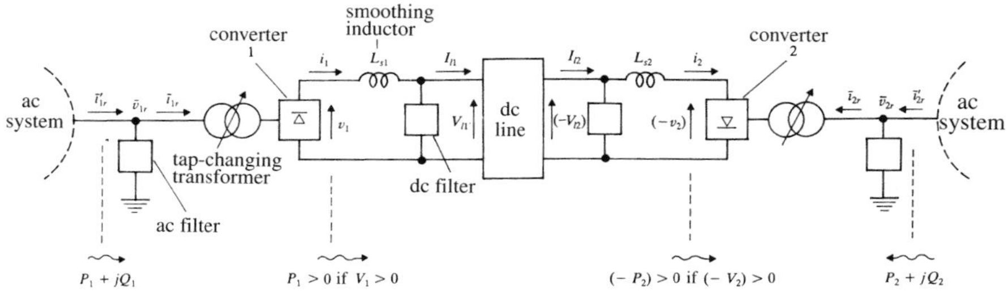

Figure 5.30. Typical diagram of a dc link.

- between each converter and the respective ac system: a tap-changing transformer and an ac filter.

The system includes protective and maneuvering devices such as surge arresters, "bypass" valves, disconnectors, etc., that will not be considered. Following a fault on the link (short-circuit in the line, malfunction of a converter, etc.) the current limitation can be left to the control systems described later on, to avoid the use of breakers.
In case of a link interconnecting networks at different frequencies, the line may even be absent (back-to-back configuration).

By using one of the converters as a *rectifier* and the other as an *inverter*, an (active) power flow from the former to the latter can be achieved. For instance, using the notation of Figure 5.30(1), if converter 1 is a rectifier and converter 2 an inverter, then $v_1 > 0$ and $v_2 < 0$, so that the power flows in the direction $1 \to 2$ (power $-P_2$ is positive like P_1 and slightly smaller because of line losses).

To reverse the direction of the power flow (it is not possible to invert that of the current), it is necessary to invert the sign of voltages by interchanging the role of converters. To obtain the desired voltage level at the dc side, each terminal may be equipped with more converters, in parallel on the ac side and in series on the dc side: however, a single (equivalent) converter will be considered at each terminal. The following considerations may be extended, although with some complication, to the case of links with more than two terminals.

The *dc line* can be represented, under the hypothesis of uniformity, by the equivalent circuit shown in Figure 5.31a (based on equations similar to those in Section 5.4), by assuming that a is the line length and:

$$n(s) \triangleq [(r + sl)(g + sc)]^{1/2}$$

$$z(s) \triangleq \frac{n(s)}{g + sc} = \frac{r + sl}{n(s)}$$

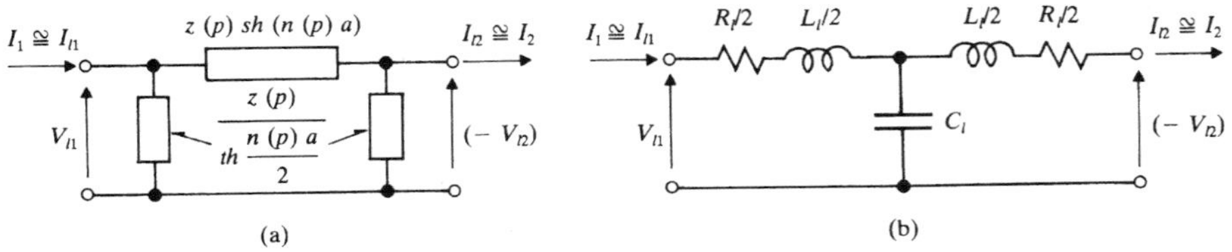

Figure 5.31. A dc line as seen by its terminals: (a) equivalent circuit; (b) simplified equivalent circuit.

(1) As the roles of the two converters can be interchanged, it is convenient to adopt the same sign convention for both of them, e.g., the generators' convention (Fig. 5.30). Similarly, it is convenient to define voltages at line terminals with the polarities indicated in Figure 5.30.

where r, l, g, c are the (series) resistance and inductance, and the (shunt) conductance and capacitance, per unit of length, respectively.

In practice, the shunt conductance can be disregarded, and the equivalent circuit may be approximated (except in the case of very fast phenomena) as a single lumped-parameter "cell" as indicated in Figure 5.31b. Specifically, the total resistance R_l and total inductance L_l of the line are shared into equal parts between the two series branches, whereas the total capacitance C_l defines the shunt branch ($R_l \triangleq ra$, $L_l \triangleq la$, $C_l \triangleq ca$).

Referring to Figure 5.30, because of the *smoothing inductors* and *dc filters*, currents and voltages at the two line terminals vary slowly, and it can be assumed that:

$$\left.\begin{aligned} I_{l1} &\cong I_1, \quad I_{l2} \cong I_2 \\ V_{l1} &\cong V_1 - sL_{s1}I_1, \quad V_{l2} \cong V_2 - sL_{s2}I_2 \end{aligned}\right\} \qquad [5.5.1]$$

where I_1, I_2 and V_1, V_2 are, respectively, the slowest components of currents i_1, i_2 and voltages v_1, v_2 at the converter outputs.

Regarding the operating characteristics of each *converter*, it is possible to refer to the typical scheme indicated in Figure 5.32, having six valves connected in the Graetz bridge configuration, and three "switching" inductances L_c (considering the presence of the transformer) on the ac side.

By assuming that:

- the voltages v_a, v_b, v_c are sinusoidal and of positive sequence at frequency ω, with a rms value V_e (the corresponding Park's vector has an amplitude $v_r = \sqrt{3}V_e$; see Appendix 2);
- the current i is constant, i.e., $i = I$ (obviously positive);
- the valves are ideal ones, controlled by "firing" at every one-sixth of a cycle ($\omega \Delta t = 60°$), in the sequence $1, 2, \ldots, 6$;

the time behaviors in Figure 5.33 can be derived for the voltages v_M, v_N and currents i_a, i_b, i_c (it is intended that voltages v_a, v_b, v_c, v_M, v_N are evaluated

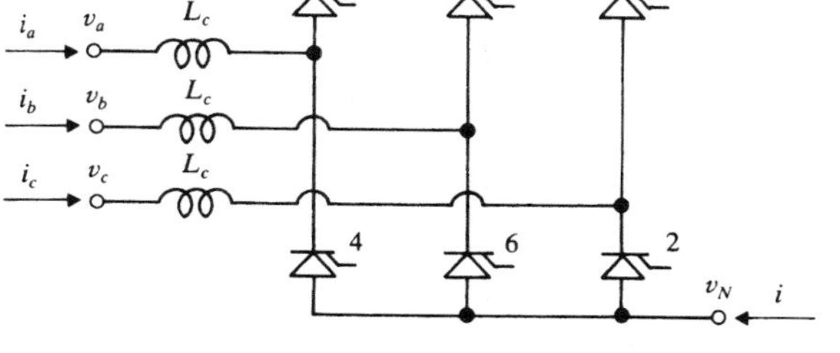

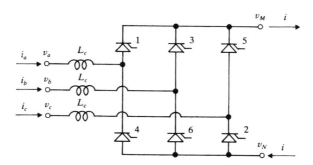

Figure 5.32. Typical scheme of a converter.

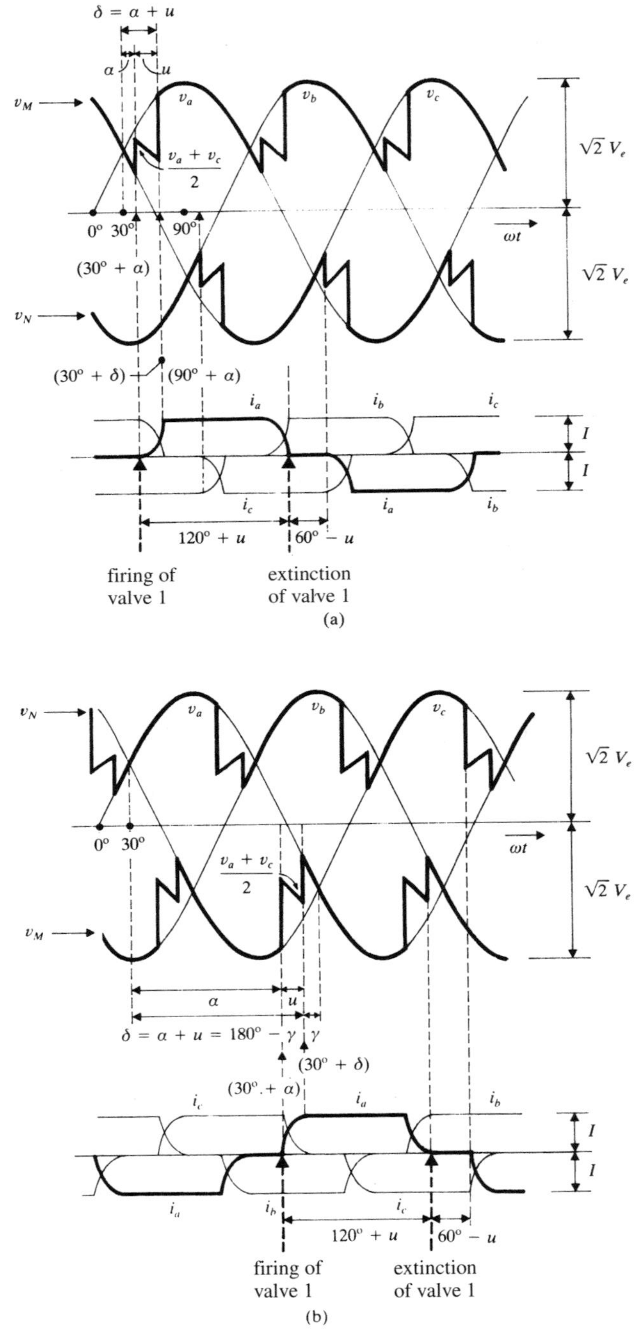

Figure 5.33. Time behaviors of voltages and currents (under the conditions specified in text): (a) operation as a rectifier; (b) operation as an inverter.

with the same reference), with:

$$0 < \alpha < \alpha + u < 180^\circ$$

where:

- the angle α, called the "*firing angle*," defines the delay at firing with respect to the natural switching (which would occur at $\omega t = 30^\circ$ for the valve 1, and so on);
- the angle u, called the "*overlapping angle*," corresponds to intervals of simultaneous conduction of two valves (it must be $u \leq 120^\circ$, and in the figure, as well as in the following, it is assumed $u < 60^\circ$; note that the value of u is usually no larger than 20°–25°).

Specifically, having posed, for brevity:

$$\delta \triangleq \alpha + u$$

it follows that:

- for $\omega t \in (30^\circ + \alpha, 30^\circ + \delta)$:

$$\begin{cases} i_b = -I \quad i_c = i - i_a \\ v_M = v_a - L_c \dfrac{\mathrm{d}i_a}{\mathrm{d}t} \\ v_M = v_c - L_c \dfrac{\mathrm{d}i_c}{\mathrm{d}t} = v_c + L_c \dfrac{\mathrm{d}i_a}{\mathrm{d}t} \\ v_N = v_b - L_c \dfrac{\mathrm{d}i_b}{\mathrm{d}t} = v_b \end{cases}$$

  from which:

$$\begin{cases} v_M = \dfrac{v_a + v_c}{2} \\ L_c \dfrac{\mathrm{d}i_a}{\mathrm{d}t} = \dfrac{v_a - v_c}{2} \\ v_N = v_b \end{cases}$$

- for $\omega t \in (30^\circ + \delta, 90^\circ + \alpha)$:

$$\begin{cases} v_M = v_a \\ v_N = v_b \end{cases}$$

and so on. Through trivial developments, by setting:

$$\left.\begin{aligned} E_o &\triangleq \frac{3\sqrt{6}}{\pi} V_e = \frac{3\sqrt{2}}{\pi} v_r \\ \lambda &\triangleq \frac{3}{\pi} \omega L_c \end{aligned}\right\} \qquad [5.5.2]$$

it is possible to derive that:

- the voltage $(v_M - v_N)$ has a mean value of:

$$V = E_o \frac{\cos\alpha + \cos\delta}{2} \qquad [5.5.3]$$

  or equivalently:

$$V = E_o \cos\left(\alpha + \frac{u}{2}\right) \cos\frac{u}{2}$$

  so it follows that:

$$\begin{cases} V > 0 \quad \text{if} \quad \alpha + \dfrac{u}{2} < 90^\circ \quad \textit{(operation as a rectifier)} \\ V < 0 \quad \text{if} \quad \alpha + \dfrac{u}{2} > 90^\circ \quad \textit{(operation as an inverter)} \end{cases}$$

- the angle δ is determined by:

$$\lambda I = E_o \frac{\cos\alpha - \cos\delta}{2} \qquad [5.5.4]$$

By eliminating δ it is possible to derive, under the above hypotheses, the converter equation:

$$V = E_o \cos\alpha - \lambda I \qquad [5.5.5]$$

which defines the mean value V of the dc voltage as a function of E_o (i.e., of the amplitude of the alternating voltage), the firing angle α, and the (dc) current I[(2)].

Moreover, the currents i_a, i_b, i_c are periodic with a zero mean value. By considering only their fundamental components (sinusoidal and of positive sequence, at frequency ω), it is possible to determine that the active and reactive powers entering into the converter from the ac side are respectively:

---

(2) Harmonic components at relatively high frequencies are superimposed to this mean value; their effects on the line voltage (and, in practice, on the current i itself) can be considered negligible, as already said, because of the presence of the smoothing inductor and the dc filter.

$$\left.\begin{aligned} P &= \frac{E_o^2}{4\lambda}(\cos^2\alpha - \cos^2\delta) = VI \\ Q &= \frac{E_o^2}{4\lambda}(u + \sin\alpha\cos\alpha - \sin\delta\cos\delta) \end{aligned}\right\} \qquad [5.5.6]$$

(where $u = \delta - \alpha$)(3).

Specifically:

- The active power $P = VI$ is a power purely "transmitted" by the converter, with no losses. This corresponds to the adopted hypotheses; note that the parameter λ, which appears in Equation [5.5.5] as an equivalent output resistance, does not actually imply any loss at all. Moreover, active power $P = VI$ has the same sign of V, so that it is positive or negative according to whether the converter operates as a rectifier or an inverter.
- The reactive power Q is positive with no counterpart from the dc side, so that it is, in any case, a power "absorbed" by the converter.

The previous equations are related to the adopted hypotheses. However, they can be accepted for a dynamic operating condition, provided it is slow for the frequency (6ω) at which valve firing meanly occurs. For a better approximation, it is possible in Equation [5.5.5] to account for the voltage drops in the valves and in the transformer resistances, and for the further drop, proportional to $\mathrm{d}I/\mathrm{d}t$, on the switching inductance during transients.

Also, by recalling Figure 5.31b and Equations [5.5.1], the whole link can be represented, using the stated approximations and not considering the control systems, by the equivalent circuit shown in Figure 5.34.

5.5.2. Operating Point

If the converter firing angles α_1 and α_2 were kept constant, the operating point of the link would be defined, based on the above, by:

$$E_{o1}\cos\alpha_1 + E_{o2}\cos\alpha_2 = (\lambda_1 + R_l + \lambda_2)I$$

(3) As a first approximation, the rms value I'_e of the fundamental components of i_a, i_b, i_c is $I'_e \cong (\sqrt{6}/\pi)I$; this corresponds to:

- an apparent power $3V_e I'_e \cong E_o I$;
- a power factor (with a lagging current) $P/(3V_e I'_e) \cong V/E_o$.

Currents include harmonic components at relatively high frequencies. However, the effects of such harmonics on the ac network (and, in practice, on the voltages v_a, v_b, v_c themselves) are modest if the considered node has a large short-circuit power, and furthermore are reduced by the ac filter (including possible shunt condensers used for reactive compensation).

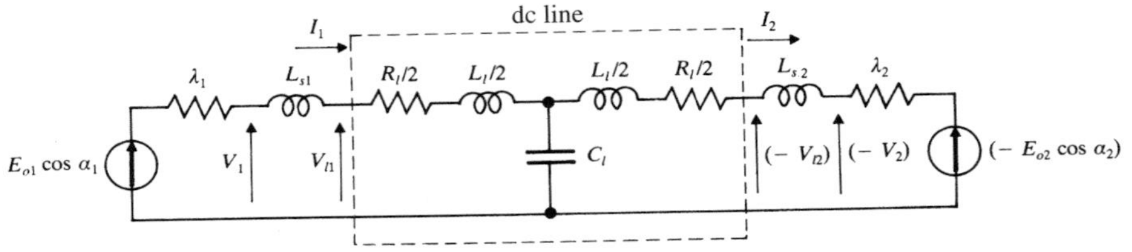

Figure 5.34. Equivalent circuit of the whole link in the absence of control.

where $I_1 = I_2 \triangleq I > 0$, and the voltage in the middle of the line (Fig. 5.34) would be:

$$E_{o1}\cos\alpha_1 - \left(\lambda_1 + \frac{R_l}{2}\right) I = -E_{o2}\cos\alpha_2 + \left(\lambda_2 + \frac{R_l}{2}\right) I$$

However, such a situation would not be acceptable in practice, as it would easily lead to large variations in current for relatively small variations in E_{o1}, E_{o2} (caused by perturbations in the ac systems).

For nominal conditions of the midline voltage and current, each of the relative voltage drops caused by λ_1, R_l, λ_2 can be, for instance, 5–10%, so that the relative current variations are approximately 3–7 times larger than those of $E_{o1}\cos\alpha_1$, $E_{o2}\cos\alpha_2$. These last variations can be limited by regulating the ac voltages, particularly by the tap-changing transformers (Fig. 5.30). However, such effects are usually negligible in short times because of the relative slowness of the considered regulations. (By varying the transformation ratio, the corresponding value of L_c and that of λ are modified. However, in the following, this fact will be disregarded.)

The most traditional solution is to achieve, for each converter (and for given E_o), a static characteristic (V, I) as shown in Figure 5.35a, in which:

- for intermediate values of voltage: $I = \text{constant} = I_{\text{rif}}$ (and thus α automatically adapted, for varying V, so *to regulate the current* to its reference value I_{rif});
- for the upper and lower voltage values:
  $\alpha = \text{constant} = \alpha_{\min}$, when operating as a rectifier $(V > 0)$;
  $\gamma = \text{constant} = \gamma_{\text{rif}} \geq \gamma_{\min}$, when operating as an inverter $(V < 0)$;

  where $\gamma \triangleq 180° - \delta = 180° - (\alpha + u)$ is the so-called *"margin angle" of the inverter* (Fig. 5.33b).

For a good operation of the converter, it is convenient that:

- $\alpha \geq \alpha_{\min} > 0$ when operating as rectifier (with $\alpha_{\min}$, e.g., 5°);

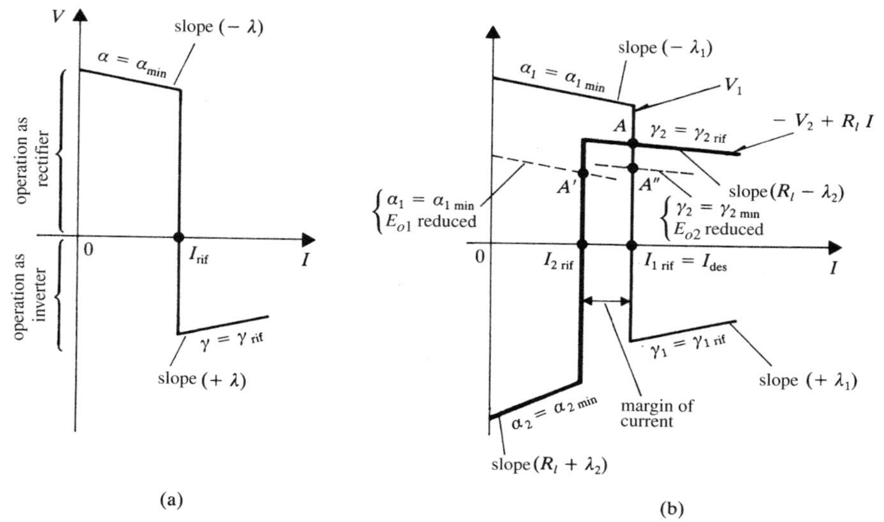

Figure 5.35. Steady-state behavior: (a) static characteristic for each converter (typical example, in the presence of control); (b) determination of the operating point (converter 1: rectifier; converter 2: inverter).

- $\gamma \geq \gamma_{\min} > 0$ when operating as inverter (with $\gamma_{\min}$, e.g., $10°-15°$; correspondingly, the reference γ_{rif} can then be, e.g., $15°-20°$).

In Figure 5.35a:

- the piece at α constant has a slope $\partial V/\partial I = -\lambda < 0$ as a result of Equation [5.5.5];
- the piece at γ constant instead has a positive slope equal to $+\lambda$; in fact, from Equations [5.5.3] and [5.5.4], by eliminating $\cos\alpha$ and recalling that $\delta = \alpha + u$, $\cos\delta = -\cos\gamma$, it follows that:

$$V = -E_o \cos\gamma + \lambda I$$

Since the operating point is defined by:

$$V_1 = -V_2 + R_l I$$

(Fig. 5.34), it becomes possible by a proper choice of the set points, to achieve the operation at the desired values of current (I_{des}) and voltages, by keeping $I = I_{\text{des}}$ even following (moderate) variations of E_{o1}, E_{o2}. Refer to Figure 5.35b, where the operating point (A) is defined by the intersection of the two characteristics (in the example, converter 1 operates as rectifier).

More precisely:

- The value of current is equal to the larger of $I_{1\text{rif}}$ and $I_{2\text{rif}}$, which must be set at the value I_{des}. The corresponding converter (converter 1 in the figure) operates as the rectifier, thus transmitting power toward the line. Therefore, representing variables related to the rectifier and the inverter by the subscripts R, I, respectively, it holds that:

$$I = I_{R\,\text{rif}} = I_{\text{des}} > I_{I\,\text{rif}}$$

The positive difference $(I_{R\,\text{rif}} - I_{I\,\text{rif}})$ is called the "*margin of current*" and it is assumed to be, e.g., 15% of the nominal current.

- For the given I, the voltage V_1 (and thus V_2, etc.) depends on $E_{oI} \cos \gamma_I$, and it can be set to the desired value, at a reasonable E_{oI}, through a proper choice of $\gamma_{I\,\text{rif}}$.

In other words:

- the current I is determined by the current reference of the rectifier;
- the values of voltages are then determined by the margin angle reference of the inverter;

and, for the example in Figure 5.35b, it is possible to relate to the operating point A, the equivalent circuit in Figure 5.36a (note that the equivalent resistance $(-\lambda_2)$ of the inverter is negative).

To interchange the roles (rectifier and inverter) of the two converters, i.e., to reverse the direction of the power, it is sufficient to invert the sign of the difference $(I_{1\,\text{rif}} - I_{2\,\text{rif}})$ between the current references, by transferring the contribution corresponding to the "margin of current" from one set-point to the other. In practice, this requires some caution, as the line capacitances must be discharged and recharged at opposite polarity.

In normal operating conditions, the variations of E_{oR} have, from one steady-state to another, the effect of modifying α_R (to keep the desired current value) within a relatively narrow range, e.g., 10°–20°. On the contrary, for $\gamma_I = \gamma_{I\,\text{rif}}$, the variations of E_{oI} directly influence the voltage steady-state of the whole link.

The control of the tap-changing transformers can neutralize such effects by returning E_{oR} and E_{oI} to their original values. However, this can be done only within given limits and

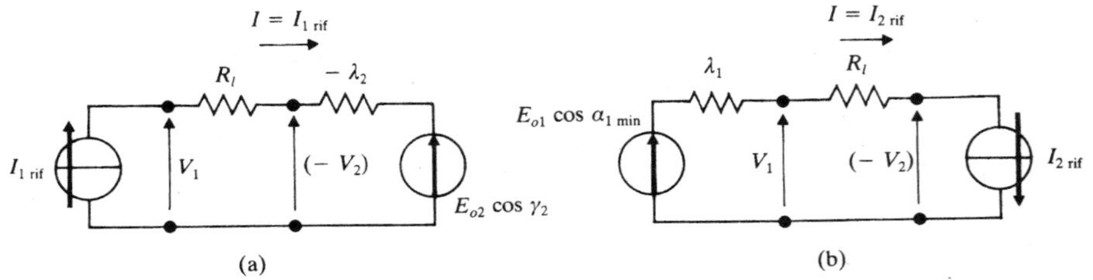

Figure 5.36. Equivalent circuit at steady-state: (a) around the point $A(\gamma_2 = \gamma_{2\text{rif}})$ or $A''(\gamma_2 = \gamma_{2\,\text{min}})$ (Fig. 5.35b); (b) around the point $A'(\alpha_1 = \alpha_{1\,\text{min}})$ (Fig. 5.35b).

in relatively long times. To improve the support of the steady-state voltages, it may be better to control the inverter by regulating the voltage instead of the "margin angle," according to Section 5.5.3 and with the obvious condition $\gamma_I \geq \gamma_{I\,\min}$. It is possible that the voltage control may be activated only when the absolute value of the voltage error becomes larger than a predefined value.

Generally, a short-circuit on one of the two ac networks can cause significant decrements in E_{oR} or in E_{oI}.

In the former case, the operating point can be changed (in the example of Fig. 5.35b) from A to A', where the rectifier is kept at the minimum firing angle, whereas the role of regulating current (to the value $I_{I\,\text{rif}} < I_{\text{des}}$) is taken by the inverter (see Fig. 5.36b).

In the latter case, a significant voltage drop can occur, which may be limited (but only for $\gamma_I \geq \gamma_{I\,\min}$) by the inverter control, according to above. For example, refer to the point A'' (Fig. 5.35b), and to the equivalent circuit in Figure 5.36a, with $\gamma_{2\,\text{rif}}$ replaced by $\gamma_{2\,\min}$.

As an alternative to the above solution, the rectifier may be used *to regulate (instead of the current) the transmitted power* at a proper point of the link, such as the power P_m at the middle of the line. Assuming, as per Figure 5.35b, that converter 1 acts as rectifier (and thus that $P_m = V_{lm} I$, with $V_{lm} = V_1 - (R_l/2)I$), the piece of its characteristic $I = I_{1\,\text{rif}} = I_{\text{des}}$ must be substituted by a piece defined by:

$$I = I_{1\,\text{rif}} = \frac{P_{m\,\text{des}}}{V_{lm}} = \frac{P_{m\,\text{des}}}{V_1 - \dfrac{R_l}{2} I}$$

or equivalently:

$$V_1 = \frac{P_{m\,\text{des}}}{I} + \frac{R_l}{2} I$$

with the condition that $I \in [I_{\min}, I_{\max}]$ (where $I_{\min}$ and $I_{\max}$ have proper values, possibly variable with voltage). Similar considerations hold for the inverter, considering the margin of current (see Fig. 5.37). The power regulation may

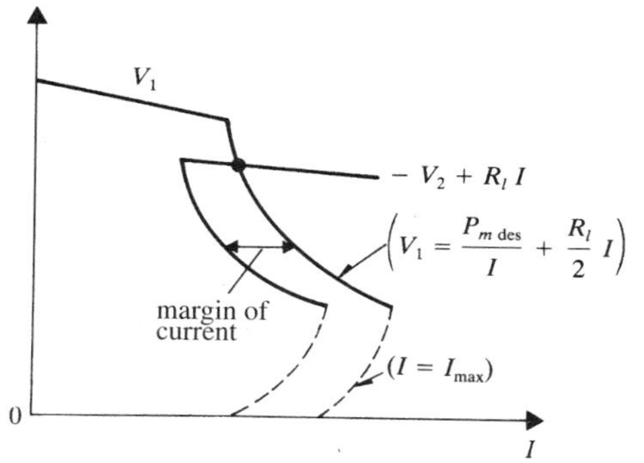

Figure 5.37. Static characteristics and operating point under power regulation (converter 1: rectifier; converter 2: inverter).

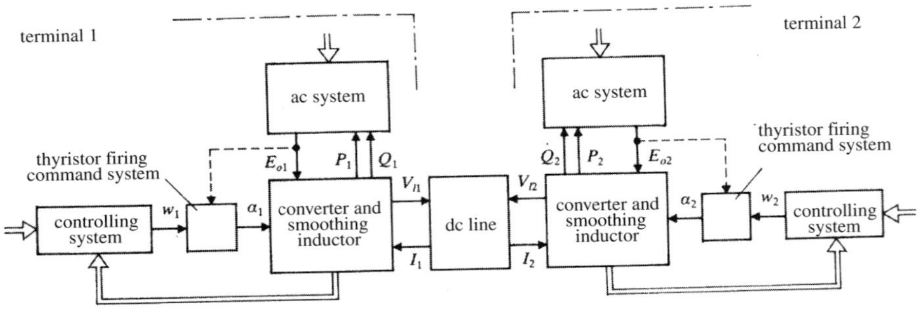

Figure 5.38. Dynamic behavior: overall block diagram.

be used in the context of the primary or secondary f/P control, according to Section 3.3.1.

5.5.3. Dynamic Behavior

The analysis of the system for any operating condition can be based on Figure 5.38 (for simplicity, frequency variations in the ac systems are disregarded).

For each block "converter and smoothing inductor," from Equations [5.5.1] and [5.5.5] it is evident that, for given E_o and I, the voltage V_l depends in a linear way not on α, but on $\cos\alpha$. To linearize the dependence of V_l on the output w of the controlling system, it is convenient that the "*thyristor firing command system*" be defined by an equation $\alpha = \text{arcos}(w/B)$ or equivalently $w = B\cos\alpha$, where B is a proper constant.

In principle, such a law can be achieved by comparing the variable w with a sinusoidal signal with amplitude B, frequency ω, and (for each valve) a proper phase. For instance, with the notation of Figure 5.33, it may be imposed that valve 1 is fired at the instant t_1 where the signal $B\cos(\omega t - 30°)$ reaches, during decreasing, the value w (it then holds $B\cos\alpha = B\cos(\omega t_1 - 30°) = w$), and so on. In the more traditional "analog" solutions, for which the reference signals are derived from the ac network, the amplitude B is proportional to E_o, and this must be considered, as specified below.

Actually, the subsequent valve firing, and thus the updating of α, does not occur continuously, but rather on the average of one-sixth of a cycle, i.e., after a time equal to $\tau = \pi/(3\omega)$ ($\tau = (1/300)$ sec at 50 Hz). For phenomena that are not varying quickly, this can be considered by approximating the actual (discontinuous) behavior of α with that of $\text{arcos}(w/B)$, delayed of the quantity $\tau/2$. It then follows that:

$$\alpha(t) = \arccos \frac{w\left(t - \dfrac{\tau}{2}\right)}{B}$$

and thus, in terms of transfer functions for small variations:

$$\Delta(\cos\alpha) \cong \epsilon^{-s\frac{\tau}{2}} \left(\frac{\Delta w}{B^o} - \sigma \frac{\Delta E_o}{E_o^o} \right) \qquad [5.5.7]$$

where $\sigma = 0$ if B is constant, and $\sigma = \cos\alpha^o$ if B is proportional to E_o (steady-state values are indicated by the superscript "o").

By linearizing Equations [5.5.1] and [5.5.5] it can be derived, for the generic block "*converter and smoothing inductor*":

$$\Delta V_l \cong \cos\alpha^o \Delta E_o + E_o^o \Delta(\cos\alpha) - (\lambda + sL_s)\Delta I$$

from which, eliminating $\Delta(\cos\alpha)$ by Equation [5.5.7] and omitting the approximation sign:

$$\Delta V_l = \frac{E_o^o}{B^o} \epsilon^{-\frac{s\tau}{2}} \Delta w + (\cos\alpha^o - \sigma\epsilon^{-\frac{s\tau}{2}})\Delta E_o - (\lambda + sL_s)\Delta I \qquad [5.5.8]$$

For the outputs P, Q, from Equations [5.5.6] it can be similarly derived:

$$\begin{cases} \Delta P = \left(\dfrac{2P}{E_o} - I\cos\delta \right)^o \Delta E_o + \left[\dfrac{E_o^2}{2\lambda}(\cos\alpha - \cos\delta) \right]^o \Delta(\cos\alpha) \\ \qquad + (E_o \cos\delta)^o \Delta I \\ \Delta Q = \left(\dfrac{2Q}{E_o} - I\sin\delta \right)^o \Delta E_o + \left[\dfrac{E_o^2}{2\lambda}(\sin\alpha - \sin\delta) \right]^o \Delta(\cos\alpha) \\ \qquad + (E_o \sin\delta)^o \Delta I \end{cases}$$

(recall that the angle δ depends on E_o, α, I, and more precisely it is, as a result of Equation [5.5.4], $\cos\delta = \cos\alpha - 2\lambda I/E_o$), from which ΔP, ΔQ can be derived as functions of Δw, ΔE_o, ΔI, by eliminating $\Delta(\cos\alpha)$ using Equation [5.5.7].

Furthermore, among the outputs of the block it is necessary to include the margin angle γ, as it may act (as well as voltage V_l and current I) on the converter control feedbacks. Since $\gamma \triangleq 180° - \delta$, it is possible to derive from Equation [5.5.4]:

$$\cos\gamma = -\cos\delta = -\cos\alpha + \frac{2\lambda I}{E_o}$$

$$-\sin\gamma^o \Delta\gamma = -\Delta(\cos\alpha) + \frac{2\lambda}{E_o^o}\left(\Delta I - I^o \frac{\Delta E_o}{E_o^o} \right)$$

and by eliminating $\Delta(\cos\alpha)$ (see Equation [5.5.7]):

$$\Delta\gamma = \frac{1}{\sin\gamma^o} \left[\epsilon^{-\frac{s\tau}{2}} \frac{\Delta w}{B^o} + \left(\frac{2\lambda I^o}{E_o^o} - \sigma\epsilon^{-\frac{s\tau}{2}} \right) \frac{\Delta E_o}{E_o^o} - \frac{2\lambda}{E_o^o}\Delta I \right] \qquad [5.5.9]$$

For each of the two blocks called "*ac system*," it may be assumed that $\overline{v}_r \overline{\imath}_r'^{*} = P + jQ$, where the Park's vectors $\overline{\imath}_r'$ and $\overline{v}_r$ are defined as per Figure 5.30, thus:

$$(P + jQ)^o \frac{\Delta \overline{v}_r}{\overline{v}_r^o} + \overline{v}_r^o \Delta \overline{\imath}_r'^{*} = \Delta P + j \Delta Q$$

and as a result of the first part of Equations [5.5.2] (aside from considering the possible control of the tap-changing transformer):

$$\frac{\Delta v_r}{v_r^o} = \frac{\Delta E_o}{E_o^o}$$

By assuming, for small variations, the following law:

$$\Delta \overline{\imath}_r' = -(G_r(s) + jB_r(s)) \Delta \overline{v}_r + \ldots$$

where the dotted term corresponds to perturbations on the ac system, it is possible to derive:

$$\Delta E_o = -K_{EP}(s) \Delta P - K_{EQ}(s) \Delta Q + \ldots \qquad [5.5.10]$$

where:

$$\begin{cases} K_{EP}(s) \triangleq \dfrac{E_o^o (P + G_r(s) v_r^2)^o}{(G_r^2(s) + B_r^2(s)) v_r^{o4} - (P^2 + Q^2)^o} \\ K_{EQ}(s) \triangleq \dfrac{E_o^o (Q - B_r(s) v_r^2)^o}{(G_r^2(s) + B_r^2(s)) v_r^{o4} - (P^2 + Q^2)^o} \end{cases}$$

For instance, in the simple case for which the ac system output impedance is purely inductive, by recalling Section 5.2.2 (with $\Omega_r = \omega$), it is possible to write:

$$\Delta \overline{\imath}_r' = -\frac{1}{(s + j\omega)L} \Delta \overline{v}_r + \ldots$$

from which $G_r(s) = s/((s^2 + \omega^2)L)$, $B_r(s) = -\omega/((s^2 + \omega^2)L)$. Therefore, in this case, each of the functions $K_{EP}(s)$ and $K_{EQ}(s)$ has two zeros and two poles.

Alternatively, if there is interest in analyzing the effects of the dc link on the ac system, it is necessary to express, based on the link equations, ΔP and ΔQ as functions of ΔE_o and the inputs that act on the link itself (set points of the converter controls, possible disturbances, etc.).

Finally, by considering Equation [5.5.10] and the expressions relating ΔP and ΔQ to Δw, ΔE_o, ΔI, it is possible to obtain ΔE_o in the form of:

$$\Delta E_o = -K_{Ew}(s) \Delta w - K_{Ei}(s) \Delta I + \ldots \qquad [5.5.11]$$

where, by omitting the indication of the argument s:

$$\begin{cases} K_{Ew} \triangleq \left(\dfrac{E_o^2}{2\lambda B}\right)^o [K_{EP}(\cos\alpha - \cos\delta)^o + K_{EQ}(\sin\alpha - \sin\delta)^o]\dfrac{\epsilon^{-s\tau/2}}{D(s)} \\ K_{Ei} \triangleq E_o^o(K_{EP}\cos\delta^o + K_{EQ}\sin\delta^o)\dfrac{1}{D(s)} \end{cases}$$

having assumed for brevity that:

$$D(s) \triangleq 1 + K_{EP}\left[\frac{2P}{E_o} - I\cos\delta - \frac{E_o}{2\lambda}(\cos\alpha - \cos\delta)\sigma\epsilon^{-\frac{s\tau}{2}}\right]^o$$
$$+ K_{EQ}\left[\frac{2Q}{E_o} - I\sin\delta - \frac{E_o}{2\lambda}(\sin\alpha - \sin\delta)\sigma\epsilon^{-\frac{s\tau}{2}}\right]^o$$

and intending that the dotted term corresponds to perturbations in the ac system.

By eliminating ΔE_o in Equations [5.5.8] and [5.5.9] it can be derived, for the set of three blocks considered up to now (i.e., for the generic *terminal without control*):

$$\left.\begin{aligned} \Delta V_l &= K(s)\Delta w - Z_n(s)\Delta I + \Delta N \\ \Delta\gamma &= K_\gamma(s)\Delta w - H_\gamma(s)\Delta I + \Delta N_\gamma \end{aligned}\right\} \quad [5.5.12]$$

with:

$$\left.\begin{aligned} K &\triangleq \left(\frac{E_o}{B}\right)^o \epsilon^{-\frac{s\tau}{2}} - K_{Ew}\left(\cos\alpha^o - \sigma\epsilon^{-\frac{s\tau}{2}}\right) \\ Z_n &\triangleq \lambda + sL_s + K_{Ei}\left(\cos\alpha^o - \sigma\epsilon^{-\frac{s\tau}{2}}\right) \\ K_\gamma &\triangleq \frac{1}{E_o^o\sin\gamma^o}\left[\left(\frac{E_o}{B}\right)^o \epsilon^{-\frac{s\tau}{2}} - K_{Ew}\left(\frac{2\lambda I^o}{E_o^o} - \sigma\epsilon^{-\frac{s\tau}{2}}\right)\right] \\ H_\gamma &\triangleq \frac{1}{E_o^o\sin\gamma^o}\left[2\lambda + K_{Ei}\left(\frac{2\lambda I^o}{E_o^o} - \sigma\epsilon^{-\frac{s\tau}{2}}\right)\right] \end{aligned}\right\} \quad [5.5.13]$$

whereas ΔN, ΔN_γ act as disturbances caused by perturbations in the ac system (see Fig. 5.39).

Specifically, note that $Z_n(s)$ is the equivalent "natural" (i.e., in the absence of control) impedance $(-\partial V_l/\partial I)$. In its expression there are three terms, the first two of which (λ and sL_s), respectively, correspond to the switching and smoothing inductances, whereas the last accounts for the effects of the ac system.

The complexities caused by the output impedance on the ac side are evident. In fact, if this impedance were zero, then $G_r = B_r = \infty$, $K_{EP} = K_{EQ} = 0$, $K_{Ew} =$

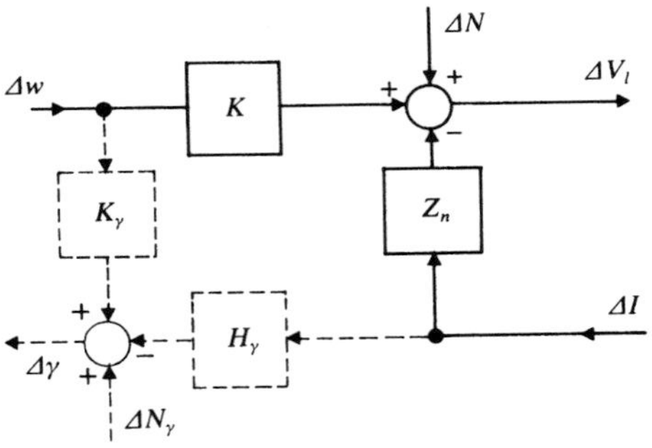

Figure 5.39. Block diagram of each terminal for small variations in the absence of control.

$K_{ei} = 0$ and thus, in a much simpler way:

$$\left.\begin{aligned} K &= \left(\frac{E_o}{B}\right)^o \epsilon^{-\frac{s\tau}{2}} \\ Z_n &= \lambda + sL_s \\ K_\gamma &= \frac{1}{(B\sin\gamma)^o}\epsilon^{-\frac{s\tau}{2}} \\ H_\gamma &= \frac{2\lambda}{(E_o \sin\gamma)^o} \end{aligned}\right\} \qquad [5.5.13']$$

For sufficiently slow phenomena, Equations [5.5.13] can be simplified by disregarding the pure delay $\tau/2$ (this holds also for Equation [5.5.13′]) and substituting $K_{Ew}(s)$, $K_{Ei}(s)$ by their respective static gains $K_{Ew}(0)$, $K_{Ei}(0)$.

For each terminal of the link, it is necessary to account for the *control actions*. By recalling Section 5.5.2, the situations of practical interest are the following;

(1) Rectifier:

- Operation at $\alpha_R = \text{constant} = \alpha_{R\min}$: see Figure 5.40a.
- Operation under current (or power) regulation: see Figure 5.40b, where $G_{R(i)} = G_{R(i)}(s)$ is the transfer function of the current regulator. (Under power regulation $I_{R\,\text{rif}} = P_{m\,\text{des}}/\hat{V}_{lm}$, where $\hat{V}_{lm} \triangleq V_{lR} - (R_l/2)I_R$ is the estimated value of the voltage at the midpoint of the line. It follows, by assuming $P_{R\,\text{rif}} = P_{m\,\text{des}}$, that $\Delta I_{R\,\text{rif}} = \Delta P_{R\,\text{rif}}/\hat{V}^o_{lm} - (P_{R\,\text{rif}}/\hat{V}^2_{lm})^o \Delta\hat{V}_{lm}$, where $\Delta\hat{V}_{lm} = \Delta V_{lR} - (R_l/2)\Delta I_R$.)

(2) Inverter:

- Operation under margin angle or voltage regulation: see Figs. 5.41a and 5.41b, where $G_{I(\gamma)} = G_{I(\gamma)}(s)$ and $G_{I(v)} = G_{I(v)}(s)$ are the regulator

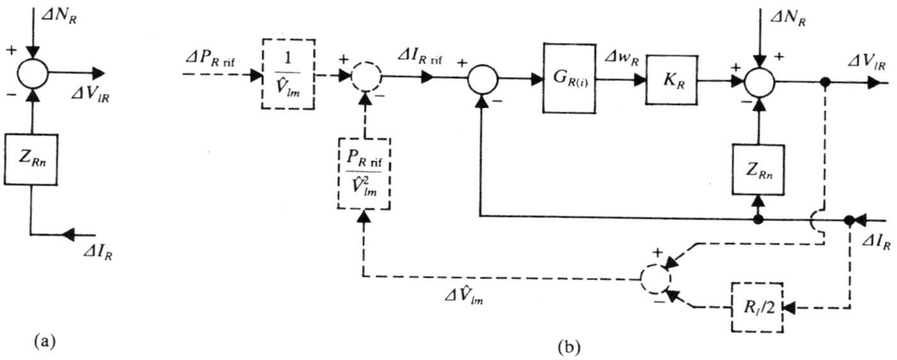

Figure 5.40. Block diagram of the terminal when operating as a rectifier, for small variations: (a) in the absence of control (α_R = constant = $\alpha_{R\,\min}$); (b) under current regulation (without dotted blocks), or under power control.

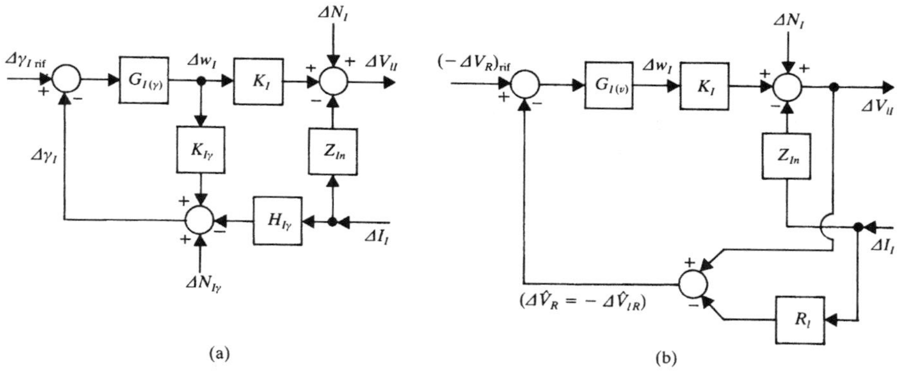

Figure 5.41. Block diagram of the terminal when operating as an inverter, for small variations: (a) under margin angle regulation; (b) under voltage regulation.

transfer functions for the two cases respectively. It is intended that in the latter case the regulated voltage is, apart from its sign, that on the rectifier side, i.e., the estimated voltage $-\hat{V}_{lR} = V_{lI} - R_l I_I$.

- (Occasional) operation under current regulation: the diagram is similar to that of the rectifier (it is sufficient to substitute the index R by I), with $I_{I\,\mathrm{rif}}$ smaller than $I_{R\ \mathrm{rif}}$ of a quantity equal to the margin of current. If the rectifier is under power regulation, its reference $I_{R\,\mathrm{rif}}$ depends not only on $P_{m\,\mathrm{des}}$ but also on the voltage. Consequently, the same happens for $I_{I\,\mathrm{rif}}$. It is worth avoiding, even during transient operation, that the margin of current drops below the desired value.

Under margin angle regulation, the measurements of γ_I for the subsequent valves are available, on average, at every one-sixth of a cycle (corresponding to time τ). To avoid unacceptable reductions of γ_I caused by abrupt current increments, it is convenient to add, for $\mathrm{d}I/\mathrm{d}t > 0$, a term proportional to $\mathrm{d}I/\mathrm{d}t$ to the measure of γ_I. If variations of E_o can be considered negligible, the margin angle may be regulated in an open-loop mode. By recalling that $\cos\alpha = 2\lambda I/E_o - \cos\gamma$, it is necessary to modify w_I (which can be considered proportional to $\cos\alpha_I$, with a delay $\tau/2$) for varying current, to maintain γ_I to the desired value. Moreover, similarly to above, a term proportional to $\mathrm{d}I/\mathrm{d}t$ can be added (when $\mathrm{d}I/\mathrm{d}t > 0$) to the measured value of I.

The *synthesis of regulators*, for different situations, must be performed accounting for the equations of the dc line. Therefore, the previous block diagrams must be integrated with the one that refers to the line. Alternatively, the equations of the two terminals can be translated into proper equivalent circuits, connected to each other through the equivalent circuit of the line, according to Figure 5.42, for which:

- Z_{Rn}, Z_{In} are the equivalent natural impedances related, respectively, to the rectifier and the inverter;
- $\Delta E_{I(\gamma)}$, $Z_{I(\gamma)}$ and $\Delta E_{I(v)}$, $Z_{I(v)}$ are the emfs and the equivalent impedances (for small variations) of the inverter, respectively, under margin angle regulation and voltage regulation. They are expressed as in the legend of Figure 5.42.

If the (relatively small) delays of the margin angle regulation are disregarded, it can be derived more simply, as if $G_{I(\gamma)} = \infty$:

$$\left.\begin{aligned} \Delta E_{I(\gamma)} &= \frac{K_I}{K_{I\gamma}}(\Delta\gamma_{I\,\mathrm{rif}} - \Delta N_{I\gamma}) + \Delta N_I \\ Z_{I(\gamma)} &= Z_{In} - \frac{K_I H_{I\gamma}}{K_{I\gamma}} \end{aligned}\right\} \qquad [5.5.14]$$

Analogously, if delays in voltage regulation are disregarded, it may be assumed that $G_{I(v)} = \infty$ and thus $\Delta E_{I(v)} = -\Delta V_{R\,\mathrm{rif}}$, $Z_{I(v)} = -R_l$.

Moreover, Equations [5.5.14] can be further simplified if the output impedance of the ac system is negligible. In such a case, recalling Equations [5.5.13′] it follows:

$$\left.\begin{aligned} \Delta E_{I(\gamma)} &= (E_{oI}\sin\gamma_I)^o(\Delta\gamma_{I\,\mathrm{rif}} - \Delta N_{I\gamma}) + \Delta N_I \\ Z_{I(\gamma)} &= Z_{In} - 2\lambda_I = -\lambda_I + sL_{sI} \end{aligned}\right\} \qquad [5.5.14']$$

(note the negative term $-\lambda_I$, which is the equivalent resistance of the inverter under the operation at E_o, γ constant; see Fig. 5.36a).

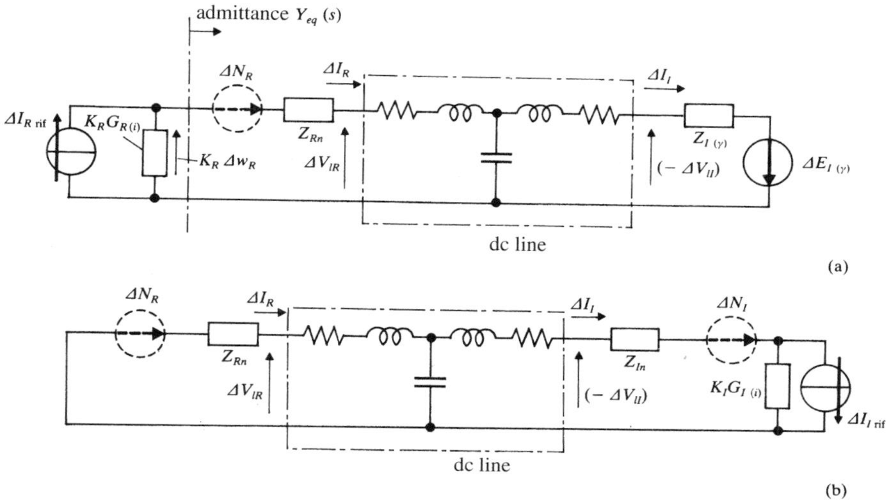

Figure 5.42. Equivalent circuit of the whole link for small variations with:

(a) rectifier under current regulation and inverter under margin angle regulation (see point A or A'' in Fig. 5.35b), with:

$$\Delta E_{I(\gamma)} = \frac{K_I G_{I(\gamma)}}{1 + K_{I\gamma} G_{I(\gamma)}} (\Delta \gamma_{I\,\text{rif}} - \Delta N_{I\gamma}) + \Delta N_I,$$

$$Z_{I(\gamma)} = Z_{In} - \frac{K_I G_{I(\gamma)} H_{I\gamma}}{1 + K_{I\gamma} G_{I(\gamma)}}$$

or rectifier under current regulation, and inverter under voltage regulation, provided $\Delta E_{I(\gamma)}$, $Z_{I(\gamma)}$ are respectively replaced by:

$$\Delta E_{I(v)} = \frac{-K_I G_{I(v)} \Delta V_{R\,\text{rif}} + \Delta N_I}{1 + K_I G_{I(v)}}, \qquad Z_{I(v)} = \frac{Z_{In} - K_I G_{I(v)} R_l}{1 + K_I G_{I(v)}}$$

(b) rectifier at minimum firing angle, and inverter under current regulation (see point A' in Fig. 5.35b).

Specifically, if:

- the rectifier is under *current regulation*,
- the inverter is under *margin angle regulation*

(see Fig. 5.42a and points A or A'' in Fig. 5.35b), the total transfer function of the current regulation loop is given by $G_{R(i)}\, K_R Y_{\text{eq}}$, where $Y_{\text{eq}} = Y_{\text{eq}}(s) \triangleq \partial I_R / \partial (K_R w_R)$ is the equivalent admittance reported in Figure 5.42a.

By accepting Equations [5.5.13′] and [5.5.14′], and representing the line by the circuit in Figure 5.31b, the function $Y_{\text{eq}}(s)$ may be written:

$$Y_{eq}(s) = Y_{eq}(0)\frac{1 + 2\zeta'\dfrac{s}{\nu'_o} + \dfrac{s^2}{\nu'^2_o}}{(1 + sT)\left(1 + 2\zeta\dfrac{s}{\nu_o} + \dfrac{s^2}{\nu_o^2}\right)}$$

with two zeros and three poles. More precisely, considering the usual values assumed by parameters:

- the zeros are complex conjugate, with a critical frequency ν'_o ("antiresonance" frequency) within 100–300 rad/sec, and damping factor ζ' (positive or negative) of small absolute value;
- the real pole $-1/T$ lies instead within a range of much lower frequencies, e.g., 3–10 rad/sec (corresponding to $T \cong 0.1$–0.3 sec);
- finally, the remaining two poles are complex conjugate, with a critical frequency ν_o ("resonance" frequency) slightly higher than ν'_o and again with a damping factor ζ of small absolute value.

For the case under examination, having set for brevity:

$$\begin{cases} R_1 \triangleq \lambda_R + \dfrac{R_l}{2} \\ L_1 \triangleq L_{sR} + \dfrac{L_l}{2} \end{cases} \qquad \begin{cases} R_2 \triangleq \dfrac{R_l}{2} - \lambda_I \\ L_2 \triangleq \dfrac{L_l}{2} + L_{sI} \end{cases}$$

and furthermore:

$$Z_1 \triangleq Z_{Rn} + \frac{R_l + sL_l}{2} = R_1 + sL_1, \qquad Z_2 \triangleq \frac{R_l + sL_l}{2} + Z_{In} = R_2 + sL_2$$

it is trivial to derive:

$$\begin{aligned} Y_{\text{eq}}(s) &= \frac{1 + sC_l Z_2}{Z_1 + Z_2 + sC_l Z_1 Z_2} \\ &= \frac{1 + sR_2C_l + s^2L_2C_l}{R_1 + R_2 + s(L_1 + L_2 + R_1R_2C_l) + s^2(R_1L_2 + R_2L_1)C_l + s^3L_1L_2C_l} \end{aligned}$$

More particularly, it can then be derived $Y_{\text{eq}}(0) = 1/(R_1 + R_2)$, and furthermore that:

$$\nu'_o = \frac{1}{\sqrt{L_2C_l}}, \qquad \zeta' = \frac{R_2}{2}\sqrt{\frac{C_l}{L_2}}$$

so that $\zeta' > 0$ if and only if:

$$R_2 > 0 \qquad [5.5.15]$$

which is equivalent to saying that $\lambda_I < R_l/2$.

Regarding the poles of $Y_{\text{eq}}(s)$, it can be first stated, by applying the Routh-Hurwitz criterion, that they all have negative real part (so that the system is asymptotically stable for $G_{R(i)} = 0$, i.e., in the absence of current regulation) if and only if, simultaneously:

$$\begin{cases} R_1 L_2 + R_2 L_1 > 0 \\ R_1 L_2^2 + R_2 L_1^2 + R_1 R_2 C_l (R_1 L_2 + R_2 L_1) > 0 \\ R_1 + R_2 > 0 \end{cases}$$

The last term in the left-hand side of the second condition is usually negligible (because of the relatively small value of C_l). It is then easy to verify that the first condition is no longer necessary, so that in practice, the following conditions hold:

$$\left.\begin{array}{l} R_1 + R_2 > 0 \\ R_1 L_2^2 + R_2 L_1^2 > 0 \end{array}\right\} \qquad [5.5.16]$$

or even only the condition $R_1 + R_2 > 0$, if $L_{sR} = L_{sI}$ and thus $L_1 = L_2$.

Finally, by similar approximations it is possible to derive:

$$\begin{cases} T \cong \dfrac{L_1 + L_2}{R_1 + R_2} \\ \nu_o \cong \sqrt{\dfrac{L_1 + L_2}{L_1 L_2 C_l}} \\ \zeta \cong \sqrt{\dfrac{C_l}{L_1 L_2 (L_1 + L_2)}} \dfrac{R_1 L_2^2 + R_2 L_1^2}{2(L_1 + L_2)} \end{cases}$$

to which conditions [5.5.16] correspond, to have T, ζ both positive. Specifically, if $L_{sR} = L_{sI}$, it holds:

$$T \cong \frac{2L_1}{R_1 + R_2}, \qquad \nu_o \cong \sqrt{\frac{2}{L_1 C_l}} = \sqrt{2}\nu_o' \qquad \zeta \cong \sqrt{\frac{C_l}{L_1}} \frac{R_1 + R_2}{4\sqrt{2}} \cong \frac{1}{2\nu_o T}$$

If the full model of the line were considered, infinite resonances and antiresonances at higher frequencies would result. Moreover, for frequencies not much lower than 6ω (the frequency at which the valve firings occur on average), it would be necessary to review also the model of converters, filters, etc.

The synthesis of $G_{R(i)}(s)$ is simpler if conditions [5.5.15] and [5.5.16] are satisfied. However, in many practical cases this does not happen for condition [5.5.15], and the zeros of $Y_{\text{eq}}(s)$ imply further undesirable phase delays. Moreover, the pure delay term $\epsilon^{-s\tau/2}$ should be considered in $K_R(s)$. As a result of stability requirements, the above practically limits the cutoff frequency of the loop, e.g., within 20–30 rad/sec. By doing so, modeling the higher frequencies may become less relevant.

Assuming a proportional-integral regulator, i.e.,

$$G_{R(i)}(s) = K_i \frac{1 + sT_i'}{s}$$

the integral gain K_i is chosen so as to obtain an acceptable cutoff frequency, whereas $1/T_i'$ can be, e.g., 20–50 $\sec^{-1}$. However, if the integral effect is omitted, thus accepting a (small) nonzero steady-state error $(\Delta I_{R\,\mathrm{rif}} - \Delta I_R)$, it is possible to assume a function $G_{R(i)}(s) = K_i T_i (1 + sT_i')/(1 + sT_i)$, with $T_i \gg T_i'$ (e.g., $1/T_i = 0.2$–$0.3\ \sec^{-1}$).

The treatment can be extended to *other situations*. Specifically, the case of Figure 5.42b (for which the rectifier operates at minimum firing angle, and the inverter is under current regulation; see point A' in Fig. 5.35b) is apparently similar to the cases above. On the contrary, this situation is actually much less critical, since the converter that does not regulate the current (i.e., in this case, the rectifier) has a positive equivalent resistance (λ_R), instead of a negative one ($-\lambda_I$) as in the previous case.

If the ac systems have output impedances that cannot be neglected, it is necessary to adopt Equations [5.5.13]. The possible use of the low-frequency approximations ($\epsilon^{-s\tau/2} \cong 1$, $K_{Ew}(s) \cong K_{Ew}(0)$, $K_{Ei}(s) \cong K_{Ei}(0)$) requires some cautions. For instance, such simplifications appear difficult to justify in the frequency range of interest for the line resonance.

5.6. SINGLE AND COMPOSITE LOADS

5.6.1. Generalities

A dynamic model overview for types of loads should first cover a wide and heterogeneous number of cases, much larger than the limits of the present work. It should make use of updated and not easily available information. On the other hand, a detailed representation of loads can be justified only for particular problems concerning, e.g., a limited number of loads supplied by a prevailing (schematically: infinite) power network and undergoing given perturbations. Vice versa, for problems which involve the whole system or a relevant part of it, the modeling simplifications can be not only less critical, but even reasonable, if uncertainty of actual load data (and the set itself) is considered.

Therefore, in the following, some illustrative information concerning "single" and "composite" loads (i.e., sets of loads, as seen from the nodes of the distribution or transmission network) will be reported. Also, for the sake of homogeneity (and of formal comparison) to what has already been presented about the synchronous machine and transformers, some deeper detail will be devoted to the asynchronous machine in Section 5.6.2.

As a *single load* we may intend a single user, with possible auxiliary equipment adopted to improve its behavior (e.g., voltage stabilizers, etc.) or for protective reasons (e.g., devices for automatic shedding on low-voltage or overcurrent conditions, etc.).

Single loads may be broadly classified into "static" and "rotating" loads. According to Section 3.1.3:

- electrical loads (lighting, heaters, arc furnaces, TV sets, etc.), as well as electrochemical loads and dc electromechanical loads (each including rectifier, dc motor, and mechanical load), can be considered "*static*" loads;
- ac electromechanical loads (each including asynchronous or, more rarely synchronous motor, and mechanical load), that depend on network frequency through their inertias, will be considered as "*rotating*" loads.

Actually, a generic load apparatus can include components of both the above-mentioned categories.

The *static characteristics* relating the absorbed active (P) and reactive (Q) powers to the voltage amplitude v and to the frequency ω, can be generally approximated, apart from possible additional constant terms, by the following functions:

$$\left.\begin{aligned} P &\propto v^{a_{pv}} \omega^{a_{p\omega}} \\ Q &\propto v^{a_{qv}} \omega^{a_{q\omega}} \end{aligned}\right\} \qquad [5.6.1]$$

at least in a proper neighborhood around the nominal operating point (e.g., in "per unit," $|\Delta v| < 20\%$, $|\Delta \omega| < 5\%$). On the other hand, the operation at abnormal values of v and ω can be considered unlikely and not allowed by the protective devices through automatic shedding of the load from the network.

If the load could be described, at any assigned ω, as a constant admittance, in Equations [5.6.1] it would evidently be that $a_{pv} = a_{qv} = 2$. However, such a situation is true only in few cases. The simplest example is the case of an electrical heater (stoves, boilers, irons, etc.), for which it can be assumed that $P \propto v^2$, $Q = 0$ and consequently $a_{pv} = 2$, as it is for a simple constant conductance. Incandescent lamps can be considered as purely resistive, with $Q = 0$, but the increase of the resistance with the temperature (i.e., with voltage) leads to a somewhat smaller value of a_{pv}, approximately 1.5–1.6. For the other load types it is necessary, more generally, to consider Equations [5.6.1] with proper and relatively diverse values of a_{pv}, a_{qv}, $a_{p\omega}$, $a_{q\omega}$. For instance, the following variation ranges have been identified for "static" loads:

$$\begin{cases} a_{pv} \in (0.9, 2.4) \\ a_{p\omega} \in (-0.55, +1.5) \\ a_{qv} \in (0.9, 4.3) \\ a_{q\omega} \in (-2.7, +1.4) \end{cases}$$

whereas the values corresponding to "rotating" loads can vary significantly from case to case, according to motor parameters and to the static characteristic that relates the resistant torque (or mechanical power) to the speed Ω_c of the mechanical load (e.g., a resistant power proportional to Ω_c^α, with α usually 1–3; see Section 3.1.3).

At any assigned ω, with $P = P(v)$, $Q = Q(v)$ (see Equations [5.6.1]), and assuming a three-phase and physically symmetrical load, it is possible to derive (in terms of Park's vectors, with $\Omega_r = \omega$ and omitting the subscript "r") $\bar{\imath} = (P - jQ)/\bar{v}^*$ and thus the equivalent admittance:

$$\overline{Y}_{\text{eq}} \triangleq \frac{\bar{\imath}}{\bar{v}} = \frac{P(v) - jQ(v)}{v^2}$$

Apart from the case for which $P(v)$ and $Q(v)$ are proportional to v^2 (and thus $a_{pv} = a_{qv} = 2$), such admittance depends on v, so that the load is called "*nonlinear*." For instance, a load is nonlinear if, at steady-state, and at assigned ω and any v, the following quantities are constant:

- the active and reactive powers (as well as the power factor) and thus $a_{pv} = a_{qv} = 0$;
- or alternatively, the current (in amplitude) and the power factor, and thus $a_{pv} = a_{qv} = 1$.

Such a definition of nonlinearity only refers to the relationship between vectors $\bar{\imath}$ and $\bar{v}$. As a particular consequence, for any generic steady-state at $\bar{v}$ constant, it follows that $\bar{\imath} = \overline{Y}_{\text{eq}}\bar{v} = \text{constant}$, so that not only phase voltages but also phase currents are sinusoidal and of positive sequence, without any current harmonics despite nonlinearity. Similar considerations apply for the synchronous machine in the presence of magnetic saturation, according to the model described in Section 4.3.3 and in Figure 4.16.

The *dynamic characteristics* of the different loads can involve the transient behavior of inductive and/or capacitive elements (see Section 5.2.1), and of asynchronous or synchronous motors, in addition to what defines the final parts of loads (chemical and mechanical parts, etc., including possible dc motors), according to very specific models. Herein we will limit ourselves to observe that, for phenomena not varying too fast, Equations [5.6.1] can appear applicable for "static" loads. However, for "rotating" loads, a dynamic dependence of P, Q on v, ω must be considered, at least for the most important electromagnetic transients in motors and the mechanical transients related to inertias. (In a dynamic operating condition, the meaning itself of ω requires further specifications; see also Section 5.6.2.)

For instance, in the case of a load driven by an asynchronous motor, if the electromagnetic transients are disregarded it is possible to derive, through considerations similar to those developed in Section 3.1.3, transfer functions such as:

- $(\partial P/\partial v)(s)$, $(\partial P/\partial \omega)(s)$, $(\partial Q/\partial \omega)(s)$ having one zero (and one pole, which corresponds however to a relatively high critical frequency; recall Equations [3.1.15] and [3.1.16]);

- $(\partial Q/\partial v)(s)$ almost constant (or alternatively, having one zero and the same pole as above, both corresponding to relatively high critical frequencies).

A *"composite" load*, defined at a node of the distribution or transmission network, can include many users, besides condensers for power factor correction, lines, transformers (step-down transformers, from high to low voltages), voltage regulation, and protection systems. The presence of these devices can significantly modify the characteristics of the composite load with respect to individual loads contributing to it.

For *static characteristics*, it is possible to refer, in terms of Park's vectors and under the hypothesis of physical symmetry, to the equivalent circuit in Figure 5.43a and to the block diagram in Figure 5.43b, with an obvious meaning of symbols (the "generic user" block can include more than one load and other devices, connected to the node at voltage v_i).

It is evident that, in the ideal case where the single v_i's can be considered assigned because of voltage regulations:

- powers P_i, Q_i can only vary with frequency ω;
- the dependence of powers P, Q (absorbed by the composite load) on v is determined by the parameters of the "interposed network";
- the dependence of P, Q on ω is determined by the above and by the relationship between P_i, Q_i, and ω;

whereas no trace is kept of the dependence of P_i, Q_i on their respective v_i, as caused by the characteristics of individual loads.

If the generic v_k is the secondary voltage of a tap-changing transformer and the regulation of v_k is done by acting on the transformer ratio, the dependence of P and Q on ω (with all the v_i's assigned) is influenced by the relationship between P_k, Q_k and ω, as noted above.

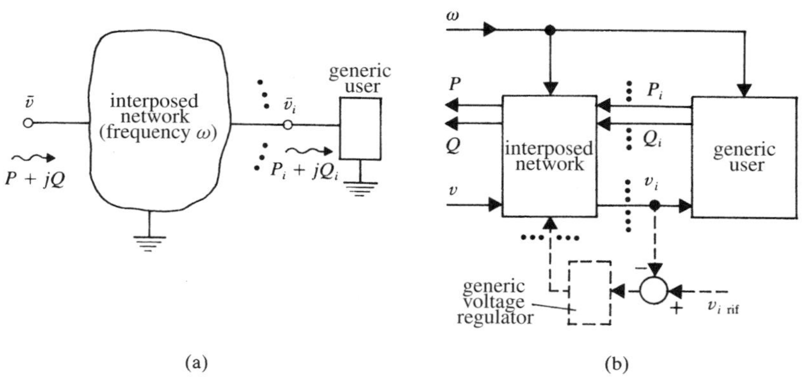

Figure 5.43. Composite load: (a) general equivalent circuit; (b) block diagram.

Instead, if the regulation of v_k is performed by means of a compensator, by acting on the reactive power that it supplies, it is easy to verify that (with all the v_i's assigned) the powers P and Q are not influenced by Q_k, so that no trace remains of the dependence of Q_k on ω. In other words, the characteristics of the k-th load influence the characteristics of the composite load only for the dependence of P_k on ω.

With the assumption of ideal voltage regulations, a further simplification can be accepted if losses in the "interposed network" are negligible. In such a case, the active power $P = \sum_i P_i$ can be practically considered independent of v.

Instead, in the opposite case for which there is no voltage regulation, it can be expected that the effect of the "interposed network" will be to reduce the nonlinearities at any given ω, thus making the dependence of P, Q on v closer to the quadratic-type relationship that corresponds to a constant admittance.

In general, it seems convenient to derive the model of composite loads based on specific evaluations and experimental measurements.

Indicatively, by assuming Equations [5.6.1] with reference to existing composite loads, it has been found that:

$$\begin{cases} a_{pv} \in (0.2, 1.8) \\ a_{p\omega} \in (0.2, 1.5) \\ a_{qv} \in (0.5, 3.0) \\ a_{q\omega} \in (-0.1, +0.5) \end{cases}$$

where the smaller values of a_{pv}, e.g., $a_{pv} \in (0.2, 1)$, and the larger of $a_{p\omega}$ correspond to industrial loads, predominantly of the rotating type.

The matter becomes more complicated for the *dynamic characteristics* of composite loads. Even if the faster phenomena are disregarded, it should be necessary to consider the (electromagnetic and mechanical) transient behavior of rotating loads, as well as voltage regulation. It becomes particularly important to group, into properly simplified dynamic equivalents, asynchronous motor loads (see Section 5.6.2), whose number can be very large. If only predominantly mechanical phenomena are of interest, as a first approximation (regarding the transfer functions $(\partial P/\partial v)(s)$, $(\partial P/\partial \omega)(s)$ etc.) it is possible to reach conclusions similar to those reported above for single loads, with the advice to consider the tap-changers' control which, because of its relative slowness, can interact with the mechanical transients in a nonnegligible way.

5.6.2. The Asynchronous Machine

With some usually acceptable approximations, consider an asynchronous machine including a three-phase symmetrical winding on the stator (indices a, b, c) and a three-phase symmetrical winding on the rotor (indices A, B, C), as indicated, for a bipolar machine, in the diagram of Figure 5.44a. For each winding, phases are wye-connected with isolated neutral (Fig. 5.44b) so that homopolar currents

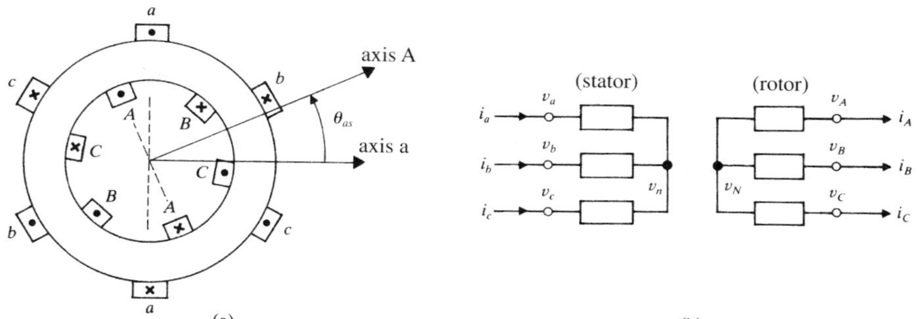

Figure 5.44. Schematic representation of the bipolar asynchronous machine with three stator circuits (subscripts a, b, c) and three rotor circuits (subscripts A, B, C).

are zero. Under these conditions, the d and q components will be considered (without considering the homopolar ones) when applying Park's transformation.

With respect to the *electrical part* of the machine, the following equations can be assumed.

(1) For stator circuits, by using the loads' convention (the most adequate for motor operation):

$$\begin{cases} v_a - v_n = Ri_a + \dfrac{d\psi_a}{dt} \\ v_b - v_n = Ri_b + \dfrac{d\psi_b}{dt} \\ v_c - v_n = Ri_c + \dfrac{d\psi_c}{dt} \end{cases}$$

intending that $(v_a - v_n)$, i_a, ψ_a are, respectively, the voltage, current, and flux of phase a, and similarly for phases b and c, whereas v_n is the neutral voltage and R is the phase resistance. By applying Park's transformation (Appendix 2) with an arbitrary angular reference θ_r, it can be derived, with $\Omega_r \triangleq d\theta_r/dt$:

$$\overline{v}_r = R\overline{\imath}_r + (p + j\Omega_r)\overline{\psi}_r \qquad [5.6.2]$$

(2) For rotor circuits, by using the generators' convention:

$$\begin{cases} v_A - v_N = -R_R i_A + \dfrac{d\psi_A}{dt} \\ v_B - v_N = -R_R i_B + \dfrac{d\psi_B}{dt} \\ v_C - v_N = -R_R i_C + \dfrac{d\psi_C}{dt} \end{cases}$$

intending that $(v_A - v_N)$, i_A, ψ_A are, respectively, the voltage, current, and flux of phase A, and similarly for phases B and C, whereas v_N is the neutral voltage and R_R is the phase resistance. Indicating the leading electrical angle of the first rotor phase axis (A) with respect to the first stator phase axis (a) by θ_{as} (Fig. 5.44a) (so that $\Omega_{as} \triangleq d\theta_{as}/dt$ is equal to the electrical speed of the rotor), and by applying Park's transformation but with an angular reference $(\theta_r - \theta_{as})$, it is possible to derive:

$$\overline{v}_{Rr} = -R_R \overline{\imath}_{Rr} + (p + j(\Omega_r - \Omega_{as}))\overline{\psi}_{Rr} \qquad [5.6.3]$$

(the subscript "R" stands for rotor variables).

For the *magnetic part* (which will be assumed as linear and conservative, excluding saturation, hysteresis, etc.), it is possible to write the following equations:

$$\begin{cases} \psi_a = L_{aa} i_a + L_{ab} i_b + L_{ac} i_c - L_{aA} i_A - L_{aB} i_B - L_{aC} i_C \\ \psi_b = \dots \\ \psi_c = \dots \end{cases}$$

$$\begin{cases} \psi_A = L_{Aa} i_a + L_{Ab} i_b + L_{Ac} i_c - L_{AA} i_A - L_{AB} i_B - L_{AC} i_C \\ \psi_B = \dots \\ \psi_C = \dots \end{cases}$$

where the expressions of the dotted terms can be found in a trivial way by "rotating" the indices a, b, c and A, B, C.

For the inductances appearing in these equations, it can be assumed (with a usually acceptable approximation, particularly under the assumption of isotropy):

$$\begin{cases} L_{aa} = L_{bb} = L_{cc} = \text{ constant} \\ L_{ab} = L_{ba} = L_{bc} = L_{cb} = L_{ca} = L_{ac} = \text{ constant} \end{cases}$$

$$\begin{cases} L_{AA} = L_{BB} = L_{CC} = \text{ constant} \\ L_{AB} = L_{BA} = L_{BC} = L_{CB} = L_{CA} = L_{AC} = \text{ constant} \end{cases}$$

whereas for the remaining inductances (which define the mutual inductances between stator and rotor windings, and consequently are surely varying with θ_{as}) it is possible to assume:

$$\begin{cases} L_{aA} = L_{Aa} = L_{bB} = L_{Bb} = L_{cC} = L_{Cc} = M_{\max} \cos\theta_{as} \\ L_{bA} = L_{Ab} = L_{cB} = L_{Bc} = L_{aC} = L_{Ca} = M_{\max} \cos(\theta_{as} - 120^\circ) \\ L_{cA} = L_{Ac} = L_{aB} = L_{Ba} = L_{bC} = L_{Cb} = M_{\max} \cos(\theta_{as} - 240^\circ) \end{cases}$$

with $M_{\max}$ constant.

By applying Park's transformation as already described (with an arbitrary reference θ_r for stator variables, and $(\theta_r - \theta_{as})$ for rotor variables), and by setting:

$$L_S \triangleq L_{aa} - L_{ab}$$
$$L_R \triangleq L_{AA} - L_{AB}$$
$$L_m \triangleq \frac{3}{2} M_{\max}$$

it then follows:

$$\left.\begin{aligned} \overline{\psi}_r &= L_S \overline{\imath}_r - L_m \overline{\imath}_{Rr} \\ \overline{\psi}_{Rr} &= -L_R \overline{\imath}_{Rr} + L_m \overline{\imath}_r \end{aligned}\right\} \qquad [5.6.4]$$

From Equations [5.6.2], [5.6.3], and [5.6.4] it is possible to derive, whatever the choice of θ_r, any one of the equivalent circuits reported in Figure 5.45. In normal operation it must be intended that the rotor winding is short-circuited, i.e., $\overline{v}_{Rr} = 0$.

Note that:

- for $\theta_r = \theta_{as}$ (and thus $\Omega_r = \Omega_{as}$) and $\overline{v}_{Rr} = 0$, there is a formal analogy — apart from the interchange between the loads' sign convention and the generators' one — between the circuits in Figure 5.45a and the equivalent circuits of the synchronous machine with damper windings (Fig. 4.8), under the hypothesis of isotropy and not considering the field circuit;
- for θ_{as} = constant (and thus $\Omega_{as} = 0$), there is a formal analogy between the circuit in Figure 5.45b and the equivalent circuit of the two-winding transformer (Fig. 5.5b).

If the rotor is of the *"squirrel-cage"* type, it is possible to obtain a better approximation by considering two rotor windings instead of one (each equivalent circuit includes a further rotor branch, similar to that in Figure 5.45). Furthermore, to account for *magnetic saturation*, the equivalent circuits may be modified similarly to the synchronous machine and the transformer.

The circuit of Figure 5.45c is formally interesting, because it is a generalization of the well-known steady-state equivalent circuit, reported (for $\overline{v}_{Rr} = 0$) in Figure 5.46a.

Finally, for the *mechanical part*, equations similar to those in Section 4.1.2 for the synchronous machine can be adopted. More precisely, it is possible to derive, for any θ_r:

$$C_e = N_p \langle j\overline{\psi}_r, \overline{\imath}_r \rangle \qquad [5.6.5]$$

where C_e is the ("electromagnetic") generated torque, and N_p the number of pole pairs. It is easy to verify that the generated mechanical power $P_e = C_e \Omega_{as}/N_p$ is equal to the total electric power absorbed by the voltage generators reported in Figure 5.45a or by that of Figure 5.45b(4). In fact, as a result of Equations [5.6.4],

(4) In the equivalent circuit of Figure 5.45c there are no voltage generators, but the operational impedance $(p + j\Omega_r)R_R/(p + j(\Omega_r - \Omega_{as}))$ varies with Ω_{as}, as well as the voltage $(p + j\Omega_r)\overline{v}_{Rr}/(p + j(\Omega_r - \Omega_{as}))$.

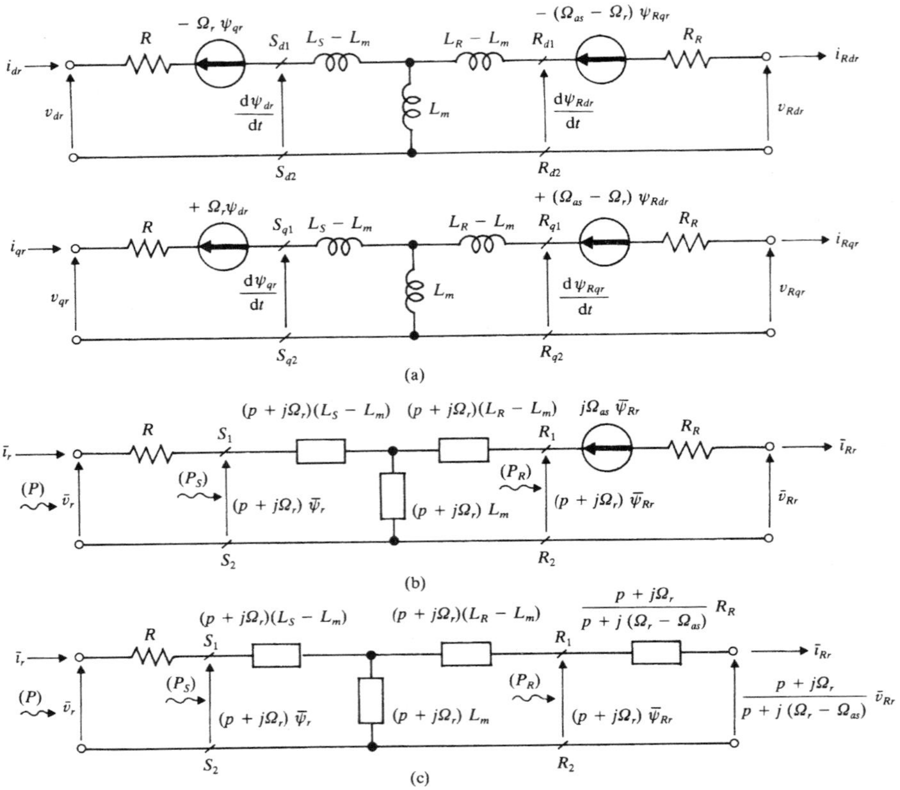

Figure 5.45. Equivalent circuits of the asynchronous machine (electrical and magnetic parts) in the absence of magnetic saturation.

it follows that:

$$\left\langle j\overline{\psi}_r, \overline{\imath}_r \right\rangle = -L_m \left\langle j\overline{\imath}_{Rr}, \overline{\imath}_r \right\rangle = \left\langle j\overline{\psi}_{Rr}, \overline{\imath}_{Rr} \right\rangle$$

and thus:

$$\begin{aligned}\left\langle j\Omega_r\overline{\psi}_r, \overline{\imath}_r \right\rangle + \left\langle j(\Omega_{as} - \Omega_r)\overline{\psi}_{Rr}, \overline{\imath}_{Rr} \right\rangle &= \left\langle j\Omega_{as}\overline{\psi}_{Rr}, \overline{\imath}_{Rr} \right\rangle \\ &= \Omega_{as}\left\langle j\overline{\psi}_r, \overline{\imath}_r \right\rangle = \Omega_{as}\frac{C_e}{N_p} = P_e\end{aligned}$$

As to active power flows in the machine:

$$P - Ri_r^2 = P_S = \mathcal{R}e\left(\frac{\mathrm{d}\overline{\psi}_r}{\mathrm{d}t}\overline{\imath}_r^*\right) + \frac{\Omega_r}{\Omega_{as}}P_e$$

(where $(\Omega_r/\Omega_{as})P_e = C_e\Omega_r/N_p$) and furthermore, for $\overline{v}_{Rr} = 0$:

$$\mathcal{R}e\left(\frac{d\overline{\psi}_{Rr}}{dt}\overline{\imath}^*_{Rr}\right) + \frac{\Omega_r}{\Omega_{as}}P_e = P_R = P_e + R_R i^2_{Rr}$$

where (Fig. 5.45b,c) P is the active power entering through the stator terminals, and P_S, P_R are the active powers flowing through the pairs of terminals (S_1, S_2) and (R_1, R_2), respectively (the difference ($P_S - P_R$) is the active power absorbed by the magnetic part of the machine).[(5)]

By *(steady-state) "equilibrium" operation* of the asynchronous machine, we refer to the operation in which:

- the rotor winding is short-circuited, i.e.,

$$\overline{v}_{Rr} = 0$$

  (if instead, each rotor phase were connected to an external impedance $\overline{Z}_{Re}$, it would be sufficient to formally substitute R_R by $R_R + \overline{Z}_{Re}$);
- stator currents are sinusoidal and of positive sequence, at frequency ω, i.e.,

$$\begin{cases} i_a = \sqrt{2}I_{(F)}\cos(\omega t + \alpha_I) \\ i_b = \sqrt{2}I_{(F)}\cos(\omega t + \alpha_I - 120°) \\ i_c = \sqrt{2}I_{(F)}\cos(\omega t + \alpha_I - 240°) \end{cases}$$

  where $I_{(F)}$, α_I are constant (more specifically, $I_{(F)}$ is the rms value of phase-currents);
- electrical speed Ω_{as} is constant but different from ω, i.e.,

$$\Omega_{as} = (1 - \sigma')\omega$$

  where $\sigma' \triangleq (\omega - \Omega_{as})/\omega$ (constant) is the "relative slip."

By assuming $\Omega_r = \omega$, i.e., $\theta_r = \int \Omega_r dt = \omega t + \theta_{ro}$ (reference of Park's transformation for stator variables), it follows that:

$$\overline{\imath}_r = \sqrt{3}I_{(F)}\epsilon^{j(\alpha_I - \theta_{ro})} = \text{constant}$$

so that the equivalent circuit is under static operation, i.e., at constant voltages, currents, and fluxes. Therefore, it can be determined that stator voltages and

(5) In Section 3.1.3, a partially different notation from the present one was adopted (more particularly, P_{cj}, P_{mcj}, Ω_{cj} instead of P, P_e, Ω_{as}).

fluxes (as well as currents) are sinusoidal of the positive sequence at frequency ω, whereas rotor currents, voltages, and fluxes are sinusoidal of the positive sequence at the "slip" frequency (equal to the slip $\omega - \Omega_{as} = \sigma'\omega$).

More precisely, from Equations [5.6.2] and [5.6.3] with $p = 0$, it follows that:

$$\begin{cases} \overline{v}_r = R\overline{\imath}_r + j\omega\overline{\psi}_r \\ \overline{\imath}_{Rr} = j\sigma'\omega\overline{\psi}_{Rr}/R_R \end{cases}$$

whereas Equations [5.6.4], which allow one to derive $\overline{\psi}_r$ and $\overline{\psi}_{Rr}$, can be rewritten as:

$$\overline{\psi}_r - (L_S - L_m)\overline{\imath}_r = L_m(\overline{\imath}_r - \overline{\imath}_{Rr}) = \overline{\psi}_{Rr} + (L_R - L_m)\overline{\imath}_{Rr}$$

Such equations correspond to the equivalent circuit in Figure 5.46a and to the vector diagram in Figure 5.46b.

If ω, Ω_{as} (or equivalently σ') and $\overline{\psi}_r$ are assumed as assigned, from Equations [5.6.3] and [5.6.4] it is possible to derive:

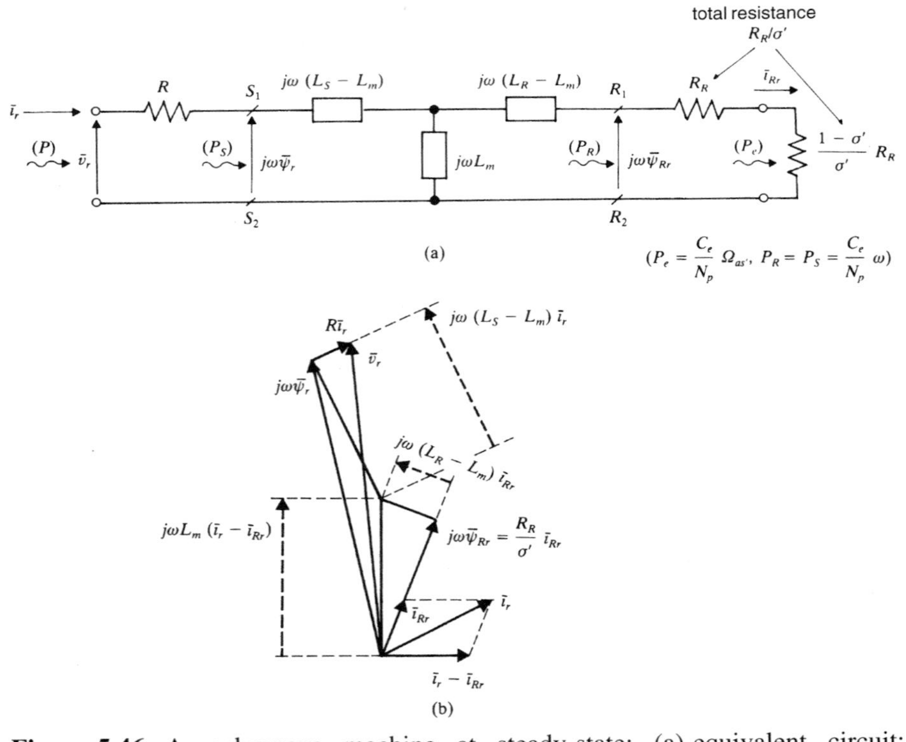

Figure 5.46. Asynchronous machine at steady-state: (a) equivalent circuit; (b) vector diagram.

$$\left.\begin{aligned}\overline{\psi}_{Rr} &= \frac{L_m R_R}{L_S(R_R + j\omega\sigma' L'_R)}\overline{\psi}_r \\ \overline{\imath}_{Rr} &= \frac{j\omega\sigma' L_m}{L_S(R_R + j\omega\sigma' L'_R)}\overline{\psi}_r \\ \overline{\imath}_r &= \frac{R_R + j\omega\sigma' L_R}{L_S(R_R + j\omega\sigma' L'_R)}\overline{\psi}_r\end{aligned}\right\} \qquad [5.6.6]$$

where $L'_R \triangleq L_R - L_m^2/L_S$ is the inductance seen from terminals (R_1, R_2) when (S_1, S_2) are short-circuited (see Figs. 5.45 and 5.46a). Furthermore, it follows that:

$$P_S = P_R = \frac{\omega C_e}{N_p} = \frac{(\omega\psi_r L_m/L_S)^2(R_R/\sigma')}{(R_R/\sigma')^2 + (\omega L'_R)^2} \qquad [5.6.7]$$

from which P_e can be derived (by multiplying for Ω_{as}/ω), whereas active and reactive powers entering through stator terminals are:

$$\begin{cases} P = P_S + Ri_r^2 \\ Q = (\omega\psi_r)^2\left(\dfrac{1}{\omega L_S} + \dfrac{(L_m/L_S)^2\omega L'_R}{(R_R/\sigma')^2 + (\omega L'_R)^2}\right) = \dfrac{\omega\psi_r^2}{L_S} + \dfrac{\omega L'_R\sigma'}{R_R}P_S \end{cases}$$

Equation [5.6.2] then gives $\overline{v}_r = R\overline{\imath}_r + j\omega\overline{\psi}_r$, and it allows the evaluation of $\overline{\psi}_r$, etc. starting from the voltage $\overline{v}_r$, when $\overline{v}_r$ is assigned instead of $\overline{\psi}_r$.

Finally, the previous equations presume, as Ω_{as} is constant, the equality between the generated torque (which is considered as driving, with reference to the case of the asynchronous motor) and the resistant torque, which generally depends on Ω_{as}. Such a condition allows the determination of Ω_{as} (or of σ'), etc. starting from ω and $\overline{\psi}_r$ (or $\overline{v}_r$).

From Equation [5.6.7] it is possible to derive, for assigned ω and ψ_r, a characteristic (C_e, Ω_{as}) as indicated in Figure 5.47a.

Regarding the dependence of P and Q on ω and ψ_r, in practice, it is often possible to assume, at least as a first approximation, $Ri_r^2 \ll P_S$, $(\omega L'_R)^2 \ll (R_R/\sigma')^2$ and thus:

$$\begin{cases} P \cong P_S = \dfrac{\omega C_e}{N_p} \cong \dfrac{\sigma'}{R_R}\left(\omega\psi_r\dfrac{L_m}{L_S}\right)^2 \\ Q \cong \omega\left(\dfrac{\psi_r^2}{L_S} + L'_R\left(\dfrac{C_e}{N_p}\right)^2\left(\dfrac{L_S}{\psi_r L_m}\right)^2\right) \end{cases}$$

where C_e is equal to the resistant torque and thus depends, generally, on $\Omega_{as} = (1 - \sigma')\omega$. In the simple case when C_e can be considered constant, it can be concluded that, with the above-mentioned approximations:

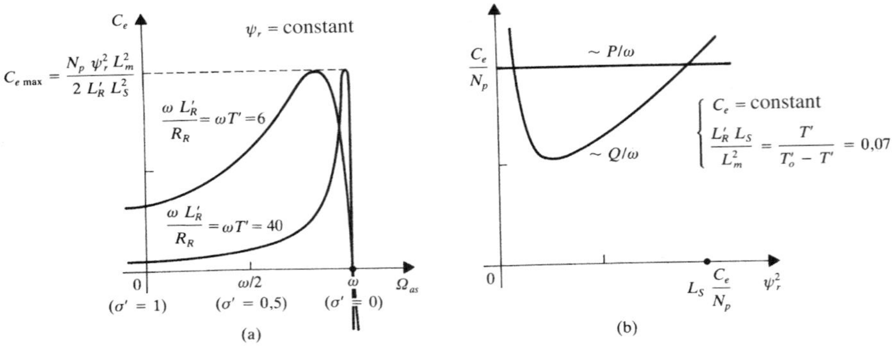

Figure 5.47. Static characteristics of the asynchronous machine: (a) dependence of the torque on speed, for an assigned amplitude of the stator flux and at an assigned frequency; (b) dependence of active and reactive powers on the amplitude of the stator flux and on the frequency, for an assigned torque (see text).

- both powers P and Q are proportional to ω;
- only Q depends on ψ_r, and it is the sum of two terms that are respectively proportional to ψ_r^2 and to $1/\psi_r^2$;

according to Figure 5.47b.

The *dynamic behavior* for small variations, around a generic operating point indicated by the superscript "o," can be analyzed in transfer functions by first linearizing Equations [5.6.3] and [5.6.4], with $\Omega_r = \omega^o$ and $\overline{v}_{Rr} = 0$ (note the convenience of assuming Ω_r = constant $= \omega^o$). By eliminating $\Delta\overline{\psi}_{Rr}$ and $\Delta\overline{\imath}_{Rr}$, it can be derived:

$$\Delta\overline{\imath}_r = \overline{A}(s)\Delta\overline{\psi}_r + \overline{B}(s)\Delta\Omega_{as} \qquad [5.6.8]$$

having posed for brevity:

$$\begin{cases} \overline{A}(s) \triangleq \dfrac{1 + sT_o' + j\omega^o\sigma'^o T_o'}{L_S(1 + sT' + j\omega^o\sigma'^o T')} \\[2ex] \overline{B}(s) \triangleq \dfrac{-jL_m\overline{\psi}_{Rr}^o}{R_R L_S(1 + sT' + j\omega^o\sigma'^o T')} \end{cases}$$

from which, expressing $\overline{\psi}_{Rr}^o$ by the first of Equations [5.6.6], it also follows that:

$$\overline{B}(s) = -j\frac{\overline{A}(s) - \overline{A}(0)}{s}\overline{\psi}_r^o$$

where $T' \triangleq L_R'/R_R$ and $T_o' \triangleq L_R/R_R$ are, respectively, the so-called *"short-circuit (transient)" and "open-circuit (transient)" time constants*. Recall the similar definitions for the synchronous machine; with reference to Figure 5.45, the specifications "short-circuit" and "open-circuit" refer to terminals (S_1, S_2). For similar reasons, the stator inductance L_S is then called "synchronous" or "open-circuit" (or "at zero slip"), whereas $L'_S \triangleq L_S - L_m^2/L_R = L_S T'/T'_o$ is the "transient" or "short-circuit" inductance.

It can be derived:

$$\begin{cases} \Delta P_S + j\Delta Q = (s + j\omega^o)\Delta\overline{\psi}_r \overline{\imath}_r^{o*} + j\omega^o \overline{\psi}_r^o \Delta\overline{\imath}_r^* \\ \Delta P = \Delta P_S + 2R\,\mathcal{R}e(\overline{\imath}_r^o \Delta\overline{\imath}_r^*) \\ \Delta(C_e/N_p) = -Im(\Delta\overline{\psi}_r \overline{\imath}_r^{o*} + \overline{\psi}_r^o \Delta\overline{\imath}_r^*) \end{cases}$$

and by applying Equation [5.6.8] it is possible to determine the dynamic dependence of ΔP_S, ΔP, ΔQ, ΔC_e on $\Delta\overline{\psi}_r$ and $\Delta\Omega_{as}$. Specifically, by observing that (as a result of the third of Equations [5.6.6]) $\overline{\imath}_r^o = \overline{A}(0)\overline{\psi}_r^o$, it follows that:

$$\begin{aligned} \Delta\left(\frac{C_e}{N_p}\right) &= Im(\overline{A}(s) + \overline{A}(0))\psi_r^o \Delta\psi_r + \mathcal{R}e(\overline{A}(s) - \overline{A}(0))\psi_r^{o2}\Delta\angle\overline{\psi}_r \\ &\quad + Im(\overline{\psi}_r^{o*}\overline{B}(s))\Delta\Omega_{as} = Im(\overline{A}(s) + \overline{A}(0))\psi_r^o\Delta\psi_r \\ &\quad + \mathcal{R}e\left(\frac{\overline{A}(s) - \overline{A}(0)}{s}\right)\psi_r^{o2}\Delta(\omega_\psi - \Omega_{as}) \end{aligned} \qquad [5.6.9]$$

where $\omega_\psi \triangleq \omega^o + \mathrm{d}\angle\overline{\psi}_r/\mathrm{d}t$ can be interpreted as the frequency associated with the stator flux (i.e., with the vector $\overline{\psi}_r$), whereas $\Omega_{as} \triangleq \mathrm{d}\theta_{as}/\mathrm{d}t$ is the electric speed of the rotor.

This last equation defines the *transfer functions* that relate $\Delta(C_e/N_p)$ to $\Delta\psi_r$, $\Delta\omega_\psi$, $\Delta\Omega_{as}$, or, more precisely, to the variation $\Delta\psi_r$ and to the slip variation $\Delta(\omega_\psi - \Omega_{as})$. Similarly, with inputs $\Delta\psi_r$, $\Delta\omega_\psi$, $\Delta\Omega_{as}$, it is possible to derive the transfer functions relative to other output variables.

One can immediately determine that the poles of such transfer functions are the poles of $\mathcal{R}e(\overline{A}(s))$ and $Im(\overline{A}(s))$, and thus satisfy the equation:

$$0 = (1 + sT')^2 + (\omega^o\sigma'^o T')^2$$

from which $s = -1/T' \pm \tilde{j}\omega^o\sigma'^o$. Therefore, there are two complex conjugate poles, that can be written in the usual form:

$$s = \left(-\zeta \pm \tilde{j}\sqrt{1-\zeta^2}\right)\nu_o \qquad [5.6.10]$$

with a "resonance frequency":

$$\nu_o = \sqrt{\left(\frac{1}{T'}\right)^2 + (\omega^o\sigma'^o)^2} \qquad [5.6.11]$$

and a "damping factor":

$$\zeta = \frac{1}{\sqrt{1 + (\omega^o \sigma'^o T')^2}} \qquad [5.6.12]$$

In conclusion, by using such a notation and recalling the expression of $\overline{A}(s)$, from Equation [5.6.9] it is possible to derive $\Delta(C_e/N_p) = G_\psi(s)\Delta\psi_r + G_\sigma(s)$ $\Delta(\omega_\psi - \Omega_{as})$, with:

$$\left.\begin{aligned} G_\psi(s) &\triangleq \frac{\partial(C_e/N_p)}{\partial\psi_r}(s) = 2\zeta^2(T'_o - T')\frac{\omega^o\sigma'^o\psi_r^o}{L_S}\frac{1 + 2\zeta'\dfrac{s}{\nu'_o} + \dfrac{s^2}{\nu'^2_o}}{1 + 2\zeta\dfrac{s}{\nu_o} + \dfrac{s^2}{\nu_o^2}} \\ G_\sigma(s) &\triangleq \frac{\partial(C_e/N_p)}{\partial(\omega_\psi - \Omega_{as})}(s) = (2\zeta^2 - 1)\zeta^2(T'_o - T')\frac{\psi_r^{o2}}{L_S}\frac{1 + s\dfrac{\zeta^2 T'}{2\zeta^2 - 1}}{1 + 2\zeta\dfrac{s}{\nu_o} + \dfrac{s^2}{\nu_o^2}} \end{aligned}\right\} \qquad [5.6.13]$$

and moreover:

$$\nu'_o = \sqrt{2}\nu_o, \quad \zeta' = \frac{\zeta}{\sqrt{2}}$$

Note that the static gains $G_\psi(0)$ and $G_\sigma(0)$ are equal, respectively, to the partial derivatives of C_e/N_P in respect to ψ_r and $\omega\sigma'$, which can be determined from Equation [5.6.7].

For many practical cases, it can be assumed as a first approximation (as already stated), that $(\omega^o\sigma'^o T')^2 \ll 1$ and thus $\nu_o \cong 1/T'$, $\zeta \cong 1$, $\nu'_o \cong \sqrt{2}/T'$, $\zeta' \cong 1/\sqrt{2}$. Through the second of Equations [5.6.13] it can be specifically derived:

$$G_\sigma(s) \cong (T'_o - T')\frac{\psi_r^{o2}}{L_S}\frac{1}{1 + sT'}$$

As a result of Equation [5.6.2], the variations $\Delta\overline{\psi}_r$ and $\Delta\overline{\imath}_r$ are related to $\Delta\overline{v}_r$ by:

$$\Delta\overline{v}_r = R\Delta\overline{\imath}_r + (s + j\omega^o)\Delta\overline{\psi}_r \qquad [5.6.14]$$

If $\Delta\overline{v}_r$ and $\Delta\Omega_{as}$ are considered as the inputs (instead of $\Delta\overline{\psi}_r$ and $\Delta\Omega_{as}$), this equation allows one to derive (from Equation [5.6.8]):

$$\Delta\overline{\psi}_r = \frac{\Delta\overline{v}_r - R\overline{B}(s)\Delta\Omega_{as}}{s + j\omega^o + R\overline{A}(s)} \qquad [5.6.15]$$

and thus all the transfer functions that relate $\Delta\bar{\imath}_r$, $\Delta(C_e/N_P)$ etc. to $\Delta\bar{v}_r$ and $\Delta\Omega_{as}$. These new transfer functions have four poles which, as a result of the expressions of $\overline{A}(s)$ and $\overline{B}(s)$, are the solutions of:

$$0 = \left[s(1+sT') + \frac{R}{L_S}(1+sT_o') - \omega^{o2}\sigma'^o T'\right]^2 + \omega^{o2}\left[1 + s(1+\sigma'^o)T' + \frac{R}{L_S}\sigma'^o T_o'\right]^2$$

As a first approximation, by neglecting the stator resistance R, Equation [5.6.15] becomes more simply $\Delta\overline{\psi}_r \cong \Delta\bar{v}_r/(s + j\omega^o)$ (whereas $\partial\bar{\imath}_r/\partial\bar{v}_r \cong (\partial\bar{\imath}_r/\partial\overline{\psi}_r)/(s + j\omega^o)$, etc.), and the four poles are given by the Equation [5.6.10] and by the solutions ($s = \pm\tilde{j}\omega^o$) of equation $0 = s^2 + \omega^{o2}$. Specifically, with $R = 0$, it follows that:

$$\left.\begin{aligned} \Delta\psi_r &= \frac{\omega^o \Delta v_r - v_r^o \Delta\omega_v}{s^2 + \omega^{o2}} \\ \Delta\omega_\psi &= \frac{\omega^o\left(s^2 \dfrac{\Delta v_r}{v_r^o} + \omega^o \Delta\omega_v\right)}{s^2 + \omega^{o2}} \end{aligned}\right\} \qquad [5.6.16]$$

where $\omega_v \triangleq \omega^o + \mathrm{d}\angle\bar{v}_r/\mathrm{d}t$ is the frequency associated with the stator voltage. Therefore, contrary to $\Delta\omega_\psi$, the transient effects of $\Delta\omega_v$ on $\Delta(C_e/N_p)$ cannot be added to those of $(-\Delta\Omega_{as})$. This means that, for transient operation, it is not correct to assume the slip variation $\Delta(\omega_v - \Omega_{as})$ as input.

Finally, it is necessary to add to the previous equations the equation of the whole mechanical part, including that associated with the machine.

For an asynchronous machine operating as a motor, assuming that a variation $\Delta\Omega_{as}/N_p$ of the mechanical speed causes a variation $K_c\Delta\Omega_{as}/N_p$ of the resistant torque, and indicating by J the total moment of inertia, it is possible to write:

$$\Delta\left(\frac{C_e}{N_p}\right) = (K_c' + sJ')\Delta\Omega_{as}$$

with $K_c' \triangleq K_c/N_p^2$, $J' \triangleq J/N_p^2$. By also considering Equations [5.6.13] (and [5.6.16]), we can then derive the block diagram in Figure 5.48 and, specifically,

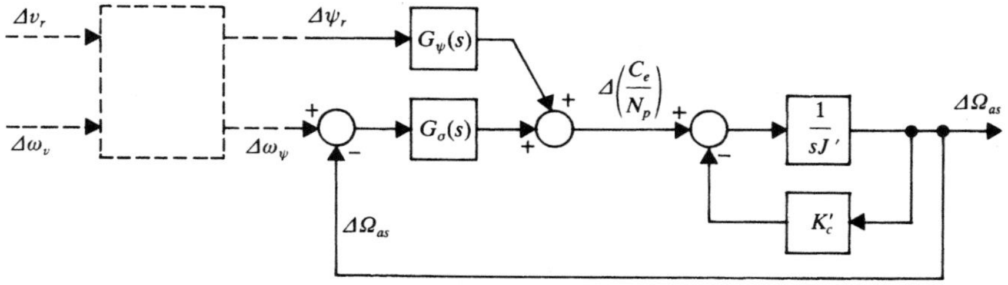

Figure 5.48. Block diagram of the asynchronous machine and the mechanical system connected to it.

the equation:

$$\Delta\Omega_{as} = \frac{G_\psi(s)\Delta\psi_r + G_\sigma(s)\Delta\omega_\psi}{G_\sigma(s) + K'_c + sJ'}$$

which expresses the dynamic dependence of $\Delta\Omega_{as}$ on $\Delta\psi_r$ and $\Delta\omega_\psi$, and allows, with the equations already written, one to derive the transfer functions concerning different output variables, by assuming $\Delta\psi_r$ and $\Delta\omega_\psi$ (or similarly Δv_r and $\Delta\omega_v$) as the inputs.

It can be easily recognized that:

- if $\Delta\psi_r$ and $\Delta\omega_\psi$ are assumed as the inputs, the poles of the transfer functions are the three roots of the equation:

$$0 = G_\sigma(s) + K'_c + sJ' \qquad [5.6.17]$$

- if, instead, Δv_r and $\Delta\omega_v$ are assumed as the inputs and the stator resistance is disregarded, the two poles $s = \pm\tilde{j}\omega^o$ are added to the above three roots.

The single parameters of the asynchronous machine—considering the mechanical load that it drives during the operation as a motor—exhibit different values depending on the situation. This is indicatively illustrated in Table 5.1[(6)], for different orders of nominal active power P_{nom}.

[(6)] Stator parameters R, L_S, L'_S are reported in per unit, by intending, as for the synchronous machine,

- $V_{(F)\text{nom}}$ = nominal rms value of the phase voltages;
- A_{nom} = nominal apparent power;
- $Z_{(F)\text{nom}} = 3V^2_{(F)\text{nom}}/A_{\text{nom}}$ = nominal value of the per-phase impedance;
- ω_{nom} = nominal (network) frequency.

If the stator inductances are formally augmented, according to the following, by the addition of an external inductance (x_e, in pu), the value corresponding to the time constant T'_o does not change, whereas x_S, x'_S, $T' \triangleq x'_S T'_o / x_S$ must be substituted, by $(x_S + x_e)$, $(x'_S + x_e)$, $(x'_S + x_e)T'_o/(x_S + x_e)$, respectively.

Furthermore, for the parameters of the mechanical part it can be observed that:

- by assuming a resistant torque proportional to Ω^β_{as} (i.e., a resistant power proportional to Ω^α_{as}, with $\alpha \triangleq \beta + 1$; see also Section 3.1.3), it follows that:

$$\frac{K'_c\omega^2_{\text{nom}}}{A_{\text{nom}}} = \beta\frac{P^o_e}{P_{\text{nom}}}\cos\varphi_{\text{nom}}\left(\frac{\omega_{\text{nom}}}{\Omega_{as}}\right)^2$$

- by defining the "start-up time" as for the synchronous machine, i.e., $T_a \triangleq (J'\omega^2_{\text{nom}})/P_{\text{nom}}$ (see T_{ac}, in Section 3.1.3):

$$\frac{J'\omega^2_{\text{nom}}}{A_{\text{nom}}} = T_a\cos\varphi_{\text{nom}}$$

where $\cos\varphi_{\text{nom}} \triangleq P_{\text{nom}}/A_{\text{nom}}$ is the nominal power factor.

It must be indicated that the value of J' is particularly subject to changes for what is shown in Table 5.1, depending on the contribution of the mechanical load.

Table 5.1 Indicative Values of Asynchronous Machine Parameters

P_{nom}	$<\sim$ 50 kW	$>\sim$ 50 kW
$r \triangleq \dfrac{R}{Z_{(F)\text{nom}}}$	0.02–0.06	0.01–0.02
$x_S \triangleq \dfrac{\omega_{\text{nom}} L_S}{Z_{(F)\text{nom}}}$	1.5–3.0	3.0–5.5
$x'_S \triangleq \dfrac{\omega_{\text{nom}} L'_S}{Z_{(F)\text{nom}}}$	0.1–0.2	0.15–0.25
T'_o	0.15–3.00 sec	2.0–6.0 sec
$T'(x_e = 0)$	0.01–0.20 sec	0.15–0.30 sec
$\dfrac{K'_c \omega^2_{\text{nom}}}{A_{\text{nom}}}$	0.2–1.5	
$\dfrac{J' \omega^2_{\text{nom}}}{A_{\text{nom}}}$	0.04–0.50 sec	0.2–6.0 sec

In the case of several asynchronous loads, supplied in a parallel way (with the same $\overline{v}_r$, i.e., the same v_r and ω_v), it is important to define, for at least small variations, a proper *"equivalent load"* (or equivalent loads, in the smallest possible number), corresponding to transfer functions close to actual ones, and of small dynamic order. The equivalence must be referred, in particular, to the relationship between the total absorbed powers (active and reactive) and the inputs v_r and ω_v.

On the other hand, the possibility of grouping more loads into an equivalent load of a minimal dynamic order, equal to that of a generic single load, is obvious if and only if the transfer function poles have the same values for all considered loads.

By disregarding, as is usually acceptable, the stator resistances, the problem becomes that of checking for which loads Equation [5.6.17] has approximately the same set of solutions.

Apart from the opportunity of specific, more accurate evaluations (and other criteria, considering the behavior for large variations):

(1) if $(\omega^o \sigma'^o T')^2 \ll 1$, Equation [5.6.17] can be substituted by:

$$\left.\begin{aligned} 0 &= 1 + sT' \\ 0 &= (T'_o - T')\frac{\psi_r^{o2}}{L_S} + K'_c + s(K'_c T' + J') + s^2 J' T' \end{aligned}\right\} \qquad [5.6.17']$$

(2) the first of Equations [5.6.17′] leads to the solution $s = -1/T'$;

(3) in the second of such equations, the "known term," which (at nominal voltage and frequency) is:

$$\left(\frac{(T_o' - T')A_{\text{nom}}}{x_S\omega_{\text{nom}}} + K_c'\right)$$

and the coefficient of s, which is $(K_c'T' + J')$, may be often approximated by $T_o'A_{\text{nom}}/(x_S\omega_{\text{nom}})$ and J' respectively, so that the solutions become:

$$s_{1,2} \cong \frac{-1 \pm \sqrt{1-a}}{2T'} \qquad [5.6.18]$$

where:

$$a \triangleq \frac{4T_o'T'A_{\text{nom}}}{x_S J'\omega_{\text{nom}}}$$

For varying a, the behavior of $s_{1,2}$ in the complex plane is reported in Figure 5.49 (accepting the Equation [5.6.18]). Such solutions are associated with a damping factor equal to $1/\sqrt{a}$, and are real if $a < 1$, and complex conjugate, with a magnitude $\sqrt{a}/2T'$, if $a > 1$ (for $a > 4$ it holds $|s_1| = |s_2| > 1/T'$).
As a conclusion, within the acceptable limits of the above-mentioned approximations:

- for the generic load, the different transfer functions include, in the denominator, a product:

$$D(s) \triangleq (1 + sT')\left(1 + \frac{s}{-s_1}\right)\left(1 + \frac{s}{-s_2}\right)\left(1 + \frac{s^2}{\omega^{o2}}\right)$$

  where s_1, s_2 are functions of a and T' according to Equation [5.6.18];
- it is possible to group those loads with equal values of both $T'$and a.

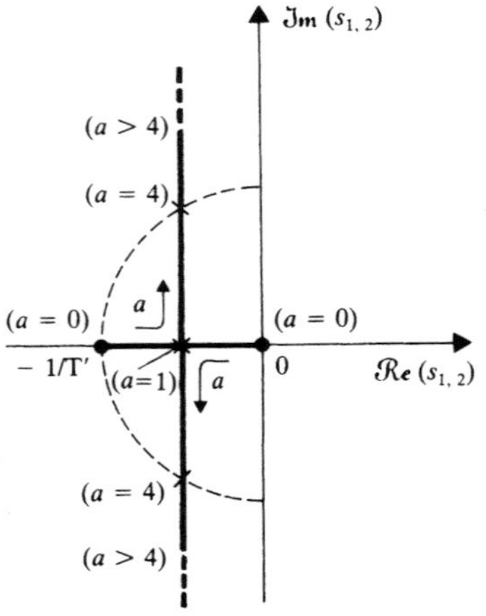

Figure 5.49. Dependence of the characteristic roots s_1 and s_2 on the parameter a (see text).

However, if the examined phenomena are slow, the adopted criterion can be applied by not considering poles with a relatively large magnitude. For frequency response ($s = \tilde{j}\nu$), if we are interested in accurately reproducing the behavior for $\nu < 1/T'$ (neglecting the delays associated with the poles $-1/T', \pm \tilde{j}\omega^o$):

- for the generic load, it can be assumed that $D(s) \cong (1 + s/ - s_1)(1 + s/-s_2)$ or even $D(s) \cong 1$, according to the value of a (analogous approximations may concern the numerators of the transfer functions, according to the values of respective zeros);
- loads with a relatively large a (for instance $a > 4$; see Fig. 5.49) can be grouped together;
- other loads can be grouped based on the values of s_1, s_2 ($a < 4$).

A similar problem is posed if the generic load is connected to the common supply node, through a link that can be viewed (in terms of equivalent circuit) as a proper external inductance. In such a case, however, new values must be considered, as in footnote [(6)], instead of x_S and T', so that the approximations indicated above in (1) and (3) may no longer be applicable. Further complications can arise from the interaction with voltage regulation. However, for predominantly mechanical phenomena, the simplified analysis in Section 3.1.3 should be recalled.

5.7. PURELY ELECTRICAL PHENOMENA

5.7.1. Preliminaries

According to Section 1.8.2, an important category of dynamic phenomena is that of purely electrical ones:

- caused by perturbations on the electrical part of the system;
- sufficiently *fast* so that it is possible to assume as constant not only the speed of the rotating machines, but also the amplitudes and electrical rotating speeds of the rotor fluxes (the operation of control systems is not considered, because it is unavoidably delayed).

More particularly, we will assume that perturbations consist of opening and/or closing of connections, caused by:

- *maneuvers* (typically a breaker operation);
- *faults* (typically a short-circuit, i.e., the closing of an anomalous connection between two points which should be isolated);

by disregarding:

- the effects of possible voltage and/or current *harmonics*, which can be analyzed in a relatively obvious way (such effects can be enhanced by electrical resonances, as illustrated in the following, with frequencies close to those of the harmonics);

- the effects of *"externally originated" perturbations* (lightning etc.), which should be considered through proper models (more or less detailed) based on specific hypotheses and/or experimental indications.

The analysis of the perturbed system implies the determination of voltages and currents at the most interesting points, referring to the choice or check of:

- insulation conditions, based on time behavior of overvoltages;
- conditions avoiding the arc restriking after breaker opening, based on the time behavior of the voltage at breaker terminals;
- protection setting (to be coordinated to guarantee a quick, secure, and selective intervention), based on voltage and current time behaviors;
- breakers' capacity, based on the value of interrupted currents (even in relatively short times; the automatic opening of a breaker may be performed with a delay of few tenths of second);

and so on. Therefore, the phenomena to be analyzed are, at most, relatively fast, so that the above-stated hypotheses (constancy of the speed of the machines and amplitude and speed of rotor fluxes) may be accepted.

For the generic *synchronous machine*, remembering Chapter 4 (and not considering magnetic saturation), it is easy to determine that the hypothesis of constant rotor fluxes can be translated into an equation:

$$\overline{v} = \overline{e}_{\psi R} - R\overline{\imath} - (p + j\Omega)\big((L_{\mathrm{d}})_{\psi R} i_d + (L_q)_{\psi R} j i_q\big) \qquad [5.7.1]$$

where:

- the emf $\overline{e}_{\psi R}$ is proportional to the electrical speed Ω ($\overline{e}_{\psi R}/\Omega$ depends only on rotor fluxes, and thus is constant in amplitude and phase);
- the inductances $(L_d)_{\psi R}$ and $(L_q)_{\psi R}$ are those seen from the stator (along the d and q axes, respectively) under operation at constant rotor fluxes (in pu they are $l_d(\infty)$ and $l_q(\infty)$, respectively; see Sections 4.1.3 and 4.3.2).

In practice, such inductances differ slightly from one another, so that they can both be substituted by a single value $L_{\psi R}$ (typically $L_{\psi R} = L_2$, where L_2 is the negative sequence inductance; see Section 4.4.3). The previous equation can be then written as:

$$\overline{v} \cong \overline{e}_{\psi R} - \big(R + (p + j\Omega)L_{\psi R}\big)\overline{\imath} \qquad [5.7.1']$$

The approximation $(L_d)_{\psi R} \cong (L_q)_{\psi R}$, i.e., $l_d(\infty) \cong l_q(\infty)$, actually results from the presence of additional rotor circuits, according to Section 4.3.2. If these circuits were not considered and if, therefore, the model described in Sections 4.1.2 and 4.1.3 was adopted, it would result in, at constant ψ_f:

$$\overline{v} = -R\overline{\imath} + (p + j\Omega)\overline{\psi} \qquad \text{[4.1.5$''$ repeated]}$$

$$\overline{\psi} = \frac{L_{md}}{L_f}\psi_f - (\hat{L}'_d i_d + L_q j i_q) \qquad \text{[4.1.7$'''$ repeated]}$$

from which Equation [5.7.1] with $\overline{e}_{\psi R} = j\Omega(L_{md}/L_f)\psi_f$ (recall the "transient emf" $\overline{e}'$ defined by Equation [4.2.9], at $\Omega = \omega_{\text{nom}}$ and in pu values), and $(L_d)_{\psi R} = \hat{L}'_d$, $(L_q)_{\psi R} = L_q$ where $\hat{L}'_d$ and L_q can be considerably different.

Furthermore, Equation [5.7.1] is derived from the application of Park's transformation with an angular reference θ defined by the electrical angular position of the rotor, whereas when analyzing the whole system, it is necessary to refer to a common angular reference θ_r. If it is assumed that $\delta \triangleq \theta - \theta_r$ it follows $\overline{v}_r = \overline{v}\epsilon^{j\delta}$, $\overline{\imath}_r = \overline{\imath}\epsilon^{j\delta}$, and Equation [5.7.1$'$] becomes:

$$\overline{v}_r \cong (\overline{e}_{\psi R})_r - \big(R + (p + j\Omega_r)L_{\psi R}\big)\,\overline{\imath}_r \qquad [5.7.2]$$

where:

$$(\overline{e}_{\psi R})_r = \overline{e}_{\psi R}\epsilon^{j\delta} \qquad [5.7.3]$$

and $\Omega_r \triangleq \mathrm{d}\theta_r/\mathrm{d}t$ (note that $(p\overline{\imath})\epsilon^{j\delta} = p\overline{\imath}_r - j(\Omega - \Omega_r)\overline{\imath}_r$).

For the generic *asynchronous machine*, applying the model described in Section 5.6.2 and assuming that the resulting rotor flux has a constant magnitude ψ_R and a phase $\theta_{\psi R}$ (which corresponds to the electrical rotating speed $\Omega_{\psi R} \triangleq \mathrm{d}\theta_{\psi R}/\mathrm{d}t$), it is possible to write, by adopting the loads' convention, the following equation:

$$\overline{v}_r = (\overline{e}_{\psi R})_r - (R + (p + j\Omega_r)L'_S)\overline{\imath}_r \qquad [5.7.4]$$

similar to [5.7.2], in which:

$$(\overline{e}_{\psi R})_r = j\Omega_{\psi R}\frac{L_m}{L_R}\psi_R\epsilon^{j\delta} \qquad [5.7.5]$$

with $\delta \triangleq \theta_{\psi R} - \theta_r$. Similar conclusions can be obtained with several rotor windings.

If, in terms of Park's variables, the models of network elements and "static" loads are considered, it is possible to determine that, for the system under examination:

- the *input variables* are constituted by the emfs (of the motional type) of synchronous and asynchronous machines (see Equations [5.7.3] and [5.7.5], respectively);
- the *state variables* are constituted by currents into inductive elements and voltages on capacitive elements.

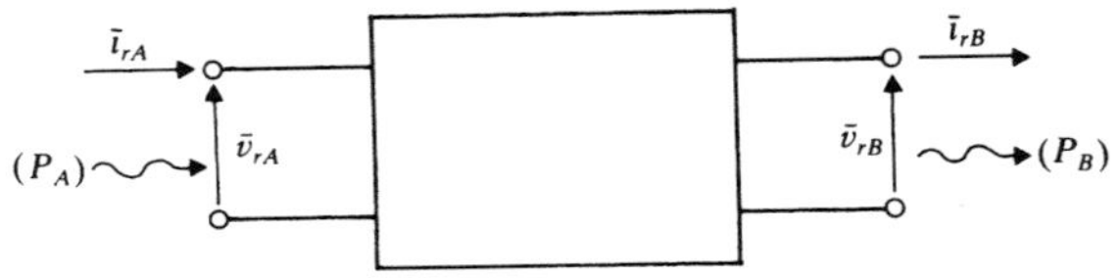

Figure 5.50. Two-port element in terms of Park's vectors.

During the transients considered, the inductive and capacitive elements can deliver or absorb active power, with consequent variation in their stored energy. As a simple example, consider a symmetrical two-port element defined (see also Fig. 5.50) by:

$$\begin{cases} \bar{\imath}_{rA} = (G_{rAA}(s) + jB_{rAA}(s))\bar{v}_{rA} + (G_{rAB}(s) + jB_{rAB}(s))\bar{v}_{rB} \\ \bar{\imath}_{rB} = -(G_{rAB}(s) + jB_{rAB}(s))\bar{v}_{rA} - (G_{rAA}(s) + jB_{rAA}(s))\bar{v}_{rB} \end{cases}$$

and assume that it is nondissipative (consequently, it only includes inductive and/or capacitive elements), so that:

$$G_{rAA}(0) = G_{rAB}(0) = 0$$

By linearizing the equations around a generic operating point (denoted by the superscript "o"), and intending $\bar{v}_{rA} = v_A \epsilon^{j\alpha_{rA}}$, $P_A = \mathcal{R}e(\bar{v}_{rA}\bar{\imath}^*_{rA})$ etc., it follows that:

$$\begin{aligned} \Delta P_A = {} & [B_{rAB}(0)v_B^o \sin(\alpha_{rA} - \alpha_{rB})^o + G_{rAA}(s)v_A^o]\Delta v_A \\ & + [B_{rAB}(s)v_A^o \sin(\alpha_{rA} - \alpha_{rB})^o + G_{rAB}(s)v_A^o \cos(\alpha_{rA} - \alpha_{rB})^o]\Delta v_B \\ & + [(B_{rAA}(0) - B_{rAA}(s))v_A^{o2} + B_{rAB}(0)v_A^o v_B^o \cos(\alpha_{rA} - \alpha_{rB})^o]\Delta\alpha_{rA} \\ & + [-B_{rAB}(s)\cos(\alpha_{rA} - \alpha_{rB})^o + G_{rAB}(s)\sin(\alpha_{rA} - \alpha_{rB})^o]v_A^o v_B^o \Delta\alpha_{rB} \end{aligned}$$

and similarly for $(-\Delta P_B)$. It is easily determined that ΔP_A and ΔP_B are not equal (apart from the condition $s = 0$, i.e., at the end of transients).
Moreover, the dependence of ΔP_A and ΔP_B on the phase variations $\Delta\alpha_{rA}$, $\Delta\alpha_{rB}$ cannot be related (apart from the condition $s = 0$) to a dependence on the phase-shift variation $\Delta(\alpha_{rA} - \alpha_{rB})$ only.

The following *simplifying assumptions* will be used in the development of the analysis:

(1) the system is linear;
(2) for $t < 0$, the system is at steady-state and thus, more particularly:
- the system is physically symmetrical;
- the electrical speeds related to the rotor fluxes of the synchronous ($\Omega = \mathrm{d}\theta/\mathrm{d}t$) and asynchronous ($\Omega_{\psi R} = \mathrm{d}\theta_{\psi R}/\mathrm{d}t$) machines are all equal;
- for Park's variables, by assuming $\Omega_r = \Omega$ (equal to the above-mentioned electrical speed), all the emfs $(\bar{e}_{\psi R})_r$ (see Equations [5.7.3] and [5.7.5]) acting on the system are constant;

- for phase variables, the operating condition is sinusoidal and of the positive sequence at the frequency Ω;

(3) the configuration of the system is subject to a perturbation of the type mentioned (i.e., opening or closing), which is actuated instantaneously at $t = 0$.

Within this concern, it can be observed that:

- the hypothesis (1) can be considered verified, if nonlinear static loads have little effect on the phenomenon under examination, and can be accounted for by proper linear composite loads (in practical cases, it may be sufficient that nonlinear loads are far enough both from the point of application of the perturbation, and from the system part where the response to the perturbation is of more interest for the analysis);
- the hypotheses (2) and (3) exclude the case of multiple perturbations, such as a short-circuit followed by the opening (and possible reclosing) of a breaker, or an arc restriking after the opening of a breaker, etc.;
- the hypothesis (3) implies the ideal operation of breakers (instantaneous arc extinction during the opening, etc.).

For greater generality, both “symmetrical” and “nonsymmetrical” perturbations will be considered. The former (three-phase symmetrical opening or closing) are the simplest to analyze. However, it is not always true, as in the following, that they are the most severe type (even if they affect all the three phases). Furthermore, they are less likely to happen and can be considered somewhat unrealistic, because of the nonsimultaneous opening of the three poles of breakers, or the nonsimultaneous short-circuit on the three phases.

Moreover, the typical nonsymmetrical perturbations are:

- opening of a single phase, e.g., as the result of breaker operation (activated by the protection system) following a single-phase short-circuit;
- opening of two phases, because of the failed opening of the third phase by the breaker;
- single-phase (phase-to-earth) short-circuit, which is the most frequent fault, caused by insulation failure or accidental contact;
- two-phase to earth short-circuit (phase-to-phase-to-earth), which is quite frequent, also as a consequence of a single-phase short-circuit;
- two-phase isolated short-circuit (phase-to-phase), which is rare and, e.g., may occur because of the action of wind on two windings.

5.7.2. Response to Symmetrical Perturbations

If the perturbation is symmetrical, the system remains physically symmetrical even for $t > 0$. Consequently it can be examined by considering the Park’s vectors, avoiding the consideration of homopolar components.

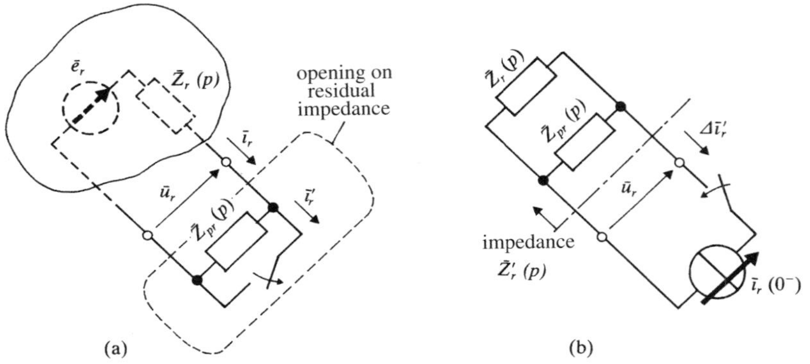

Figure 5.51. Opening of a link: (a) circuit under examination; (b) equivalent circuit, for the determination of the voltage $\overline{u}_r(t)$ caused by the opening.

Furthermore, since the system is at steady-state for $t < 0$ and inputs (that is vectors $(\overline{e}_{\psi R})_r$, with $\Omega_r = \Omega$) remain constant, the "response" to the assigned perturbation is determined by the initial ($t = 0$) values of the state variables, which no longer correspond to a steady-state condition. As a consequence, the problem becomes evaluation of the "free" response of the system in its new configuration.

For the opening of a link, it is possible to refer to Figure 5.51a, where it is intended (for greater generality) that the opening occurs on a residual impedance $\overline{Z}_{pr}(p)$ (in parallel to the terminals), so that:

$$\overline{\imath}'_r = \overline{\imath}_r - \frac{\overline{u}_r}{\overline{Z}_{pr}(p)}$$

whereas the rest of the system is represented by:

- a constant emf $\overline{e}_r$ (vector) resulting from the emfs of the machines;
- an output impedance $\overline{Z}_r(p)$, equal to the impedance seen from the examined terminals.

It is then possible to derive:

$$\overline{u}_r = \overline{e}'_r - \overline{Z}'_r(p)\overline{\imath}'_r$$

where $\overline{e}'_r \triangleq \overline{Z}_{pr}\overline{e}_r/(\overline{Z}_r + \overline{Z}_{pr})$, $\overline{Z}'_r \triangleq \overline{Z}_r\overline{Z}_{pr}/(\overline{Z}_r + \overline{Z}_{pr})$. (At $t = 0^-$ it is $\overline{u}_r(0^-) = 0$ and the system is at equilibrium; thus: $\overline{\imath}'_r(0^-) = \overline{\imath}_r(0^-) = \overline{e}'_r/\overline{Z}'_r(0) = \overline{e}_r/\overline{Z}_r(0)$.)

Similarly, when closing a link (on a residual impedance $\overline{Z}_{sr}(p)$, in series to the terminals) it is possible to write, according to Figure 5.52a:

$$\overline{v}'_r = \overline{v}_r - \overline{Z}_{sr}(p)\overline{\jmath}_r$$

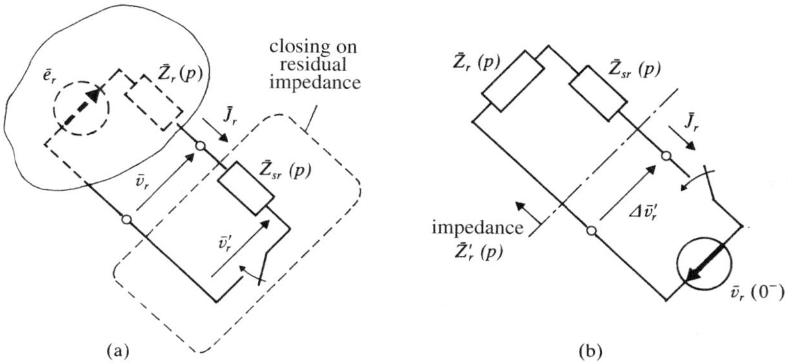

Figure 5.52. Closing of a link: (a) circuit under examination; (b) equivalent circuit, for the determination of the current $\overline{\jmath}_r(t)$ caused by the closing.

and furthermore:

$$\overline{\jmath}_r = \frac{\overline{e}_r - \overline{v}'_r}{\overline{Z}'_r(p)}$$

where $\overline{Z}'_r \triangleq \overline{Z}_r + \overline{Z}_{sr}$. (At $t = 0^-$ it is $\overline{\jmath}_r(0^-) = 0$ and the system is at equilibrium; thus: $\overline{v}'_r(0^-) = \overline{v}_r(0^-) = \overline{e}_r$.)

The voltage $\overline{u}_r(t)$ caused by the opening, with $\overline{u}_r(0^-) = 0$, can be determined as the "forced" response of the only impedance $\overline{Z}'_r(p)$ to a current step ("equivalent injection") equal to $-\overline{\imath}_r(0^-)$, according to Figure 5.51b. Similarly, the current $\overline{\jmath}_r(t)$ caused by the closing, with $\overline{\jmath}_r(0^-) = 0$, can be determined as the forced response of $1/\overline{Z}'_r(p)$ to a voltage step equal to $-\overline{v}_r(0^-)$, according to Figure 5.52b.

In both cases it must be assumed that:

- the emfs are short-circuited;
- the state variables have a zero initial value (i.e., the inductive and capacitive elements are initially "discharged").

On the other hand, in terms of Laplace transforms:

- at opening (Fig. 5.51):

$$(\mathcal{L}\overline{u}_r)(s) = (\mathcal{L}\Delta\overline{u}_r)(s) = -\overline{Z}'_r(s)(\mathcal{L}\Delta\overline{\imath}'_r)(s) = \overline{Z}'_r(s)\frac{\overline{\imath}_r(0^-)}{s} \qquad [5.7.6]$$

- at closing (Fig. 5.52):

$$(\mathcal{L}\overline{\jmath}_r)(s) = (\mathcal{L}\Delta\overline{\jmath}_r)(s) = -\frac{1}{\overline{Z}'_r(s)}(\mathcal{L}\Delta\overline{v}'_r)(s) = \frac{1}{\overline{Z}'_r(s)}\frac{\overline{v}_r(0^-)}{s} \qquad [5.7.7]$$

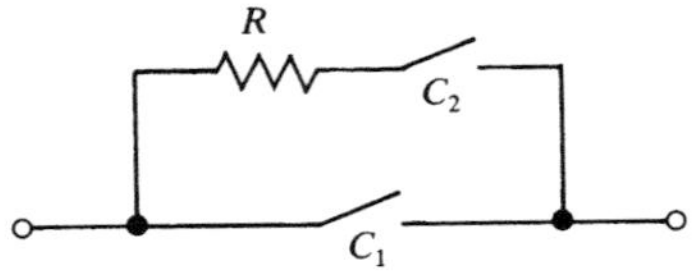

Figure 5.53. Schematic representation of a breaker with preinsertion resistor.

The above can be applied to determine the voltage on a breaker after opening or the current after closing or after a short-circuit.

For a short-circuit, it is evident that a *"fault" impedance* $\overline{Z}_{sr}(p)$ (Fig. 5.52) can modify the current behavior $\overline{\jmath}_r(t)$ and reduce its values. For similar reasons, it can be useful to make breakers in accordance with Figure 5.53, i.e., equipped with a pair of contacts C_1 and C_2 and a *"preinsertion resistor"* R, so that manoeuvres can be performed more gradually, and less cumbersome. When opening, the breaker C_1 is operated before C_2; when closing, C_2 is operated before C_1. In both cases, the resistor R remains temporarily in series in the circuit. Therefore it can smooth the voltage or current transients. The case under examination is outside of the present treatment, because it does not imply a single perturbation, but the effect of R is apparent from a qualitative point of view.

Steady-State Response

For phase variables, the steady-state response is constituted by a sinusoidal steady-state of the positive sequence at frequency Ω, which can be derived based on the new (constant) values assumed by Park's vectors. Specifically, at the point where the perturbation is applied:

- at opening (see Fig. 5.51 and Equation [5.7.6]): $\overline{u}_r = \overline{e}'_r = \overline{Z}'_r(0)\overline{\imath}_r(0^-)$;
- at closing (see Fig. 5.52 and Equation [5.7.7]): $\overline{\jmath}_r = \overline{e}_r/\overline{Z}'_r(0) = \overline{v}_r(0^-)/\overline{Z}'_r(0)$;

where $\overline{Z}'_r(0)$ is the impedance evaluated at the frequency $\Omega_r = \Omega$, of the type $\overline{Z}'_r(0) = Z'(j\Omega)$ (recall Equation [5.2.8]). The deduction of all the other variables at different points of the network is similarly obvious.

However, the above data imply that the characteristic roots defined by Equation [5.7.6] or [5.7.7] (i.e., the poles of $|\overline{Z}'_r(s)|$ or of $|1/\overline{Z}'_r(s)|$), respectively in the two cases, have a negative real part, so that they contribute to the transient by damped components which tend to zero. Such a condition is actually satisfied because of resistances present in the network (or in the residual impedance: preinsertion resistance, fault resistance, etc.).

For a three-phase short-circuit with zero fault impedance (see Fig. 5.52, with $\overline{Z}_{sr} = 0$) $\overline{\jmath}_r = \overline{v}_r(0^-)/\overline{Z}_r(0)$. If at $t = 0^-$ the voltage amplitude had its nominal value, i.e., $v_r(0^-) = \sqrt{3}V_{(F)\text{nom}}$, where $V_{(F)\text{nom}}$ is the nominal rms value of the phase voltages, it would follow that $|\overline{v}_r(0^-)\overline{\jmath}_r^*| = 3V_{(F)\text{nom}}^2/|\overline{Z}_r(0)|$. This last quantity is the so-called "*short-circuit power*" of the node under consideration, referring to the (relatively fast) purely electrical phenomena considered, with machines simply represented by Equations [5.7.2] and [5.7.4]. For (longer term) steady-state response, similar definitions can be adopted (even considering zero output impedances of the machines because of voltage regulation etc.). It is possible to determine, with reference to practical cases (predominantly inductive network), that:

- The addition of links, i.e., a more meshed network, causes an increase in the short-circuit power at different nodes.
- If the short-circuit power at a given node is increased, short-circuits at the other nodes have a smaller effect on the voltage at the node under examination (the node becomes "*stronger*"). However, a fault at this node implies a larger short-circuit current and greater effects on other node voltages.

As an elementary example, also used in the following:

- consider the system defined in Figure 5.54a, which corresponds (in terms of Park's vectors and homopolar components, respectively), to the equivalent circuits in Figure 5.54b, assuming that the emfs e_{ga}, e_{gb}, e_{gc} are sinusoidal and of positive sequence, with:

$$e_{ga}(t) = E_{gM}\cos(\Omega t + \alpha), \quad \text{etc.} \qquad [5.7.8]$$

- as a perturbation, assume the opening or short-circuit at the indicated nodes.

As evident, the case under examination can correspond to a line terminated with a reactive load or short-circuited (with parameters $L = L_1 + L_2, C$), and supplied by a node of infinite short-circuit power (with emf $\overline{e}_{gr}$).

The perturbations then correspond, respectively, to the opening of the supply-side breaker (with preinsertion resistances ρ_p) and to a line short-circuit (with fault resistances ρ_s, ρ_{st}). Furthermore, the emfs are wye-connected, with Z_{nt1} the (scalar) operational impedance between neutral and earth. The (scalar) impedances Z_{t12} and Z_{t23} define the earth circuit, according to Figure 5.54a. However, these last three impedances, as well as ρ_{st}, are related only to the homopolar components and thus do not intervene in the present analysis.

For the (three-phase) opening it follows, after minor developments:

$$\overline{u}_r = \frac{\overline{e}_{gr}}{1 + \dfrac{j\Omega L}{\rho_p(1 - \Omega^2 LC)}}$$

where the vector $\overline{u}_r$ is the voltage across the breaker. For the (three-phase) short-circuit:

$$\overline{\jmath}_r = \frac{\overline{e}_{gr}}{\dfrac{\rho_s L}{L_2} + j\Omega L_1}$$

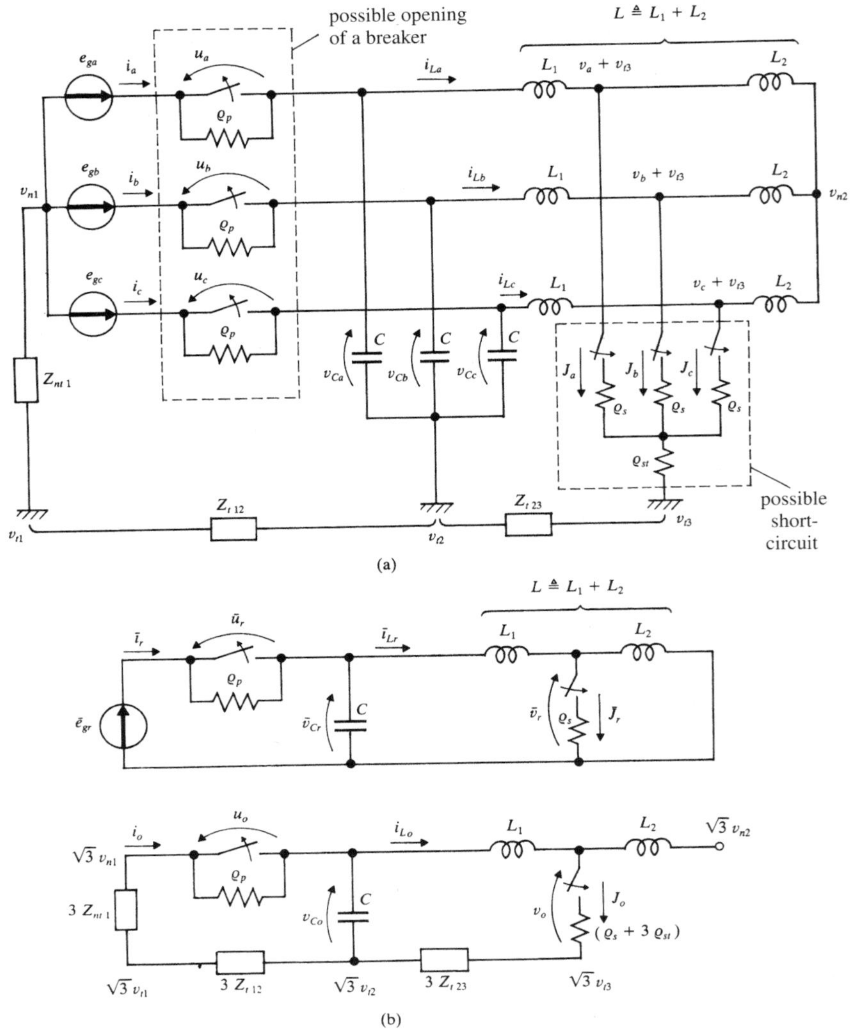

Figure 5.54. Elementary example of three-phase system: (a) circuit under examination; (b) equivalent circuits.

where the vector $\overline{\jmath}_r$ is the short-circuit current. (Here and in the following, other network variables will not be evaluated.) More particularly, as evident, the resistance ρ_p reduces the amplitude of $\overline{u}_r$ (with respect to the case $\rho_p = \infty$). Similarly, ρ_s reduces the amplitude of $\overline{\jmath}_r$ (with respect to the case $\rho_s = 0$).

Transient Response

The transient response includes, in addition to the steady-state response, the purely transient (i.e., going to zero) components related to the characteristic roots of the system (it has been assumed that these roots have negative real parts). By using the notation of Figures 5.51 and 5.52, the response $\overline{u}_r(t)$ (to link opening) or $\overline{\jmath}_r(t)$ (to link closing) can be obtained by inverse transformation of Equation [5.7.6] or [5.7.7], respectively.

In Figure 5.54, the damping of the transient components is clearly the result of (noninfinite) resistance ρ_p after opening and (nonzero) resistance ρ_s after short-circuit. However, to simplify the analytical deduction of the transient responses, it will be assumed here and in the following that $\rho_p = \infty$ and $\rho_s = 0$, which imply purely imaginary characteristic roots. In practice, the above-mentioned resistances (and those of lines etc.) can result in reduction of the amplitude of the transient response and damping of the transient components.

For the (three-phase) *opening* with $\rho_p = \infty$, Equation [5.7.6] becomes:

$$(\mathcal{L}\overline{u}_r)(s) = \overline{Z}_r(s)\frac{\overline{\imath}_r(0^-)}{s}$$

where $\overline{Z}_r(s)$ (impedance seen from the breaker terminals) is, by adopting Figure 5.54:

$$\overline{Z}_r(s) = \frac{(s + j\Omega)L}{1 + (s + j\Omega)^2 LC}$$

whereas $\overline{\imath}_r(0^-) = (1 - \Omega^2 LC)\overline{e}_{gr}/(j\Omega L)$; by inverse transformation of the expression for $(\mathcal{L}\overline{u}_r)(s)$ and setting $\Omega' \triangleq \Omega - 1/\sqrt{LC}$, $\Omega'' \triangleq \Omega + 1/\sqrt{LC}$:

$$\overline{u}_r(t) = \overline{e}_{gr}\left(1 - \frac{1}{2\Omega}(\Omega''\epsilon^{-j\Omega' t} + \Omega'\epsilon^{-j\Omega'' t})\right)$$

where the first term, equal to $\overline{e}_{gr}$, corresponds to the steady-state response previously derived for $\rho_p \to \infty$.

For phase variables, by using the equations in Appendix 2 and recalling Equation [5.7.8], we can derive:

$$u_a(t) = E_{gM}\cos(\Omega t + \alpha) - E_{gM}\left(\cos\alpha\cos(\nu_o t) - \frac{\nu_o}{\Omega}\sin\alpha\sin(\nu_o t)\right)$$

where:

$$\nu_o \triangleq \frac{1}{\sqrt{LC}}$$

and further similar expressions for $u_b(t)$ and $u_c(t)$.

In the expression of $u_a(t)$, the first term (equal to $e_{ga}(t)$), is the so-called "recovery voltage," which corresponds to the steady-state response (for $\rho_p \to \infty$), at frequency Ω and with amplitude E_{gM}.

The remaining term, which defines the voltage v_{Ca} on the condenser of the phase a, is sinusoidal at the "resonance frequency" ν_o. Its amplitude depends on

α, i.e., or on the instant at which the opening occurs (recall Equation [5.7.8]). Indicating such amplitude by U'_M, it is trivial to derive:

$$U'_M = E_{gM}\sqrt{(\cos\alpha)^2 + \left(\frac{\nu_o}{\Omega}\sin\alpha\right)^2}$$

so that U'_M is maximum at $\alpha = 90°$ ($U'_M = E_{gM}\nu_o/\Omega$) and minimum at $\alpha = 0°$ ($U'_M = E_{gM}$), if $\nu_o > \Omega$ (as usually occurs), and vice versa in the opposite case (see also Figure 5.60b). The variations with time of the voltage v_{Ca} on the condenser and current i_{La} on the inductor (both for the phase a) are indicated in Figure 5.55, under the hypothesis $\nu_o > \Omega$ and:

- for $\alpha = 0°$, i.e., for an opening at an instant of maximum voltage $v_{Ca}(t)$;
- for $\alpha = 90°$, i.e., for an opening at an instant of maximum current $i_{La}(t)$.

The maximum value instantaneously assumed by the voltage $u_a(t)$ (equal to that assumed by $u_b(t)$ and $u_c(t)$) also can be determined. If the frequency ν_o is different enough from Ω, such a value approximately equals $(E_{gM} + U'_M)$, which is intermediate between $2E_{gM}$ and $(1 + \nu_o/\Omega)E_{gM}$.

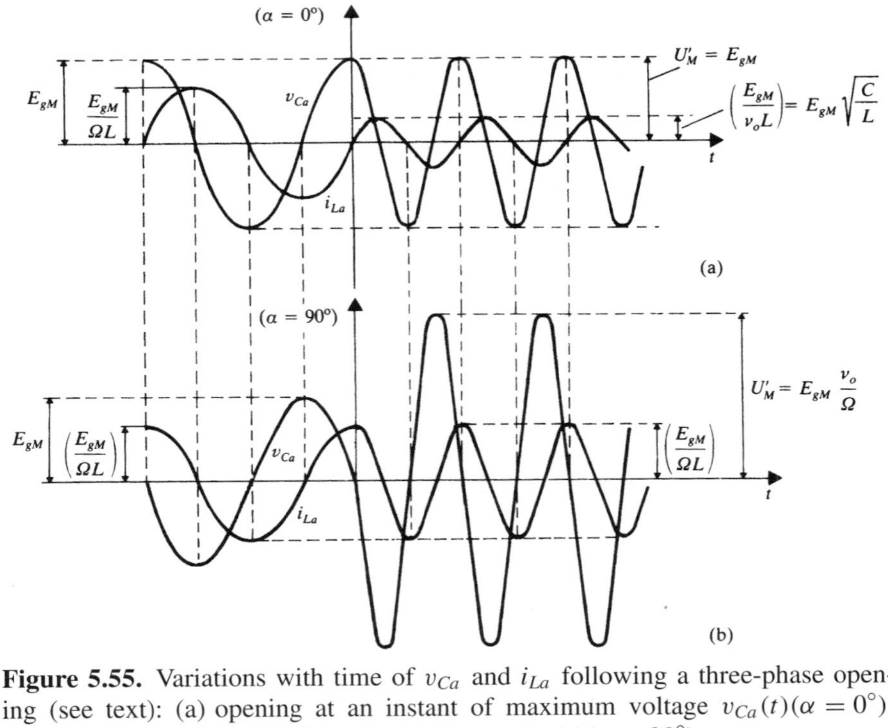

Figure 5.55. Variations with time of v_{Ca} and i_{La} following a three-phase opening (see text): (a) opening at an instant of maximum voltage $v_{Ca}(t)(\alpha = 0°)$; (b) opening at an instant of maximum current $i_{La}(t)(\alpha = 90°)$.

For the (three-phase) *short-circuit* with $\rho_s = 0$, Equation [5.7.7] becomes:

$$(\mathcal{L}\overline{\mathcal{J}}_r)(s) = \frac{1}{\overline{Z}_r(s)} \frac{\overline{v}_r(0^-)}{s}$$

where now it is:

$$\overline{Z}_r(s) = (s + j\Omega)\frac{L_1 L_2}{L_1 + L_2}$$

whereas $\overline{v}_r(0^-) = L_2 \overline{e}_{gr}/(L_1 + L_2)$; by inverse transformation of the expression for $(\mathcal{L}\overline{\mathcal{J}}_r)(s)$ it follows that:

$$\overline{\mathcal{J}}_r(t) = \frac{\overline{e}_{gr}}{j\Omega L_1}(1 - \epsilon^{-j\Omega t})$$

where the term $\overline{e}_{gr}/(j\Omega L_1)$ corresponds to the steady-state response previously determined for $\rho_s \rightarrow 0$.

For phase variables, it is possible to derive:

$$\mathcal{J}_a(t) = \frac{E_{gM}}{\Omega L_1}(\sin(\Omega t + \alpha) - \sin\alpha)$$

and similarly $\mathcal{J}_b(t)$, $\mathcal{J}_c(t)$.

In the expression of $\mathcal{J}_a(t)$, the term $(E_{gM}/(\Omega L_1))\sin(\Omega t + \alpha)$ corresponds to the steady-state response (for $\rho_s \rightarrow 0$), at frequency Ω and with amplitude $E_{gM}/(\Omega L_1)$.

The remaining term is constant[7], and its absolute value J'_M depends on α (i.e., on the instant at which the short-circuit occurs), being zero for $\alpha = 0°$ and maximum for $\alpha = \pm 90°(J'_M = E_{gM}/(\Omega L_1))$ (see Figures 5.61b and 5.62b). Correspondingly, the maximum absolute value instantaneously assumed by $\mathcal{J}_a(t)$ (and by $\mathcal{J}_b(t)$, $\mathcal{J}_c(t)$) can vary between $E_{gM}/(\Omega L_1)$ and $2E_{gM}/(\Omega L_1)$.

5.7.3. Response to Nonsymmetrical Perturbations

If the perturbation is nonsymmetrical, the system loses (for $t > 0$) its physical symmetry, and the analysis must also consider, in general, the homopolar components.

On the other hand, any assigned perturbation can be translated into precise conditions on phase variables and thus Park's variables, based on the equations in Appendix 2. For instance, the zero-impedance short-circuit between phases b and c (isolated from earth) can be translated into the conditions $\mathcal{J}_a = \mathcal{J}_b + \mathcal{J}_c = 0$,

---

[7] The constant term represents, for the phase a, the "unidirectional" component of the short-circuit transient, similarly to that in Section 4.4.4 with reference to the synchronous machine. In practice, such a component is damped because of circuit resistances.

$v_b - v_c = 0$, from which it can be derived:

$$\begin{cases} \mathcal{I}_o = 0 \\ \overline{\mathcal{I}}_r \epsilon^{j\theta_r} + (\overline{\mathcal{I}}_r \epsilon^{j\theta_r})^* = 0 \\ \overline{v}_r \epsilon^{j\theta_r} - (\overline{v}_r \epsilon^{j\theta_r})^* = 0 \end{cases}$$

where θ_r is linearly varying with time, with $\mathrm{d}\theta_r/\mathrm{d}t = \Omega_r = \Omega$. The first of these conditions is a simplification (as it is possible not to consider the homopolar components), whereas the two remaining ones can be used together with an equation:

$$\overline{v}_r = \overline{e}_r - \overline{Z}_r(p)\overline{\mathcal{I}}_r$$

(Fig. 5.52a) to determine $\overline{v}_r(t)$, $\overline{\mathcal{I}}_r(t)$ and consequently the behavior of different phase variables.

However, the procedure indicated here is cumbersome, as is evident from the previous example (which is a simplified one), because of the dependence of θ_r on time. Therefore, it may be convenient to use other procedures, based for instance on the definition of "symmetrical components" (vectors, or phasors) for the determination of steady-state response, and on direct analysis (in terms of phase variables) for the determination of transient response.

Steady-State Response

For phase variables, the steady-state response is constituted, under the usual hypothesis of damped transient components, by a sinusoidal operation at frequency Ω.

However, each three-phase set generally includes components of the three sequences (positive, negative, and zero). Therefore, the generic three-phase set can be posed in the following form:

$$\left.\begin{aligned} w_a &= W_{M(0)}\cos(\Omega t + \alpha_{(0)}) + W_{M(1)}\cos(\Omega t + \alpha_{(1)}) + W_{M(2)}\cos(\Omega t + \alpha_{(2)}) \\ w_b &= W_{M(0)}\cos(\Omega t + \alpha_{(0)}) + W_{M(1)}\cos(\Omega t + \alpha_{(1)} - 120^\circ) \\ &\quad + W_{M(2)}\cos(\Omega t + \alpha_{(2)} - 240^\circ) \\ w_c &= W_{M(0)}\cos(\Omega t + \alpha_{(0)}) + W_{M(1)}\cos(\Omega t + \alpha_{(1)} - 240^\circ) \\ &\quad + W_{M(2)}\cos(\Omega t + \alpha_{(2)} - 120^\circ) \end{aligned}\right\} \quad [5.7.9]$$

If the following constant vectors are defined (which correspond to the "*symmetrical components*"; see Appendices 1 and 2):

$$\left.\begin{aligned} \overline{w}_{(0)} &\triangleq \sqrt{\frac{3}{2}} W_{M(0)} \epsilon^{j\alpha_{(0)}} \\ \overline{w}_{(1)} &\triangleq \sqrt{\frac{3}{2}} W_{M(1)} \epsilon^{j\alpha_{(1)}} \\ \overline{w}_{(2)} &\triangleq \sqrt{\frac{3}{2}} W_{M(2)} \epsilon^{j\alpha_{(2)}} \end{aligned}\right\} \quad [5.7.10]$$

the previous equations can be rewritten as:

$$\left.\begin{aligned} w_a &= \sqrt{\frac{2}{3}}\mathcal{R}e[(\overline{w}_{(0)} + \overline{w}_{(1)} + \overline{w}_{(2)})\epsilon^{j\Omega t}] \\ w_b &= \sqrt{\frac{2}{3}}\mathcal{R}e[(\overline{w}_{(0)} + \overline{w}_{(1)}\epsilon^{-j120^\circ} + \overline{w}_{(2)}\epsilon^{-j240^\circ})\epsilon^{j\Omega t}] \\ w_c &= \sqrt{\frac{2}{3}}\mathcal{R}e[(\overline{w}_{(0)} + \overline{w}_{(1)}\epsilon^{-j240^\circ} + \overline{w}_{(2)}\epsilon^{-j120^\circ})\epsilon^{j\Omega t}] \end{aligned}\right\} \quad [5.7.11]$$

(see Equations [A2.9] and [A2.11]).

The procedure based on the Equations [5.7.10], for the determination of the steady-state response, may be illustrated by considering the systems in Figures 5.56a (breaker opening) and 5.57a (short-circuit).

In both cases, the rest of the system (including the earth circuit) can be represented by equivalent circuits as shown, for the three sequences, in Figures 5.56b and 5.57b. More precisely:

- $\overline{e}_{(1)}$ is the constant emf (vector) resulting from different emfs present in the system, and it acts on the positive sequence circuit (because of the adopted hypotheses, all the single emfs are of the positive sequence, so that in the other two circuits there is no emf);

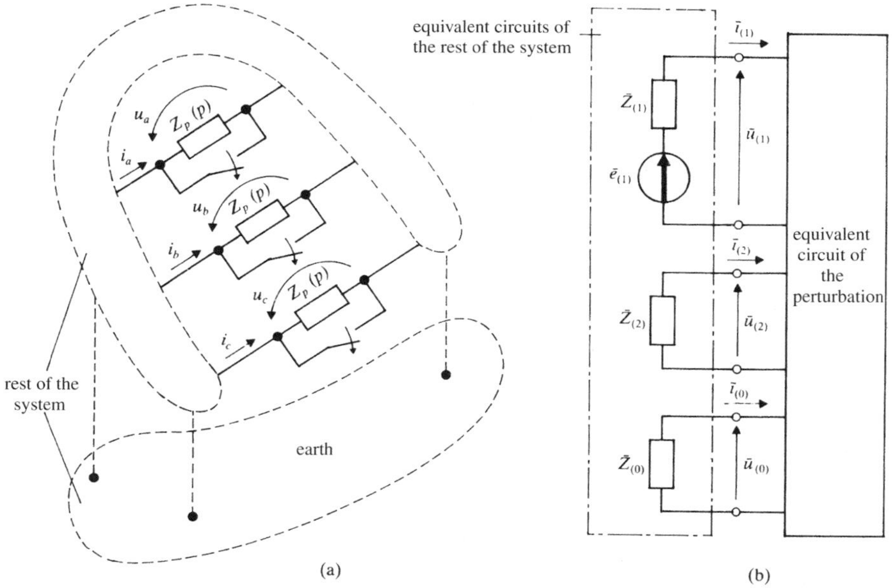

Figure 5.56. Opening of one or more phases: (a) system under examination; (b) equivalent circuit for the determination of the steady-state response.

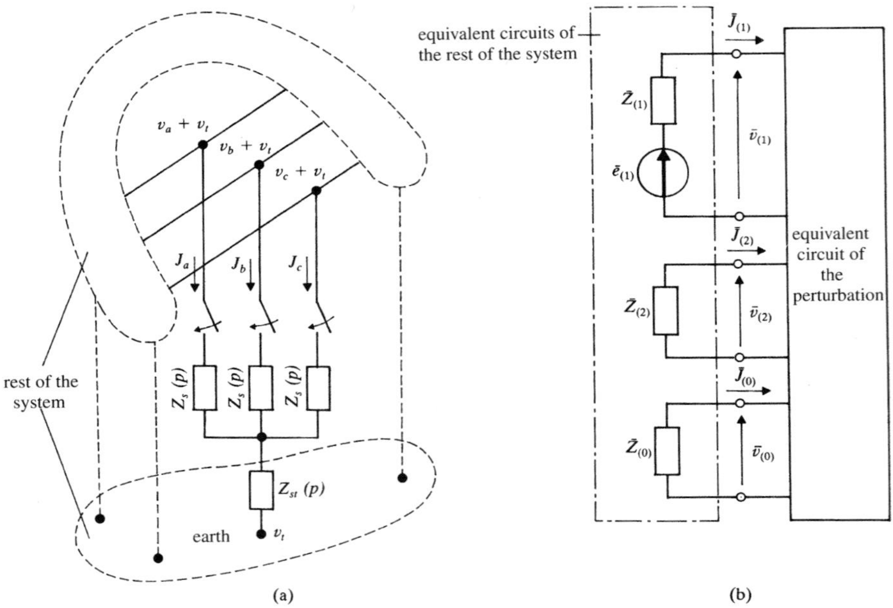

Figure 5.57. Short-circuit on one or more phases: (a) system under examination; (b) equivalent circuit for the determination of the steady-state response.

- $\overline{Z}_{(1)}, \overline{Z}_{(2)}, \overline{Z}_{(0)}$ are the output impedances of the system, under operation at frequency Ω, respectively for the positive, negative, and zero sequences.

It is easy to determine that, if $\overline{e}_r$ and $\overline{Z}_r(p)$ are the emf and the operational impedance relative to the Park's vectors (see Figs. 5.51a and 5.52a, respectively, for the opening and short-circuit), it follows that:

$$\overline{e}_{(1)} = \overline{e}_r$$
$$\overline{Z}_{(1)} = \overline{Z}_{(2)} = \overline{Z}_r(0)$$

whereas, if $Z_o(p)$ is the (scalar) operational impedance relative to the homopolar components:

$$\overline{Z}_{(0)} = Z_o(j\Omega)$$

In Figures 5.56b and 5.57b, the same symbols $\overline{e}_{(1)}, \overline{Z}_{(1)}$ etc. are used for simplicity, although the respective values are different in the two cases.
Moreover, the definitions of $\overline{Z}_{(1)}$ and $\overline{Z}_{(2)}$ and the equality $\overline{Z}_{(1)} = \overline{Z}_{(2)}$, are strictly related to the adoption of Equations [5.7.2] and [5.7.4] for the machines, as a consequence of the hypotheses already illustrated.
For the actual behavior of the synchronous machine, see Section 4.4.3.

Furthermore, by recalling Equations [5.7.11], each perturbation can be translated into an equivalent circuit in terms of symmetrical components, according to the following examples. For brevity, it is assumed that:

$$\overline{Z}_p \triangleq Z_p(j\Omega), \quad \overline{Z}_s \triangleq Z_s(j\Omega), \quad \overline{Z}_{st} \triangleq Z_{st}(j\Omega)$$

where the scalar impedances Z_p and Z_s, Z_{st} are defined in Figures 5.56a and 5.57a, respectively, and furthermore:

$$h \triangleq \epsilon^{+j120^\circ} = \epsilon^{-j240^\circ}$$

For two-phase perturbations, a simplification is obtained by assuming the (possible) residual impedances on the phases to be equal, i.e., Z_p after the opening or Z_s after the short-circuit.

- *Single-phase opening* (phase a):

$$\begin{cases} u_a = Z_p i_a \\ u_b = 0 \\ u_c = 0 \end{cases} \qquad \begin{cases} \overline{u}_{(0)} + \overline{u}_{(1)} + \overline{u}_{(2)} = \overline{Z}_p(\overline{\imath}_{(0)} + \overline{\imath}_{(1)} + \overline{\imath}_{(2)}) \\ \overline{u}_{(0)} + h^2\overline{u}_{(1)} + h\overline{u}_{(2)} = 0 \\ \overline{u}_{(0)} + h\overline{u}_{(1)} + h^2\overline{u}_{(2)} = 0 \end{cases}$$

from which (Fig. 5.58a):

$$\overline{u}_{(0)} = \overline{u}_{(1)} = \overline{u}_{(2)} = \frac{\overline{Z}_p}{3}(\overline{\imath}_{(0)} + \overline{\imath}_{(1)} + \overline{\imath}_{(2)})$$

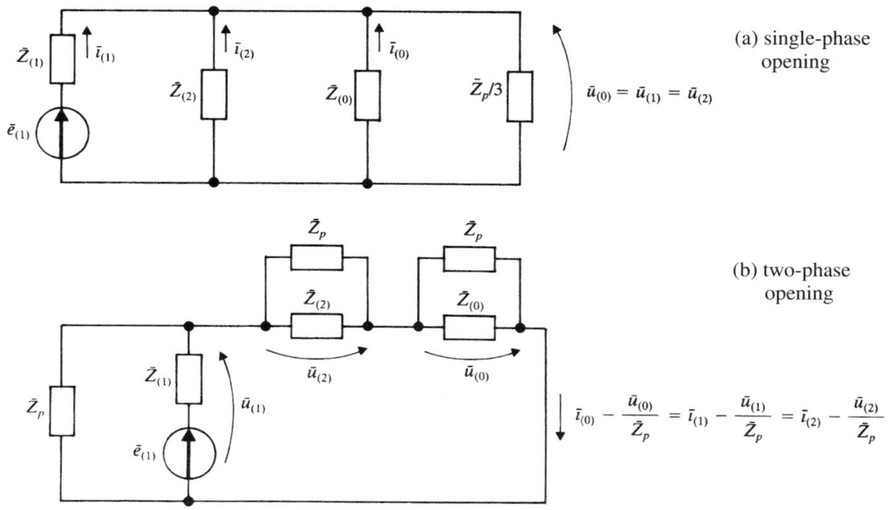

Figure 5.58. Determination of the steady-state response for the opening (a) single-phase, (b) two-phase.

- *Two-phase opening* (phases b, c):

$$\begin{cases} u_a = 0 \\ u_b = Z_p i_b \\ u_c = Z_p i_c \end{cases} \qquad \begin{cases} \overline{u}_{(0)} + \overline{u}_{(1)} + \overline{u}_{(2)} = 0 \\ \overline{u}_{(0)} + h^2\overline{u}_{(1)} + h\overline{u}_{(2)} = \overline{Z}_p(\bar{\imath}_{(0)} + h^2\bar{\imath}_{(1)} + h\bar{\imath}_{(2)}) \\ \overline{u}_{(0)} + h\overline{u}_{(1)} + h^2\overline{u}_{(2)} = \overline{Z}_p(\bar{\imath}_{(0)} + h\bar{\imath}_{(1)} + h^2\bar{\imath}_{(2)}) \end{cases}$$

from which (Fig. 5.58b):

$$\overline{u}_{(0)} + \overline{u}_{(1)} + \overline{u}_{(2)} = 0$$

$$\bar{\imath}_{(0)} - \frac{\overline{u}_{(0)}}{\overline{Z}_p} = \bar{\imath}_{(1)} - \frac{\overline{u}_{(1)}}{\overline{Z}_p} = \bar{\imath}_{(2)} - \frac{\overline{u}_{(2)}}{\overline{Z}_p}$$

- *Single-phase short-circuit* (phase a):

$$\begin{cases} v_a = (Z_s + Z_{st})\mathcal{J}_a \\ \mathcal{J}_b = 0 \\ \mathcal{J}_c = 0 \end{cases} \qquad \begin{cases} \overline{v}_{(0)} + \overline{v}_{(1)} + \overline{v}_{(2)} = (\overline{Z}_s + \overline{Z}_{st})(\overline{\mathcal{J}}_{(0)} + \overline{\mathcal{J}}_{(1)} + \overline{\mathcal{J}}_{(2)}) \\ \overline{\mathcal{J}}_{(0)} + h^2\overline{\mathcal{J}}_{(1)} + h\overline{\mathcal{J}}_{(2)} = 0 \\ \overline{\mathcal{J}}_{(0)} + h\overline{\mathcal{J}}_{(1)} + h^2\overline{\mathcal{J}}_{(2)} = 0 \end{cases}$$

from which (Fig. 5.59a):

$$\overline{\mathcal{J}}_{(0)} = \overline{\mathcal{J}}_{(1)} = \overline{\mathcal{J}}_{(2)} = \frac{1}{3(\overline{Z}_s + \overline{Z}_{st})}(\overline{v}_{(0)} + \overline{v}_{(1)} + \overline{v}_{(2)})$$

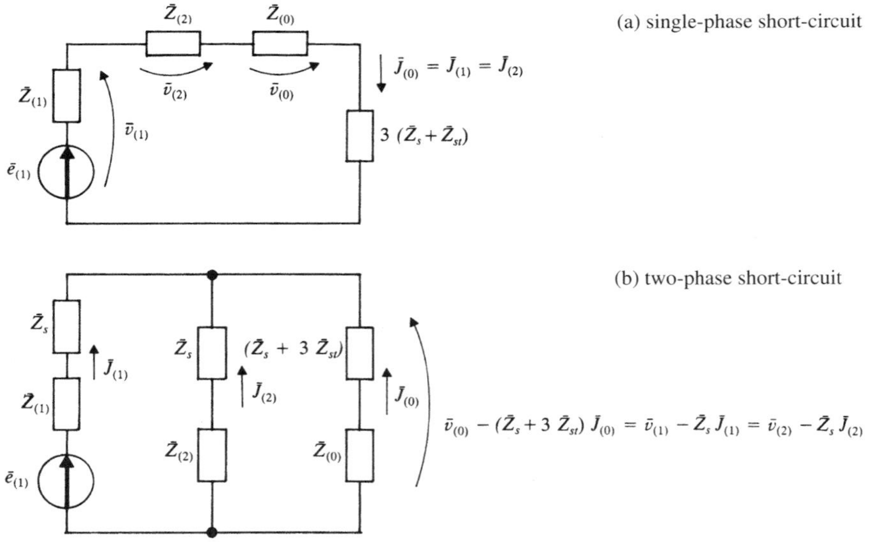

Figure 5.59. Determination of the steady-state response for the short-circuit (a) single-phase, (b) two-phase. For two-phase isolated short-circuit, it holds $\overline{Z}_{st} = \infty$, $\overline{\mathcal{J}}_{(0)} = 0$, so that the branch at the extreme right side can be eliminated.

- *Two-phase to earth short-circuit* (phases b, c, and earth):

$$\begin{cases} \mathcal{I}_a = 0 \\ v_b = Z_s \mathcal{I}_b + Z_{st}(\mathcal{I}_b + \mathcal{I}_c) \\ v_c = Z_s \mathcal{I}_c + Z_{st}(\mathcal{I}_b + \mathcal{I}_c) \end{cases} \quad \begin{cases} \overline{\mathcal{I}}_{(0)} + \overline{\mathcal{I}}_{(1)} + \overline{\mathcal{I}}_{(2)} = 0 \\ \overline{v}_{(0)} + h^2\overline{v}_{(1)} + h\overline{v}_{(2)} = \overline{Z}_s(\overline{\mathcal{I}}_{(0)} + h^2\overline{\mathcal{I}}_{(1)} + h\overline{\mathcal{I}}_{(2)}) \\ \qquad + \overline{Z}_{st}(2\overline{\mathcal{I}}_{(0)} - \overline{\mathcal{I}}_{(1)} - \overline{\mathcal{I}}_{(2)}) \\ \overline{v}_{(0)} + h\overline{v}_{(1)} + h^2\overline{v}_{(2)} = \overline{Z}_s(\overline{\mathcal{I}}_{(0)} + h\overline{\mathcal{I}}_{(1)} + h^2\overline{\mathcal{I}}_{(2)}) \\ \qquad + \overline{Z}_{st}(2\overline{\mathcal{I}}_{(0)} - \overline{\mathcal{I}}_{(1)} - \overline{\mathcal{I}}_{(2)}) \end{cases}$$

from which (Fig. 5.59b):

$$\overline{\mathcal{I}}_{(0)} + \overline{\mathcal{I}}_{(1)} + \overline{\mathcal{I}}_{(2)} = 0$$

$$\overline{v}_{(0)} - (\overline{Z}_s + 3\overline{Z}_{st})\overline{\mathcal{I}}_{(0)} = \overline{v}_{(1)} - \overline{Z}_s\overline{\mathcal{I}}_{(1)} = \overline{v}_{(2)} - \overline{Z}_s\overline{\mathcal{I}}_{(2)}$$

- *Two-phase isolated short-circuit* (phases b, c); it is equivalent to the previous case with $Z_{st} \to \infty$ ($\mathcal{I}_a = 0$, $\mathcal{I}_b + \mathcal{I}_c = 0$, $v_b - v_c = Z_s(\mathcal{I}_b - \mathcal{I}_c)$), so that:

$$\overline{\mathcal{I}}_{(0)} = \overline{\mathcal{I}}_{(1)} + \overline{\mathcal{I}}_{(2)} = 0$$

$$\overline{v}_{(1)} - \overline{Z}_s\overline{\mathcal{I}}_{(1)} = \overline{v}_{(2)} - \overline{Z}_s\overline{\mathcal{I}}_{(2)}$$

The above allows the determination for any assigned case of different symmetrical components, from which it is possible to derive the phase variables by applying Equations [5.7.11]. For instance, for single-phase opening (Fig. 5.58a), by assuming $Z_p = \infty$ and accounting $\overline{Z}_{(1)} = \overline{Z}_{(2)}$, it follows that:

$$\overline{u}_{(0)} = \overline{u}_{(1)} = \overline{u}_{(2)} = \frac{\overline{e}_{(1)}}{2 + \dfrac{\overline{Z}_{(1)}}{\overline{Z}_{(0)}}}$$

and because of the first part of Equations [5.7.11] we can conclude that the amplitude of the sinusoid $u_a(t)$ is:

$$U_{aM} = \sqrt{\frac{2}{3}} \cdot 3 \left| \frac{\overline{e}_1}{2 + \overline{Z}_{(1)}/\overline{Z}_{(0)}} \right|$$

or equivalently, if $\overline{e}_{(1)}$ corresponds (for each single phase) to an emf of amplitude E_M, so that $|\overline{e}_{(1)}| = \sqrt{3/2}E_M$:

$$U_{aM} = \frac{3E_M}{\left| 2 + \dfrac{\overline{Z}_{(1)}}{\overline{Z}_{(0)}} \right|}$$

and so on.

Going back to the example of Figure 5.54, it is easily determined that the quantities $\overline{e}_{(1)}$, $\overline{Z}_{(1)}$, $\overline{Z}_{(2)}$, $\overline{Z}_{(0)}$ assume the following values:

- for the breaker opening:

$$\overline{e}_{(1)} = \overline{e}_{gr}$$

$$\overline{Z}_{(1)} = \overline{Z}_{(2)} = \frac{j\Omega L}{1 - \Omega^2 LC}$$

$$\overline{Z}_{(0)} = \frac{-j}{\Omega C} + 3(\overline{Z}_{nt1} + \overline{Z}_{t12})$$

- for the short-circuit:

$$\overline{e}_{(1)} = \frac{L_2 \overline{e}_{gr}}{L_1 + L_2}$$

$$\overline{Z}_{(1)} = \overline{Z}_{(2)} = j\Omega \frac{L_1 L_2}{L_1 + L_2}$$

$$\overline{Z}_{(0)} = j\Omega L_1 + \frac{3(\overline{Z}_{nt1} + \overline{Z}_{t12})}{1 + 3(\overline{Z}_{nt1} + \overline{Z}_{t12}) j\Omega C} + 3\overline{Z}_{t23}$$

where $\overline{Z}_{nt1} \triangleq Z_{nt1}(j\Omega)$, and so on.

Assuming that $Z_{t23} = 0$, and $Z_{nt1} + Z_{t12} = 0$ or $Z_{nt1} + Z_{t12} = \infty$ (more particularly, neutral connected to earth or isolated), and disregarding resistances ρ_p, ρ_s, ρ_{st}, the results summarized in Figures 5.60a, 5.61a, and 5.62a can be obtained for the different cases, where:

- U_M^o = voltage amplitude at breaker terminals, on the opened phase or on each of the opened phases;
- J_M^o = current amplitude, in the short-circuited phase or in each of the short-circuited phases;

whereas E_{gM} is the amplitude of the emfs e_{ga}, e_{gb}, e_{gc} (see Equation [5.7.8]).

For any given value of $\Omega^2 LC$, it is possible to identify the most severe perturbation, in terms of U_M^o or J_M^o (which may not be the symmetrical one). Moreover, the effect of the "status" of the neutral (connected to earth, or isolated) is evident, for single-phase or two-phase opening, and for single-phase or two-phase to earth short-circuit.

Transient Response

The determination of transient response is much more complicated. If results are requested in short times, limiting analysis to the determination of the steady-state response may be convenient, then applying proper incremental coefficients (on

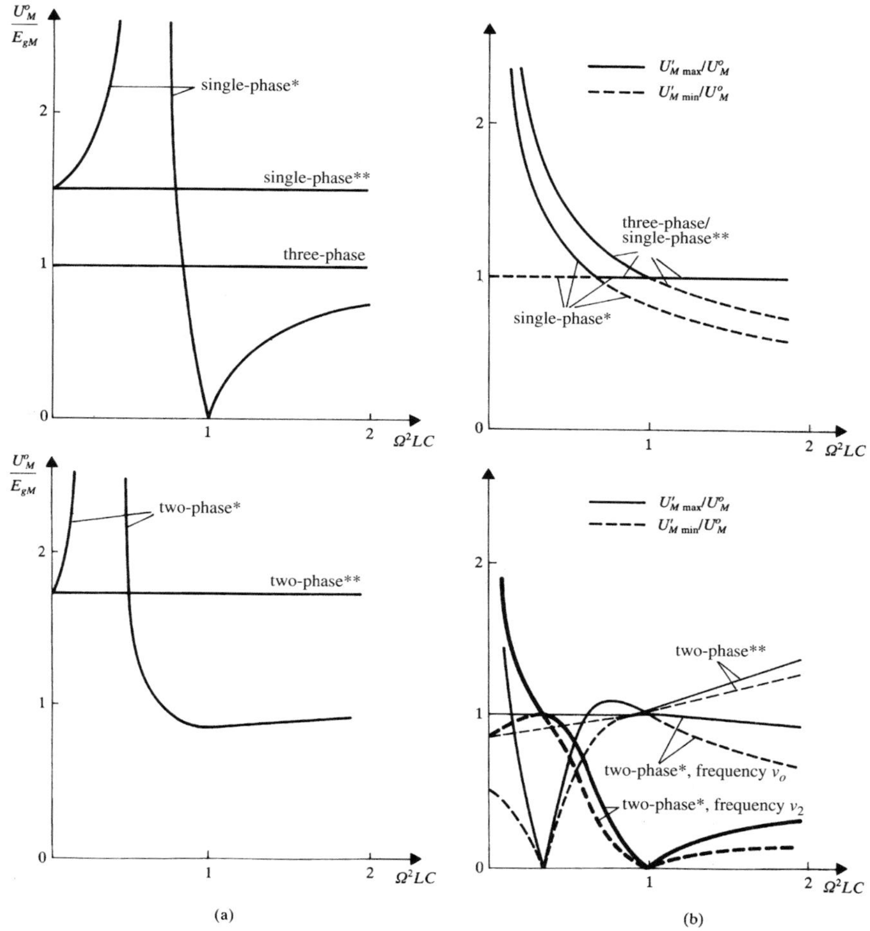

Figure 5.60. Opening of the breaker in Figure 5.54a, on one or more phases, with $\rho_p = \infty$: (a) amplitude of the steady-state response; (b) minimum and maximum relative amplitude of the transient-response components (see text).

$$^*Z_{nt1} + Z_{t12} = 0 \qquad ^{**}Z_{nt1} + Z_{t12} = \infty$$

the safety side) to estimate the maximum values assumed by variables during the transient.

For phase variables, under the hypothesis of nonzero damping, the purely transient components are related to the characteristic roots of the system, in its new configuration for the three phases.

In Figure 5.54, by disregarding resistances ρ_p, ρ_s, ρ_{st} (and thus damping), assuming $Z_{t23} = 0$, and recalling Equation [5.7.8], the following equations can be obtained.

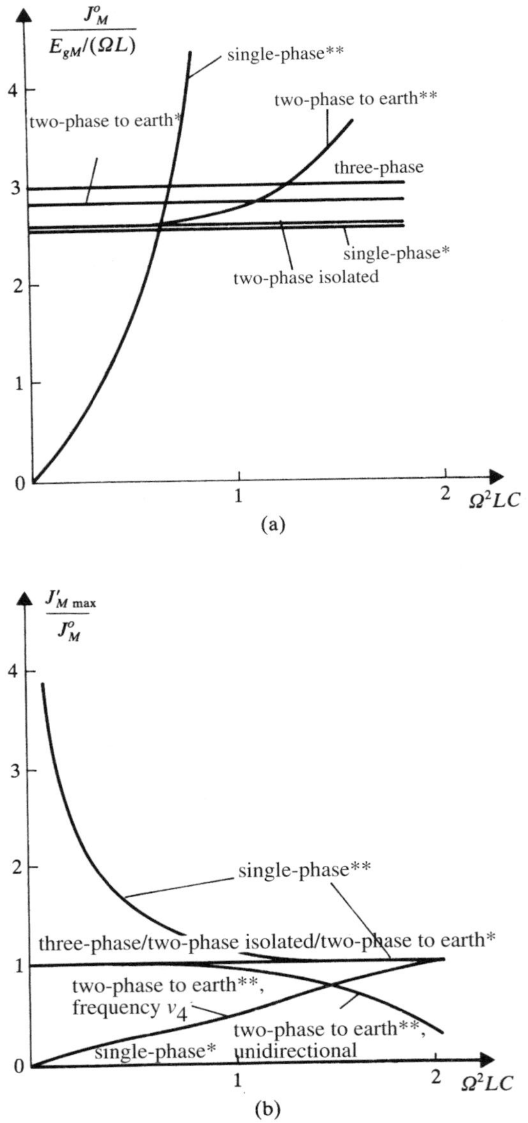

Figure 5.61. Short-circuit as in Figure 5.54a, on one or more phases, with $\rho_s = \rho_{st} = Z_{t23} = 0$, $L_1 = L_2/2$: (a) amplitude of the steady-state response; (b) maximum relative amplitude of the transient-response components (see text).

$^{*}Z_{nt1} + Z_{t12} = 0 \qquad ^{**}Z_{nt1} + Z_{t12} = \infty$

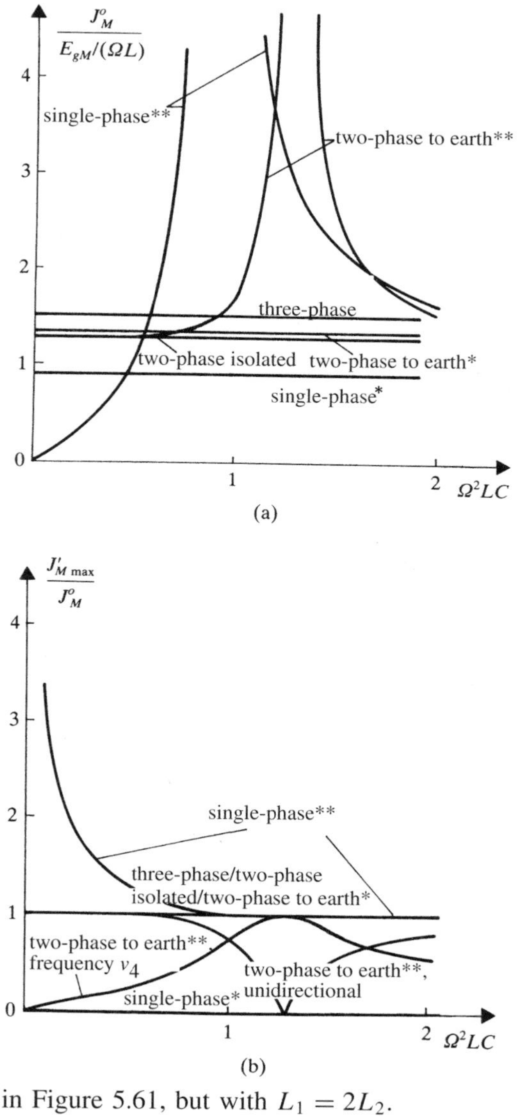

Figure 5.62. As in Figure 5.61, but with $L_1 = 2L_2$.

$^{*}Z_{nt1} + Z_{t12} = 0 \qquad ^{**}Z_{nt1} + Z_{t12} = \infty$

- *Single-phase opening* (phase a), with $Z_{nt1} + Z_{t12} = 0$:

$$\frac{u_a(t)}{E_{gM}} = \frac{\frac{3}{2}\nu_1^2 - \Omega^2}{\nu_1^2 - \Omega^2}\left[\cos(\Omega t + \alpha) - \left(\cos\alpha\cos(\nu_1 t) - \frac{\nu_1}{\Omega}\sin\alpha\sin(\nu_1 t)\right)\right]$$

with a "resonance frequency" of:

$$\nu_1 \triangleq \sqrt{\frac{2}{3LC}}$$

- *Single-phase opening* (phase a), with $Z_{nt1} + Z_{t12} = \infty$:

$$\frac{u_a(t)}{E_{gM}} = \frac{3}{2}\left[\cos(\Omega t + \alpha) - \left(\cos\alpha\cos(\nu_o t) - \frac{\nu_o}{\Omega}\sin\alpha\sin(\nu_o t)\right)\right]$$

with a resonance frequency of:

$$\nu_o \triangleq \frac{1}{\sqrt{LC}}$$

- *Two-phase opening* (phases b, c), with $Z_{nt1} + Z_{t12} = 0$:

$$\left.\begin{matrix}\dfrac{u_b(t)}{E_{gM}} \\ \dfrac{u_c(t)}{E_{gM}}\end{matrix}\right\} = \frac{\nu_o^2 - \Omega^2}{2(\nu_2^2 - \Omega^2)}\left[-\cos(\Omega t + \alpha) + \cos\alpha\cos(\nu_2 t) - \frac{\nu_2}{\Omega}\sin\alpha\sin(\nu_2 t)\right]$$

$$\pm \frac{\sqrt{3}}{2}\left[\sin(\Omega t + \alpha) - \sin\alpha\cos(\nu_o t) - \frac{\nu_o}{\Omega}\cos\alpha\sin(\nu_o t)\right]$$

with two resonance frequencies, ν_o (defined as above) and:

$$\nu_2 \triangleq \frac{1}{\sqrt{3LC}}$$

- *Two-phase opening* (phases b, c), with $Z_{nt1} + Z_{t12} = \infty$:

$$\left.\begin{matrix}\dfrac{u_b(t)}{E_{gM}} \\ \dfrac{u_c(t)}{E_{gM}}\end{matrix}\right\} = \frac{3}{2}\left[-\cos(\Omega t + \alpha) + \cos\alpha\cos(\nu_o t) - \frac{\Omega}{\nu_o}\sin\alpha\sin(\nu_o t)\right]$$

$$\pm \frac{\sqrt{3}}{2}\left[\sin(\Omega t + \alpha) - \sin\alpha\cos(\nu_o t) - \frac{\Omega}{\nu_o}\cos\alpha\sin(\nu_o t)\right]$$

with the only resonance frequency ν_o.

- *Single-phase short-circuit* (phase a), with $Z_{nt1} + Z_{t12} = 0$:

$$\frac{\jmath_a(t)}{E_{gM}} = \frac{3L_2}{\Omega L_1(L_1 + 3L_2)}\sin(\Omega t + \alpha)$$

as in the steady-state response.

- *Single-phase short-circuit* (phase a), with $Z_{nt1} + Z_{t12} = \infty$:

$$\frac{\mathcal{I}_a(t)}{E_{gM}} = \frac{3\Omega \nu_3^2 L_2 C}{(\nu_3^2 - \Omega^2)(L_1 + L_2)} \Big[-\sin(\Omega t + \alpha) + \sin\alpha \cos(\nu_3 t) + \frac{\nu_3}{\Omega} \cos\alpha \sin(\nu_3 t) \Big]$$

with a resonance frequency of:

$$\nu_3 \triangleq \sqrt{\frac{L_1 + L_2}{(L_1 + 3L_2)L_1 C}}$$

- *Two-phase to earth short-circuit* (phases b, c, and earth), with $Z_{nt1} + Z_{t12} = 0$:

$$\left.\begin{array}{l} \dfrac{\mathcal{I}_b(t)}{E_{gM}} \\ \dfrac{\mathcal{I}_c(t)}{E_{gM}} \end{array}\right\} = \frac{3}{2} \frac{L_2}{\Omega(2L_1 + 3L_2)L_1} [\sin\alpha - \sin(\Omega t + \alpha)] \pm \frac{\sqrt{3}}{2\Omega L_1} [\cos\alpha - \cos(\Omega t + \alpha)]$$

with a "unidirectional" component.

- *Two-phase to earth short-circuit* (phases b, c, and earth), with $Z_{nt1} + Z_{t12} = \infty$:

$$\left.\begin{array}{l} \dfrac{\mathcal{I}_b(t)}{E_{gM}} \\ \dfrac{\mathcal{I}_c(t)}{E_{gM}} \end{array}\right\} = \frac{3}{2} \frac{\Omega L_2 C}{(2L_1 + 3L_2)L_1 C(\nu_4^2 - \Omega^2)} \cdot \Big[\sin(\Omega t + \alpha) - \sin\alpha \cos(\nu_4 t) - \frac{\nu_4}{\Omega} \cos\alpha \sin(\nu_4 t) \Big] \pm \frac{\sqrt{3}}{2\Omega L_1} [\cos\alpha - \cos(\Omega t + \alpha)]$$

with a unidirectional component, and a resonance frequency equal to:

$$\nu_4 \triangleq \sqrt{\frac{2(L_1 + L_2)}{(2L_1 + 3L_2)L_1 C}}$$

- *Two-phase isolated short-circuit* (phases b, c), for any $Z_{nt1} + Z_{t12}$:

$$\frac{\mathcal{I}_b(t)}{E_{gM}} = -\frac{\mathcal{I}_c(t)}{E_{gM}} = \frac{\sqrt{3}}{2\Omega L_1} [\cos\alpha - \cos(\Omega t + \alpha)]$$

with a unidirectional component.

The analytical difficulties are evident, even for the simple example considered. However, the results obtained can be considered typical, as they imply the presence of:

- *unidirectional components*;
- *oscillatory components*, not only at the forced frequency Ω, but also resulting from *resonances between inductive and capacitive elements* (if the line were represented by a distributed parameter model, also *propagation phenomena* as in Section 5.4 would appear).

The importance of the value of α (i.e., of the instant of application of the perturbation) should be particularly underlined for its effect on the amplitudes of the different sinusoidal components (at the indicated resonance frequencies) or unidirectional ones, and the maximum instantaneous values assumed by the different quantities.

Figures 5.60b, 5.61b, and 5.62b summarize previous results, where:

- $U'_{M\,\min}$ = minimum amplitude (obtainable at a proper α) of the generic component of the voltage $u_a(t)$ (single-phase opening), or both voltages $u_b(t)$, $u_c(t)$ (two-phase opening), or all the three voltages $u_a(t)$, $u_b(t)$, $u_c(t)$ (three-phase opening);
- $U'_{M\,\max}$ = maximum amplitude etc., but related to one of the voltages $u_b(t)$ and $u_c(t)$ in the case of two-phase opening;
- $J'_{M\,\max}$ = similarly, with reference to the current $\mathcal{J}_a(t)$ (single-phase short-circuit), or one of the currents $\mathcal{J}_b(t)$, $\mathcal{J}_c(t)$ (two-phase short-circuit), or all the three currents $\mathcal{J}_a(t)$, $\mathcal{J}_b(t)$, $\mathcal{J}_c(t)$ (three-phase short-circuit);

whereas U^o_M and J^o_M are the amplitudes of the (steady-state) components at frequency Ω. The knowledge of the ratios $U'_{M\,\min}/U^o_M$, $U'_{M\,\max}/U^o_M$, $J'_{M\,\max}/J^o_M$ can give useful indications (particularly if the frequencies involved are rather different) on the incremental coefficients to be applied, starting from the steady-state response only.

Moreover, the analysis can be extended to other variables of particular interest, such as the currents in the unopened phases, the voltages in the unfaulted phases, and so on.

As a *concluding remark*, it is useful to indicate:

- The phenomena considered here are assumed to be fast, so that the simplifications of the machines' models may appear credible. Thus, the "steady-state" responses described above hold in the short term, i.e., in accordance to the mentioned simplifications.
- For longer times, it is necessary to consider the variations in amplitude of rotor fluxes (recall the short-circuit transient of the synchronous machine for the conditions described in Section 4.4.4), and speed variations. If the

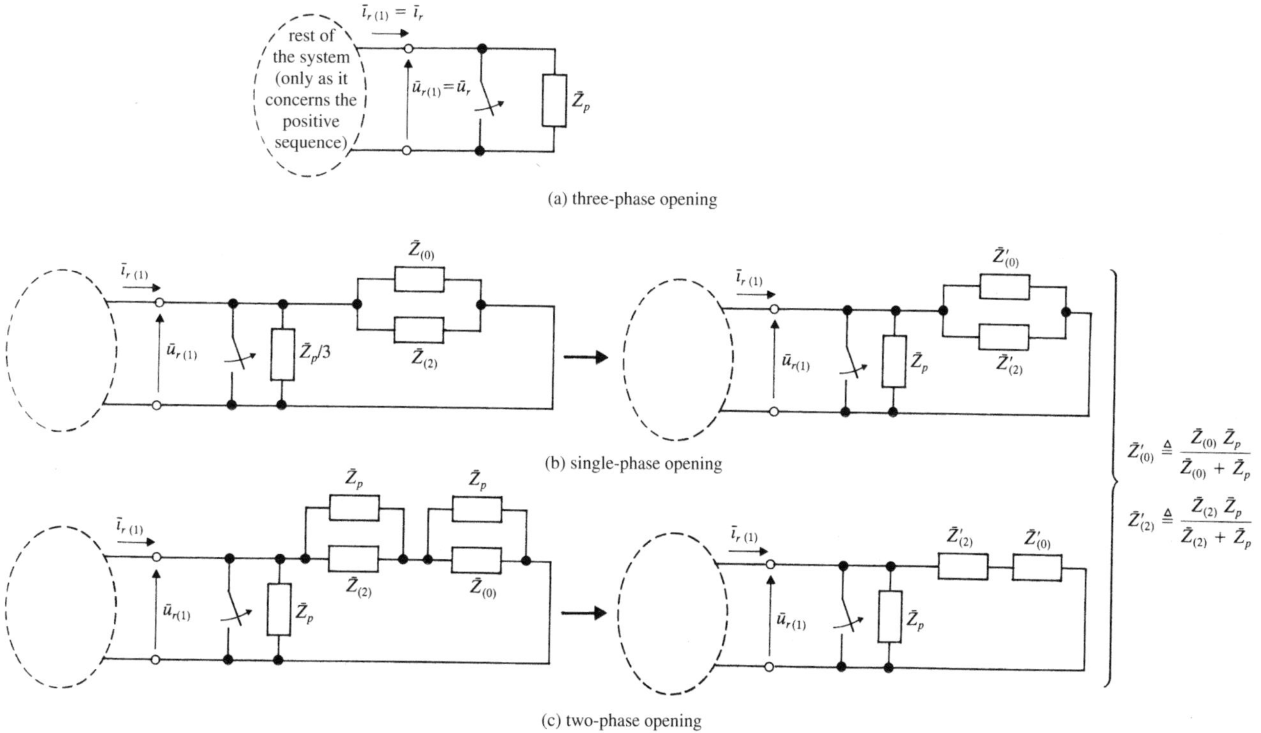

Figure 5.63. Equivalent impedances in terms of the positive sequence in the case of the opening (a) three-phase, (b) single-phase, (c) two-phase.

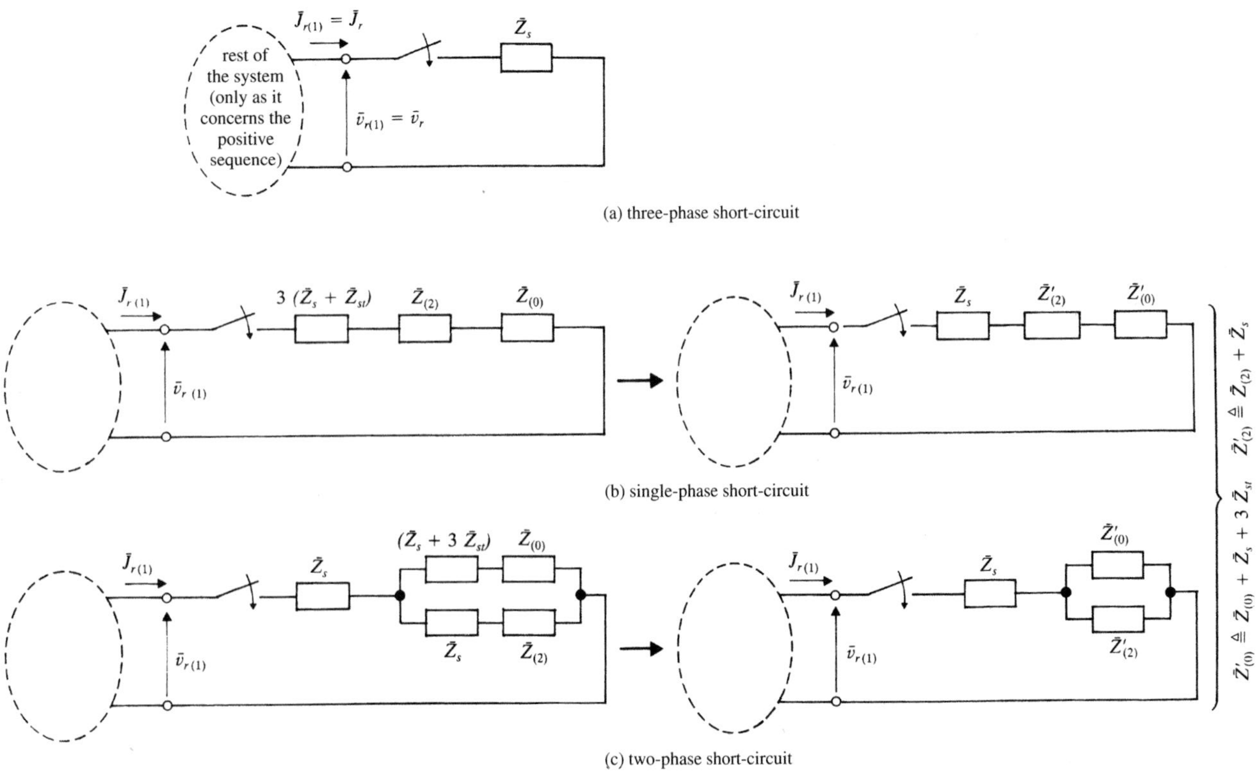

Figure 5.64. Equivalent impedances in terms of the positive sequence in the case of the short-circuit (a) three-phase, (b) single-phase, (c) two-phase. For the two-phase isolated short-circuit, it is $\overline{Z}_{st} = \infty$, $\overline{Z}'_{(0)} = \infty$.

transient components considered here are slightly damped, overlapping of different phenomena makes use of *simulations* practically essential.

- If instead such transient components are quickly damped, the remaining phenomena (related to the variations in amplitude of rotor fluxes and speed) can be analyzed by considering only the "steady-state" responses, as if the electrical part of the system were in a sinusoidal operation (in general, at the three sequences), *slowly varying* in amplitude and frequency.

In this last case (assuming that the three sequences are present):

- The behavior of the phase variables can be derived, starting from the corresponding symmetrical components.
- The analysis of the electromechanical phenomena can be simplified by considering, for each machine, the torque resulting only from the positive sequence (recall Section 4.4.3). Therefore, it is possible to refer to Park's vectors corresponding only to the positive sequence, by considering perturbations on the electrical part by *equivalent impedances* derived from Figs. 5.58 and 5.59, according to Figures 5.63 and 5.64.

ANNOTATED REFERENCES

Among references of more general interest, the following ones are mentioned: 5, 6, 23, 25, 26, 42, 47, 50, 64.

Furthermore, regarding

- network elements (in general): 99;
- regulating transformers: 184;
- ac lines: 188, 193, 196, 306, 319;
- dc links: 24, 139, 304;
- loads: 112, 153, 164, 255, 295, 296, 297, 318, 328, 334, in addition to some notes prepared by the author in view of the writing of 53;
- short-circuit currents: 29.

CHAPTER 6

VOLTAGE AND REACTIVE POWER CONTROL

6.1. GENERALITIES

Based on Chapter 2, the close relationship between the voltage magnitudes and reactive powers at steady-state is evident. The approximations defined in Figure 2.7, for the case of the "dc model," should be recalled. (A similar interrelation holds between voltage phase-shifts and active powers, as shown also in Figure 2.6.)

Within this concern it must be noted that, under the most usual operating conditions, the network is predominantly inductive and thus[(1)]:

- reactive powers absorbed by the network itself (e.g., lines, transformers, terminals of dc links, etc.) must be considered, in addition to reactive powers demanded by loads;
- the transportation of reactive powers from generators to loads would usually imply, in addition to the previously mentioned absorption, unacceptable voltage drops (in the same sense, i.e., from generators to loads) if proper actions are not undertaken within the network as specified in the following.

[(1)] The situation also can be inverted at least partially in some cases, e.g., during light load hours, with reactive power generated by lines (which are relatively unloaded) and increasing voltages from generators to loads. Furthermore, what has been discussed so far can be emphasized in the case where long lines are present, e.g., constructed to exploit hydraulic energy sources that are far from the rest of the system.

On the other hand, reactive power injections that can be achieved by *generators* are generally not sufficient for:

- the matching of the total reactive power demand, because of the limits on generators themselves (Fig. 2.9);
- the accomplishment of an acceptable voltage steady-state, because of the concentration of generators in relatively few "sites" (consistent with technical, environmental constraints etc.).

It is then convenient to intervene also within the transmission and distribution networks and near to loads, for instance by:

- injecting reactive power (usually positive, or possibly negative in those cases mentioned in footnote[(1)]) using shunt condensers or inductors or, for more general control functions, static or synchronous compensators;
- reducing the total absorbed reactive power using series condensers;
- adjusting voltage levels using tap-changing transformers.

The locations at which such remedies are installed can be generally chosen without particular constraints of the environmental type, etc. Therefore, the situation under examination can be considered as a complementary one with respect to that concerning the active power steady-state. This holds as far as both power generation and transportation are of concern, because active power transportation implies relatively small losses and usually acceptable phase-shifts (without prejudices for stability), so that the active power generation can be effectively concentrated at few proper sites as previously mentioned.

The voltage "support" along a *line*, aside from that which is intrinsically due to series inductances and shunt capacitances of the line itself:

- may make it possible to have acceptable voltage values at any location (also matching the requirements of possible intermediate loads);
- also can be important in relation to the transfer limit of the active power (see Section 6.4c).

It also must be remembered that, along a uniform and nondissipative line (see Section 5.4):

- the voltage amplitude is constant (assuming of course that such a value is achieved at the line terminals) if and only if the transmitted active power is equal to the "characteristic" power $P^{(o)}$;
- in such conditions, the transmitted reactive power is zero at any line location (whereas the phase-shift δ between terminal voltages is βa, where β is the "phase constant" and a is the line length); thus, in practice, *a good voltage profile implies also a low value of transmitted reactive power* at

different locations, and thus low current and low losses at equal transmitted active power.

Broadly it can be thought that, regarding the steady-state:

- the addition of shunt condensers or reactors or static compensators is equivalent to a modification of the parameter c of the line (capacitance for unit of line length);
- the addition of series condensers is equivalent to a modification of the parameter l (inductance for unit of line length);

so that, if such equivalencies were exact, a variation of $P^{(o)}$ and β would result.
In such conditions, the constant voltage profile also could be viewed as the result of an "adjustment" of $P^{(o)}$ to the value P of active power to be transmitted (to which would automatically correspond $\delta = \beta a$).
However, the above cannot be generally extended to the transient operation (recall the discussion in Section 5.2 regarding the dynamic behavior of condensers and inductors). Furthermore, note that an excessive value of $\delta = \beta a$ can be unacceptable for stability reasons.

Regarding the generic *load*, the addition of a shunt reactive element to absorb a reactive power Q_R, and zero active power, obviously provides the capability to adjust the load voltage and/or the overall required reactive power. As an important additional effect, harmonic and/or flicker filtering can be obtained.

In the typical example of Figure 6.1 (in which Park's transformation is applied with a reference speed $\Omega_r = \omega$, equal to the network frequency, and $(R + jX)$ defines the impedance of the line going to the load):

$$\overline{e} = \overline{v} + (R + jX)\overline{\imath} = \overline{v} + (R + jX)\frac{P - jQ}{\overline{v}^*}$$

from which:

$$\frac{\overline{e}}{\overline{v}} = 1 + \frac{(RP + XQ) + j(XP - RQ)}{v^2}$$

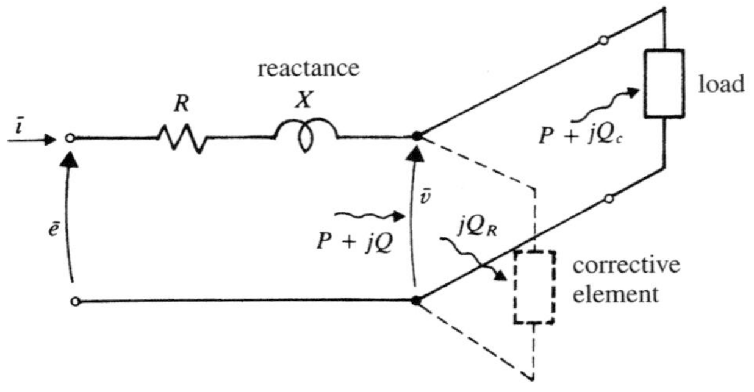

Figure 6.1. Addition of a shunt reactive element at a load node.

Specifically:

- if the goal is $v = e$, it must be imposed that:

$$1 = \left(1 + \frac{RP + XQ}{v^2}\right)^2 + \left(\frac{XP - RQ}{v^2}\right)^2$$

  from which $Q(=Q_c + Q_R)$ and thus Q_R, as functions of P and e[(2)];
- if, instead, the goal is $Q = 0$, i.e., total *"power factor correction,"* it is necessary to set $Q_R = -Q_c$, whereas v varies with P and e.

By posing $u_R \triangleq (RP + XQ)/v^2$, $u_X \triangleq (XP - RQ)/v^2$, usually:

$$u_X \ll 1 + u_R$$

so that:

$$\frac{e}{v} = \sqrt{(1 + u_R)^2 + u_X^2} \cong 1 + u_R$$

$$\frac{e - v}{v} \cong u_R$$

Furthermore, if $R \ll X$ and $v \cong e$, it can be written that $u_R \cong XQ/e^2 \ll 1$, and thus:

$$v \cong e - \frac{X}{e} Q \qquad [6.1.1]$$

However, in practice, the minimum and maximum limits on Q_R and Q_c also should be considered. Therefore,

- to have $v = e$ (from which the value of Q is derived) for any Q_c, with:

$$Q_R = (Q - Q_c) \in [Q - Q_{c\max}, Q - Q_{c\min}]$$

it is necessary and sufficient that:

$$\begin{cases} Q_{R\min} \le Q - Q_{c\max} \\ Q_{R\max} \ge Q - Q_{c\min} \end{cases}$$

- if, instead, a voltage $v \neq e$ is accepted, it also can be accepted that $Q_{R\min} > Q - Q_{c\max}$ and/or $Q_{R\max} < Q - Q_{c\min}$.

---

(2) Note that the solution for Q is given, for any desired value of v, by:

$$Q = \frac{X(ve\cos\alpha - v^2) - Rve\sin\alpha}{R^2 + X^2}$$

where the phase shift $\alpha \triangleq \angle\overline{e} - \angle\overline{v}$ depends on the active power P, according to:

$$P = \frac{Xve\sin\alpha + R(ve\cos\alpha - v^2)}{R^2 + X^2}$$

The addition of shunt reactive elements also may permit the *"balancing" of load on the three phases.* As an example, let us assume that:

- the (composite) load is linear and can be represented by three delta-connected admittances (in the case of a wye connection with isolated neutral, one can apply the well-known wye-delta transformation);
- voltages v_a, v_b, v_c (at the delta terminals) are sinusoidal of the positive sequence and at frequency ω.

By generically referring to Figure 6.2, it follows, in terms of symmetrical components (phasors), and for brevity setting $h \triangleq e^{\tilde{j}120^\circ} = e^{-\tilde{j}240^\circ}$, $\tilde{Y}_{ab} \triangleq Y_{ab}(\tilde{j}\omega)$ etc. (see Appendix 1):

$$\tilde{v}_a = \frac{1}{\sqrt{3}}\tilde{v}_{(1)}, \quad \tilde{v}_b = \frac{1}{\sqrt{3}}h^2\tilde{v}_{(1)}, \quad \tilde{v}_c = \frac{1}{\sqrt{3}}h\tilde{v}_{(1)}$$

$$\tilde{\imath}_{ab} = \tilde{Y}_{ab}(\tilde{v}_a - \tilde{v}_b), \quad \tilde{\imath}_{bc} = \tilde{Y}_{bc}(\tilde{v}_b - \tilde{v}_c), \quad \tilde{\imath}_{ca} = \tilde{Y}_{ca}(\tilde{v}_c - \tilde{v}_a)$$

$$\tilde{\imath}_a = \tilde{\imath}_{ab} - \tilde{\imath}_{ca}, \quad \tilde{\imath}_b = \tilde{\imath}_{bc} - \tilde{\imath}_{ab}, \quad \tilde{\imath}_c = \tilde{\imath}_{ca} - \tilde{\imath}_{bc}$$

and finally:

$$\begin{cases} \tilde{\imath}_{(0)} = \dfrac{1}{\sqrt{3}}(\tilde{\imath}_a + \tilde{\imath}_b + \tilde{\imath}_c) \\ \tilde{\imath}_{(1)} = \dfrac{1}{\sqrt{3}}(\tilde{\imath}_a + h\tilde{\imath}_b + h^2\tilde{\imath}_c) \\ \tilde{\imath}_{(2)} = \dfrac{1}{\sqrt{3}}(\tilde{\imath}_a + h^2\tilde{\imath}_b + h\tilde{\imath}_c) \end{cases}$$

from which it follows:

$$\left.\begin{aligned} \tilde{\imath}_{(0)} &= 0 \\ \tilde{\imath}_{(1)} &= \tilde{v}_{(1)}(\tilde{Y}_{ab} + \tilde{Y}_{bc} + \tilde{Y}_{ca}) \\ \tilde{\imath}_{(2)} &= -h^2\tilde{v}_{(1)}(\tilde{Y}_{ab} + h\tilde{Y}_{bc} + h^2\tilde{Y}_{ca}) \end{aligned}\right\} \qquad [6.1.2]$$

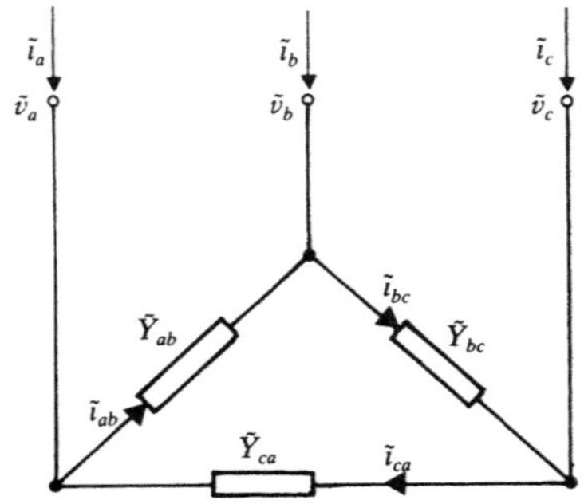

Figure 6.2. Example of three-phase load (the symbol $\tilde{Y}$ represents admittance in phasor terms).

To balance the load on the three phases, it is necessary to set $\tilde{\imath}_{(2)} = 0$ (so that there are only currents i_a, i_b, i_c of the positive sequence) and thus, due to the last part of Equations [6.1.2]:

$$0 = \tilde{Y}_{ab} + h\tilde{Y}_{bc} + h^2\tilde{Y}_{ca} \qquad [6.1.3]$$

The reactive power Q absorbed, which can be set to a desired value (e.g., $Q = 0$), is then:

$$Q = Im(\tilde{v}_{(1)}\tilde{\imath}^*_{(1)}) = -3V_e^2\, Im\left(\frac{\tilde{\imath}_{(1)}}{\tilde{v}_{(1)}}\right) = -3V_e^2\, Im(\tilde{Y}_{ab} + \tilde{Y}_{bc} + \tilde{Y}_{ca}) \qquad [6.1.4]$$

where V_e is the rms value of v_a, v_b, v_c. In particular, the condition $Q = 0$ corresponds to:

$$0 = Im(\tilde{Y}_{ab} + \tilde{Y}_{bc} + \tilde{Y}_{ca}) \qquad [6.1.4']$$

On the other hand, by means of Equations [6.1.2] (with $\tilde{Y}_{ab} = G_{ab} + \tilde{\jmath}B_{ab}$, etc.) it is possible to derive:

$$\left.\begin{aligned}
B_{ab} &= \frac{1}{3}\, Im\left(\frac{\tilde{\imath}_{(1)} + \tilde{\imath}_{(2)}}{\tilde{v}_{(1)}}\right) - \frac{1}{\sqrt{3}}\mathcal{R}e\left(\frac{\tilde{\imath}_{(2)}}{\tilde{v}_{(1)}}\right) + \frac{1}{\sqrt{3}}(G_{ca} - G_{bc})\\
B_{bc} &= \frac{1}{3}\, Im\left(\frac{\tilde{\imath}_{(1)} - 2\tilde{\imath}_{(2)}}{\tilde{v}_{(1)}}\right) + \frac{1}{\sqrt{3}}(G_{ab} - G_{ca})\\
B_{ca} &= \frac{1}{3}\, Im\left(\frac{\tilde{\imath}_{(1)} + \tilde{\imath}_{(2)}}{\tilde{v}_{(1)}}\right) + \frac{1}{\sqrt{3}}\mathcal{R}e\left(\frac{\tilde{\imath}_{(2)}}{\tilde{v}_{(1)}}\right) + \frac{1}{\sqrt{3}}(G_{bc} - G_{ab})
\end{aligned}\right\} \qquad [6.1.5]$$

from which, by imposing $\tilde{\imath}_{(2)} = 0$, and Q to the desired value, the following conditions result:

$$\left.\begin{aligned}
B_{ab} &= -\frac{1}{9}\frac{Q}{V_e^2} + \frac{1}{\sqrt{3}}(G_{ca} - G_{bc})\\
B_{bc} &= -\frac{1}{9}\frac{Q}{V_e^2} + \frac{1}{\sqrt{3}}(G_{ab} - G_{ca})\\
B_{ca} &= -\frac{1}{9}\frac{Q}{V_e^2} + \frac{1}{\sqrt{3}}(G_{bc} - G_{ab})
\end{aligned}\right\} \qquad [6.1.6]$$

Assuming that such conditions are not already satisfied by the load admittances, it is possible to achieve the desired values B_{ab}, B_{bc}, B_{ca} by adding the susceptances $B_{ab(R)}$, $B_{bc(R)}$, $B_{ca(R)}$ as indicated in Figure 6.3, with:

$$\left.\begin{aligned}
B_{ab(R)} &= B_{ab} - B_{ab(L)}\\
B_{bc(R)} &= B_{bc} - B_{bc(L)}\\
B_{ca(R)} &= B_{ca} - B_{ca(L)}
\end{aligned}\right\} \qquad [6.1.7]$$

Therefore, by doing so, the positive sequence voltages v_a, v_b, v_c lead to currents i_a, i_b, i_c of only positive sequence. However, in general, possible voltages of the negative sequence would not lead to currents of only negative sequence.

Under operation, the deduction of the required values $B_{ab(R)}$ etc. requires—based on Equations [6.1.6] and [6.1.7]—the knowledge of load conductances (G_{ab} etc.) and

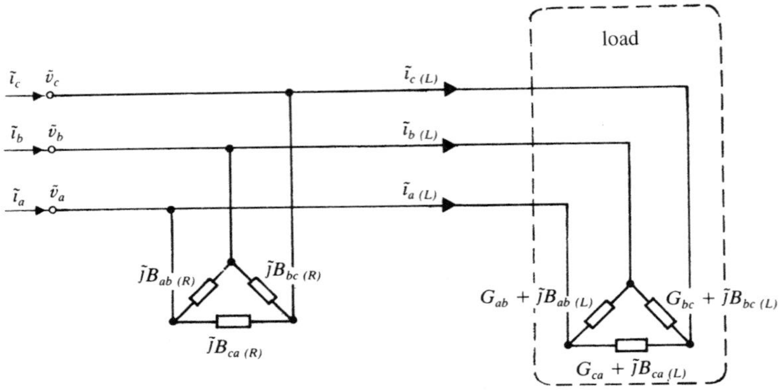

Figure 6.3. Balancing of load on the three phases (for the example of Fig. 6.2).

susceptances ($B_{ab(L)}$ etc.) which must be estimated from measurements of $\tilde{v}_a$, $\tilde{v}_b$, $\tilde{v}_c$, $\tilde{\imath}_{a(L)}$, $\tilde{\imath}_{b(L)}$, $\tilde{\imath}_{c(L)}$ (Fig. 6.3).
Alternatively, by writing the load susceptances in a form similar to Equations [6.1.5], it can be derived:

$$\begin{cases} B_{ab(R)} = -\dfrac{1}{9}\dfrac{Q}{V_e^2} - \dfrac{1}{3}\,Im\left(\dfrac{\tilde{\imath}_{(L)(1)} + \tilde{\imath}_{(L)(2)}}{\tilde{v}_{(1)}}\right) + \dfrac{1}{\sqrt{3}}\,\mathcal{R}e\left(\dfrac{\tilde{\imath}_{(L)(2)}}{\tilde{v}_{(1)}}\right) \\ B_{bc(R)} = -\dfrac{1}{9}\dfrac{Q}{V_e^2} - \dfrac{1}{3}\,Im\left(\dfrac{\tilde{\imath}_{(L)(1)} - 2\tilde{\imath}_{(L)(2)}}{\tilde{v}_{(1)}}\right) \\ B_{ca(R)} = -\dfrac{1}{9}\dfrac{Q}{V_e^2} - \dfrac{1}{3}\,Im\left(\dfrac{\tilde{\imath}_{(L)(1)} + \tilde{\imath}_{(L)(2)}}{\tilde{v}_{(1)}}\right) - \dfrac{1}{\sqrt{3}}\,\mathcal{R}e\left(\dfrac{\tilde{\imath}_{(L)(2)}}{\tilde{v}_{(1)}}\right) \end{cases}$$

where:

$$\begin{cases} \tilde{\imath}_{(L)(1)} = \dfrac{1}{\sqrt{3}}(\tilde{\imath}_{a(L)} + h\tilde{\imath}_{b(L)} + h^2\tilde{\imath}_{c(L)}) \\ \tilde{\imath}_{(L)(2)} = \dfrac{1}{\sqrt{3}}(\tilde{\imath}_{a(L)} + h^2\tilde{\imath}_{b(L)} + h\tilde{\imath}_{c(L)}) \\ \tilde{v}_{(1)} = \sqrt{3}\,\tilde{v}_a \end{cases}$$

so that the values $B_{ab(R)}$ etc. also can be derived more directly, by measurements of $\tilde{v}_a$, $\tilde{\imath}_{a(L)}$, $\tilde{\imath}_{b(L)}$, $\tilde{\imath}_{c(L)}$.

To obtain satisfactory working points for the different loading conditions, and to control the set of voltages and reactive powers during transients, it is possible to use[(3)]:

(3) Some of these actions can be performed to avoid unstable situations following large perturbations (see specifically Section 7.3).

- discontinuous variations (by switching) of condenser capacitances and "reactor" inductances, at least at the most important nodes of the network;
- continuous variations of the equivalent reactance (positive or negative) of static compensators, and of the excitation voltage of synchronous compensators and, more generally, of synchronous machines;
- quasi-continuous variations (i.e., switching, but through relatively small steps) of the transformation ratio of tap-changing transformers.

With condensers and reactors, control may be achieved through suitable "thresholds." In the other cases, the control can be performed, exactly or almost so, with continuity.

In the rest of this chapter, these last types of control (continuous or almost so) — including synchronous machine excitation, tap-changing transformers, and static compensators — will be managed by considering some simple (ideal) cases in which they:

- appear separately (this will allow the examination of their respective characteristics; in real cases, the control actions can instead interact with one another to various degrees);
- do not interact with any electromechanical phenomenon (in real cases, these interactions can be significant; see Sections 7.2.2, 7.2.3, and 8.5.2).

6.2. CONTROL OF SYNCHRONOUS MACHINE EXCITATION

6.2.1. The Synchronous Machine in Isolated Operation

To examine the characteristics of the v/Q control, by disregarding the interactions with electromechanical phenomena, consider a synchronous machine in isolated operation as shown in Figure 6.4. The interactions with electromechanical phenomena will be considered later in Sections 7.2.2, 7.2.3, and 8.5.2.

In a schematic way, to obtain a more direct comparison between different situations, let us furthermore assume that the load is "purely" inductive or resistive or capacitive. In practice:

- the case of a purely reactive load may, for example, correspond to the operation on an "unloaded" line (i.e., not connected, at the end terminal, to any load);
- the case of a purely active load may correspond to a limit condition in (occasional) isolated operation, with the load located very close to the synchronous generator.

For simplicity, it can be further assumed that the load is directly connected to the generator terminals. If however, the connection is made through an inductive

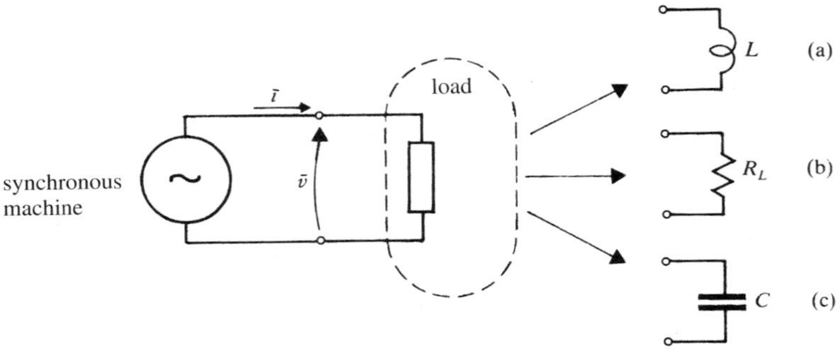

Figure 6.4. Synchronous machine in isolated operation on (a) inductive, (b) resistive, (c) capacitive load.

element (e.g., the step-up transformer), it may be considered, without any formal change, by simply adding its inductance to the machine's inductances.

Finally, it will be assumed that the voltage to be regulated is the load voltage, and thus the one downstream of this (possible) inductive element.

(a) Operation on Inductive Load (Fig. 6.4a)

Let us assume Park's transformation with an angular reference (θ_r) rotating at the same electrical speed (Ω) as the machine, so that $\mathrm{d}\theta_r/\mathrm{d}t \triangleq \Omega_r = \Omega$, and further assume that this speed remains constant.

By the notation already defined in Chapter 4, it then follows that:

$$(s + j\Omega)\overline{\psi} - R\overline{\imath} = \overline{v} = (s + j\Omega)L\overline{\imath} \tag{6.2.1}$$

or equivalently:

$$\left.\begin{aligned} s\psi_d - \Omega\psi_q - Ri_d = v_d = sLi_d - \Omega Li_q \\ \Omega\psi_d + s\psi_q - Ri_q = v_q = \Omega Li_d + sLi_q \end{aligned}\right\} \qquad [6.2.1']$$

and further, disregarding saturation:

$$\left.\begin{aligned} \psi_d &= A(s)v_f - \mathcal{L}_d(s)i_d \\ \psi_q &= -\mathcal{L}_q(s)i_q \end{aligned}\right\} \qquad [6.2.2]$$

where $\mathcal{L}_d(s)$ and $\mathcal{L}_q(s)$ are, for example, of the type given by Equations [4.1.26] or of a higher order (recall Equation [4.3.5] and also, in "per unit", [4.3.6] etc.)[(4)].

---

(4) The hypothesis that Ω = constant would imply an inertia coefficient $M = \infty$, as the electromagnetic torque C_e is not zero because of both the active power dissipated in the armature resistance R and that absorbed or delivered by the machine and load inductances (caused by changes in the stored energy) during transients. However, the values of C_e can be disregarded.

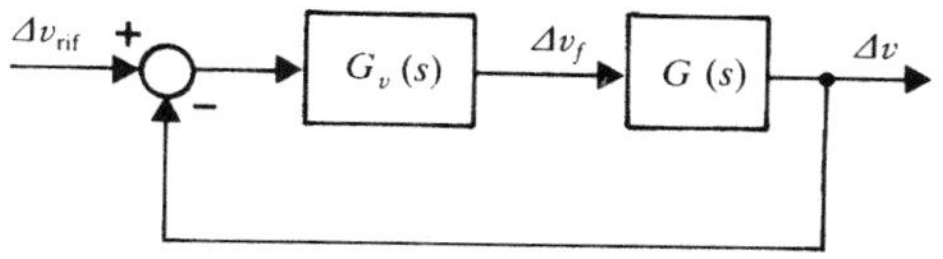

Figure 6.5. Voltage regulation by excitation control: broad block diagram.

For the transfer function $G(s) \triangleq (\Delta v/\Delta v_f)(s)$ reported in Figure 6.5, it should be noted that:

- it is $v^2 = v_d^2 + v_q^2$, and thus $\Delta v = (v_d^o \Delta v_d + v_q^o \Delta v_q)/v^o$, where v^o is positive by definition, whereas v_d^o and v_q^o can assume either sign;
- from Equations [6.2.1′] and [6.2.2], it is possible to derive equations such as:

$$\begin{cases} \dfrac{v_d}{v_f} = \Omega^2 L N_d(s) \dfrac{A(s)}{\Psi(s)} \\ \dfrac{v_q}{v_f} = \Omega^2 L N_q(s) \dfrac{A(s)}{\Psi(s)} \end{cases}$$

where $N_d(s)$ and $N_q(s)$ are proper functions, and:

$$\Psi(s) \triangleq (s^2 + \Omega^2)(\mathcal{L}_d(s) + L)(\mathcal{L}_q(s) + L) + sR(\mathcal{L}_d(s) + \mathcal{L}_q(s) + 2L) + R^2 \qquad [6.2.3]$$

- it then follows, because $A(0)$, $\Psi(0) > 0$, and assuming that $v_f^o > 0$:

$$G(s) = \frac{F(s)}{\sqrt{F(0)}} \Omega^2 L \frac{A(s)}{\Psi(s)}$$

where:

$$F(s) \triangleq N_d(0) N_d(s) + N_q(0) N_q(s)$$

(if, instead, we would assume $v_f^o < 0$, $G(s)$ would change its sign).

Similar conclusions formally hold for the following cases (b), (c) also.

As an example, Figure 6.6a shows the frequency response diagrams of $G(\tilde{\jmath}\nu)$, assuming $A(s) = A(0)/(1 + s\hat{T}'_{do})$, $\mathcal{L}_d(s) = (L_d + s\hat{T}'_{do}\hat{L}'_d)/(1 + s\hat{T}'_{do})$, with $\hat{L}'_d = 0.3L_d$, $\hat{T}'_{do} = 2000/\Omega$, $\mathcal{L}_q(s) = L_q = 0.6L_d$ (salient pole machine), $L = L_d$, and $R = 2 \cdot 10^{-3}\Omega L_d$. Such diagrams remain practically unchanged when assuming $R = 0$ and/or disregarding the dynamic behavior of inductances.

In the absence of control (i.e., if the excitation voltage v_f is constant), the characteristic equation is given by $\Psi(s) = 0$ (recall Equation [6.2.3]).

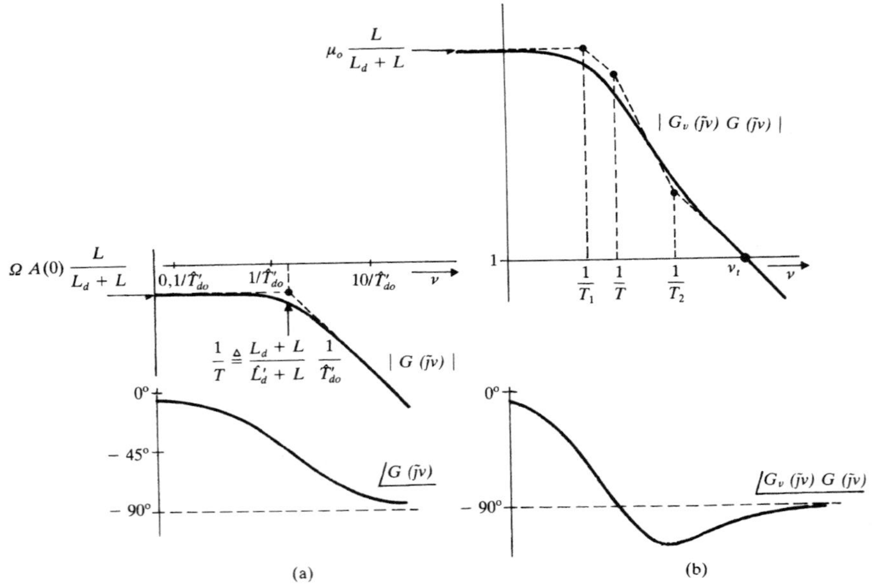

Figure 6.6. Operation on inductive load: frequency response: (a) of the transfer function $(\Delta v/\Delta v_f)(s) = G(s)$; (b) of the overall transfer function $G_v(s)G(s)$ of the loop (see Fig. 6.5 and Equation [6.2.7]).

If the dynamic behavior of inductances is considered, but it is assumed that $R = 0$, such an equation simply leads to:

$$\left.\begin{aligned} 0 &= s^2 + \Omega^2 \\ 0 &= \mathcal{L}_d(s) + L \\ 0 &= \mathcal{L}_q(s) + L \end{aligned}\right\} \qquad [6.2.4]$$

from which it is possible to derive:

- two purely imaginary characteristic roots ($\pm \tilde{\jmath}\Omega$) that correspond to a relatively high resonance frequency (Ω);
- other characteristic roots that are real negative and can be derived from the last two parts of Equations [6.2.4]; specifically, by assuming $\mathcal{L}_d(s) = (L_d + s\hat{T}'_{do}\hat{L}'_d)/(1 + s\hat{T}'_{do})$, $\mathcal{L}_q(s) = L_q$, such roots are reduced to a single real negative root $(-(L_d + L)/(\hat{L}'_d + L)\hat{T}'_{do})$ at low frequency.

Furthermore it can be derived:

$$G(s) \triangleq \frac{\Delta v}{\Delta v_f}(s) = \Omega L \frac{A(s)}{\mathcal{L}_d(s) + L} \qquad [6.2.5]$$

the poles of which are only the solutions of the second of Equations [6.2.4], and thus they are real and negative.

Note that Equations [6.2.4], respectively, correspond to the dynamic behaviors of:

- inductances (of machine and load);
- rotor windings on the d axis;
- rotor windings on the q axis;

which therefore, for $R = 0$, are noninteracting.

In practice, the presence of the armature resistance R causes some interactions which, however, are usually negligible because of the relatively small value of R. The most important effect of R is modifying the two high-frequency roots, the real part of which becomes negative[(5)]. As a consequence, the transients related to the dynamic behavior of inductances become damped.

Considering this, as well as the relative slowness of the phenomena related to excitation control, in the following we will also assume $R = 0$ and disregard the dynamic behavior of inductances.

Still in the absence of control, *if the dynamic behavior of inductances is not considered* and $R = 0$ is assumed, then:

- the characteristic equation leads to the second and the third of Equations [6.2.4];
- the transfer function [6.2.5] applies again, which is not affected by the dynamic behavior of rotor windings on the q axis, nor by that of inductances (whereas it can be intended that ψ_q, v_d, and i_q are zero).

For $R \neq 0$, Equations [6.2.1] and [6.2.1′] must be substituted by:

$$j\Omega\overline{\psi} - R\overline{\imath} = \overline{v} = j\Omega L\overline{\imath} \qquad [6.2.6]$$

$$\left.\begin{aligned} -\Omega\psi_q - Ri_d &= v_d = -\Omega L i_q \\ \Omega\psi_d - Ri_q &= v_q = \Omega L i_d \end{aligned}\right\} \qquad [6.2.6']$$

so that, considering Equations [6.2.2], the following characteristic equation is derived:

$$0 = \Omega^2(\mathcal{L}_d(s) + L)(\mathcal{L}_q(s) + L) + R^2 \triangleq \phi(s)$$

whereas it holds that:

$$G(s) = \frac{F(s)}{\sqrt{F(0)}}\Omega^2 L \frac{A(s)}{\phi(s)}$$

---

(5) For the usual values of R, which are relatively small, such roots can be approximated by $\sim (\alpha \pm \tilde{\jmath}\Omega)$, intending:

$$\alpha \triangleq -\frac{R}{2}\mathcal{R}e\left(\frac{1}{\mathcal{L}_d(\tilde{\jmath}\Omega) + L} + \frac{1}{\mathcal{L}_q(\tilde{\jmath}\Omega) + L}\right)$$

with:

$$F(s) = \Omega^2(\mathcal{L}_q(s) + L)(L_q + L) + R^2$$

For $\mathcal{L}_q(s) = L_q$, the characteristic equation is simply given by:

$$0 = \mathcal{L}_d(s) + L + \frac{R^2}{\Omega^2(L_q + L)}$$

whereas:

$$G(s) = \Omega L \frac{A(s)\sqrt{1 + \left(\dfrac{R}{\Omega(L_q + L)}\right)^2}}{\mathcal{L}_d(s) + L + \dfrac{R^2}{\Omega^2(L_q + L)}}$$

and such expressions are similar to the second of Equations [6.2.4] and to [6.2.5], respectively. For example, in Equation [6.2.4], it is sufficient to substitute L by $L + R^2/(\Omega^2(L_q + L))$.

The criteria for the *synthesis of the voltage regulation loop* (Fig. 6.5) should, as usual, account for (1) the static effect of disturbances, and (2) the response speed and stability.

(1) Regarding the static effect of disturbances, it is insightful to consider the voltage drop from no-load to on-load conditions. By assuming for simplicity $R = 0$:

- in the absence of regulation it holds that:

$$v = G(0)v_f = \Omega L \frac{A(0)}{L_d + L} v_f = \frac{L}{L_d + L} v^{(o)}$$

  where $v^{(o)}$ is the no-load voltage ($L = \infty$), and therefore, in relative terms:

$$a \triangleq \frac{v^{(o)} - v}{v^{(o)}} = \frac{L_d}{L_d + L}$$

  (e.g., $a = 50\%$ for $L = L_d$, $a = 67\%$ for $L = L_d/2$);

- in the presence of regulation and assuming $\mu_o \triangleq G_v(0)\Omega A(0)$[6], it instead holds that:

$$v = \frac{G_v(0)G(0)}{1 + G_v(0)G(0)} v_f = \frac{\mu_o L}{L_d + L + \mu_o L} v_f$$

$$v^{(o)} = \frac{\mu_o}{1 + \mu_o} v_f$$

(6) Note that the static gain of the regulation loop (Fig. 6.5) is given by $G_v(0)G(0)$, where $G(0)$ is equal to $\Omega A(0)$ at "no-load" operation ($L = \infty$); thus μ_o is, by definition, the static gain of the loop *at no-load operation*.

so that:

$$a = \frac{L_d}{L_d + (1 + \mu_o)L}$$

For any desired a, one can then derive μ_o (and thus also $G_v(0)$), i.e.,

$$\mu_o = \frac{L_d}{aL} - \frac{L_d + L}{L}$$

(where the last term is usually negligible). For example, by assuming that $a = 0.5\%$, it follows that $\mu_o \cong 200$ for $L = L_d$, $\mu_o \cong 400$ for $L = L_d/2$. As a result of the regulation, the value of a is reduced approximately $\mu_o L/(L_d + L)$ times.

(2) Regarding the response speed and stability, the case considered above does not imply any particular problem, with the simplifications adopted. If for instance, it is assumed that $R = 0$, $A(s) = A(0)/(1 + s\hat{T}'_{do})$, and $\mathcal{L}_d(s) = (L_d + s\hat{T}'_{do}\hat{L}'_d)/(1 + s\hat{T}'_{do})$, the response delay of $G(s)$ is, in fact, defined only by the time constant $T \triangleq ((\hat{L}'_d + L)/(L_d + L))\hat{T}'_{do}$ which, when L is not too small, is slightly smaller than $\hat{T}'_{do}$. It is possible then, to think of a transfer function simply of the proportional type, i.e., $G_v(s) \triangleq (\Delta v_f/\Delta(v_{\text{rif}} - v))(s) = \text{constant} = G_v(0)$, with a sufficiently high value (at least as high as in (1)), and corresponding to a cutoff frequency:

$$\nu_t \cong \frac{G_v(0)G(0)}{T} = \frac{\mu_o L}{T(L_d + L)} = \frac{\mu_o L}{\hat{T}'_{do}(\hat{L}'_d + L)}$$

This value is (slightly) smaller than $\mu_o/\hat{T}'_{do}$, which, for example, is 50 sec^{-1} for $\hat{T}'_{do} = 7$ sec and $\mu_o = 350$ (recall footnote[(4)], Chapter 3).

However, this would not be practically acceptable for various reasons, including:

- other delays disregarded here, that would cause instability;
- saturation (hitting the *"ceiling"*; see Section 6.2.2a) of the excitation system during transients, even for relatively small disturbances (this would make the response speed, evaluated with reference only to the linear behavior, unrealistic);
- inconsistency with other situations, such as operation when connected to the network (for which the interactions with the electromechanical phenomena must be considered; see Section 7.2), and so on.

Therefore, it is convenient to accept a relatively small cutoff frequency (e.g., 5 rad/sec, about 10 times lower than what is seen above), that corresponds to a "medium frequency" gain sufficiently smaller than μ_o.

The above requirements may be satisfied, apart from further unavoidable delays (e.g., time constants of some hundredth of a second, caused by the v transducer in the feedback path, etc.), by a transfer function:

$$G_v(s) = \frac{1}{\Omega A(0)}\mu_o \frac{1+sT_2}{1+sT_1} \quad [6.2.7]$$

In fact, by assuming $T_1 \gg T_2$, the low-pass effect of this function permits $G_v(0)$ to have the desired magnitude, and the cutoff frequency $\nu_t \cong (\mu_o T_2/T_1)L/(\hat{T}'_{do}(\hat{L}'_d + L))$ to be not too large; see Figure 6.6b.

(b) Operation on Resistive Load (Fig. 6.4b)

Under the usual assumptions, the following equations now hold:

$$(s + j\Omega)\overline{\psi} - R\overline{\imath} = \overline{v} = R_L\overline{\imath} \quad [6.2.8]$$

$$\left.\begin{aligned} s\psi_d - \Omega\psi_q - Ri_d &= v_d = R_L i_d \\ \Omega\psi_d + s\psi_q - Ri_q &= v_q = R_L i_q \end{aligned}\right\} \quad [6.2.8']$$

In the absence of control ($v_f =$ constant) and assuming $\Omega =$ constant, the characteristic equation is again of the type $\Psi(s) = 0$ (recall Equation [6.2.3]) with R replaced by $(R + R_L)$, and $L = 0$. Furthermore, it is:

$$G(s) \triangleq \frac{\Delta v}{\Delta v_f}(s)$$

$$= \frac{F(s)}{\sqrt{F(0)}}\Omega R_L \frac{A(s)}{(s^2+\Omega^2)\mathcal{L}_d\mathcal{L}_q + s(R+R_L)(\mathcal{L}_d+\mathcal{L}_q) + (R+R_L)^2}$$

where:

$$F(s) \triangleq (s^2+\Omega^2)\mathcal{L}_q(s)L_q + s(R+R_L)L_q + (R+R_L)^2$$

For $s = \tilde{\jmath}\nu$, the continuous line diagrams of Figure 6.7a are obtained, by assuming $A(s) = A(0)/(1+s\hat{T}'_{do})$, $\mathcal{L}_d(s) = (L_d + s\hat{T}'_{do}\hat{L}'_d)/(1+s\hat{T}'_{do})$, $\hat{L}'_d = 0.3L_d$, $\hat{T}'_{do} = 2000/\Omega$, $\mathcal{L}_q(s) = L_q = 0.6L_d$, and $R_L = \Omega L_d$, $R = 0$.

Disregarding, for reasons similar to those given above, the dynamic behavior of inductances, it follows that:

$$\left.\begin{aligned} -\Omega\psi_q - Ri_d &= v_d = R_L i_d \\ \Omega\psi_d - Ri_q &= v_q = R_L i_q \end{aligned}\right\} \quad [6.2.9]$$

and the characteristic equation becomes, without control:

$$0 = \Omega^2\mathcal{L}_d(s)\mathcal{L}_q(s) + (R+R_L)^2$$

whereas:

$$G(s) = \frac{F(s)}{\sqrt{F(0)}}\Omega R_L \frac{A(s)}{\Omega^2\mathcal{L}_d(s)\mathcal{L}_q(s) + (R+R_L)^2}$$

with $F(s) = \Omega^2\mathcal{L}_q(s)L_q + (R+R_L)^2$.

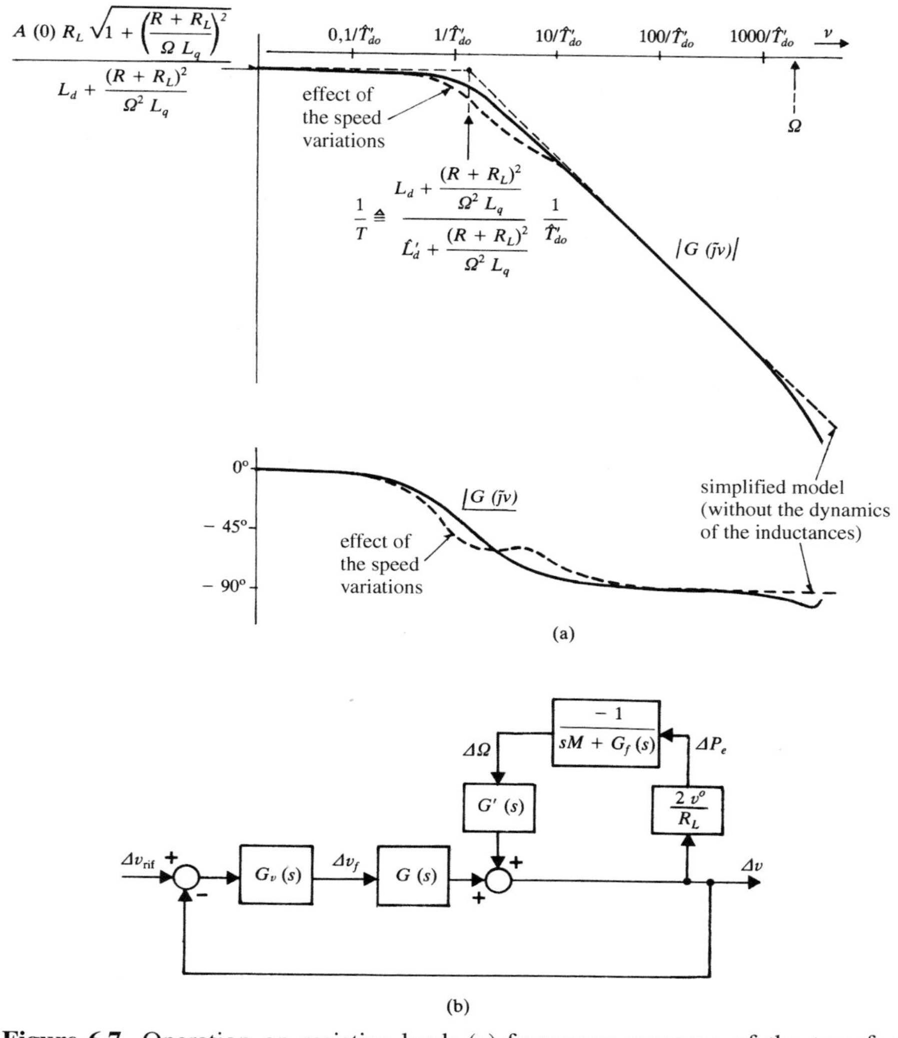

Figure 6.7. Operation on resistive load: (a) frequency response of the transfer function $(\Delta v/\Delta v_f)(s) = G(s)$; (b) block diagram (including speed variations).

For $\mathcal{L}_q(s) = L_q$ the characteristic equation is simply given by:

$$0 = \mathcal{L}_d(s) + \frac{(R + R_L)^2}{\Omega^2 L_q}$$

whereas:

$$G(s) = R_L \sqrt{1 + \left(\frac{R + R_L}{\Omega L_q}\right)^2} \frac{A(s)}{\mathcal{L}_d(s) + \dfrac{(R + R_L)^2}{\Omega^2 L_q}}$$

and such expressions are formally similar to the second of Equations [6.2.4] (with the inductance L substituted by $(R + R_L)^2/(\Omega^2 L_q)$) and to Equation [6.2.5] (with L substituted by $R_L\sqrt{1 + ((R + R_L)/(\Omega L_q))^2}/\Omega$ for the numerator and, again, by $(R + R_L)^2/(\Omega^2 L_q)$ for the denominator. For $R = 0$ and R_L in the order of ΩL_q, such values are, respectively, approximately $\sqrt{2}L_q$ and L_q).

For a better approximation, it is possible to also account for the *speed variations* Ω caused by the variations in the electromagnetic torque (and in the driving torque, due to the effect of the speed governor).

To this purpose, using the same notation (and approximations) seen in Chapter 3, it is possible to write an equation:

$$\Delta\Omega = -\frac{\Delta P_e}{sM + G_f(s)}$$

where, disregarding R:

$$\Delta P_e = \Delta\left(\frac{v^2}{R_L}\right) = \frac{2v^o}{R_L}\Delta v$$

Furthermore, by linearizing Equations [6.2.9] (for $R = 0$) and with some development, it can be derived:

$$\Delta v = G(s)\Delta v_f + G'(s)\Delta\Omega$$

(see Figure 6.7b), where:

$$G'(s) \triangleq R_L \frac{R_L \dfrac{v^o}{\Omega^o} + \dfrac{v_d^o v_q^o}{v^o}(\mathcal{L}_q(s) - \mathcal{L}_d(s))}{\Omega^{o2}\mathcal{L}_d(s)\mathcal{L}_q(s) + R_L^2}$$

$$= b\frac{\Omega^{o2} L_q(L_q + \mathcal{L}_q(s) - \mathcal{L}_d(s)) + R_L^2}{\Omega^{o2}\mathcal{L}_d(s)\mathcal{L}_q(s) + R_L^2}$$

having set, for brevity:

$$b \triangleq \frac{v^o/\Omega^o}{1 + (\Omega^o L_q/R_L)^2}$$

By eliminating $\Delta\Omega$ and ΔP_e, it can be finally derived:

$$\Delta v = \frac{G(s)}{1 + \dfrac{2v^o}{R_L}\dfrac{G'(s)}{sM + G_f(s)}}\Delta v_f$$

which however, in practice, differs significantly from $\Delta v = G(s)\Delta v_f$ only in a range of relatively low frequencies (see the qualitative example in Figure 6.7a), in which the gain of the regulation loop can be made large enough (recall Equation [6.2.7]) to noticeably reduce the speed variation effects on voltage.

In any case, for the same reasons previously given, the suitability of a $G_v(s)$ similar to Equation [6.2.7] is confirmed.

Regarding the variations of $\Delta v/v$ at steady-state without control, if we assume, for simplicity, $R = 0$ and $L_d = L_q$, then:

- in the case of a purely inductive load, we found that $v = \Omega L(A(0)/(L_d + L))v_f$, and therefore:

$$\frac{\Delta v/v}{\Delta L/L} = \frac{1}{1 + L/L_d}$$

- instead, in the present case (purely resistive load), it holds that:

$$v = R_L \frac{A(0)}{L_d\sqrt{1 + \left(\dfrac{R_L}{\Omega L_d}\right)^2}} v_f$$

and thus:

$$\frac{\Delta v/v}{\Delta R_L/R_L} = \frac{1}{1 + (R_L/\Omega L_d)^2}$$

which is larger or smaller than the similar ratio for the previous case, according to whether $R_L^2 \lessgtr \Omega^2 L L_d$.

(c) Operation on Capacitive Load (Fig. 6.4c)

Unlike the previous cases, the dynamic behavior of the load capacitance now must be considered. By assuming again that $\Omega_r = \Omega =$ constant, it follows that:

$$(s + j\Omega)\overline{\psi} - R\overline{\imath} = \overline{v} = \frac{1}{(s + j\Omega)C}\overline{\imath} \qquad [6.2.10]$$

or equivalently:

$$\left.\begin{aligned} (s^2 - \Omega^2)\psi_d - 2s\Omega\psi_q - sRi_d + \Omega Ri_q = sv_d - \Omega v_q = \frac{1}{C}i_d \\ 2s\Omega\psi_d + (s^2 - \Omega^2)\psi_q - \Omega Ri_d - sRi_q = \Omega v_d + sv_q = \frac{1}{C}i_q \end{aligned}\right\} \qquad [6.2.10']$$

in addition to Equations [6.2.2]. With the hypothesis $\Omega =$ constant, recall footnote[(4)] referring to the case of inductive load.

Based on what has been observed in previous cases, one could be inclined to disregard the dynamic behavior of machine inductances and of load capacitance (shortly, the "LC dynamics"). However, as we will see, this is not always an acceptable approximation, especially for small values of $1/(\Omega C)$[7].

If the LC dynamics are not considered, the explicit terms in s in the previous equations simply must be disregarded, which is equivalent to substitution of the reactance ΩL by $-1/(\Omega C)$ in Equations [6.2.6] and [6.2.6′]. By also considering Equations [6.2.2] the following characteristic equation then holds, in the absence of control (v_f = constant):

$$0 = \left(\Omega \mathcal{L}_d(s) - \frac{1}{\Omega C}\right)\left(\Omega L_q(s) - \frac{1}{\Omega C}\right) + R^2 \triangleq \phi(s) \qquad [6.2.11]$$

and it is:

$$G(s) \triangleq \frac{\Delta v}{\Delta v_f}(s) = \pm \frac{F(s)}{\sqrt{F(0)}} \frac{1}{C} \frac{A(s)}{\phi(s)}$$

according to whether $v_f^o/\phi(0) \gtrless 0$, with;

$$F(s) \triangleq \left(\Omega \mathcal{L}_q(s) - \frac{1}{\Omega C}\right)\left(\Omega L_q - \frac{1}{\Omega C}\right) + R^2$$

If R is disregarded, Equation [6.2.11] simply leads to:

$$\left.\begin{aligned} 0 &= \Omega^2 \mathcal{L}_d(s) C - 1 \\ 0 &= \Omega^2 \mathcal{L}_q(s) C - 1 \end{aligned}\right\} \qquad [6.2.12]$$

without any interaction between the dynamic behavior of rotor windings on the d axis and that of rotor windings on the q axis. It further results that $\sqrt{F(0)} = |\Omega L_q - 1/(\Omega C)|$, and we can derive:

$$G(s) = \pm \Omega \frac{A(s)}{1 - \Omega^2 \mathcal{L}_d(s) C}$$

according to whether $v_f^o/(1 - \Omega^2 L_d C) \gtrless 0$. Therefore, it simply holds:

$$G(s) = \Omega \frac{A(s)}{1 - \Omega^2 L_d(s) C}$$

by assuming that v_f^o has the same sign as $(1 - \Omega^2 L_d C)$.

(7) In the following analysis, relatively large values of C also will be considered, which correspond to small load impedances of little practical interest. This will be done not only for reasons of formal completeness, but also to illustrate the effects of the LC dynamics for capacitance values comparable with those adopted in the "series" line compensation (see Section 7.2.4).

Note that, in terms of transfer functions:

$$\begin{cases} \dfrac{i_d}{v_f}(s) = -\Omega^2 C \dfrac{A(s)}{1-\Omega^2 \mathcal{L}_d(s) C} \\ \dfrac{i_q}{v_f}(s) = 0 \end{cases} \qquad \begin{cases} \dfrac{v_d}{v_f}(s) = 0 \\ \dfrac{v_q}{v_f}(s) = \Omega \dfrac{A(s)}{1-\Omega^2 \mathcal{L}_d(s) C} \end{cases}$$

Based on the above assumptions, it then follows that $(v_q/v_f)(s) = G(s)$, which means that $\Delta v_q = +\Delta v$, and at steady-state $v_q^o = v^o > 0$(8).

Assuming that $A(s) = A(0)/(1 + s\hat{T}'_{do})$, $\mathcal{L}_d(s) = (L_d + s\hat{T}'_{do}\hat{L}'_d)/(1 + s\hat{T}'_{do})$, $\mathcal{L}_q(s) = L_q$, the characteristic equation in the absence of control is reduced to:

$$0 = 1 - \Omega^2 L_d C + s\hat{T}'_{do}(1 - \Omega^2 \hat{L}'_d C)$$

and it is:

$$G(s) = \Omega \frac{A(0)}{1 - \Omega^2 L_d C + s\hat{T}'_{do}(1 - \Omega^2 \hat{L}'_d C)}$$

with a real pole, which is either positive or negative according to whether $1/(\Omega C)$ is internal or external to the interval $(\Omega \hat{L}'_d, \Omega L_d)$; refer to the Nyquist diagrams in Figure 6.8 for the numerical example with $\hat{L}'_d = 0.3L_d$, $\hat{T}'_{do} = 2000/\Omega$, $L_q = 0.6L_d$. Therefore, without control, the (asymptotic) stability condition can be written as:

$$\frac{1}{\Omega C} \notin [\Omega \hat{L}'_d, \Omega L_d]$$

More generally, but still for $R = 0$, $\hat{L}'_d$ can be substituted by $\mathcal{L}_d(\infty)$.

However, if the LC dynamics also are considered, the stability condition may be more restrictive (according to condition [6.2.15], it must hold $1/(\Omega C) > \Omega L_d$).

In any case, the possible instability constitutes the well-known phenomenon called *"self-excitation,"* which also involves the magnetic characteristic of the machine, according to Figure 6.9 (the notations of which are the same as those in Fig. 4.10). More precisely, if $1/(\Omega C)$ is within $\Omega \hat{L}'_d$ and ΩL_d, and more in general if $1/(\Omega C) < \Omega L_d$, the working point A is unstable, and the system reaches point A' (or A'') (with relatively large values of current, flux, and voltage), in which:

$$\frac{d\psi_{sd}^o}{d(-i_d^o)} < \frac{1}{\Omega^2 C} - L_l$$

(8) Therefore, differently from what was reported in the footnote(6) of Chapter 4, Section 4.4.1, it is here assumed that $v_q^o > 0$ (instead of $u_q^o \triangleq v_q^o + \Omega L_q i_d^o > 0$), so that it results in $v_f^o < 0$ for $1/(\Omega C) < \Omega L_d$, or equivalently for $Q^o/v^{o2} = -\Omega C < -1/(\Omega L_d)$ (instead of $Q^o/v^{o2} \in (-1/(\Omega L_q), -1/(\Omega L_d))$).

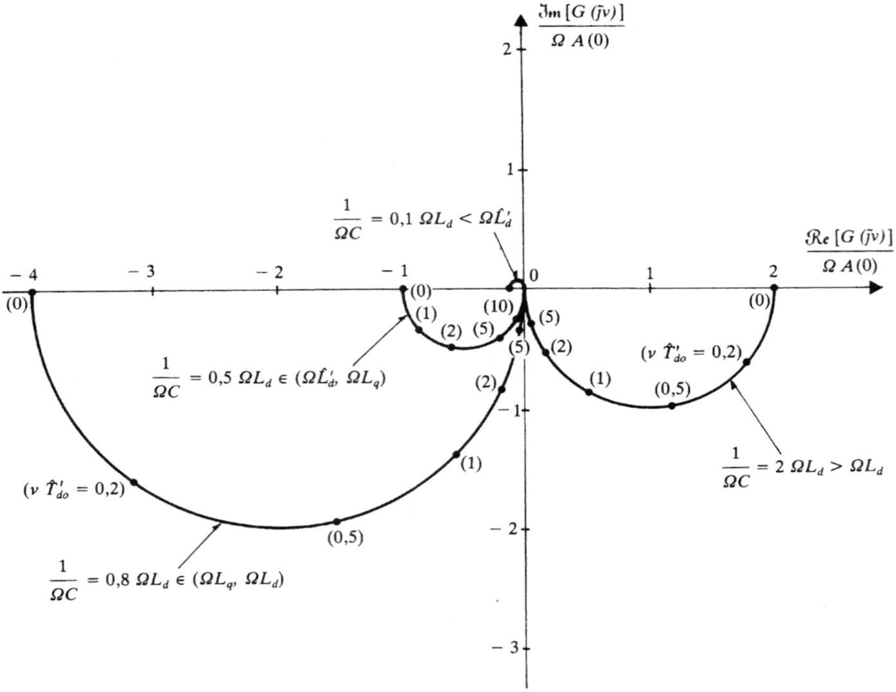

Figure 6.8. Operation on capacitive load: frequency response of the transfer function $(\Delta v/\Delta v_f)(s)/(\Omega A(0)) = G(s)/(\Omega A(0))$, with $R = 0$.

so that $1/(\Omega C)$ is larger than the equivalent reactance given by:

$$(\Omega L_d)_{\text{eq}} \triangleq \Omega \left(\frac{\mathrm{d}\psi_{sd}^o}{\mathrm{d}(-i_d^o)} + L_l \right)$$

The self-excitation phenomenon may be complicated by the presence of hysteresis in the magnetic characteristic. However, it must be underlined that, to explain the self-excitation behavior, the consideration of hysteresis (as usually done) *is both insufficient* (unless a dynamic model is also defined) *and unnecessary* (according to the discussion above).

For $R \neq 0$, and again assuming $\mathcal{L}_d(s) = (L_d + s\hat{T}'_{do}\hat{L}'_d)/(1 + s\hat{T}'_{do})$, $\mathcal{L}_q(s) = L_q$, the characteristic Equation [6.2.11] can be written as:

$$0 = \left[\Omega^2 (L_d + s\hat{T}'_{do}\hat{L}'_d) - \frac{1 + s\hat{T}'_{do}}{C} \right] \left(\Omega^2 L_q - \frac{1}{C} \right) + R^2 \Omega^2$$

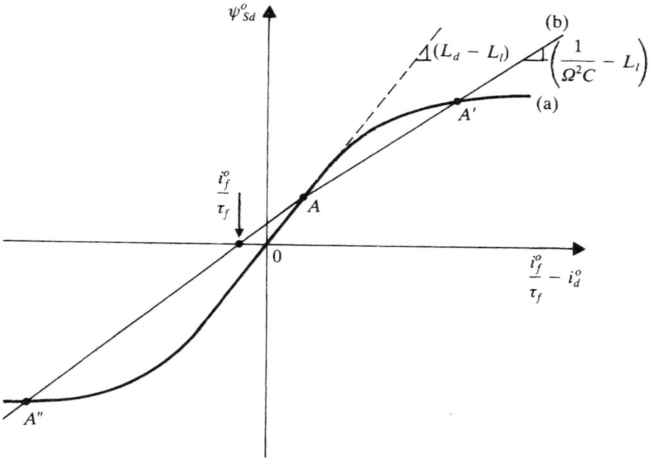

Figure 6.9. Self-excitation behavior, for $1/(\Omega C) < \Omega L_d$, i.e., for $1/(\Omega^2 C) - L_l < L_d - L_l$, with (see also Fig. 4.10):

$$\psi^o_{Sd} = \begin{cases} (L_d - L_l)\left(\dfrac{i^o_f}{\tau_f} - i^o_d - i^o_{Sd}\right) & \text{(characteristic (a))} \\ \psi^o_d + L_l i^o_d = -\left(\dfrac{1}{\Omega^2 C} - L_l\right) i^o_d & \text{(straight-line (b))} \end{cases}$$

from which:

$$s = -\frac{\Omega L_d - \dfrac{1}{\Omega C} + \dfrac{R^2}{\Omega L_q - \dfrac{1}{\Omega C}}}{\left(\Omega \hat{L}'_d - \dfrac{1}{\Omega C}\right) \hat{T}'_{do}}$$

Therefore, a condition $R > R_{\min}$ can be derived for the (asymptotic) stability, where $R_{\min}$ depends on $1/(\Omega C)$ according to Figure 6.10.

Regarding the effect of the voltage regulation, by assuming for simplicity $G_v(s) = \text{constant} = \mu/(\Omega A(0)) > 0$ (and $R = 0$), and indicating by $P_{(+)}$ the number of poles of $G(s)$ having positive real part, the following stability conditions can be derived, in terms of μ:

- case 1: $\dfrac{1}{\Omega C} > \Omega L_d \quad (P_{(+)} = 0)$ any μ
- case 2: $\dfrac{1}{\Omega C} \in (\Omega \hat{L}'_d, \Omega L_d) \quad (P_{(+)} = 1)$ $\mu > \Omega^2 L_d C - 1$
- case 3: $\dfrac{1}{\Omega C} < \Omega \hat{L}'_d \quad (P_{(+)} = 0)$ $\mu < \Omega^2 L_d C - 1$

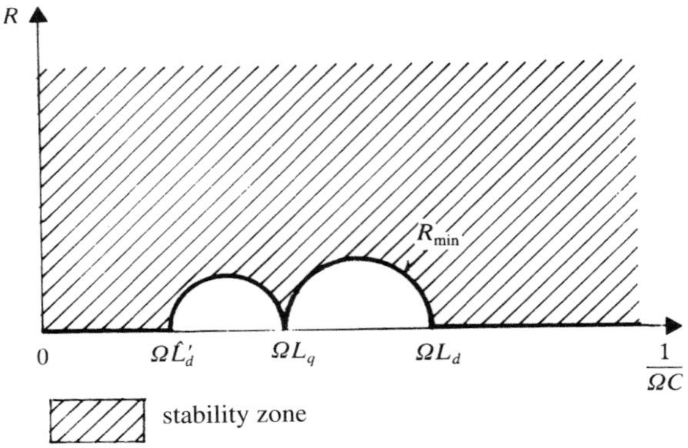

Figure 6.10. Operation on capacitive load: minimum value of the armature resistance R, for stability (without considering the LC dynamics).

according to Figure 6.11a (recall also Fig. 6.8 and the Nyquist criterion). Therefore, the regulation can also provide a stabilizing effect (case 2) or a destabilizing one (case 3).

By the Nyquist criterion, it is easy to extend the analysis to the case for which $G_v(s)$ is not constant. For instance, if $G_v(s)$ is given by Equation [6.2.7] (with the usual values of T_1, T_2), it is possible to again find the previous conditions by assuming $\mu = \mu_o$.

However, the interactions with the LC dynamics may be important. To determine if this is the case, at least qualitatively (without making the analysis too burdensome), it may be sufficient to again assume that $R = 0$. *If the LC dynamics are considered*, the following characteristic equation is obtained in the absence of control:

$$0 = (s^2 + \Omega^2)^2 \mathcal{L}_d(s)\mathcal{L}_q(s)C^2 + (s^2 - \Omega^2)(\mathcal{L}_d(s) + \mathcal{L}_q(s))C + 1 \triangleq \Psi'(s) \quad [6.2.13]$$

and it is:

$$G(s) = \Omega \frac{A(s)(1 - (s^2 + \Omega^2)\mathcal{L}_q(s)C)}{\Psi'(s)}$$

Furthermore, by assuming for simplicity $A(s) = A(0)/(1 + s\hat{T}'_{do})$, $\mathcal{L}_d(s) = (L_d + s\hat{T}'_{do}\hat{L}'_d)/(1 + s\hat{T}'_{do})$, $\mathcal{L}_q(s) = L_q$, Equation [6.2.13] can be written:

$$0 = a_5 s^5 + a_4 s^4 + a_3 s^3 + a_2 s^2 + a_1 s + a_0 \triangleq \Psi''(s) \quad [6.2.14]$$

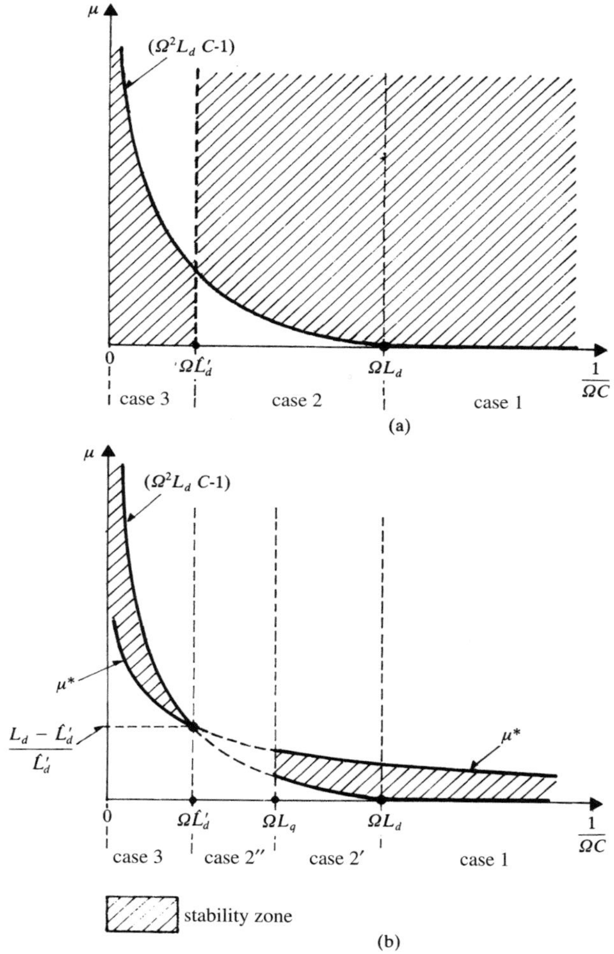

Figure 6.11. Operation on capacitive load, with $R = 0$ and $G_v(s) = \text{constant} = \mu/(\Omega A(0))$: admissible values of μ, for stability: (a) not considering the LC dynamics; (b) considering the LC dynamics.

where:

$$a_5 = \hat{T}'_{do}\hat{L}'_d L_q C^2, \quad a_4 = L_d L_q C^2, \quad a_3 = \hat{T}'_{do}(\hat{L}'_d + L_q + 2\Omega^2 \hat{L}'_d L_q C)C$$
$$a_2 = (L_d + L_q + 2\Omega^2 L_d L_q C)C, \quad a_1 = \hat{T}'_{do}(1 - \Omega^2 \hat{L}'_d C)(1 - \Omega^2 L_q C)$$
$$a_0 = (1 - \Omega^2 L_d C)(1 - \Omega^2 L_q C)$$

whereas:

$$G(s) = \Omega \frac{A(0)(1 - (s^2 + \Omega^2)L_q C)}{\Psi''(s)}$$

By applying, for instance, the Routh-Hurwitz criterion to Equation [6.2.14], the following (asymptotic) stability conditions can be derived:

$$\frac{1}{\Omega C} > \Omega L_q, \quad \frac{1}{\Omega C} > \Omega L_d$$

or equivalently (since $L_q \leq L_d$) the only condition:

$$\frac{1}{\Omega C} > \Omega L_d \qquad [6.2.15]$$

Note that:

- for $1/(\Omega C) = \Omega \hat{L}'_d$ and for $1/(\Omega C) = \Omega L_d$, the coefficients a_1 and a_0, respectively, become zero;
- for $1/(\Omega C) = \Omega L_q$, both a_1 and a_0 become zero.

Furthermore, the number of poles of $G(s)$ having positive real part is:

$$P_{(+)} = \begin{cases} 0 \quad \text{for } \dfrac{1}{\Omega C} > \Omega L_d & \text{(case 1)} \\ 1 \quad \text{for } \dfrac{1}{\Omega C} \in (\Omega L_q, \Omega L_d) & \text{(case 2}'\text{)} \\ 2 \quad \text{for } \dfrac{1}{\Omega C} \in (\Omega \hat{L}'_d, \Omega L_q) & \text{(case 2}''\text{)} \\ \quad \text{and for } \dfrac{1}{\Omega C} < \Omega \hat{L}'_d & \text{(case 3)} \end{cases}$$

so that case 2 previously defined ($1/(\Omega C) \in (\Omega \hat{L}'_d, \Omega L_d)$) now corresponds to the two (very distinct) cases $2'$ and $2''$.

To evaluate the effect of the voltage regulation, let us first assume $G_v(s) =$ constant $= \mu/(\Omega A(0)) > 0$. The Nyquist diagrams of the function $G(\tilde{j}\nu)$ for the different cases 1, $2'$, $2''$, 3 are similar to those of Figure 6.8, for values of ν which are not too high.

Therefore, without considering further details for such diagrams, one may be tempted to conclude, based on Nyquist criterion, that the LC dynamics do not modify the stability conditions in cases 1 and $2'$, and that there is no value $\mu > 0$ which can stabilize the (closed-loop) system in cases $2''$ and 3.

However, by carefully examining the different situations (also recalling the Routh-Hurwitz criterion) and by writing:

$$\mu^* \triangleq (L_d - \hat{L}'_d) \frac{\sqrt{(L_q - \hat{L}'_d)^2 + 8\hat{L}'_d L_q(\hat{L}'_d + L_q)\Omega^2 C} + \hat{L}'_d - L_q}{2\hat{L}'_d(\hat{L}'_d + L_q)}$$

the following (asymptotic) stability conditions can be deduced:

- case 1: $\mu < \mu^*$
- case 2′: $\mu \in (\Omega^2 L_d C - 1, \mu^*)$
- case 2″: no value $\mu > 0$
- case 3: $\mu \in (\mu^*, \Omega^2 L_d C - 1)$

which are summarized in Figure 6.11b. Therefore, in cases 1 and 2′ there is also the upper limitation $\mu < \mu^*$, and in case 3 there is the possibility of stabilizing the system by a proper choice of μ.

The above conditions are indeed confirmed by the Nyquist criterion, if the actual high-frequency behavior of $G(\tilde{\jmath}\nu)$ is considered. This can be seen, from a qualitative point of view (that is, not to scale), from the diagrams of Figure 6.12.

For $1/(\Omega C)$ decreasing from ΩL_d to 0, the function $G(s)$ presents, at first, a positive real pole (case 2′), then two positive real poles, and finally a pair of complex conjugate poles with a positive real part.

However, the hypothesis that $G_v(s) =$ constant may not appear to be reasonable, since the conditions derived above correspond to the behavior of $G(s)$ in frequency ranges that are very different from one another. In fact, in Figure 6.12, the point $1/(1 - \Omega^2 L_d C)$ corresponds to $\nu = 0$, whereas the point $-1/\mu^*$ corresponds to $\nu = \nu^*$, where:

$$\nu^{*2} \triangleq \frac{\hat{L}'_d + L_q + 2\Omega^2 \hat{L}'_d L_q C - \sqrt{(L_q - \hat{L}'_d)^2 + 8\hat{L}'_d L_q (\hat{L}'_d + L_q)\Omega^2 C}}{2\hat{L}'_d L_q C}$$

which is relatively large (in the cases of Fig. 6.8, with $1/(\Omega^2 L_d C) = 2;\ 0.8;\ 0.1$, it results in $\nu^*/\Omega \cong 1.11;\ 0.31;\ 0.5$, respectively).

If $G_v(s)$ is of the type given in Equation [6.2.7], instead of the above-reported conditions, it is possible to obtain the following ones:

- case 1: $\mu_o T_2/T_1 < \mu^*$
- case 2′: $\mu_o > \Omega^2 L_d C - 1, \quad \mu_o T_2/T_1 < \mu^*$
- case 2″: still no solutions
- case 3: $\mu_o T_2/T_1 > \mu^*, \quad \mu_o < \Omega^2 L_d C - 1$

and these conditions appear (with $\mu = \mu_o$, and $T_2 < T_1$):

- less restrictive in cases 1 and 2′, which incidentally are the only meaningful ones for the practical situations of isolated operation, as they do not imply excessive currents;
- more restrictive in case 3.

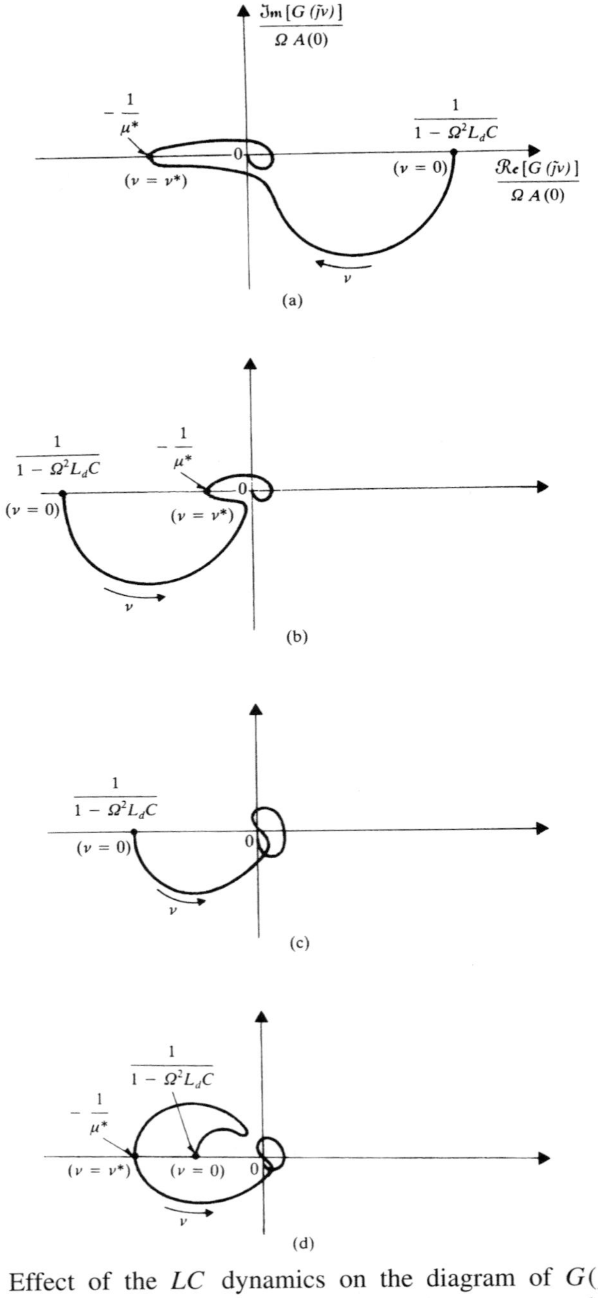

Figure 6.12. Effect of the LC dynamics on the diagram of $G(\tilde{j}\nu)/(\Omega A(0))$ (see Fig. 6.8) at high frequencies (qualitative behaviors, not to scale): (a) case 1: $1/(\Omega C) > \Omega L_d$; (b) case 2′: $1/(\Omega C) \in (\Omega L_q, \Omega L_d)$; (c) case 2″: $1/(\Omega C) \in (\Omega \hat{L}'_d, \Omega L_q)$; (d) case 3: $1/(\Omega C) < \Omega \hat{L}'_d$.

Moreover, such effects can be similarly enhanced because of further delays in $G_v(s)$, not considered in Equation [6.2.7].

The relatively small value of μ^* (note that certainly $\mu^* < (L_d - \hat{L}'_d)/\hat{L}'_d$) is the result of the presence of a poorly damped resonance in $G(s)$, which is responsible for increases in magnitude (relatively large, as shown in Fig. 6.12) at the point $-1/\mu^*$. In practice, such a resonance may instead be damped because of the effects of resistances (machine armature and network resistances), and it may happen that, in previous conditions, μ^* should be substituted by a much higher value.

Moreover, for a greater realism, functions $\mathcal{L}_d(s)$ and $\mathcal{L}_q(s)$ different (especially at high frequencies) from the ones previously considered also should be considered, as well as the nonlinearities caused by the magnetic characteristic of the machine etc. The hypothesis itself of purely capacitive load should be reviewed. However, the analysis developed here is still useful, at least from a qualitative perspective.

6.2.2. Typical Control Schemes

(a) Primary Control

(a1) Machine Voltage Regulation

As seen in Section 6.2.1, a transfer function $G_v(s) \triangleq (\Delta v_f/\Delta(v_{\text{rif}} - v))(s)$ of the type given in Equation [6.2.7] (neglecting small delays), i.e.,

$$G_v(s) = \frac{1}{\Omega A(0)}\mu_o \frac{1 + sT_2}{1 + sT_1} \qquad [6.2.16]$$

is acceptable in operation for inductive or resistive load and, with some caution, for capacitive load. A possible integral action, with $G_v(0) = \infty$, can be accomplished by assuming (at unchanged μ_o/T_1) μ_o, $T_1 \to \infty$.

The consequent control scheme is partially conditioned by the "excitation system," which may be basically of the following types (Fig. 6.13):

- "rotating" excitation:
  - by "dc exciter" (Fig. 6.13a),
  - by "(synchronous) ac exciter" and rotating diodes (Fig. 6.13b, where the exciter is itself excited through the stator, so that slip rings and brushes are completely avoided);
- "static" excitation:
  - by "thyristor exciter," with an "independent" or "dependent" supply (Fig. 6.13c).

The typical control schemes are correspondingly illustrated in Figure 6.14 (without considering delay time constants of some hundredths of a second, such as in feedback transducers and amplifier).

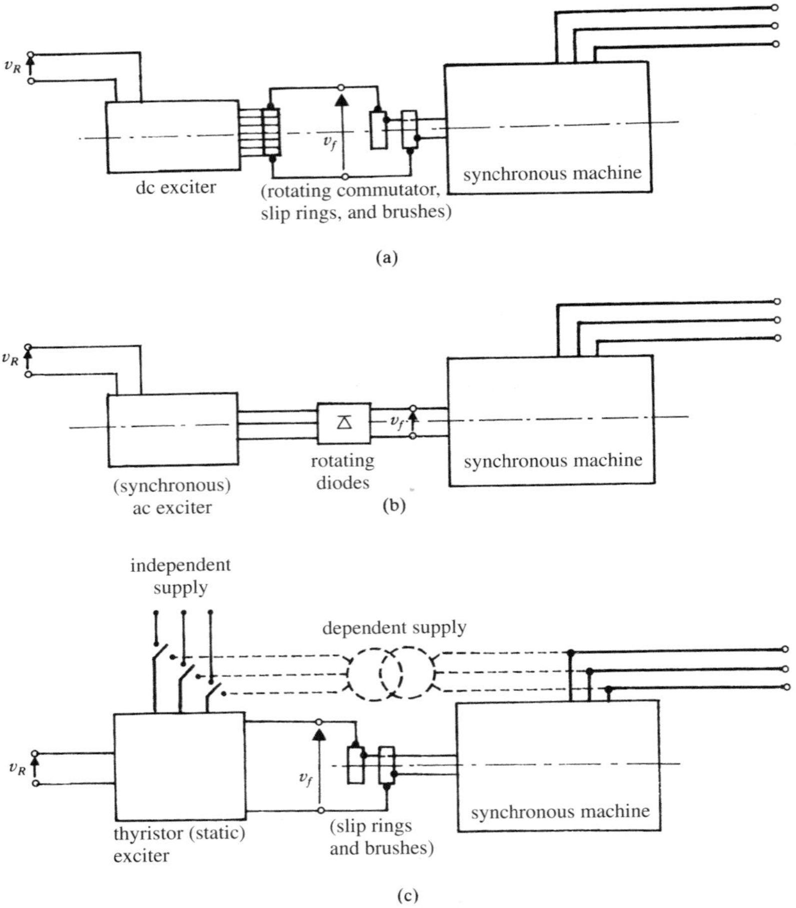

Figure 6.13. Excitation systems: (a) rotating excitation, by "dc exciter"; (b) as above, by "(synchronous) ac exciter" and rotating diodes; (c) static excitation, by "thyristor exciter."

More precisely, by *rotating excitation*:

- the excitation system may be represented, apart from magnetic saturation, by a transfer function $K_e/(1 + sT_e)$ (where T_e is typically 0.5–1 sec), as indicated in Figures 6.14a,b;
- the control system may be of the type shown in Figure 6.14a, with "transient feedback," or of the type shown in Figure 6.14b.

In the case of Figure 6.14a:

- with a dc exciter, there is an amplifier with a gain K_A, and a transient feedback having a transfer function $H(s) = K_t sT_d/(1 + sT_d)$, where the

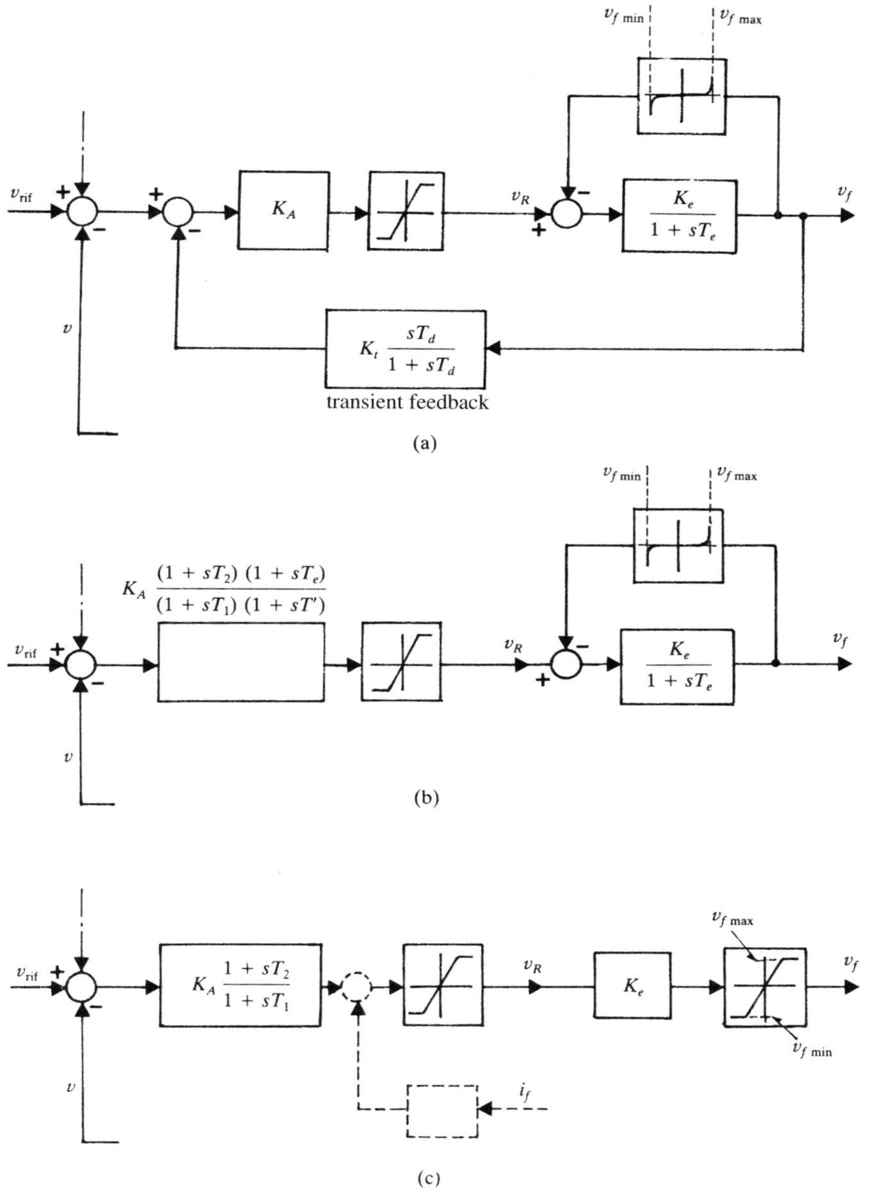

Figure 6.14. Excitation control: block diagrams for the case of rotating excitation (a) with and (b) without transient feedback; (c) for the case of static excitation. The input signals indicated by a dashed-dotted line may be representative of those caused by over- and underexcitation limiters, and of further possible ones.

product $K_A K_e K_t$ is large enough to obtain:

$$G_v(s) = K_A K_e \frac{1 + sT_d}{(1 + sT_e)(1 + sT_d) + sK_A K_e K_t T_d}$$
$$= K_A K_e \frac{1 + sT_d}{(1 + sT_1')(1 + sT_1'')}$$

with $T_1' \cong K_A K_e K_t T_d$ and $T_1'' \cong T_e/(K_A K_e K_t)$, i.e., neglecting T_1'' (which can be actually 0.07 sec or even smaller), similar to Equation [6.2.16], with:

$$\begin{cases} G_v(0) = \dfrac{1}{\Omega A(0)} \mu_o = K_A K_e \\ T_2 = T_d \\ T_1 = T_1' \gg T_2 \end{cases}$$

- with an ac exciter and rotating diodes, the feedback must be performed (to avoid sliding contacts), from the voltage v_R at the output of the amplifier (and not from v_f); to obtain a result similar to the previous one, it is sufficient to accomplish a feedback transfer function $H(s)K_e/(1 + sT_e)$.

In the case of Figure 6.14b, the control system instead acts only on the forward chain, by a transfer function (electronically achieved):

$$G_A(s) = K_A \frac{(1 + sT_2)(1 + sT_e)}{(1 + sT_1)(1 + sT')} = G_v(s) \frac{1 + sT_e}{K_e(1 + sT')}$$

where again $K_A = G_v(0)/K_e$, and the time constant T' can be made negligible.
Finally, by *static excitation* (Fig. 6.14c):

- the excitation system may be represented, apart from a small response delay approximately of a cycle (i.e., 0.02 sec for a 50-Hz power supply), by a simple gain K_e proportional to the supply voltage (and thus to the machine voltage v, in the case of "dependent" supply);
- the control system may be similar to that in Figure 6.14b, with $T_e = T' = 0$ and thus:

$$G_A(s) = K_A \frac{1 + sT_2}{1 + sT_1} = G_v(s) \frac{1}{K_e}$$

- because of the high response speed of the excitation system, it may be convenient to add a feedback from the field current i_f, to accelerate the response of i_f itself and obtain a faster regulation.

According to Figure 6.14, also the limits on v_R and v_f must be considered for large variations. The limits on v_f are the results of:

- magnetic saturation in the case of rotating excitation;

- the thyristor bridge characteristics in the case of static excitation (if the bridge is "half-controlled" it must be intended $v_{f\,\min} = 0$).

The values $v_{f\,\max}$ and $v_{f\,\min}$ constitute the so-called *"ceiling"* voltages, respectively positive and negative (or zero). Typical values are:

$$v_{f\,\max} \cong (2\text{–}2.5)V_{f(n)}, \qquad v_{f\,\min} \cong -(1.5\text{–}2.5)V_{f(n)}$$

(or $v_{f\,\min} = 0$, in the case just discussed), where $V_{f(n)}$ is the field voltage required at nominal conditions[(9)]. Large values of $v_{f\,\max}$ or of $|v_{f\,\min}|$ may be suitable in the cases of a short-circuit close to the machine or of a rejection from parallel operation, respectively.

One should also remember the *over- and underexcitation "limiters"* that are threshold devices intervening on the control system to keep (although not in short time intervals) the reactive power Q within a proper interval $[Q_{\min}, Q_{\max}]$ (see Section 2.2.1 and Fig. 2.9).

Finally, the control system can be requested to accept *further input signals*, particularly for the "compound" and/or secondary control (discussed below), or to stabilize the electromechanical oscillations ("additional" signals; see Sections 7.2.2 and 8.5.2).

(a2) Voltage Regulation at a Downstream Node with Possible "Compound" For greater generality, assume (Fig. 6.15a) that:

- the machine is connected to the rest of the system by a purely inductive element (i.e., step-up transformer and part of a line);
- the amplitude v' of the voltage downstream of this element must be regulated.

Disregarding the dynamic behavior of inductances and indicating by X the reactance of the considered element (evaluated at the frequency $\Omega_r = \Omega^o$, equal to the electrical speed of the machine at steady-state), it is possible to write;

$$\frac{Q}{v} \cong \frac{v - v'}{X} \qquad [6.2.17]$$

(see Fig. 2.7 and, by similar approximation, Equation [6.1.1]), from which:

$$v' \cong v - X\frac{Q}{v}$$

(9) Indicating by V_f^* (see Section 4.1.2) the value of v_f that corresponds to $v = v_{\text{nom}}$ at no-load operation, with $\Omega = \Omega_{\text{nom}}$ and without saturation, $V_{f(n)}$ can be typically (1.5–2) V_f^* for hydrounits and $(2\text{–}3)V_f^*$ for thermal units.

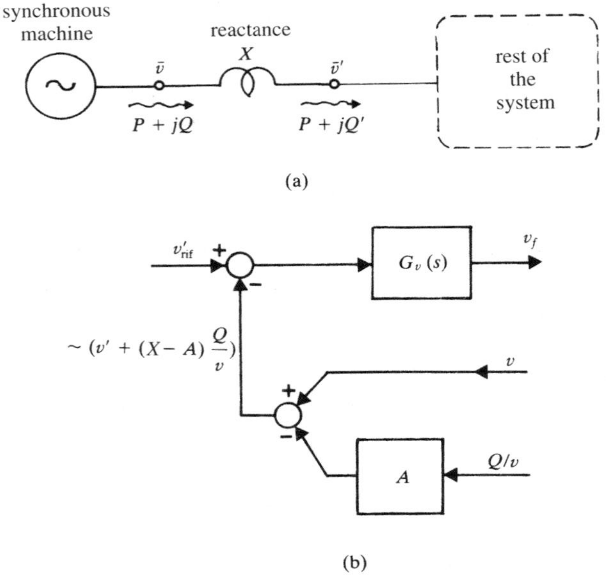

Figure 6.15. Voltage regulation at a downstream node by a single machine: (a) system under examination; (b) block diagram.

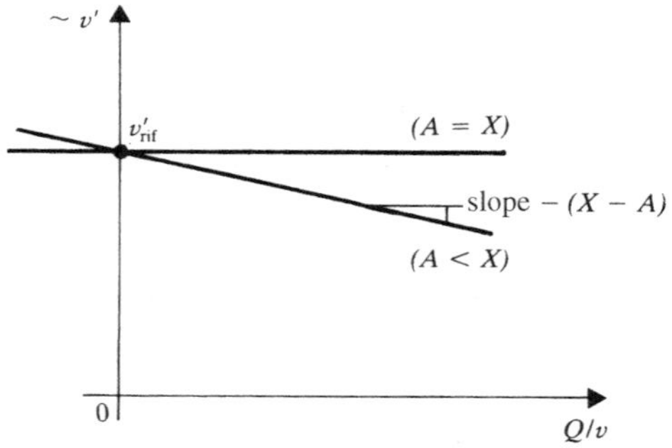

Figure 6.16. Regulation static characteristic for the system in Figure 6.15 (with $G_v(0) \to \infty$).

To regulate v' to the value v'_{rif} it is sufficient, with the stated approximations, to substitute v_{rif} by v'_{rif} and v by $v - XQ/v$ (see Fig. 6.15b, with $A = X$). In fact, by assuming μ_o (and thus $G_v(0)$) to be very large, we obtain $v' \cong v'_{\text{rif}}$ at steady-state, independently of the supplied reactive current Q/v (see Fig. 6.16).

Similarly to what was said for the speed regulation of a single unit (see Section 3.1.2), it can be stated that the static characteristic (v', Q/v) corresponds, at the conditions considered, to a zero-"droop" operation.

Generally, if v is substituted by $v - AQ/v$, at steady-state it is possible to obtain $v - AQ/v \cong v'_{\text{rif}}$, $v' \cong v'_{\text{rif}} + (A - X)Q/v$. This corresponds to an operation at nonzero (positive or negative) droop. Usually, the system seen by the machine has an inductive behavior, with $\partial(Q/v)/\partial v \in (0, 1/X)$, so that the term $-AQ/v$ in the feedback path causes a positive feedback (obviously if $A > 0$), opposite to the negative feedback which starts from the voltage v (Fig. 6.15b). For this reason, it is usually convenient to assume a value $A < X$, as if the aim were to regulate the voltage at an intermediate point of the interposed inductive element.

In the case of more machines connected to the node at voltage v' (Fig. 6.17a), it would not make sense to regulate v' by assuming a zero droop for all the machines (and neither for two of them), since:

- at equal $v'_{\text{rif}\,i}$'s, the sharing of Q_i's would remain undetermined;
- at different $v'_{\text{rif}\,i}$'s, some machines would operate at $Q_{i\,\text{min}}$ and others at $Q_{i\,\text{max}}$, whereas a single machine would operate at an intermediate Q_i value.

On the other hand, if only one machine was regulated at zero droop, this alone would be required to respond to (within its own limits) the reactive power variations demanded when passing from one steady-state to another.

Therefore, it is necessary to set nonzero droops for all regulators, i.e., $A_i \neq X_i$, as in Figure 6.17b.

Assuming, with usual approximations:

$$\frac{Q'}{v'} \cong \sum_i \frac{Q_i}{v_i} \cong \sum_i \frac{v_i - v'}{X_i}$$

and again assuming that $G_{vi}(0)$'s are very large, for brevity using the notation $B_i \triangleq 1/(X_i - A_i)$, $B \triangleq \sum_i B_i$, it then follows at steady-state[10]:

$$v' \cong \frac{1}{B}\left(\sum_i B_i v'_{\text{rif}\,i} - \frac{Q'}{v'}\right)$$

$$\frac{Q_k}{v_k} \cong \frac{B_k}{B}\left[\frac{Q'}{v'} + \sum_{i \neq k} B_i (v'_{\text{rif}\,k} - v'_{\text{rif}\,i})\right]$$

$$v_k \cong v'_{\text{rif}\,k} + \frac{A_k B_k}{B}\left(\frac{Q'}{v'} - \sum_i B_i v'_{\text{rif}\,i}\right)$$

[10] In particular, if the $v'_{\text{rif}\,i}$'s are all the same ($\triangleq v'_{\text{rif}}$), at steady-state:

$$v' \cong v'_{\text{rif}} - \frac{1}{B}\frac{Q'}{v'}, \qquad \frac{Q_k}{v_k} \cong \frac{B_k}{B}\frac{Q'}{v'}$$

with $v_k \cong (1 - A_k B_k) v'_{\text{rif}} + \dfrac{A_k B_k}{B}\dfrac{Q'}{v'}$.

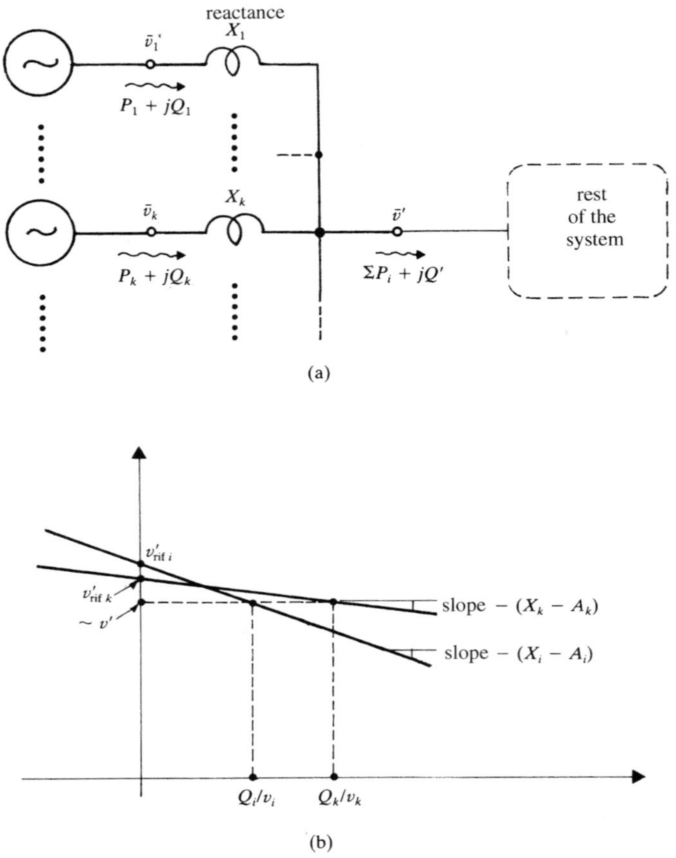

Figure 6.17. Voltage regulation at a downstream node by more machines: (a) system under examination; (b) regulation static characteristics and sharing of reactive currents (or equivalently powers).

By doing so:

- it is possible to determine the steady-state sharing, between the different machines, of the reactive currents Q_k/v_k for any total demanded current Q'/v', or equivalently of the reactive powers Q_k (the sharing may be done in such a way as to make uniform the reactive power margins of the machines; see particularly Sections 2.2.1 and 2.2.6);
- the value v' nevertheless varies with Q', even if it is assumed $G_{vi}(0) \to \infty$.

(b) Secondary Control

The secondary control (if any) is slower than the primary one and acts on the set points of the machine voltage regulators (or of some of them) to achieve a satisfactory voltage and reactive power steady-state in the whole network.

In this regard, the network may be assimilated to one or more "areas," in each of which:

- the nodes are "electrically" close one another (high sensitivity of the voltage at each node, with respect to voltages of other nodes);
- there is at least one machine of sufficient power (to make it possible to effectively control the voltage values);
- the mean voltage level is strictly conditioned by the voltage at a given node (sufficiently "strong"; see also Section 5.7.2), called *"pilot node."*

Furthermore, in the case of more areas, it is assumed that:

- the nodes of different areas are electrically far from one another.

In the case of *a single area*, the general diagram of Figure 6.18a can be applied, for which the "secondary" regulator, possibly realized through the computer of the central dispatching office (Fig. 2.43) with, for instance, a sampling period $\Delta t = 5{-}10$ sec:

- regulates the voltage v_p at the pilot node (telemetered or obtained through "state estimation"; see Section 2.5), with the set point $v_{p\,\mathrm{rif}}$ determined at the reactive dispatching stage:
- carries out the desired dispatching of the Q_i's between the different machines (in particular, by making uniform the respective reactive power margins).

The regulation of the voltage v_p and the control of the reactive powers imply actions of the integral type, so that at steady-state it holds $v_p = v_{p\,\mathrm{rif}}$ and $Q_i = Q_{\mathrm{d}i}$, where $Q_{\mathrm{d}i}$ is the desired value of Q_i.

With *more areas* that interact slightly, such a scheme could be repeated for each area. Generally, it is instead convenient that each set of notably interacting areas be controlled by the same secondary regulator (having as many inputs and outputs as the number of involved areas), as shown in Figure 6.18b, where the first set includes n_1 areas, and so on.

As far as the synthesis of the secondary control loops is concerned, considering the difficulties in the availability of an (updated) knowledge of parameters etc., it is generally convenient to resort to simple criteria, similar to those used for the f/P control. For example:

- the control of reactive powers is made sufficiently slower than the primary control (perhaps 10 times slower), so that, as a first approximation, the response delays of the latter can be disregarded;
- similarly, the voltage regulation at the pilot node (or nodes) is made slower (perhaps again 10 times) of the reactive power control, so that, as a first approximation, the response delays of the latter control also can be disregarded.

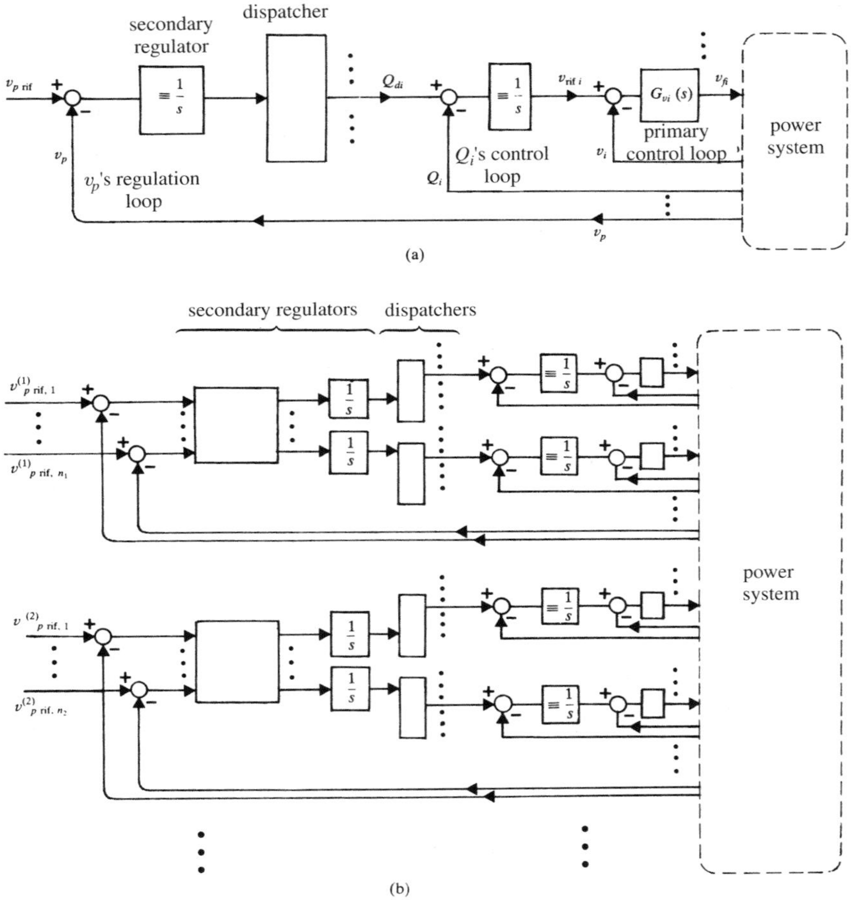

Figure 6.18. Block diagram of the secondary v/Q control: (a) case of a single area; (b) case of more areas.

6.3. CONTROL OF TAP-CHANGERS

Consider the system in Figure 6.19, which includes:

- an infinite power network (the frequency and voltage of which can be considered as "inputs" of the system; see also Section 7.1);
- a connection line, defined by the reactance X evaluated at the network frequency (because of the relative slowness of the control, the dynamics of inductances and possible capacitances can be disregarded);

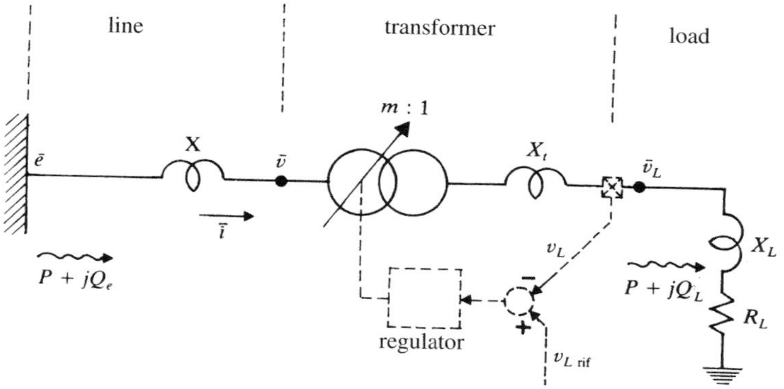

Figure 6.19. Control of an on-load tap-changing transformer: system under examination.

- a transformer equipped with an on-load tap-changer (with variable ratio m), represented by an ideal transformer and a constant reactance X_t on the secondary side (see Fig. 5.7a, assuming that the number of coils varies at the primary side).
- a load (as seen by the secondary side of the transformer) of the linear "static" type, represented by the impedance $R_L + jX_L$.

By applying Park's transformation with Ω_r equal to the network frequency (and, for simplicity reasons, not repeating the index "r" in the Park's vectors), it follows that:

$$\begin{cases} \bar{v}_L = \bar{e}\dfrac{m(R_L + jX_L)}{m^2 R_L + j(X + m^2 X'_L)} \\ \bar{v} = \bar{e}\dfrac{m^2(R_L + jX'_L)}{m^2 R_L + j(X + m^2 X'_L)} \end{cases}$$

and furthermore:

$$\frac{P - jQ_e}{\bar{e}^*} = \bar{\imath} = \frac{\bar{e}}{m^2 R_L + j(X + m^2 X'_L)}$$

where:

$$X'_L \triangleq X_t + X_L$$

By writing:

$$Z_e \triangleq \sqrt{m^4 R_L^2 + (X + m^2 X'_L)^2}$$

which is the magnitude of the impedance as seen from the network, it is then possible to derive:

$$\begin{cases} v_L = \dfrac{em\sqrt{R_L^2 + X_L^2}}{Z_e} \\ v = \dfrac{em^2\sqrt{R_L^2 + X_L'^2}}{Z_e} \\ P = \dfrac{e^2 m^2 R_L}{Z_e^2} \\ Q_e = \dfrac{e^2(X + m^2 X_L')}{Z_e^2} \end{cases}$$

(note also that $v = mv_L\sqrt{(R_L^2 + X_L'^2)/(R_L^2 + X_L^2)}$ and $P = v_L^2 R_L/(R_L^2 + X_L^2)$).

The transformation ratio may be adequately varied by an on-load switching to regulate the voltage magnitude v_L to the desired value $v_{L\,\mathrm{rif}}$. In this concern, v_L (as well as P) exhibits, for varying m, a maximum at $m = m^*$, with:

$$m^{*2} \triangleq \frac{X}{\sqrt{R_L^2 + X_L'^2}}$$

and, more precisely, it results in (see also Fig. 6.20)

$$v_{L\max}^2 = e^2 a \frac{R_L}{2X}$$

where for brevity, the following notation has been used:

$$a \triangleq (1 + b^2)(\sqrt{1 + b'^2} - b'), \quad b \triangleq \frac{X_L}{R_L}, \quad b' \triangleq \frac{X_L'}{R_L} = b + \frac{X_t}{R_L}$$

Furthermore, for each desired value of v_L, obviously smaller than $v_{L\max}$, there are two possible solutions for m (points 1 and 2 in Fig. 6.20). It is easy to determine that the solution point 2 (for a larger m) corresponds to a relatively large voltage v and a relatively small current i at the primary side of the transformer, whereas the opposite occurs with the other solution.

By assuming for simplicity that $X_L = 0$ and disregarding X_t, the vector diagram of Figure 6.21 can be derived, by which the two solutions in m for $v_L < v_{L\max}$ are evident. For the mentioned conditions, the values m^*, $v_{L\max}$ and $P_{\max}$ correspond to a 45° phase-shift between vectors $\overline{e}$ and $\overline{v}$.

Under normal operation, it is convenient to regulate the voltage v_L at the solution point (2) with $m > m^*$, which implies a smaller line current and consequently

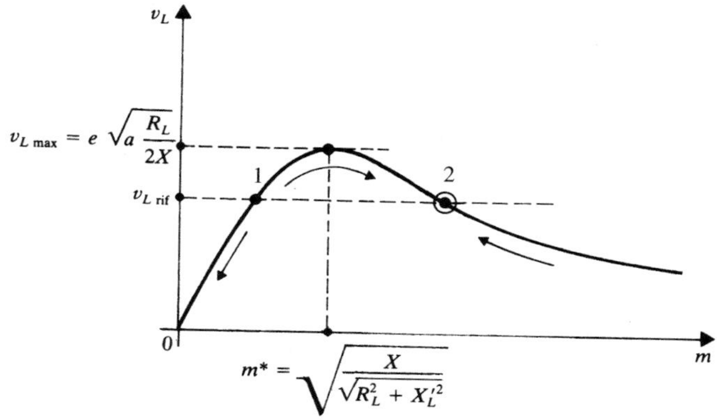

Figure 6.20. Dependence of the voltage v_L on the transformation ratio m, in the case of a linear static load (see Fig. 6.19).

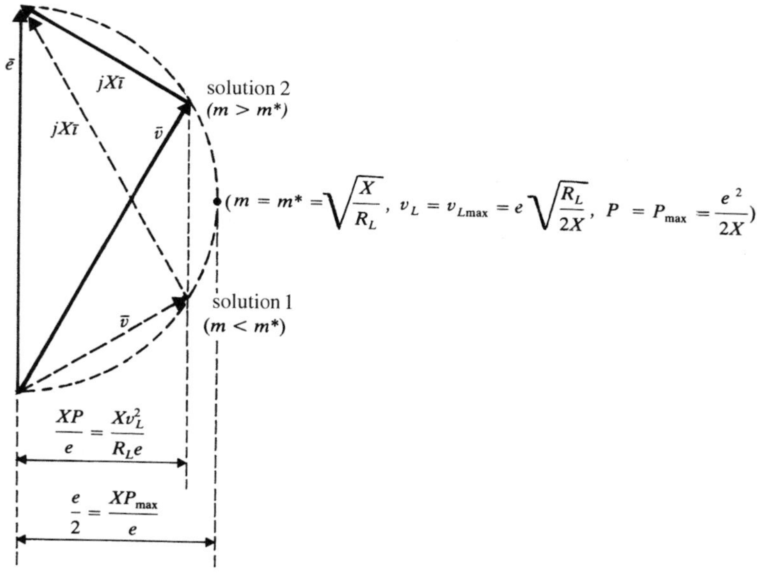

Figure 6.21. Vector diagram for $X_L = X_t = 0$.

a smaller voltage drop (i.e., an acceptable v), and smaller losses because of the unavoidable resistances. Around this solution, however, $\partial v_L/\partial m < 0$ (Fig. 6.20), so that it must be assumed:

$$\frac{m}{v_{L\,\text{rif}} - v_L}(s) = -G_v(s)$$

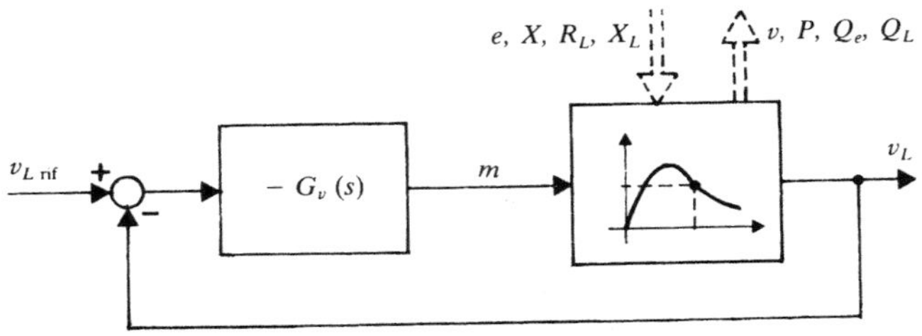

Figure 6.22. Block diagram of the regulation loop.

as indicated in Figure 6.22, where the transfer function $G_v(s)$ must be chosen by the usual criteria[(11)].

For each assigned operating point, the synthesis of $G_v(s)$ does not imply any particular problems, at least for the simplifying conditions considered here (infinite power network, "static" load etc.). In particular, if the $G_v(s)$ includes an integral effect, it is possible to obtain $v_L = v_{L\,\text{rif}}$ at steady-state. If it is simply assumed that:

$$G_v(s) = \frac{K_v}{s} \qquad [6.3.1]$$

the cutoff frequency of the regulation loop is $\nu_t = -K_v \partial v_L / \partial m$[(12)], and the closed-loop response delay is represented by the time constant $1/\nu_t$ (in practical cases, ν_t can be, for instance, approximately 0.07 rad/sec, corresponding to a time constant of 15 sec).

If Equation [6.3.1] is adopted, the effect of the regulator is moving the generic point (m, v_L) in the sense indicated by the arrows in Figure 6.20.

However, some possible drawbacks must be identified:

(1) For decreasing values of R_L (caused by load insertion) or for increasing X (e.g., because of a line opening of two lines initially in a parallel configuration), the characteristic (v_L, m) is lowered as indicated in Figure 6.23a, and the same occurs (without that m^* increases) for decreasing e. In such cases:

- before the intervention of the regulator, voltages v_L and v decrease:
- the regulator forces m to decrease so as to take v_L back to its desired value (Fig. 6.23a), but in doing so it causes a further reduction of the impedance as seen by the primary side of the transformer, and then of v;

(11) For simplicity, minimum and maximum limits on m are not considered (for instance, $m/m_{\text{nom}} = 0.9-1.1$), as well as the discontinuities in the variations of m (e.g., an elementary variation Δm of 0.5% of m_{nom}).

(12) For simplicity here and in the following, the superscript, "o" relative to values computed at the operating point, will not be indicated.

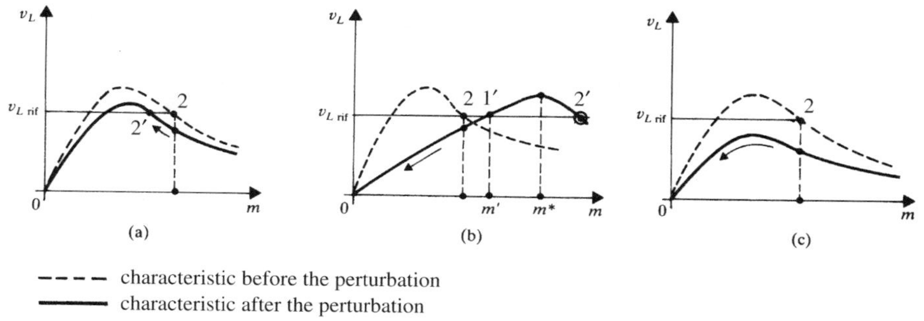

Figure 6.23. Behavior of the controlled system following perturbations (see text).

(similar conclusions hold for variations of the opposite sign). Therefore, the regulation of v_L *makes the "support" of the voltage worse at the primary side* of the transformer.

(2) If, after some relatively significant perturbation, the initial values are $m < m'$ and $v_L < v_{L\,\text{rif}}$ (Fig. 6.23b), the regulator forces m, as well as v_L and v, to decrease further (voltage collapse caused by regulation instability at point 1′; note that, if initially $m \in (m', m^*)$, such an instability would not cause the collapse, but it simply would lead the operating point to 2′). Remedies may include stopping the regulation (by blocking the value of m) or changing the sign of the regulator gain so as to reach the point 1′, even if it is not desired in normal situations.

(3) The value $v_{L\max}$ can decrease for several reasons, particularly because of a reduction of e or R_L, or for an increment of X. For a reduction of $v_{L\max}$ (or even for an incorrect setting of $v_{L\,\text{rif}}$) it may happen that $v_{L\max} < v_{L\,\text{rif}}$, in which case the regulator continues forcing m to decrease (according to Fig. 6.23c), thus causing a *voltage instability* (see also Section 1.5). To avoid the voltage collapse, it is necessary to perform actions such as a timely load-shedding, reduction of $v_{L\,\text{rif}}$, or alternatively to inhibit the regulation (in this case there is no more use in changing the sign of the regulator gain).

(4) A possible variation Δe causes variations ΔP and ΔQ_e which, in the absence of regulation, would have the same sign as Δe. Because of the regulation effects, it instead results in, for small variations:

$$G_{pe}(s) \triangleq \frac{\mathcal{L}\Delta P}{\mathcal{L}\Delta e} = \frac{\partial P}{\partial e} + \frac{\partial P}{\partial m}\,\frac{G_v(s)\dfrac{\partial v_L}{\partial e}}{1 - G_v(s)\dfrac{\partial v_L}{\partial m}}$$

$$G_{qe}(s) \triangleq \frac{\mathcal{L}\Delta Q_e}{\mathcal{L}\Delta e} = \frac{\partial Q_e}{\partial e} + \frac{\partial Q_e}{\partial m}\,\frac{G_v(s)\dfrac{\partial v_L}{\partial e}}{1 - G_v(s)\dfrac{\partial v_L}{\partial m}}$$

Assuming for simplicity that Equation [6.3.1] holds, the above transfer functions have a pole ($-\nu_t$, real and negative) and a zero (respectively $-\nu_t G_{pe}(0)/G_{pe}(\infty)$, $-\nu_t G_{qe}(0)/G_{qe}(\infty)$). It finally results in, around a generic operating point with $m > m^*$:

$$G_{pe}(0) = \left(\frac{\Delta P}{\Delta e}\right)_{\Delta v_L=0} = 0, \quad G_{pe}(\infty) = \left(\frac{\Delta P}{\Delta e}\right)_{\Delta m=0} = \frac{\partial P}{\partial e} > 0$$

$$G_{qe}(0) = \left(\frac{\Delta Q_e}{\Delta e}\right)_{\Delta v_L=0} < 0, \quad G_{qe}(\infty) = \left(\frac{\Delta Q_e}{\Delta e}\right)_{\Delta m=0} = \frac{\partial Q_e}{\partial e} > 0$$

so that $G_{pe}(s)$ has a zero static gain (and a zero at the origin), whereas $G_{qe}(s)$ has a negative static gain (and a positive real zero). This last circumstance, for actual cases with a noninfinite power network, can be a drawback for the network voltage regulation.

(5) In relatively realistic terms, if the *limits on the reactive power* (Q_e) that the network can supply are reached, they may be considered by imposing $\Delta Q_e = 0$ instead of $\Delta e = 0$ (and assuming that the variations Δe are generated within the network itself, so as to maintain constant Q_e; recall the operation of the over- and underexcitation limiters of synchronous machines). Under such conditions, we must consider:

$$\left(\frac{\Delta v_L}{\Delta m}\right)_{\Delta Q_e=0} = \frac{\partial v_L}{\partial m} - \frac{\dfrac{\partial v_L}{\partial e}\dfrac{\partial Q_e}{\partial m}}{\dfrac{\partial Q_e}{\partial e}}$$

instead of $\partial v_L/\partial m$, and since:

$$\left(\frac{\Delta v_L}{\Delta m}\right)_{\Delta Q_e=0} = \frac{eX\sqrt{R_L^2 + X_L^2}}{(X + m^2 X_L')Z_e}$$

it can be concluded that the regulation becomes unstable, unless $X + m^2 X_L' < 0$, i.e., $Q_e < 0$ (underexcitation limit, with $X_L' < 0$, $X_L < -X_t$, $m^2 > X/(-X_L')$, $v_{L\,\mathrm{rif}} = v_L < (e/R_L)\sqrt{(-X_L'/X)(R_L^2 + X_L^2)}$).

Note that:

$$\left(\frac{\Delta v_L}{\Delta m}\right)_{\Delta Q_e=0} = \frac{\partial v_L}{\partial m}\frac{G_{qe}(0)}{G_{qe}(\infty)} \qquad [6.3.2]$$

so that the zero of the function $G_{qe}(s)$ is $-\nu_t G_{qe}(0)/G_{qe}(\infty) = K_v(\Delta v_L/\Delta m)_{\Delta Q_e=0}$. The reported results are confirmed, as a particular case, by the following treatment, which is extended to nonlinear static loads.

More generally, in the case of a *nonlinear* static load (with P and Q_L known functions of v_L), it is convenient to separate the equations of line and transformer, and the equations of load.

To this aim, it should be observed that, by putting:

$$X' \triangleq X + m^2 X_t \qquad [6.3.3]$$

it results:

$$\frac{P - jQ_e}{\overline{e}^*} = \overline{\imath} = \frac{\overline{e} - m\overline{v}_L}{jX'} = \frac{P - jQ_L}{m\overline{v}_L^*}$$

from which the following equations for line and transformer can be obtained:

$$X'(P^2 + Q_e^2) + e^2(Q_L - Q_e) = 0 \qquad [6.3.4]$$

$$X'(Q_e + Q_L) - e^2 + m^2 v_L^2 = 0 \qquad [6.3.5]$$

Moreover, the following equations may be generically assumed for the load:

$$\left.\begin{aligned} P &= P(v_L) \\ Q_L &= Q_L(v_L) \end{aligned}\right\} \qquad [6.3.6]$$

From Equation [6.3.5], it particularly follows that:

$$Q_e = \frac{e^2 - m^2 v_L^2}{X'} - Q_L \qquad [6.3.7]$$

so that, substituting into Equation [6.3.4] and considering [6.3.3], the following equation can be derived:

$$0 = (X + m^2 X_t)^2 (P^2 + Q_L^2) + 2(X + m^2 X_t) m^2 v_L^2 Q_L + m^2 v_L^2 (m^2 v_L^2 - e^2) \qquad [6.3.8]$$

which is of the second degree in m^2 (for given e, v_L). It is possible to determine that the two solutions in m^2 are real and positive if and only if:

$$e^4 v_L^2 - 4X\left[P^2(v_L)(X_t e^2 + X v_L^2) + e^2 Q_L(v_L)(v_L^2 + X_t Q_L(v_L))\right] \geq 0 \qquad [6.3.9]$$

This condition, for any assigned e, defines the admissible values of v_L and thus of $v_{L\,\mathrm{rif}}$. For a qualitative example, refer to Figure 6.24, in which it is assumed $Q_L = 0$ and P proportional to $(v_L)^{a_p}$, with $a_p = 0, 1, 2$. Of the two possible solutions in m^2, it appears more convenient to use the larger one, which implies a smaller value for the line current. In fact, for any (admissible) value

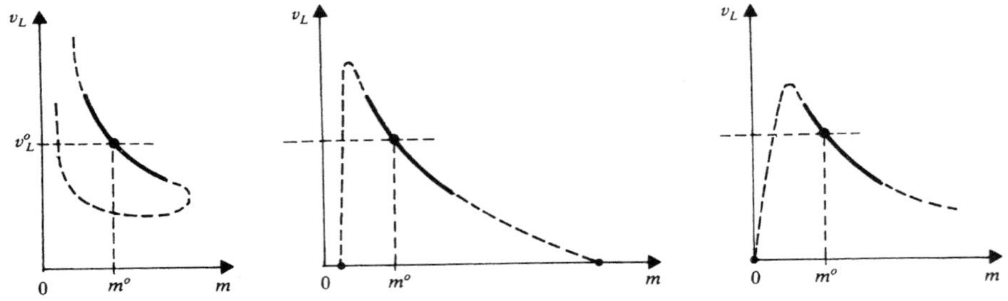

Figure 6.24. Dependence of the voltage v_L on the transformer ratio m (qualitative outlines), in the case of a purely active static load, with: (a) $P = P^o(a_p = 0)$; (b) $P = \gamma_L v_L(a_p = 1)$; (c) $P = v_L^2/R_L(a_p = 2$: see Fig. 6.20). The bolded line sections are those useful for normal operation, i.e., for operation at $\partial v_L/\partial m < 0$ and, for the same v_L, at smaller i.

of v_L, the value of the load current is given, so that i is inversely proportional to m(13).

For small deviations and having set:

$$\begin{cases} a_p \triangleq \dfrac{\mathrm{d}P/\mathrm{d}v_L}{P/v_L} \\ a_q \triangleq \dfrac{\mathrm{d}Q_L/\mathrm{d}v_L}{Q_L/v_L} \end{cases}$$

(recall a_{pv} and a_{qv} in Equations [5.6.1]), it specifically follows:

(13) The treatment would be simplified if X' were not dependent on m, or equivalently if the number of coils varied at the secondary side of the transformer. In such a case, a constant reactance should be considered at the primary side (Fig. 5.7b), which it would result in $X' = X + \text{constant} = \text{constant}$. In practice, the approximation of $X' = \text{constant}$ can become acceptable only if, within the variation range of m, $m^2 X_t \ll X$.

In particular, under this approximation:

- Equation [6.3.4] no longer contains m and permits—for any assigned e and by expressing Q_L as a function of P—the determination of a curve in the plane (P, Q_e), which defines the network load conditions for varying P, or equivalently for varying v_L (and $v_{L\,\text{rif}}$) (if for instance Q_L were proportional to P, and thus the load power factor were constant, such a curve would be a circle).
- Equation [6.3.5] then allows the determination, for any assigned situation, of:

$$m^2 = \frac{e^2 - X'(Q_e + Q_L)}{v_L^2}$$

$$\left(\frac{\Delta v_L}{\Delta m}\right)_{\Delta e=0} = \frac{\partial v_L}{\partial m}$$

$$= \frac{m v_L^3 (2X Q_e - e^2)}{a_p X'^2 P^2 + a_q X'(m^2 v_L^2 Q_e - X'P^2) + m^2 v_L^2 (e^2 - 2X' Q_e)} \quad [6.3.10]$$

$$\left(\frac{\Delta Q_e}{\Delta e}\right)_{\Delta v_L=0} = \frac{2Xe}{X'} \frac{Q_e - Q_L}{2X Q_e - e^2} \quad [6.3.11]$$

$$\left(\frac{\Delta v_L}{\Delta m}\right)_{\Delta Q_e=0} = \frac{m X v_L^3}{X'} \frac{Q_e - Q_L}{a_p X' P^2 + a_q \left(m^2 v_L^2 \left(Q_e - \frac{Q_L}{2}\right) - X'P^2\right) - m^2 v_L^2 (Q_e - Q_L)} \quad [6.3.12]$$

where it is intended that the different quantities are evaluated at the operating point.

The value of Q_e is obtainable from e, v_L, m^2, based on Equation [6.3.7]. Note that:

$$Q_e = X' \frac{P^2 + Q_L^2}{m^2 v_L^2} + Q_L$$

and thus $Q_e > Q_L$. On the other hand, the difference $(Q_e - Q_L)$ is the (positive) reactive power absorbed by the line and the transformer.

Similar to the discussion above, related to the linear load, and again assuming (Fig. 6.22) a transfer function $G_v(s)$ given by [6.3.1]:

- Equation [6.3.10] allows the assessment of the stability under normal operation (for which e is intended as an "input" of the system), and it is necessary that $(\Delta v_L/\Delta m)_{\Delta e=0} < 0$.
- Equation [6.3.11] defines the static gain (usually negative) of the function $G_{qe}(s)$, whereas Equation [6.3.12] allows the determination of its zero, which is equal to $K_v(\Delta v_L/\Delta m)_{\Delta Q_e=0}$, and is usually positive (the value $G_{qe}(\infty)$ can be then deduced from Equation [6.3.2]); such knowledge may be of some importance, as already indicated, for network voltage regulation.
- Finally, Equation [6.3.12] itself permits the assessment of the stability in the possible operation at constant Q_e, where it is necessary that $(\Delta v_L/\Delta m)_{\Delta Q_e=0} < 0$.

Equations [6.3.10] and [6.3.12] depend on the values of a_p and a_q, i.e., on the relationships by which P and Q_L depend on v_L. This does not happen for Equation [6.3.11] since it refers to the condition $\Delta v_L = 0$.

Some simplifications can be obtained *when the load power factor is constant*, and therefore $Q_L/P =$ constant. In such a case $a_p = a_q$, so that, in addition to

Equation [6.3.11], it holds:

$$\left(\frac{\Delta v_L}{\Delta m}\right)_{\Delta e=0} = \frac{v_L}{m}\frac{2XQ_e - e^2}{(a_p-2)X'Q_e + e^2} \qquad [6.3.10']$$

$$\left(\frac{\Delta v_L}{\Delta m}\right)_{\Delta Q_e=0} = \frac{Xv_L}{mX'}\frac{Q_e - Q_L}{a_p\left(Q_e - \dfrac{Q_L}{2}\right) + Q_L - Q_e} \qquad [6.3.12']$$

If $Q_L \geq 0$, as is usually the case, it follows (by Equation [6.3.4], with $P > 0$) that $Q_e \in (0, e^2/X')$, and furthermore $Q_e > Q_L$. It is then easy to verify that:

- the condition $(\Delta v_L/\Delta m)_{\Delta e=0} < 0$ (stability under normal operation) also can be turned, for $a_p \geq 1$, into:

$$Q_e < \frac{e^2}{2X}$$

- the condition $(\Delta v_L/\Delta m)_{\Delta Q_e=0} < 0$ (stability under the operation at constant Q_e) is always satisfied if $a_p \leq 0$, and not satisfied if $a_p \geq 1$, whereas it becomes:

$$Q_e > \frac{2-a_p}{2(1-a_p)}Q_L$$

if $a_p \in (0, 1)$.

Finally, more drastic simplifications can be obtained *if the load is purely active*, i.e., for a unity power factor of the load. In this case, $Q_L = a_q Q_L = 0$, and it is possible to derive:

$$\left(\frac{\Delta v_L}{\Delta m}\right)_{\Delta e=0} = \frac{v_L}{m}\frac{2XQ_e - e^2}{(a_p-2)X'Q_e + e^2} \qquad [6.3.10' \text{ rep.}]$$

$$\left(\frac{\Delta Q_e}{\Delta e}\right)_{\Delta v_L=0} = \frac{2Xe}{X'}\frac{Q_e}{2XQ_e - e^2} \qquad [6.3.11']$$

$$\left(\frac{\Delta v_L}{\Delta m}\right)_{\Delta Q_e=0} = \frac{Xv_L}{mX'}\frac{1}{a_p - 1} \qquad [6.3.12'']$$

Regarding the stability assessment under normal operation, it can be observed based on Equation [6.3.10′] that, for $Q_L = 0$:

$$Q_e = \frac{X'P^2}{m^2 v_L^2}$$

(with $X' \triangleq X + m^2 X_t$), where m^2 is the larger of the two solutions of:

$$0 = (X + m^2 X_t)^2 P^2 + m^2 v_L^2 (m^2 v_L^2 - e^2)$$

and where it must be:

$$P \leq \frac{e^2}{2X\sqrt{1+\dfrac{X_t e^2}{X v_L^2}}} \triangleq P_{\max}$$

(recall Equations [6.3.8] and [6.3.9]). By considering the above, the condition $(\Delta v_L/\Delta m)_{\Delta e=0} < 0$ can be usefully expressed in terms of XP/e^2 and a_p, for any assigned value of the ratio $X_t e^2/(X v_L^2)$, as indicated in Figure 6.25a.

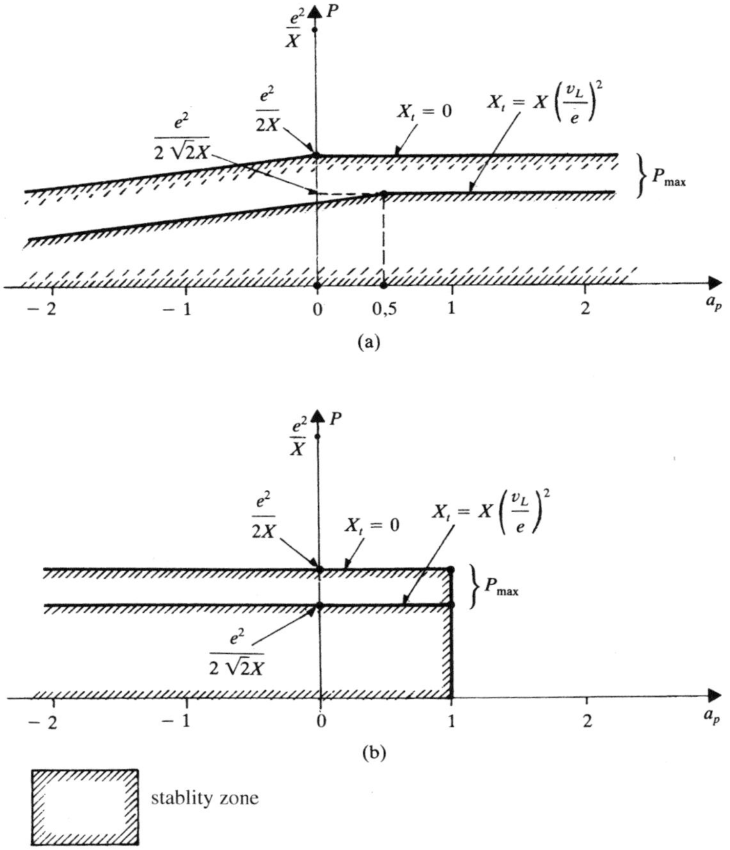

Figure 6.25. Stability zones in the plane (P, a_p), for $Q_L = 0$: (a) under normal operation (having set $\alpha \triangleq X_t e^2/(X v_L^2)$, $\beta \triangleq \sqrt{1-4(1+\alpha)(XP/e^2)^2} \in [0, 1]$, the stability condition can be written as $a_p > (\alpha - 2\beta/(1-\beta))/(1+\alpha))$; (b) under operation at constant Q_e, refer to the text (the stability condition is $a_p < 1$).

The stability assessment in the operation at constant Q_e is instead immediate. Equation [6.3.12″] is extremely simple, and based on this equation, it is possible to conclude that such operation is stable or unstable according to $a_p \lessgtr 1$, as indicated in Figure 6.25b.

6.4. CONTROL OF STATIC COMPENSATORS

(a) Generalities

Static compensators can be viewed as variable shunt-connected susceptances. Less-recent solutions are based on *"saturated (nonlinear) reactors,"* with series and parallel condensers, as shown in Figure 6.26a (there are different types of devices, rated at various, relatively small power levels). The magnetic characteristic of the reactors is schematically shown in Figure 6.26b, with:

- a large reactance for small currents;
- a small residual (under saturation) reactance;

where ψ_r and i_r are, respectively, the magnetic flux in the generic reactor, and the current entering into it.

If the current i_r is sinusoidal and with an amplitude that is not too small, the flux ψ_r is similar to a square wave, and the fundamental component of the voltage $v_r = \mathrm{d}\psi_r/\mathrm{d}t$ is almost constant in amplitude (Fig. 6.26c).

If it is assumed that the voltage harmonics are sufficiently filtered, so that their presence can be disregarded, it is possible to deduce, in the absence of condensers, a characteristic (v', i) of the type (1) indicated in Figure 6.26d. In this case, v' and i are proportional, with the same proportionality factor, to the amplitude of the fundamental component of v_r and to the amplitude of i_r, respectively. However, in the following we will assume $i \gtrless 0$, based on whether the current is inductive or capacitive; the absorbed reactive power Q' is then proportional to the product $v'i$.

The aim of the series condensers (see C_s in Fig. 6.26a) is to compensate for the residual reactances of reactors, to further reduce the slope of the characteristic (v', i) under saturation. Refer to the characteristic (2) in Figure 6.26d, in which v' accounts for the contribution of the series condensers, whereas it still holds that $i > 0$.

Because of the effect of the parallel condensers (see C_p in Fig. 6.26a), capacitive operation becomes possible (with $i < 0$, which corresponds to an absorbed reactive power $Q' < 0$). Refer to the characteristic (3) in Figure 6.26d, in which i accounts for the (capacitive) current absorbed by the parallel condensers.

A similar characteristic (v, i) (where $i \gtrless 0$ depending on whether the overall current is inductive or capacitive) can be obtained by compensating the reactance of the (possible) transformer with a condenser, as shown as C_t in Figure 6.26a.

In this way it is possible to automatically achieve (however, without the typical flexibility of control systems) the operation at constant v within a given interval

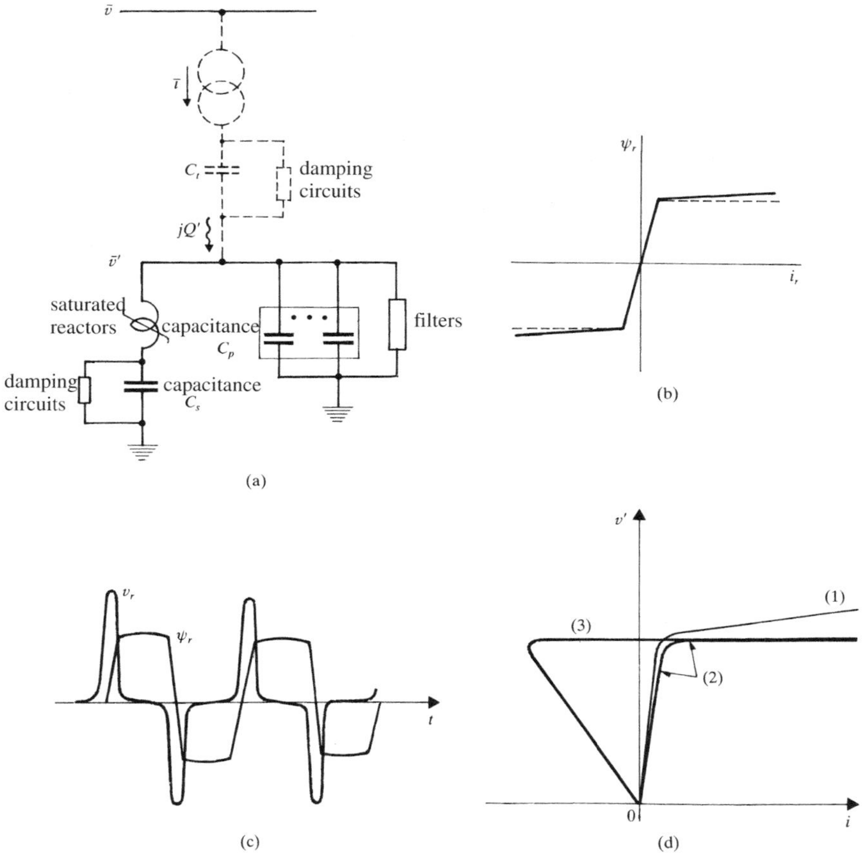

Figure 6.26. Static compensation by saturated reactors: (a) schematic representation; (b) magnetic characteristic of the reactors; (c) time behaviors of flux and voltage on the generic reactor, under the hypothesis of sinusoidal current; (d) voltage-current characteristic ((1) without condensers, (2) with series condensers, (3) with series and parallel condensers).

of admissible values for Q'. Such an interval may be modified by realizing C_p by several parallel condensers and by varying, through switching, such a set. Also considering the characteristics of the rest of the system, the response delay is generally small, e.g., a cycle (i.e., 0.02 sec at 50 Hz).

However:

- harmonics actually must be filtered (their effect may be further amplified by possible LC resonances, at frequencies close to those of the harmonics themselves), for instance, by adding proper LC filters in parallel to C_p (Fig. 6.26a);

- the simultaneous presence of inductive and capacitive elements may cause, independently of the presence of harmonics, undesirable *LC* resonance phenomena, similar to those seen in Sections 5.7.2 and 6.2.1c, so that it is convenient to add proper damping circuits (including resistive elements) in parallel to C_s, etc. (see again Fig. 6.26a).

The most recent solutions are based on *"controlled (linear) reactors"* (thyristor-controlled reactors), with parallel condensers such as in Figure 6.27a[(14)].

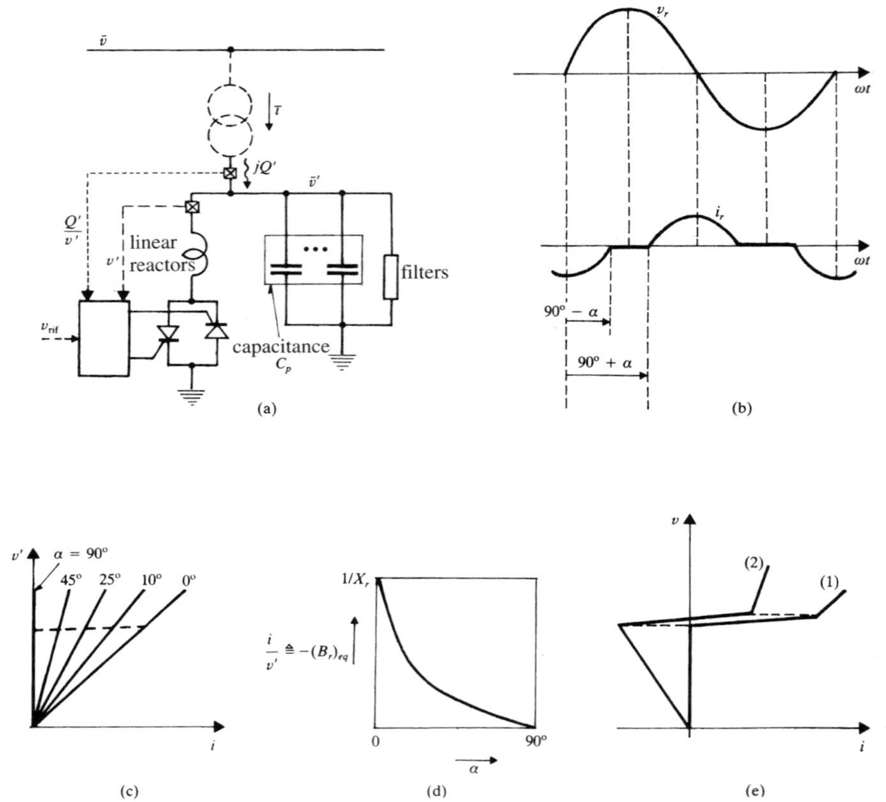

Figure 6.27. Static compensation by controlled reactors: (a) schematic representation; (b) time behavior of the current in a generic reactor, for a given firing angle α and under the hypothesis of sinusoidal voltage; (c) voltage-current characteristics, for different values of the firing angle α and in the absence of condensers; (d) equivalent susceptance of the reactor, as function of the firing angle; (e) regulation characteristic (1) without and (2) with parallel condensers.

(14) It is assumed, for simplicity, that each phase has a reactor in series with two thyristors. Actually, there are circuit solutions different from this one, which are specifically designed to reduce harmonic generation.

The time behavior of the (inductive) current i_r in a reactor, for an assigned "firing angle" α of its thyristors and with a sinusoidal voltage v_r at frequency ω, is schematically indicated in Figure 6.27b.

If it is assumed that the current harmonics are sufficiently filtered so that their presence can be disregarded, it is possible to determine — for different values of α, and not considering parallel condensers — (v', i) characteristics as indicated by continuous lines in Figure 6.27c. In these characteristics, v' and i are proportional to the amplitude of v_r and the amplitude of the fundamental component of i_r, respectively (the absorbed reactive power Q' is then proportional to the product $v'i$). More precisely, if X_r is the reactor reactance, by applying the Fourier analysis, it is possible to derive:

$$\frac{i}{v'} = \frac{\pi - 2\alpha - \sin 2\alpha}{\pi X_r} \triangleq -(B_r)_{\text{eq}} > 0 \qquad [6.4.1]$$

where $(B_r)_{\text{eq}}$ may be viewed as the equivalent (negative) susceptance of the reactor (Fig. 6.27d).

As already said for saturated reactors, it is convenient that harmonics are adequately filtered (see filters indicated in Fig. 6.27a.)

The typical goal of control is to vary α so as to regulate v at its desired value v_{rif}, with a steady-state zero error (regulator with an integral action) or sufficiently small error; see Fig. 6.28. If the transformer is not considered ($v = v'$), it is possible to realize a characteristic like the one represented by a dashed line in Figure 6.27c, i.e., the characteristic (1) in Figure 6.27e, where the slope $\Delta v/\Delta i$ in the intermediate zone is zero if the regulator has an integral action[(15)].

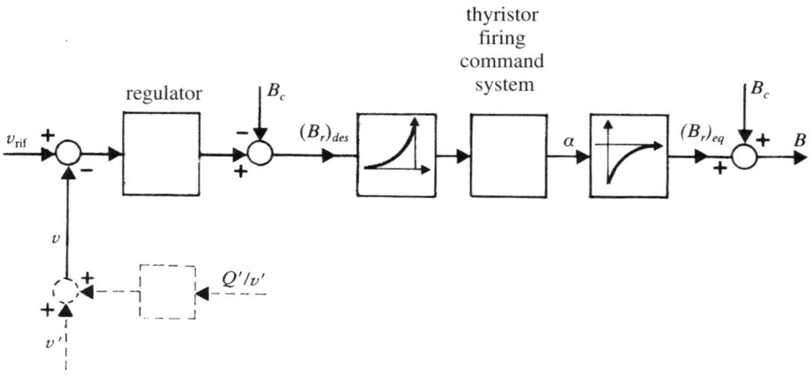

Figure 6.28. Static compensator control: block diagram.

(15) The control of reactors also can be used, alternatively, to regulate the power factor at a load node, or it can be included in the secondary v/Q control, etc.

The parallel condensers (see C_p in Fig. 6.27a) also allow capacitive operation, i.e., at $Q' < 0$ or equivalently $i < 0$ (see characteristic (2) in Fig. 6.27e), within a given range of admissible values for Q'. This range can be variable according to the reactive power demands for each specific situation, by modifying the condenser set through switching, possibly performed by other thyristors.

If it is posed that $B_c \triangleq \omega C_p$, the (equivalent) overall susceptance is equal to:

$$B = B_c + (B_r)_{\text{eq}} \qquad [6.4.2]$$

(smaller than B_c), where $(B_r)_{\text{eq}}$ is given by Equation [6.4.1] and thus depends on the thyristor firing angle α according to a nonlinear relationship (Fig. 6.27d). To compensate this nonlinearity between α and $(B_r)_{\text{eq}}$, it is possible to realize the inverse nonlinearity between $(B_r)_{\text{des}}$ and α, as indicated in Figure 6.28.

The presence of the transformer can suggest the addition of a reactive current feedback (see the signal Q'/v' in Figs. 6.27a and 6.28) to regulate v instead of v', similarly to what was seen in Section 6.2.2(a2)[16].

The response delay of the controlled reactors may be considered as a pure delay in the range from 1/6 to 1/2 cycle, and therefore it can be generally considered negligible. The cutoff frequency of the control loop, then, depends also on the characteristics of the rest of the system, as shown in the following examples.

The control of reactors also can be useful in improving the damping of electromechanical oscillations (by "additional" signals, having functions similar to those described in Sections 7.2.2 and 8.5.2), or in avoiding the loss of synchronism following large perturbations (see Section 7.3).

(b) Voltage Regulation at a Load Node

Through approximations similar to those in Section 6.3, consider the system in Figure 6.29, which includes[17]:

- an infinite power network;
- a connection line defined by the reactance X (evaluated at the network frequency);
- a controlled reactor static compensator represented by a variable susceptance B (recall Equations [6.4.1] and [6.4.2]);
- a load of the linear "static" type represented by the impedance $R_L + jX_L$ (in general, this means the equivalent load, as "seen" by the node to which the compensator is connected).

---

(16) There is also a "controlled transformer" version, without reactors, for which the reactance X_r is constituted by the leakage reactance of the transformer itself (with this reactance intentionally designed to be large), whereas the capacitors are directly connected to the node at voltage v.

(17) The present case is a generalization of Example 1 reported in Section 1.5, and it is, for many aspects, similar to the case in Section 6.3.

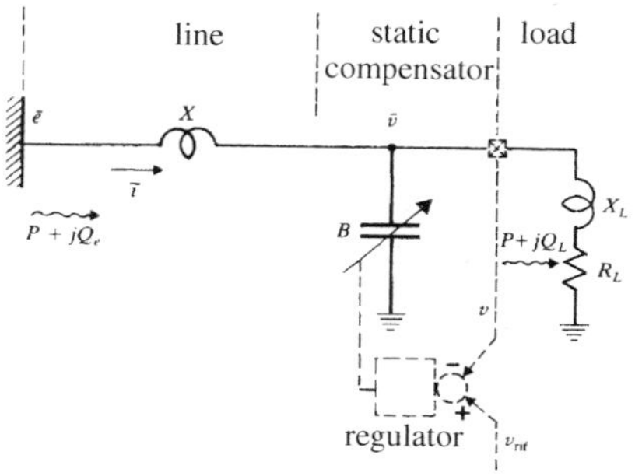

Figure 6.29. Voltage regulation at a load node by a static compensator: system under examination.

By applying Park's transformation with Ω_r equal to the network frequency (disregarding for simplicity the index "r" in the Park's vectors), by the notation of the figure it follows:

$$\begin{cases} \overline{v} = \overline{e}\dfrac{R_L + jX_L}{R_L(1 - XB) + j(X + X_L - XX_L B)} \\ \dfrac{P - jQ_e}{\overline{e}^*} = \overline{\imath} = \overline{e}\dfrac{(1 - X_L B) + jR_L B}{R_L(1 - XB) + j(X + X_L - XX_L B)} \end{cases}$$

from which:

$$\begin{cases} v = \dfrac{e\sqrt{R_L^2 + X_L^2}}{D} \\ P = \dfrac{e^2 R_L}{D^2} \\ Q_e = \dfrac{e^2 A}{D^2} \end{cases}$$

having used for brevity:

$$\begin{cases} D \triangleq \sqrt{R_L^2(1 - XB)^2 + (X + X_L - XX_L B)^2} \\ A \triangleq X + X_L - B(R_L^2 + X_L^2 + 2XX_L) + B^2 X(R_L^2 + X_L^2) \end{cases}$$

Specifically, $P = v^2 R_L/(R_L^2 + X_L^2)$.

The susceptance B can be properly varied, by the reactor control, to regulate the voltage v at the desired value v_{rif} (Fig. 6.28). For varying B, v (as well as P) exhibits a maximum value at $B = B^*$, with:

$$B^* \triangleq \frac{R_L^2 + X_L^2 + XX_L}{(R_L^2 + X_L^2)X} = \frac{1}{X} + \frac{X_L}{R_L^2 + X_L^2}$$

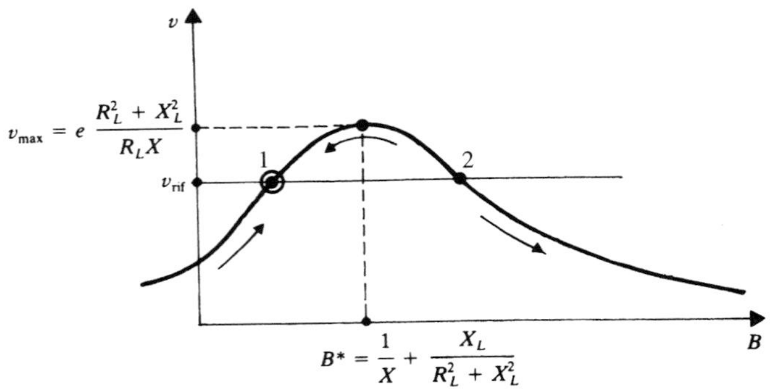

Figure 6.30. Dependence of the voltage v on the susceptance B, in the case of a linear static load (see Fig. 6.29).

which corresponds to (Fig. 6.30):

$$v_{\max} = e\frac{R_L^2 + X_L^2}{R_L X}$$

Furthermore, for each desired value of v (smaller than $v_{\max}$) there are two possible solutions in B (points 1 and 2 in Fig. 6.30). In practical cases, with $v \cong e$, the solution (1) at a smaller B appears preferable, as it results in a smaller current i(18).

Around this solution $\partial v/\partial B > 0$ (Fig. 6.30), so that the regulation loop may be as shown in Figure 6.31, where the transfer function $G_v(s)$ must be chosen based on the usual criteria. It is assumed that the nonlinearity between α and $(B_r)_{\text{eq}}$ (Fig. 6.28) is adequately compensated. The minimum and maximum limits on B are not considered for simplicity.

For each assigned operating point, the synthesis of $G_v(s)$ does not imply any particular problems, at least for the simplifying conditions considered here (infinite power network, "static" load, etc.). In particular, by an integral effect in $G_v(s)$, it is possible to obtain at steady-state $v = v_{\text{rif}}$. If it is simply assumed that:

$$G_v(s) = \frac{K_v}{s} \tag{6.4.3}$$

(18) To determine this, it is sufficient to note that at $v = e$ the reactive power (Xi^2) absorbed by the line is supplied equally by the two terminals, so that it holds:

$$Q_e = \frac{Xi^2}{2} = Be^2 - Q_L(e)$$

where $Q_L(e) \triangleq e^2 X_L/(R_L^2 + X_L^2)$ is preassigned; therefore, i^2 is smaller for the solution at smaller B. The assumption $v = e$ implies the condition $e < v_{\max}$, i.e., $R_L X < R_L^2 + X_L^2$, which is usually satisfied.

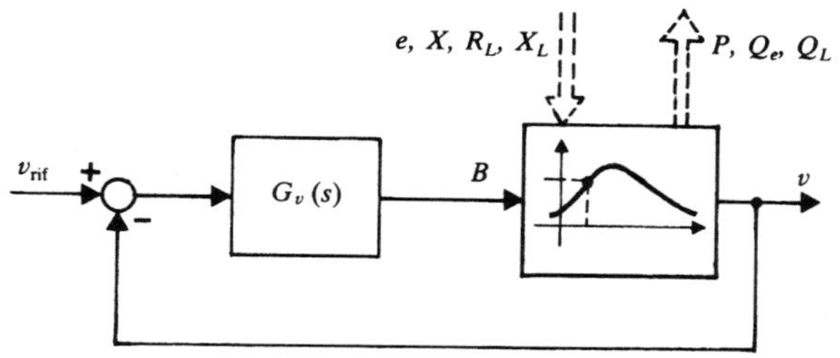

Figure 6.31. Block diagram of the regulation loop.

the cutoff frequency of the regulation loop is $\nu_t = K_v \partial v/\partial B$[19], and the closed-loop response delay is represented by the time constant $1/\nu_t$.

If the cutoff frequency is relatively high, the analysis should consider the dynamic behavior of the various inductive and capacitive elements.

Furthermore, if Equation [6.4.3] is adopted, the effect of the regulator is to move the generic point (B, v) in the sense indicated by the arrows in Figure 6.30.

However, some drawbacks should be indicated, in a partial analogy to Section 6.3.

(1) If, after some relatively significant perturbation, the initial values are $B > B''$ and $v < v_{\text{rif}}$ (Fig. 6.32a), the regulator causes B to increase further, causing a voltage collapse. This is caused by the instability of regulation at

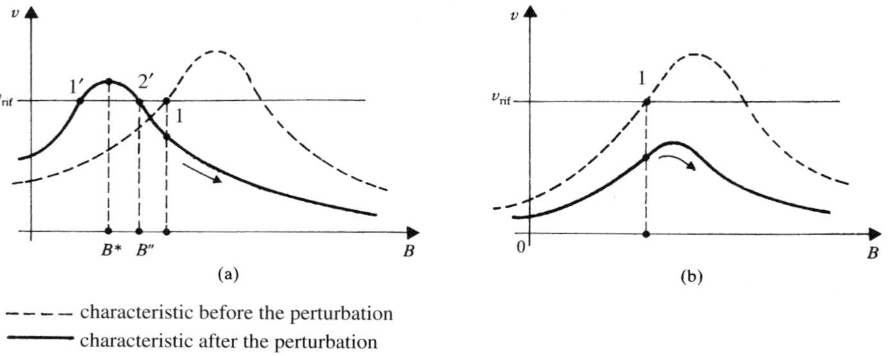

Figure 6.32. Behavior of the controlled system following perturbations (see text).

(19) For simplicity, the indication of the superscript, "o" relative to values computed at the operating point, will be omitted here and in the following.

point 2′. (On the other hand, if initially $B \in (B^*, B'')$, such an instability would simply move the operating point to 1′.) Remedies may include stopping the regulation (by blocking the value of B) or even changing the sign of the regulator gain so as to reach the point 2′, even if it is not desired in normal situations.

(2) The value $v_{\max}$ can diminish for different reasons, particularly because of a reduction in e, or in R_L (e.g., caused by a load connection), or an increase in X (e.g., caused by opening of a line out of two lines, initially in a parallel configuration). For a reduction of $v_{\max}$, or even for an erroneous setting of v_{rif}, it may happen that $v_{\max} < v_{\text{rif}}$, in which case the regulator continues increasing B according to Figure 6.32b, thus causing a *voltage instability* (recall Section 1.5). To avoid the voltage collapse, it is necessary to realize within a short time, a load-shedding or a reduction of v_{rif}, or alternatively to inhibit the regulation (in this case, there is no use in changing the sign of the regulator gain).

(3) A variation Δe causes variations ΔP and ΔQ_e which, in the absence of regulation, would have the same sign as Δe. Because of the effects of regulation, it instead results in, for small variations:

$$G_{pe}(s) \triangleq \frac{\mathcal{L}\Delta P}{\mathcal{L}\Delta e} = \frac{\partial P}{\partial e} - \frac{\partial P}{\partial B}\frac{G_v(s)\dfrac{\partial v}{\partial e}}{1 + G_v(s)\dfrac{\partial v}{\partial B}}$$

$$G_{qe}(s) \triangleq \frac{\mathcal{L}\Delta Q_e}{\mathcal{L}\Delta e} = \frac{\partial Q_e}{\partial e} - \frac{\partial Q_e}{\partial B}\frac{G_v(s)\dfrac{\partial v}{\partial e}}{1 + G_v(s)\dfrac{\partial v}{\partial B}}$$

Assuming, for simplicity, that Equation [6.4.3] holds, the above-mentioned transfer functions have a pole ($-\nu_t$, real and negative) and a zero (respectively: $-\nu_t G_{pe}(0)/G_{pe}(\infty)$, $-\nu_t G_{qe}(0)/G_{qe}(\infty)$). Around a generic operating point with $B < B^*$, it finally results in:

$$G_{pe}(0) = \left(\frac{\Delta P}{\Delta e}\right)_{\Delta v=0} = 0, \quad G_{pe}(\infty) = \left(\frac{\Delta P}{\Delta e}\right)_{\Delta B=0} = \frac{\partial P}{\partial e} > 0$$

so that $G_{pe}(s)$ has a zero static gain (and a zero at the origin). Moreover, the values:

$$G_{qe}(0) = \left(\frac{\Delta Q_e}{\Delta e}\right)_{\Delta v=0} = \frac{e\dfrac{\mathrm{d}A}{\mathrm{d}B}}{D\dfrac{\mathrm{d}D}{\mathrm{d}B}}$$

$$G_{qe}(\infty) = \left(\frac{\Delta Q_e}{\Delta e}\right)_{\Delta B=0} = \frac{\partial Q_e}{\partial B} = \frac{2eA}{D^2}$$

(as well as the zero of $G_{qe}(s)$) can assume both signs, depending on the operating point. In practice, with a noninfinite power network, this can cause problems in network voltage regulation.

(4) In relatively realistic terms, reaching the *limits on the reactive power* (Q_e) that can be supplied by the network may be considered by imposing $\Delta Q_e = 0$ instead of $\Delta e = 0$ (recall Section 6.3). In such conditions, it must be imposed, for the stability, that $(\Delta v/\Delta B)_{\Delta Q_e=0} > 0$ instead of $\partial v/\partial B > 0$, and since:

$$\left(\frac{\Delta v}{\Delta B}\right)_{\Delta Q_e=0} = -\frac{e\sqrt{R_L^2 + X_L^2}\dfrac{\mathrm{d}A}{\mathrm{d}B}}{2AD}$$

it can be concluded that the regulation remains stable if and only if, at the given operating point, it is $(\mathrm{d}A/\mathrm{d}B)/A > 0$.

The analysis of the cases, with respect to the sign of the above-mentioned quantities, will be included as a particular case in the next treatment, extended to nonlinear static loads. Note that:

$$\left(\frac{\Delta v}{\Delta B}\right)_{\Delta Q_e=0} = \frac{\partial v}{\partial B}\frac{G_{qe}(0)}{G_{qe}(\infty)} \qquad [6.4.4]$$

so that the zero of $G_{qe}(s)$ is also $-K_v(\Delta v/\Delta B)_{\Delta Q_e=0}$.

Generally, in the case of a *nonlinear* static load (with P and Q_L known functions of v), it is convenient to separate the equations of line and compensator, and the equations of load.

In this context:

$$\frac{P - jQ_e}{\overline{e}^*} = \overline{\imath} = \frac{\overline{e} - \overline{v}}{jX} = jB\overline{v} + \frac{P - jQ_L}{\overline{v}^*}$$

from which, for the line and the compensator:

$$X(P^2 + Q_e^2) + e^2(Q_L - Q_e - Bv^2) = 0 \qquad [6.4.5]$$

$$X(Q_e + Q_L) - e^2 + (1 - XB)v^2 = 0 \qquad [6.4.6]$$

whereas for the load, it may be generically written that:

$$\left.\begin{aligned} P &= P(v) \\ Q_L &= Q_L(v) \end{aligned}\right\} \qquad [6.4.7]$$

From Equation [6.4.6], it can be specifically derived:

$$Q_e = \frac{e^2 - (1 - XB)v^2}{X} - Q_L \qquad [6.4.8]$$

so that, by substituting into Equation [6.4.5], the following results:

$$0 = X^2(P^2 + Q_L^2) + v^2[2X(1 - XB)Q_L + (1 - XB)^2 v^2 - e^2] \qquad [6.4.9]$$

which is of the second degree in B (for given e, v), with two real solutions if and only if:

$$P \leq \frac{ev}{X} \triangleq P_{\max} \qquad [6.4.10]$$

($P_{\max}$ is the "transmissibility limit" of the active power; see Section 1.5). Of the two possible solutions in B, it is convenient to choose, for the reasons already stated, the one with the smaller value, which is:

$$B = \frac{1}{X} + \frac{Q_L - \sqrt{P_{\max}^2 - P^2}}{v^2}$$

For small variations, having set:

$$\begin{cases} a_p \triangleq \dfrac{\mathrm{d}P/\mathrm{d}v}{P/v} \\ a_q \triangleq \dfrac{\mathrm{d}Q_L/\mathrm{d}v}{Q_L/v} \end{cases}$$

(recall a_{pv}, a_{qv} in Equation [5.6.1]) and furthermore, for brevity:

$$p' \triangleq \sqrt{P_{\max}^2 - P^2}$$

the following can be obtained:

$$\left(\frac{\Delta v}{\Delta B}\right)_{\Delta e=0} = \frac{\partial v}{\partial B} = \frac{v^3 p'}{a_p P^2 + a_q Q_L p' + p'^2 - P^2 - 2Q_L p'} \qquad [6.4.11]$$

$$\left(\frac{\Delta Q_e}{\Delta e}\right)_{\Delta v=0} = \frac{e}{X}\frac{2p' - \dfrac{v^2}{X}}{p'} \qquad [6.4.12]$$

$$\left(\frac{\Delta v}{\Delta B}\right)_{\Delta Q_e=0} = \frac{v^3\left(p' - \dfrac{v^2}{2X}\right)}{a_p P^2 + a_q Q_L\left(p' - \dfrac{v^2}{2X}\right) + \dfrac{v^2}{X}(Q_L - p') + p'^2 - P^2 - 2Q_L p'} \qquad [6.4.13]$$

where it is intended that the quantities are evaluated at the operating point.

In analogy to the case of a linear load and again adopting Equation [6.4.3]:

- Equation [6.4.11] allows the assessment of the stability under normal operation, and it must be that $(\Delta v/\Delta B)_{\Delta e=0} > 0$.

- Equation [6.4.12] defines the static gain of the function $G_{qe}(s)$, whereas Equation [6.4.13] allows the evaluation of its zero (the value $G_{qe}(\infty)$ can be then derived from Equation [6.4.4]), and such information may be important for network voltage regulation.
- Finally, Equation [6.4.13] itself allows the assessment of the stability for operation at constant Q_e, where it must be that $(\Delta v/\Delta B)_{\Delta Q_e=0} > 0$.

If the load is purely active (i.e., with a unity power factor), $Q_L = a_q Q_L = 0$, and it can be more simply derived:

$$\left(\frac{\Delta v}{\Delta B}\right)_{\Delta e=0} = \frac{v^3 p'}{a_p P^2 + p'^2 - P^2} \qquad [6.4.11']$$

$$\left(\frac{\Delta Q_e}{\Delta e}\right)_{\Delta v=0} = \frac{e}{X}\frac{2p' - \dfrac{v^2}{X}}{p'} \qquad [6.4.12 \text{ rep.}]$$

$$\left(\frac{\Delta v}{\Delta B}\right)_{\Delta Q_e=0} = \frac{v^3\left(p' - \dfrac{v^2}{2X}\right)}{a_p P^2 - \dfrac{v^2}{X}P' + p'^2 - P^2} \qquad [6.4.13']$$

where, as already said, $p' \triangleq \sqrt{P^2_{\max} - P^2} \triangleq \sqrt{(ev/X)^2 - P^2}$. In particular, the stability condition under normal operation, or under that at constant Q_e, can be usefully expressed in terms of XP/e^2 and a_p, for any assigned value of the ratio v/e. Refer, for $v/e = 1$, to Figure 6.33, which permits a direct comparison with the already treated case in which the load voltage is regulated by an on-load tap-changing transformer (recall Fig. 6.25).

(c) Voltage Regulation at One or Several Nodes of a Line

Now consider the system in Figure 6.34, which includes;

- two infinite power networks operating at the same frequency ω and having voltages $\overline{e}_A$ and $\overline{e}_B$ of equal magnitude;
- a connection line defined by the reactance $X_{\text{tot}} \triangleq X_1 + X_2$ evaluated at the network frequency[(20)];
- a controlled reactor static compensator represented (as in the previous example) by a variable susceptance B.

[(20)] The effect of line capacitances can be considered, for each section of the line, by two shunt capacitors at the terminal nodes of the section itself. The capacitors at the infinite power nodes can be disregarded, whereas the susceptance of the other two capacitors can be included in that of the compensator.

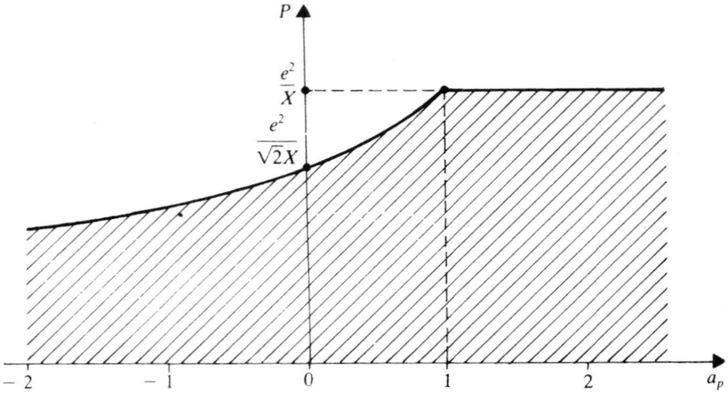

(a)

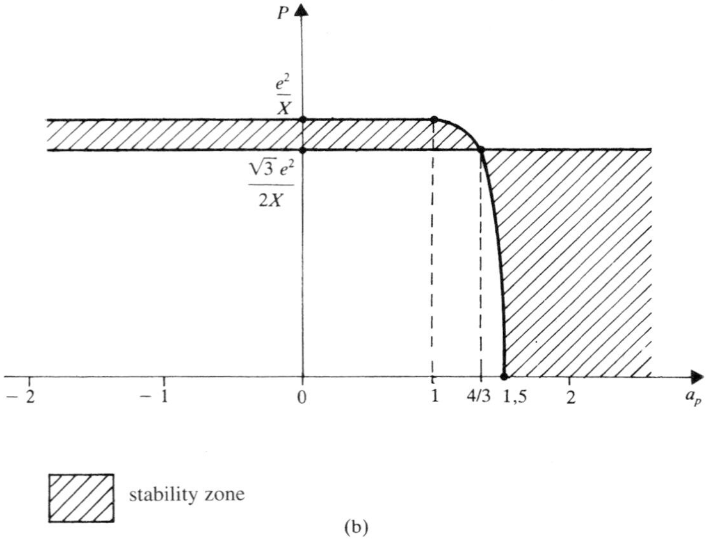

(b)

Figure 6.33. Stability zones in the plane (P, a_p), for $Q_L = 0$ and (at the operating point) $v = e$: (a) under normal operation (the stability condition is $a_p > 2 - (ev/(XP))^2$); (b) under the operation at constant Q_e, see text (having set $\alpha \triangleq e/v$, $\beta \triangleq \sqrt{1 - \alpha^2(XP/e^2)^2} \in [0, 1]$, the stability condition is $a_p \lessgtr 1 + \beta(1 - \alpha\beta)/(\alpha(1 - \beta^2))$ according to whether $\beta \lessgtr 1/(2\alpha)$).

By applying Park's transformation with $\Omega_r = \omega$ (and again disregarding, for simplicity, the index "r" in the Park's vectors), by the notation of the figure, it follows:

$$\overline{v} = \frac{X_2\overline{e}_A + X_1\overline{e}_B}{X_1 + X_2 - X_1X_2B}$$

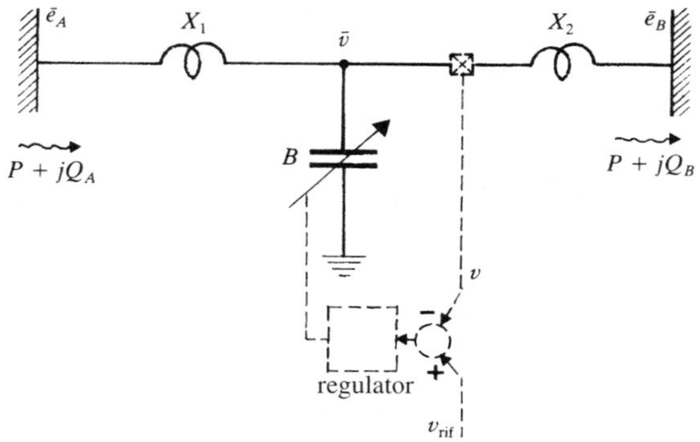

Figure 6.34. Voltage regulation at an intermediate node of a line by a static compensator: system under examination.

from which, as a result of the assumption $e_A = e_B \triangleq e$ and setting $\delta \triangleq \angle\bar{e}_A - \angle\bar{e}_B$:

$$v = \frac{e\sqrt{X_1^2 + X_2^2 + 2X_1X_2\cos\delta}}{|X_1 + X_2 - X_1X_2B|} \qquad [6.4.14]$$

whereas the active power P transmitted by the system (in the sense indicated in the figure) is:

$$P = \frac{e^2 \sin\delta}{X_1 + X_2 - X_1X_2B} \qquad [6.4.15]$$

Based on Equation [6.4.14], the dependence of v on B is as indicated in Figure 6.35, with $v = \infty$ (in the ideal case, without any resistance, considered here) for $B = B^*$, where:

$$B^* \triangleq \frac{1}{X_1} + \frac{1}{X_2}$$

In Equation [6.4.14], the numerator becomes zero if and only if $X_1 = X_2$ and $\cos\delta = -1$. The diagram of Figure 6.35 refers to a generic situation in which such conditions do not hold.

By the reactor control, the susceptance B can be changed so as to regulate the voltage v at the desired value v_{rif} ("voltage support" along the line). On the other hand, for each value of v there are two possible solutions in B (points 1 and 2 in Fig. 6.35), of which the solution 1, at a smaller B, is usually the preferred one, as it implies a smaller current in the compensator and in the two sections of the line.

Around this solution, it holds that $\partial v/\partial B > 0$ (Fig. 6.35), so that the regulation loop can be as in Figure 6.36, which is similar to that of Figure 6.31. Also, the synthesis of $G_v(s)$ does not imply, for each assigned operating point, particular problems with the adopted simplifications. In particular, by an integral

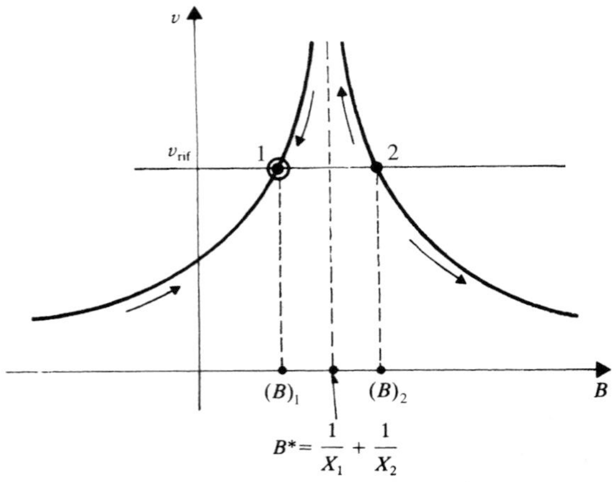

Figure 6.35. Dependence of the voltage v on the susceptance B (see Fig. 6.34).

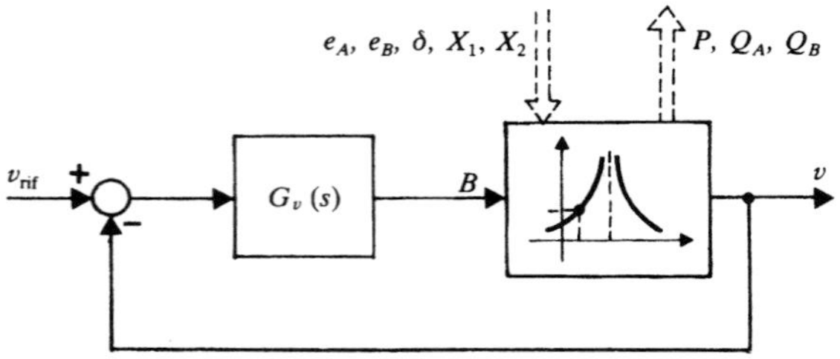

Figure 6.36. Block diagram of the regulation loop.

effect in $G_v(s)$ it is possible to obtain $v = v_{\text{rif}}$ at steady-state. If it is simply assumed that:

$$G_v(s) = \frac{K_v}{s} \qquad [6.4.16]$$

the cutoff frequency of the regulation loop is $\nu_t = K_v \partial v / \partial B$, whereas the closed-loop response delay is represented by the time constant $1/\nu_t$. If the cutoff frequency is relatively high, the analysis should also consider the dynamic behavior of the inductive and capacitive elements.

Furthermore, if Equation [6.4.16] is adopted, the effect of the regulator is to move the generic point (B, v) in the sense indicated by the arrows in Figure 6.35. Therefore, if for some reasons it were true that $B > B^*$, the regulator would cause (unless proper countermeasures were taken):

- the voltage collapse, for $B > (B)_2$;
- the reaching of point 1, but at the cost of unacceptable overvoltages, for $B \in (B^*, (B)_2)$.

In practical cases, with noninfinite power networks, it also may be useful to evaluate the transfer functions that relate the variations ΔP, ΔQ_A, ΔQ_B to variations Δe_A, Δe_B, $\Delta\delta$. In particular, from Equations [6.4.14] and [6.4.15] it is possible to derive the transfer function:

$$K(s) \triangleq \frac{\mathcal{L}\Delta P}{\mathcal{L}\Delta\delta} = \frac{\partial P}{\partial \delta} - \frac{\partial P}{\partial B}\frac{G_v(s)\dfrac{\partial v}{\partial \delta}}{1 + G_v(s)\dfrac{\partial v}{\partial B}} \qquad [6.4.17]$$

the behavior of which can be considered important for keeping synchronism between the two networks, as specified in the following.

The regulation under examination limits voltage variations during transients, thereby allowing a better operation of the line (without regulation, the overvoltages might find some limitation only because of the "corona effect," caused by themselves) and a better supply to possible loads located along the line. However, it is easy to determine that the regulation also can increase the transmittable active power, according to the following.

Assuming, as it looks obvious, $v_{\text{rif}} = e$, based on the previous equations the equilibrium points 1 and 2 are, respectively, defined by:

$$(B^o)_{1,2} = B^* \mp \frac{\sqrt{X_1^2 + X_2^2 + 2X_1X_2\cos\delta^o}}{X_1X_2}$$

and:

$$(P^o)_{1,2} = \pm\frac{e^2\sin\delta^o}{\sqrt{X_1^2 + X_2^2 + 2X_1X_2\cos\delta^o}}$$

Therefore, the active power P^o has at the stable solution point 1, the same sign as $\sin\delta^o$ and depends on δ^o as indicated by the bold line in Figure 6.37a, whereas $(P^o)_2 = -(P^o)_1$.

If the regulation were blocked, with $B = \text{constant}$, because of Equation [6.4.15] there would be a single characteristic (P^o, δ^o), as indicated by the thinner line in Figure 6.37a, having a slope:

$$K(\infty) = \left(\frac{\Delta P}{\Delta\delta}\right)_{B=\text{constant}} \gtrless 0 \quad \text{according to whether } \delta^o \lessgtr 90^\circ$$

The operation at constant B automatically occurs if B reaches its minimum or maximum limit. If for instance, the maximum limit is reached at a power P smaller than the value P_M indicated in the figure, the characteristic must be modified in an obvious way.

For $B = 0$, i.e., without compensator, it would be:

$$P^o = \frac{e^2\sin\delta^o}{X_1 + X_2}$$

as indicated by the dashed line in Figure 6.37a.

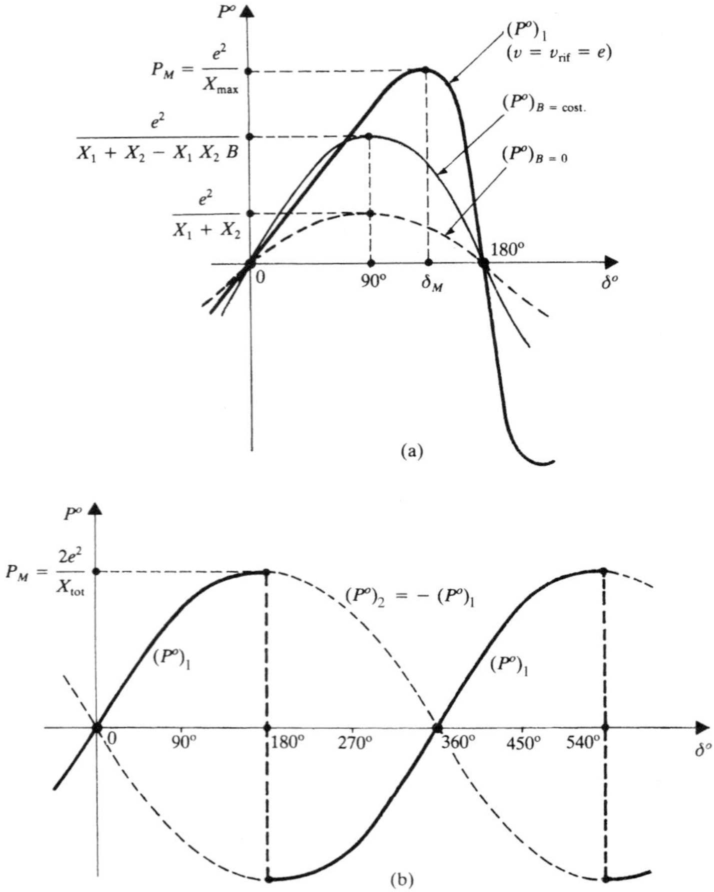

Figure 6.37. Dependence of the transmitted active power (P^o) on the phase-shift (δ^o) between the voltage vectors at the line terminals, in steady-state conditions: possible solutions, with a compensator (see text; figure (b) corresponds to the case $X_1 = X_2$).

With the regulation such that $v^o = v_{rif} = e$, the slope of the characteristic is instead:

$$K(0) = \left(\frac{\Delta P}{\Delta \delta}\right)_{v=e} \gtrless 0 \quad \text{according to whether } \delta^o \lessgtr \delta_M$$

where δ_M is defined by:

$$\cos \delta_M \triangleq -\frac{X_{min}}{X_{max}}$$

where $X_{\min}$ and $X_{\max}$ are respectively the smaller and the larger of the two reactances X_1, X_2.

Correspondingly, the *maximum active power* P_M *which can be transmitted* in a stable way is increased from $e^2/(X_1 + X_2)$ (without compensator) to $e^2/(X_1 + X_2 - X_1 X_2 B)$ (with $B =$ constant), or, in the presence of regulation, to:

$$P_M = e^2 \sqrt{\frac{1 - \left(\dfrac{X_{\min}}{X_{\max}}\right)^2}{X_{\max}^2 - X_{\min}^2}} = \frac{e^2}{X_{\max}}$$

Assuming, for simplicity, that Equation [6.4.16] holds and that $v_{\text{rif}} = e$, it can be further deduced, around the generic stable operating point 1, that:

$$K(s) = \frac{K(0) + sTK(\infty)}{1 + sT}$$

where $T \triangleq 1/\nu_t$, and $K(0)$ and $K(\infty)$ (equal to the slope of the characteristic (P^o, δ^o), respectively, at $v = v_{\text{rif}} = e$ and at $B =$ constant $= (B^o)_1$) can be of either sign, as already seen, depending on the value of δ^o. Furthermore, it results that $K(0) > K(\infty)$ except for $\delta^o = 0$.

Assuming as a first approximation that the phases of $\overline{e}_A$, $\overline{e}_B$ are associated to two non-infinite inertia coefficients M_A and M_B, so that it can be written (see also Sections 1.6 and 3.3.1):

$$\frac{\mathrm{d}^2 \Delta\delta}{\mathrm{d}t^2} = -\left(\frac{1}{M_A} + \frac{1}{M_B}\right) \Delta P \qquad [6.4.18]$$

the following characteristic equation is derived:

$$0 = s^2 M_{AB} + K(s)$$

where $M_{AB} \triangleq M_A M_B/(M_A + M_B)$, i.e.,

$$0 = s^3 T M_{AB} + s^2 M_{AB} + sTK(\infty) + K(0)$$

For stability, the following conditions must then hold:

$$K(\infty) > K(0) > 0 \qquad [6.4.19]$$

the former of which is never satisfied. Therefore, under the adopted hypotheses, the operation is unstable (the two networks lose their synchronism).

However, it is possible to reach different conclusions by substituting Equation [6.4.18] by a model with better approximations (see also Section 7.2.3.).

If $X_1 = X_2 = X_{tot}/2$, the result is more simply (with $B^* = 4/X_{tot}$):

$$(B^o)_{1,2} = B^* \mp \frac{4\sqrt{\dfrac{1+\cos\delta^o}{2}}}{X_{tot}} = \frac{4}{X_{tot}}\left(1 \mp \left|\cos\frac{\delta^o}{2}\right|\right)$$

$$(P^o)_{1,2} = \pm\frac{e^2\sin\delta^o}{X_{tot}\sqrt{\dfrac{1+\cos\delta^o}{2}}} = \pm\frac{e^2}{X_{tot}}\frac{\sin\delta^o}{\left|\cos\dfrac{\delta^o}{2}\right|}$$

(i.e., $(P^o)_1 = \pm 2(e^2/X_{tot})\sin(\delta^o/2)$ according to whether $\cos(\delta^o/2) \gtrless 0$, whereas $(P^o)_2 = -(P^o)_1$), and the diagrams are changed to those in Figure 6.37b, with:

$$P_M = 2\frac{e^2}{X_{tot}}$$

At this point, however, some explanations should be given to *avoid commonly made and dangerous misunderstandings*.

(1) For $X_1 = X_2$ one might spontaneously write, referring to each section of the line, $P^o = e^2\sin(\delta^o/2)/(X_{tot}/2)$, and thus $P^o_{max} = 2e^2/X_{tot}$. Instead, it is necessary to consider the existence of the opposite solution. Recall that the operating points 1 (stable) and 2 (unstable) pertain partially to one solution and partially to the other.

(2) For $X_1 = X_2$, the stability limit is at $\delta^o = 180°$ (so that $P_M = 2e^2/X_{tot} = P^o_{max}$), where $\partial P^o/\partial\delta^o = 0$. This last condition has all the appearance of a condition for maintaining synchronism between the two networks, but actually it is a *purely casual coincidence*. Actually:

- Under the hypothesis that the two networks are infinite power networks, the problem of maintaining synchronism does not exist, and the stability to be considered is only that of the regulation (if, for instance, it were assumed that $K_v < 0$, the stability and the instability conditions, which are based on the sign of $\partial v/\partial B$, would reciprocally change).
- The problem of maintaining synchronism might arise only if the hypothesis of infinite power networks were removed. But in such a case, the whole function $K(s)$ should be considered, as seen above.
- The above-mentioned coincidence is clearly missing for $X_1 \neq X_2$ (recall the bold-line characteristic in Figure 6.37a corresponding to a stable operation, for which $\partial P^o/\partial\delta^o$ can be negative), and as illustrated below, for the case of two or more compensators.

In the case of *n compensators* (with $n > 1$), the analysis implies greater difficulties. Useful insight, at least from the qualitative point of view, may be obtained by more simply assuming that:

- the line is "uniform" at the operating point, i.e.,

$$\left.\begin{aligned} X_1 = X_2 = \cdots = X_{(n+1)} \triangleq \frac{X_{\text{tot}}}{n+1} \triangleq X \\ B_1^o = B_2^o = \cdots = B_n^o \triangleq B^o \end{aligned}\right\} \qquad [6.4.20]$$

  where B_k^o is the value of the generic susceptance B_k at the operating point (see Fig. 6.38);

- all the n regulators have set-points equal to each other (and to $e_A = e_B \triangleq e$) and equal transfer functions, i.e.,

$$\left.\begin{aligned} v_{\text{rif }1} = v_{\text{rif }2} = \cdots = v_{\text{rif }n} \triangleq v_{\text{rif}} = e \\ G_{v1}(s) = G_{v2}(s) = \cdots = G_{vn}(s) \triangleq G_v(s) \end{aligned}\right\} \qquad [6.4.21]$$

  (where $G_{vk}(s) \triangleq \mathcal{L}\Delta B_k/\mathcal{L}(\Delta v_{\text{rif }k} - \Delta v_k)$: see Fig. 6.39).

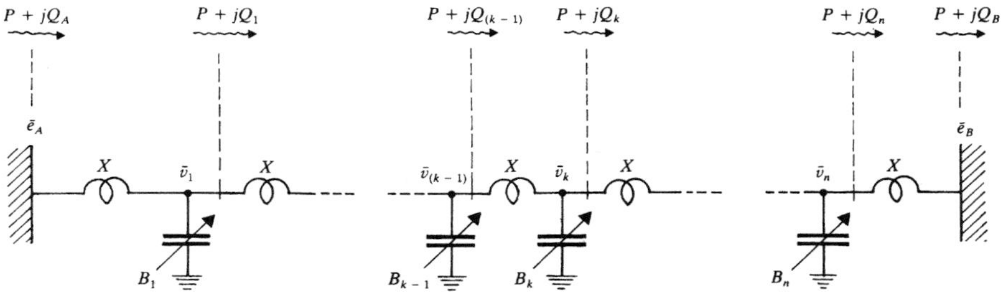

Figure 6.38. Voltage regulation along a line by more than one static compensator: system under examination.

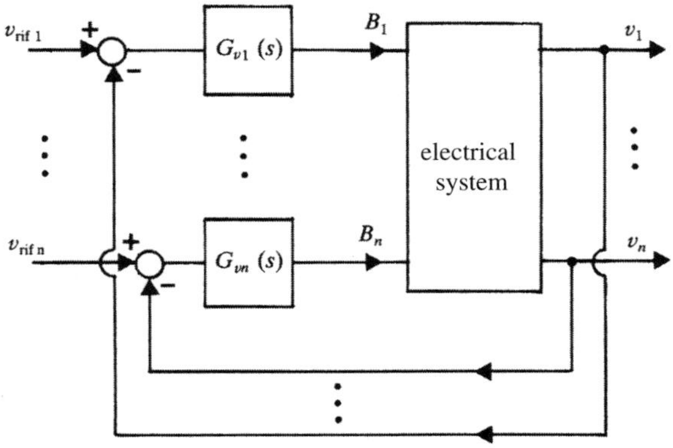

Figure 6.39. Overall block diagram of the regulations.

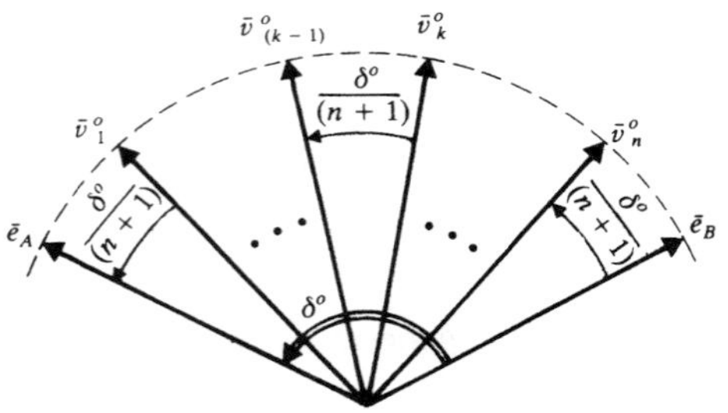

Figure 6.40. Diagram of voltage vectors at the considered steady-state (see text).

At the operating point, it results that (by assuming regulators with integral effect):

$$v_1^o = v_2^o = \cdots = v_n^o \triangleq v^o = v_{\text{rif}} = e$$

and the subsequent phase-shifts between voltages are equal one another, i.e., with $\alpha_0 \triangleq \angle \bar{e}_A$, $\alpha_k \triangleq \angle \bar{v}_k (k = 1, \ldots, n)$, $\alpha_{(n+1)} \triangleq \angle \bar{e}_B = \alpha_0 - \delta$:

$$(\alpha_0 - \alpha_1)^o = (\alpha_1 - \alpha_2)^o = \cdots = (\alpha_n - \alpha_{(n+1)})^o = \frac{\delta^o}{n+1}$$

as indicated in Figure 6.40.

By generalizing what has already been seen for a single compensator, with $X_1 = X_2$, one might be tempted:

- to simply write, referring to each section of the line, that:

$$P^o = \frac{v^{o2}}{X} \sin \frac{\delta^o}{n+1} = \frac{(n+1)e^2}{X_{\text{tot}}} \sin \frac{\delta^o}{n+1} \qquad [6.4.22]$$

  which has a maximum value of:

$$P^o_{\max} = (n+1) \frac{e^2}{X_{\text{tot}}}$$

  at $\delta^o = (n+1)90^\circ$, apart from multiples of 360° (see the thin-line characteristics in Figure 6.41);

- to assume that the maximum power which can be transmitted in a stable way (P_M), is given by $P^o_{\max}$ (as it happens in the case of a single compensator) and thus can be incremented *at will* by simply increasing the number of compensators.

The situation is actually very different, both because of the existence of other solutions (P^o, δ^o) and because of the limitations on P_M as a result of stability, as indicated in the following analysis.

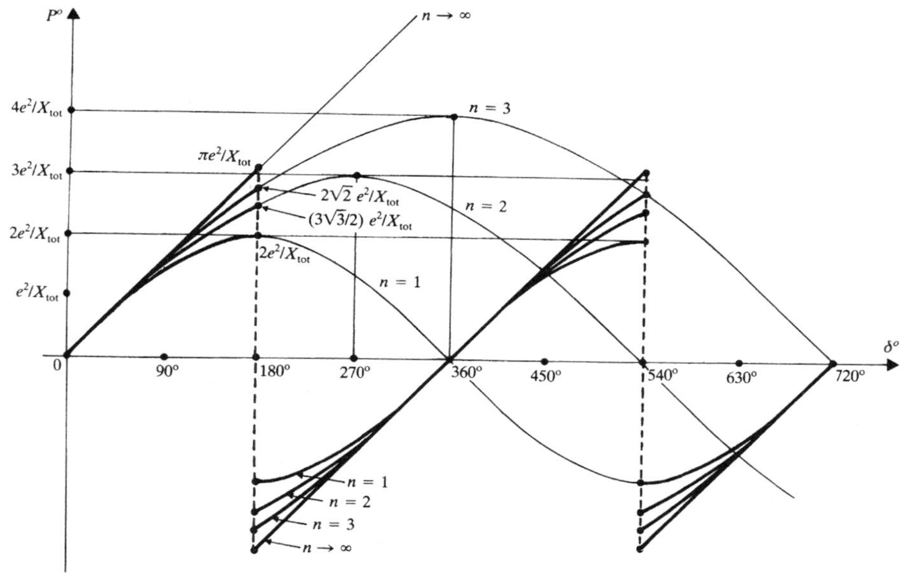

Figure 6.41. Dependence of the transmitted active power (P^o) on the phase-shift (δ^o) between the voltage vectors at the line terminals, in steady-state conditions: some solutions, with one or more compensators (see text; the bold-line characteristics are those of practical interest).

Regarding the value B^o of the susceptance, there are $(n+1)$ possible solutions, which, by considering any possible $|\delta^o|$ (even larger than 180°), can be expressed in the form:

$$(B^o)_r = \frac{2}{X}(1 - \cos\gamma_r) \quad (r = 0, 1, \ldots, n) \qquad [6.4.23]$$

where:

$$\gamma_r \triangleq \frac{\delta^o - r360^\circ}{n+1} \qquad [6.4.24]$$

according to Figure 6.42. Among such $(n+1)$ solutions, the one with the smallest value (indicated by the bold line in the figure), is usually the preferable solution, as it implies smaller currents in the compensators and in the different sections of the line.

Correspondingly, there are $(n+1)$ solutions also in P^o (for any given δ^o), defined by:

$$(P^o)_r = \frac{e^2 \sin\gamma_r}{X} \quad (r = 0, 1, \ldots, n-1) \qquad [6.4.25]$$

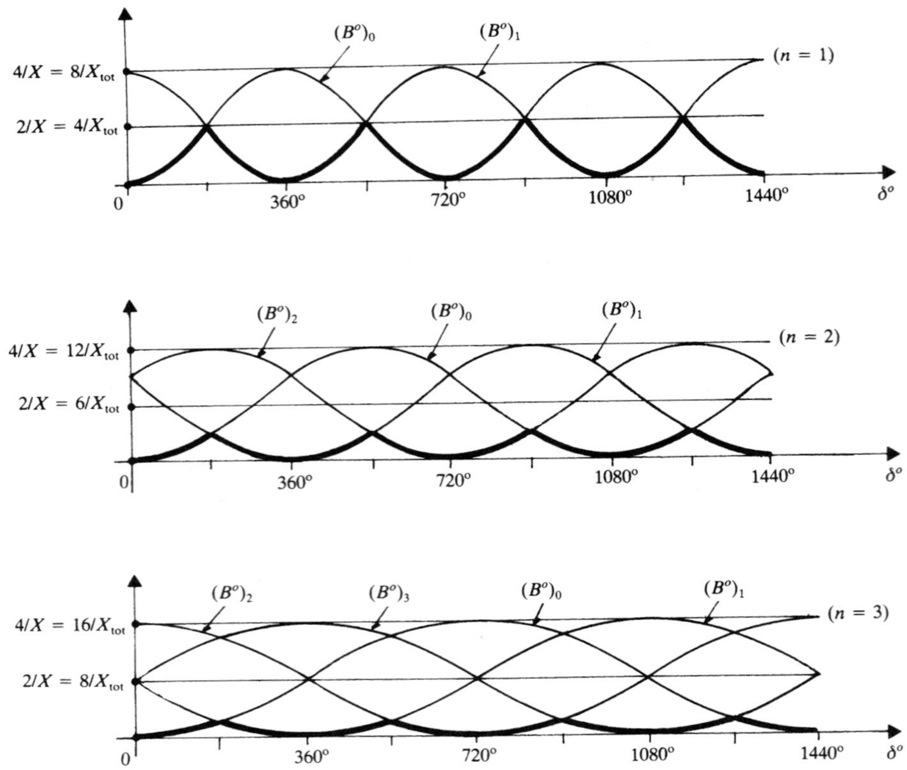

Figure 6.42. Dependence of the susceptance (B^o) on the phase-shift (δ^o) between the voltage vectors at the line terminals, in steady-state conditions: possible solutions, with one or more compensators (see text; the bold-line characteristics are those of practical interest).

Note that [6.4.22] is only the particular solution $(P^o)_0$, corresponding to $r = 0$, $\gamma_r = \gamma_0 = \delta^o/(n+1)$.

Furthermore, for $n \to \infty$, $\sin(\delta^o/(n+1)) \to \delta^o/(n+1)$, and thus:

$$(P^o)_0 \to \frac{e^2 \delta^o}{X_{\text{tot}}}$$

whereas the total susceptance is:

$$B_{\text{tot}} = (n+1)(B^o)_0 \to \frac{n+1}{X}\gamma_o^2 = \frac{\delta^{o2}}{X_{\text{tot}}}$$

In such conditions, indicating the line length by a, it also can be stated that the "phase constant" (see Section 5.4.2 and the second of Equations [5.4.15′]) is, at steady-state:

$$\beta = \frac{\sqrt{X_{\text{tot}} B_{\text{tot}}}}{a} = \frac{\delta^o}{a}$$

so that $\beta a = \delta^o$, whereas the "characteristic power" of the line is automatically equal to the transmitted active power, whichever its value (recall Equation [5.4.27] and Fig. 5.25). However, such conclusions hold *only at steady-state*, and may be of practical interest only if the operation is stable.

Moreover, for any operating conditions, it can be written:

$$\left.\begin{aligned} v_k &= \left(v_{(k-1)} - \frac{XQ_{(k-1)}}{v_{(k-1)}}\right)\cos(\alpha_{(k-1)} - \alpha_k) + \frac{XP}{v_{(k-1)}}\sin(\alpha_{(k-1)} - \alpha_k) \\ 0 &= \left(v_{(k-1)} - \frac{XQ_{(k-1)}}{v_{(k-1)}}\right)\sin(\alpha_{(k-1)} - \alpha_k) - \frac{XP}{v_{(k-1)}}\cos(\alpha_{(k-1)} - \alpha_k) \\ Q_k &= \frac{v_k}{v_{(k-1)}}[Q_{(k-1)}\cos(\alpha_{(k-1)} - \alpha_k) - P\sin(\alpha_{(k-1)} - \alpha_k)] + B_k v_k^2 \end{aligned}\right\} \quad [6.4.26]$$

($k = 1, \ldots, n$, with $v_0 = e_A$, $Q_0 = Q_A$), and furthermore:

$$e_B = \left(v_n - \frac{XQ_n}{v_n}\right)\cos(\alpha_n - \alpha_{(n+1)}) + \frac{XP}{v_n}\sin(\alpha_n - \alpha_{(n+1)}) \quad [6.4.27]$$

with $\alpha_{(n+1)} = \alpha_0 - \delta$.

Consider now the behavior *for small variations* around the former of the solutions given by [6.4.23], which corresponds to $r = 0$, i.e.,

$$B^o = (B^o)_0 = \frac{2}{X}(1 - \cos\gamma_0)$$

where:

$$\gamma_0 = \frac{\delta^o}{n+1} \quad [6.4.28]$$

By linearizing the Equations [6.4.26] and [6.4.27] and considering that:

$$\Delta B_k = G_v(s)(\Delta v_{\text{rif } k} - \Delta v_k)$$

it is possible to derive Δe_B and $\Delta\delta$ starting from Δe_A, ΔP, ΔQ_A, $\Delta v_{\text{rif } 1}, \ldots,$ $\Delta v_{\text{rif } n}$, according to equations:

$$\left.\begin{aligned} \Delta e_B &= \frac{\partial e_B}{\partial P}\Delta P + \frac{\partial e_B}{\partial Q_A}\Delta Q_A + \cdots \\ \Delta\delta &= \frac{\partial\delta}{\partial P}\Delta P + \frac{\partial\delta}{\partial Q_A}\Delta Q_A + \cdots \end{aligned}\right\} \quad [6.4.29]$$

where the dotted terms depend on Δe_A, $\Delta v_{\text{rif } 1}, \ldots, \Delta v_{\text{rif } n}$.

Since Δe_A, Δe_B, $\Delta\delta$, $\Delta v_{\text{rif }1}, \ldots, \Delta v_{\text{rif }n}$ are assumed as "inputs" of the system, the characteristic equation can be derived by setting to zero the determinant of the coefficients (functions of s) of ΔP and ΔQ_A in these equations, i.e.,

$$0 = \frac{\partial e_B}{\partial P}\frac{\partial \delta}{\partial Q_A} - \frac{\partial e_B}{\partial Q_A}\frac{\partial \delta}{\partial P}$$

If it is formally posed that:

$$\cos(2\sigma) \triangleq \cos(2\gamma_0) + \frac{XeG_v(s)}{2}\cos\gamma_0 \qquad [6.4.30]$$

it can be derived:

$$\left.\begin{aligned}
\frac{\partial e_B}{\partial P} &= e\frac{\partial \delta}{\partial Q_A} = \frac{X\sin\gamma_0}{e}\left(\frac{\sin((n+1)\sigma)}{\sin\sigma}\right)^2\\
\frac{\partial e_B}{\partial Q_A} &= -\frac{X\cos\gamma_0}{e}\frac{\sin(2(n+1)\sigma)}{\sin(2\sigma)}\\
\frac{\partial \delta}{\partial P} &= \frac{X}{e^2\cos\gamma_0}\left[(n+1)\left(1-\left(\frac{\sin\gamma_0}{\sin\sigma}\right)^2\right)+\left(\frac{\sin\gamma_0}{\sin\sigma}\right)^2\frac{\sin(2(n+1)\sigma)}{2\tan\sigma}\right]
\end{aligned}\right\} \qquad [6.4.31]$$

and the characteristic equation can be translated into:

$$0 = \frac{\sin((n+1)\sigma)}{\sin(2\sigma)} \qquad [6.4.32]$$

$$\cos(2\gamma_0) = \frac{\cos(2\sigma) - F}{1 - F} \qquad [6.4.33]$$

having assumed, for brevity:

$$F \triangleq \frac{\tan((n+1)\sigma)}{(n+1)\tan\sigma}$$

It is easy to determine that, in Equations [6.4.32] and [6.4.33], the dependence on σ is actually a dependence on $\cos(2\sigma)$. More precisely:

- from Equation [6.4.32], it follows that:

$$\sigma = h\frac{180^\circ}{n+1}, \qquad h = 1, \ldots, h_{\max}$$

(where $h_{\max} = (n-1)/2$ if n is odd, and $h_{\max} = n/2$ if n is even), from which the following $h_{\max}$ real solutions (that we will call *"type 1" solutions*) result:

$$\cos(2\sigma)_h = \cos\left(h\frac{360^\circ}{n+1}\right)$$

(for $n = 1$ there is no solution; for $n = 2$ there is the solution $\cos(2\sigma) = -1/2$; for $n = 3$ there is the solution $\cos(2\sigma) = 0$; for $n = 4$ instead there are the solutions $\cos 72^\circ$ and $\cos 144^\circ$; for $n = 5$ there are the solutions $1/2$ and $-1/2$; and so on);

- Equation [6.4.33] gives then, for any assigned γ_0, other $(n - h_{\max})$ real solutions in $\cos(2\sigma)$ (*"type 2" solutions*),

with a total of n solutions (see Fig. 6.43).

If Equation [6.4.30] is considered, it can be concluded that the characteristic roots of the system are deducible from the equations:

$$0 = 1 + \frac{X_e}{2} \frac{\cos\gamma_0}{\cos(2\gamma_0) - \cos(2\sigma)_i} G_v(s) \qquad [6.4.34]$$

$(i = 1, \ldots, n)$, where $\cos(2\sigma)_1, \ldots, \cos(2\sigma)_n$ are the solutions (of the types 1 and 2) defined above.

By assuming, for instance, that Equation [6.4.16] holds, i.e., assuming a $G_v(s)$ of the purely integral type with a positive gain K_v, the following stability conditions can be derived:

$$\frac{\cos\gamma_0}{\cos(2\gamma_0) - \cos(2\sigma)_i} > 0 \quad \forall i = 1, \ldots, n$$

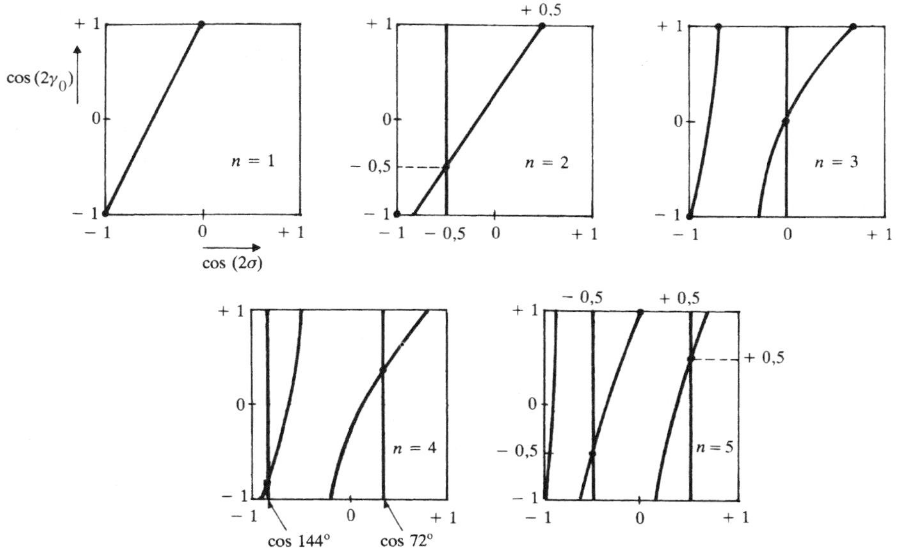

Figure 6.43. "Type 1" and "type 2" solutions, for the deduction of the characteristic roots of the system in Figure 6.38 (with n controlled compensators).

which define the admissible ranges of values for γ_0 and thus for $\delta^o = (n+1)\gamma_0$. More precisely, *the sign of* $(\cos(2\gamma_0) - \cos(2\sigma)_i)$ *must remain unchanged for all* $i = 1, \ldots, n$, *and it must be the same sign of* $\cos\gamma_0$.

By assuming $\gamma_0 \in (-180°, +180°]$, the above value ranges are defined as follows (see also Fig. 6.43):

- for $n = 1$: $|\gamma_0| < 90°$ and thus $|\delta^o| < 180°$
- for $n = 2$: $|\gamma_0| \begin{cases} < 60° \\ \in (90°, 120°) \end{cases}$ and thus $|\delta_0| \begin{cases} < 180° \\ \in (270°, 360°) \end{cases}$
- for $n = 3$: $|\gamma_0| < 45°$ and thus $|\delta^o| < 180°$
- for $n = 4$: $|\gamma_0| \begin{cases} < 36° \\ \in (90°, 108°) \end{cases}$ and thus $|\delta_0| \begin{cases} < 180° \\ \in (450°, 540°) \end{cases}$
- for $n = 5$: $|\gamma_0| < 30°$ and thus $|\delta^o| < 180°$

and so on. Note that, for $n > 1$, the limitation $|\delta^o| < 180°$ is the result of the "type 1" solution corresponding to $h = 1$.

Up to now, reference has been made only to the solution $(B^o)_0$, i.e., to the former of the solutions given by [6.4.23]. However, the treatment presented can be repeated with reference to the other solutions $(B^o)_1, \ldots, (B^o)_n$, provided δ^o is increased by the quantities $360°, \ldots, n\,360°$, respectively (recall Equation [6.4.24] and Fig. 6.42).

Therefore, based on previous results, it can be finally determined that for practical purposes, and with $\delta^o > 0$:

- the meaningful solution corresponds to $(B^o)_0$ for $\delta^o < 180°$, to $(B^o)_1$ for $\delta^o \in (180°, 540°)$, and so on, i.e., in any case, to the smaller susceptance (Fig. 6.42);
- correspondingly, the characteristic (P^o, δ^o), for the reported values of n, results in the solution represented by the bold line in Figure 6.41;
- the *maximum active power that can be transmitted* in a stable way is given (see Equation [6.4.25]) by:

$$P_M = \frac{e^2 \sin \dfrac{180°}{n+1}}{X} = \frac{(n+1)e^2}{X_{\text{tot}}} \sin \frac{180°}{n+1}$$

  and it is then within $2e^2/X_{\text{tot}}$ (for $n = 1$) and $\pi e^2/X_{\text{tot}}$ (for $n \to \infty$), with *a small increment* for increasing n.

Actually, if if for instance $n = 2$, the solution $(B^o)_0$ is possible also for $\delta^o \in (270°, 360°)$. But, for this range of values, the solution $(B^o)_1$ is smaller and thus appears preferable. Similar considerations apply for $n = 4$ and $\delta^o \in (450°, 540°)$, and so on.

Based on Equation [6.4.34], the previous considerations can be extended to the case of functions $G_v(s)$ of a more general type, for instance:

$$G_v(s) = K_v \frac{1 + sT'}{s(1 + sT)}$$

i.e., regulators of the proportional-integral type, with a delay defined by a time constant $T \geq 0$.

Moreover, regarding the dependence of ΔP on $\Delta\delta$, which in practical cases may be of interest in maintaining synchronism between the two networks, it can be derived from Equations [6.4.29] that:

$$K(s) \triangleq \frac{\mathcal{L}\Delta P}{\mathcal{L}\Delta\delta} = \frac{\dfrac{\partial e_B}{\partial Q_A}}{\dfrac{\partial e_B}{\partial Q_A}\dfrac{\partial \delta}{\partial P} - \dfrac{\partial e_B}{\partial P}\dfrac{\partial \delta}{\partial Q_A}}$$

and then, because of Equations [6.4.31] (with reference to the solution $B^o = (B^o)_0$):

$$K(s) = \frac{e^2 \cos\gamma_0}{X_{\text{tot}}} \frac{1 - \cos(2\sigma)}{(1 - F)\cos(2\gamma_0) + F - \cos(2\sigma)}$$

This equation also shows that the poles of $K(s)$ are constituted only by the characteristic roots that correspond to the "type 2" solutions (see Equation [6.4.33]). Then, if signals sensitive to the active power P were added as inputs to the regulators, the characteristic roots that correspond to the type 1 solutions would remain unchanged. Therefore, *nothing would change* with respect to the limitation $|\delta^o| < 180°$ and the maximum transmittable power P_M.

Such a conclusion can be reached by considering also the other solutions $(B^o)_1, \ldots, (B^o)_n$, similarly to what was shown above.

ANNOTATED REFERENCES

Among the works of more various or general interest, it can be made reference to: 25, 37, 42, 45, 47, 71, 74, 79, 130, 188, 193, 196, 217, 224, 225, 229, 232, 239, 244, 245, 259, 275, 280, 286, 288, 292, 293, 294, 306, 319, 326, 331.

More specifically, for what concerns the excitation control of the synchronous machines: 9, 51, 68, 72, 77, 81, 88, 114, 115, 125, 126 (for terminological aspects), 146, 212.

Moreover, for what concerns

- the control of tap-changers: 194, 322;
- the control of static compensators: 63, 228, 242, 274, 325;

further than some notes (particularly regarding the stability conditions) prepared by the author in view of the writing of 53.

CHAPTER 7

THE SYNCHRONOUS MACHINE CONNECTED TO AN INFINITE BUS

7.1. PRELIMINARIES

For a better understanding of different dynamic phenomena, it is useful to consider the case of a single synchronous machine connected to an infinite power network (infinite bus). On the other hand, such a schematization can be considered acceptable in several cases, when one or more machines relatively "close" to one another (e.g., machines of the same power plant, that may be suitably thought of as a single equivalent machine) are connected to a node of a much larger power.

For a greater generality, reference can be made to Figure 7.1, which includes:

- a synchronous machine with its excitation control and its turbine valve control;
- a transformer;
- a line;
- an infinite power network (at constant frequency equal to the "synchronous" speed Ω_s);
- an intermediate load, possibly voltage-controlled by an on-load tap-changing transformer or a static compensator.

The vectors indicated in the figure are obtained by applying the Park's transformation with the "machine" angular reference $\theta_r = \theta$ ($\Omega_r = \Omega$), with θ the electrical angular position (and $\Omega = d\theta/dt$ the speed) of the rotor of the machine.

The choice $\theta_r = \theta$ allows the machine equations to remain in their original form (Chapter 4). Nevertheless, if it is necessary to account for the dynamics of inductive and capacitive elements, with a dynamic order higher than 1

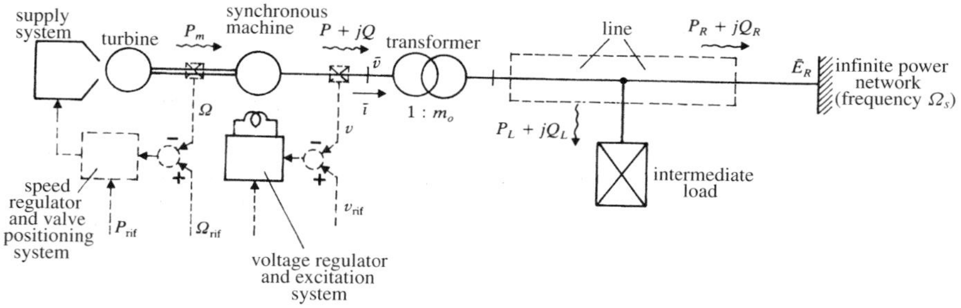

Figure 7.1. A synchronous machine connected to an infinite power network: system under examination.

(Appendix 2), it is convenient to apply the Park's transformation with a "network" angular reference, equal to the electrical angular position θ_s of a fictitious rotor rotating at the (constant) synchronous speed $\Omega_s = \mathrm{d}\theta_s/\mathrm{d}t$.

In the former case, the symbol of the Park's vectors will have no index, whereas in the latter case it will have the index s. The generic vector $\overline{y}$ (with the machine reference) is related to the vector $\overline{y}_s$ (with the network reference) through $\overline{y}\epsilon^{j\theta} = \overline{y}_s\epsilon^{j\theta_s}$.

With regard to the definition *of "infinite power" network*:

- the first assumption may be $\overline{E}_{Rs}$ = constant, for instance:

$$\overline{E}_{Rs} = jE_R$$

  with E_R constant, as if it were the emf of an ideal machine, with constant excitation, zero output impedances, and electrical angular position θ_s, and thus rotating at a constant electrical speed Ω_s (infinite inertia);
- more generally, it may be assumed that $\overline{E}_{Rs}$ is:

$$\overline{E}_{Rs} = jE_R\epsilon^{j\alpha_e} \qquad [7.1.1]$$

  where E_R and α_e can vary, but independently of what happens on the machine link, so they can be considered "inputs" of the system.

  As to this last consideration, if we assume that:
- the upward system can be approximated by a circuit as in Figure 7.2, with a constant "internal" emf $\overline{E}_{os}$ (having a magnitude E_o), a load, and a set of delta-connected constant impedances (evaluated at the network frequency Ω_s);
- the effects of $\overline{I}_{Rs}$ on $\overline{E}_{Rs}$ are negligible;

we can deduce:

$$\overline{E}_{Rs} \cong \overline{E}_{os} - \overline{ZI}_{os}$$

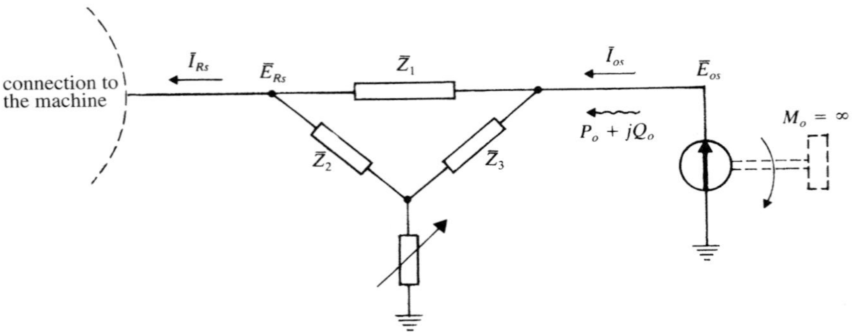

Figure 7.2. Schematic interpretation of the "infinite power network" (see text).

with:

$$\overline{Z} \triangleq \frac{\overline{Z}_1\overline{Z}_3}{\overline{Z}_1 + \overline{Z}_2 + \overline{Z}_3}$$

$$\overline{I}_{os} = \frac{P_o - jQ_o}{\overline{E}^*_{os}}$$

from which, following a load variation:

$$\Delta\overline{E}_{Rs} \cong -\overline{Z}\Delta\overline{I}_{os} = -\overline{Z}\frac{\Delta P_o - j\Delta Q_o}{\overline{E}^*_{os}}$$

If $\overline{Z}_1$, $\overline{Z}_2$, $\overline{Z}_3$ are purely imaginary, $\overline{Z}$ is imaginary as well, so that it is possible to set $\overline{Z} = jX$; by further assuming $X \to 0$, it can be derived:

$$\overline{E}^o_{Rs} \cong \overline{E}_{os}$$

$$\frac{\Delta\overline{E}_{Rs}}{\overline{E}^o_{Rs}} \cong -X\frac{j\Delta P_o + \Delta Q_o}{E_o^2}$$

and thus, being $\Delta\overline{E}_{Rs}/\overline{E}^o_{Rs} = \Delta E_R/E^o_R + j\Delta\alpha_e$:

$$\frac{\Delta E_R}{E^o_R} \cong -X\frac{\Delta Q_o}{E_o^2}$$

$$\Delta\alpha_e \cong -X\frac{\Delta P_o}{E_o^2}$$

Therefore, with the adopted schematizations:

- a variation $\Delta E_R/E^o_R$ (relative variation in magnitude) can be interpreted in terms of a variation in the reactive power delivered by the internal emf (and referred to the power E_o^2/X that is infinite), following a reactive load perturbation;

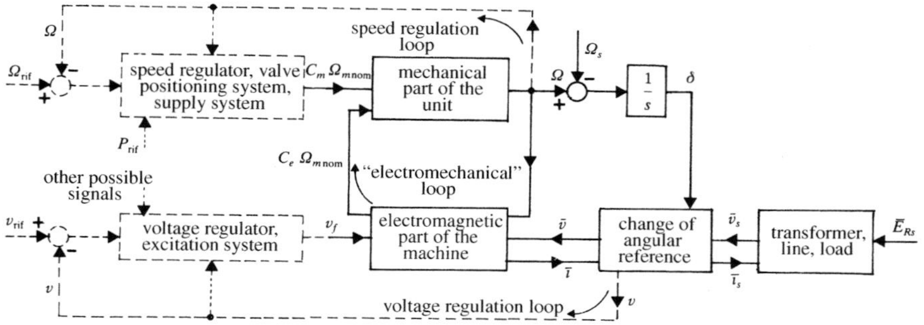

Figure 7.3. Block diagram of the system in Figure 7.1.

- a variation $\Delta\alpha_e$ (phase variation) can be interpreted in terms of a variation in the active power delivered by the internal emf (again referred to the power E_o^2/X), following an active load perturbation.

Such schematizations allow nonzero-phase variations, despite the assumption of infinite inertia ($M_o = \infty$).

The system in Figure 7.1 can be represented by the block diagram of Figure 7.3 (see also Figure 4.5) by particularly intending:

$$\delta \triangleq \theta - \theta_s \qquad [7.1.2]$$

and thus $d\delta/dt = \Omega - \Omega_s$ (it then follows $\overline{v} = \overline{v}_s \epsilon^{-j\delta}$, $\overline{\imath}_s = \overline{\imath}\epsilon^{j\delta}$, etc.); for small variations, the block diagram of Figure 7.4a can be further deduced.

To analyze the electromechanical phenomena (relative motion between the machine and the infinite bus), it is useful to write out the transfer functions:

$$D_m(s) \triangleq \frac{-\mathcal{L}(\Delta C_m \Omega_{m\,\text{nom}})}{\mathcal{L}(\Delta\Omega)} \qquad [7.1.3]$$

$$K_e(s) \triangleq \frac{\mathcal{L}(\Delta C_e \Omega_{m\,\text{nom}})}{\mathcal{L}(\Delta\delta)} \qquad [7.1.4]$$

(Fig. 7.4b), which respectively account for:

- driving torque variations caused by speed variations, through the turbine valve control (the torque variations related to the natural characteristics of the unit and the mechanical losses usually can be disregarded or possibly considered into $D_m(s)$; then, by using the notation of Chapter 3, it can be assumed that $D_m(s) = G_f(s)$ or, more generally, $D_m(s) = G_f(s) + G_g(s)$);
- electromagnetic torque variations caused by phase-shift variations ($\Delta\delta$) and speed variations ($\Delta\Omega = d\Delta\delta/dt$), according to the characteristics of:

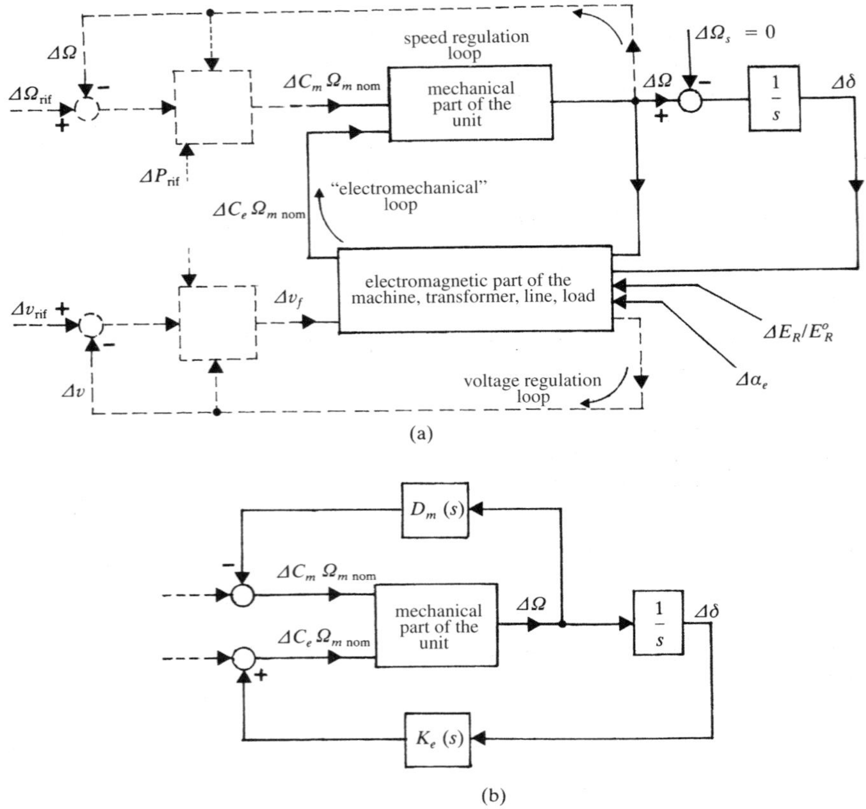

Figure 7.4. Linearized system: (a) block diagram; (b) definition of the functions $D_m(s)$ and $K_e(s)$.

- the electromagnetic part of the machine;
- the excitation control;
- transformer, line, and intermediate load (with possible control).

Finally, for the mechanical part of the machine, the inertia of the unit must be considered, as well as torsional phenomena (see Section 4.3.4). If these last phenomena may be disregarded, it simply follows:

$$\Delta\Omega = \frac{1}{sM}(\Delta C_m - \Delta C_e)\ \Omega_{m\,\mathrm{nom}} \qquad [7.1.5]$$

with M the inertia coefficient defined in Section 3.1.1.

In the following, we will assume as "*basic case*" (to which other situations are referred) the one for which:

- the transformer and line can be simply considered — at the machine side of the transformer — as inductances L_t, L_l in series between the machine and the infinite bus (disregarding the line capacitances);
- there is no intermediate load.

Furthermore, for simplicity, we will assume $\Omega_s = \Omega_{\text{nom}}$, and disregard:

- the armature resistance;
- the magnetic saturation.

For the basic case, we may then assume:

- the diagram of Figure 7.5a with $L_e \triangleq L_t + L_l$, $\overline{e}_R \triangleq \overline{E}_R/m_o$ (m_o is the transformation ratio of the transformer in Figure 7.1);
- the vector diagram of Figure 7.5b with reference to the steady-state (with $\alpha_e^o = 0$);

With notations used in Chapter 4, the following equations hold:

- for the mechanical part of the machine, disregarding the torsional phenomena (otherwise, see Section 4.3.4) and mechanical losses:

$$C_m - C_e = \frac{M}{\Omega_{m\,\text{nom}}} \frac{\mathrm{d}\Omega}{\mathrm{d}t} \tag{7.1.6}$$

further:

$$\Omega - \Omega_s = \Omega - \Omega_{\text{nom}} = \frac{\mathrm{d}\delta}{\mathrm{d}t} \tag{7.1.7}$$

- for the rest of the system, by applying the Park's transformation with the "machine" angular reference:

$$\left.\begin{aligned} v_d &= \frac{\mathrm{d}\psi_d}{\mathrm{d}t} - \Omega\psi_q \\ v_q &= \Omega\psi_d + \frac{\mathrm{d}\psi_q}{\mathrm{d}t} \end{aligned}\right\} \tag{7.1.8}$$

where:

$$\left.\begin{aligned} \psi_d &= A(s)v_f - \mathcal{L}_d(s)\, i_d \\ \psi_q &= -\mathcal{L}_q(s)\, i_q \end{aligned}\right\} \tag{7.1.9}$$

and moreover:

$$\left.\begin{aligned} e_R \sin(\delta - \alpha_e) &= v_d - L_e\left(\frac{\mathrm{d}i_d}{\mathrm{d}t} - \Omega i_q\right) \\ e_R \cos(\delta - \alpha_e) &= v_q - L_e\left(\Omega i_d + \frac{\mathrm{d}i_q}{\mathrm{d}t}\right) \end{aligned}\right\} \tag{7.1.10}$$

$$C_e = \frac{\Omega_{\text{nom}}}{\Omega_{m\,\text{nom}}}(\psi_d i_q - \psi_q i_d) \tag{7.1.11}$$

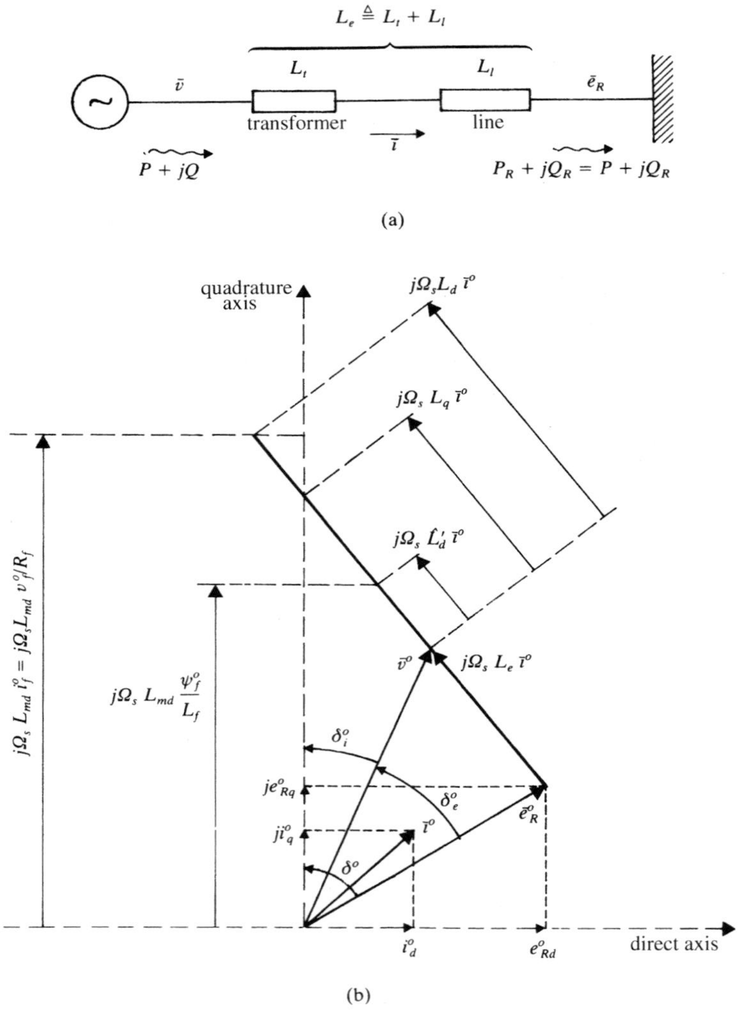

Figure 7.5. "Basic case": (a) system under examination; (b) vector diagram at steady-state (with $\alpha_e^o = 0$).

whereas C_m and v_f, respectively, depend on the turbine valve control and on the excitation control.

7.2. SMALL PERTURBATION BEHAVIOR

7.2.1. "Basic Case" in the Absence of Control

By linearizing Equations [7.1.6] to [7.1.11] around a generic operating point (indicated by the superscript "o"), the block diagram of Figure 7.6 can be derived

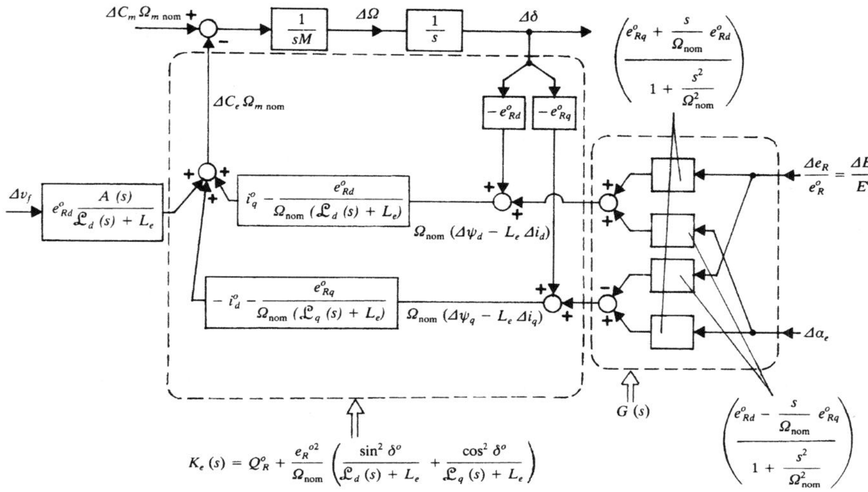

Figure 7.6. Block diagram of the linearized system ("basic case").

for which (in the absence of control) $\Delta C_m = 0$, $\Delta v_f = 0$ (apart from the small torque variations caused by the "natural" characteristics of the unit etc., and the variations in v_f caused by voltage variations in the case of exciter with dependent supply).

In particular, it results:

$$D_m(s) = 0$$

By eliminating Δv_d, Δv_q, $\Delta\psi_d$, $\Delta\psi_q$, Δi_d, Δi_q it follows:

$$K_e(s) = Q_R^o + \frac{e_R^{o2}}{\Omega_{\text{nom}}}\left(\frac{\sin^2\delta^o}{\mathcal{L}_d(s) + L_e} + \frac{\cos^2\delta^o}{\mathcal{L}_q(s) + L_e}\right) \qquad [7.2.1]$$

according to what is specified in Figure 7.6, where:

$$Q_R^o = e_{Rq}^o i_d^o - e_{Rd}^o i_q^o$$

(with $e_{Rd}^o = e_R^o \sin\delta^o$, $e_{Rq}^o = e_R^o \cos\delta^o$) is the reactive power absorbed by the network (Fig. 7.5).

Equation [7.2.1] shows the effect of the operating point on $K_e(s)$, in terms of Q_R^o, e_R^o, δ^o. As an alternative, the angle δ^o also can be expressed as a function of P^o, Q_R^o, e_R^o:

$$\tan\delta^o = \frac{P^o}{Q_R^o + \dfrac{e_R^{o2}}{\Omega_{\text{nom}}(L_q + L_e)}} \qquad [7.2.2]$$

(with $L_q = \mathcal{L}_q(0)$: see also Fig. 7.5b).

In more synthetic terms, having set for brevity:

$$q \triangleq \frac{Q_R^o}{e_R^o i^o}$$

$$l_{dt}(s) \triangleq \frac{\Omega_{\text{nom}}(\mathcal{L}_d(s) + L_e)\, i^o}{e_R^o}$$

$$l_{qt}(s) \triangleq \frac{\Omega_{\text{nom}}(\mathcal{L}_q(s) + L_e)\, i^o}{e_R^o}$$

$$x_{qt} \triangleq l_{qt}(0) = \frac{\Omega_{\text{nom}}(L_q + L_e)\, i^o}{e_R^o}$$

(with i^o the current at the operating point, and $P^{o2} + Q_R^{o2} = (e_R^2 i^o)^2$), it can be written:

$$\frac{K_e(s)}{e_R^o i^o} = q + \frac{1}{l_{dt}(s)} + \cos^2\delta^o \left(\frac{1}{l_{qt}(s)} - \frac{1}{l_{dt}(s)}\right)$$

where:

$$\cos^2\delta^o = \frac{\left(q + \dfrac{1}{x_{qt}}\right)^2}{1 + \dfrac{2q}{x_{qt}} + \dfrac{1}{x_{qt}^2}}$$

depends on q according to Figure 7.7, for different values of x_{qt}. The effects of q (or even of the "power factor" at the infinite bus) can be simply evaluated, once $l_{dt}(s)$ and $l_{qt}(s)$ are known.

Moreover, the present results can be expressed in terms of P^o, Q^o, v^o, by considering that:

$$\left.\begin{aligned} Q_R^o &= Q^o - \Omega_{\text{nom}} L_e i^{o2} \\ e_R^{o2} &= v^{o2} + (\Omega_{\text{nom}} L_e i^o)^2 - 2\Omega_{\text{nom}} L_e Q^o \end{aligned}\right\} \qquad [7.2.3]$$

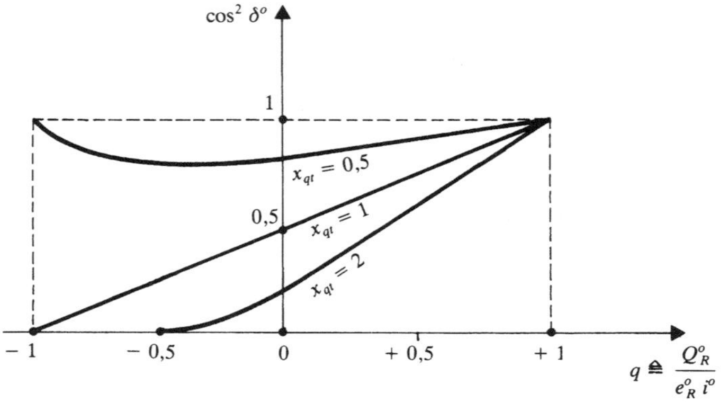

Figure 7.7. Dependence of $\cos^2\delta^o$ on q, for different values of x_{qt}.

where:

$$i^{o2} = \frac{P^{o2} + Q^{o2}}{v^{o2}} \qquad [7.2.4]$$

Note that Equation [7.2.1] can be also obtained:

- by disregarding the dynamic behavior of the inductive elements (which are present in the machine, transformer, and line);
- by approximating Ω by Ω_{nom} in Equations [7.1.8] and [7.1.10] (and similarly by replacing P_m with $C_m \Omega_{m\,\text{nom}}$, and P_e by $C_e \Omega_{m\,\text{nom}}$).

This is equivalent to replacing Equations [7.1.6], [7.1.8], [7.1.10], and [7.1.11] by the following simpler equations:

$$P_m - P_e = M \frac{\mathrm{d}\Omega}{\mathrm{d}t} \qquad [7.2.5]$$

$$\left.\begin{aligned} v_d &= -\Omega_{\text{nom}} \psi_q \\ v_q &= \Omega_{\text{nom}} \psi_d \end{aligned}\right\} \qquad [7.2.6]$$

$$\left.\begin{aligned} e_R \sin(\delta - \alpha_e) &= v_d + \Omega_{\text{nom}} L_e i_q \\ e_R \cos(\delta - \alpha_e) &= v_q - \Omega_{\text{nom}} L_e i_d \end{aligned}\right\} \qquad [7.2.7]$$

$$P_e = \Omega_{\text{nom}} (\psi_d i_q - \psi_q i_d) = v_d i_d + v_q i_q = P \qquad [7.2.8]$$

which correspond to the simplifications illustrated in Chapters 3 and 4.

Such approximations can, therefore, be adopted, as *they have effect neither on the transfer function* $K_e(s)$, nor, consequently, on the total transfer function of the electromechanical loop ($K_e(s)/(s^2 M)$ in Fig. 7.6).

The considered approximations also have no influence on the transfer function $\mathcal{L}(\Delta C_e \Omega_{m\,\text{nom}})/\mathcal{L}(\Delta v_f)$. The only impact they have is to replace the matrix $G(s)$ (indicated in Figure 7.6) by $G(0)$, with consequences on the response to variations Δe_R, $\Delta \alpha_e$. In particular, the effects of $\Delta \alpha_e$ turn out to be considered equivalent to those of variation $\Delta\delta = -\Delta\alpha_e$, so that it is sufficient to consider the variation $(\Delta\delta - \Delta\alpha_e)$ (see also Fig. 7.13).

Nevertheless these conclusions are related to the hypothesis that the armature resistance is zero. Otherwise, or generally in the presence of *resistances* in the electrical system, the dynamics of the inductive elements can somewhat affect $K_e(s)$, but only at relatively high frequencies (with a good approximation level: a damped resonance at the frequency Ω_{nom}, and an antiresonance at frequency $\sim \Omega_{\text{nom}}$), and therefore with negligible effects on the "electromechanical" oscillation, the frequency of which (as it will be seen) is usually much lower than Ω_{nom}.

Similarly, the behavior of the *line capacitances* also may be approximated by representing the line with simple inductive (series) and capacitive (shunt) reactances. By applying the Thevenin theorem to the set line-network, an equivalent scheme like in Figure 7.5a can again be found. However, some cautions may be necessary if the line is very long, with some resonance frequencies relatively low (refer to Fig. 5.27).

Furthermore, the dynamics of the inductive and capacitive elements deserve particular attention in the case of a series-compensated line, according to Section 7.2.4.

If the presence of additional rotor circuits is not considered, in Equations [7.1.9] it results (see Section 4.1.3):

$$\begin{cases} A(s) = \dfrac{A(0)}{1 + s\hat{T}'_{do}} \\ \mathcal{L}_d(s) = \dfrac{L_d + s\hat{T}'_{do}\hat{L}'_d}{1 + s\hat{T}'_{do}} \\ \mathcal{L}_q(s) = L_q \end{cases}$$

with $A(0) = L_{md}/R_f$. Through Equation [7.2.1] it follows:

$$K_e(s) = \frac{K_1 + sTK'_1}{1 + sT} \qquad [7.2.9]$$

with:

$$\left.\begin{aligned} K_1 &\triangleq Q_R^o + e_R^{o2}\left(\frac{\sin^2\delta^o}{X_d + X_e} + \frac{\cos^2\delta^o}{X_q + X_e}\right) \\ K'_1 &\triangleq Q_R^o + e_R^{o2}\left(\frac{\sin^2\delta^o}{\hat{X}'_d + X_e} + \frac{\cos^2\delta^o}{X_q + X_e}\right) \\ T &\triangleq \frac{(\hat{X}'_d + X_e)\hat{T}'_{do}}{X_d + X_e} \end{aligned}\right\} \qquad [7.2.9']$$

where $X_d \triangleq \Omega_{\text{nom}} L_d$, etc.

In particular:

- It can be determined that at the operating point, K_1 and K'_1 are equal to the slope of the characteristics (P_e^o, δ^o), respectively, for constant i_f^o and constant ψ_f^o (Fig. 7.8), corresponding to the equations:

$$\left.\begin{aligned} P_e &= \frac{X_{md} i_f e_R}{X_d + X_e}\sin\delta + \frac{(X_d - X_q)e_R^2}{2X_d X_q}\sin 2\delta \\ P_e &= \frac{X_{md}\psi_f e_R}{(\hat{X}'_d + X_e)L_f}\sin\delta - \frac{(X_q - \hat{X}'_d)e_R^2}{2\hat{X}'_d X_q}\sin 2\delta \end{aligned}\right\} \qquad [7.2.10]$$

  which are valid (for the adopted model, and for $\alpha_e = 0$) at any operating condition.

Both characteristics (P_e^o, δ^o) are static in the usual sense. However, only the characteristic for constant i_f^o is actually called "static," whereas that for constant ψ_f^o is called "transient."

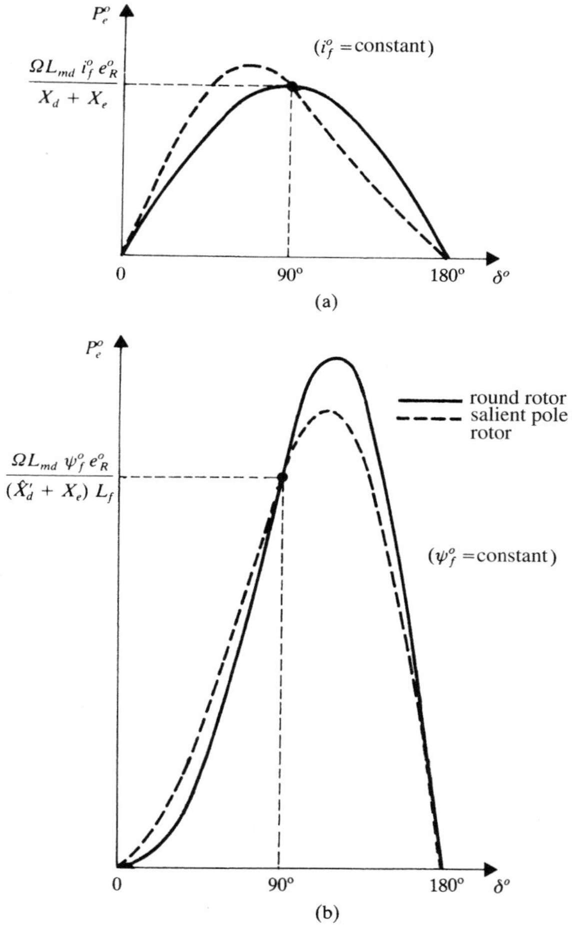

Figure 7.8. Dependence of P_e^o on δ^o for: (a) i_f^o constant; (b) ψ_f^o constant.

- The time constant T is defined similarly to the time constant $\hat{T}_d'$ ("short-circuit transient" time constant; see Section 4.1.4) of the machine, but with reference to the network node instead of the machine node.

For the system under examination, the characteristic equation is:

$$0 = 1 + \frac{K_e(s)}{s^2 M} \qquad [7.2.11]$$

i.e., because of Equation [7.2.9]:

$$0 = s^3 MT + s^2 M + sTK_1' + K_1 \qquad [7.2.11']$$

from which the following (asymptotic) stability conditions can be derived:

$$\left.\begin{aligned} T &> 0 \\ K_1 &> 0 \\ K_1' &> K_1 \end{aligned}\right\} \qquad [7.2.12]$$

Usually, such conditions are all met, and:

- the total transfer function of the electromechanical loop $K_e(s)/(s^2M)$ is represented by Bode diagrams as indicated in Figure 7.9;
- in the closed-loop operation, the system exhibits:

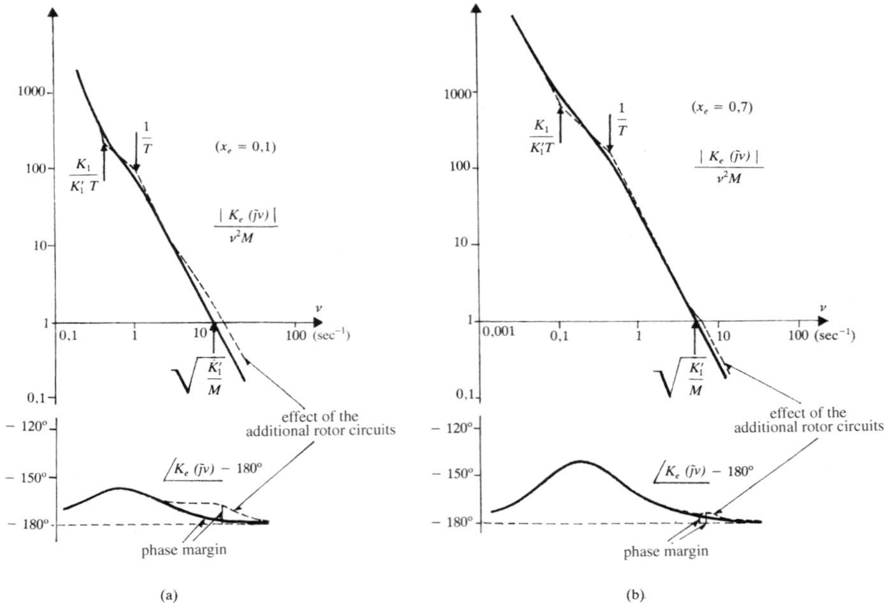

Figure 7.9. Frequency response of the total transfer function of the electromechanical loop (in the absence of control) for: (a) $x_e = 0.1$; (b) $x_e = 0.7$; and furthermore assuming that, for the bold line diagrams:

$$\begin{aligned} &\Omega_{nom} = 314 \text{ rad/sec} && x_d = x_q = 2 \\ &v^o = v_{nom} && \hat{x}_d' = 0.2 \\ &P^o = P_{nom} = 0.8A_{nom} && \hat{T}_{do}' = 7 \text{ sec} \\ &Q^o = 0.6A_{nom} && T_a = \frac{\Omega_{nom} M}{P_{nom}} = 8 \text{ sec} \end{aligned}$$

where x_d, x_q, $\hat{x}_d'$, x_e are the reactances in "machine" pu, i.e., referred to v_{nom}^2/A_{nom}, and T_a is the start-up time.

- a real characteristic root: the higher the gain of the loop, the closer is the root to $-K_1/(K_1'T)$;
- a pair of complex conjugate characteristic roots, which correspond to the ("electromechanical") oscillation between the machine and the infinite bus, with a resonance frequency:

$$\nu_o \cong \nu_t \cong \sqrt{\frac{K_1'}{M}} \gg \frac{1}{T}$$

(ν_t is the cutoff frequency of the loop), and with a (positive) damping factor ζ that increases with the "phase margin" $\angle K_e(\tilde{\jmath}\nu_t)$ and that also may be approximated by (see Equation [7.2.15]):

$$\zeta \cong \frac{1}{2} \tan \angle K_e(\tilde{\jmath}\nu_o)$$

As it will be seen, the characteristic equation remains similar for cases different from the basic one considered here, and its solutions (i.e., the characteristic roots) are usually constituted by a real root and a pair of complex conjugate roots.
Furthermore:

- if $K_1 < 0$: the real root would be positive (instability of the aperiodic type, caused by insufficient synchronizing actions);
- if $K_1 > 0$, but $K_1' < K_1$: the pair of complex conjugate roots would have a positive real part (instability of the oscillatory type, corresponding to a negative damping of the electromechanical oscillation).

By referring to Equations [7.2.9′], it is clear that in the present case:

- the first of conditions [7.2.12] is surely satisfied;
- it results:

$$K_1' - K_1 = (e_R^o \sin \delta^o)^2 \frac{X_d - \hat{X}_d'}{(\hat{X}_d' + X_e)(X_d + X_e)}$$

where $X_d > \hat{X}_d'$; thus, the third of conditions [7.2.12] is also satisfied (except for $\sin \delta^o = 0$, i.e., $P^o = P_e^o = 0$);

so that the complex conjugate roots have negative real part (or zero when $P^o = 0$), and *the instability can only be of the aperiodical type*, for $K_1 < 0$. (It is possible to draw the same conclusion when accounting for the additional rotor circuits.)
Therefore, the stability conditions are reduced only to $K_1 > 0$, i.e., $\delta^o < \delta_m^o$, by assuming that δ_m^o is the value of δ^o corresponding, for constant i_f^o, to maximum P_e^o. It results $\delta_m^o = 90°$ for the round rotor and $\delta_m^o < 90°$ for the salient pole rotor (see Fig. 7.8a).

If we define:

$$\left.\begin{aligned} Q' &\triangleq Q^o + \frac{v^{o2}}{X_q} \\ Q'' &\triangleq \frac{v^{o2}}{X_e} - Q^o \end{aligned}\right\} \qquad [7.2.13]$$

and moreover $\delta^o \triangleq \delta_i^o + \delta_e^o$, where δ_i^o and δ_e^o, respectively, represent the "internal" and the "external" phase-shifts (Fig. 7.5b), we may write, in terms of P^o, Q^o, v^o:

$$\tan\delta_i^o = \frac{P^o}{Q'}$$

and consequently, with obvious sign convention for the field circuit:

$$\sin\delta_i^o = \frac{P^o}{\sqrt{P^{o2} + Q'^2}}, \qquad \cos\delta_i^o = \frac{Q'}{\sqrt{P^{o2} + Q'^2}}$$

$$\begin{cases} e_R^o = \dfrac{X_e}{v^o}\sqrt{P^{o2} + Q''^2} \\ \sin\delta_e^o = \dfrac{P^o}{\sqrt{P^{o2} + Q''^2}}, \qquad \cos\delta_e^o = \dfrac{Q''}{\sqrt{P^{o2} + Q''^2}} \end{cases}$$

from which:

$$\tan\delta_e^o = \frac{P^o}{Q''}$$

according to Figure 7.10a.

By proper developments and referring to Equations [7.2.3] and [7.2.4], the first part of Equations [7.2.9′] also can be rewritten as:

$$K_1 = \frac{X_q X_e}{(X_q + X_e)v^{o2}}(Q'Q'' - P^{o2}) - \frac{(X_d - X_q)(X_q + X_e)v^{o2}}{(X_d + X_e)X_q^2}\frac{P^{o2}}{P^{o2} + Q'^2} \qquad [7.2.14]$$

whereas for K_1', a similar expression holds with $\hat{X}_d'$ instead of X_d.

Therefore, the condition $K_1 > 0$ can be translated, in the plane $((P/v^2)^o, (Q/v^2)^o)$, into a stability zone as indicated in Figures 7.10b,c, which is symmetrical with the abscissa axis. This zone decreases as X_e increases.

In the case of round rotor ($X_d = X_q$), the stability limit $K_1 = 0$ corresponds to:

$$P^{o2} = Q'Q'' = \left(Q^o + \frac{v^{o2}}{X_q}\right)\left(\frac{v^{o2}}{X_e} - Q^o\right)$$

which represents the equation of a circle as in Figure 7.10b (whereas the stability zone is constituted by the points internal to the circle). Furthermore:

- for $X_e = 0$: the limit curve degenerates into a vertical line, and the stability condition becomes $(Q/v^2)^o > -1/X_q$;
- for large X_e (e.g., 0.5 or larger in "machine" pu), the limit curve may cross the one corresponding to the capability limits (Figure 2.9), thus implying a further reduction (obviously not desired) of the zone in which the machine can be operated.

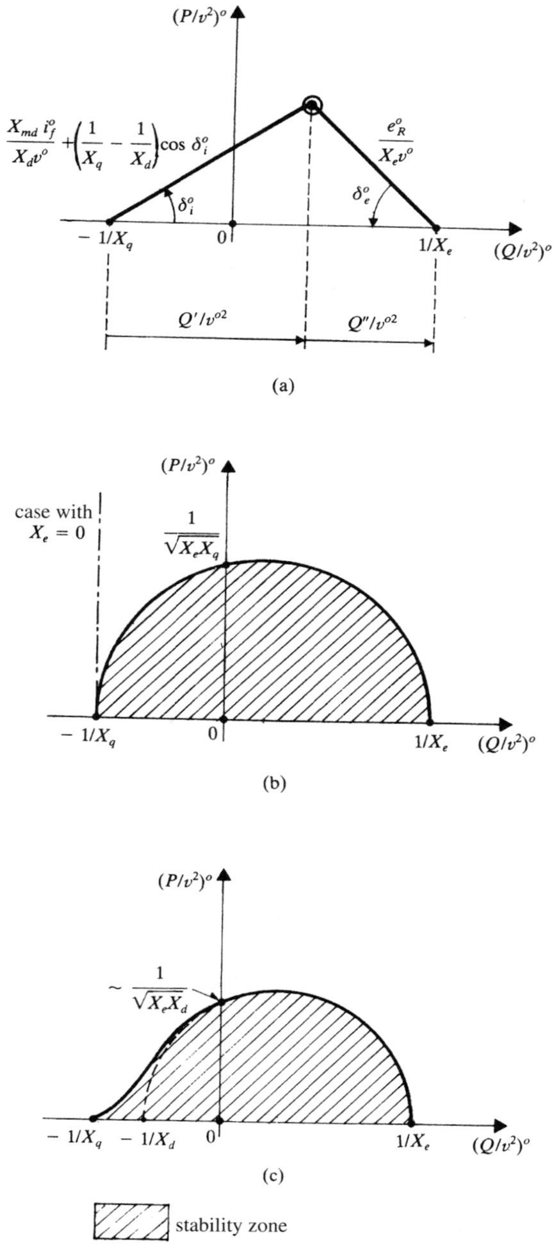

Figure 7.10. Representation in the plane $((P/v^2)^o, (Q/v^2)^o)$: (a) deduction of the "internal" (δ_i^o) and "external" (δ_e^o) phase-shifts; (b) stability zone in the absence of control for the round rotor; (c) stability zone in the absence of control for the salient pole rotor.

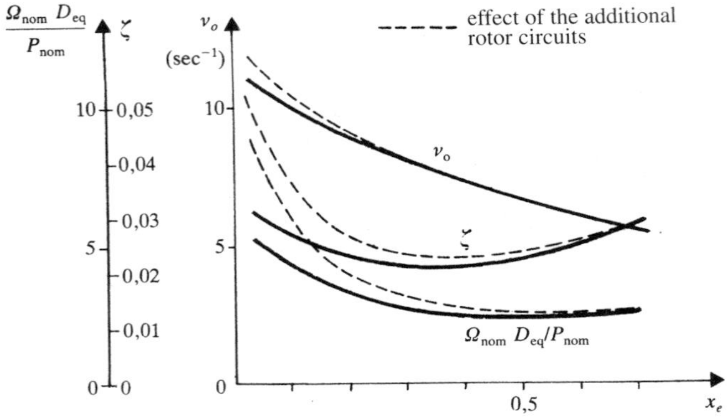

Figure 7.11. Frequency and damping of the electromechanical oscillation for the same numerical example as in Figure 7.9 (without additional rotor circuits: bold line diagrams).

Usually the electromechanical oscillation, lightly damped, has a frequency ν_o within 5–10 rad/sec (approximately 1–2 Hz) or more, as indicated in Figure 7.11 (*see footnote*[(4)] *in Chapter 3*). It therefore indicates, as already assumed, little interaction with the much faster dynamics of the inductive elements (even in the presence of resistances) and (although with some caution) with the torsional phenomena.

With $(a \pm \tilde{j}b)$ denoting the complex conjugate roots that correspond to $\nu_o = \sqrt{a^2 + b^2}$, $\zeta = -a/\nu_o$, because of Equation [7.2.11] it is possible to write, in general:

$$0 = 1 + \frac{(K_e(a + \tilde{j}b))}{(a + \tilde{j}b)^2 M}$$

from which:

$$\begin{cases} b^2 - a^2 = \dfrac{\mathcal{R}e(K_e(a + \tilde{j}b))}{M} \\ -2ab = \dfrac{Im(K_e(a + \tilde{j}b))}{M} \end{cases}$$

If the damping of the electromechanical oscillation is small, it then follows:

$$0 < -a \ll b \cong \nu_o$$

and thus:

$$\nu_o \cong \sqrt{\frac{\mathcal{R}e(K_e(\tilde{j}\nu_o))}{M}}$$

(usually, $\nu_o \cong \nu_t \cong \sqrt{K_1'/M}$, $\mathcal{R}e(K_e(\tilde{j}\nu_o)) \cong K_1'$), whereas:

$$\zeta \cong \frac{1}{2} \tan \angle K_e(\tilde{j}\nu_o) \cong \frac{Im(K_e(\tilde{j}\nu_o))}{2\nu_o^2 M} \qquad [7.2.15]$$

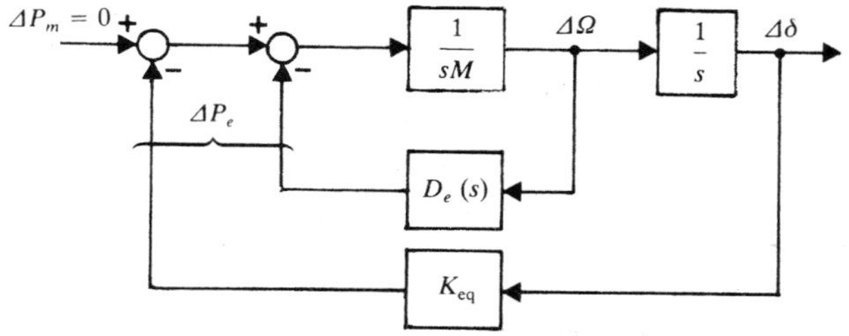

Figure 7.12. Block diagram of the linearized system with definition of K_{eq} and $D_e(s)$.

In the present conditions, the damping characteristics also can be shown by considering the block diagram of Figure 7.12, with:

$$\begin{cases} K_{\text{eq}} \triangleq \nu_o^2 M \\ D_e(s) \triangleq \dfrac{K_e(s) - K_{\text{eq}}}{s} \end{cases}$$

Substituting Equation [7.2.15] by:

$$\zeta \cong \frac{Im(\tilde{j}\nu_o D_e(\tilde{j}\nu_o))}{2\nu_o^2 M} = \frac{\mathcal{R}e(D_e(\tilde{j}\nu_o))}{2\nu_o M} \qquad [7.2.15']$$

the function $D_e(s)$ also can be replaced, as far as only the electromechanical oscillation is concerned, by the constant gain:

$$D_{\text{eq}} \triangleq 2\zeta\nu_o M \cong \mathcal{R}e(D_e(\tilde{j}\nu_o)) \qquad [7.2.16]$$

Noting that usually $\nu_o \cong \sqrt{K_1'/M} \gg 1/T$, the following equations can be finally derived:

$$\begin{cases} K_{\text{eq}} \cong K_1' \\ D_e(s) \cong \dfrac{-(K_1' - K_1)}{s(1 + sT)} \end{cases}$$

from which, because of Equations [7.2.15′] and [7.2.16] (see Fig. 7.11):

$$\left. \begin{aligned} \zeta &\cong \frac{1}{2\nu_o M}\frac{K_1' - K_1}{\nu_o^2 T} \cong \frac{1}{2\nu_o T}\frac{K_1' - K_1}{K_1'} \\ D_{\text{eq}} &\cong \frac{M}{T}\frac{K_1' - K_1}{K_1'} \end{aligned} \right\} \qquad [7.2.17]$$

As shown in Figures 7.9 and 7.11, a decrease of X_e implies an increasing of the frequency ν_o. However, particularly for small X_e, and for ν_o at least in the order of 8–10 rad/sec, the effect of *additional rotor circuits* — and specifically of the so-called "dampers" — should be considered. Such effect can be qualitatively translated into the dashed characteristics in Figure 7.9, which correspond (particularly for small X_e) to a significant increase of the phase margin and thus of the damping factor (see also Fig. 7.11).

Finally, the damping contribution of the additional rotor circuits becomes essential at $P^o = 0$—as it occurs for a synchronous compensator—because, without them, the damping would be zero (at $P^o = 0$, it holds $K_1' = K_1$), and it would remain so, even when considering the excitation control (see Section 7.2.2).

7.2.2. "Basic Case" in the Presence of Control

Based on previous considerations (Section 7.2.1), let us disregard, for simplicity:

- the torsional phenomena and the inductive element dynamics (thus accepting the approximations of Equations [7.2.5],..., [7.2.8]);
- the effects of additional rotor circuits.

For small variations, it is possible to derive the block diagram of Figure 7.13a, where:

- the time constant T is still defined by the third of Equations [7.2.9′];
- the gains K_1, K_1' (defined by the first two of Equations [7.2.9′], or equivalently by Equation [7.2.14] and the similar one) and K_2, K_2', K_3, K_4, H_1, H_1', H_2, H_2' depend on the operating point;
- the transfer functions $G_f(s)$ and $G_v(s)$, respectively, consider the turbine valve control (for speed regulation) and the excitation control (for voltage regulation), according to Sections 3.1, 3.2, and 6.2.

With reference to Figure 7.4b (and to Equations [7.1.3] and [7.1.4]), it can be assumed $D_m(s) = G_f(s)$, whereas $K_e(s)$ depends also on $G_v(s)$. Generally, it also could be assumed, as already specified, $D_m(s) = G_f(s) + G_g(s)$, and furthermore account could be taken, in $K_e(s)$, of the possible variations of v_f directly depending on Δv, in the case of exciter with dependent supply.

In terms of P^o, Q^o, v^o, the dependence of the gain K_1 on the operating point is defined by Equation [7.2.14], and it further results:

$$\left.\begin{aligned}
K_2 &= \frac{X_q X_e}{(X_q + X_e)v^o} P^o + \frac{(X_d - X_q)X_e v^o}{(X_d + X_e)X_q} \frac{P^o Q'}{P^{o2} + Q'^2} \\
K_3 &= \frac{X_{md}}{R_f} \frac{(X_q + X_e)v_d^o}{(X_d + X_e)X_q} \\
K_4 &= \frac{X_{md}}{R_f} \frac{X_e v_q^o}{(X_d + X_e)v^o} \\
H_1 &= P^o \left(1 + \frac{X_d - X_q}{(X_d + X_e)X_q} \left(\frac{X_q + X_e}{X_q} \frac{v^{o2} Q'}{P^{o2} + Q'^2} - X_e\right)\right) \\
H_2 &= \frac{X_d X_e}{X_d + X_e} \left(\frac{Q''}{v^o} - \frac{(X_d - X_q)v^o}{X_d X_q} \frac{P^{o2}}{P^{o2} + Q'^2}\right)
\end{aligned}\right\} \qquad [7.2.18]$$

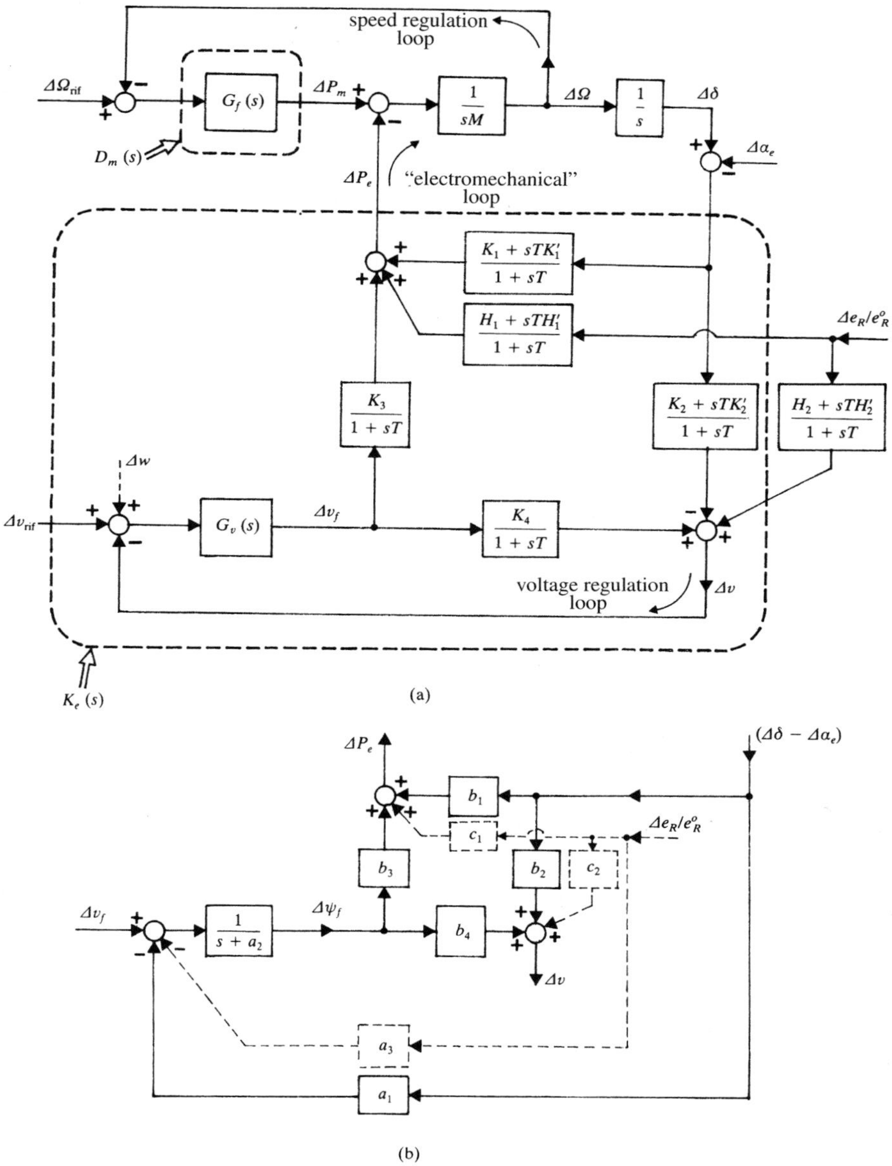

Figure 7.13. Linearized system in the presence of control (without additional rotor circuits): (a) block diagram (Δw is the sum of other possible signals, acting on the voltage regulator); (b) detailed representation including the state variable $\Delta\psi_f$.

(whereas the expressions of K_1', K_2', H_1', H_2' can be obtained from those of K_1, K_2, H_1, H_2, by simply substituting $\hat{X}_d'$ for X_d), where Q' and Q'' are defined by Equations [7.2.13], and furthermore:

$$\begin{cases} v_d^o = v^o \sin \delta_i^o = \dfrac{v^o P^o}{\sqrt{P^{o2} + Q'^2}} \\ v_q^o = v^o \cos \delta_i^o = \dfrac{v^o Q'}{\sqrt{P^{o2} + Q'^2}} \end{cases}$$

where δ_i^o is the internal phase-shift defined in Figure 7.5b.

If P^o, e_R^o, v^o are assigned (instead of P^o, Q^o, v^o), in Equations [7.2.14] and [7.2.18] we must consider that:

$$\begin{cases} Q''^2 = \left(\dfrac{e_R^o v^o}{X_e}\right)^2 - P^{o2} \\ Q' = v^{o2}\left(\dfrac{1}{X_e} + \dfrac{1}{X_q}\right) - Q'' \end{cases}$$

whereas:

$$Q^o = Q' - \frac{v^{o2}}{X_q} = \frac{v^{o2}}{X_e} - Q''$$

Specifically:

- for assigned Q^o and v^o — or even for assigned e_R^o and v^o — the gains K_1, K_1' and K_4 are "even" functions of P^o (i.e., they only depend on $|P^o|$), whereas K_2, K_2' and K_3 are "odd" functions of P^o (i.e., they simply change their sign if P^o does so);
- for assigned e_R^o and v^o, the gains K_1, K_1' and K_4 are usually positive (apart from $K_1 < 0$ or even $K_1' < 0$, for large P^o), as well as (for $P^o > 0$) K_2 and K_3, whereas K_2' becomes positive only for sufficiently large P^o (see Figure 7.14);
- instead, for $P^o = 0$ (e.g., in the case of a synchronous compensator):

  $$K_2 = K_2' = K_3 = 0$$

  so that, in the absence of additional rotor circuits, the electromechanical loop is decoupled with respect to the voltage regulation loop, whereas, as already indicated:

  $$K_1' = K_1$$

  (furthermore, $H_1 = H_1' = 0$).

Finally, with the adopted model:

$$s\Delta\psi_f = \Delta v_f - R_f \Delta i_f \qquad [7.2.19]$$

and Δi_f, ΔP_e, and Δv are linear combinations of $(\Delta\delta - \Delta\alpha_e)$, $\Delta\psi_f$ and $\Delta e_R/e_R^o$.

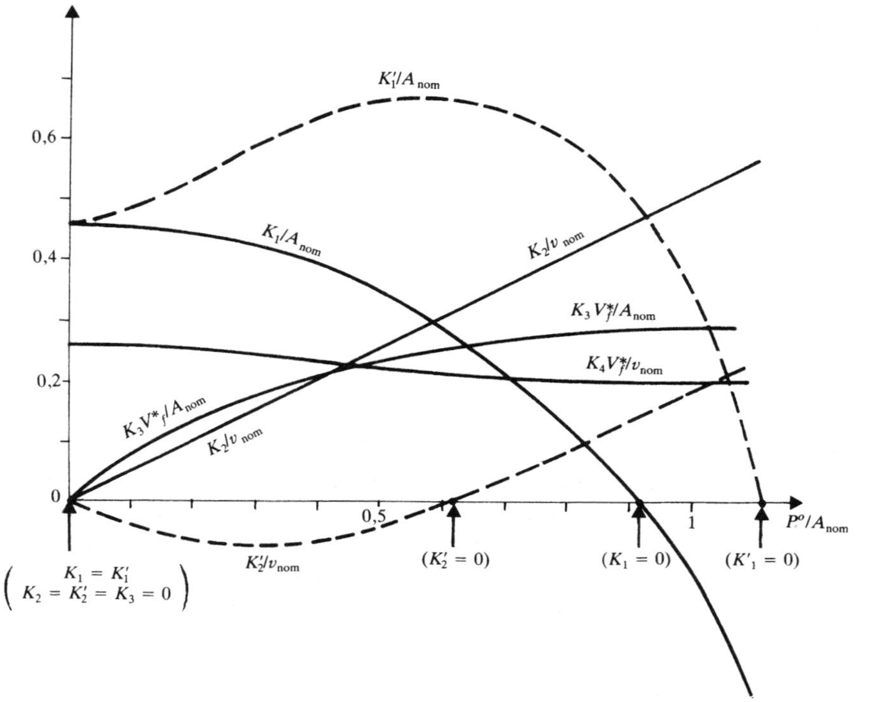

Figure 7.14. Dependence of the parameters K_1, K_1', K_2, etc. on P^o, assuming $e_R^o = 0.8v_{\text{nom}}$, $v^o = v_{\text{nom}}$ and furthermore (in "machine" pu) $x_d = x_q = 2$, $\hat{x}_d' = 0.2$, $x_e = 0.7$, whereas $V_f^* = R_f v_{\text{nom}}/X_{md}$. The diagrams for $P^o < 0$ can be immediately derived. In fact, if P^o changes only its sign, K_1, K_1', and K_4 (which are "even" functions of P^o) do not vary, whereas K_2, K_2', and K_3 (which are "odd" functions of P^o) simply change their sign.

By generically setting:

$$\left.\begin{aligned} R_f \Delta i_f &= a_1(\Delta\delta - \Delta\alpha_e) + a_2 \Delta\psi_f + a_3 \frac{\Delta e_R}{e_R^o} \\ \Delta P_e &= b_1(\Delta\delta - \Delta\alpha_e) + b_3 \Delta\psi_f + c_1 \frac{\Delta e_R}{e_R^o} \\ \Delta v &= b_2(\Delta\delta - \Delta\alpha_e) + b_4 \Delta\psi_f + c_2 \frac{\Delta e_R}{e_R^o} \end{aligned}\right\} \qquad [7.2.20]$$

and eliminating $R_f \Delta i_f$, it can be derived:

$$\Delta\psi_f = \frac{\Delta v_f - a_1(\Delta\delta - \Delta\alpha_e) - a_3 \dfrac{\Delta e_R}{e_R^o}}{s + a_2}$$

as shown in the block diagram of Figure 7.13b.

Eliminating $\Delta\psi_f$, it is possible to derive the transfer functions that relate ΔP_e and Δv to $(\Delta\delta - \Delta\alpha_e)$, Δv_f, $\Delta e_R/e_R^o$ (Fig. 7.13a), with:

$$\left.\begin{array}{llll} T = \dfrac{1}{a_2} & & & \\ K_1 = b_1 - a_1 b_3 T, & K_1' = b_1, & H_1 = c_1 - a_3 b_3 T, & H_1' = c_1 \\ -K_2 = b_2 - a_1 b_4 T, & -K_2' = b_2, & H_2 = c_2 - a_3 b_4 T, & H_2' = c_2 \\ K_3 = b_3 T & & & \\ K_4 = b_4 T & & & \end{array}\right\} \quad [7.2.21]$$

so that it results:

$$\frac{K_1' - K_1}{K_2 - K_2'} = \frac{H_1' - H_1}{H_2' - H_2} = \frac{K_3}{K_4} \quad [7.2.22]$$

(a) Effects of Speed Regulation

For the electromechanical loop, the effect of speed regulation is to substitute the transfer function $1/(sM)$ with $1/(D_m(s) + sM)$ (see Fig. 7.13).

As in Sections 3.1 and 3.2, such an effect is appreciable particularly at low frequencies (not much higher than the cutoff frequency of the regulation loop, e.g., 0.3 rad/sec). Thus, the speed regulation usually has a small effect on the electromechanical oscillation, and:

- the frequency ν_o of the electromechanical oscillation remains practically unchanged;
- the corresponding damping factor ζ undergoes a small variation.

For the damping variation, by examining Figures 7.12 and 7.13 it is clear that $D_m(s) = G_f(s)$ has effects similar to those of $D_e(s)$; by referring to Equation [7.2.16] it is then possible to write:

$$\Delta D_{\text{eq}} \triangleq 2\Delta\zeta\nu_o M \cong \mathcal{Re}(G_f(\tilde{j}\nu_o))$$

from which:

$$\Delta\zeta \cong \frac{\mathcal{Re}(G_f(\tilde{j}\nu_o))}{2\nu_o M}$$

Actually, the damping contribution of the speed regulation also can be negative—and not negligible, for relatively low ν_o—especially with hydroelectric units at full load (see Figure 7.15). Moreover, the effects of a possible "backlash" in the regulator and/or the valve-positioning system can be regarded, at first approximation, as an additional phase delay (increasing up to 90°, for diminishing oscillation amplitude) in $G_f(\tilde{j}\nu)$, with consequences on ΔD_{eq} and $\Delta\zeta$.

(b) Effects of Voltage Regulation

For the electromechanical loop, the voltage regulation modifies the function $K_e(s)$ (as well as $D_e(s)$).

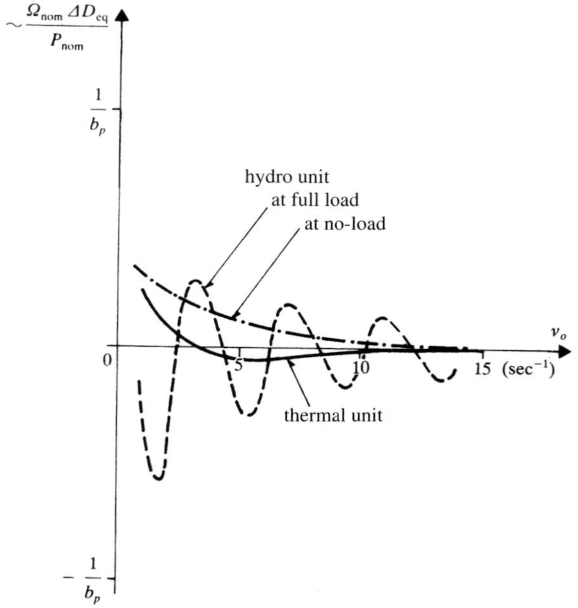

Figure 7.15. Damping contribution of the speed regulation (as a function of oscillation frequency), assuming:

$$G_f(s) = \frac{P_{nom}}{\Omega_{nom} b_p} \frac{(1+3.5s)}{(1+10s)(1+0.3s)} g(s)$$

with:

$$g(s) = \begin{cases} 1 & \text{for a hydrounit at no load} \\ \dfrac{1 - 1.28\tanh(0.785s)}{1 + 0.64\tanh(0.785s)} & \text{for a hydrounit at full load} \\ \dfrac{1}{1+0.3s} & \text{for a thermal unit} \end{cases}$$

Based on the diagram in Figure 7.13a, it can be easily derived:

$$K_e(s) = \frac{K_1 + sTK_1'}{1+sT} + \Delta K_e(s) \quad [7.2.23]$$

with:

$$\Delta K_e(s) \triangleq \frac{K_3}{K_4} F_v^{\cdot}(s) \frac{K_2 + sTK_2'}{1+sT} \quad [7.2.24]$$

where:

$$F_v(s) \triangleq \frac{G_v(s)K_4}{1 + sT + G_v(s)K_4} \quad [7.2.25]$$

is the transfer function which relates (at closed loop) Δv to Δv_{rif}, without considering the interactions with the electromechanical loop (as if $K_3 = 0$, or $K_2 = K_2' = 0$, or $\Delta\delta = 0$).

By developing and referring to Equations [7.2.22], it also can be deduced:

$$K_e(s) = \frac{K_1 + G_v(s)(K_1 K_4 + K_2 K_3) + sTK_1'}{1 + G_v(s)K_4 + sT} \qquad [7.2.26]$$

and therefore, as can be directly determined from Figure 7.13a, the static gain:

$$K_e(0) = K_1 + \frac{G_v(0)K_2 K_3}{1 + G_v(0)K_4}$$

also can significantly increase because of the regulation, whereas the gain $K(\infty) = K_1'$ remains unchanged.

Quite often, the *increase in* $K_e(0)$ is accepted in a favorable way without any reserve, as—according to the new static characteristic (P_e^o, δ^o), for v_{rif} constant and $G_v(s) = G_v(0)$—it allows operation at larger values of phase-shift and active power, without aperiodic instability (see Fig. 7.16, where P_e^o is maximum for $\delta^o = \delta_m'^o > 90°$)(1).

However, such a conclusion is often illusory, as it does not consider the further stability conditions, more strictly related to the dynamic effects of regulation. To account for such effects, the function $G_v(s)$ can be substituted, with usually acceptable approximation, by a proper gain K_v that estimates its behavior (for $s = \tilde{j}\nu$) around the electromechanical oscillation frequency. Referring to Equation [6.2.7], and considering usual values of the time constants T_1 and T_2, it can be assumed $K_v = G_v(0)T_2/T_1$.

Correspondingly, considering that usually $K_v K_4 \gg 1$, Equation [7.2.25] can be rewritten as:

$$F_v(s) \cong \frac{K_v K_4}{(1 + K_v K_4)\left(1 + \dfrac{s}{\nu_v^o}\right)} \cong \frac{1}{1 + \dfrac{s}{\nu_v^o}} \qquad [7.2.25']$$

where:

$$\nu_v^o \cong \frac{1 + K_v K_4}{T} \cong \frac{K_v K_4}{T}$$

The frequency $\nu_v^\circ (\gg 1/T)$ constitutes the cutoff frequency of the regulation loop, evaluated by ignoring the interactions with the electromechanical loop.

(1) For $G_v(0) = \infty$, i.e., $v^o = v_{\text{rif}} =$ constant, the maximum of P_e^o corresponds to a 90° lead of $\overline{v}^o$ on $\overline{e}_R^o$, and thus it holds that $P_{e\max}^o = v_{\text{rif}}\, e_R^o / X_e$. Furthermore, for $X_d = X_q$:

$$\tan\delta_m'^o = -\frac{X_d + X_e}{X_d}\frac{v_{\text{rif}}}{e_R^o}, \qquad \frac{i_f^o}{I_f^*} = \frac{X_d + X_e}{X_e \sin\delta_m'^o}$$

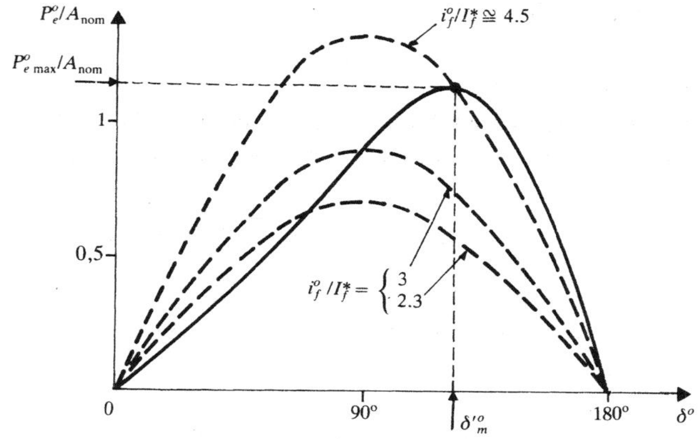

Figure 7.16. Dependence of P_e^o on δ^o for constant v^o ($G_v(0) = \infty$, bold line characteristic) and for constant i_f^o (dashed line characteristics), assuming $e_R^o = 0.8 v_{\text{nom}}$, $v^o = v_{\text{rif}} = v_{\text{nom}}$, $x_d = x_q = 2$, $x_e = 0.7$ (see Fig. 7.14), whereas $I_f^* = v_{\text{nom}}/X_{md}$.

In practice, ν_v° can be approximately 1–5 rad/sec, slightly lower than the ν_t defined in Section 6.2.1a, because of the factor $v_q^o/v^o < 1$ in K_4 (see the third of Equations [7.2.18]) and possibly the smaller value of the external reactance.

Furthermore, from Equation [7.2.26] the following expression can be derived:

$$K_e(s) \cong \frac{\hat{K}_1 + s\hat{T}\hat{K}_1'}{1 + s\hat{T}} \qquad [7.2.26']$$

which is similar to [7.2.9], with:

$$\left.\begin{aligned} \hat{K}_1 &\triangleq K_1 + \frac{K_v K_2 K_3}{1 + K_v K_4} \\ \hat{T} &\triangleq \frac{T}{1 + K_v K_4} \cong \frac{1}{\nu_v^o} \end{aligned}\right\} \qquad [7.2.27]$$

whereas $\hat{K}_1' = K_1'$. Note that, for $K_v K_4 \to \infty$, it results $\hat{K}_1 \to K_1 + K_2 K_3 / K_4 = K_1' + K_2' K_3 / K_4$.

From Equation [7.2.26′] the following stability conditions can be derived, which are similar to conditions [7.2.12]:

$$\left.\begin{aligned} &\hat{T} > 0 \\ &\hat{K}_1 > 0 \\ &K_1' = \hat{K}_1' > \hat{K}_1 \end{aligned}\right\} \qquad [7.2.28]$$

Of the three above conditions:

- the first two are usually satisfied;
- the third, associated with the damping of electromechanical oscillation, may instead not be satisfied if the increase in $\hat{K}_1$ is excessive.

Considering Equations [7.2.22] and [7.2.27], the above conditions also can be translated into:

$$\left.\begin{array}{l} 1 + K_v K_4 > 0 \\ K_1 + K_v(K_1 K_4 + K_2 K_3) > 0 \\ K_1' - K_1 - K_v K_2' K_3 > 0 \end{array}\right\} \qquad [7.2.28']$$

Correspondingly, a stability zone (in the plane (K_v, P^o), for given e_R^o, v^o) like that indicated in Figure 7.17 can be derived. In particular, any positive value can be accepted for K_v only for sufficiently small values of $|P^o|$, for which $K_2' K_3 < 0$.

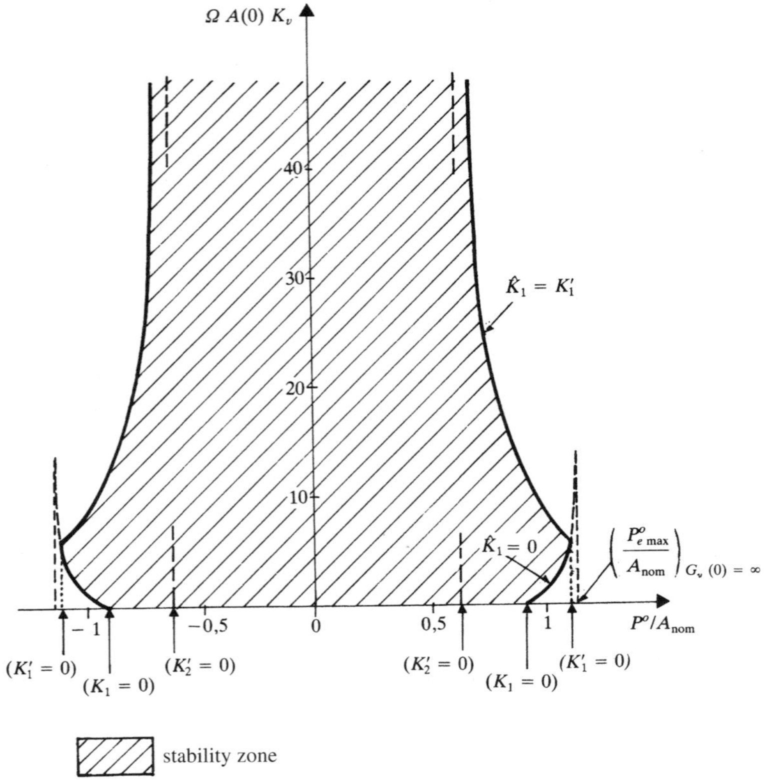

Figure 7.17. Stability zone in the plane (K_v, P^o) for the numerical example in Figure 7.14 and for $K_v > 0$.

Therefore, the voltage regulation — although improving the situation for what concerns the aperiodical-type instability — can significantly reduce the damping of electromechanical oscillation, or even cause (if $\hat{K}_1 > K_1'$) negative damping, i.e., *oscillatory-type instability*.

Conversely, the electromechanical phenomenon can cause voltage regulation instability. For this regulation it is possible, for small variations, to refer to a block diagram like that shown in Figure 6.5, where (with the adopted model) the transfer function $G(s) \triangleq \mathcal{L}(\Delta v)/\mathcal{L}(\Delta v_f)$ now includes:

- three poles, given by the solutions of Equation [7.2.11′] (characteristic roots in the absence of control);
- two imaginary conjugate zeros, corresponding to an undamped antiresonance, at the frequency $\nu = \sqrt{(K_1 + K_2K_3/K_4)/M} \cong \sqrt{\hat{K}_1/M}$;

and by imposing the regulation stability, conditions [7.2.28] can be derived.

If it is assumed that the resulting damping is small, it is possible to apply an approximation similar to Equation [7.2.15] and derive:

$$\hat{\zeta} \cong \frac{Im(K_e(\tilde{j}\hat{\nu}_o))}{2\hat{\nu}_o^2 M}$$

where $\hat{\nu}_o$ and $\hat{\zeta}$ are the electromechanical oscillation frequency and damping factor, in the presence of regulation.

From Equation [7.2.26′], it follows that:

$$\hat{\zeta} \cong \frac{\hat{T}}{2\hat{\nu}_o M(1 + \hat{\nu}_o^2 \hat{T}^2)}(K_1' - \hat{K}_1) \qquad [7.2.29]$$

Because of Equations [7.2.22] and the first of Equations [7.2.27], we also have:

$$\hat{\zeta} \cong \frac{\hat{T}}{2\hat{\nu}_o M(1 + \hat{\nu}_o^2 \hat{T}^2)} \frac{K_3}{K_4} \left(\frac{K_2}{1 + K_v K_4} - K_2' \right) \qquad [7.2.29']$$

where the last factor (depending on K_v) can become negative (at least for sufficiently large P^o) because of the regulation.

Similarly, by recalling Equation [7.2.16] it can be written:

$$\hat{D}_{eq} \triangleq 2\hat{\zeta}\hat{\nu}_o M \cong \mathcal{R}e(D_e(\tilde{j}\hat{\nu}_o))$$

with $D_e(s) = (K_e(s) - K_{eq})/s$ and $K_{eq} = \hat{\nu}_o^2 M$.

As it can be checked based on the Bode diagrams, in practice the frequency $\hat{\nu}_0$ differs slightly from $\nu_o \cong \sqrt{K_1'/M}$, which holds in the absence of regulation, provided that ν_v^0 is sufficiently smaller than ν_o (otherwise, $\hat{\nu}_o$ can be greater than ν_o). By assuming:

$$\frac{1}{\hat{T}} \cong \nu_v^o \ll \nu_o \cong \sqrt{\frac{K_1'}{M}} \cong \hat{\nu}_o$$

from Equation [7.2.29] it can be derived, similarly to the first of [7.2.17]:

$$\hat{\zeta} \cong \frac{1}{2\nu_o \hat{T}} \frac{K_1' - \hat{K}_1}{K_1'} = \frac{1}{2\nu_o T} \frac{K_1' - K_1 - K_v K_2' K_3}{K_1'} \tag{7.2.30}$$

or even, because of Equations [7.2.21]:

$$\hat{\zeta} \cong \frac{1}{2\nu_o} \frac{b_3(a_1 + K_v b_2)}{b_1} \tag{7.2.30'}$$

The variation in the damping factor, caused by regulation, is then:

$$\Delta\zeta \cong -\frac{1}{2\nu_o T} \frac{K_v K_2' K_3}{K_1'} = \frac{1}{2\nu_o} \frac{K_v b_2 b_3}{b_1} \tag{7.2.31}$$

and *is usually negative*. Similarly, it holds[(2)]:

$$\hat{D}_{eq} \cong \frac{M}{\hat{T}} \frac{K_1' - \hat{K}_1}{K_1'} = \frac{M}{T} \frac{K_1' - K_1 - K_v K_2' K_3}{K_1'}$$

$$= M \frac{b_3(a_1 + K_v b_2)}{b_1} \tag{7.2.32}$$

$$\Delta D_{eq} \cong -\frac{M}{T} \frac{K_v K_2' K_3}{K_1'} = M \frac{K_v b_2 b_3}{b_1} \tag{7.2.33}$$

It also should be remembered that at $P^o = 0$ — as it occurs for a synchronous compensator — we have instead $K_2 = K_2' = K_3 = \Delta K_e(s) = 0$, so that the voltage regulation changes neither the damping nor the frequency of the oscillation.

---

(2) Because of the hypothesis that the oscillation frequency remains unchanged, the expressions of $\Delta\zeta$ and D_{eq} also can be directly derived based on:

$$\Delta D_{eq} = 2\Delta\zeta\,\hat{\nu}_o M \cong \mathcal{R}e\left(\frac{\Delta K_e(\tilde{j}\hat{\nu}_o)}{\tilde{j}\hat{\nu}_o}\right) = \frac{Im(\Delta K_e(\tilde{j}\hat{\nu}_o))}{\hat{\nu}_o}$$

where $\Delta K_e(s)$ is given by Equation [7.2.24].

With *machine representation*, it can be observed that:

- the *third-order* model, defined in Section 4.2.2 and adopted here, can be acceptable at least for sufficiently large X_e (in this case, the effect of additional rotor circuits may be disregarded);
- with particular reference to the electromechanical oscillation, if $\hat{\nu}_0 T \gg 1$ we may have the approximations:

$$\begin{cases} \Delta P_e \cong K_1'(\Delta\delta - \Delta\alpha_e) + \dfrac{K_3}{sT}\Delta v_f + H_1' \dfrac{\Delta e_R}{e_R^o} \\ \Delta v \cong -K_2'(\Delta\delta - \Delta\alpha_e) + \dfrac{K_4}{sT}\Delta v_f + H_2' \dfrac{\Delta e_R}{e_R^o} \end{cases}$$

  according to Equations [7.2.20] and [7.2.21], with $\Delta\psi_f = \Delta v_f / s$; such approximations correspond to a simplified model:

  - again of the third order, with zero damping in the absence of control;
  - consistent with the model defined by Equations [4.2.9] and [4.2.10] (in pu, with machine equivalent reactances $\hat{x}_d'$ and x_q, respectively on the direct and quadrature axes), but with a possibly varying ψ_f ($\mathrm{d}\psi_f/\mathrm{d}t = v_f - v_f^o$);

- the further passage to a *second-order* model (see also Section 4.2.3), with proper equivalent reactances ($X_{d\,\mathrm{eq}}$, $X_{q\,\mathrm{eq}}$) on the machine direct and quadrature axes, can be somewhat justified for approximated analyses when the damping is negligible, e.g., in the cases of the:

  - synchronous compensator, for large X_e;
  - generator, when the positive damping effect of the field circuit (and other rotor circuits) is cancelled by the negative effect of regulations;

  whereas the possible assumption of the same equivalent reactance on the two axes of the machine, can appear acceptable only if the difference between $X_{d\,\mathrm{eq}}$ and $X_{q\,\mathrm{eq}}$ is actually small (also with respect to the external reactance X_e);

- furthermore, as a first approximation, the damping is sometimes considered as if it were of mechanical origin, by assuming (see also Fig. 7.4b) $D_m(s) = \text{constant} = D_{\mathrm{eq}}$; by doing so, the model remains of the second order.

(c) Effects of Additional Signals in Excitation Control

As seen, the damping of the electromechanical oscillation can become too small and even negative, above all because of the voltage regulation. On the other hand, for large values of the reactance X_e, the oscillation frequency $\hat{\nu}_o$ value is relatively small, and therefore the damping effect of the additional rotor circuits cannot be relied on.

To guarantee a satisfactory damping factor (e.g., not smaller than ~0.1) it can be thought to make use of "additional signals" in the excitation control, so as to realize a further dependence of ΔP_e on $\Delta\Omega$, as follows:

$$(\Delta P_e)_{\mathrm{add}} = \Delta D_{\mathrm{eq}} \Delta\Omega \qquad [7.2.34]$$

where ΔD_{eq} is a proper constant. According to Equation [7.2.16], it is possible to obtain an increase in the damping factor equal to $\Delta\zeta = \Delta D_{eq}/2\hat{\nu}_o M$, if the oscillation frequency $\hat{\nu}_o$ remains unchanged.

On the other hand, based on the diagram of Figure 7.13a, it is easy to show that, in response to the signal Δw, it results:

$$(\Delta P_e)_{add} = \frac{K_3}{K_4} F_v(s) \Delta w \quad [7.2.35]$$

(where $F_v(s)$ is the closed-loop transfer function defined by [7.2.25]), or further, if we accept Equation [7.2.25′]:

$$(\Delta P_e)_{add} \cong \frac{K_3/K_4}{1 + s/\nu_v^o} \Delta w \quad [7.2.35']$$

As a consequence, Equation [7.2.34] can be satisfied by assuming:

$$\Delta w = \frac{K_4}{K_3 F_v(s)} \Delta D_{eq} \Delta\Omega \cong \frac{K_4}{K_3} \Delta D_{eq} \left(1 + \frac{s}{\nu_v^o}\right) \Delta\Omega \quad [7.2.36]$$

i.e., by deriving a signal Δw sensitive both to the speed variation $\Delta\Omega$ and the acceleration $d\Omega/dt$.

However, it must be observed that usually, because of the small value of $|D_m(\tilde{j}\hat{\nu}_o)|$, the electromechanical oscillation is accompanied by negligible variations ΔP_m of the driving power. Thus, it is possible to assume, around the frequency $\hat{\nu}_0$:

$$-\Delta P_e \cong sM\Delta\Omega \quad [7.2.37]$$

By recalling Equation [7.2.36] (and without the approximation marks), it is possible to derive the following relationship:

$$\Delta w = K_\omega \Delta\Omega - K_p \Delta P_e \quad [7.2.36']$$

with:

$$\left.\begin{aligned} K_\omega &= \frac{K_4}{K_3} \Delta D_{eq} \\ K_p &= \frac{K_4}{K_3} \frac{\Delta D_{eq}}{M \nu_v^o} \end{aligned}\right\} \quad [7.2.38]$$

($K_\omega = M\nu_v^0 K_p$), according to the diagram of Figure 7.18.

The following are some possibilities for simplification:

- if $\hat{\nu}_o \gg \nu_v^o$, Equation [7.2.25′] can be substituted around the oscillation frequency by $F_v(s) \cong \nu_v^o/s$, so that Equation [7.2.36] reduces to:

$$\Delta w \cong \frac{K_4}{K_3} \Delta D_{eq} \frac{s}{\nu_v^o} \Delta\Omega$$

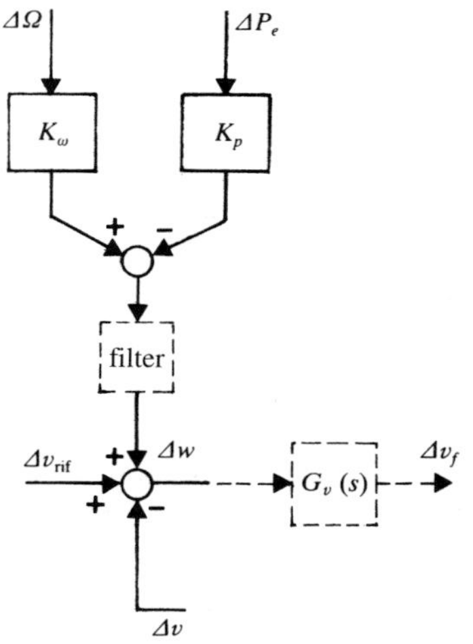

Figure 7.18. Block diagram of the linearized system: definition of the additional signals in excitation control (see also Fig. 7.13a).

and thus $K_\omega = 0$ in [7.2.36′] (that is a single additional signal, sensitive to active power);

- if, instead, $\hat{v}_o \ll v_v^o$, it can be similarly assumed $F_v(s) \cong 1$:

$$\Delta w \cong \frac{K_4}{K_3} \Delta D_{\text{eq}} \Delta \Omega$$

and thus $K_p = 0$ (a single additional signal that is sensitive to speed).

Before concluding, it is useful to point out the following.

(1) If the approximation $G_v(s) \cong K_v$ is accepted, the system is of the third order, with state variables $\Delta\Omega$, $\Delta\delta$, and $\Delta\psi_f$ (ψ_f is the machine field flux). In general terms, the system dynamic behavior can be improved (particularly by "shifting" its characteristic roots) by proper feedbacks starting from the state variables. On the other hand, in the absence of disturbances on the network it results:

$$\Delta P_e = K_1' \Delta\delta + \frac{K_3}{T} \Delta\psi_f$$

(recall Equations [7.2.20] and [7.2.21]), so that in practice (as $\Delta\psi_f$ is not directly measurable) the feedbacks can be conveniently achieved from $\Delta\Omega$, $\Delta\delta$ and ΔP_e.

After these premises, it can be recognized (based on the diagram of Figure 7.13a, assuming K_3 and K_4 positive, and recalling Equation [7.2.35′]) that:

- feedback of the type $\Delta w = K_\omega \Delta \Omega$, with $K_\omega > 0$, increases $\hat{K}_1'$;
- feedback of the type $\Delta w = -K_p \Delta P_e$, with $K_p > 0$, reduces $\hat{K}_1$ and $\hat{T}$;
- feedback of the type $\Delta w = K_\delta \Delta \delta$, with $K_\delta > 0$, increases $\hat{K}_1$.

Therefore the convenience of using additional signals sensitive to speed and active power is confirmed, particularly to shift the pair of complex conjugate roots that correspond to the electromechanical oscillation. Instead, a signal sensitive to the phase-shift δ, apart from the difficulties for measuring it, would not appear as useful, except for the possibility of properly adjusting the shift of the remaining (real) characteristic root.

(2) It is also necessary to consider that the parameters of the linearized system, as considered so far, actually vary with the operating point.
However, even assuming that K_ω and K_p are constants, it is possible to obtain a significant increase of the stability zone in the plane $((P/v^2)^o, (Q/v^2)^o)$, as illustrated in the qualitative example of Figure 7.19.
On the other hand, at low loads:
- the gain K_3/K_4 that appears in Equation [7.2.35] is relatively small, so that the effect of Δw is reduced;
- the gain K_2' that appears in Equations [7.2.29′] and [7.2.31] is small and can even be negative, so that the destabilizing effect of voltage regulation may disappear;

as a consequence, the additional signals are often disconnected when P^o falls below a given value (e.g., 30% of the nominal power).
Furthermore, the gain K_3 has the same sign as P^o (as well as v_d^o; note the second of Equations [7.2.18]), so that:
- if $P^o = 0$ $(K_3 = 0)$—as in the case of a *synchronous compensator*—the additional signals cannot have effect (on the other hand, it is already seen that the voltage regulation does not modify the damping);

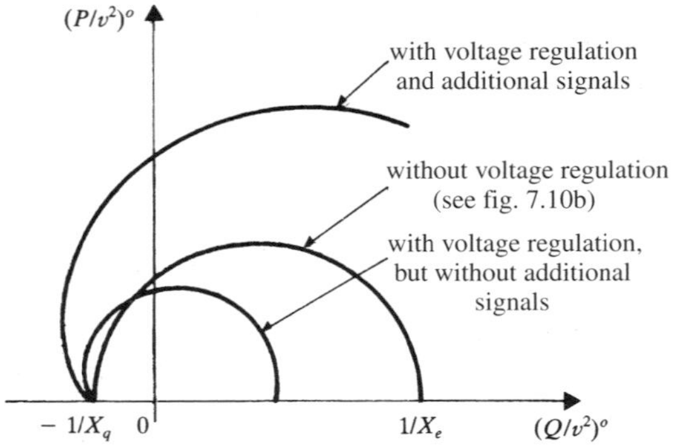

Figure 7.19. Stability zone in the plane $((P/v^2)^o, (Q/v^2)^o)$ (qualitative example for the round rotor).

- if $P^o < 0$ ($K_3 < 0$)—as in the case of a *pumping hydroelectric unit*—it is necessary to change the sign of K_ω, K_p.

(3) If some interaction with the torsional phenomena is possible, it is convenient that the speed measurement (for the feedback) is performed at a place in the shaft for which the torsional speed oscillations are sufficiently small. In general, the speed measurement can be avoided by recalling Equation [7.2.37] and approximating [7.2.36′] by a relationship of the form:

$$\Delta w = -\left(\frac{K_\omega T_\omega}{M(1 + sT_\omega)} + K_p\right)\Delta P_e$$

with $T_\omega \gg 1/\hat{\nu}_o$.

(4) The additional signals cause a disturbance to the voltage regulation, as Δw is equivalent (Fig. 7.13a) to a variation in the reference Δv_{rif} (specifically, the dependency of Δw on Δv_f, through $\Delta\Omega$ and ΔP_e, is equivalent to a modification in the transfer function $G_v(s)$, as far as the voltage regulation is concerned). By interposing a proper band-pass filter (Fig. 7.18), it is possible to operate so that Δw is practically active only for a narrow band of frequencies around the electromechanical oscillation frequency $\hat{\nu}_o$ (where it is possible to realize the desired damping effect). Considering the low-pass effect of $F_v(s)$, it can be sufficient to use a filter with a transfer function of the type $s\tau/(1 + s\tau)$, i.e., a high-pass filter, with $\hat{\nu}_o\tau \gg 1$ (e.g., τ in the order of seconds).

(5) Usually, the additional signals are limited in magnitude to reduce possible destabilizing effects in case of large perturbations (see Section 7.3).

(6) In real cases with several machines, the additional signals can be usefully adopted in a similar way, to improve the damping of one or more oscillations (see Section 8.5.2).

7.2.3. Effects of an Intermediate Load

With reference to the system in Figure 7.1, again disregard, for simplicity:

- the line capacitances (Fig. 7.20);
- the armature resistance;
- the magnetic saturation;
- the additional rotor circuits;
- the torsional phenomena.
- the dynamics of the inductive and capacitive elements (in Fig. 7.20, the reactance X_{e1} represents the transformer and the first section of the line, whereas the reactance X_{e2} represents the second section of the line; furthermore, G and B are the load conductance and susceptance, possibly variable as specified below);

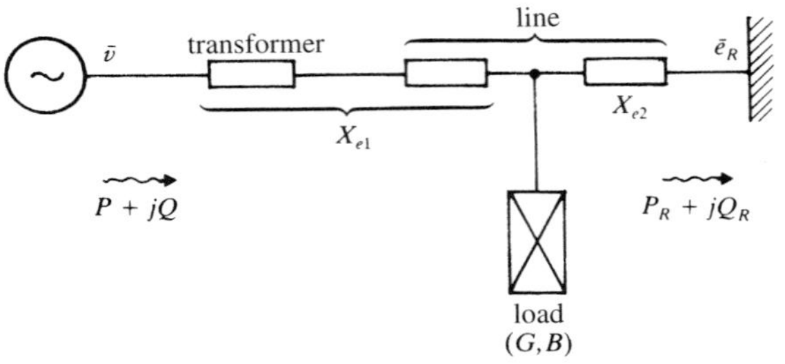

Figure 7.20. Case with intermediate load: system under examination.

and consider the following cases:

- system without regulations;
- load voltage regulated through the machine excitation;
- load voltage regulated through an on-load tap-changing transformer or through a static compensator.

For small variations, the block diagram of Figure 7.21 can be generically derived (with respect to the diagram of Figure 7.13a, the inputs ΔG and ΔB are added; further, the presence of the load modifies the different parameters T, K_1, K_2, etc., as it will be shown).

More precisely:

- Equation [7.2.19] still holds,
- Equations [7.2.20] can be generalized by the following equations:

$$\left.\begin{aligned} R_f \Delta i_f &= a_1(\Delta\delta - \Delta\alpha_e) + a_2 \Delta\psi_f + a_3 \frac{\Delta e_R}{e_R^o} + a_4 \Delta G + a_5 \Delta B \\ \Delta P_e &= b_1(\Delta\delta - \Delta\alpha_e) + b_3 \Delta\psi_f + c_1 \frac{\Delta e_R}{e_R^o} + d_1 \Delta G + e_1 \Delta B \\ \Delta v &= b_2(\Delta\delta - \Delta\alpha_e) + b_4 \Delta\psi_f + c_2 \frac{\Delta e_R}{e_R^o} + d_2 \Delta G + e_2 \Delta B \end{aligned}\right\} \quad [7.2.39]$$

- by eliminating $R_f \Delta i_f$ and $\Delta\psi_f$, relationships like [7.2.21] can be obtained and similarly:

$$\left.\begin{aligned} D_1 &= d_1 - a_4 b_3 T, \quad D_1' = d_1, \quad E_1 = e_1 - a_5 b_3 T, \quad E_1' = e_1 \\ D_2 &= d_2 - a_4 b_4 T, \quad D_2' = d_2, \quad E_2 = e_2 - a_5 b_4 T, \quad E_2' = e_2 \end{aligned}\right\} \quad [7.2.40]$$

whereas we have the following identities (corresponding to Equations [7.2.22]):

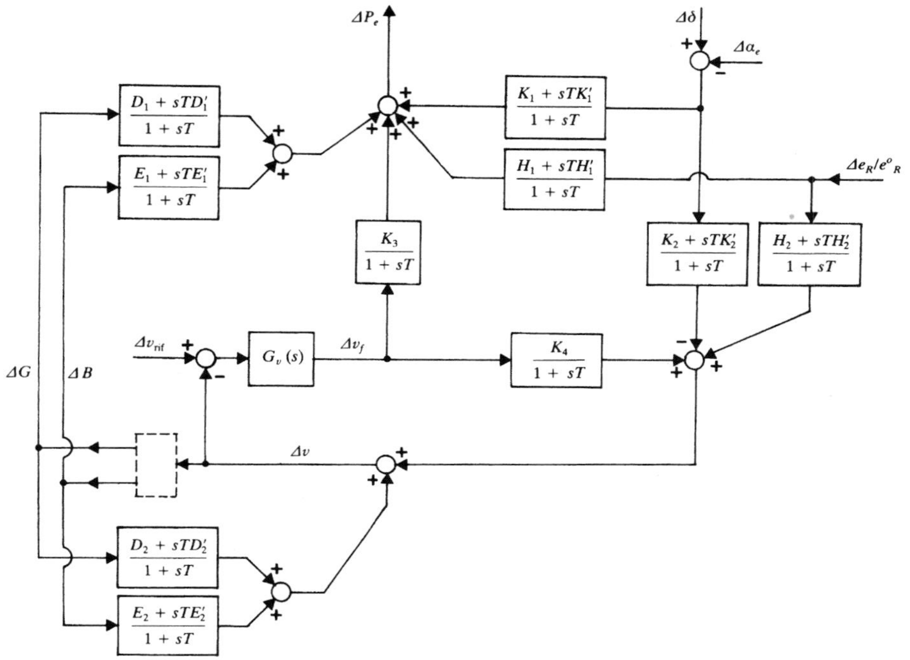

Figure 7.21. Block diagram of the linearized system in the presence of control and with an intermediate load (and without additional rotor circuits).

$$\frac{K_1' - K_1}{K_2 - K_2'} = \frac{H_1' - H_1}{H_2' - H_2} = \frac{D_1' - D_1}{D_2' - D_2} = \frac{E_1' - E_1}{E_2' - E_2} = \frac{K_3}{K_4} \quad [7.2.41]$$

For simpler notation, in the following it will be assumed that $X_{e1} = 0$, $X_{e2} = X_e$, as in the case of load at machine terminals. Generally, it would be necessary to add X_{e1} to the machine reactances and to substitute X_e by X_{e2}.

For the time constant T, it can be deduced:

$$T = \frac{\hat{X}_d' N' \hat{T}_{do}'}{X_d N} \quad [7.2.42]$$

where:

$$\left.\begin{aligned} N &\triangleq \left(\frac{1}{X_e} + \frac{1}{X_d} - B^o\right)\left(\frac{1}{X_e} + \frac{1}{X_q} - B^o\right) + G^{o2} \\ N' &\triangleq \left(\frac{1}{X_e} + \frac{1}{\hat{X}_d'} - B^o\right)\left(\frac{1}{X_e} + \frac{1}{X_q} - B^o\right) + G^{o2} \end{aligned}\right\} \quad [7.2.43]$$

Furthermore, in terms of P^o, Q^o, v^o (and of G^o, B^o), instead of Equations [7.2.14] and [7.2.18] it can be shown that:

$$\left.\begin{aligned}
K_1 &= \frac{1}{N}\left\{\left(\frac{1}{X_e}+\frac{1}{X_d}-B^o\right)\frac{Q'Q''-P^{o2}}{v^{o2}}+G^{o2}Q'\right.\\
&\quad\left.-\left(\frac{1}{X_q}-\frac{1}{X_d}\right)\left[\left(\frac{1}{X_e}+\frac{1}{X_q}-B^o\right)^2+G^{o2}\right]\frac{v^{o2}P^{o2}}{P^{o2}+Q'^2}\right\}\\
K_2 &= \frac{1}{N}\left\{\left(\frac{1}{X_e}+\frac{1}{X_d}-B^o\right)\frac{P^o}{v^o}-G^o\frac{Q'}{v^o}\right.\\
&\quad\left.+\left(\frac{1}{X_q}-\frac{1}{X_d}\right)\left[\left(\frac{1}{X_e}+\frac{1}{X_q}-B^o\right)Q'+G^oP^o\right]\frac{v^oP^o}{P^{o2}+Q'^2}\right\}\\
K_3 &= \frac{X_{md}}{R_fX_dN}\left\{\left[\left(\frac{1}{X_e}+\frac{1}{X_q}-B^o\right)^2+G^{o2}\right]v_d^o+G^o\frac{\sqrt{P^{o2}+Q'^2}}{v^o}\right\}\\
K_4 &= \frac{X_{md}}{R_fX_dN}\left[\left(\frac{1}{X_e}+\frac{1}{X_q}-B^o\right)v_q^o+G^ov_d^o\right]\frac{1}{v^o}\\
H_1 &= P^o-\frac{1}{N}\left\{G^o\frac{P^{o2}+Q'^2}{v^{o2}}+\left(\frac{1}{X_q}-\frac{1}{X_d}\right)\left[\left(\frac{1}{X_e}+\frac{1}{X_q}-B^o\right)P^o-G^oQ'\right.\right.\\
&\quad\left.\left.-\left(\left(\frac{1}{X_e}+\frac{1}{X_q}-B^o\right)^2+G^{o2}\right)\frac{v^{o2}P^oQ'}{P^{o2}+Q'^2}\right]\right\}\\
H_2 &= v^o-\frac{1}{N}\left[\left(\frac{1}{X_e}+\frac{1}{X_q}-B^o\right)Q'+G^oP^o\right]\left[\frac{1}{v^o}-\left(\frac{1}{X_q}-\frac{1}{X_d}\right)\frac{v^oQ'}{P^{o2}+Q'^2}\right]
\end{aligned}\right\}\quad [7.2.44]$$

and moreover:

$$D_1 = \frac{1}{N}\left\{\left(\frac{1}{X_e}+\frac{1}{X_d}-B^o\right)Q'-G^oP^o-\left(\frac{1}{X_q}-\frac{1}{X_d}\right)\left[\left(\frac{1}{X_e}+\frac{1}{X_q}-B^o\right)P^o+G^oQ'\right]\frac{v^{o2}P^o}{P^{o2}+Q'^2}\right\}$$

$$D_2 = -\frac{v^o}{N}\left[\left(\frac{1}{X_q}-\frac{1}{X_d}\right)\frac{P^oQ'}{P^{o2}+Q'^2}+G^o\right]$$

$$E_1 = \frac{1}{N}\left\{\left(\frac{1}{X_e}+\frac{1}{X_d}-B^o\right)P^o+G^oQ'+\left(\frac{1}{X_q}-\frac{1}{X_d}\right)\left[\left(\frac{1}{X_e}+\frac{1}{X_q}-B^o\right)Q'-G^oP^o\right]\frac{v^{o2}P^o}{P^{o2}+Q'^2}\right\}$$

$$E_2 = \frac{1}{N}\left[\left(\frac{1}{X_e}+\frac{1}{X_d}-B^o\right)P^{o2}+\left(\frac{1}{X_e}+\frac{1}{X_q}-B^o\right)Q'^2\right]\frac{v^o}{P^{o2}+Q'^2}$$

(the expressions of K_1', K_2', H_1', H_2', D_1', D_2', E_1', E_2' can be obtained from those of K_1, K_2, etc., by simply replacing X_d by $\hat{X}_d'$ and N by N'). In these equations it is assumed:

$$\begin{cases} Q' \triangleq Q^o + \dfrac{v^{o2}}{X_q} \\ Q'' \triangleq v^{o2}\left(\dfrac{1}{X_e} - B^o\right) - Q^o \end{cases}$$

and further:

$$\begin{cases} v_d^o = v^o \sin\delta_i^o = \dfrac{v^o P^o}{\sqrt{P^{o2} + Q'^2}} \\ v_q^o = v^o \cos\delta_i^o = \dfrac{v^o Q'}{\sqrt{P^{o2} + Q'^2}} \end{cases}$$

Note that, with respect to the case without load (Fig. 7.10a), it now holds:

$$\begin{cases} \dfrac{v^o e_R^o}{X_e} \sin\delta_e^o = P^o - G^o v^{o2} \\ \dfrac{v^o e_R^o}{X_e} \cos\delta_e^o = v^{o2}\left(\dfrac{1}{X_e} - B^o\right) - Q^o = Q'' \end{cases}$$

and thus:

$$\begin{cases} e_R^o = \dfrac{X_e}{v^o}\sqrt{(P^o - G^o v^{o2})^2 + Q''^2} \\ \tan\delta_e^o = \dfrac{P^o - G^o v^{o2}}{Q''} \end{cases}$$

whereas it still holds:

$$\tan\delta_i^o = \frac{P^o}{Q^o + \dfrac{v^{o2}}{X_q}} = \frac{P^o}{Q'}$$

If, instead, P^o, e_R^o, v^o are assigned, in the previous expressions we have to note that:

$$\begin{cases} Q''^2 = \left(\dfrac{e_R^o v^o}{X_e}\right)^2 - (P^o - G^o v^{o2})^2 \\ Q' = v^{o2}\left(\dfrac{1}{X_e} + \dfrac{1}{X_q} - B^o\right) - Q'' \end{cases}$$

whereas:

$$Q^o = Q' - \frac{v^{o2}}{X_q} = v^{o2}\left(\frac{1}{X_e} - B^o\right) - Q''$$

Moreover, for given e_R^o and v^o:

- K_1, K_1', and K_4 are no longer "even" functions of P^o, and K_2, K_2', and K_3 are no longer "odd" functions of P^o;

- for $P^o = 0$ we no longer have $K_2 = K_2' = K_3 = 0$, so that the voltage regulation loop remains coupled with the electromechanical loop; furthermore, K_1' is no longer equal to K_1 (for $P^o = 0$, we have $K_1' - K_1 = (-1/(NN'))(1/\hat{X}_d' - 1/X_d)(G^o Q'/v^o)^2$).

Specifically, from the diagrams in Figure 7.22 (compared with those without load, in Fig. 7.14) it can be concluded that:

- for very small $|P^o|$, $K_1' < K_1$ holds;
- for $P^o > 0$: K_2 is negative for small P^o, and K_2' changes its sign for a value of P^o larger than that without load;
- for $P^o < 0$: K_2' can be always negative, whereas K_3 remains positive for sufficiently small $|P^o|$.

As far as the operating point is concerned, note that:

- at the same values of P^o, e_R^o, v^o, an increase in G^o implies a reduction in $P_R^o = P^o - G^o v^{o2}$ (Fig. 7.20) and possibly $P_R^o < 0$ (i.e., the load is also supplied by the network);

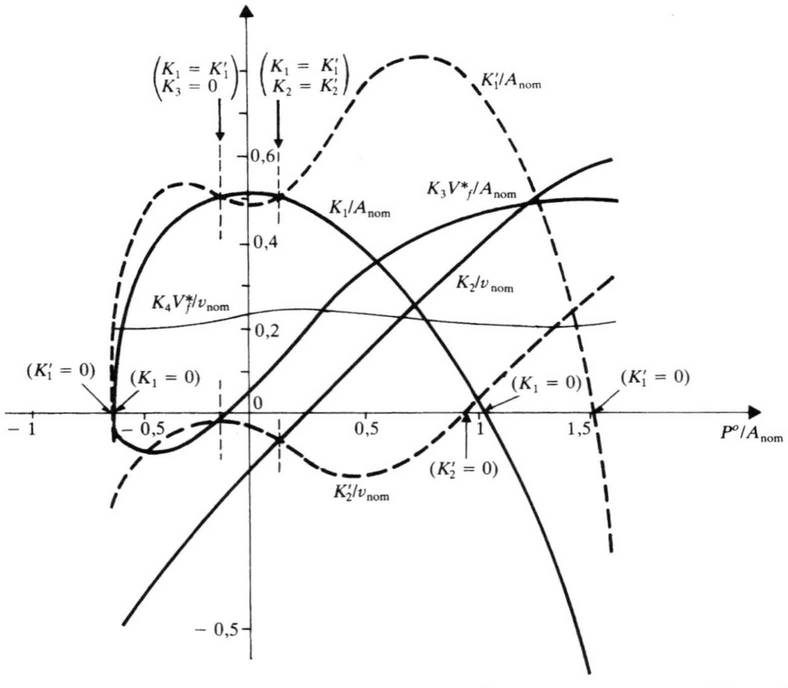

Figure 7.22. Dependence of parameters K_1, K_1', K_2, etc. on P^o, with an intermediate load, assuming that $e_R^o = 0.8v_{nom}$, $v^o = v_{nom}$, $x_d = x_q = 2$, $\hat{x}_d' = 0.2$, $x_e = 0.7$ as in Figure 7.14, and $B^o = 0$, $G^o = 0.5A_{nom}/v_{nom}^2$.

- without the load, a large value of X_e has no practical interest, because it would imply an underutilization of the machine; instead, in the presence of the load, X_e can be large (in the case $P_R^o = 0$, the connection to the network may be of interest as a reserve to supply the load) and thus *the electromechanical oscillation frequency may be low*.

In the absence of regulation, the characteristic Equation [7.2.11′] and the stability conditions [7.2.12] can be found again.

The condition $T > 0$ is usually satisfied (recall Equations [7.2.42] and [7.2.43], where both N and N' are usually positive). Furthermore, except for very small $|P^o|$, the condition $K_1' > K_1$ is also satisfied, so that the only condition to consider, as in the "basic case" without load, is $K_1 > 0$ (otherwise, there is an aperiodical-type instability), for which the slope of the "static" characteristic (P_e^o, δ^o) at the operating point must be positive (Fig. 7.23).

If the voltage at the load is regulated by the machine excitation (see also Section 6.2.2a2, with voltage regulation at a downstream node), the stability implies conditions [7.2.28] (or equivalently [7.2.28′]), of which $K_1' > \hat{K}_1$ is usually the more stringent one (otherwise, there is an oscillatory-type instability).

In comparison to the case without load, the stability zone in the plane (K_v, P^o) is modified according to Figure 7.24 (recall Fig. 7.17). Specifically, the operation as generator ($P^o > 0$) with arbitrarily large K_v becomes possible for a range of P^o larger than that without load (with $K_2' < 0$ and $K_3 > 0$).

If the voltage at the load is regulated by an on-load tap-changing transformer or by a static compensator (Fig. 7.25a and b, respectively), the dependence of ΔG and ΔB on Δv also should be considered (see the dashed-line block in the diagram of Fig. 7.21).

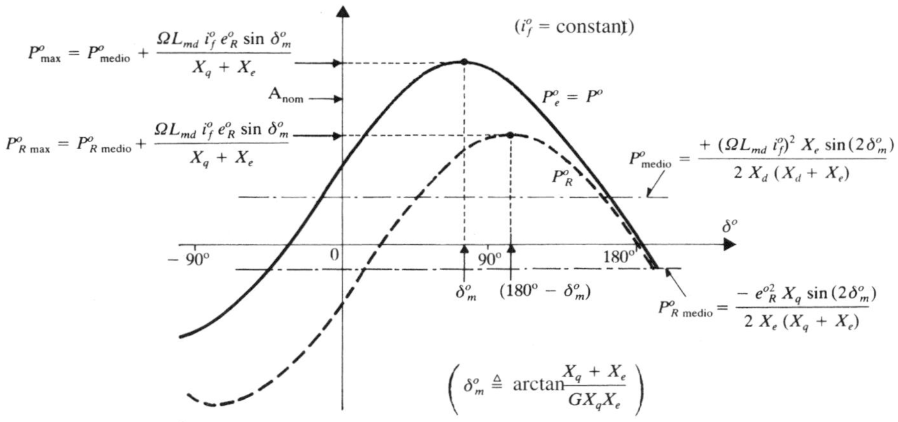

Figure 7.23. Dependence of P_e^o and P_R^o on δ^o, for constant i_f^o (assuming $e_R^o = 0.8v_{\text{nom}}$, $x_d = x_q = 2$, $x_e = 0.7$, $B^o = 0$, $G^o = 0.5A_{\text{nom}}/v_{\text{nom}}^2$, and $i_f^o = 3I_f^* = 3v_{\text{nom}}/X_{md}$).

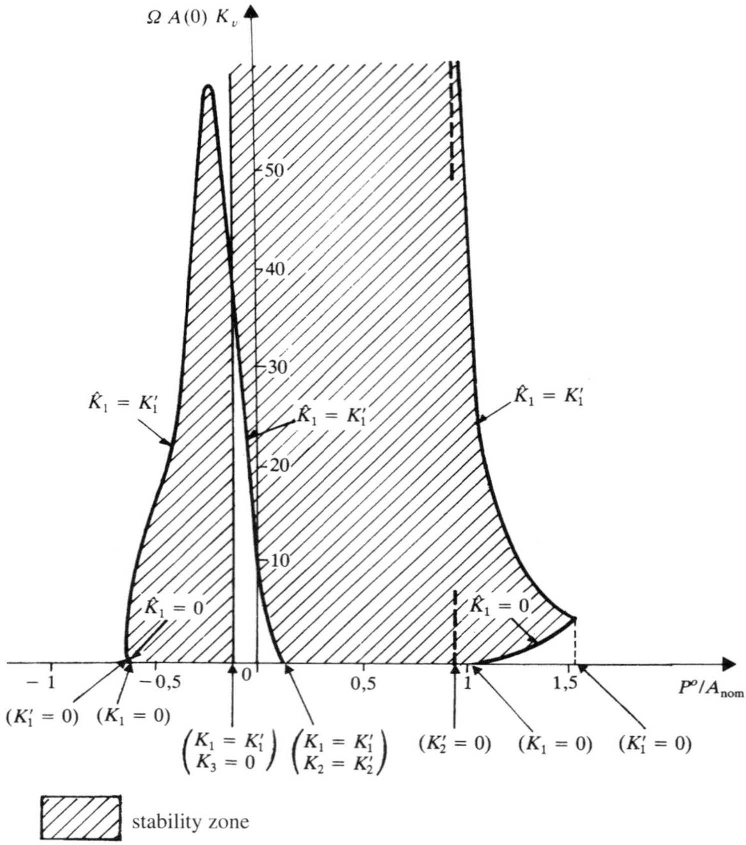

Figure 7.24. Stability zone in the plane (K_v, P^o), with an intermediate load and $K_v > 0$ (numerical example as in Fig. 7.22).

By letting:

$$\left.\begin{aligned} \Delta G &= -R_G(s)\Delta v \\ \Delta B &= -R_B(s)\Delta v \end{aligned}\right\} \qquad [7.2.45]$$

and recalling Equations [7.2.41], it can be derived that:

$$\begin{aligned} K_e(s) &\triangleq \frac{\mathcal{L}(\Delta P_e)}{\mathcal{L}(\Delta\delta)} = \\ &= \frac{[K_1(1 + K_4 G_v + D_2 R_G + E_2 R_B) + K_2(K_3 G_v + D_1 R_G + E_1 R_B)]}{(1 + K_4 G_v + D_2 R_G + E_2 R_B) + sT(1 + D_2' R_G + E_2' R_B)} \\ &\quad + \frac{sT[K_1'(1 + D_2' R_G + E_2' R_B) + K_2'(D_1' R_G + E_1' R_B)]}{(1 + K_4 G_v + D_2 R_G + E_2 R_B) + sT(1 + D_2' R_G + E_2' R_B)} \qquad [7.2.46] \end{aligned}$$

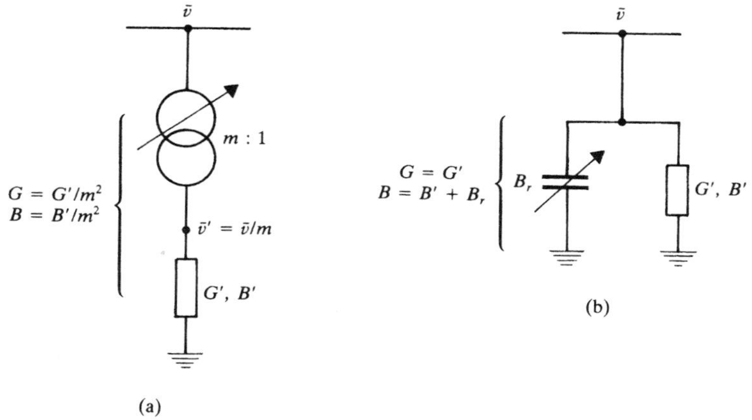

Figure 7.25. Load voltage regulation: (a) by an on-load tap-changing transformer; (b) by a static compensator.

(If the load voltage is regulated only as specified above, i.e., if the machine excitation is kept constant, it must be assumed that $G_v = 0$.)

Assuming, as specified, $G_v(s) \cong K_v$ and supposing for simplicity that the electromechanical oscillation frequency remains practically unchanged with respect to the case $R_G = R_B = 0$, and equal to:

$$\hat{\nu}_o \cong \sqrt{\frac{K_1'}{M}} \gg \frac{1}{\hat{T}}$$

(with $\hat{T} \triangleq T/(1 + K_v K_4)$), the effect that the above regulations have on the damping can be translated at a first approximation into:

$$\Delta D_{\text{eq}} = 2\Delta\zeta\,\hat{\nu}_o M \cong \mathcal{Re}(\Delta D_e(\tilde{j}\hat{\nu}_o)) = \frac{Im(\Delta K_e(\tilde{j}\hat{\nu}_o))}{\hat{\nu}_o} \qquad [7.2.47]$$

where:

$$\Delta K_e(s) \triangleq K_e(s) - (K_e(s))_{R_G = R_B = 0}$$

With the *on-load tap-changing transformer* (Fig. 7.25a), by neglecting the transformer reactance it follows:

$$v = mv', \quad G = \frac{G'}{m^2}, \quad B = \frac{B'}{m^2}$$

and for small variations, by assuming $\Delta m = (K_m/s)\Delta v'$ (see Section 6.3), it can be derived that:

$$R_G(s) = \frac{2G^o/v^o}{1 + s\tau}, \quad R_B(s) = \frac{2B^o/v^o}{1 + s\tau}$$

with $\tau \triangleq m^{o2}/(K_m v^o)$.

With the *static compensator* (Fig. 7.25b) it holds:

$$G = G', \quad B = B' + B_r$$

and, by assuming $\Delta B_r = -(K_B/s)\Delta v$ (see Section 6.4), it can be directly derived:

$$R_G(s) = 0, \quad R_B(s) = \frac{K_B}{s}$$

If $G' = 0$, we are again in the case for which the compensator has only the task of "voltage support."

By applying Equation [7.2.47], the diagrams of Figure 7.26 can be obtained under the indicated assumptions.

7.2.4. Case with Series-Compensated Line

If the line is series-compensated by the interposition of a capacitive element (see Fig. 7.27a), the "LC dynamics" cannot generally be neglected, primarily because of its interaction with the torsional phenomena on the unit shaft.

To account for the LC dynamics, it is convenient (as already specified) to apply the Park's transformation with a "network" angular reference, equal to the electrical angular position θ_s of a fictitious rotor, rotating at the constant speed $d\theta_s/dt = \Omega_s$ (and we assume $\Omega_s = \Omega_{nom}$).

On the other hand, useful indications can be obtained — avoiding to burden the analytical treatment too much — representing the electromagnetic part of the

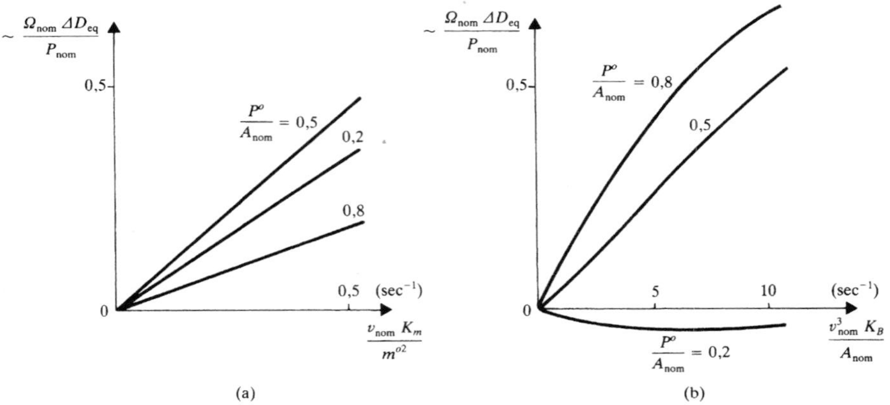

Figure 7.26. Damping contribution of the load voltage regulation (see text), obtained (a) by an on-load tap-changing transformer, (b) by a static compensator, for the numerical example in Figure 7.22, and by further assuming $\Omega_{nom} = 314$ rad/sec, $\hat{T}'_{do} = 7$ sec, $T_a = \Omega_{nom}M/P_{nom} = 8$ sec, $G_v = 0$. (Note that $P^o/A_{nom} = 0.2, 0.5, 0.8$, respectively, means $P^o_R/A_{nom} = -0.3, 0, +0.3$).

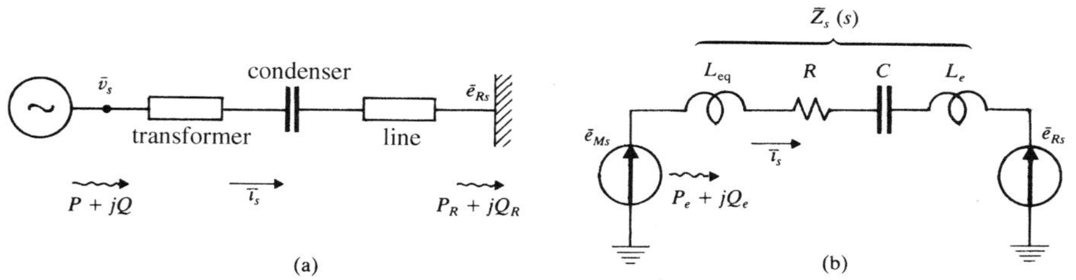

Figure 7.27. Case with series-compensated line: (a) system under examination; (b) equivalent circuit (with a simplified model of the machine; see text).

machine (and its voltage regulation) by a purely algebraic model, e.g., substituting Equations [7.1.9] by equations like:

$$\left.\begin{aligned} \psi_d &= (\psi_d)_{i_d=0} - L_{eq} i_d \\ \psi_q &= -L_{eq} i_q \end{aligned}\right\} \qquad [7.2.48]$$

where $(\psi_d)_{i_d=0}$ and L_{eq} (equivalent inductance) have proper constant values. With such assumption, in the "basic case" the damping of the electromechanical oscillation would be zero; to counteract this, as a first approximation it is possible to introduce a proper coefficient D_{eq} in the model of the mechanical part of the machine, according to the end of Section 7.2.2b.

By adopting the subscript "s" for the Park's vectors obtained by the "network" reference, and by accounting in general terms for a total series impedance $\overline{Z}_s(s)$ (Fig. 7.27b), it can be derived:

$$\bar{e}_{Ms} = \overline{Z}_s(s)\bar{\imath}_s + \bar{e}_{Rs}$$

where, by recalling Equations [7.1.1] and [7.1.2] and assuming $e_R = E_R/m_o$ (m_o is the transformation ratio of the transformer in Fig. 7.1):

$$\begin{aligned} \bar{e}_{Ms} &= j\Omega(\psi_d)_{i_d=0}\,\epsilon^{j\delta} \\ \bar{e}_{Rs} &= j e_R \epsilon^{j\alpha_e} \end{aligned}$$

and furthermore:

$$C_e \Omega_{m\,\text{nom}} = \Omega_{\text{nom}}(\psi_d)_{i_d=0}\, i_q = e_M^o(-i_{ds}\sin\delta + i_{qs}\cos\delta)$$

with $e_M^0 \triangleq \Omega_{\text{nom}}(\psi_d)_{i_d=0}$, whereas it results $\bar{\imath} = \bar{\imath}_s \epsilon^{-j\delta}$, and thus:

$$\begin{cases} i_d = i_{ds}\cos\delta + i_{qs}\sin\delta \\ i_q = -i_{ds}\sin\delta + i_{qs}\cos\delta \end{cases}$$

For small variations, by letting:

$$\overline{Z}_s(s) \triangleq r(s) + jx(s)$$

it can then be derived, for $\alpha_e^0 = 0$:

$$\begin{aligned}\Delta C_e \Omega_{m\,\text{nom}} = {} & \frac{e_M^o e_R^o}{r^2(s) + x^2(s)} \\ & \cdot \left[(-r(s)\cos\delta^o + x(s)\sin\delta^o)\frac{\Delta e_r}{e_R^o} - (r(s)\sin\delta^o + x(s)\cos\delta^o)\Delta\alpha_e \right] \\ & + \left[-Q_e^o + \frac{e_M^{o2}}{r^2(s) + x^2(s)} \left(\frac{sr(s)}{\Omega_{\text{nom}}} + x(s) \right) \right] \Delta\delta \end{aligned} \quad [7.2.49]$$

where:

$$Q_e^o \triangleq e_M^o i_d^o$$

is the reactive power delivered by the emf. $\overline{e}_{Ms}$ (Fig. 7.27b).

In the present case, by denoting $L \triangleq L_{\text{eq}} + L_e$, and considering more generally a series resistance R (possibly including the machine armature resistance) it results (see Fig. 7.27b and Section 5.2):

$$\overline{Z}_s(s) = R + (s + j\Omega_{\text{nom}})L + \frac{1}{(s + j\Omega_{\text{nom}})C}$$

and thus:

$$\begin{cases} r(s) = R + sL + \dfrac{s}{(s^2 + \Omega_{\text{nom}}^2)C} \\ x(s) = \Omega_{\text{nom}}L - \dfrac{\Omega_{\text{nom}}}{(s^2 + \Omega_{\text{nom}}^2)C} \end{cases}$$

Substituting the above into Equation [7.2.49], it follows:

$$K_e(s) \triangleq \frac{\mathcal{L}(\Delta C_e \Omega_{m\,\text{nom}})}{\mathcal{L}(\Delta\delta)} = -Q_e^o + \frac{e_M^{o2}}{\Omega_{\text{nom}}L}\frac{n(s)}{d(s)} \quad [7.2.50]$$

where $n(s)$ and $d(s)$ are polynomials of the fourth degree and precisely:

$$\left.\begin{aligned} n(s) &\triangleq s^4 + \frac{R}{L}s^3 + \left(2\Omega_{\text{nom}}^2 + \frac{1}{LC}\right)s^2 + \Omega_{\text{nom}}^2\frac{R}{L}s + \Omega_{\text{nom}}^2\left(\Omega_{\text{nom}}^2 - \frac{1}{LC}\right) \\ d(s) &\triangleq s^4 + \frac{2R}{L}s^3 + \left(2\left(\Omega_{\text{nom}}^2 + \frac{1}{LC}\right) + \frac{R^2}{L^2}\right)s^2 + \frac{2R}{L}\left(\Omega_{\text{nom}}^2 + \frac{1}{LC}\right)s \\ &\quad + \left(\Omega_{\text{nom}}^2 - \frac{1}{LC}\right)^2 + \left(\frac{\Omega_{\text{nom}}R}{L}\right)^2 \end{aligned}\right\} \quad [7.2.51]$$

For $R = 0$, the results are simply:

$$\left.\begin{aligned} n(s) &= (s^2 + \nu_1'^2)(s^2 + \nu_2'^2) \\ d(s) &= (s^2 + \nu_1^2)(s^2 + \nu_2^2) \end{aligned}\right\} \qquad [7.2.51']$$

with:

$$\nu_1'^2, \nu_2'^2 = \Omega_{\text{nom}}^2 + \frac{1}{2LC}\left(1 \mp \sqrt{1 + 8\Omega_{\text{nom}}^2 LC}\right) \qquad [7.2.52]$$

$$\nu_1^2, \nu_2^2 = \left(\Omega_{\text{nom}} \mp \frac{1}{\sqrt{LC}}\right)^2 \qquad [7.2.53]$$

Thus, for $R = 0$, the transfer function $K_e(s)$ exhibits two pairs of imaginary conjugate poles, which correspond to *two* undamped *resonances*, respectively, at the frequencies ν_1 and ν_2 (positive solutions of Equation [7.2.53]).

In particular, the former of these resonances — at the frequency ν_1, lower than the synchronous frequency Ω_{nom} — defines a "subsynchronous" oscillation (it is assumed that $1/\sqrt{LC} < 2\Omega_{\text{nom}}$, so that $\nu_1 = |\Omega_{\text{nom}} - 1/\sqrt{LC}| < \Omega_{\text{nom}}$).

Furthermore, the zeros of $K_e(s)$ are given by the equation:

$$0 = -Q_e^o d(s) + \frac{e_M^{o2}}{\Omega_{\text{nom}} L} n(s) \qquad [7.2.54]$$

thus, they depend on $\alpha \triangleq Q_e^o \Omega_{\text{nom}} L / e_M^{o2}$, according to Figure 7.28. It is assumed that $1/\sqrt{LC} < \Omega_{\text{nom}}$, thus:

$$\frac{1}{\Omega_{\text{nom}} C} < \Omega_{\text{nom}} L \qquad [7.2.55]$$

This means that the capacitive reactance is not sufficient to compensate the overall inductive reactance.

In particular, note that:

- it results:

$$\nu_1 < \nu_1' < \Omega_{\text{nom}} < \nu_2' < \nu_2$$

  (where ν_1' and ν_2' are the positive solutions of Equation [7.2.52]);
- for $Q_e^o = 0$, there are two pairs of conjugate imaginary zeros that correspond to two undamped antiresonances, at the frequencies ν_1' and ν_2';
- for $Q_e^o < 0$, there are again two undamped antiresonances (at frequencies, respectively, between ν_1' and ν_1, and between ν_2' and ν_2);
- instead, for $Q_e^o > 0$, the undamped antiresonances are kept only for values of Q_e^o sufficiently small (the corresponding frequencies are between ν_1' and ν_2') or relatively large (with frequencies beyond the interval (ν_1, ν_2));

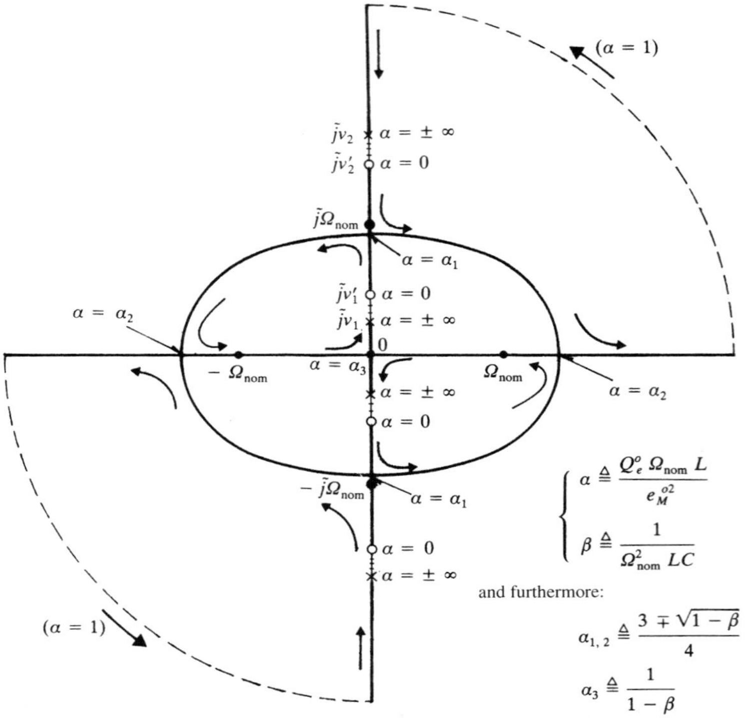

Figure 7.28. Shifting of the zeros of $K_e(s)$ in the complex plane, when varying $\alpha \triangleq Q_e^o \Omega_{nom} L / e_M^{o2}$, for $R = 0$ and $\beta \triangleq 1/(\Omega_{nom}^2 LC) = 0.5$. (Arrows correspond to shifting for $\alpha > 0$.)

- for intermediate values of $Q_e^o > 0$, there are one or two zeros having a positive real part, i.e., the function $K_e(s)$ becomes a "nonminimum phase" function, with significant phase delays in the frequency response (see also Fig. 7.29).

Note, finally, that:

- for $1/(\Omega_{nom} C) = 0$ (i.e., in the absence of compensation), $\nu_1 = \nu_1' = \nu_2' = \nu_2 = \Omega_{nom}$, and thus $K_e(s) = -Q_e^o + e_M^{o2}/(\Omega_{nom} L) =$ constant, independent of the inductive element dynamics (this confirms what is already pointed out about Equation [7.2.1]);
- for $1/(\Omega_{nom} C) \to \Omega_{nom} L$ (almost total compensation), we have instead $\nu_1 \to 0$, $\nu_1' \to 0$, $\nu_2' \to \sqrt{3}\Omega_{nom}$, $\nu_2 \to 2\Omega_{nom}$.

In practice, for $R > 0$, the above-mentioned resonances (and antiresonances) become slightly damped, but the situation basically remains similar; see the diagrams in Figure 7.29.

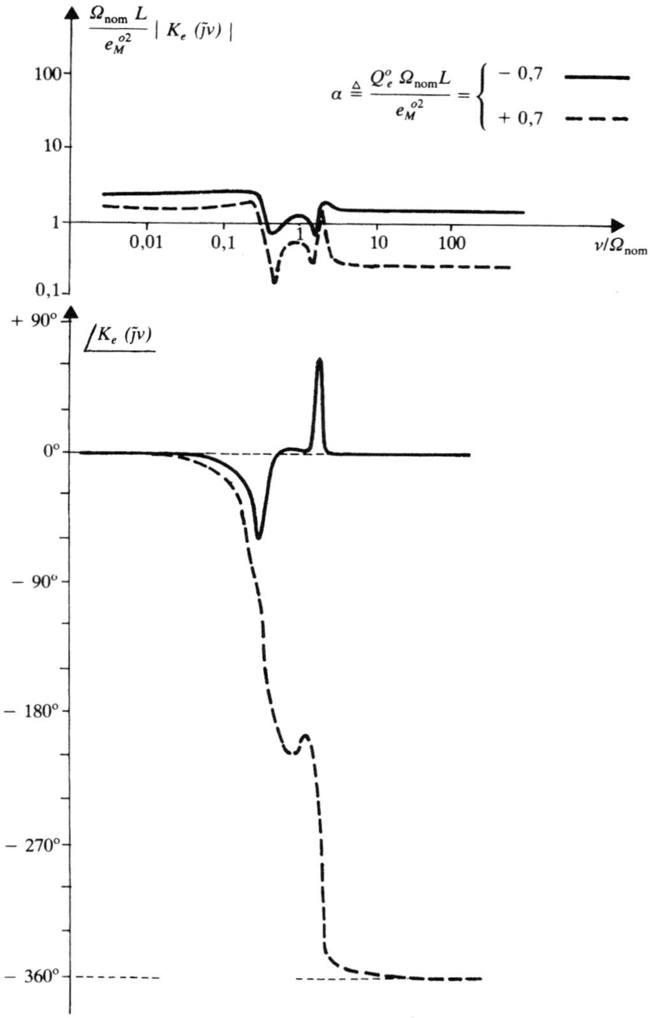

Figure 7.29. Frequency response of the function $K_e(s)$, for $R/(\Omega_{\text{nom}}L) = 0.2$ and $\beta \triangleq 1/(\Omega_{\text{nom}}^2 LC) = 0.5$.

With the electromechanical oscillation, the compensation under examination appears to be destabilizing, as (with the adopted model) it usually results:

$$Im(K_e(\tilde{j}\nu_o)) < 0$$

(remember Equation [7.2.15] and Figure 7.29, with $\nu_o/\Omega_{\text{nom}}$ within the range of 0.005–0.05). Nevertheless, the resulting damping may be positive (because of the actual behavior of the machine), or made positive through proper additional signals.

It is instead important to note that the frequency ν_l of the subsynchronous oscillation may easily be within the frequency range of the torsional oscillations of the unit shaft, according to Section 4.3.4 (e.g., for $1/(\Omega_{\text{nom}}C) = \Omega_{\text{nom}}L/2$, it follows $\nu_l \cong 0.3\Omega_{\text{nom}}$, i.e., approximately 15 Hz for a system operating at 50 Hz). Consequently:

- the variations ΔC_e of electromagnetic torque may contain *subsynchronous oscillatory components*, particularly heavy for the torsional deformation of the shaft, even up to its breaking (conversely, the $\Delta\delta$ oscillations may cause large torque variations);
- the stability of the electromechanical loop can be compromised by these interactions between the phenomena of electrical resonance (subsynchronous oscillations) and those of mechanical resonance (torsional oscillations).

Therefore, because of their importance, such interactions must be analyzed with particular attention, by adopting an adequate model (also for the dependence of $\Delta\Omega$ on ΔC_e; see Section 4.3.4).

As seen, for $R = 0$ the function $K_e(s)$ exhibits, under the simplified Equations [7.2.48], four purely imaginary poles (conjugate in pairs) resulting from electrical resonance phenomena between the different inductances and capacitances.

If the electromagnetic part of the machine is instead correctly considered by using Equations [7.1.9], the poles of $K_e(s)$ are defined by the equation:

$$0 = (s^2 + \Omega_{\text{nom}}^2)^2(\mathcal{L}_d(s) + L_e)(\mathcal{L}_q(s) + L_e)C^2 + (s^2 - \Omega_{\text{nom}}^2)(\mathcal{L}_d(s) + \mathcal{L}_q(s) + 2L_e)C + 1$$

(which can be derived similarly to [6.2.13]). By assuming for simplicity $\mathcal{L}_d(s) = (L_d + s\hat{T}'_{do}\hat{L}'_d)/(1 + s\hat{T}'_{do})$, $\mathcal{L}_q(s) = L_q$, it can be shown that:

- the function $K_e(s)$ has five poles;
- the number of poles of $K_e(s)$ having positive real part, is:

$$P_{(+)} = \begin{cases} 0 \\ 1 \\ 2 \end{cases} \text{according to whether } \frac{1}{\Omega_{\text{nom}}C} \begin{cases} > \Omega_{\text{nom}}(L_d + L_e) \\ \in (\Omega_{\text{nom}}(L_q + L_e), \Omega_{\text{nom}}(L_d + L_e)) \\ < \Omega_{\text{nom}}(L_q + L_e) \end{cases}$$

(see, in Section 6.2.1c, the condition [6.2.15] and the following considerations).

Therefore, if $1/(\Omega_{\text{nom}}C)$ is relatively small as it may occur in practical cases, the actual behavior of the electromagnetic part of the machine may be responsible (for what concerns the electrical resonances) for a destabilizing effect, similar to that of a negative resistance and thus, actually, more or less compensated by the presence of the resistance R.

Anyway, the electrical resonances can be conveniently attenuated by inserting proper series filters on the line or damping circuits in parallel to the condensers, whereas the subsynchronous torque oscillations can be further reduced (and damped) through the use of additional signals into the machine excitation.

7.3. LARGE PERTURBATION BEHAVIOR

The behavior for large variations is affected by different nonlinearities of the system, particularly by the nonlinear relationship between the resistant torque C_e and the state variables ψ_f (field flux) and δ (electrical angular position of the rotor, with respect to the "network" synchronous reference; see Equation [7.1.2]).

The greater concern on the system operation usually comes from the *risk of "loss of synchronism"* between the machine and the infinite bus, as described in Section 1.6. Therefore, in this section particular attention will be given to the analysis of relative motion between the machine and the infinite bus, and to the means that can be adopted to reduce such a risk.

The loss of synchronism is an undesired phenomenon, as explained in Section 1.6. Particularly, the excessive increase in current (which can be viewed as produced by a short-circuit at the so-called "electrical center") can cause protection intervention, and definitive unit rejection. By considering the probability of events that can cause the loss of synchronism, the set of acceptable operating points may be significantly reduced, below what can be allowed — at steady-state — by the stability conditions for small variations, by the machine capability limits, by the transmissibility limit on the link, etc.

If the machine maintains the synchronism, the angle δ and the other quantities of the system (voltage and currents magnitudes, etc.) exhibit, as seen, electromechanical oscillations. Instead, during the loss of synchronism, there is a substantially "aperiodic" increase of $\delta(t)$, following which the magnitudes of voltages, currents, etc. again assume an oscillatory-type time behavior. However, such oscillations are only the result of the dependence of these quantities on $\sin\delta$ and $\cos\delta$ (and thus of their periodic dependence on δ), and therefore are not at all related with the "electromechanical" oscillations that accompany the keeping of synchronism.

Following large perturbations, a rigorous analysis should require the use of simulations, because of the complexities of the model. However, some simplifications may appear reasonable, permitting a first-level approximate analysis, such as that reported in Section 1.6.

Usually, the possibility of keeping synchronism becomes evident within tenths of a second (time after which the angle δ diminishes again; see $\delta_{ab}(t)$ in Fig. 1.6). Consequently, to check whether or not synchronism is kept, it may be thought, as a first approximation, that in such a time interval:

(1) the purely electrical transients (associated with inductive and capacitive elements) can be considered vanished, as well as those of the "subtransient" type of the machine dynamics (associated with the additional rotor circuits);

(2) the field flux ψ_f, instead, remains practically constant.

Accordingly, we can:

- write:

$$P_m - P_e(\delta) = M\frac{\mathrm{d}^2\delta}{\mathrm{d}t^2} \qquad [7.3.1]$$

  where the dependence of P_e on δ is defined by the "transient" characteristic, at constant ψ_f (for the "basic case", see the second of Equations [7.2.10] or the characteristic in Figure 7.8b)(3);
- apply the analysis described in Section 1.6 (with $M_b = \infty$) based on the *"equal area" criterion*: see the examples in Figure 7.30, which refer to:
  - disconnection of an intermediate active load (at the instant $t = t_i$);
  - removal of a line section (at the instant $t = t_i$);
  - short-circuit on a line section (at the instant $t = t_i$) and subsequent removal of the section itself (at the instant $t = t_e$, by the protection system);

  under the hypothesis that the system is initially at a steady-state condition and that the driving power P_m remains constant(4).

With the approximations specified in Section 7.1:

- an active load disconnection in the network could be assimilated to a step $\Delta\alpha_e > 0$, and thus to a sudden shift to right (by the quantity $\Delta\alpha_e$) of the characteristic $P_e(\delta)$;
- a reactive load connection in the network, instead, could be assimilated to a step $\Delta e_R < 0$, and thus to a sudden reduction in $P_{e\max}$, similar to that in Figure 7.30b.

---

(3) More generally, it must be assumed $P_e = P_e(e_R, \delta - \alpha_e)$, where e_R and α_e can vary because of load perturbations in the network (see Section 7.1).

The analysis also can be extended, by similar approximations, to the case of two machines (see Section 1.6). Moreover, each machine may represent a whole subsystem (including machines, loads etc.) within which the angular slips could be disregarded (see also Section 8.4).

(4) In Figures 7.30b and 7.30c, it is assumed, for greater generality, that the opening or the short-circuit occurs with a proper equivalent impedance, which accounts for:

- the possible residual impedance;
- the negative and zero sequence impedances;

according to Figures 5.63 and 5.64. In fact, in the present analysis, only the mechanical balance expressed by Equation [7.3.1] is of interest, and, in the cases of unsymmetrical opening or short-circuit, P_e can be evaluated by considering only the contribution related to the positive sequence (see the conclusion of Section 5.7.3).

Moreover, with the only aim of a graphic simplicity, in the figures it is assumed that, for $t > t_i$ (i.e., after the initial time), the dependence of P_e on δ is $P_e = P_{e\max}\sin\delta$, where $P_{e\max}$ varies in a discontinuous way corresponding to possible perturbations. This is equivalent to assume that it is $\hat{X}'_d = X_q$ (see the second of Equations [7.2.10]), $\alpha_e = 0$, and that the system does not include any resistive element for $t > t_i$.

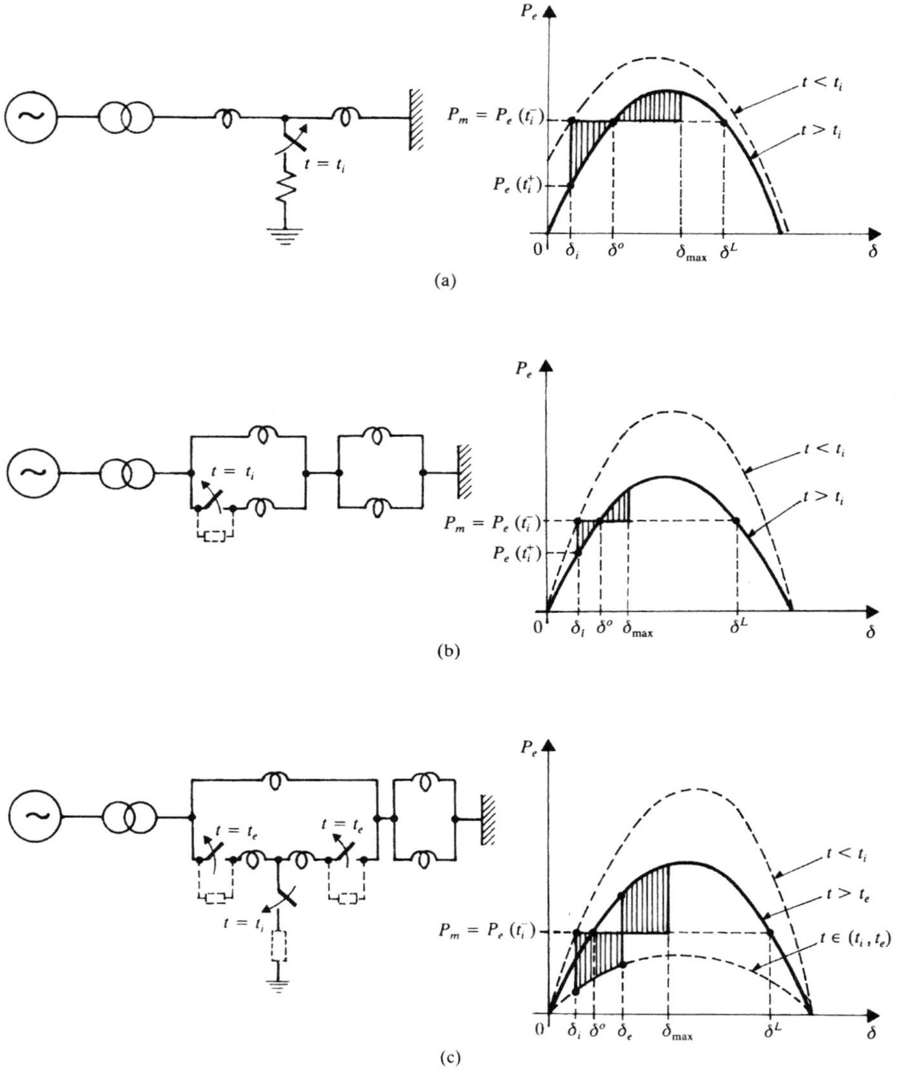

Figure 7.30. Application of the "equal area" criterion, in the case of:

(a) disconnection of an intermediate active load;

(b) removal of a line section;

(c) short-circuit on a line section and subsequent removal of the section itself.

It is assumed that $\delta_i \triangleq \delta(t_i)$, $\delta_e \triangleq \delta(t_e)$, whereas δ_{max} is the maximum value reached by δ (which must be smaller than δ^L, to keep the synchronism), and δ^o is the value of δ corresponding to the final steady-state condition.

By the examination of different cases, the necessity of properly limiting the value $P_m = P_e(t_i^-)$ appears evident, so that synchronism can be maintained ($\delta_{\max} < \delta^L$). In Figure 7.30c, with a double perturbation (short-circuit at the instant t_i, and removal of the faulted line section at the instant t_e), it becomes important to greatly reduce the duration $(t_e - t_i)$ (typical minimum values: 0.1 sec or smaller).

In Figures 7.30b and 7.30c we should emphasize the convenience of having:

- more lines in parallel (otherwise, a three-phase perturbation might cause $P_e = 0$ for $t > t_i$);
- an intermediate station or more such stations (to avoid marked reductions of the characteristic $P_e(\delta)$ after removal of the line section).

For a better approximation, it would be necessary to consider the variations in P_e caused by variations in the field flux ψ_f (and in the fluxes associated to the additional rotor circuits), which themselves depend on:

- the perturbation type (e.g., ψ_f can increase for the load disconnection in Fig. 7.30a and decrease for the short-circuit in Fig. 7.30c);
- the excitation control;

whereas the variations in P_m caused by speed regulation usually can be considered negligible.

In particular, the presence of the additional rotor circuits can lead to a beneficial effect in the case of Figure 7.30a (because of the increase in the rotor fluxes) and to a disadvantage in the case of Figure 7.30c (for the opposite reason).

Instead, the excitation control has generally a beneficial effect, which can be translated (as in Figure 7.31, with a successful reclosure) into an increase of the maximum acceptable power $P_e(t_i^-)$, e.g., 5% for rotating excitation and 10% for static excitation with

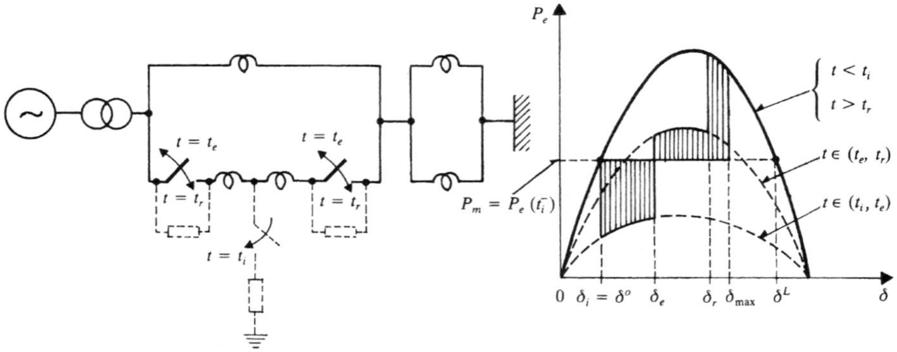

Figure 7.31. The case of Figure 7.30c, followed by a "fast reclosure" of the section under examination (assuming $\delta_r \triangleq \delta(t_r)$, etc.).

independent supply (with a dependent supply, such an effect can be significantly reduced because of the voltage drop). Similar considerations can further apply for other voltage regulation types, the benefits of which also can be considerable if the regulation is fast, e.g., performed by means of a shunt-connected static compensator.

To reduce the risk of losing synchronism, it is possible to adopt specific countermeasures[(5)], such to reduce (e.g., within a few tenths of a second) the accelerating power ($P_m - P_e$) through rather sudden variations of P_e and/or P_m (however, it is necessary to check that such variations do not result in unacceptable torsional stresses).

The countermeasures under examination can be achieved based on different signals:

- associated with protections or sensitive to the status of line breakers;
- sensitive — in a way possibly coordinated with the control systems — to the speed variation $\Delta\Omega$, the acceleration $\mathrm{d}\Omega/\mathrm{d}t$, the power variation ΔP_e, the voltage variation Δv, etc. (in particular, the acceleration measurement may require proper filtering of torsional oscillation effects).

Following a short-circuit and the subsequent removal of the faulted line section (Fig. 7.30c), the typical countermeasure consists of the "*fast reclosure*" (at the instant $t = t_r$) of this line section, which is automatically performed by the protections, so that the original configuration is restored as well as (under the above-recalled approximations (1) and (2)) the original characteristic $P_e(\delta)$ (see Fig. 7.31).

Obviously, it is necessary to reclose after a "waiting" time ($t_r - t_e$) sufficient for the extinction of the arc that accompanies the short-circuit (e.g., 0.3–0.5 sec for a three-phase short-circuit, and approximately 2 sec for a single-phase short-circuit), hoping that the causes for the short-circuit have been removed in the meantime. Possibly, the reclosure also can occur after more attempts; however, the arc persistence may lead to an "unsuccessful" reclosure, particularly for a single-phase short-circuit and a relatively long line.

More generally, other countermeasures — not all of which are commonly adopted — also can be temporarily used for increasing P_e or diminishing P_m.

[(5)] Moreover, preventive-type countermeasures may concern:

- the machine parameters; specifically, to reduce the acceleration (at given accelerating power) the inertia coefficient M could be increased by adding a fly wheel;
- the parameters and the structure of the link, e.g., by arranging more parallel lines (also to reduce the longitudinal reactance of the link) and intermediate stations;
- the operating point; specifically, to reduce δ_i at the same initial power $P_e(t_i^-)$ it is convenient to operate the system at a large e_R (e.g., 1.1 pu) and large ψ_f (i.e., $Q > 0$ and sufficiently large).

Such countermeasures must be based on local-type strategies, accounting for the nature of the perturbation and the possible occurrence of the reclosure, as described above.

More precisely, *to increase* P_e at the same δ, it is possible to:

(1) amplify the characteristic $P_e(\delta)$, in the sense of the ordinates (Fig. 7.32a):

   (1a) by reducing the longitudinal reactance of the link, through the insertion of series condensers[(6)];

   (1b) increasing the voltage magnitude by "forcing" the excitation control (or the control of the possible static compensator or in-phase regulating transformers), or by the insertion of shunt-connected condensers or the disconnection of reactors[(7)].

(2) move the characteristic $P_e(\delta)$, predominantly toward the left side (Fig. 7.32b), by increasing the phase difference between the link terminals; this can be performed by "forcing" quadrature-regulating transformers

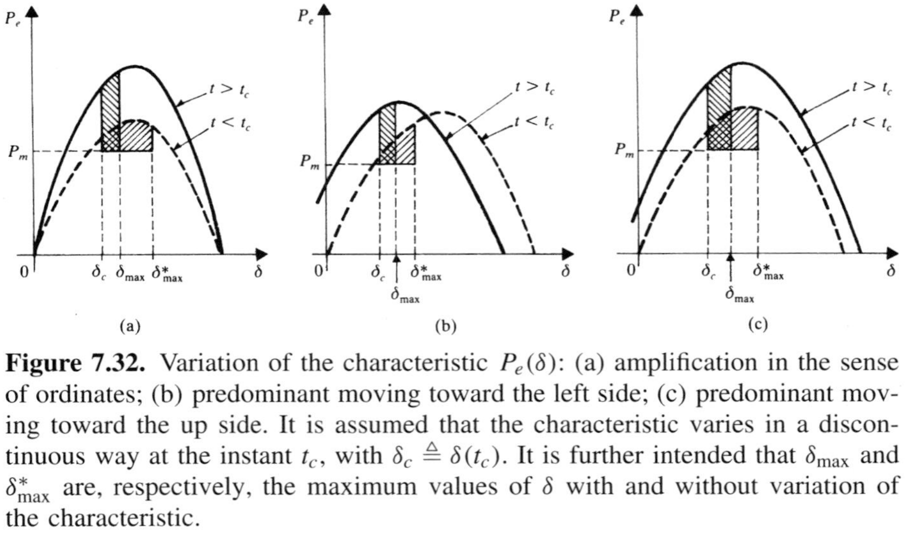

Figure 7.32. Variation of the characteristic $P_e(\delta)$: (a) amplification in the sense of ordinates; (b) predominant moving toward the left side; (c) predominant moving toward the up side. It is assumed that the characteristic varies in a discontinuous way at the instant t_c, with $\delta_c \triangleq \delta(t_c)$. It is further intended that δ_{max} and δ^*_{max} are, respectively, the maximum values of δ with and without variation of the characteristic.

---

(6) For instance, if the link consists of four line sections, two-by-two parallel connected (see the previous figures), each having a reactance X:

- the resulting reactance initially (before the short-circuit) is X;
- after a short-circuit and the (complete) removal of the faulted section, such a reactance can be returned to the value X (to restore the original characteristic $P_e(\delta)$, as in the case of successful reclosure) by means of a series capacitive reactance $-X/2$.

(7) In the presence of loads, the effects of the voltage increase can be different, according to the characteristics that relate the (active and reactive) absorbed powers to the voltage magnitude.

(similarly, a shift in the opposite sense can be useful, whenever δ reaches relatively large values);

(3) move the characteristic $P_e(\delta)$, predominantly toward the up side (Fig. 7.32c), by inserting a shunt-connected fictitious load, e.g., constituted by "*braking resistors*"[(8)].

Finally, *to decrease* P_m it is possible, for a thermal unit, to stop the steam flow to the turbine by opening proper "bypass" valves ("*fast-valving*"; see Fig. 7.33), based on signals sensitive to the acceleration or the delivered active power.

If the machine considered so far actually represents a multimachine subsystem (see footnote[(3)]), the disconnection of hydroelectric units also may be provided.

Normally, the above-reported countermeasures should be used temporarily. On the other hand, once the loss of synchronism has been prevented, they:

- generally are not convenient in the subsequent stage at $\Delta\Omega < 0$, for which it would be preferable to decrease P_e and/or increase P_m (e.g., by removing the braking resistors and/or restoring the steam to the turbine);
- may correspond, at steady-state conditions, to unacceptable situations (e.g., because of an excessive value of the machine excitation current).

Moreover, by switching the characteristic $P_e(\delta)$ (as described above, but in both senses alternatively) it is possible to improve the subsequent transient, until reaching the final steady-state condition.

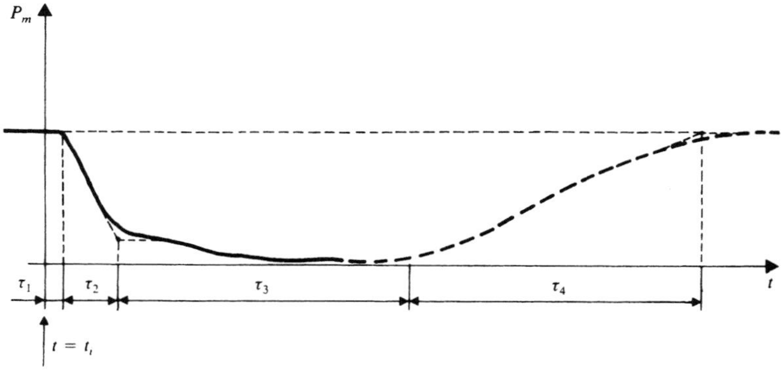

Figure 7.33. Example of behavior of $P_m(t)$ under "fast-valving" (e.g., $\tau_1 = 0.1$–0.15 sec; $\tau_2 = 0.2$–0.4 sec; $\tau_3 = 1$–3 sec; $\tau_4 = 1.5$–2 sec).

(8) By doing so, the energy absorbed by the artificial load is dissipated. Alternatively, it can be thought to store this energy in the form of magnetic energy (which can be successively reused), by realizing the fictitious load by means of inductive circuits of negligible resistance (i.e., superconductors), operating (through the interposition of a converter) in a "direct current (dc) mode."

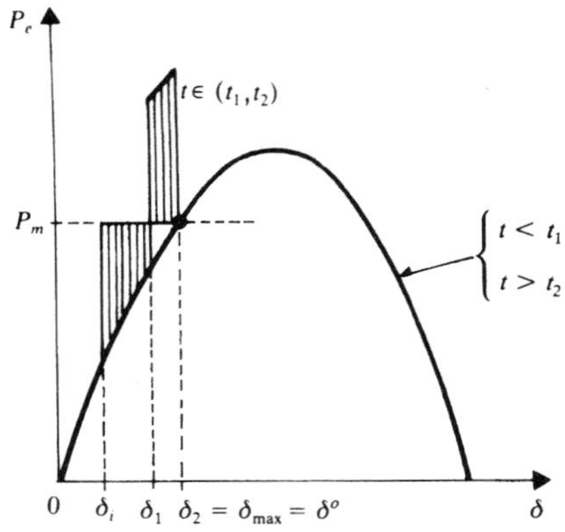

Figure 7.34. Reaching the final steady-state condition, by switching the characteristic $P_e(\delta)$: ideal case, with only two switchings ($\delta_1 \triangleq \delta(t_1)$, $\delta_2 \triangleq \delta(t_2)$ $= \delta_{\text{max}} = \delta^o$).

According to Figure 7.34, such a steady-state condition ($\Delta\Omega = 0$, $\delta = \delta^o$) could be reached even after only two switchings—at proper instants $t = t_1$ and $t = t_2$—of the characteristic $P_e(\delta)$. In practice this is not achievable, both because of the complexities and uncertainties in modelling (and in diagnosis), and because of the discontinuities in the tuning of the switching effects. However, reaching the final steady-state condition can clearly be accelerated by switchings performed at proper instants.

ANNOTATED REFERENCES

Among the works of more general interest, it can be made reference to: 4, 6, 7, 10, 19, 37, 42, 50, 51, 68, 77, 81, 88, 98, 103, 106, 109, 111, 115, 125, 135, 147, 191, 214, 292, 293.

More specifically, for what concerns

- the effects of an intermediate load: notes prepared by the author, in view of the writing of 53;
- the case of a series-compensated line: 36, 208, 209, 223;
- the large perturbations behavior: 98, 118, 161, 173, 262, 264, 275, 279, 283, 287, 302, 315.

CHAPTER 8

ELECTROMECHANICAL PHENOMENA IN A MULTIMACHINE SYSTEM

8.1. PRELIMINARIES

To completely represent the dynamic behavior of a power system, a nonlinear model of a very high dynamic order should be used, which would certainly be prohibitive for an analysis not supported by simulation. The complications of the model also would cause problems within the simulations, even if a powerful computer is available.

Therefore it is necessary, in practice, to simplify the model according to the specific problem to be solved.

A commonly used simplification is assuming a synchronous machine connected to a network of infinite power (see Chapter 7). This scenario is likely acceptable for particular problems of "local" character, i.e., when the phenomenon mainly concerns a single machine (e.g., by the effect of a disturbance near it) and is primarily determined by the dynamic behavior of this machine.

It must be indicated, however, that even in this case use of a complete model (for the machine and its control systems) could be avoided, because details of the model can actually be disregarded without compromising the accuracy of the results.

In the general case of a multimachine system, the model complications also depend on the number of machines (which may be very large), on the characteristics of the network, and so on.

On the other hand, if the problem is not of a "local" character as mentioned above, it may be presumed that a model which specifically considers interactions between the components, can give reasonable results (at least for a first approximation analysis), even if the representation of each component is drastically simplified.

In this chapter, analysis will be developed referring to the machine motions (transients of angular position and speed of the rotors, with respect to a given synchronous steady-state operation), assuming simplified models for the machines, loads, and network elements.

In qualitative terms, these assumptions appear very useful to show, relatively simply, the fundamental characteristics of the machine motions, considering the interactions caused by the rest of the system. It must not be forgotten that, notwithstanding the simplifications, the results of this analysis can be useful also from a quantitative point of view — subject to a suitable choice of the value of parameters — as it has been confirmed by some experimental results achieved on actual systems.

In equilibrium steady-state conditions, all the (synchronous) machines in a given system have the same electrical angular speed (equal to the "synchronous" speed Ω_s), whereas their electrical angular positions differ from each other, with constant shifts depending on the loading conditions of individual machines.

In any dynamic condition, the mechanical balance equations of an N-machine system can generally be written in the following form:

$$\left.\begin{aligned} \frac{d\Omega_i}{dt} &= \frac{1}{M_i}(P_{mi} - P_{ei}) \\ \frac{d\delta_i}{dt} &= \Omega_i - \Omega_s \end{aligned}\right\} \quad (i = 1, \ldots, N) \qquad [8.1.1]$$

where, for the i-th machine:

- Ω_i = electrical angular speed;
- δ_i = electrical angular position, evaluated with respect to that (θ_s) of a fictitious rotor, rotating at the synchronous speed $d\theta_s/dt = \Omega_s$;
- P_{mi} = mechanical driving power;
- P_{ei} = active electrical power generated;
- M_i = coefficient of inertia;

(see Equation [3.1.2], with mechanical losses neglected).

If it is assumed that the powers P_{mi} and P_{ei} are only functions of the speeds and angular positions of the machines, and of possible input variables (represented by a "column matrix" u), the motion of the machines turns out to be fully represented by the Equations [8.1.1], which define a model of $2N$ dynamic order, with state variables $\Omega_1, \ldots, \Omega_N$, $\delta_1, \ldots, \delta_N$, and input u. This first approximation model will be detailed later in this chapter. The input variables may, for example, correspond to external actions on the turbines (as the set-points of governors), capable of varying the P_{mi}'s independently of the speed variations, or to perturbations on the structure of the electric part of the system (opening of lines, load switching, etc.), capable of directly varying the P_{ei}'s. The synchronous speed Ω_s may, on the other hand, be regarded as a parameter, with an arbitrary constant value.

Actually, for a more complete dynamic representation of the system, other differential equations should be added to [8.1.1] to take into account the dynamic behaviour of the electromagnetic parts of the machines, of control systems, etc. Therefore, more generally, a better approximation model can be obtained by adding to [8.1.1] a matrix equation of the type

$$\frac{\mathrm{d}x''}{\mathrm{d}t} = f''(\Omega, \delta, x'', u) \qquad [8.1.2]$$

where:

- Ω = column matrix of the speeds $\Omega_1, \ldots, \Omega_N$;
- δ = column matrix of the positions $\delta_1, \ldots, \delta_N$;
- x'' = column matrix of the additional state variables (e.g., magnetic flux linkages in the machines, state variables of the control systems etc.);
- u = input column matrix, appropriately defined (including, for example, the set points of the regulators, etc.);

and assuming that the powers P_{mi} and P_{ei}, as well as the vector f'', in general depend on Ω, δ, x'', u.

Then, the first approximation model can be considered a simplified version of the more complete model, obtained by replacing the differential Equation [8.1.2] with an algebraic equation that makes it possible to eliminate x''.

As a first approximation, applying the Park's transformation with a synchronous angular reference θ_s and neglecting, for simplicity, the subscript "s" in the Park's vectors, it can be assumed that (Fig. 8.1):

- the generic i-th machine can be represented, possibly considering its voltage regulation, by an "equivalent" electromotive force (vector):

$$\overline{e}_i = j e_i \epsilon^{j\delta_i} \qquad (i = 1, \ldots, N) \qquad [8.1.3]$$

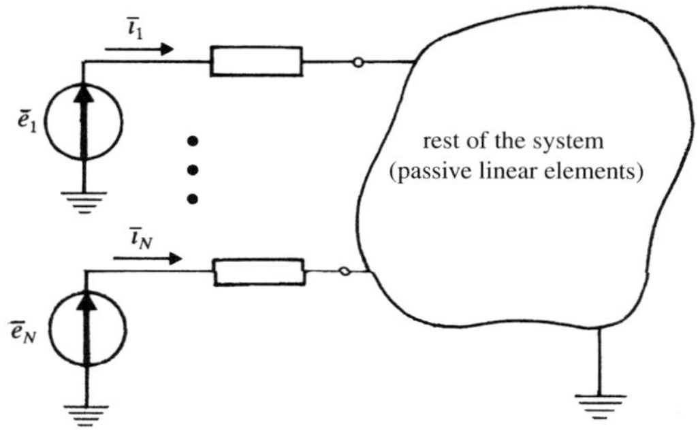

Figure 8.1. Multimachine system under examination.

behind a suitable output impedance (as a result of Equation [8.1.3], the phases of the electromotive forces are given by $\delta_i + 90^\circ$, which corresponds to the assumption that the electromotive forces act along the "quadrature axes" of the respective machines);
- the rest of the system (network and loads) consists of passive linear elements that can be represented by suitable impedances or admittances evaluated at frequency Ω_s.

Under these hypotheses, the currents $\bar{\imath}_1, \ldots, \bar{\imath}_N$ (vectors) indicated in Figure 8.1 are linked to the electromotive forces $\bar{e}_1, \ldots, \bar{e}_N$ by means of linear equations:

$$\bar{\imath}_i = \sum_1^N {}_j \overline{Y}_{ij} \bar{e}_j \qquad (i = 1, \ldots, N) \qquad [8.1.4]$$

where the $\overline{Y}_{ij}$'s are constant complex admittances ($i, j = 1, \ldots, N$), with $\overline{Y}_{ij} = \overline{Y}_{ji}$, whereas the active electrical powers are expressed by:

$$P_{ei} = \mathcal{R}e\,(\bar{e}_i \bar{\imath}_i^*) \qquad (i = 1, \ldots, N) \qquad [8.1.5]$$

From Equations [8.1.3], [8.1.4], and [8.1.5], by letting $\overline{Y}_{ij} \triangleq G_{ij} + jB_{ij}$, $\delta_{ij} \triangleq \delta_i - \delta_j$, it can be derived:

$$P_{ei} = e_i^2 G_{ii} + e_i \sum_1^N {}_{j \neq i}\, e_j (G_{ij} \cos \delta_{ij} + B_{ij} \sin \delta_{ij}) \qquad (i = 1, \ldots, N) \qquad [8.1.6]$$

so that the powers P_{ei} are functions of:

- the magnitudes $e_1, \ldots, e_N$ of the electromotive forces;
- the phase differences between the electromotive forces (i.e., the differences between the electrical angular positions of the rotors) or equivalently the $(N-1)$ differences $\delta_{1N}, \ldots, \delta_{(N-1)N}$, having taken as phase reference the phase of the N-th electromotive force;
- the conductances G_{ij} and the susceptances B_{ij} that characterize the system interposed between the electromotive forces.

By substituting Equation [8.1.6] in [8.1.1], we can derive the simplified model:

$$\left.\begin{aligned} \frac{\mathrm{d}\Omega_i}{\mathrm{d}t} &= \frac{1}{M_i}(P_{mi} - P_{ei}(e, \delta, \overline{Y}_{(i)})) \\ \frac{\mathrm{d}\delta_i}{\mathrm{d}t} &= \Omega_i - \Omega_s \end{aligned}\right\} \qquad (i = 1, \ldots, N) \qquad [8.1.7]$$

where e, δ, $\overline{Y}_{(i)}$ are column matrices, respectively, consisting of the magnitudes $e_1, \ldots, e_N$ of the emfs, the angular shifts $\delta_1, \ldots, \delta_N$, and the admittances $\overline{Y}_{i1}, \ldots, \overline{Y}_{iN}$.

For a correct formulation of the analysis, it seems appropriate to present some *clarifications about the model* in Equations [8.1.7] and about other models that may be deduced from it.

If it is assumed that the driving powers P_{mi} are directly controllable (by suitable actions on the turbines), as well as the magnitudes e_i of the electromotive forces (by the excitation of the machines) and the admittances $\overline{Y}_{ij} = G_{ij} + jB_{ij}$ (by structure perturbations), Equations [8.1.7] constitute a model of $2N$ *dynamic order*, with state variables Ω_i, δ_i, and input variables P_{mi}, e_i, G_{ij}, B_{ij} $(i, j = 1, \ldots, N)$.

The equilibrium points of the system in Equations [8.1.7] are defined by the conditions:

$$\frac{d\Omega_i}{dt} = 0, \quad \frac{d\delta_i}{dt} = 0 \quad \forall i = 1, \ldots, N$$

i.e.,

$$\left.\begin{aligned} &P_{mi} - P_{ei}(e, \delta, \overline{Y}_{(i)}) = 0 \quad &[8.1.8]\\ &\Omega_i - \Omega_s = 0 \quad &[8.1.9]\end{aligned}\right\} \quad \forall i = 1, \ldots, N$$

Conditions [8.1.9] lead to $\Omega_1 = \cdots = \Omega_N = \Omega_s$, implying that speeds of all machines are equal (zero slips) to the synchronous speed. Conditions [8.1.8] constitute, for given values of the input variables, N equations in $\delta_1, \ldots, \delta_N$. It is, however, clear that these equations only have solutions for particular choices of the input variables, because the powers P_{ei} actually depend on $(N-1)$ phase differences, rather than on the N phases $\delta_1, \ldots, \delta_N$ independently.

In other words, to determine possible solutions in the $(N-1)$ unknown $\delta_{1N}, \ldots, \delta_{(N-1)N}$, only $(N-1)$ of the Equations [8.1.8] are sufficient, after which the remainder of [8.1.8]—with the δ_{iN}'s found $(i = 1, \ldots, N-1)$—defines a very precise *constraint* on the input variables (see also Section 2.1.5). In practice, this constraint is automatically satisfied, by suitable distribution of the P_{mi}'s caused by the f/P control.

For each solution $\delta_{1N}, \ldots, \delta_{(N-1)N}$ it is then derived $\delta_i = \delta_{iN} + \delta_N$ $(i = 1, \ldots, N)$ with arbitrary δ_N. Because of this arbitrariness, every solution corresponds to a given equilibrium point in the space $\{\delta_{1N}, \ldots, \delta_{(N-1)N}\}$, and to a *straight-line* of possible equilibrium points in the space $\{\delta_1, \ldots, \delta_N\}$. This must be considered, for example, to avoid an incorrect application of the "Lyapunov-like" theorems for the stability analysis (see Section 8.3.)

For this reason, it may be more convenient to:

- rewrite Equations [8.1.7] in the form:

$$\left.\begin{aligned} \frac{d\Omega_i}{dt} &= \frac{1}{M_i}(P_{mi} - P_{ei}(e, \delta', \overline{Y}_{(i)})) \qquad (i = 1, \ldots, N)\\ \frac{d\delta_{kN}}{dt} &= \Omega_k - \Omega_N \qquad (k = 1, \ldots, N-1)\end{aligned}\right\} \quad [8.1.10]$$

$$\frac{d\delta_N}{dt} = \Omega_N - \Omega_s \qquad [8.1.11]$$

where $\delta' \triangleq [\delta_{1N}, \ldots, \delta_{(N-1)N}]^T$;

- pay attention only to Equations [8.1.10], which define a model of $(2N - 1)$ *dynamic order*, with state variables $\Omega_1, \ldots, \Omega_N$, $\delta_{1N}, \ldots, \delta_{(N-1)N}$ (with the exception of considering also Equation [8.1.11] in rare cases where the behavior of δ_N, and thus of the individual angular positions, may be of some interest).

By imposing the equilibrium conditions of the system [8.1.10], a constraint is found again on the input variables, as already specified (in fact, the N equations $P_{mi} - P_{ei} = 0$, in the $(N - 1)$ unknowns $\delta_{1N}, \ldots, \delta_{(N-1)N}$, must be consistent with each other). If this constraint is satisfied, every equilibrium point is characterized by constant values of the angular differences (obtained from the $P_{mi} - P_{ei} = 0$) and zero values of the slips, whereas the constant value of the speed (common to all the machines) is arbitrary.

Moreover, Equations [8.1.10] can be rewritten as:

$$\left.\begin{aligned} \frac{d\Omega_{kN}}{dt} &= \left(\frac{P_{mk}}{M_k} - \frac{P_{mN}}{M_N}\right) - \left(\frac{P_{ek}}{M_k} - \frac{P_{eN}}{M_N}\right) \\ \frac{d\delta_{kN}}{dt} &= \Omega_{kN} \end{aligned}\right\} \quad (k = 1, \ldots, N - 1) \qquad [8.1.12]$$

$$\frac{d\Omega_N}{dt} = \frac{1}{M_N}(P_{mN} - P_{eN}) \qquad [8.1.13]$$

with $\Omega_{kN} \triangleq \Omega_k - \Omega_N$, $P_{ek} = P_{ek}(e, \delta', \overline{Y}_{(k)})$, $P_{eN} = P_{eN}(e, \delta', \overline{Y}_{(N)})$. Equations [8.1.12] define a $(2N - 2)$ *dynamic order* model (with state variables Ω_{kN}, δ_{kN}, $k = 1, \ldots, N - 1$) that fully describes the relative machine motion with respect to the motion of the N-th machine. Note that the evolution of system [8.1.12] influences, by the dependence of P_{eN} on the δ_{kN}'s, Equation [8.1.13], but not vice versa (see also, for the linearized case, Fig. 8.3).

Imposing the equilibrium conditions for Equations [8.1.12]:

- the (constant) values of the $(N - 1)$ unknowns $\delta_{1N}, \ldots, \delta_{(N-1)N}$ must be derived from the $(N - 1)$ equations:

$$0 = \frac{1}{M_k}(P_{mk} - P_{ek}) - \frac{1}{M_N}(P_{mN} - P_{eN}) \qquad (k = 1, \ldots, N - 1)$$

  without any constraint on the input variables;
- one may then derive $\Omega_{kN} = 0$ $(k = 1, \ldots, N - 1)$, i.e., zero slips, whereas the individual speeds, equal to each other, may vary with time.

This corresponds to an equilibrium *point* in the space $\{\Omega_{1N}, \ldots, \Omega_{(N-1)N}, \delta_{1N}, \ldots, \delta_{(N-1)N}\}$.

8.2. LINEARIZED ANALYSIS

8.2.1. Basic Equations

Linearizing Equations [8.1.7] around a given equilibrium point (characterized by the superscript "o"), we may obtain:

$$\left.\begin{aligned} \frac{\mathrm{d}\Delta\Omega_i}{\mathrm{d}t} &= \frac{1}{M_i}\left(\Delta u_i - \sum_1^N {}_j K_{ij}\Delta\delta_j\right) \\ \frac{\mathrm{d}\Delta\delta_i}{\mathrm{d}t} &= \Delta\Omega_i \end{aligned}\right\} \qquad (i = 1, \ldots, N) \qquad [8.2.1]$$

with:

$$\left.\begin{aligned} \Delta u_i &\triangleq \Delta P_{mi} - \sum_1^N {}_j \left(\frac{\partial P_{ei}}{\partial e_j}\right)^o \Delta e_j - \Delta P_{ei}^S \\ K_{ij} &\triangleq \left(\frac{\partial P_{ei}}{\partial \delta_j}\right)^o \end{aligned}\right\} \qquad [8.2.2]$$

by denoting ΔP_{ei}^S as the variation of P_{ei} caused by possible variations of the G_{ij}, B_{ij} (structure perturbations; see for example Section 3.3.3).

In agreement with what is already said, Equations [8.2.1] define a linear model of $2N$ dynamic order, with state variables $\Delta\Omega_i$, $\Delta\delta_i$, and input variables Δu_i $(i = 1, \ldots, N)$ (Δu_i can be interpreted as an equivalent variation of the driving power on the i-th machine).

The K_{ij} coefficients, which we will call *"synchronizing" coefficients*, satisfy the conditions:

$$\sum_1^N {}_j K_{ij} = 0 \qquad \forall i = 1, \ldots, N \qquad [8.2.3]$$

because the power P_{ei}'s depend on the phase differences (rather than on the phases $\delta_1, \ldots, \delta_N$ independently), so that $\Delta P_{ei} = 0$ if $\Delta\delta_1 = \cdots = \Delta\delta_N$. Therefore, in the matrix $K \triangleq \{K_{ij}\}$, the sum of the elements in each row is zero.

Particularly, from Equation [8.1.6] we may derive:

$$K_{ij} = \begin{cases} [e_i e_j (G_{ij}\sin\delta_{ij} - B_{ij}\cos\delta_{ij})]^o & \text{if } j \neq i \\ \displaystyle\sum_1^N {}_{h\neq i}[e_i e_h(-G_{ih}\sin\delta_{ih} + B_{ih}\cos\delta_{ih})]^o & \text{if } j = i \end{cases} \qquad [8.2.4]$$

from which Equation [8.2.3].

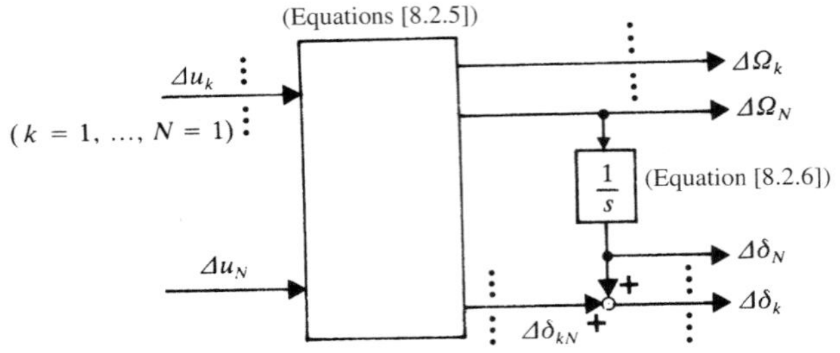

Figure 8.2. Block diagram of the linearized system.

Therefore, in the first of Equations [8.2.1] it also results:

$$\sum_{1}^{N}{}_j\ K_{ij}\Delta\delta_j = \sum_{1}^{N}{}_{j\neq i}\ K_{ij}\Delta\delta_{ji} = \sum_{1}^{N-1}{}_j\ K_{ij}\Delta\delta_{jN}$$

Moreover, Equations [8.2.1] can be rewritten as (see Fig. 8.2):

$$\left.\begin{aligned} \frac{d\Delta\Omega_i}{dt} &= \frac{1}{M_i}\left(\Delta u_i - \sum_{1}^{N-1}{}_j K_{ij}\Delta\delta_{jN}\right) && (i = 1, \ldots, N) \\ \frac{d\Delta\delta_{kN}}{dt} &= \Delta\Omega_k - \Delta\Omega_N && (k = 1, \ldots, N-1) \end{aligned}\right\} \qquad [8.2.5]$$

$$\frac{d\Delta\delta_N}{dt} = \Delta\Omega_N \qquad [8.2.6]$$

as can be derived by linearizing, respectively, Equations [8.1.10] and [8.1.11]. For what is already said with regard to Equations [8.1.10], the analysis will be carried out with particular regard to the linearized model [8.2.5], of $(2N - 1)$ dynamic order, with state variables $\Delta\Omega_1, \ldots, \Delta\Omega_N$, $\Delta\delta_{1N}, \ldots, \Delta\delta_{(N-1)N}$ and input variables $\Delta u_1, \ldots, \Delta u_N$.

Finally, to show the relative machine motion, it can be useful to rewrite Equations [8.2.5] as follows:

$$\left.\begin{aligned} \frac{d\Delta\Omega_{kN}}{dt} &= \frac{1}{M_k}\left(\Delta u'_k - \sum_{1}^{N-1}{}_j K'_{kj}\Delta\delta_{jN}\right) \\ \frac{d\Delta\delta_{kN}}{dt} &= \Delta\Omega_{kN} \end{aligned}\right\} \quad (k = 1, \ldots, N-1) \qquad [8.2.7]$$

$$\frac{d\Delta\Omega_N}{dt} = \frac{1}{M_N}\left(\Delta u_N - \sum_{1}^{N-1}{}_j K_{Nj}\Delta\delta_{jN}\right) \qquad [8.2.8]$$

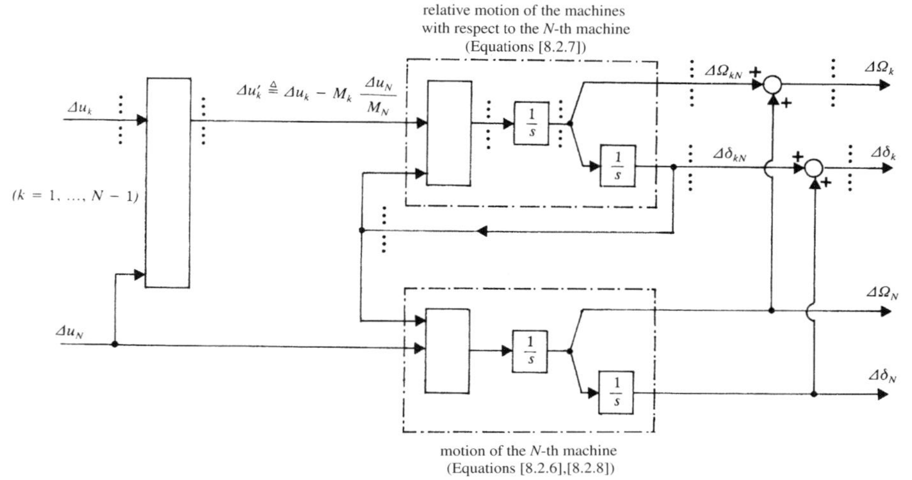

Figure 8.3. Variant of the block diagram to show the relative motion of the machines with respect to one of them.

as can be derived by linearizing Equations [8.1.12] and [8.1.13], respectively, and denoting:

$$\begin{cases} \Delta u'_k \triangleq \Delta u_k - M_k \dfrac{\Delta u_N}{M_N} \\ K'_{kj} \triangleq K_{kj} - M_k \dfrac{K_{Nj}}{M_N} \end{cases}$$

according to Figure 8.3.

8.2.2. Characteristic Roots and Stability

With matrix notations, Equation [8.2.1] can be rewritten as:

$$\left.\begin{aligned} \frac{\mathrm{d}\Delta\Omega}{\mathrm{d}t} &= -M^{-1}K\Delta\delta + M^{-1}\Delta u \\ \frac{\mathrm{d}\Delta\delta}{\mathrm{d}t} &= \Delta\Omega \end{aligned}\right\} \qquad [8.2.9]$$

with $M \triangleq \mathrm{diag}\{M_1, \ldots, M_N\}$, $K \triangleq \{K_{ij}\}(i, j = 1, \ldots, N)$, whereas $\Delta\Omega$, $\Delta\delta$, Δu are the column matrices consisting of $\Delta\Omega_i$, $\Delta\delta_i$, Δu_i $(i = 1, \ldots, N)$, respectively.

From these equations, it also follows:

$$\frac{\mathrm{d}^2\Delta\delta}{\mathrm{d}t^2} = -M^{-1}K\Delta\delta + M^{-1}\Delta u$$

so that the characteristic roots $\Lambda_1, \ldots, \Lambda_{2N}$ of the system in question can be derived by solving for the unknown Λ in the equation:

$$0 = \det(\Lambda^2 I_{(N)} + M^{-1}K) \qquad [8.2.10]$$

On the other hand, Equation [8.2.10] has solutions:

$$\Lambda^2 = L_1, \ldots, L_N$$

where $L_1, \ldots, L_N$ denote the N eigenvalues of the matrix $[-M^{-1}K]$; therefore, the characteristic roots $\Lambda_1, \ldots, \Lambda_{2N}$ are given by the square roots of $L_1, \ldots, L_N$.

If the properties defined by [8.2.3] are considered, it can be seen that the "synchronizing" matrix K is singular, and therefore also the $[-M^{-1}K]$ matrix. Consequently, (at least) one of the eigenvalues $L_i (i = 1, \ldots, N)$ is equal to zero.

By means of a suitable choice of indices, assuming:

$$L_N = 0 \qquad [8.2.11]$$

it can be written:

$$\left.\begin{array}{l} \Lambda_h = -\Lambda_{(h+N-1)} = (L_h)^{1/2} \quad (h = 1, \ldots, N-1) \\ \Lambda_{(2N-1)} = 0 \end{array}\right\} \qquad [8.2.12]$$

and furthermore:

$$\Lambda_{2N} = 0$$

The $(2N-1)$ characteristic roots of the system [8.2.5]—for which the analysis will be particularly developed here—are the same as those of the system [8.2.1] except for one at the origin, corresponding to Equation [8.2.6] (see also Fig. 8.2); more precisely, these characteristic roots are expressed by Equations [8.2.12].

It is easy to prove that, for the stability of the system [8.2.5], it is necessary (and sufficient) that $L_1, \ldots, L_{N-1}$ are distinct, real, and negative (as we shall assume later on). In fact, in this case, it results, from Equations [8.2.12]:

$$\left.\begin{array}{l} \Lambda_h = -\Lambda_{(h+N-1)} = \tilde{j}\nu_h \quad (h = 1, \ldots, N-1) \\ \Lambda_{(2N-1)} = 0 \end{array}\right\} \qquad [8.2.13]$$

with:

$$\nu_h \triangleq \sqrt{-L_h} \qquad [8.2.14]$$

(i.e., all simple and imaginary characteristic roots), which corresponds to "weak" stability of the linearized system [8.2.5]. In any other case, there would be at least one characteristic root with a positive real part, or multiple with real part equal to zero, and the system in question would be unstable.

As expressed by Equations [8.2.13], one of the characteristic roots lies at the origin, whereas the others make up $(N-1)$ pairs of imaginary conjugate roots $\pm \tilde{j}\nu_h$, corresponding to just as many oscillatory "modes" with zero damping and resonance frequencies $\nu_1, \ldots, \nu_{(N-1)}$ given by Equation [8.2.14].

In principle, the weak stability of the linearized system does not guarantee anything about the actual stability properties of the (nonlinear) system around the considered equilibrium point. To obtain exhaustive indications, it is necessary to refer to an analysis of the Lyapunov type, such as that illustrated in Section 8.3. However, the present results may appear useful, as they correspond (because of the approximations adopted) to a limit situation, with respect to what may be obtained by adopting more realistic models with nonimaginary characteristic roots (see Section 8.5).

8.2.3. Modal Analysis: "Mean Motion" and Electromechanical Oscillations

Assuming, as said, that the matrix $[-M^{-1}K]$ has distinct eigenvalues, it can be written in the form:

$$-M^{-1}K = ALA^{-1} \tag{8.2.15}$$

with $L \triangleq \text{diag}\{L_1, \ldots, L_N\}$, whereas A is the matrix consisting of the "(column) eigenvectors" corresponding to $L_1, \ldots, L_N$, respectively.

Note that it results:

$$\sum_1^N {}_j (A^{-1})_{hj} A_{jN} = \sum_1^N {}_i (A^{-1})_{Ni} A_{ih} = 0 \qquad \forall h = 1, \ldots, N-1 \tag{8.2.16}$$

whereas:

$$\sum_1^N {}_i (A^{-1})_{Ni} A_{iN} = 1 \tag{8.2.17}$$

Moreover, as $L_N = 0$, the elements of the last column of A (i.e., the elements of the eigenvector corresponding to L_N) are equal to each other, i.e.,

$$A_{1N} = \cdots = A_{NN} \tag{8.2.18}$$

so that it can be derived, by accounting for Equations [8.2.16] and [8.2.17], respectively:

$$\sum_1^N {}_j (A^{-1})_{hj} = 0 \qquad \forall h = 1, \ldots, N-1 \tag{8.2.19}$$

$$\sum_1^N {}_i (A^{-1})_{Ni} = (A_{1N})^{-1} \tag{8.2.20}$$

In terms of Laplace transforms, from Equations [8.2.9] and recalling [8.2.15] it is possible to derive:

$$\Delta\Omega_i(s) = \sum_1^{N-1} {}_h \frac{s}{s^2+\nu_h^2} \phi_{\Omega_i}^{(h)}(s) + \Delta\Omega_o(s) \tag{8.2.21}$$

$$\Delta\delta_i(s) = \sum_1^{N-1} {}_h \frac{s}{s^2+\nu_h^2} \phi_{\delta_i}^{(h)}(s) + \Delta\delta_o(s) \tag{8.2.21'}$$

$(i = 1, \ldots, N)$, where:

$$\phi_{\Omega_i}^{(h)}(s) \triangleq A_{ih} \sum_1^N {}_j (A^{-1})_{hj} \left(\frac{\Delta u_j(s)}{M_j} + \Delta\Omega_j^{(o)} - \nu_h^2 \frac{\Delta\delta_j^{(o)}}{s} \right) \tag{8.2.22}$$

$$\phi_{\delta_i}^{(h)}(s) \triangleq A_{ih} \sum_1^N {}_j (A^{-1})_{hj} \left(\frac{1}{s} \left(\frac{\Delta u_j(s)}{M_j} + \Delta\Omega_j^{(o)} \right) + \Delta\delta_j^{(o)} \right) \tag{8.2.22'}$$

$(h = 1, \ldots, N-1)$, whereas:

$$\Delta\Omega_o(s) = \frac{1}{s} A_{1N} \sum_1^N {}_j (A^{-1})_{Nj} \left(\frac{\Delta u_j(s)}{M_j} + \Delta\Omega_j^{(o)} \right) \tag{8.2.23}$$

$$\Delta\delta_o(s) = \frac{1}{s} \left(\Delta\Omega_o(s) + A_{1N} \sum_1^N {}_j (A^{-1})_{Nj} \Delta\delta_j^{(o)} \right) \tag{8.2.23'}$$

In the above, the superscript "(o)" denotes initial conditions, i.e., the values at $t = 0$.

For any given h, the functions $\phi_{\Omega_i}^{(h)}(s)$ are all similar to each other in that they depend on the index i only through the proportionality coefficients A_{ih}. The same holds for their inverse transforms. (This is also valid for the functions $\phi_{\delta_i}^{(h)}(s)$.)

These results are worth some important comments, as follows:

(1) In Equations [8.2.21] and [8.2.21′], the terms $\Delta\Omega_o$ and $\Delta\delta_o$ do not depend on the index i, and therefore form a common characteristic for machine motion. It is natural to assume these terms as representatives of the "mean motion" of the set of the machines. In particular, the common concept of "*mean frequency*" of the network, generally used in an empirical way, gets a precise definition here. Then, with the model adopted, the term $\Delta\Omega_o$ represents the "mean frequency" variation.
On the contrary, the other terms in Equations [8.2.21] and [8.2.21′] depend on the index i, and therefore characterize the relative motion of the machines between themselves, and with respect to the mean motion. These terms correspond to the oscillatory modes already mentioned, i.e.,

to the "*electromechanical oscillations*" of the machines, at frequencies $\nu_1, \ldots, \nu_{(N-1)}$.

(2) By recalling Equations [8.2.16] and [8.2.20], it is easy to verify that it results (both in the s and the t domains):

$$\Delta\Omega_o = A_{1N} \sum_1^N {}_i (A^{-1})_{Ni} \Delta\Omega_i \tag{8.2.24}$$

and similarly:

$$\Delta\delta_o = A_{1N} \sum_1^N {}_i (A^{-1})_{Ni} \Delta\delta_i \tag{8.2.24'}$$

so that $\Delta\Omega_o$ may be interpreted as an "*ensemble average*" of the $\Delta\Omega_i$'s and more precisely (recalling Equation [8.2.20]) as a "weighted" average with a unitary sum of the weights. The same applies for $\Delta\delta_o$, for what concerns the $\Delta\delta_i$'s. Moreover, using the following notation for brevity:

$$\left(\frac{\Delta u}{M}\right)_o \triangleq A_{1N} \sum_1^N {}_j (A^{-1})_{Nj} \frac{\Delta u_j}{M_j} \tag{8.2.24''}$$

the *mean motion* may be represented by the differential equations:

$$\left.\begin{aligned} \frac{\mathrm{d}\Delta\Omega_o}{\mathrm{d}t} &= \left(\frac{\Delta u}{M}\right)_o \\ \frac{\mathrm{d}\Delta\delta_o}{\mathrm{d}t} &= \Delta\Omega_o \end{aligned}\right\} \tag{8.2.25}$$

i.e., by a second-order system, with both the characteristic roots at the origin.

Furthermore it can be checked that, for every $j = 1, \ldots, N$, it results:

$$(A^{-1})_{Nj} = \frac{\gamma_j M_j}{\sum_1^N {}_k \gamma_k M_k} (A_{1N})^{-1} \tag{8.2.26}$$

where $\gamma_1, \ldots, \gamma_N$ are the cofactors of the $K_{1i}, \ldots, K_{Ni}$ elements of any i-th column of the K matrix (the arbitrariness of the column derives from [8.2.3], i.e., from the fact that, in the K matrix, each column is the opposite of the sum of the others). In fact, for any given $i = 1, \ldots, N$ it holds:

$$\sum_1^N {}_j \gamma_j K_{ji} = \det K = 0 \tag{8.2.27}$$

and similarly (since $A^{-1}M^{-1}K = -LA^{-1}$, $L_N = 0$):

$$\sum_1^N {}_j \frac{(A^{-1})_{Nj}}{M_j} K_{ji} = -L_N(A^{-1})_{Ni} = 0 \qquad [8.2.27']$$

so that the ratios $(A^{-1})_{Nj}/(M_j\gamma_j)$ $(j = 1, \ldots, N)$ are independent of the index j; recalling [8.2.20], the equation [8.2.26] can be then derived.
Alternatively it is possible to write, denoting by adj (...) the "adjoint" matrix:

$$\gamma_j = [\mathrm{adj}(K)]_{ij} \qquad \forall i = 1, \ldots, N$$

where (since $K = -MALA^{-1}$):

$$\mathrm{adj}(K) = A\,\mathrm{adj}(L)A^{-1}\mathrm{adj}(M)(-1)^{N-1}$$

so that, by developing, it can be derived:

$$\gamma_j = \frac{(A^{-1})_{Nj}}{M_j} A_{1N}\nu_1^2 \ldots \nu_{(N-1)}^2 M_1 \ldots M_N$$

and thus again Equation [8.2.26], as a result of [8.2.20]; in particular, note that:

$$\sum_1^N {}_k \gamma_k M_k = \nu_1^2 \ldots \nu_{(N-1)}^2 M_1 \ldots M_N$$

As a consequence, Equations [8.2.24], [8.2.24′], and [8.2.24″] may be finally rewritten as:

$$\Delta\Omega_o = \frac{\sum_1^N {}_i \gamma_i M_i \Delta\Omega_i}{\sum_1^N {}_k \gamma_k M_k} \qquad [8.2.28]$$

$$\Delta\delta_o = \frac{\sum_1^N {}_i \gamma_i M_i \Delta\delta_i}{\sum_1^N {}_k \gamma_k M_k} \qquad [8.2.28']$$

$$\left(\frac{\Delta u}{M}\right)_o = \frac{\sum_1^N {}_i \gamma_i \Delta u_i}{\sum_1^N {}_k \gamma_k M_k} \qquad [8.2.28'']$$

so that, in the definition of the mean motion:

- the variations of the angular positions and speeds must be combined linearly, with *weights proportional to* $\gamma_i M_i$, *and not simply* M_i, as it is frequently assumed without a precise justification;
- for the computation of $(\Delta u/M)_o$ (i.e., of the equivalent input, for what concerns the mean motion), the variations $\Delta u_1, \ldots, \Delta u_N$ and the inertia coefficients $M_1, \ldots, M_N$ must be *linearly combined*, with weights respectively proportional to $\gamma_1, \ldots, \gamma_N$, *and not simply added.*

Usually, however, the values of $\gamma_1, \ldots, \gamma_N$ are rather close to each other, so that it may be reasonably assumed $(\Delta u/M)_o \cong \sum_1^N {}_i \Delta u_i / \sum_1^N {}_k M_k$, etc. (see Section 8.2.4.)

(3) If we define:

$$\left.\begin{aligned} \Delta\Omega_{io} &\triangleq \Delta\Omega_i - \Delta\Omega_o \\ \Delta\delta_{io} &\triangleq \Delta\delta_i - \Delta\delta_o \\ \Delta u_{i(o)} &\triangleq \Delta u_i - M_i \left(\frac{\Delta u}{M}\right)_o \end{aligned}\right\} \qquad [8.2.29]$$

Equations [8.2.1] or [8.2.9], also may be substituted by:

- Equations [8.2.25], which define a system of second dynamic order, corresponding to the mean motion;
- the following additional equations:

$$\left.\begin{aligned} \frac{\mathrm{d}\Delta\Omega_{ko}}{\mathrm{d}t} &= \frac{1}{M_k}\left(\Delta u_{k(o)} - \sum_1^{N-1} {}_j K_{kj}(\Delta\delta_{jo} - \Delta\delta_{No})\right) \\ \frac{\mathrm{d}\Delta\delta_{ko}}{\mathrm{d}t} &= \Delta\Omega_{ko} \end{aligned}\right\} \quad (k = 1, \ldots, N-1) \qquad [8.2.30]$$

which define a system of $(2N-2)$ dynamic order, corresponding to the *motion of the machines with respect to the mean motion.*

Note that $\Delta\Omega_{No}$ can be derived starting from the values $\Delta\Omega_{io}$ for $i = 1, \ldots, N-1$, as it results (because of Equations [8.2.20] and [8.2.24]):

$$0 = \sum_1^N {}_i (A^{-1})_{Ni} \Delta\Omega_{io}$$

and a similar conclusion holds for $\Delta\delta_{No}$, $\Delta u_{N(o)}/M_N$, starting from $\Delta\delta_{io}$, $\Delta u_{i(o)}/M_i$, $i = 1, \ldots, N-1$.

It is interesting to observe that the systems [8.2.25] and [8.2.30] are *noninteracting*, as indicated in Figure 8.4 (this, in contrast, does not occur for the

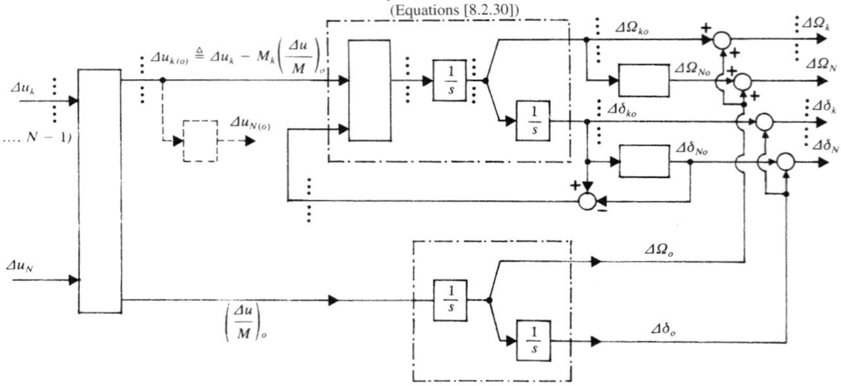

Figure 8.4. Variant of the block diagram to show the mean motion and the machine motion with respect to it.

systems in Fig. 8.3). Furthermore, the characteristic roots of the system [8.2.30] (machine motion with respect to the mean motion) coincide with those of the system [8.2.7] (relative machine motion between themselves), and consist of the $(N - 1)$ pairs of imaginary conjugate roots $\pm\tilde{j}\nu_1, \ldots, \pm\tilde{j}\nu_{(N-1)}$.

It is possible to conclude this both by considering Equations [8.2.21] and [8.2.21′], and by observing that such characteristic roots must be the same as for the full Equations [8.2.1], except for two roots at the origin which correspond to [8.2.25] (mean motion; see also Fig. 8.4) or to [8.2.6] and [8.2.8] (Fig. 8.3).

From the present formulations, which correspond to the "*modal*" *analysis* of the system [8.2.1], it is also possible to derive, in quite expressive terms, the scheme of Figure 8.5, where, in accordance with Equations [8.2.21] and [8.2.22]:

- the generic coefficient $(A^{-1})_{hj}/M_j$, which, for brevity, will be called "*excitance*," constitutes a measure of how much the h-th "mode" is excited by an equivalent variation (Δu_j) of driving power on the j-th machine $(h, j = 1, \ldots, N)$;
- the generic coefficient A_{ih}, which will be called "*accipiency*," constitutes a measure of how much the h-th "mode" is present in the variation of speed $(\Delta\Omega_i)$ of the i-th machine $(h, i = 1, \ldots, N)$;

whereas:

$$\frac{\mathrm{d}\Delta\delta_i}{\mathrm{d}t} = \Delta\Omega_i, \qquad \Delta\delta_i(s) = \frac{\Delta\Omega_i(s) + \Delta\delta_i^{(o)}}{s}$$

Specifically, as a result of Equations [8.2.18], *the accipiencies relative to the mode* $h = N$, which corresponds to the mean motion, *are all equal*.

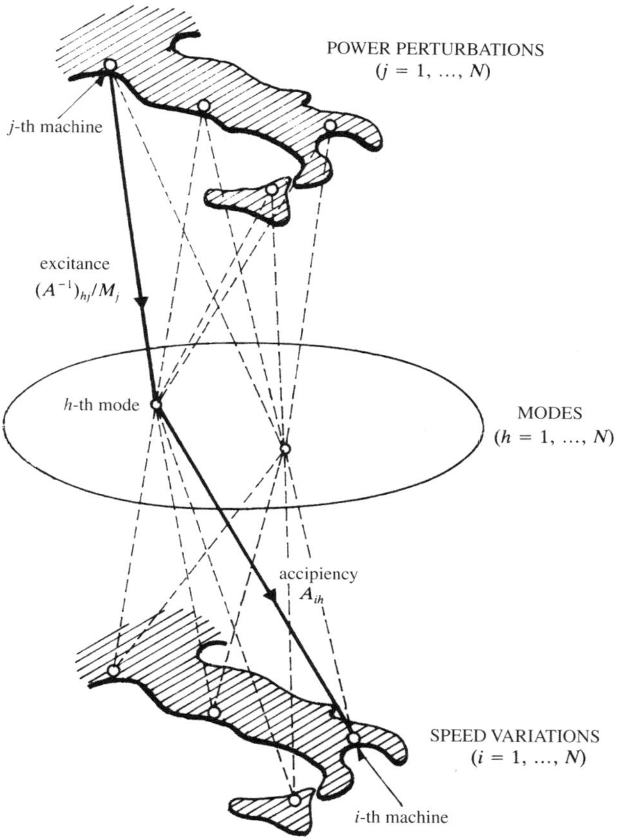

Figure 8.5. Effects of power perturbations on speed variations: interpretation in "modal" terms.

Actually, the "measures" expressed by the above-mentioned coefficients must be interpreted in relative terms among the different machines. On the other hand, it must be recalled that A is an eigenvector matrix, and thus, for any given h, all the accipiencies A_{ih} are defined with an arbitrary proportionality constant. This is also true for the excitances $(A^{-1})_{hj}/M_j$, with a proportionality constant that is the inverse of the previous one. To overcome this arbitrariness, such coefficients may be properly "normalized," e.g., according to Equations [8.2.38] described in the next section.

8.2.4. Particular Cases

If the synchronizing coefficients K_{ij} satisfy, in addition to [8.2.3], the conditions:

$$\sum_{1}^{N} {}_i K_{ij} = 0 \qquad \forall j = 1, \ldots, N \qquad [8.2.31]$$

i.e., *in the K matrix, the sum of the elements in each column is also equal to zero*, just as the case for the elements of each row, from Equations [8.2.27] and [8.2.27′] we may derive:

$$\gamma_1 = \cdots = \gamma_N \tag{8.2.32}$$

as well as:

$$\frac{(A^{-1})_{N1}}{M_1} = \cdots = \frac{(A^{-1})_{NN}}{M_N} \tag{8.2.32′}$$

(and vice versa, these last equations imply [8.2.31]). By letting $M_T \triangleq \sum_1^N {}_i M_i$ (total inertia coefficient), from Equation [8.2.20] or [8.2.26] it can be derived:

$$\frac{(A^{-1})_{Nj}}{M_j} = \frac{1}{A_{1N} M_T} \qquad \forall j = 1, \ldots, N \tag{8.2.33}$$

Therefore, in this case, even *the excitances relative to the mode* $h = N$ *are* (as the accipiencies) *all equal*. Furthermore:

$$\left.\begin{aligned} \Delta\Omega_o &= \sum_1^N {}_i M_i \Delta\Omega_i / M_T \\ \Delta\delta_o &= \sum_1^N {}_i M_i \Delta\delta_i / M_T \\ \left(\frac{\Delta u}{M}\right)_o &= \sum_1^N {}_i \Delta u_i / M_T \end{aligned}\right\} \tag{8.2.34}$$

according to what is usually adopted in the definition of the mean motion.

Finally, from Equation [8.2.15] it can be derived that $MAL = -KA$, and thus, for $h = 1, \ldots, N-1$:

$$M_i A_{ih} = -\sum_1^N {}_j K_{ij} A_{jh} / L_h$$

from which, under the present assumptions:

$$\sum_1^N {}_i M_i A_{ih} = 0 \qquad \forall h = 1, \ldots, N-1$$

The conditions [8.2.31] are equivalent to:

$$\frac{\partial \left(\sum_1^N {}_i P_{ei}\right)}{\partial \delta_j} = 0 \qquad \forall j = 1, \ldots, N \tag{8.2.35}$$

for which the total generated power $\left(\sum_1^N {}_i P_{ei}\right)$ is invariant to possible variations of the angular positions of the machines. This might occur in each of the following cases:

- the system supplied by the emfs consists of only reactive elements and, removing the linearity hypothesis, by loads at constant active power;
- as above but with loads (with active and reactive powers dependent on the voltage magnitude) perfectly voltage-regulated by a suitable action on the magnitudes $e_1, \ldots, e_N$ of the emfs;
- the matrix K is symmetric (see the following).

In fact, in the first two cases, the total power $\sum_1^N {}_i P_{ei}$ is constant and thus Equation [8.2.35] would apply (with K nonsymmetric in general), whereas in the third case [8.2.3] is directly translated into [8.2.31].

If *the matrix K is symmetrical*, the conditions [8.2.31] hold, also implying all the properties just described. Moreover, by recalling Equation [8.2.15] it can be written:

$$MALA^{-1} = -K = -K^T = (A^{-1})^T L A^T M$$
$$A^T MAL = LA^T MA$$

so that (because L is diagonal, having distinct elements) the matrix $[A^T MA]$ is diagonal. If:

$$\beta_h \triangleq ([A^T MA]_{hh})^{-1} = \left(\sum_1^N {}_k M_k A_{kh}^2\right)^{-1} \qquad (h = 1, \ldots, N) \qquad [8.2.36]$$

from $A^{-1} = [A^T MA]^{-1} A^T M$ it follows that:

$$\frac{(A^{-1})_{hj}}{M_j} = \beta_h A_{jh} \qquad (h, j = 1, \ldots, N) \qquad [8.2.37]$$

i.e., for each index h, *the excitances relative to the different machines are simply proportional to the respective accipiencies.*

Equation [8.2.37] may also suggest the following definition of "normalized" (nonadimensional) accipiencies and excitances:

$$\left.\begin{aligned} a_{ih} &\triangleq A_{ih}\sqrt{\beta_h} \\ e_{hj} &\triangleq \frac{(A^{-1})_{hj}}{M_j\sqrt{\beta_h}} \end{aligned}\right\} \qquad [8.2.38]$$

so that $a_{ih}e_{hj} = A_{ih}(A^{-1})_{hj}/M_j$ and moreover, for symmetrical K, $e_{hj} = a_{jh}$.

Finally, it is interesting to observe that the eigenvalues $L_1, \ldots, L_N$ of the matrix $[-M^{-1}K]$ are real. In fact, the equation:

$$0 = \det(-M^{-1}K - \lambda I_{(N)})$$

is equivalent to:

$$0 = \det(M^{-1/2}KM^{-1/2} + \lambda I_{(N)})$$

and thus the eigenvalues under consideration are also those of $[-M^{-1/2}KM^{-1/2}]$, which under the present assumption are real, because of the symmetry of this last matrix.

The assumption of a symmetric K may correspond to each of the following cases:

- the system supplied by the emfs consists of only reactive elements, and of loads of the synchronous type, or located at emf's terminals (this implies $G_{ij} = 0\ \forall i, j$, and from Equations [8.2.4] it follows $K_{ij} = K_{ji}$);
- as above, but with loads at constant active power and voltage magnitude (this situation also may be seen as a limit case of the previous one, by assimilating the loads under consideration to synchronous loads having negligible inertia);
- at the considered operating point, $\delta^o_{ij} = 0\ \forall i, j$, i.e., all the rotors are in phase with each other (from Equations [8.2.4] it follows $K_{ij} = K_{ji}$).

Such examples obviously constitute ideal cases. However, the treatment based on the assumption of a symmetrical K retains its precise interest in that—for many practical cases, because of the concomitance of situations not far from the ones described—the dissymmetries of K are actually relatively modest and, above all, Equations [8.2.34] and [8.2.37] appear applicable as a good approximation.

If one of the emfs (let us assume the N-th) corresponds to an "*infinite power*" *network*, the model defined by Equations [8.2.1] is translated into the following one, with a dynamic order $(2N - 2)$:

$$\begin{cases} \dfrac{\mathrm{d}\Delta\Omega_i}{\mathrm{d}t} = \dfrac{1}{M_i}\left(\Delta\hat{u}_i - \displaystyle\sum_1^{N-1}{}_j K_{ij}\Delta\delta_j\right) \\ \dfrac{\mathrm{d}\Delta\delta_i}{\mathrm{d}t} = \Delta\Omega_i \end{cases} \qquad (i = 1, \ldots, N-1)$$

where the inputs $\Delta\hat{u}_i \triangleq \Delta u_i - K_{iN}\Delta\delta_N$ also consider, for a greater generality (see Section 7.1), possible variations in the magnitude and the phase of the emf $\overline{e}_N$ (remember that the effect of Δe_N is already included in the Δu_i's; see the first of Equations [8.2.2]).

Through considerations similar to those already developed, it can be shown that:

- the $(2N-2)$ characteristic roots are given by the square roots of the eigenvalues $\hat{L}_1, \ldots, \hat{L}_{(N-1)}$ of the matrix $[-\hat{M}^{-1}\hat{K}]$, where:

$$\begin{cases} \hat{M} \triangleq \mathrm{diag}\{M_1, \ldots, M_{(N-1)}\} \\ \hat{K} \triangleq \{K_{ij}\} \qquad \text{with } i, j = 1, \ldots, N-1 \end{cases}$$

- for the (weak) stability of the linearized system, it is necessary and sufficient that the above-mentioned eigenvalues are distinct, real, and negative;
- the system then exhibits $(N-1)$ pairs of imaginary conjugate characteristic roots $\pm\tilde{\jmath}\hat{\nu}_h$, corresponding to as many oscillatory modes, having zero damping and resonance frequencies $\hat{\nu}_h = \sqrt{-\hat{L}_h}$ $(h = 1, \ldots, N-1)$;
- expressions similar to Equations [8.2.21] and [8.2.21′], etc. can be derived, without the terms corresponding to the mean motion;
- if the matrix $\hat{K}$ is symmetrical, the proportionality between the excitances and accipiencies is found again for each oscillatory mode (moreover, the eigenvalues $\hat{L}_h$ are real).

Finally, the extension to the case of several infinite power networks is obvious.

8.2.5. Time-Domain Response

From Equations [8.2.21] and [8.2.23] it can be derived, in general, that:

$$\Delta\Omega_i(t) = \sum_1^{N-1}{}_h\, \mathcal{L}^{-1}\left(\frac{s}{s^2+\nu_h^2}\phi_{\Omega_i}^{(h)}(s)\right) + \Delta\Omega_o(t) \qquad [8.2.39]$$

$(i = 1, \ldots, N)$, where, by recalling the first of Equations [8.2.25]:

$$\begin{aligned} \Delta\Omega_o(t) &= A_{1N}\sum_1^N{}_j (A^{-1})_{Nj}\left(\int_0^t \frac{\Delta u_j(t)}{M_j}\,\mathrm{d}t + \Delta\Omega_j^{(o)}\right) \\ &= \int_0^t \left(\frac{\Delta u}{M}\right)_o (t)\,\mathrm{d}t + \Delta\Omega_o^{(o)} \end{aligned} \qquad [8.2.40]$$

In the following, the analysis of the variations $\Delta\delta_i(t) (i = 1, \ldots, N)$ and $\Delta\delta_o(t)$ is omitted for simplicity; note that $\Delta\delta_i(t) = \Delta\delta_i^{(o)} + \int_0^t \Delta\Omega_i(t)\,\mathrm{d}t$, $\Delta\delta_o(t) = \Delta\delta_o^{(o)} + \int_0^t \Delta\Omega_o(t)\,\mathrm{d}t$.

In Equation [8.2.39], the generic function $\mathcal{L}^{-1}(\ldots)$ includes a sinusoidal term equal to:

$$|\phi_{\Omega_i}^{(h)}(\tilde{\jmath}\nu_h)| \cos\left(\nu_h t + \angle\phi_{\Omega_i}^{(h)}(\tilde{\jmath}\nu_h)\right)$$

the magnitude of which is proportional to $|A_{ih}|$ as a result of Equation [8.2.22]. (It is assumed that the functions $\Delta u_j(s)$, and thus $\phi_{\Omega_i}^{(h)}(s)$, have no poles at $\tilde{j}\nu_h$.)

Therefore:

- $(N-1)$ sinusoidal components at the frequencies $\nu_1, \ldots, \nu_{(N-1)}$ (electromechanical oscillations) are present in each of the $\Delta\Omega_i(t)$.
- For any given h, *the magnitude of the oscillation varies from one machine to the other, proportionally to the magnitude of the respective accipiency* $A_{ih} (i = 1, \ldots, N)$. This is true *independently of the nature, the size, and the point of application of the perturbations.*

The sign of A_{ih} influences instead the oscillation phase, as the machines with accipiencies of the same sign oscillate in phase with each other, and in opposite phase with respect to the remaining ones.

In particular, the "free" response (i.e., the response with zero inputs Δu_i, $i = 1, \ldots, N$) is defined by:

$$\Delta\Omega_i(t) = \sum_1^{N-1}{}_h A_{ih} \sum_1^{N}{}_j (A^{-1})_{hj} (\Delta\Omega_j^{(o)} \cos \nu_h t - \nu_h \Delta\delta_j^{(o)} \sin \nu_h t) + \Delta\Omega_o(t) \tag{8.2.41}$$

$$\Delta\Omega_o(t) = A_{1N} \sum_1^{N}{}_j (A^{-1})_{Nj} \Delta\Omega_j^{(o)} = \Delta\Omega_o^{(o)} = \text{constant} \tag{8.2.42}$$

Regarding the "forced" response (i.e., the response at zero initial conditions $\Delta\Omega_j^{(o)}$, $\Delta\delta_j^{(o)}$), if:

$$\begin{cases} \Delta u_j(t) = 1(t) & \text{(unit step)} \\ \Delta u_k(t) = 0 & \forall k \neq j;\ k = 1, \ldots, N \end{cases}$$

it can be derived:

$$\Delta\Omega_i(t) = \sum_1^{N-1}{}_h \frac{A_{ih}(A^{-1})_{hj}}{\nu_h M_j} \sin \nu_h t + \Delta\Omega_o(t) \tag{8.2.43}$$

$$\Delta\Omega_o(t) = \frac{A_{1N}(A^{-1})_{Nj}}{M_j} t \tag{8.2.44}$$

In particular, from Equation [8.2.43] it is possible to derive that, for any given h, the machines oscillate at zero phase or at 180° phase, and at different magnitudes, according to the value of $A_{ih}(A^{-1})_{hj}/M_j$, where the accipiency A_{ih} depends on the machine considered, whereas the excitance $(A^{-1})_{hj}/M_j$ depends on the point of step application.

The initial accelerations are:

$$\frac{d\Delta\Omega_i}{dt}(0^+) = \sum_{1}^{N-1}{}_h \frac{A_{ih}(A^{-1})_{hj}}{M_j} + \frac{A_{1N}(A^{-1})_{Nj}}{M_j} = \begin{cases} 1/M_j & \text{for } i = j \\ 0 & \text{for } i \neq j \end{cases}$$

i.e., as it is obvious, the unit step in Δu_j causes a nonzero initial acceleration (equal to $1/M_j$) only on the j-th machine to which it is applied.

The effects of a sudden structure perturbation (e.g., a line opening, a load rejection, a generator tripping, etc.) may be assimilated to those of a set of steps $\Delta u_i = -\Delta P_{ei}^S$ (see the first of Equations [8.2.2]), where ΔP_{ei}^S is the variation of P_{ei} caused by the perturbation considered, at the instant the perturbation occurs.

In Equations [8.2.41] and [8.2.43], the term $\Delta\Omega_o(t)$ represents not only an "ensemble average" (as already specified) of $\Delta\Omega_1(t), \ldots, \Delta\Omega_N(t)$ at each instant, but also, with its time behavior, the "*time average*" of each of the $\Delta\Omega_i(t)$. Such a property is important as it may allow to experimentally derive the means frequency starting from only one of the machine speeds, or from the frequency at a generic network location. In particular, the response to a sudden structure perturbation is (with the present model):

$$\Delta\Omega_o(t) = -A_{1N}\left(\sum_{1}^{N}{}_j (A^{-1})_{Nj}\frac{\Delta P_{ej}^S}{M_j}\right)t = -\frac{\sum_{1}^{N}{}_j \gamma_j \Delta P_{ej}^S}{\sum_{1}^{N}{}_k \gamma_k M_k}t \qquad [8.2.45]$$

which, by assuming that the coefficients $(A^{-1})_{Nj}/M_j$ or equivalently $\gamma_j (j = 1, \ldots, N)$ are slightly different from each other, may be approximated by:

$$\Delta\Omega_o(t) \cong -\frac{\sum_{1}^{N}{}_j \Delta P_{ej}^S}{M_T}t \qquad [8.2.45']$$

and this allows one to experimentally derive (with some approximations and with the condition that the actual total perturbation $\sum_1^N{}_j \Delta P_{ej}^S$ is evaluated with the required care; see Section 3.3.3) the total inertia coefficient M_T, after estimating $\Delta\Omega_o(t)$ from the frequency recording at a generic network location.

8.3. STABILITY OF THE RELATIVE MOTION

To study the stability without explicitly resolving the system equations by simulation, the methods based on the Lyapunov theory may be (at least in principle) used.

This approach may generally seem particularly attractive, not only as an alternative to the simulation methods for stability checks (in effect, the calculation of the transients may be found to be rather heavy, and may also be superfluous when the only information required concerns the stability), but also as a rapid mean for the detection of critical situations, as can be required in planning studies (with many cases to analyze) or for checking the system during operation (with very little time available).

However, the practical applicability of the methods under consideration is radically limited by various orders of difficulty, and in particular:

- the difficulty to find a suitable "Lyapunov function" (on which the analysis is based), unless the model of the system is sufficiently simple;
- the difficulty, of both an analytical and computational nature, to determine the "stability region" (in the state space) around a given stable equilibrium point[(1)].

Therefore, in the following we will limit ourselves to outline a possible procedure, illustrating also the limitations and difficulties that it may imply.

In the field of multimachine systems, an example of application regarding the stability of the relative motion (and in particular, for large variations, the "keeping in step" of the machines) is to use the first approximation model defined by Equations [8.1.12], with the further hypotheses:

$$P_{mi(o)} \triangleq P_{mi} - M_i \frac{\sum_1^N {}_k P_{mk}}{M_T} = \text{constant} \qquad \forall i = 1, \ldots, N$$

where $M_T \triangleq \sum_1^N {}_k M_k$ is the total inertia coefficient, or equivalently (for variations of any magnitude):

$$\frac{\Delta P_{m1}}{M_1} = \cdots = \frac{\Delta P_{mN}}{M_N} \qquad [8.3.1]$$

and furthermore:

$$e_i = \text{constant} \qquad \forall i = 1, \ldots, N$$

$$P_{ei(o)} \triangleq P_{ei} - M_i \frac{\sum_1^N {}_k P_{ek}}{M_T} = P_{ei(o)}(\delta_{1N}, \ldots, \delta_{(N-1)N})$$

---

(1) It is assumed that the system configuration does not change from the instant t^i (referred to as the initial instant). In the presence of structure perturbations, e.g., caused by a short-circuit and/or intervention of protections, it must be assumed that all perturbations occur within $t < t^i$, and their effect on the initial state of the system ($t = t^i$) must be evaluated in advance.

with:

$$\frac{\partial P_{ei(o)}}{\partial \delta_{jN}} = \frac{\partial P_{ej(o)}}{\partial \delta_{iN}} \qquad \forall i, j = 1, \ldots, N-1 \tag{8.3.2}$$

In practice, Equations [8.3.1] do not appear to be particularly limiting. In fact, the variations ΔP_{mi} caused by the speed regulators (relatively slow, and thus often negligible within the scope of the considered problems) as a first approximation can likely be considered proportional to the nominal powers, or even to the inertia coefficients M_i of the respective units.

Equation [8.3.2] must then apply for any set of values δ_{iN} and is verified if the loads are synchronous, or directly supplied by the emfs, or characterized by constant active power and voltage magnitude, whereas all the other network elements are purely reactive (symmetrical matrix K).

Note that $\sum_1^N {}_k P_{mk} \neq \sum_1^N {}_k P_{ek}$ may hold; therefore, it also is possible to analyze the stability, e.g., after a generator tripping, load-shedding, or after the separation from other systems.

Under the above stated hypotheses, the following "*Lyapunov function*" can be assumed:

$$\mathcal{V}(\delta', \Omega') = W(\delta') + C(\Omega') \tag{8.3.3}$$

where:

$$W(\delta') \triangleq \sum_1^{N-1} {}_i \int_{\delta_{iN}^{(R)}}^{\delta_{iN}} P_{ei(o)}(\delta_{1N}, \ldots, \delta_{iN}, \delta_{(i+1)N}^{(R)}, \ldots, \delta_{(N-1)N}^{(R)})\, \mathrm{d}\delta_{iN}$$

$$+ \sum_1^{N-1} {}_i P_{mi(o)}(\delta_{iN}^{(R)} - \delta_{iN}) \tag{8.3.3$'$}$$

$$C(\Omega') \triangleq \frac{1}{2M_T} \sum_1^{N-1} {}_i M_i \sum_{i+1}^{N} {}_j M_j (\Omega_{iN} - \Omega_{jN})^2 \tag{8.3.3$''$}$$

where $\delta' \triangleq [\delta_{1N} \ldots \delta_{(N-1)N}]^T$, $\Omega' \triangleq [\Omega_{1N} \ldots \Omega_{(N-1)N}]^T$, while $\delta_{1N}^{(R)}, \ldots, \delta_{(N-1)N}^{(R)}$ may take arbitrary values. Equation [8.3.3] constitutes a generalization of [1.6.2], which is derived from the case of two machines.

For the function $\mathcal{V}(\delta', \Omega')$ we have, because of Equations [8.1.12]:

$$\dot{\mathcal{V}} \triangleq \sum_1^{N-1} {}_i \left(\frac{\partial \mathcal{V}}{\partial \delta_{iN}} \frac{\mathrm{d}\delta_{iN}}{\mathrm{d}t} + \frac{\partial \mathcal{V}}{\partial \Omega_{iN}} \frac{\mathrm{d}\Omega_{iN}}{\mathrm{d}t} \right) = 0 \tag{8.3.4}$$

in the whole state space $\{\delta', \Omega'\}$. Therefore, it can be stated that, along every generic trajectory of the system, the $\mathcal{V}$ function remains constant (an analogy can be seen with a conservative system characterized by a constant $\mathcal{V}$ energy).

It can be further verified that:

- the equilibrium points may at most be "weakly" stable (recall the conditions in Section 8.2.2);
- in the (weakly) stable equilibrium points (and only in them), the function $\mathcal{V}$ exhibits a minimum;
- consequently, around the mentioned points, the surfaces $\mathcal{V} =$ constant are closed.

(Moreover, recall that, at each equilibrium point, $\Omega' = 0$, $C(\Omega') = 0$.)

The "*stability region*" $\mathcal{R}(\delta'^o, 0)$, corresponding to a given stable equilibrium point $(\delta'^o, 0)$, can be defined as the state space region within which the surfaces $\mathcal{V} =$ constant maintain themselves closed around $(\delta'^o, 0)$ (recall e.g., for $N = 2$, the dashed region in Figure 1.7, around the point $(\delta^o_{ab}, 0)$).

To check maintainance of machine synchronism starting from an assigned initial point (δ'^i, Ω'^i) of the state space, it is sufficient to determine that a point $(\delta'^o, 0)$ exists such that (δ'^i, Ω'^i) pertains to $\mathcal{R}(\delta'^o, 0)$. In the presence of damping, which is ignored here, the point $(\delta'^i, 0)$ might define the final equilibrium point to which reference must be made in the "static security" checks (see Section 2.2.5c).

As in Section 1.6 for the case $N = 2$, the maintenance of synchronism may be checked through a procedure of the following type:

- starting from (δ'^i, Ω'^i), let us move in the state space with continually decreasing values of $\mathcal{V}$, down until a minimum of the function $\mathcal{V}$ is reached, i.e., a stable equilibrium point $(\delta'^o, 0)$;
- determine the (unstable) equilibrium points, which can be connected to $(\delta'^o, 0)$ through a path for which $\mathcal{V}$ is always decreasing;
- check that it holds that:
$$\mathcal{V}(\delta'^i, \Omega'^i) < \mathcal{V}^L \qquad [8.3.5]$$
where $\mathcal{V}^L$is the minimum among the values assumed by $\mathcal{V}$ in the above-mentioned points ($\mathcal{V}^L$is the upper limit of the values of $\mathcal{V}$ in $\mathcal{R}(\delta'^o, 0)$).

It is clear, according to what has been already stated, that the procedure described can imply considerable practical difficulties (particularly for the evaluation of $\mathcal{V}^L$) if the number of machines is not small. In particular, it may be necessary to calculate in advance an enormous number of equilibrium points, to check their connection to $(\delta'^o, 0)$, as specified above.

Some modest simplifications may derive from the fact that, at the equilibrium points, it holds $\Omega' = 0$, $\mathcal{V} = W$, so that the determination of $(\delta'^o, 0)$ and of $\mathcal{V}^L$ can be performed based on the behavior only of the function W in the subspace $\{\delta'\}$, whereas the term $C(\Omega')$ of Equation [8.3.3] must be considered only for evaluating $\mathcal{V}(\delta'^i, \Omega'^i)$.

However, for particular classes of problems, it is possible to obtain, through a supplementary theoretical investigation, an analytical formulation of the stability properties. Assume, for example, that the dependence of the P_{ei}'s on the phase shifts δ_{kN} can be expressed by equations:

$$P_{ei} = b_i \sum_{1}^{N} {}_j b_j \sin \delta_{ij} \qquad (i = 1, \ldots, N) \tag{[8.3.6]}$$

with $b_1, \ldots, b_N$ constant and all being positive, $\delta_{ij} \triangleq \delta_i - \delta_j$, and moreover that:

$$P_{mi(o)} = 0 \qquad \forall i = 1, \ldots, N-1 \tag{[8.3.7]}$$

Equation [8.3.6] may be verified, e.g., in the case of a system like the one indicated in Figure 8.6a, as specified in the following, or also generally in the case of Figure 8.6b. Equation [8.3.7], which also implies $P_{mN(o)} = 0$, then corresponds to the conditions $P_{m1}/M_1 = \cdots = P_{mN}/M_N$, by which the powers P_{mi} (which may possibly vary) must practically be, with a good approximation, proportional to the respective nominal values. The analysis can be extended, without excessive difficulties, to the rather realistic case of $P_{mi(o)}$ with relatively small absolute value.

Under the adopted hypotheses, and by proper development, it can be derived:

$$W(\delta') = \frac{1}{2}\left[\sum_{1}^{N} {}_i b_i^2 - \left(\sum_{1}^{N} {}_i b_i \cos \delta_{iN}\right)^2 - \left(\sum_{1}^{N-1} {}_i b_i \sin \delta_{iN}\right)^2\right] \tag{[8.3.8]}$$

apart from a constant additional term that depends on the (arbitrary) choice of $\delta_{1N}^{(R)}, \ldots, \delta_{(N-1)N}^{(R)}$, and that therefore may be assumed to be zero, in a simpler way and without any consequence on the results of the analysis.

Furthermore, it can be demonstrated that:

- the equilibrium points can be of the following three types:
  - solutions of the "type 0" (stable): those solutions with all the rotors in phase (apart from multiples of 360°);
  - solutions of the "type 1" (unstable): those solutions with one or more rotors in phase with each other and in opposite phase with respect to the remaining ones;
  - solutions of the "type 2" (unstable), with notzero $\sin \delta_{iN}$ at least for a value of the index i;
- by assuming $\mathcal{V}^L \triangleq \mathcal{V}(\delta'^L, 0) = W(\delta'^L)$, the search for δ'^L may be limited to the "type 1" solutions;

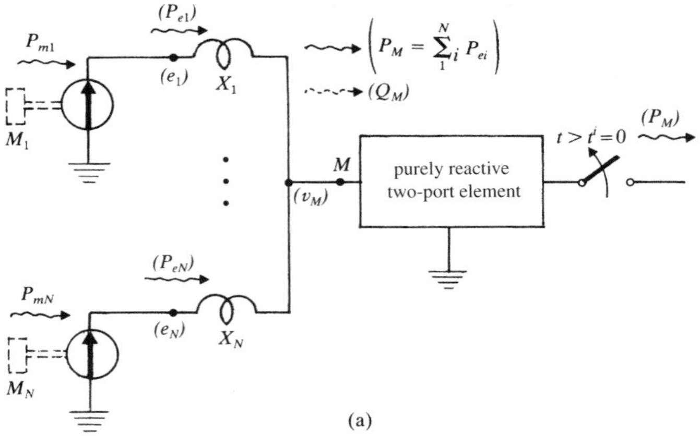

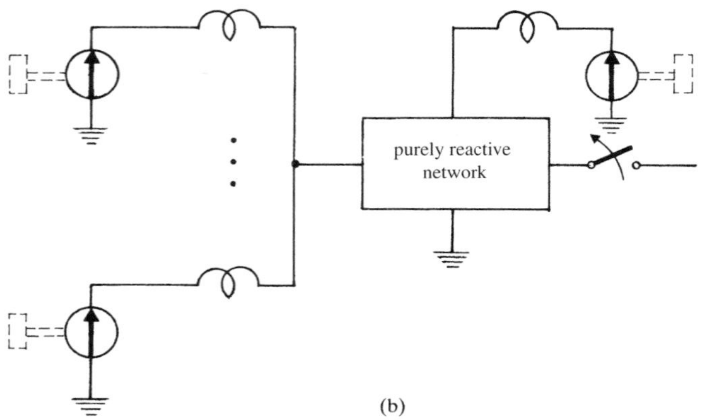

Figure 8.6. Stability analysis for large variations: (a) system under examination; (b) possible generalization of the system under examination.

- by examining the values of W at the “type 1” solutions, it then follows:

$$\mathcal{V}^L = \frac{1}{2}\left[\sum_1^N {}_h b_h^2 - \left(\sum_1^N {}_h b_h - 2b_{\min}\right)^2\right] \qquad [8.3.9]$$

with:

$$b_{\min} \triangleq \min(b_1, \ldots, b_N)$$

(note that, if $b_{\min} = b_s$, the solution δ'^L is the one for which the s-th rotor is in opposite phase with respect to all the other ones).

If the example of Figure 8.6a (with $t^i = 0$) is considered, assuming that X_M is the reactance of the two-port element as seen from the node M at $t > 0$, and moreover:

$$X_p \triangleq \left(\frac{1}{X_M} + \sum_1^N {}_i \frac{1}{X_i}\right)^{-1} > 0$$

it can be derived (see Equation [8.3.6]):

$$b_i = \frac{e_i}{X_i}\sqrt{X_p} \qquad (i = 1, \ldots, N)$$

Furthermore, the value $W(\delta'^i)$ that corresponds to the initial point, can be expressed as a function of the values assumed by P_M, Q_M and v_M (see Fig. 8.6a) at $t = 0^-$. Finally, the stability condition [8.3.5] may be rewritten in the form:

$$\left[\left(\frac{Q_M}{v_M} + v_M \sum_1^N {}_i \frac{1}{X_i}\right)^2 + \left(\frac{P_M}{v_M}\right)^2\right]_{t=0^-} > \left(\sum_1^N {}_h \frac{e_h}{X_h} - 2\left(\frac{e}{X}\right)_{\min}\right)^2 + \frac{2C(\Omega'^i)}{X_p} \qquad [8.3.10]$$

with:

$$\left(\frac{e}{X}\right)_{\min} \triangleq \min\left(\frac{e_1}{X_1}, \ldots, \frac{e_N}{X_N}\right)$$

This corresponds to the stability zone indicated in Figure 8.7.

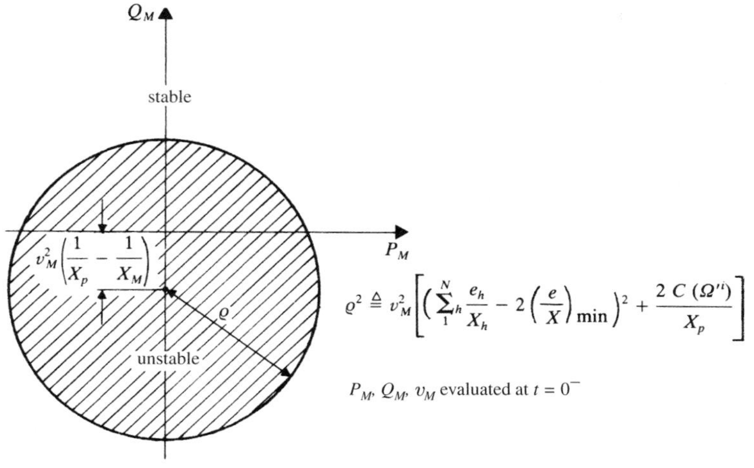

Figure 8.7. Stability zone for large variations.

In particular, if the system is at equilibrium for $t < 0$, in condition [8.3.10] it results that $C(\Omega'^i) = 0$, whereas it can be demonstrated that, for $t < 0$:

$$\frac{Q_M}{v_M} + v_M \sum_1^N {}_i \frac{1}{X_i} = \sum_1^N {}_h \sqrt{\left(\frac{e_h}{X_h}\right)^2 - \left(\frac{M_h}{M_T}\frac{P_M}{v_M}\right)^2}$$

so that [8.3.10] is finally translated only into a condition on the value assumed by $(P_M/v_M)^2$ for $t < 0$.

It is possible to determine that such a value cannot be larger than $[M_T(e/XM)_{\min}]^2$, with:

$$\left(\frac{e}{XM}\right)_{\min} \triangleq \min\left(\frac{e_1}{X_1M_1}, \ldots, \frac{e_N}{X_NM_N}\right)$$

Therefore, it appears significant to check if the stability is guaranteed within the whole interval of admissibility for $[(P_M/v_M)^2]_{t<0}$ (this case corresponds to a "total security" condition) or only within a part of it ("conditioned security"). Note that the stability condition is certainly verified for some values of $[(P_M/v_M)^2]_{t<0}$; e.g., when such a value is zero.

Referring to the problem considered, it is useful to notice that the left-hand side of condition [8.3.10] is a nonincreasing function of $[(P_M/v_M)^2]_{t<0}$. As a consequence, by putting into [8.3.10]:

$$\left[\left(\frac{P_M}{v_M}\right)^2\right]_{t<0} = \left[M_T\left(\frac{e}{XM}\right)_{\min}\right]^2$$

which is the maximum admissible value, the following *total security condition* can be derived:

$$\left[\sum_1^N {}_h \sqrt{\left(\frac{e_h}{X_h}\right)^2 - \left(M_h\left(\frac{e}{XM}\right)_{\min}\right)^2}\right]^2 + \left[M_T\left(\frac{e}{XM}\right)_{\min}\right]^2$$
$$> \left[\sum_1^N {}_h \frac{e_h}{X_h} - 2\left(\frac{e}{X}\right)_{\min}\right]^2 \qquad (8.3.11]$$

which constitutes a condition on $(2N-2)$ adimensional parameters (e.g., the $(N-1)$ ratios between the e_i/X_i's, and those between the M_i's, whereas the reactance X_M does not appear in the condition itself).

For example, if:

$$\left(\frac{e}{X}\right)_{\min} = \frac{e_N}{X_N}$$

and moreover:

$$\begin{cases} \dfrac{e_1}{X_1M_1} = \cdots = \dfrac{e_r}{X_rM_r} \\ \dfrac{e_{(r+1)}}{X_{(r+1)}M_{(r+1)}} = \cdots = \dfrac{e_N}{X_NM_N} \end{cases}$$

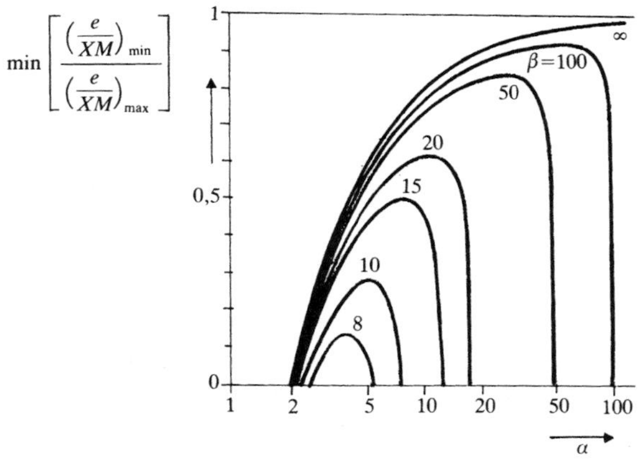

Figure 8.8. "Total" security condition (see text).

(with $0 \le r < N$), assuming:

$$\begin{cases} \alpha \triangleq \sum_1^r {}_i \dfrac{e_i/X_i}{e_N/X_N} \\ \beta \triangleq \sum_1^N {}_h \dfrac{e_h/X_h}{e_N/X_N} \end{cases}$$

the condition [8.3.11], through proper developments, can be translated into:

$$\frac{\left(\dfrac{e}{XM}\right)_{\min}}{\left(\dfrac{e}{XM}\right)_{\max}} > 1 - \frac{2(\beta - 1)}{\alpha(\beta - \alpha)} \qquad [8.3.12]$$

The above is related to the diagrams of Figure 8.8. (Note that $\alpha \ge r$, $\beta \ge N$, $\beta - \alpha \ge N - r > 0$ and, furthermore, that condition [8.3.12] is certainly satisfied if $\alpha \le 2$, or if $\beta \le 4 + 2\sqrt{2} \cong 6.83$.)

8.4. SIMPLIFICATION OF THE OVERALL MODEL

8.4.1. Generalities

In the dynamic analysis of a multimachine system it is often necessary to make adequate simplifications; in fact, the use of an overdetailed model may be:

- *burdensome*, not only for the analysis itself, but also for other aspects such as data collection (and consequent setting up of the model), simulations by computer, etc.

- *inadequate*, above all if one considers the practical difficulty that exists in defining a complete model for all system parts, and in obtaining well-founded information on all the necessary data (especially on external systems, interconnected with that of more direct interest), as well as the minor importance of many model details, for the specific problem under examination.

In particular, then, the simplified model should be realizable based on *essential and actually available data*.

The simplifications may concern not only the individual components, as already underlined in Section 8.1, but also the whole system. Even if the individual components are represented in a very simplified manner, the overall model may, in fact, be unacceptable especially because of the great number of machines and the nonlinear characteristics of interaction between them.

Moreover, the simplifications may be required for the entire system or only for some parts of it. For instance, given parts of the system (with or without the respective controlling systems) must be represented by a "simplified equivalent" as viewed from the rest of the system; or the whole set of machines, network, and loads, must be represented by a simplified equivalent as viewed from the controlling systems (see Fig. 8.9).

The validity of the simplifications, within the requested approximations and with regard to the particular problem examined, may be supported by:

- intuitive considerations: for instance, two or more machines may likely be considered "coherent" (i.e., with equal electrical speeds) if they are electrically close enough to each other;
- experimental or simulation tests: for instance, the "spectral" analysis of the interesting variables may reveal, in linearized terms, the existence of only a few "dominant modes," as in the case of a system of dynamic order much smaller than the actual one;
- theoretical justifications: for instance, if the transients to be analyzed have a short duration, only the "transient" characteristics of the machines etc. may seem essential, whereas the analysis of the relatively slow transients (such as those related to the f/P control) can make use of significant simplifications for the faster transient components; moreover, possible system parts that are subject to small perturbations, can be represented by linearized models.

Finally, the degree of approximation of the simplified model can be defined on the basis of various objectives, among which the satisfactory reproduction of the behavior of some selected variables (voltages, speeds, power flows at given points of the network, etc.), for a given time interval (for instance, up to a few seconds, or to several minutes) and after given perturbations; or the

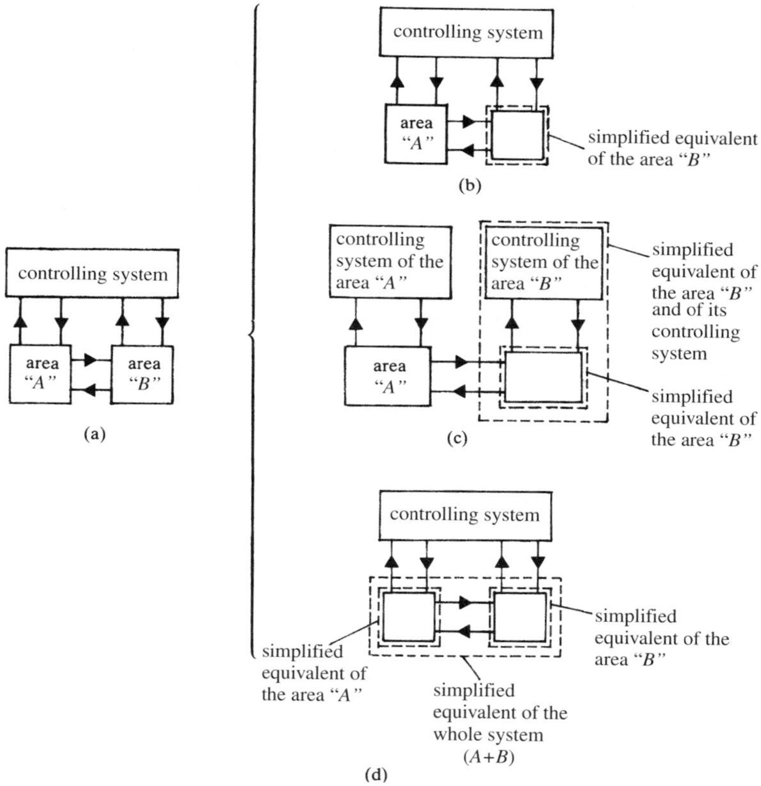

Figure 8.9. Simplification of the overall model: (a) original model; (b), (c), (d) typical examples of simplified models. The model (c) assumes that the controlling systems of areas "A" and "B" can be considered separately.

satisfactory reproduction of some properties of the real system (for instance, stability limits)(2).

8.4.2. "Area" Models

The common simplified models are based on the definition of one or more (geographically bounded) "areas", within each of which some simplification criteria *independent of the rest of the system* are applied. As already pointed out, the area(s) under study may cover the whole or part of the system.

(2) It also must be remembered that the required degree of approximation may be very different according to the type of problem, e.g., previsional scheduling, reconstruction (through off-line simulation) of events actually occurred, and so on.

Clearly, the wide use of these models is justified by their flexibility and the relative simplicity in setup, because the simplification of the overall model is the result of one or several partial simplifications made independently of each other.

The quality of a multiarea model depends on the choice of the areas themselves and on the simplification criteria applied to each. Once the choice of the areas has been determined, the model setup is generally quite simple; however, the choice of areas may imply some uncertainties or perplexities for its possible consequences on the resulting approximation. The boundaries of one or several areas are sometimes suggested by considerations of practical convenience, based, for instance, on strictly geographical criteria. Moreover, an external system not well-known in detail is often assumed as a single area, since rough representation of such a system is presumably acceptable, also considering the uncertainties of the corresponding data.

The more usual simplification criterion is based on the assumption that, *within each area, the rotors of the machines are "coherent" to each other*. Therefore, if for instance the considered area includes the machines $1, \ldots, r$, it is assumed that the phase displacements $(\delta_i - \delta_j)$ remain constant in time for all the $i, j = 1, \ldots, r$, and that the electrical angular speeds $\Omega_1, \ldots, \Omega_r$ are equal to each other at any time instant.

For a reasonable application of such criterion, it is appropriate to make preliminary checks about the practical "coherency" of the machines assigned to each area, for example by analyzing the behavior of the system for small variations, through simulations or, less empirically, by means of modal analysis techniques of the type described in Section 8.4.3. Ignoring such checks, the choice of areas may be based on more or less empirical evaluations of the "electrical distance" between the machines. A rather used procedure, for the case in which the perturbation is constituted by a zero-impedance, three-phase short-circuit (at the instant $t = 0$), is for example the following one:

- Based on the model represented in Figure 8.1, the relative voltage drops, caused (by the short-circuit) at the terminals of the different machines, are evaluated for $t = 0^+$.
- Two machine subsets (A, B) are defined, respectively, constituted by the machines that have relative voltage drops larger than a predetermined value (subset A) and by the remaining ones (subset B); each machine pertaining to the subset A will be kept in the final model, whereas those pertaining to the subset B will be grouped into one or more areas.
- The machines $i \in B$ which have, with respect to the subset A, values of the "*electrical distance*" $d(i)$ not much different, are grouped into the same area (i.e., they are considered as "coherent" with each other); the generic $d(i)$ may be, for example, simply defined by:

$$d(i) = \max_{j \in A} |\overline{Y}_{ij}| \qquad (i \in B)$$

where $\overline{Y}_{ij}$ is the mutual admittance between the emfs $\overline{e}_i$, $\overline{e}_j$ (see Equation [8.1.4]), or by the so-called "reflection coefficient":

$$d(i) = \max_{j \in A} |a_{ji}| \qquad (i \in B)$$

where a_{ji} is the acceleration of the j-th machine caused by the variation $\Delta\delta_i$ after a sufficiently small time τ (following the short-circuit). By obvious notation, the accelerations a_{ji} are evaluated based on $(\mathrm{d}\Omega_i/\mathrm{d}t)(0^+) = (P_{ei}(0^-) - P_{ei}(0^+))/M_i$, $\Delta\delta_i(\tau) \cong (\mathrm{d}\Omega_i/\mathrm{d}t)(0^+)\tau^2/2$, $a_{ji} \cong K_{ji}\Delta\delta_i(\tau)/M_i$.

Through such a procedure, it is possible to obtain relatively wide areas, however based on an assumed "coherency" between machines that actually can be far from each other.

The equations of mechanical balance for the individual machines of the generic area (i.e., Equations [8.1.1] with $i = 1, \ldots, r$) are then simply replaced by:

$$\left.\begin{aligned} \frac{\mathrm{d}\Omega_a}{\mathrm{d}t} &= \frac{1}{M_a}(P_{ma} - P_{ea}) \\ \frac{\mathrm{d}\delta_a}{\mathrm{d}t} &= \Omega_a - \Omega_s \end{aligned}\right\} \qquad [8.4.1]$$

by letting:

$$\left.\begin{aligned} \Omega_a &\triangleq \Omega_1 = \cdots = \Omega_r \\ \Delta\delta_a &\triangleq \Delta\delta_1 = \cdots = \Delta\delta_r \end{aligned}\right\} \qquad [8.4.2]$$

$$P_{ma} \triangleq \sum_1^r {}_i P_{mi}$$

$$P_{ea} \triangleq \sum_1^r {}_i P_{ei}$$

$$M_a \triangleq \sum_1^r {}_i M_i$$

so that the dynamic order of the model, for the given area, is reduced by $(2r - 2)$.

However, no appreciable computational savings may be expected from this reduction of the dynamic order, if the dynamic order of the original model (for the whole system) is relatively high. The greatest advantages are obtained when the original model of each machine considered is already the simplified second-order model, and the dynamic order of the area model is therefore reduced from $2r$ (with state variables Ω_i, δ_i, $i = 1, \ldots, r$) to 2 (with state variables Ω_a, δ_a).

However, also in this case, the advantages in question could still be moderate, since the computation work depends not only on the dynamic order of the model, but also (if not mostly) on the complexity of the algebraic-type operations to be effected. In particular, the computation of:

$$P_{ea} = \sum_1^r {}_i P_{ei}(\delta_1(\delta_a), \ldots, \delta_r(\delta_a), \delta_{(r+1)}, \ldots) \triangleq P_{ea}(\delta_a, \delta_{(r+1)}, \ldots)$$

may still imply a considerable work — with no practical advantage with respect to the original model — if the dependence of P_{ea} on δ_a cannot be expressed directly, i.e., without passing through the intermediary of the variables $\delta_1, \ldots, \delta_r$. A similar conclusion can be drawn regarding the powers delivered by the machines outside the area and other variables depending on δ_a.

In this respect, it is possible to demonstrate that, under suitable conditions usually adopted as valid within the considered area (machines represented by emfs of constant magnitude, connected between them and to the boundary nodes by a linear network, including loads), the generic area can be represented by a single "*equivalent machine*" (for which Equations [8.4.1] can be interpreted as the equations of mechanical balance), connected to the boundary nodes through a suitable "*equivalent connection network*." The above-mentioned requirement of expressing directly the dependence of P_{ea} etc. on δ_a is therefore satisfied.

If the area has a single boundary node (Fig. 8.10a), with:

$$\begin{cases} \bar{\imath} = \bar{y}_{00}\bar{v} + \sum_1^r {}_k \bar{y}_{0k}\bar{e}_k \\ \bar{\imath}_k = \bar{y}_{k0}\bar{v} + \sum_1^r {}_i \bar{y}_{ki}\bar{e}_i \qquad (k = 1, \ldots, r) \end{cases}$$

imposing the "coherency" between the machines, the equivalent system indicated in Figure 8.10b can be derived, through proper developments, with:

$$\begin{cases} \overline{Y}_s \triangleq \bar{y}_{00} \dfrac{v_v}{e_{\text{eq}}} \epsilon^{-j\beta/2} \\ \overline{Y}_{d1} \triangleq \bar{y}_{\text{eq}} - \overline{Y}_s \\ \overline{Y}_{d2} \triangleq \bar{y}_{00} - \overline{Y}_s \end{cases}$$

$$\angle \bar{e}_{\text{eq}} \triangleq \angle \bar{v}_v + \frac{\beta}{2}$$

$$M_{\text{eq}} \triangleq \frac{M_a}{m}$$

$$P_{m\ \text{eq}} \triangleq \frac{P_{ma}}{m}$$

where:

- m and β are the parameters defined by Equation [3.3.25], referring to the boundary node (in particular, as described in Section 3.3.3, it commonly results $m \cong 1$, whereas β can be $10°$–$40°$);
- v_v and $\angle \bar{v}_v$ are the magnitude and the phase of the open-circuit voltage:

$$\bar{v}_v = -\frac{1}{\bar{y}_{00}} \sum_1^r {}_k\ \bar{y}_{0k}\bar{e}_k$$

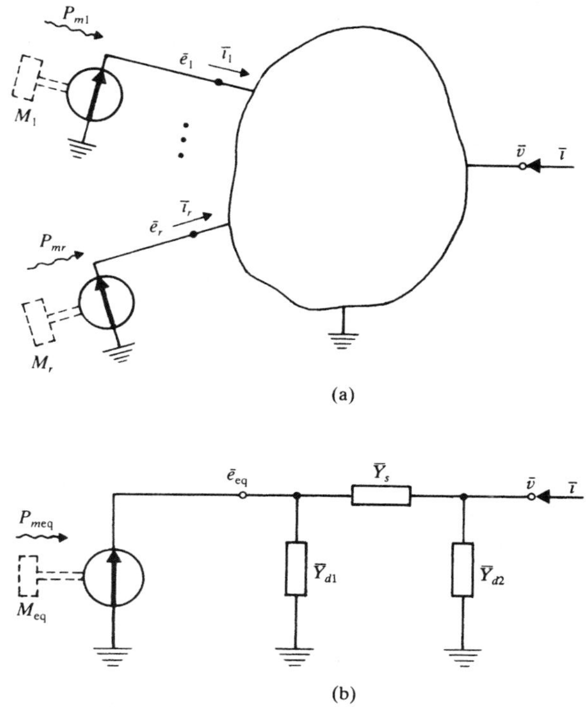

Figure 8.10. Area with a single boundary node: (a) system under examination; (b) equivalent system under the assumption of "coherency" between the machines ($\overline{Y}_s$, $\overline{Y}_{d1}$, $\overline{Y}_{d2}$ represent the admittances of the different branches).

and furthermore:

$$\bar{y}_{eq} \triangleq \sum_1^r {}_k e_k \, \epsilon^{-j\delta^o_{kr}} \sum_1^r {}_i \bar{y}_{ki} e_i \, \epsilon^{j\delta^o_{ir}} \Big/ (m e_{eq}^2)$$

whereas e_{eq}, magnitude of $\bar{e}_{eq}$, is arbitrary (in particular it can be assumed, for simplicity, $e_{eq} = v_v$). When passing from the original system to the equivalent, the relationship between $\bar{v}$ and $\bar{\imath}$ does not change; instead, the total mechanical power and the total complex power (i.e., the active and the reactive powers) supplied by the emfs are divided by m. However, with $M_{eq} = M_a/m$, the electromechanical phenomena remain unchanged.

The situation becomes more complicated, in the case of more than one boundary node. In particular, the network connecting the equivalent machine and the boundary nodes is generally characterized by a nonsymmetrical admittance matrix.

If the whole system is regarded as a single area (e.g., for problems of f/P control), all angular differences $\delta_{1N}, \ldots, \delta_{(N-1)N}$ are assumed to be constant. If, in the original model, the machines are represented by emfs with a constant magnitude and connected between them by a linear network, then the powers P_{ei}, and therefore P_{ea}, are constant (unless there are perturbations in the network), without any dependence on δ_a. For small variations around a steady-state operating point, Equations [8.4.1] then give:

$$\begin{cases} \dfrac{\mathrm{d}\Delta\Omega_a}{\mathrm{d}t} = \dfrac{1}{M_a}(\Delta P_{ma} - \Delta P_{ea}) = \sum_1^N {}_i\, \Delta u_i / M_T \\ \dfrac{\mathrm{d}\Delta\delta_a}{\mathrm{d}t} = \Delta\Omega_a \end{cases}$$

and the motion of the equivalent machine constitutes a generally acceptable approximation of the mean motion of the machines (see Equations [8.2.25] and the last of Equations [8.2.34])(3).

The previous statements can be extended to several areas, provided that the motion, common (by hypothesis) to all the machines in the same area, may be considered as a significant mean of the motions of such machines, also for what concerns the phase-shifts with respect to the machines of the other areas, and the consequent power flows.

In the most usual practice, the model of each area is based not only on the assumption of "coherency" between the machines, but also on *further simplifying assumptions* (more or less intuitive, and often empirical). This allows to define, for a given area, an equivalent machine (with an order even greater than the second one, equipped with a possible "equivalent voltage regulator", etc.), connected to the boundary nodes through a network with a symmetrical admittance matrix. A simplified equivalent of the area is then obtained, consistent with the usual models and simulation programs.

However, as a particular consequence of these further simplifications, in Equations [8.4.1] it is possible to have, under generic dynamic conditions:

$$P_{ea} \neq \sum_1^r {}_i\, P_{ei}$$

and therefore the mechanical balance of the system may not be exactly reproduced, given the assumption of coherency within each area.

On the other hand, the only hypothesis of "coherency" is sufficient to define for each area, without further approximations, an "equivalent speed governor," by simply adding

(3) $M_T \triangleq \sum_1^N {}_i\, M_i$ is the total inertia coefficient, whereas it holds $\Delta u_i \triangleq \Delta P_{mi} - \Delta P_{ei}^S$, where ΔP_{ei}^S is the variation of P_{ei} due to possible structure perturbations in the network. More generally, if the magnitudes e_i of the emfs can be considered as inputs — possibly varying — of the system, it can be assumed:

$$\Delta u_i \triangleq \Delta P_{mi} - \sum_1^N {}_j \left(\frac{\partial P_{ei}}{\partial e_j}\right)^o \Delta e_j - \Delta P_{ei}^S$$

in accordance with the first of Equations [8.2.2].

up the effects of the single governors, i.e., assuming:

$$\frac{\partial P_{ma}}{\partial \Omega_a}(s) = \sum_1^r {}_i \frac{\partial P_{mi}}{\partial \Omega_i}(s)$$

(see also Section 3.3.1).

Sometimes — in particular for problems of f/P control of interconnected systems — the boundary nodes between the areas are not explicitly taken into evidence. The effect of the electrical links is directly translated into proper (approximate) equations between the active powers delivered by the equivalent machines (or "exported" by the single areas, apart from losses) and the respective angular positions (see also Sections 3.3.1 and 3.4.1).

The assumption of machine "coherency" within each area (or the whole system), even when not accompanied by further simplifying assumptions, may generally appear rather rough and restrictive, as, in the actual system, the machines are electrically and not mechanically coupled to each other. In particular, such an assumption prevents consideration of variations of the machine relative positions within the same area, and has similar consequences on these variables that depend substantially on the above relative positions, such as powers exchanged within the area, etc.

For an improved approximation a new model has been then proposed, based on the assumption that, *within each area, the accelerating powers of the machines are, at every time instant, proportional to the respective inertia coefficients.* Such an assumption is evidently contained in the assumption of "coherency." However, the acceleration value (unique for each area) is now ascribed not to the machine rotors, but to the so-called area "*center of inertia*." The variation of position of the latter is suitably defined as function of the variations of angular position of the rotors themselves. In other words, if the considered area includes the machines $1, \ldots, r$, in Equations [8.1.1] the accelerations $\mathrm{d}\Omega_1/\mathrm{d}t, \ldots, \mathrm{d}\Omega_r/\mathrm{d}t$ of the individual area machines are all formally *replaced* by the acceleration $\mathrm{d}\Omega_a/\mathrm{d}t$ of the center of inertia. By doing so, we still have Equations [8.4.1], i.e.,

$$\left.\begin{aligned} \frac{\mathrm{d}\Omega_a}{\mathrm{d}t} &= \frac{1}{M_a}(P_{ma} - P_{ea}) \\ \frac{\mathrm{d}\delta_a}{\mathrm{d}t} &= \Omega_a - \Omega_s \end{aligned}\right\} \qquad \text{[8.4.1 rep.]}$$

(with $P_{ma} \triangleq \sum_1^r {}_i P_{mi}$, $P_{ea} \triangleq \sum_1^r {}_i P_{ei}$, $M_a \triangleq \sum_1^r {}_i M_i$), to which the following equations, in place of Equations [8.4.2], must now be added:

$$\left\{\begin{aligned} &\frac{P_{m1} - P_{e1}}{M_1} = \cdots = \frac{P_{mr} - P_{er}}{M_r} \left(= \frac{\mathrm{d}\Omega_a}{\mathrm{d}t}\right) \\ &\Delta\delta_a = \Delta\delta_a(\Delta\delta_1, \ldots, \Delta\delta_r) \end{aligned}\right. \qquad \text{[8.4.3]}$$

$$\Omega_i = \Omega_s + \frac{\mathrm{d}\delta_i}{\mathrm{d}t} \qquad (i = 1, \ldots, r) \qquad \text{[8.4.4]}$$

where the r Equations [8.4.3] make it possible to deduce behaviors of $\Delta\delta_1, \ldots, \Delta\delta_r$. In fact, for given initial values $\delta_1^{(o)}, \ldots, \delta_r^{(o)}$, the P_{ei}'s are functions of $\Delta\delta_1, \ldots, \Delta\delta_r$, in addition to other variables relative to the electrical part of the system. Regarding the choice of the function $\Delta\delta_a(\Delta\delta_1, \ldots, \Delta\delta_r)$, it may appear reasonable enough to assume:

$$\Delta\delta_a = \frac{\sum_{1}^{r}{}_i\, M_i \Delta\delta_i}{M_a}$$

even if (referring to Section 8.2) this expression only appears justifiable with reference to the mean motion of an area *not* interconnected with others, for small perturbations, and under particular assumptions (see the second part of Equations [8.2.34]).

Therefore, compared with those previously described, the present model admits variations of the relative positions for the machines of the same area and, therefore, even in all the quantities that depend on them. However, these variations are not accompanied by electromechanical oscillations between the machines themselves, and may be evaluated based on variations of the powers P_{ei} generated by the machines, according to "static"-type relationships similar to those involved in usual "load-flow" computations.

From the point of view of the resulting degree of approximation, considerable improvements may be obtained in comparison with previous models. However it is not possible, for the general case of several areas, to derive a simple equivalent of the generic area as "viewed" by the boundary nodes. Furthermore, the computation work for simulations may be considerable.

In the case of a single area ($r = N$), if the definition of δ_a and the deduction of the single δ_i's and Ω_i's are not considered, the preceding equations can be reduced to the following N ones:

$$\frac{P_{m1} - P_{e1}}{M_1} = \cdots = \frac{P_{mN} - P_{eN}}{M_N} = \frac{d\Omega_o}{dt} \qquad [8.4.5]$$

in the N unknowns $\delta_{1N}, \ldots, \delta_{(N-1)N}$, $d\Omega_o/dt$ (in fact, P_{ei}'s depend on $\delta_{1N}, \ldots, \delta_{(N-1)N}$; see the scheme of Fig. 8.11), where Ω_o can be assumed as the "mean frequency"; see the conclusion of Section 8.4.3.

Equations [8.4.5] allow the development of a particularly useful model to analyze the slowest dynamic phenomena ("*long-term dynamics*"):

- also accounting for the evolution of the (mean) regime of the different electrical quantities — voltages, power fluxes, etc. — further than the mechanical ones, for example with the aim of ("quasistatic") security assessments more credible than those of static security (see Sections 2.2.5 and 2.3.1),
- assuming that the f/P control is sensitive to the mean frequency Ω_o (instead of to the single Ω_i's).

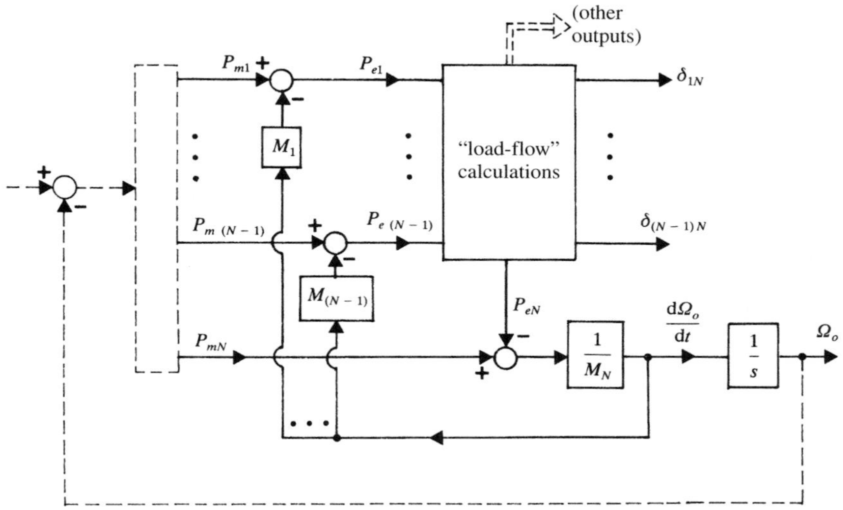

Figure 8.11. Block diagram for the analysis of the slowest phenomena ("long-term dynamics").

The assumptions described above, independent of their acceptability for given applications, are suggested by rather empirical simplification criteria (as well as the choice itself of areas) that are strictly oriented to mechanical type phenomena.

Rational criteria, based on the "modal" analysis for small variations, have been proposed for the model simplification of any given area, as viewed by its boundary nodes, considering jointly the control systems acting on the area itself.

These criteria imply, for each area, the *simplification of the modes* that may be regarded as "*nondominant*," in response to small perturbations of a given kind. Such a procedure may be applied in general, independently of the original model of the area. The simplified equivalent of the generic area is not consistent, generally, with the usual models and simulation programs, but may allow satisfactory approximations, provided that:

- the area is actually submitted to relatively small perturbations;
- the nondominant modes are identified with reference to likely meaningful input functions, with regard to the operation of the whole system (for any given area, the set of input variables also includes electrical variables, such as voltages or powers, at the boundary nodes).

Finally, the computation work depends on the desired degree of approximation.

8.4.3. Models of More General Type

Some improvement in simplifying procedures may be suggested by various considerations.

First, the simplified model of any given area is developed, according to the above, independently of the rest of the system, although it should generally depend on the characteristics of interconnected operation and therefore (in different measures) on the value of all parameters. The definition itself of "areas" in the above sense (referred to as subsystems, each simplified according to criteria independent of the rest of the system) may be considered forced and arbitrary even from a formal point of view, that is regardless of the difficulties which could be met for a satisfactory criterion for the choice of the areas themselves (unless this last problem is not particularly "well-conditioned," taking into account the structure of the network etc.).

Specifically, for small variations, the resulting "modes" of a multiarea model do not correspond, neither in frequency nor in magnitude and phase, to any subset of original model modes. However, it is important that the dominant modes (i.e., those contributing mostly to the observed variable response, under the considered operating conditions) are reproduced with a satisfactory approximation. To improve the approximation, the number of areas could be increased, but this measure would lead to a dynamic order of the simplified model rather higher than that required (which may be very small in case of few dominant modes). For instance, n areas characterized by second-order equivalent machines may account for $(n - 1)$ electromechanical oscillations, but a number of areas rather higher than 2 may be necessary to satisfactorily reproduce only one oscillatory mode of the original model. Even when not considering such an inconvenience, increasing the number of areas would be a rather empirical remedy, without the possibility of easily foreseeing the most suitable configuration to adopt for the simplified model.

In practice, the feasibility of an area subdivision also must be evaluated with respect to the particular type of problem. If the considered perturbations are concentrated in a relatively small region (e.g., after a fault at a given point in the network), as well as the variables whose behavior must be examined, it may be reasonable to simplify those system parts that are far enough from that region. This can be accomplished by treating them like "areas" in the above-mentioned sense, without expecting substantial detriment to the resulting approximation. However, this concept may not appear reliable in more general cases, when the perturbations and the variables to be observed are actually distributed (e.g., as a result of control actions) throughout the system, so that no parts of it are likely more suitable for a subdivision into areas.

Finally, it should be recalled that the traditional simplification criteria are strictly oriented to mechanical phenomena. This means that if the state variables of the original model are not of an exclusively mechanical nature (specifically, motor speeds and positions), these criteria result in a minor role in the simplification process.

It may therefore be useful to identify simplifying criteria:

- with efficiency not dependent on the particular nature (i.e., mechanical etc.) of the model to simplify;

- with a "global" character, i.e., complying with the operating characteristics of the whole system.

Such requirements may be satisfied by *"modal reduction" criteria* (of the type already mentioned at the end of Section 8.4.2, but applied to the whole system), which, even if based on the linearized behavior, prove useful also for nonsmall variations within reasonable value ranges.

By modal reduction, *the dominant modes can be exactly retained* in the simplified model, with their contributions to the response to any perturbation type. Obviously, the requirement of exactly reproducing the dominant mode contributions also may be limited to reduced sets of output and input variables. This also may lead to a reduction of the set of modes to be assumed as dominant.

Leaving out, for the sake of simplicity, the analytic (rather elementary) developments required for making a model of "modal" type for small variations, it is important to indicate that, if the original model (of n_0 dynamic order) is defined in the matrix form:

$$\left.\begin{aligned} \frac{\mathrm{d}x}{\mathrm{d}t} &= Ax + Bu \\ y &= Cx + Du \end{aligned}\right\} \qquad [8.4.6]$$

(where x, u, y are column matrices, consisting of the n_0 state variables, and of the input and output variables, respectively; see Equations [A3.1.3] and footnote(3) in Appendix 3), the simplified model (of n_r dynamic order) may be written as follows:

$$\left.\begin{aligned} \frac{\mathrm{d}z}{\mathrm{d}t} &= R(Ax + Bu) \\ x &= Sz + Tu \\ y &= Cx + Du \end{aligned}\right\} \qquad [8.4.7]$$

where:

- z is the "reduced" state (n_r variables);
- S and T are suitable matrices, depending on the choice of the dominant modes and on the simplification of the remainders;
- R is any arbitrary matrix (n_r, n_0) satisfying the condition $RS = I_{(n_r)}$.

The first of Equations [8.4.7] exhibits an interesting formal analogy with the original model, as it explicitly contains the same term $(Ax + Bu)$ appearing in the first of Equations [8.4.6]. Therefore, it may be spontaneous to extend the previous result to the *nonlinear case*, assuming, if the original model is in the form:

$$\left.\begin{aligned} \frac{\mathrm{d}x}{\mathrm{d}t} &= f(x, u) \\ y &= g(x, u) \end{aligned}\right\} \qquad [8.4.6']$$

a simplified model of the type:

$$\left.\begin{aligned} \frac{dz}{dt} &= Rf(x, u) \\ \Delta x &= Sz + T\Delta u \\ y &= g(x, u) \end{aligned}\right\} \qquad [8.4.7']$$

where it is understood that the variations Δx, Δu are evaluated with respect to the equilibrium operating point assumed for calculating the matrices S and T.

For small variations, the simplified Equations [8.4.7′] satisfy the modal reduction criterion, whatever the choice of R is (provided that $RS = I_{(nr)}$). On the other hand, for large variations, the quality of such model depends on R and can therefore be optimized to some extent (for given operating conditions) by suitably taking advantage of the degrees of freedom in the choice of R.

The modal-type model is not consistent with the most common models and simulation programs for multimachine systems. However, considerable advantages may be obtained in this connection, because of its formal analogy in respect to the original model, as above outlined.

In the case of a relatively complex system, to obtain a less-demanding setup of the simplified model, it may be advisable to apply in advance any reasonable simplification (e.g., by replacing some parts of the system with equivalent machines, so as to reduce the order n_0 before applying the modal reduction). The computational savings may be, especially in the field of small variations, of the same order—and greater, for a given degree of approximation—than those obtained by using the traditional "area" models.

Moreover, it is important to emphasize that an "area" model too can generally be led to the Equations [8.4.7′], with matrices R, S, and T replaced by suitable matrices. Therefore, it may be considered a simplified version of the modal-type model. Consequently, the knowledge of this last one also may lead to important indications regarding the best choice of possible areas, which may be used to derive an "area" model or (as a *compromise solution*) a modal-type model, developed starting from equivalent machines as mentioned above.

To better explain these considerations, we can refer to the case of a system including N machines of the second order; as specified in Section 8.2, for small variations it is possible to write (assuming initial conditions equal to zero, for simplicity):

$$\Delta\Omega_i(s) = \sum_1^{N-1}{}_h \frac{s}{s^2 + \nu_h^2} A_{ih} \sum_1^{N}{}_j (A^{-1})_{hj} \frac{\Delta u_j(s)}{M_j} + \frac{1}{s} A_{1N} \sum_1^{N}{}_j (A^{-1})_{Nj} \frac{\Delta u_j(s)}{M_j}$$

(see Equation [8.2.21] and subsequent ones), where the terms in the first summation correspond to the electromechanical oscillations, whereas the last term corresponds to the machine "mean motion."

In practice, it often occurs that in the $\Delta\Omega_i$'s of greater interest, only a few oscillatory terms are "dominant" (e.g., those for $h = 1, \ldots, q$, with q equal to 3, 4, or slightly more), whereas the other modes (obviously not unstable) can be disregarded or properly simplified, since the contributions of the respective terms $A_{ih}(A^{-1})_{hj}\Delta u_j(s)/M_j$ are small. In practice, many oscillatory modes also may be adequately simplified or disregarded, because they are characterized by high damping factors.

In such conditions, the modal reduction may lead to a simplified model, for which:

$$\Delta\Omega_i(s) = \sum_1^q {}_h \frac{s}{s^2 + \nu_h^2} A_{ih} \sum_1^N {}_j (A^{-1})_{hj} \frac{\Delta u_j(s)}{M_j} + \frac{1}{s} A_{1N} \sum_1^N {}_j (A^{-1})_{Nj} \frac{\Delta u_j(s)}{M_j} + \varepsilon_i(s) \qquad [8.4.8]$$

where $\varepsilon_i(s)$ constitutes a suitable approximation of the (minor) terms corresponding to $h = q+1, \ldots, N-1$.

With Equation [8.4.8], it is obvious that, for instance, the machines 1, 2 can be considered "coherent" to each other, for small variations at least, if:

$$A_{1h} \cong A_{2h} \qquad [8.4.9]$$

$$\frac{(A^{-1})_{h1}}{M_1} \cong \frac{(A^{-1})_{h2}}{M_2} \qquad [8.4.10]$$

$$\forall h = 1, \ldots, q, \quad \text{and moreover for } h = N$$

In fact, Equation [8.4.9] leads to $\Delta\Omega_1 \cong \Delta\Omega_2$, and [8.4.10] permits to add up the perturbations Δu_1, Δu_2 on the two machines, obtaining an equivalent total perturbation $(\Delta u_1 + \Delta u_2)$, whereas the contribution of the difference between $\varepsilon_1(s)$ and $\varepsilon_2(s)$ can even (by hypothesis) be disregarded. Note that the coherency implies not only the equality of the speeds, but also that the perturbations can be summed up, as specified above.

A comparison, for the different machines ($i = 1, \ldots, N$), between the values of the accipiencies A_{ih} and between the values of the excitances $(A^{-1})_{hj}/M_j$, only for the dominant modes ($h = 1, \ldots, q$), can therefore give some plausible practical indications for the choice of possible areas.

Therefore, the knowledge of the accipiencies and excitances may seem helpful not only for the *selection of the dominant oscillations* (based on the product values $(A_{ih}(A^{-1})_{hj}/M_j)$ for the different h's, and on the perturbation size Δu_j), but also for the *identification of possible areas*[(4)].

It should also be remembered that:

- for $h = N$, the accipiencies are equal to each other, either do the excitances at a good approximation level (see Equations [8.2.32′]);

[(4)] A particularly convenient graphic representation can be created by using geographical maps, one for each oscillation, where the values of the accipiencies and excitances are indicated for the respective machines. This may possibly allow (qualitative) geographical interpretations of the oscillations themselves, e.g., in terms of oscillations between areas, etc.

- for $h = 1, \ldots, N-1$ (oscillatory modes), the excitances are often roughly proportional to the accipiencies (see [8.2.37]);

so that the present considerations may be limited only to the knowledge of the accipiencies A_{ih} for $h = 1, \ldots, N-1$.

Finally, by linearizing the (single area) "long-term" model defined by Equations [8.4.5], it follows:

$$\begin{bmatrix} \Delta u_1 \\ \vdots \\ \Delta u_N \end{bmatrix} = \begin{bmatrix} K_{11} & \cdots & K_{1(N-1)} & M_1 \\ \cdot & \cdots & \cdot & \cdot \\ \cdot & \cdots & \cdot & \cdot \\ \cdot & \cdots & \cdot & \cdot \\ K_{N1} & \cdots & K_{N(N-1)} & M_N \end{bmatrix} \begin{bmatrix} \Delta\delta_{1N} \\ \vdots \\ \Delta\delta_{(N-1)N} \\ \dfrac{\mathrm{d}\Delta\Omega_o}{\mathrm{d}t} \end{bmatrix}$$

from which, in particular:

$$\frac{\mathrm{d}\Delta\Omega_o}{\mathrm{d}t} = \frac{\sum_{1}^{N}{}_i\, \gamma_i \Delta u_i}{\sum_{1}^{N}{}_k\, \gamma_k M_k}$$

This is in full agreement with the equations of the mean motion (see the first of Equations [8.2.25], and Equation [8.2.28″]). Then, the simplified model defined by Equations [8.4.5] satisfies, for small variations, the modal reduction criterion for the case $q = 0$. Consequently, the mean motion is then exactly reproduced, as well as the mean variations of the phase-shifts between the machines, etc.

8.5. EFFECT OF FEEDBACKS SENSITIVE TO THE MOTION OF MACHINES

8.5.1. Effect of Speed Regulations

From the results reported in Section 8.2, we can understand the importance of accounting for the speed regulations, at least for the behavior of the mean frequency (in particular, Equations [8.2.42], [8.2.44], [8.2.45], and [8.2.45′] are not practically acceptable).

To this aim it can be assumed, in terms of Laplace transforms:

$$\Delta u_i(s) \triangleq \Delta\hat{u}_i(s) - G_{fi}(s)\Delta\Omega_i(s) \qquad (i = 1, \ldots, N) \qquad [8.5.1]$$

where the last term considers the speed regulation of the single units; similarly to Equations [8.2.24″], [8.2.28″], and the last of Equations [8.2.29], it can be further assumed:

$$\left(\frac{\Delta \hat{u}}{M}\right)_o \triangleq A_{1N} \sum_1^N {}_j (A^{-1})_{Nj} \frac{\Delta \hat{u}_j}{M_j} = \frac{\sum_1^N {}_j \ \gamma_j \Delta \hat{u}_j}{\sum_1^N {}_k \ \gamma_k M_k}$$

$$\Delta \hat{u}_{i(o)} \triangleq \Delta \hat{u}_i - M_i \left(\frac{\Delta \hat{u}}{M}\right)_o \qquad (i = 1, \ldots, N)$$

From the equations in Section 8.2, it can be then derived, assuming zero initial conditions for simplicity:

$$\begin{aligned} \Delta \Omega_{io}(s) = {} & \sum_1^{N-1} {}_h \frac{s}{s^2 + \nu_h^2} A_{ih} \sum_1^N {}_j (A^{-1})_{hj} \left(\frac{\Delta \hat{u}_{j(o)}(s)}{M_j} - \frac{G_{fj}(s)}{M_j} \Delta \Omega_{jo}(s) \right) \\ & - \left[\sum_1^{N-1} {}_h \frac{s}{s^2 + \nu_h^2} A_{ih} \sum_1^N {}_j (A^{-1})_{hj} \frac{G_{fj}(s)}{M_j} \right] \Delta \Omega_o(s) \end{aligned} \qquad [8.5.2]$$

$(i = 1, \ldots, N)$, and furthermore:

$$\begin{aligned} \Delta \Omega_o(s) = {} & \frac{1}{s} \left(\left(\frac{\Delta \hat{u}}{M} \right)_o - \left(\frac{G_f(s)}{M} \right)_o \Delta \Omega_o(s) \right) \\ & - \frac{1}{s} A_{1N} \sum_1^N {}_j (A^{-1})_{Nj} \frac{G_{fj}(s)}{M_j} \Delta \Omega_{jo}(s) \end{aligned} \qquad [8.5.3]$$

with:

$$\left(\frac{G_f(s)}{M}\right)_o \triangleq A_{1N} \sum_1^N {}_j (A^{-1})_{Nj} \frac{G_{fj}(s)}{M_j} = \frac{\sum_1^N {}_j \gamma_j G_{fj}(s)}{\sum_1^N {}_k \gamma_k M_k}$$

The last terms in Equations [8.5.2] and [8.5.3] define the *interactions between the mean motion, and the machine motion relative to the mean motion itself.* However, the electromechanical oscillations usually have rather high frequencies compared to the "low-pass" characteristics of the regulators. Therefore, it can be generally assumed that the terms $G_{fj}(s)\Delta\Omega_{jo}(s)$ give negligible contributions, as if the regulators are sensitive only to the mean frequency, thus approximating [8.5.1] by:

$$\Delta u_i(s) \cong \Delta \hat{u}_i(s) - G_{fi}(s) \Delta \Omega_o(s) \qquad (i = 1, \ldots, N)$$

Under such an approximation:

- The effect of $\Delta\Omega_o$ on the $\Delta\Omega_{io}$'s remains evident in Equation [8.5.2]: i.e., the mean motion still influences the relative motion.
- Equation [8.5.3] becomes instead:

$$\Delta\Omega_o(s) \cong \frac{1}{s}\left(\left(\frac{\Delta\hat{u}}{M}\right)_o - \left(\frac{G_f(s)}{M}\right)_o \Delta\Omega_o(s)\right) \qquad [8.5.3']$$

from which, independently of the relative motion:

$$\Delta\Omega_o(s) \cong \frac{\left(\dfrac{\Delta\hat{u}}{M}\right)_o}{s + \left(\dfrac{G_f(s)}{M}\right)_o}$$

or more explicitly:

$$\Delta\Omega_o(s) \cong \frac{\sum_1^N {}_j\gamma_j \Delta\hat{u}_j(s)}{\sum_1^N {}_k\gamma_k(sM_k + G_{fk}(s))}$$

which can be simplified according to the assumptions made in Chapter 3, if $\gamma_1, \ldots, \gamma_N$ are slightly different from each other.

The interaction between the mean motion and the relative motion results in, instead, a total lack in the case (even not too far from the reality) of "similar" regulators, with:

$$\frac{G_{f1}(s)}{M_1} = \cdots = \frac{G_{fN}(s)}{M_N}$$

Remember in fact that $\sum_1^N {}_j(A^{-1})_{hj} = 0$ $(h = 1, \ldots, N-1)$, $\sum_1^N {}_j(A^{-1})_{Nj}\Delta\Omega_{jo} = 0$, so that the last terms in Equations [8.5.2] and [8.5.3] become both zero. Note in particular that, under the present hypothesis, Equation [8.5.3′] holds without the approximation sign and, inside it, it holds:

$$\left(\frac{G_f(s)}{M}\right)_o = \frac{G_{f1}(s)}{M_1} = \cdots$$

The presence of the terms $G_{fj}(s)\Delta\Omega_{jo}(s)$ in Equation [8.5.2] causes a (usually modest) variation in the characteristic roots which correspond to the relative motion. According to Equation [A3.6.1] in Appendix 3, and in analogy to Section 7.2.2a, it can be generally stated with a good approximation that, because of the j-th regulator:

- the generic oscillation frequency ν_h undergoes a variation:

$$\sim -\frac{1}{2} A_{jh} \frac{(A^{-1})_{hj}}{M_j} Im(G_{fj}(\tilde{j}\nu_h))$$

- the corresponding damping factor is no longer zero, and becomes:

$$\zeta_h = \Delta\zeta_h \cong \frac{1}{2\nu_h} A_{jh} \frac{(A^{-1})_{hj}}{M_j} \mathcal{Re}(G_{fj}(\tilde{j}\nu_h))$$

Therefore, it appears convenient that $\mathcal{Re}(G_{fj}(\tilde{j}\nu_h)) > 0$, at least for those units which correspond to the largest values of accipiencies and excitances.

A more accurate analysis, based on better-approximated models, is instead necessary in the case of very slow oscillations, such that they interact in a non-negligible way with the frequency regulation, e.g., oscillations at a frequency $\nu_h \cong 0.6-1$ rad/sec, which have been found in very large systems with a predominantly "longitudinal" structure.

8.5.2. Effect of Additional Signals in Excitation Controls

The connection with:

- large external areas, e.g., pertaining to different states, which were previously operated separately;
- new generating plants, located at very long distances, for the exploitation of energy remote resources (particularly, hydraulic ones);

has led to the development of larger and larger systems which often have a "longitudinal" and scarcely meshed structure and which are characterized by relatively slow electromechanical oscillations, e.g., oscillations at frequencies ν_h in the order of 1–3 rad/sec (which approximately correspond to 0.15–0.50 Hz, i.e., periods of 2–7 sec). Note that the extension itself of the system (not necessarily the existence of "weak" links between the machines) can be sufficient cause for the presence of slow oscillations.

For such low oscillation frequencies, according to Sections 7.2.1 and 7.2.2, it is easy to recognize that:

- damper windings of the machines are practically inactive;
- voltage regulation may have destabilizing effects;

- speed regulation has modest effects (apart from the case of even lower ν_h, e.g., $\nu_h \cong 0.85$ rad/sec, with a stabilizing effect experimentally verified of the thermal units' regulation);
- therefore the most reliable stabilizing effect (although it may be modest) is the one caused by the field windings,

so that, in all, such oscillations also may be unstable (negative damping); see for example the experimental diagrams reported in Figure 8.12a.

The damping of such oscillations can be guaranteed in a relatively simple and economic way by using "additional signals" in the excitation control of the machines (or of some of them, properly chosen), similarly to what is described in Section 7.2.2c; see for example the experimental results reported in Fig. 8.12b.

For an illustrative analysis, we can use the simplifications adopted in the previous sections (machines, real and/or equivalent ones, characterized by suitable emfs, etc.), assuming that the additional signals cause variations of the emf's magnitudes (Δe_j) depending on the respective speed variations ($\Delta\Omega_j$). In such a way, a dependence of the generated powers P_{ei} on the $\Delta\Omega_j$'s ($i, j = 1, \ldots, N$) is defined, which must be considered into Equations [8.1.1] *without any increase in the dynamic order* of the model.

For small variations, it is possible to derive (instead of Equations [8.2.1] and [8.2.2], and not considering the speed regulators):

$$\left.\begin{aligned} \frac{\mathrm{d}\Delta\Omega_i}{\mathrm{d}t} &= \frac{1}{M_i}\left(\Delta w_i - \sum_1^N{}_j D_{ij}\Delta\Omega_j - \sum_1^N{}_j K_{ij}\Delta\delta_j\right) \\ \frac{\mathrm{d}\Delta\delta_i}{\mathrm{d}t} &= \Delta\Omega_i \end{aligned}\right\} \qquad (i = 1, \ldots, N) \qquad [8.5.4]$$

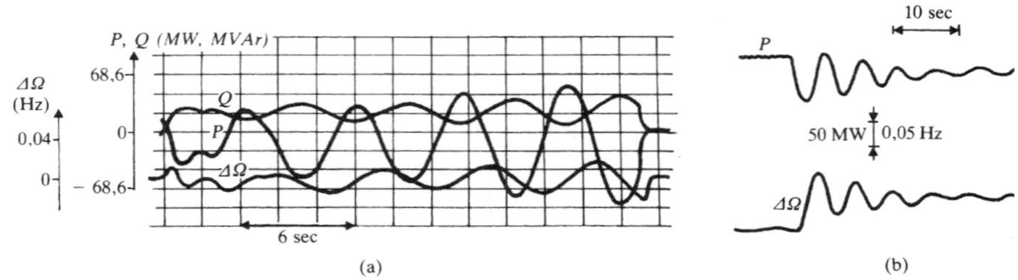

Figure 8.12. Connection of the Yugoslavian network to the UCPTE (Western Europe) network: (a) unsuccessful attempt because of unstable operation (June 19, 1973); (b) successful attempt because of the insertion of "additional signals" at the Djerdap power station (April 19, 1974). P and Q are the active and reactive powers recorded on the line Divača-Padriciano; Ω is the frequency measured on that line in the case (a), and at Belgrad in case (b) (see reference 203).

where:

$$\begin{cases} \Delta w_i \triangleq \Delta u_i + \sum_1^N {}_j \left(\dfrac{\partial P_{ei}}{\partial e_j}\right)^o \Delta e_j = \Delta P_{mi} - \Delta P_{ei}^S \\ D_{ij} \triangleq \left(\dfrac{\partial P_{ei}}{\partial \Omega_j}\right)^o = \left(\dfrac{\partial P_{ei}}{\partial e_j}\right)^o \dfrac{\Delta e_j}{\Delta \Omega_j} \qquad (\text{"braking" coefficients}) \\ K_{ij} \triangleq \left(\dfrac{\partial P_{ei}}{\partial \delta_j}\right)^o \qquad (\text{"synchronizing" coefficients}) \end{cases}$$

For a greater generality, the effect (on the ΔP_{ei}'s) of possible variations Δe_j not related to the speed variations, also can be accounted for in the Δw_i's.

Equations [8.5.4] define a linear model of $2N$ dynamic order; from this, it is possible to derive, similar to what is illustrated in Section 8.2.1, models of the $(2N-1)$ or $(2N-2)$ dynamic order, with state variables, respectively, $\Delta\Omega_i$, $\Delta\delta_{kN}$ or $\Delta\Omega_{kN}$, $\Delta\delta_{kN}$ $(i = 1, \ldots, N;\ k = 1, \ldots, N-1)$.

The relative motion of the machines $1, \ldots, N-1$ with respect to the N-th machine, now interacts with the motion of the N-th machine itself; however, if:

$$\frac{\sum_1^N {}_j D_{1j}}{M_1} = \cdots = \frac{\sum_1^N {}_j D_{Nj}}{M_N} \qquad [8.5.5]$$

then the motion of the N-th machine no longer influences the relative motion, as in the case of zero braking coefficients (see Fig. 8.3).

In matrix terms, the Equations [8.5.4] can be written as:

$$\left.\begin{aligned} \frac{d\Delta\Omega}{dt} &= -M^{-1}D\Delta\Omega - M^{-1}K\Delta\delta + M^{-1}\Delta w \\ \frac{d\Delta\delta}{dt} &= \Delta\Omega \end{aligned}\right\} \qquad [8.5.6]$$

in accordance to the block diagram of Figure 8.13, assuming that $M \triangleq \text{diag}\{M_i\}$, $D \triangleq \{D_{ij}\}$, $K \triangleq \{K_{ij}\} (i, j = 1, \ldots, N)$, and denoting by $\Delta\Omega$, $\Delta\delta$ and Δw the column matrices constituted by $\Delta\Omega_i$, $\Delta\delta_i$, and $\Delta w_i (i = 1, \ldots, N)$ respectively.

From these equations it can also be derived:

$$\frac{d^2\Delta\delta}{dt^2} = -M^{-1}D\frac{d\Delta\delta}{dt} - M^{-1}K\Delta\delta + M^{-1}\Delta w$$

so that the $2N$ characteristic roots $\Lambda_1, \ldots, \Lambda_{2N}$ of the system considered are given by the solutions Λ of the equation:

$$0 = \det(\Lambda^2 I_{(N)} + \Lambda M^{-1}D + M^{-1}K) \qquad [8.5.7]$$

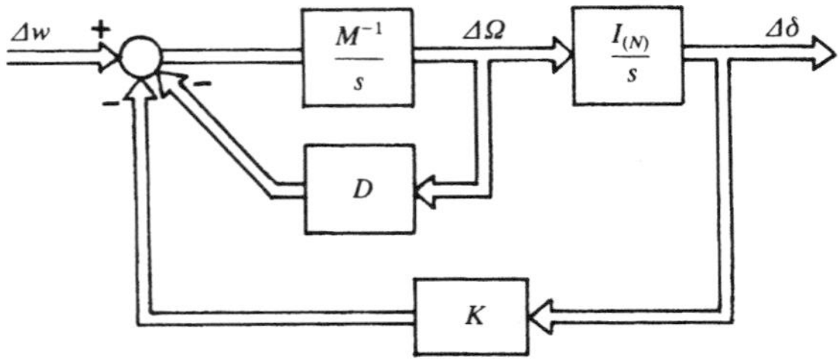

Figure 8.13. Block diagram of the linearized system in the presence of "braking" actions.

In particular, for braking coefficients of moderate value (and an obvious ordering of the indices), the first $(2N-2)$ characteristic roots are constituted by $(N-1)$ pairs of complex conjugate values, generically of the type:

$$\left\{ \begin{array}{l} \Lambda_h \\ \Lambda_{(h+N-1)} \end{array} \right\} = \left(-\zeta_h \pm \tilde{j}\sqrt{1-\zeta_h^2}\right)\nu_h \qquad (h = 1, \ldots, N-1) \qquad [8.5.8]$$

with suitable ζ_h, ν_h, whereas $\Lambda_{(2N-1)}$ and Λ_{2N} are real (and more precisely, because of the singularity of K, it results $\Lambda_{2N} = 0$).

Incidentally, if, in analogy to Equation [8.2.15]:

$$M^{-1}D = A\ d\ A^{-1}$$

with d diagonal matrix, from Equation [8.5.7] one can derive the N scalar equations:

$$0 = \Lambda^2 + \Lambda d_h - L_h \qquad (h = 1, \ldots, N)$$

from which, assuming $L_h < -d_h^2/4$ for $h = 1, \ldots, N-1$:

$$\begin{cases} \zeta_h = \dfrac{d_h}{2\sqrt{-L_h}} \\ \nu_h = \sqrt{-L_h} \end{cases}$$

whereas:

$$\begin{cases} \Lambda_{(2N-1)} = -d_N \\ \Lambda_{2N} = 0 \end{cases}$$

Under the adopted assumption, the oscillation frequencies ν_h are unchanged with respect to the case of zero braking coefficients. Therefore, as braking coefficients vary, the corresponding characteristic roots move in the complex plane along circles having their centers at the origin and radius $\sqrt{-L_h}$.

Furthermore, by recalling Equations [8.2.18], [8.2.19], and [8.2.20], it can be derived:

$$\frac{\sum_{1}^{N} {}_j D_{kj}}{M_k} = d_N \qquad \forall k = 1, \ldots, N$$

so that Equations [8.5.5] are verified.

Now examine the effect of the braking coefficients on the characteristic roots, assuming that these coefficients all have a *small size*.

To this aim, it is useful to indicate (recalling the block diagram in Fig. 8.13) that the effect of the generic coefficient D_{ji} corresponds to the feedback in Figure 8.14, where, resulting from Equation [8.2.21] and subsequent equations:

$$\mathcal{G}_{ij}(s) \triangleq \left(\frac{\partial \Omega_i}{\partial w_j}(s)\right)_{D=0} = \sum_{1}^{N-1} {}_h \frac{s}{s^2 + \nu_h^{o2}} A_{ih} \frac{(A^{-1})_{hj}}{M_j} + \frac{1}{s} A_{1N} \frac{(A^{-1})_{Nj}}{M_j}$$

with $\nu_h^o \triangleq \sqrt{-L_h}$.

By applying what is reported in Appendix 3 (Equation [A3.6.1]), the following "*sensitivity coefficients*" can be derived:

$$\left.\begin{aligned} \frac{\partial \zeta_h}{\partial D_{ji}} &= \frac{A_{ih}(A^{-1})_{hj}}{2M_j \nu_h^o} \\ \frac{\partial \nu_h}{\partial D_{ji}} &= 0 \end{aligned}\right\} \quad (h = 1, \ldots, N-1; \quad j, i = 1, \ldots, N) \qquad [8.5.9]$$

(where ζ_h and ν_h are generically defined by Equation [8.5.8]), and moreover:

$$\frac{\partial \Lambda_{(2N-1)}}{\partial D_{ji}} = -A_{1N} \frac{(A^{-1})_{Nj}}{M_j} \qquad (j, i = 1, \ldots, N) \qquad [8.5.10]$$

whereas it still holds, as already said, $\Lambda_{2N} = 0$.

Because of Equations [8.5.9], the braking coefficients (of small size) introduce a damping in the electromechanical oscillations, without practically altering their

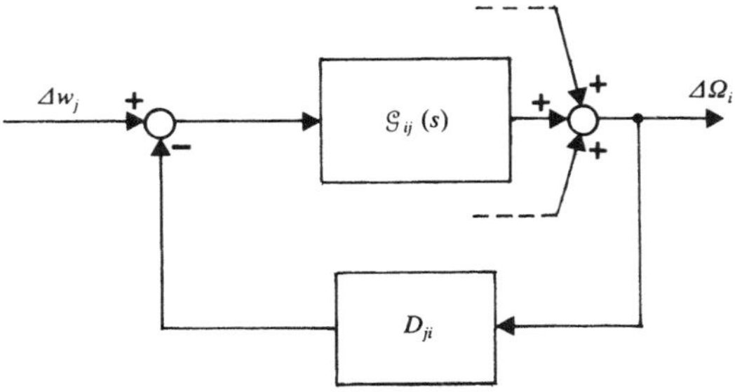

Figure 8.14. Detail of the block diagram with the "braking" coefficient D_{ji} represented.

frequency. For any given D_{ji}, the effect on ζ_h can be considered proportional to the product of the accipiency A_{ih} and the excitance $(A^{-1})_{hj}/M_j$.

Such a correlation with accipiencies and excitances seems very useful for the *synthesis of the feedbacks by additional signals*, and for the *choice of the machines to which these feedbacks should be applied* (so as to guarantee a sufficient damping for all the oscillations).

Moreover, the sensitivity coefficients [8.5.9] can be derived *experimentally*, e.g., by the response of the original system (without additional signals) to a step $\Delta u_j(t) = 1(t)$, as it simply results:

$$\Delta\Omega_i(t) = \sum_1^{N-1} {}_h 2\frac{\partial \zeta_h}{\partial D_{ji}} \sin \nu_h^o t + \Delta\Omega_o(t)$$

(recall [8.2.43]). More generally, through Equation [8.2.22] it follows:

$$\frac{\partial \zeta_h / \partial D_{ji}}{\partial \zeta_h / \partial D_{jr}} = \frac{A_{ih}}{A_{rh}} = \frac{\phi_{\Omega_i}^{(h)}}{\phi_{\Omega_r}^{(h)}} \qquad [8.5.11]$$

$(h = 1, \ldots, N-1;\ j, i, r = 1, \ldots, N)$, so that the ratios, for any given j, between the above-mentioned sensitivity coefficients can be deduced by experimentally measuring the magnitudes (and the phases) of oscillation of the machines i, r, at *any* operating conditions. Similar considerations apply also in the presence of one or more "infinite power" networks.

Finally, for Equation [8.5.10] it can be observed that the right-hand member does not actually depend on i, so that it can be substituted by:

$$\frac{\partial \Lambda_{(2N-1)}}{\partial \left(\sum_1^N {}_k D_{jk} \right)} = -A_{1N} \frac{(A^{-1})_{Nj}}{M_j} \qquad (j = 1, \ldots, N) \qquad [8.5.10']$$

Based on such equations, it can be stated that the braking coefficients move the real root $\Lambda_{(2N-1)}$ from the origin. However, such a result has a modest practical interest, if one considers that the additional signals act through proper band-pass filters to reduce the consequent disturbances on the voltage regulations (see Section 7.2.2), so that their low-frequency effects are disregarded.

If Equation [8.2.31] holds true (i.e., if in the matrix K the sum of the elements of each column is equal to zero), the right-hand member in Equation [8.5.10′] is equal to $-1/M_T$, independently of j, so that such equations can be substituted by a single one (which has a scarce practical interest, as said), i.e.,

$$\frac{\partial \Lambda_{(2N-1)}}{\partial D_T} = -\frac{1}{M_T} \qquad [8.5.10'']$$

with $D_T \triangleq \sum_1^N {}_j \sum_1^N {}_k D_{jk}$.

Moreover, if the matrix K is symmetrical:

- Because of Equation [8.2.37] it can be derived:

$$\frac{\partial \zeta_h}{\partial D_{ji}} = \frac{\beta_h}{2\nu_h^o} A_{ih} A_{jh}$$

  (in particular, then, the coefficients $\partial \zeta_h / \partial D_{ii}$ are nonnegative).

- It generically results:

$$\frac{\partial \zeta_h}{\partial D_{ir}} = \frac{\partial \zeta_h}{\partial D_{ri}} = \frac{\dfrac{\partial \zeta_h}{\partial D_{ji}} \dfrac{\partial \zeta_h}{\partial D_{jr}}}{\dfrac{\partial \zeta_h}{\partial D_{jj}}}$$

  so that, if $\partial \zeta_h / \partial D_{j1}, \ldots, \partial \zeta_h / \partial D_{jN}$ have been (experimentally) identified by a *single* test at known Δu_j, it is possible to derive *all* the other sensitivity coefficients.

- Equation [8.5.11] can be generalized into:

$$\frac{\partial \zeta_h / \partial D_{ji}}{\partial \zeta_h / \partial D_{kr}} = \frac{\phi_{\Omega_j}^{(h)} \phi_{\Omega_i}^{(h)}}{\phi_{\Omega_k}^{(h)} \phi_{\Omega_r}^{(h)}}$$

  ($h = 1, \ldots, N-1$; $j, i, k, r = 1, \ldots, N$), from which in particular:

$$\frac{\partial \zeta_h / \partial D_{ii}}{\partial \zeta_h / \partial D_{rr}} = \left(\frac{\phi_{\Omega_i}^{(h)}}{\phi_{\Omega_r}^{(h)}} \right)^2$$

  equal to the *square* of the ratio between the oscillation magnitudes of the machines i, r, under any operating conditions. This last equation is particularly indicative for the choice of the machines to which the additional signals should be applied, as usually each e_i has more influence on the respective P_{ei} than on the P_{ej}'s of the other machines, and therefore the matrix D can be considered a *dominant diagonal* matrix.

For a better evaluation of the functions $\phi_{\Omega_i}^{(h)}$ (for any given h), it is suitable that the oscillation magnitudes are not too small.

To this aim, because of Equation [8.2.22] it appears convenient to perturb the machines j in correspondence to which the excitance $(A^{-1})_{hj}/M_j$ —or equivalently the accipiency A_{jh}, within the approximation limits of Equation [8.2.37]— assumes the largest (absolute) values.

As the accipiencies are (for varying index j) proportional to the respective $\phi_{\Omega_j}^{(h)}$'s, i.e., to the oscillation magnitudes, the machines to which perturbations should be applied are those which largely oscillate. Therefore, to identify such machines it is possible to use experimental estimations under *normal* operating conditions.

Generally, if variations ΔK_{ji} also are considered (because of network parameter variations etc., or to additional signals sensitive to the positions δ_i, in which case the matching of Equation [8.2.3] may even fail), it can be derived:

$$\begin{cases} \dfrac{\partial \zeta_h}{\partial D_{ji}} = \dfrac{\partial \nu_h}{\partial K_{ji}} = \dfrac{A_{ih}(A^{-1})_{hj}}{2M_j \nu_h^o} \\ \dfrac{\partial \nu_h}{\partial D_{ji}} = \dfrac{\partial \zeta_h}{\partial K_{ji}} = 0 \end{cases}$$

$$\frac{\partial \Lambda_{(2N-1)}}{\partial D_{ji}} = \frac{\partial(\Lambda_{(2N-1)})^2}{\partial K_{ji}} = \frac{\partial(\Lambda_{2N})^2}{\partial K_{ji}} = -A_{1N}\frac{(A^{-1})_{Nj}}{M_j}$$

from which it results that (small) variations of the synchronizing coefficients can modify the oscillation frequencies (without practically introducing any damping), and move from the origin both the real roots $\Lambda_{(2N-1)}$ and Λ_{2N} (provided that Equation [8.2.3] does not hold).

If in the original system $D \neq 0$, the expressions of the sensitivity coefficients (obviously more complicated) can be deduced similarly.

In any case, the knowledge of the sensitivity coefficients may be useful also for *identifying corrections ΔD_{ji} and ΔK_{ji} to be applied to the model parameters* , starting from (small) deviations between the characteristic roots experimentally obtained and those originally assumed.

Before concluding, the following should be indicated.

- Despite different approximations, the analysis reported here proved particularly useful in practical cases (large systems, with very small or even negative damping), at least for a first-attempt selection of the machines to which the additional signals should be applied.
- For a better approximation, the response delays of the voltage regulation loops, etc. also should be considered, e.g., by replacing the braking coefficients with proper transfer functions $D_{ji}(s)$. In practice, the existence of such delays may suggest the implementation of additional signals sensitive also to the delivered active powers, similarly to what is illustrated in Section 7.2.2c.
- *In general*, and beyond any model approximation, it is important to remember (see Appendix 3) that the characteristic root variations, as a consequence of a given feedback with a moderate gain, can be estimated based on the respective open-loop "residuals," which also can be determined experimentally, at least in principle. Therefore, to estimate the effects obtainable by a feedback sensitive to the speed Ω_i (or to the power P_{ei}) of a given unit, and acting on the set point of the respective voltage regulator, it is important to know the response of Ω_i (or of P_{ei}) to perturbations on this set-point; and so on, for any possible pair of input and output variables of the original system.

ANNOTATED REFERENCES

Among the works of more general interest, the following are mentioned: 19, 37, 142, 238. More particularly, for what concerns

- the modal analysis: 102 (including the definition of the mean motion), 183, 198, 202;
- the stability of the relative motion: 27, 52, 56, 61, 69, 80, 93, 105, 108, 117, 118, 119, 127 (with the example of a complete analysis, reported in the text), 134, 140, 145, 171 (with a rich bibliography), 270, 273, 336;
- the simplification of the overall model: 28, 122, 124, 144, 155, 156, 158, 170, 172, 175, 176, 180, 206, 221, 233, 241 (with a rich bibliography), 251, 256, 258, 272, 277, 290, 291, 301, 309, 310, 316, 317, 327 (with a rich bibliography);
- the effect of feedbacks sensitive to the motion of machines: 51, 95, 113, 146, 181, 183, 191, 192, 201, 202, 203, 214, 215, 216, 254.

APPENDIX 1

TRANSFORMATION TO SYMMETRICAL COMPONENTS

For any given frequency ν, a sinusoidal variable:

$$w(t) = W_M \cos(\nu t + \varphi_w)$$

can be represented by a "*phasor*" of the type:

$$\widetilde{w} \triangleq W \epsilon^{\tilde{j}\varphi_w} \quad \text{[A1.1]}$$

where:

$W \triangleq W_M/\sqrt{2}$ is the rms value of $w(t)$,
$\tilde{j}$ is the imaginary unit in the "phasor plane";

with obvious notation it conversely results:

$$w(t) = \sqrt{2}\mathcal{Re}(\widetilde{w}\epsilon^{\tilde{j}\nu t}) \quad \text{[A1.2]}$$

It then holds, with α_1 and α_2 arbitrary constants:

$$\widetilde{(\alpha_1 w_1 + \alpha_2 w_2)} = \alpha_1 \widetilde{w}_1 + \alpha_2 \widetilde{w}_2 \quad \text{[A1.3]}$$

and moreover:

$$\left(\widetilde{\frac{\mathrm{d}w}{\mathrm{d}t}}\right) = \tilde{j}\nu\widetilde{w} \quad [A1.4]$$

In the case of a linear and stationary (continuous-time) system, operating at a sinusoidal regime at frequency ν, the relationships among the different phasors can be then directly deduced from the differential equations of the system, by substituting:

- the single variables (or their Laplace transforms) with their respective phasors;
- the operator $p = \mathrm{d}/\mathrm{d}t$ (or the Laplace variable s, in the transfer functions) by the multiplication factor $\tilde{\jmath}\nu$.

In the case of three-phase electrical systems it is possible to define, for any given point, a set of phase voltages and a set of phase currents, with indices a, b, c (according to a predetermined order).

Under the *(ideal) equilibrium steady-state* at frequency ω, such sets constitute positive sequence sinusoidal sets at frequency ω, generically of the type:

$$\left.\begin{aligned} w_a(t) &= W_{M(1)}\cos(\omega t + \alpha_{(1)}) \\ w_b(t) &= W_{M(1)}\cos(\omega t + \alpha_{(1)} - 120^\circ) \\ w_c(t) &= W_{M(1)}\cos(\omega t + \alpha_{(1)} - 240^\circ) \end{aligned}\right\} \qquad [\text{A1.5}]$$

Because of [A1.1], the phase variable $w_a(t)$ is defined, for the given frequency ω, by the following phasor:

$$\widetilde{w}_a \triangleq W_{(1)}\epsilon^{\tilde{\jmath}\alpha_{(1)}}$$

where $W_{(1)} \triangleq W_{M(1)}/\sqrt{2}$, whereas it can be similarly derived:

$$\begin{aligned} \widetilde{w}_b &= \widetilde{w}_a\epsilon^{-\tilde{\jmath}120^\circ} \\ \widetilde{w}_c &= \widetilde{w}_a\epsilon^{-\tilde{\jmath}240^\circ} \end{aligned}$$

Under the mentioned conditions, the entire set is then defined by $\widetilde{w}_a$, or equivalently by the phasor:

$$\widetilde{w}_{(1)} \triangleq \sqrt{3}\widetilde{w}_a \qquad [\text{A1.6}]$$

named "symmetrical component of the positive sequence;" it also results:

$$\widetilde{w}_{(1)} = \frac{1}{\sqrt{3}}(\widetilde{w}_a + \widetilde{w}_b\epsilon^{-\tilde{\jmath}240^\circ} + \widetilde{w}_c\epsilon^{-\tilde{\jmath}120^\circ}) \qquad [\text{A1.6}']$$

and furthermore, accounting for the previous expressions:

$$\widetilde{w}_{(1)} = \sqrt{3}W_{(1)}\epsilon^{\tilde{\jmath}\alpha_{(1)}} = \sqrt{\frac{3}{2}}W_{M(1)}\epsilon^{\tilde{\jmath}\alpha_{(1)}} \qquad [\text{A1.7}]$$

The inverse transformation is then defined by:

$$\left.\begin{aligned} \widetilde{w}_a &= \frac{1}{\sqrt{3}}\widetilde{w}_{(1)} \\ \widetilde{w}_b &= \frac{1}{\sqrt{3}}\widetilde{w}_{(1)}\epsilon^{-\tilde{\jmath}120^\circ} \\ \widetilde{w}_c &= \frac{1}{\sqrt{3}}\widetilde{w}_{(1)}\epsilon^{-\tilde{\jmath}240^\circ} \end{aligned}\right\} \qquad [\text{A1.8}]$$

whereas:

$$\begin{cases} w_a(t) = \sqrt{2}\mathcal{R}e(\widetilde{w}_a \epsilon^{\tilde{j}\omega t}) = \sqrt{\frac{2}{3}}\mathcal{R}e(\widetilde{w}_{(1)}\epsilon^{\tilde{j}\omega t}) \\ w_b(t) = \sqrt{2}\mathcal{R}e(\widetilde{w}_b \epsilon^{\tilde{j}\omega t}) = \sqrt{\frac{2}{3}}\mathcal{R}e(\widetilde{w}_{(1)}\epsilon^{\tilde{j}(\omega t-120^\circ)}) \\ w_c(t) = \sqrt{2}\mathcal{R}e(\widetilde{w}_c \epsilon^{\tilde{j}\omega t}) = \sqrt{\frac{2}{3}}\mathcal{R}e(\widetilde{w}_{(1)}\epsilon^{\tilde{j}(\omega t-240^\circ)}) \end{cases}$$

The choice of the coefficient $\sqrt{3}$ in the Equation [A1.6], or equivalently of the coefficient $1/\sqrt{3}$ in [A1.6′], appears suitable as it allows to obtain, for the voltages v_a, v_b, v_c and the currents i_a, i_b, i_c at a given point (indicating by $\tilde{\imath}^*_{(1)}$ the conjugate of $\tilde{\imath}_{(1)}$):

$$\tilde{v}_{(1)}\tilde{\imath}^*_{(1)} = 3\tilde{v}_a\tilde{\imath}^*_a = \tilde{v}_a\tilde{\imath}^*_a + \tilde{v}_b\tilde{\imath}^*_b + \tilde{v}_c\tilde{\imath}^*_c = P + \tilde{j}Q \qquad \text{[A1.9]}$$

with P and Q the corresponding active and reactive powers.

In general, under *sinusoidal steady-state* conditions at frequency ω, each set of phase variables may include components of the positive, negative, and zero sequence. More precisely, it can be recognized that a generic sinusoidal set, corresponding to the phasors $\widetilde{w}_a$, $\widetilde{w}_b$, and $\widetilde{w}_c$, can be written in the form:

$$\left.\begin{aligned} w_a(t) &= W_{M(0)}\cos(\omega t + \alpha_{(0)}) + W_{M(1)}\cos(\omega t + \alpha_{(1)}) \\ &\quad + W_{M(2)}\cos(\omega t + \alpha_{(2)}) \\ w_b(t) &= W_{M(0)}\cos(\omega t + \alpha_{(0)}) + W_{M(1)}\cos(\omega t + \alpha_{(1)} - 120^\circ) \\ &\quad + W_{M(2)}\cos(\omega t + \alpha_{(2)} - 240^\circ) \\ w_c(t) &= W_{M(0)}\cos(\omega t + \alpha_{(0)}) + W_{M(1)}\cos(\omega t + \alpha_{(1)} - 240^\circ) \\ &\quad + W_{M(2)}\cos(\omega t + \alpha_{(2)} - 120^\circ) \end{aligned}\right\} \qquad \text{[A1.10]}$$

By such positions it in fact follows:

$$\begin{cases} \widetilde{w}_a = W_{(0)}\epsilon^{\tilde{j}\alpha_{(0)}} + W_{(1)}\epsilon^{\tilde{j}\alpha_{(1)}} + W_{(2)}\epsilon^{\tilde{j}\alpha_{(2)}} \\ \widetilde{w}_b = W_{(0)}\epsilon^{\tilde{j}\alpha_{(0)}} + W_{(1)}\epsilon^{\tilde{j}(\alpha_{(1)}-120^\circ)} + W_{(2)}\epsilon^{\tilde{j}(\alpha_{(2)}-240^\circ)} \\ \widetilde{w}_c = W_{(0)}\epsilon^{\tilde{j}\alpha_{(0)}} + W_{(1)}\epsilon^{\tilde{j}(\alpha_{(1)}-240^\circ)} + W_{(2)}\epsilon^{\tilde{j}(\alpha_{(2)}-120^\circ)} \end{cases}$$

with $W_{(0)} \triangleq W_{M(0)}/\sqrt{2}$, $W_{(1)} \triangleq W_{M(1)}/\sqrt{2}$, $W_{(2)} \triangleq W_{M(2)}/\sqrt{2}$, so that $W_{M(0)}$, $\alpha_{(0)}$, $W_{M(1)}$, $\alpha_{(1)}$, $W_{M(2)}$, $\alpha_{(2)}$ can be derived starting from $\widetilde{w}_a$, $\widetilde{w}_b$, and $\widetilde{w}_c$.

The entire set can be then defined by the phasors:

$$\left.\begin{aligned} \widetilde{w}_{(0)} &\triangleq \frac{1}{\sqrt{3}}(\widetilde{w}_a + \widetilde{w}_b + \widetilde{w}_c) \\ \widetilde{w}_{(1)} &\triangleq \frac{1}{\sqrt{3}}(\widetilde{w}_a + \widetilde{w}_b\epsilon^{-\tilde{j}240^\circ} + \widetilde{w}_c\epsilon^{-\tilde{j}120^\circ}) \\ \widetilde{w}_{(2)} &\triangleq \frac{1}{\sqrt{3}}(\widetilde{w}_a + \widetilde{w}_b\epsilon^{-\tilde{j}120^\circ} + \widetilde{w}_c\epsilon^{-\tilde{j}240^\circ}) \end{aligned}\right\} \qquad \text{[A1.11]}$$

which constitute the so-called "*symmetrical components*," respectively of the zero, positive, and negative sequence; accounting for the previous expressions it can be furthermore derived, similarly to Equation [A1.7]:

$$\left.\begin{aligned}\widetilde{w}_{(0)} &= \sqrt{3}W_{(0)}\epsilon^{\tilde{j}\alpha_{(0)}} = \sqrt{\tfrac{3}{2}}W_{M(0)}\epsilon^{\tilde{j}\alpha_{(0)}}\\ \widetilde{w}_{(1)} &= \sqrt{3}W_{(1)}\epsilon^{\tilde{j}\alpha_{(1)}} = \sqrt{\tfrac{3}{2}}W_{M(1)}\epsilon^{\tilde{j}\alpha_{(1)}}\\ \widetilde{w}_{(2)} &= \sqrt{3}W_{(2)}\epsilon^{\tilde{j}\alpha_{(2)}} = \sqrt{\tfrac{3}{2}}W_{M(2)}\epsilon^{\tilde{j}\alpha_{(2)}}\end{aligned}\right\} \quad [A1.12]$$

Finally, the inverse transformation is given by:

$$\left.\begin{aligned}\widetilde{w}_a &= \tfrac{1}{\sqrt{3}}(\widetilde{w}_{(0)} + \widetilde{w}_{(1)} + \widetilde{w}_{(2)})\\ \widetilde{w}_b &= \tfrac{1}{\sqrt{3}}(\widetilde{w}_{(0)} + \widetilde{w}_{(1)}\epsilon^{-\tilde{j}120^\circ} + \widetilde{w}_{(2)}\epsilon^{-\tilde{j}240^\circ})\\ \widetilde{w}_c &= \tfrac{1}{\sqrt{3}}(\widetilde{w}_{(0)} + \widetilde{w}_{(1)}\epsilon^{-\tilde{j}240^\circ} + \widetilde{w}_{(2)}\epsilon^{-\tilde{j}120^\circ})\end{aligned}\right\} \quad [A1.13]$$

which generalize Equations [A1.8], whereas:

$$\left.\begin{aligned}w_a(t) &= \sqrt{2}\mathcal{R}e(\widetilde{w}_a\epsilon^{\tilde{j}\omega t}) = \sqrt{\tfrac{2}{3}}\mathcal{R}e[(\widetilde{w}_{(0)} + \widetilde{w}_{(1)} + \widetilde{w}_{(2)})\epsilon^{\tilde{j}\omega t}]\\ w_b(t) &= \sqrt{2}\mathcal{R}e(\widetilde{w}_b\epsilon^{\tilde{j}\omega t}) = \sqrt{\tfrac{2}{3}}\mathcal{R}e[(\widetilde{w}_{(0)} + \widetilde{w}_{(1)}\epsilon^{-\tilde{j}120^\circ} + \widetilde{w}_{(2)}\epsilon^{-\tilde{j}240^\circ})\epsilon^{\tilde{j}\omega t}]\\ w_c(t) &= \sqrt{2}\mathcal{R}e(\widetilde{w}_c\epsilon^{\tilde{j}\omega t}) = \sqrt{\tfrac{2}{3}}\mathcal{R}e[(\widetilde{w}_{(0)} + \widetilde{w}_{(1)}\epsilon^{-\tilde{j}240^\circ} + \widetilde{w}_{(2)}\epsilon^{-\tilde{j}120^\circ})\epsilon^{\tilde{j}\omega t}]\end{aligned}\right\} \quad [A1.14]$$

The second of Equations [A1.11] coincides with Equation [A1.6′]. Similar to what was noted above, the choice of the coefficient $1/\sqrt{3}$ in [A1.11] allows to obtain, for the voltages v_a, v_b, v_c and the currents i_a, i_b, i_c at a given point:

$$\tilde{v}_{(0)}\tilde{\imath}^*_{(0)} + \tilde{v}_{(1)}\tilde{\imath}^*_{(1)} + \tilde{v}_{(2)}\tilde{\imath}^*_{(2)} = \tilde{v}_a\tilde{\imath}^*_a + \tilde{v}_b\tilde{\imath}^*_b + \tilde{v}_c\tilde{\imath}^*_c = P + \tilde{j}Q \quad [A1.15]$$

with P and Q the corresponding active and reactive powers. This last relationship, clearly, generalizes Equation [A1.9]. In the presence of zero sequence currents, it may be intended that the return of the total current $(i_a + i_b + i_c)$ is through an "earth" circuit, and that the voltages v_a, v_b, v_c are evaluated with respect to earth.

ANNOTATED REFERENCES

The following references are particularly indicated: 11, 33.

APPENDIX 2

PARK'S TRANSFORMATION

In a three-phase electrical system at *any possible operating condition* (even under transient conditions), to the generic set of phase variables w_a, w_b, w_c it is possible to associate the following "Park's variables":

$$\left.\begin{aligned} w_{dr} &\triangleq K_{dq}(w_a \cos\theta_r + w_b \cos\theta'_r + w_c \cos\theta''_r) \\ w_{qr} &\triangleq -K_{dq}(w_a \sin\theta_r + w_b \sin\theta'_r + w_c \sin\theta''_r) \\ w_o &\triangleq K_o(w_a + w_b + w_c) \end{aligned}\right\} \quad \text{[A2.1]}$$

where:

K_{dq} and K_o are arbitrary constants[(1)];
θ_r is the "reference angular position";

whereas it is intended, for brevity:

$$\theta'_r \triangleq \theta_r - 120°, \qquad \theta''_r \triangleq \theta_r - 240°$$

The variables w_{dr} and w_{qr} also can be interpreted as the components of the "*Park's vector*":

$$\overline{w}_r \triangleq w_{dr} + j w_{qr} = K_{dq}(w_a + w_b \epsilon^{j120°} + w_c \epsilon^{j240°})\epsilon^{-j\theta_r} \quad \text{[A2.1']}$$

[(1)] In the text chapters it is intended $K_{dq} = \sqrt{2/3}$, $K_o = 1/\sqrt{3}$ (see Equations [A2.4]).

where j is the imaginary unit in the plane of the Park's vectors (*which must not be confused* with the analog unit $\tilde{j}$, relative to the phasors), whereas the variable w_o constitutes the "*homopolar*" (*Park's*) *variable*.

In the symbol of the Park's vector and of its components, the subscript "r" is indicative of the dependence on the reference θ_r; the homopolar variable instead is independent of this reference.

Observe that:

- since $1 + \epsilon^{j120^\circ} + \epsilon^{j240^\circ} = 0$, the Park's vector $\overline{w}_r$ depends (apart from θ_r) only on the *differences* between w_a, w_b, w_c; in particular, having set $w_{ab} \triangleq w_a - w_b$, etc., it follows:

$$\begin{aligned}\overline{w}_r\epsilon^{j\theta_r} &= K_{dq}(w_a + w_b\epsilon^{j120^\circ} + w_c\epsilon^{j240^\circ}) \\ &= \begin{Bmatrix} \dfrac{\epsilon^{-j30^\circ}}{\sqrt{3}}K_{dq}(w_{ab} + w_{bc}\epsilon^{j120^\circ} + w_{ca}\epsilon^{j240^\circ}) \\ \dfrac{\epsilon^{+j30^\circ}}{\sqrt{3}}K_{dq}(w_{ac} + w_{ba}\epsilon^{j120^\circ} + w_{cb}\epsilon^{j240^\circ}) \end{Bmatrix} \\ &= K_{dq}(w_{ab}\epsilon^{-j60^\circ} + w_{ac}\epsilon^{+j60^\circ})\end{aligned}$$

- the homopolar variable w_o depends instead only on the *sum* $(w_a + w_b + w_c)$.

Finally, the inverse transformation (from the Park's variables to the phase variables) is given by:

$$\left.\begin{aligned} w_a &= \frac{w_o}{3K_o} + \frac{2}{3K_{dq}}(w_{dr}\cos\theta_r - w_{qr}\sin\theta_r) = \frac{w_o}{3K_o} + \frac{2}{3K_{dq}}\mathcal{Re}(\overline{w}_r\epsilon^{j\theta_r}) \\ w_b &= \frac{w_o}{3K_o} + \frac{2}{3K_{dq}}(w_{dr}\cos\theta_r' - w_{qr}\sin\theta_r') = \frac{w_o}{3K_o} + \frac{2}{3K_{dq}}\mathcal{Re}(\overline{w}_r\epsilon^{j\theta_r'}) \\ w_c &= \frac{w_o}{3K_o} + \frac{2}{3K_{dq}}(w_{dr}\cos\theta_r'' - w_{qr}\sin\theta_r'') = \frac{w_o}{3K_o} + \frac{2}{3K_{dq}}\mathcal{Re}(\overline{w}_r\epsilon^{j\theta_r''}) \end{aligned}\right\} \quad \text{[A2.2]}$$

The *principal properties* of the Park's transformation, defined by Equations [A2.1], are as follows:

(1) If, with α_1 and α_2 arbitrary constants:

$$\begin{cases} w_a = \alpha_1 w_{a1} + \alpha_2 w_{a2} \\ w_b = \alpha_1 w_{b1} + \alpha_2 w_{b2} \\ w_c = \alpha_1 w_{c1} + \alpha_2 w_{c2} \end{cases}$$

then:

$$\left\{\begin{array}{l} \left.\begin{array}{l} w_{dr} = \alpha_1 w_{dr1} + \alpha_2 w_{dr2} \\ w_{qr} = \alpha_1 w_{qr1} + \alpha_2 w_{qr2} \end{array}\right\} \overline{w}_r = \alpha_1 \overline{w}_{r1} + \alpha_2 \overline{w}_{r2} \\ w_o = \alpha_1 w_{o1} + \alpha_2 w_{o2} \end{array}\right.$$

(2) If w_{a1}, w_{b1}, w_{c1} and w_{a2}, w_{b2}, w_{c2} are two generic sets of phase variables, it results:

$$\begin{aligned} w_{a1} w_{a2} + w_{b1} w_{b2} + w_{c1} w_{c2} &= \frac{w_{o1} w_{o2}}{3K_o^2} + \frac{w_{dr1} w_{dr2} + w_{qr1} w_{qr2}}{\frac{3}{2} K_{dq}^2} \\ &= \frac{w_{o1} w_{o2}}{3K_o^2} + \frac{\langle \overline{w}_{r1}, \overline{w}_{r2} \rangle}{\frac{3}{2} K_{dq}^2} \end{aligned} \qquad [A2.3]$$

where $\langle \overline{w}_{r1}, \overline{w}_{r2} \rangle = \mathcal{R}e(\overline{w}_{r1} \overline{w}_{r2}^*)$ represents the scalar product of $\overline{w}_{r1}$ and $\overline{w}_{r2}$ (in such a product, the dependence on the reference θ_r vanishes).

Therefore, if it is assumed:

$$\left\{\begin{array}{l} K_{dq} = \sqrt{\frac{2}{3}} \\ K_o = \frac{1}{\sqrt{3}} \end{array}\right. \qquad [A2.4]$$

it follows in a simpler way:

$$\begin{aligned} w_{a1} w_{a2} + w_{b1} w_{b2} + w_{c1} w_{c2} &= w_{o1} w_{o2} + w_{dr1} w_{dr2} + w_{qr1} w_{qr2} \\ &= w_{o1} w_{o2} + \langle \overline{w}_{r1}, \overline{w}_{r2} \rangle \end{aligned} \qquad [A2.3']$$

which may be more convenient to have an *energy equivalence* between the phase variable system and the Park's variable system, if the terms that appear in Equation [A2.3′] have the dimension of an energy or of a power (e.g., if the two sets, respectively, correspond to the voltages and the currents at a given point of the system, so that the left-hand member in Equation [A2.3′] represents the instantaneous active power).

(3) Applying Equations [A2.1] to the set $\mathrm{d}w_a/\mathrm{d}t$, $\mathrm{d}w_b/\mathrm{d}t$, $\mathrm{d}w_c/\mathrm{d}t$, and denoting $p \triangleq \mathrm{d}/\mathrm{d}t$, $\Omega_r \triangleq \mathrm{d}\theta_r/\mathrm{d}t$, it can be derived:

$$\left.\begin{array}{l} \left.\begin{array}{l} \left(\dfrac{\mathrm{d}w}{\mathrm{d}t}\right)_{dr} = \dfrac{\mathrm{d}w_{dr}}{\mathrm{d}t} - \Omega_r w_{qr} \\ \left(\dfrac{\mathrm{d}w}{\mathrm{d}t}\right)_{qr} = \dfrac{\mathrm{d}w_{qr}}{\mathrm{d}t} + \Omega_r w_{dr} \end{array}\right\} \overline{\left(\dfrac{\mathrm{d}w}{\mathrm{d}t}\right)_r} = (p + j\Omega_r)\overline{w}_r \\ \left(\dfrac{\mathrm{d}w}{\mathrm{d}t}\right)_o = \dfrac{\mathrm{d}w_o}{\mathrm{d}t} \end{array}\right\} \qquad [A2.5]$$

In fact it results, because of Equations [A2.1]:

$$\frac{\mathrm{d}w_{dr}}{\mathrm{d}t} = K_{dq}\left(\frac{\mathrm{d}w_a}{\mathrm{d}t}\cos\theta_r + \frac{\mathrm{d}w_b}{\mathrm{d}t}\cos\theta_r' + \frac{\mathrm{d}w_c}{\mathrm{d}t}\cos\theta_r''\right)$$
$$- K_{dq}(w_a\sin\theta_r + w_b\sin\theta_r' + w_c\sin\theta_r'')\frac{\mathrm{d}\theta_r}{\mathrm{d}t} = \left(\frac{\mathrm{d}w}{\mathrm{d}t}\right)_{dr} + w_{qr}\Omega_r$$

and similarly $\mathrm{d}w_{qr}/\mathrm{d}t = (\mathrm{d}w/\mathrm{d}t)_{qr} - w_{dr}\Omega_r$, whereas $\mathrm{d}w_o/\mathrm{d}t = (\mathrm{d}w/\mathrm{d}t)_o$.

By a similar procedure, it is possible to derive the Park's variables corresponding to the derivatives (of phase variables) of higher order. More precisely, it can be obtained:

$$\left.\begin{aligned}\overline{\left(\frac{\mathrm{d}^2 w}{\mathrm{d}t^2}\right)_r} &= \frac{\mathrm{d}^2\overline{w}_r}{\mathrm{d}t^2} + 2j\Omega_r\frac{\mathrm{d}\overline{w}_r}{\mathrm{d}t} + j\frac{\mathrm{d}\Omega_r}{\mathrm{d}t}\overline{w}_r - \Omega_r^2\overline{w}_r\\ &= (p + j\Omega_r)[(p + j\Omega_r)\overline{w}_r]\\ \left(\frac{\mathrm{d}^2 w}{\mathrm{d}t^2}\right)_o &= p^2 w_o\end{aligned}\right\}$$

in the former of which it must be intended that the operator $p = \mathrm{d}/\mathrm{d}t$ out of the square brackets is applied also to the Ω_r included within bracket. By so intending it can be finally obtained, more in general ($k = 2, 3, \ldots$):

$$\left.\begin{aligned}\overline{\left(\frac{\mathrm{d}^k w}{\mathrm{d}t^k}\right)_r} &= \underbrace{(p + j\Omega_r)\{(p + j\Omega_r)[\ldots(p + j\Omega_r)}_{k}\overline{w}_r]\}\\ \left(\frac{\mathrm{d}^k w}{\mathrm{d}t^k}\right)_o &= p^k w_o\end{aligned}\right\} \qquad \text{[A2.6]}$$

A significant simplification results when $\Omega_r = constant$, in which case Equations [A2.6] are reduced to:

$$\left.\begin{aligned}\overline{\left(\frac{\mathrm{d}^k w}{\mathrm{d}t^k}\right)_r} &= (p + j\Omega_r)^k\overline{w}_r\\ \left(\frac{\mathrm{d}^k w}{\mathrm{d}t^k}\right)_o &= p^k w_o\end{aligned}\right\} \qquad \text{[A2.6']}$$

i.e., *the application of the operator p to the phase variables becomes the application of the operator* $(p + j\Omega_r)$ *to the Park's vectors*, and of the operator p to the homopolar variables.

(4) If we move from the reference θ_r to the reference θ_s, it follows:

$$\left.\begin{aligned}w_{ds} &= w_{dr}\cos(\theta_r - \theta_s) - w_{qr}\sin(\theta_r - \theta_s)\\ w_{qs} &= w_{dr}\sin(\theta_r - \theta_s) + w_{qr}\cos(\theta_r - \theta_s)\end{aligned}\right\}\overline{w}_s = \overline{w}_r\epsilon^{j(\theta_r - \theta_s)} \qquad \text{[A2.7]}$$

whereas the homopolar variable remains, as stated, unchanged. In fact, as a result of Equation [A2.1′], the vector $\overline{w}_r \epsilon^{j\theta_r}$ is independent of the angular reference of the transformation. Note that $|\overline{w}_s| = |\overline{w}_r|$, independent of this reference.

Under *sinusoidal steady-state* conditions (of the phase variables) at frequency ω, by recalling Equations [A1.10] it can be derived:

$$\left.\begin{aligned} w_{dr} &= \tfrac{3}{2}K_{dq}(W_{M(1)}\cos(\omega t + \alpha_{(1)} - \theta_r) + W_{M(2)}\cos(\omega t + \alpha_{(2)} + \theta_r)) \\ w_{qr} &= \tfrac{3}{2}K_{dq}(W_{M(1)}\sin(\omega t + \alpha_{(1)} - \theta_r) - W_{M(2)}\sin(\omega t + \alpha_{(2)} + \theta_r)) \\ w_o &= 3K_o W_{M(0)}\cos(\omega t + \alpha_{(0)}) \end{aligned}\right\} \quad \text{[A2.8]}$$

and therefore:

- the positive- and negative-sequence components (with their magnitudes and phases) influence only the Park's vector (with its components w_{dr} and w_{qr});
- the zero-sequence component (with its magnitude and phase) influences only the homopolar component w_o.

Similarly to the symmetrical components (phasors) in Equations [A1.12] (but obviously with a different meaning), it also is possible to define the following constant vectors:

$$\left.\begin{aligned} \overline{w}_{(0)} &\triangleq \sqrt{\tfrac{3}{2}}W_{M(0)}\epsilon^{j\alpha_{(0)}} \\ \overline{w}_{(1)} &\triangleq \sqrt{\tfrac{3}{2}}W_{M(1)}\epsilon^{j\alpha_{(1)}} \\ \overline{w}_{(2)} &\triangleq \sqrt{\tfrac{3}{2}}W_{M(2)}\epsilon^{j\alpha_{(2)}} \end{aligned}\right\} \quad \text{[A2.9]}$$

It then follows, because of Equations [A2.8]:

$$\left.\begin{aligned} \overline{w}_r &= \sqrt{\tfrac{3}{2}}K_{dq}(\overline{w}_{(1)}\epsilon^{j(\omega t - \theta_r)} + \overline{w}^*_{(2)}\epsilon^{-j(\omega t + \theta_r)}) \\ w_o &= \sqrt{6}K_o \mathcal{R}e(\overline{w}_{(0)}\epsilon^{j\omega t}) \end{aligned}\right\} \quad \text{[A2.10]}$$

where the first equation simplifies to:

$$\overline{w}_r = \sqrt{\tfrac{3}{2}}K_{dq}(\overline{w}_{(1)} + \overline{w}^*_{(2)}\epsilon^{-2j\omega t}) \quad \text{[A2.10′]}$$

if it is assumed $\theta_r = \omega t$ (it furthermore results $\sqrt{3/2}K_{dq} = 1$, $\sqrt{6}K_o = \sqrt{2}$, if [A2.4] are adopted).

In analogy to Equations [A1.14], the following expressions then hold:

$$\left.\begin{aligned} w_a(t) &= \sqrt{\tfrac{2}{3}}\mathcal{R}e[(\overline{w}_{(0)} + \overline{w}_{(1)} + \overline{w}_{(2)})\epsilon^{j\omega t}] \\ w_b(t) &= \sqrt{\tfrac{2}{3}}\mathcal{R}e[(\overline{w}_{(0)} + \overline{w}_{(1)}\epsilon^{-j120^\circ} + \overline{w}_{(2)}\epsilon^{-j240^\circ})\epsilon^{j\omega t}] \\ w_c(t) &= \sqrt{\tfrac{2}{3}}\mathcal{R}e[(\overline{w}_{(0)} + \overline{w}_{(1)}\epsilon^{-j240^\circ} + \overline{w}_{(2)}\epsilon^{-j120^\circ})\epsilon^{j\omega t}] \end{aligned}\right\} \quad \text{[A2.11]}$$

whereas Equation [A1.15], for the voltages v_a, v_b, v_c and the currents i_a, i_b, i_c at a given point, with the already specified notation, translates into:

$$\overline{v}_{(0)}\overline{\imath}^*_{(0)} + \overline{v}_{(1)}\overline{\imath}^*_{(1)} + \overline{v}_{(2)}\overline{\imath}^*_{(2)} = P + jQ \qquad \text{[A2.12]}$$

By putting into evidence, in $\overline{w}_r$, the separate effects of the positive and of the negative sequences, the first of Equations [A2.10] also can be written as:

$$\overline{w}_r = \overline{w}_{r1} + \overline{w}_{r2}$$

with:

$$\begin{cases} \overline{w}_{r1} \triangleq \sqrt{\frac{3}{2}}K_{dq}\overline{w}_{(1)}\epsilon^{j(\omega t-\theta_r)} \\ \overline{w}_{r2} \triangleq \sqrt{\frac{3}{2}}K_{dq}\overline{w}^*_{(2)}\epsilon^{-j(\omega t+\theta_r)} \end{cases}$$

It can then be derived, with $\Omega_r \triangleq \mathrm{d}\theta_r/\mathrm{d}t$:

$$\begin{cases} \dfrac{\mathrm{d}\overline{w}_{r1}}{\mathrm{d}t} = j(\omega - \Omega_r)\overline{w}_{r1} \\ \dfrac{\mathrm{d}\overline{w}_{r2}}{\mathrm{d}t} = -j(\omega + \Omega_r)\overline{w}_{r2} \end{cases}$$

and thus:

$$\frac{\mathrm{d}\overline{w}_r}{\mathrm{d}t} = j\omega(\overline{w}_{r1} - \overline{w}_{r2}) - j\Omega_r(\overline{w}_{r1} + \overline{w}_{r2}) = j\omega(\overline{w}_{r1} - \overline{w}_{r2}) - j\Omega_r\overline{w}_r$$

from which, recalling Equations [A2.5]:

$$\overline{\left(\frac{\mathrm{d}w}{\mathrm{d}t}\right)_r} = (p + j\Omega_r)\overline{w}_r = j\omega(\overline{w}_{r1} - \overline{w}_{r2})$$

In particular, if the negative sequence is absent, this last equation becomes:

$$\overline{\left(\frac{\mathrm{d}w}{\mathrm{d}t}\right)_r} = (p + j\Omega_r)\overline{w}_r = j\omega\overline{w}_r$$

and this means that, for a sinusoidal steady-state of the positive sequence at frequency ω:

- it has to be intended, for what concerns the Park's vectors:

$$p = j(\omega - \Omega_r)$$

  ($p = 0$ if it is assumed $\Omega_r = \omega$);

- the generic ratio $\overline{(\mathrm{d}w/\mathrm{d}t)}_r/\overline{w}_r$ (between the vectors corresponding to the derivatives of a set, and to the set itself) is equal to $j\omega$ independently of the reference θ_r.

Note furthermore that, if the (phase) voltages and currents at a given point are only of the positive sequence, it results, recalling Equation [A2.12]:

$$\overline{v}_r\overline{\imath}_r^* = \tfrac{3}{2}K_{dq}^2\overline{v}_{(1)}\overline{\imath}_{(1)}^* = \tfrac{3}{2}K_{dq}^2(P + jQ)$$

whereas, if the above-mentioned voltages and currents are only of the negative sequence, it instead holds:

$$\overline{v}_r\overline{\imath}_r^* = \tfrac{3}{2}K_{dq}^2\overline{v}_{(2)}^*\overline{\imath}_{(2)} = \tfrac{3}{2}K_{dq}^2(P - jQ)$$

Therefore, the product $\overline{v}_r\overline{\imath}_r^*$ does not allow, in general, deduction of the reactive power Q. Recall instead that, by Equation [A2.3], it is:

$$P = \frac{v_o i_o}{3K_o^2} + \frac{\langle\overline{v}_r, \overline{\imath}_r^*\rangle}{\frac{3}{2}K_{dq}^2} = \frac{v_o i_o}{3K_o^2} + \frac{\mathcal{Re}(\overline{v}_r\overline{\imath}_r^*)}{\frac{3}{2}K_{dq}^2}$$

for *any possible operating condition*, even under transient conditions.

The variations in the magnitude and phase of the generic Park's vector can be evaluated on the basis of the identity:

$$\frac{\mathrm{d}\overline{w}_r}{\overline{w}_r} = \frac{\mathrm{d}|\overline{w}_r|}{\overline{w}_r} + j\,\mathrm{d}\angle\overline{w}_r$$

where the left-hand term is, because of Equation [A2.1′], equal to:

$$\frac{\mathrm{d}\overline{w}_r}{\overline{w}_r} = \frac{\mathrm{d}(w_a + w_b\epsilon^{j120^\circ} + w_c\epsilon^{j240^\circ})}{w_a + w_b\epsilon^{j120^\circ} + w_c\epsilon^{j240^\circ}} - j\,\mathrm{d}\theta_r$$

Intending $w_{ab} \triangleq w_a - w_b$, etc., it is then possible to derive:

$$\begin{cases} \dfrac{\mathrm{d}|\overline{w}_r|}{|\overline{w}_r|} = \mathcal{Re}\left(\dfrac{\mathrm{d}\overline{w}_r}{\overline{w}_r}\right) = \dfrac{(w_{ab}+w_{ac})\,\mathrm{d}w_a + (w_{bc}+w_{ba})\,\mathrm{d}w_b + (w_{ca}+w_{cb})\,\mathrm{d}w_c}{w_{ab}^2 + w_{bc}^2 + w_{ca}^2} \\ \qquad = \dfrac{\left(w_{ab} - \dfrac{w_{ac}}{2}\right)\mathrm{d}w_{ab} + \left(w_{ac} - \dfrac{w_{ab}}{2}\right)\mathrm{d}w_{ac}}{w_{ab}^2 + w_{ac}^2 - w_{ab}w_{ac}} \\ \mathrm{d}\angle\overline{w}_r = Im\left(\dfrac{\mathrm{d}\overline{w}_r}{\overline{w}_r}\right) = \sqrt{3}\dfrac{w_{cb}\,\mathrm{d}w_a + w_{ac}\,\mathrm{d}w_b + w_{ba}\,\mathrm{d}w_c}{w_{ab}^2 + w_{bc}^2 + w_{ca}^2} - \mathrm{d}\theta_r \\ \qquad = \dfrac{\sqrt{3}}{2}\dfrac{-w_{ac}\,\mathrm{d}w_{ab} + w_{ab}\,\mathrm{d}w_{ac}}{w_{ab}^2 + w_{ac}^2 - w_{ab}w_{ac}} - \mathrm{d}\theta_r \end{cases}$$

(where $|\overline{w}_r| = (K_{dq}/\sqrt{2})\sqrt{w_{ab}^2 + w_{bc}^2 + w_{ca}^2} = K_{dq}\sqrt{w_{ab}^2 + w_{ac}^2 - w_{ab}w_{ac}}$).

At equilibrium steady-state (w_a, w_b, w_c sinusoidal at frequency ω, and only of the positive sequence), having set $w_a = W_{M(1)} \cos(\omega t + \alpha_{(1)})$ etc., it can be derived:

$$\begin{cases} \mathrm{d}|\overline{w}_r| = 0 \\ \mathrm{d}\angle\overline{w}_r = (\omega - \Omega_r)\,\mathrm{d}t \end{cases}$$

with $\Omega_r \triangleq \mathrm{d}\theta_r/\mathrm{d}t$, so that:

- the magnitude of $\overline{w}_r$ remains constant;
- the phase of $\overline{w}_r$ varies with a derivative $(\omega - \Omega_r)$ (which is zero if $\Omega_r = \omega$).

Based on this last result, it can be intended that the "*frequency*" associated to the set w_a, w_b, w_c is given—*at any possible operating condition, even under transient conditions*—by:

$$\omega_{(w)} \triangleq \frac{\mathrm{d}\angle\overline{w}_r}{\mathrm{d}t} + \Omega_r = \sqrt{3}\,\frac{w_{cb}\dfrac{\mathrm{d}w_a}{\mathrm{d}t} + w_{ac}\dfrac{\mathrm{d}w_b}{\mathrm{d}t} + w_{ba}\dfrac{\mathrm{d}w_c}{\mathrm{d}t}}{w_{ab}^2 + w_{bc}^2 + w_{ca}^2}$$

$$= \frac{\sqrt{3}}{2}\,\frac{-w_{ac}\dfrac{\mathrm{d}w_{ab}}{\mathrm{d}t} + w_{ab}\dfrac{\mathrm{d}w_{ac}}{\mathrm{d}t}}{w_{ab}^2 + w_{ac}^2 - w_{ab}w_{ac}} \qquad \text{[A2.13]}$$

ANNOTATED REFERENCES

The following references are particularly indicated: 11, 33, 65, 99.

APPENDIX 3

ELEMENTARY OUTLINE OF THE AUTOMATIC CONTROL THEORY

A3.1. PRELIMINARIES

Consider the (dynamic, continuous-time) system represented by the n-th order differential equation:

$$a_n \frac{d^n y}{dt^n} + \cdots + a_1 \frac{dy}{dt} + a_0 y = b_n \frac{d^n u}{dt^n} + \cdots + b_1 \frac{du}{dt} + b_0 u \qquad \text{[A3.1.1]}$$

where $u = u(t)$ and $y = y(t)$, respectively, constitute the input and the output of the system, and $a_n \neq 0$[(1)].

Further assume, for simplicity, that the coefficients $a_n, \ldots, a_0, b_n, \ldots, b_0$:

- do not depend on u, y (or on their derivatives): the system is then *linear*;
- do not depend on t: the system is then *stationary*.

The value of n constitutes the *dynamic order* of the system.

The solution $y(t)$ is the sum of two contributions:

- the *"free" response* resulting from the initial conditions;
- the *"forced" response* resulting from the input function $u(t)$.

[(1)] On the contrary, the maximum derivative order at the right-hand side (i.e., related to the input function $u(t)$) can be $n' < n$, if $b_n = \cdots = b_{(n'+1)} = 0$.

More precisely, by applying the Laplace transformation and intending $Y(s) \triangleq \mathcal{L}\{y(t)\}$, $U(s) \triangleq \mathcal{L}\{u(t)\}$, it is possible to derive:

$$Y(s) = \frac{\cdots}{\mathcal{A}(s)} + \frac{\mathcal{B}(s)}{\mathcal{A}(s)} U(s) \qquad \text{[A3.1.2]}$$

where the dotted term depends only on the initial conditions, whereas:

$$\begin{cases} \mathcal{A}(s) \triangleq a_n s^n + \cdots + a_1 s + a_0 \\ \mathcal{B}(s) \triangleq b_n s^n + \cdots + b_1 s + b_0 \end{cases}$$

Therefore, the free response and the forced response can be derived respectively by inverse-transforming the two terms at the right-hand side in Equation [A3.1.2]. The function:

$$G(s) \triangleq \frac{\mathcal{B}(s)}{\mathcal{A}(s)}$$

which defines the dependence of the forced response on the input, is called *"transfer" function* of the system. The knowledge of $G(s)$ is sufficient to return to the original Equation [A3.1.1] (and thus to perform analyses also of the free response), if and only if there are no cancellations of common factors in $\mathcal{B}(s)$ and $\mathcal{A}(s)$.

Under the mentioned conditions, the generic (linear and stationary) single-input and single-output system also can be represented by equations:

$$\left.\begin{aligned} \frac{dx}{dt} &= Ax + Bu \\ y &= Cx + Du \end{aligned}\right\} \qquad \text{[A3.1.3]}$$

(*equations in "normal" form*), where:

- x is a column matrix $(n, 1)$, called *"state"* of the system[2];
- A, B, C are suitable constant matrices (respectively (n, n), $(n, 1)$, $(1, n)$ dimensional);
- D is scalar, and equal to $D = b_n/a_n$.

---

(2) The elements of x constitute the so-called "state variables," which are n. The first of Equations [A3.1.3] is equivalent to n first-order differential equations in such variables, whereas the latter equation is purely algebraic. For any given nonimpulsive function $u(t)$:

- the solution $x(t)$ is continuous, whereas $y(t)$ is nonimpulsive (and continuous, if $b_n = 0$);
- to deduce such solutions, it is necessary and sufficient to know the initial state.

From such equations, one can derive the transfer function in the following form:

$$G(s) = C(sI_{(n)} - A)^{-1}B + D \qquad \text{[A3.1.4]}$$

and similarly, as said, a single (n-th order) equation, of the type [A3.1.1]. However, if the system has cancellations in $G(s)$, the representation by Equations [A3.1.3] is exhaustive, whereas [A3.1.1] may even correspond to different systems(3).

A3.2. STABILITY AND RESPONSE MODES

The free response is said to be "*stable*" if and only if, whichever be the set of initial conditions, it is limited (it is still assumed, here and in the following, that the system is linear and stationary). Particularly, the free response is said to be "*asymptotically* stable" if, furthermore, it tends to zero for $t \to \infty$ (otherwise, it is said to be "*weakly* stable").

To evaluate the stability properties (of the free response, or "of the system" as it is commonly said), consider the *characteristic equation*:

$$0 = \mathcal{A}(s)$$

the n solutions of which are the *characteristic roots* of the system. Alternatively, with reference to the Equations [A3.1.3], the characteristic equation can be written as:

$$0 = \det(sI_{(n)} - A)$$

so that the characteristic roots are given by the eigenvalues of the matrix A.

There is stability if and only if the characteristic roots have no positive real part, and those (possible) with a null real part are simple roots. Particularly, the stability is asymptotic if and only if all the characteristic roots have a negative real part(4).

To check the asymptotic stability without computing the characteristic roots, it is possible to apply the *Routh-Hurwitz criterion*, for which it is necessary and sufficient that:

- all the $(n + 1)$ coefficients $a_n, \ldots, a_1, a_0$ have the same sign;

(3) In general, a (linear and stationary) system with m inputs and q outputs can be represented by Equations [A3.1.3], intending that u and y are column matrices (respectively $(m, 1)$ and $(q, 1)$), and that A, B, C, D are proper constant matrices (respectively (n, n), (n, m), (q, n), (q, m)). In such a case, the dependence of the forced response on the input is defined by a *"transfer" matrix* (q, m), again expressed by Equation [A3.1.4].

(4) The present condition (of asymptotic stability) is the usually desired one, also because it guarantees that the forced response is limited, when any possible limited input function is applied.

and moreover:

- in the "Routh-Hurwitz table" (defined here below), all the $(n+1)$ elements a_n, $a_{(n-1)}$, α_1, β_1, ... of the first column have the same sign.

The Routh-Hurwitz table is defined as follows:

$$
\begin{array}{lllll}
a_n & a_{(n-2)} & a_{(n-4)} & a_{(n-6)} & \cdots \\
a_{(n-1)} & a_{(n-3)} & a_{(n-5)} & \cdots & \\
\alpha_1 & \alpha_2 & \alpha_3 & \cdots & \\
\beta_1 & \beta_2 & \cdots & & \\
\vdots & \vdots & & &
\end{array}
$$

where:

- the first two rows are given by the coefficients of the polynomial $\mathcal{A}(s)$, according to above;
- the third row is given by:

$$
\begin{cases}
\alpha_1 = \dfrac{a_{(n-1)}a_{(n-2)} - a_n a_{(n-3)}}{a_{(n-1)}} \\
\cdots \\
\alpha_k = \dfrac{a_{(n-1)}a_{(n-2k)} - a_n a_{(n-2k-1)}}{a_{(n-1)}} \\
\cdots
\end{cases}
$$

  (obviously intending $a_i = 0$ for $i < 0$), up to a total number of elements equal to that of the first row, less one;
- each one of the next rows is similarly derived, starting from the elements of the two preceding rows ($\beta_1 = (\alpha_1 a_{(n-3)} - a_{(n-1)}\alpha_2)/\alpha_1$, etc.); in such a way a pseudotriangular table is obtained, having $(n+1)$ significant rows (the last of which contains only one element).

In general, the characteristic roots define (further than the stability properties) the response *"modes"* of the system.

In particular:

- a real (nonnull) characteristic root defines an *aperiodic* mode, and, in $\mathcal{A}(s)$, a factor of the following type can be associated to it:

$$(1 + sT) \qquad \text{[A3.2.1]}$$

  where T is called *"time constant"*;

- a pair of complex conjugate characteristic roots defines an *oscillatory* mode, and, in $\mathcal{A}(s)$, a factor of the following type can be associated to it:

$$\left(1 + 2\zeta\frac{s}{\nu_o} + \frac{s^2}{\nu_o^2}\right) \quad \text{[A3.2.2]}$$

  (with $|\zeta| < 1$, $\nu_o > 0$), where ν_o and ζ are, respectively, called *"resonance frequency"* and *"damping factor."*

For such cases, the condition of asymptotic stability implies, respectively, that $T > 0$ and $\zeta > 0$; if, on the contrary, it were $T < 0$ or $\zeta < 0$, there would be, respectively, an instability of the "aperiodic" or of the "oscillatory" type.

By a similar formalism, and accounting for possible null roots, the generic transfer function can be posed in the form:

$$G(s) = \frac{K}{s^h}\frac{n(s)}{d(s)} \quad \text{[A3.2.3]}$$

with h integer, whereas $n(s)$ and $d(s)$ are products of factors of the type [A3.2.1] and/or [A3.2.2] (with $n(0) = d(0) = 1$), and K is a suitable constant. In particular, the value $G(0)$ is called *"static gain"* ($G(0) = K$ if $h = 0$, i.e., if $G(s)$ has no null zeros nor poles; $G(0) = 0, \infty$ if $h \lessgtr 0$, respectively).

Furthermore, the function $G(s)$ is said to be a *"minimum-phase"* function if, and only if, in Equation [A3.2.3], the gain K, all the time constants and all the damping factors (both in $n(s)$ and $d(s)$) are positive.

A3.3. FREQUENCY RESPONSE

Now assume that the (linear and stationary) system is asymptotically stable, so that its free response tends to zero for $t \to \infty$. Then, in Equation [A3.2.3] it is $h \leq 0$, and all the factors in $d(s)$ correspond to positive time constants and/or damping factors.

The response to a sinusoid $u(t) = U_M \sin(\nu t + \varphi_u)$, tends, for long times, to a sinusoidal steady-state (or *stationary*) component, itself at the frequency ν, equal to:

$$Y_M \sin(\nu t + \varphi_y)$$

with, $Y_M = |G(\tilde{\jmath}\nu)|U_M$, $\varphi_y = \varphi_u + \angle G(\tilde{\jmath}\nu)$[5]. In terms of phasors, the ratio between the output and the input results therefore given by the complex quantity

(5) Similarly, the response to a step with amplitude U tends, for $t \to \infty$, to a constant stationary component, equal to $G(0)U$. Moreover, y undergoes, for $t = 0$, a discontinuous variation equal to $G(\infty)U$.

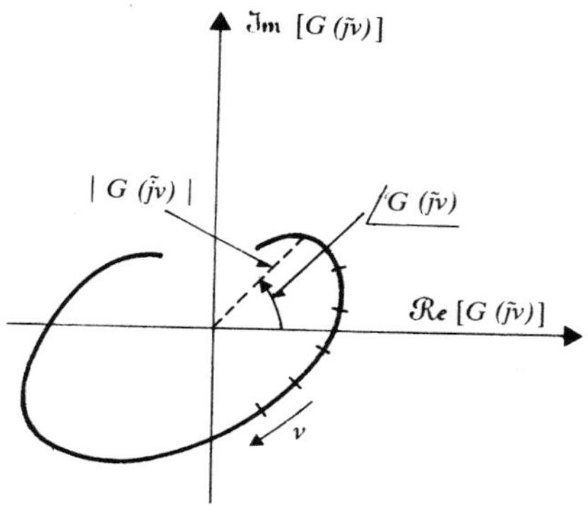

Figure A.1. Nyquist diagram.

$G(\tilde{j}\nu)$, which defines, for varying ν, the so-called *frequency response* properties of the system.

The most used diagrams for representing the dependence of $G(\tilde{j}\nu)$ on the frequency ν are of the following two types:

- the *Nyquist diagram* (or polar diagram): it is constituted by the line, in the complex plane, described by $G(\tilde{j}\nu)$ for varying ν (Fig. A.1);
- the *Bode diagrams* (or "logarithmic" diagrams): these are two diagrams that represent, respectively, the magnitude and the phase of $G(\tilde{j}\nu)$ for varying ν; in them, $|G(\tilde{j}\nu)|$ and ν are reported in a logarithmic scale, whereas $\angle G(\tilde{j}\nu)$ is reported in a linear scale (Fig. A.2).

If $G(s)$ is known in the form [A3.2.3], the Bode diagrams can be easily traced: in fact, by the above-mentioned choice of the scales, such diagrams are the *graphic* sum and/or difference of (well-known) elementary diagrams, corresponding to functions of the type K, s, $(1 + sT)$, $(1 + 2\zeta s/\nu_o + s^2/\nu_o^2)$.

Furthermore, if $G(s)$ is a minimum-phase function — as usually occurs — the knowledge of the magnitude diagram is by itself sufficient to derive the phase (and thus the whole function $G(\tilde{j}\nu)$) for any given ν.

Useful indications on the behavior of the magnitude diagram can be obtained by means of the so-called *"asymptotic" diagram*, the tracing of which is extremely easy.
More precisely, by naming *"critical" frequencies* the values of the type $|1/T|$, ν_o:

- starting from the low frequencies, the "first" asymptote corresponds to the function K/s^h, and therefore it pertains to the straight-line passing for the point $(1, |K|)$ with a slope $-h$;
- the passage from the first to the "second" asymptote occurs in correspondence to the lowest critical frequency, and is characterized by a variation in the slope equal

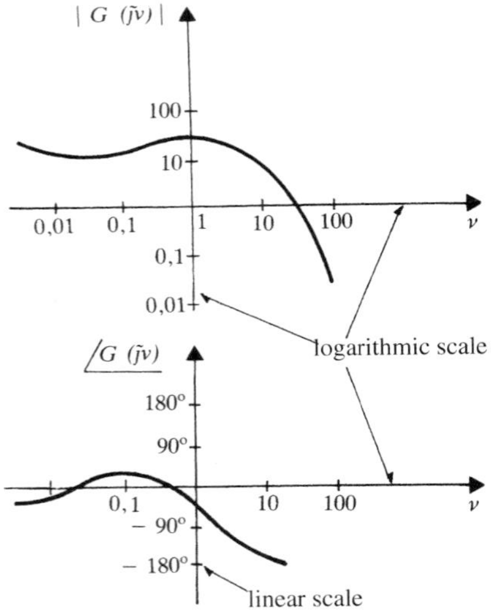

Figure A.2. Bode diagrams.

to ± 1 or ± 2, as specified here below; and the same happens in the passage to the subsequent asymptotes, in correspondence to the other critical frequencies;

- each term of the type $(1 + sT)$ causes a slope variation (at the corresponding critical frequency $|1/T|$) equal to ± 1, according to whether this term is at the numerator or at the denominator of $G(s)$;
- each term of the type $(1 + 2\zeta s/\nu_o + s^2/\nu_o^2)$ causes a slope variation (at the corresponding critical frequency ν_o) equal to ± 2, according to whether this term is at the numerator or at the denominator of $G(s)$.

A3.4. ELEMENTARY CONTROL SYSTEMS: GENERALITIES

Figure A.3a represents a typical elementary control system. According to such a scheme, the "controlled" system has for simplicity:

- a single output, that is the controlled variable y (the time behavior of which must be as close as possible to the desired one);
- two inputs, that is the "controlling" variable u (the time behavior of which is adapted to the control goals) and the "disturbance" d (which is instead the result of external causes, independent of the above-mentioned goals).

The behavior of u is moreover dependent, through the "controlling" system, on:

- the so-called "reference" variable r, which takes into proper account the desired behavior of y;

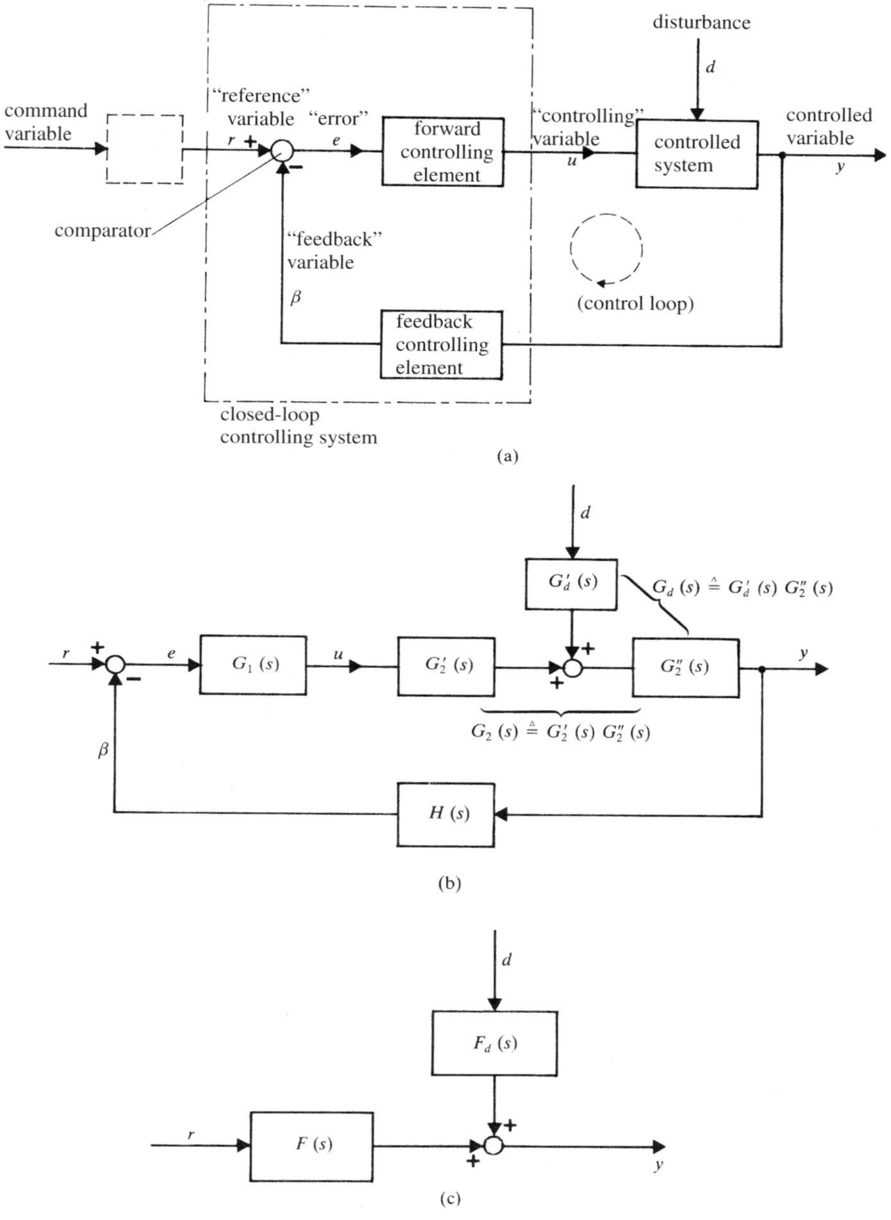

Figure A.3. Elementary control system: (a) typical configuration; (b) transfer functions of different parts; (c) resulting transfer functions.

- the actual behavior of y, through the *"feedback"* element: in such a way a *"closed-loop" control* is obtained, with a "control loop" like that defined in the Figure A.3a.

Assuming that all the elements are linear and stationary, it is possible to derive the block diagram of Figure A.3b, where each element is defined by its respective transfer function.

The "closed-loop" transfer functions, indicated in Figure A.3c, can be easily derived; they are expressed by:

$$\left.\begin{aligned} F(s) &= \frac{G_1 G_2}{1 + G_1 G_2 H}(s) \\ F_d(s) &= \frac{G_d}{1 + G_1 G_2 H}(s) \end{aligned}\right\} \qquad \text{[A3.4.1]}$$

where the product $G_1 G_2 H(s)$, present at the denominator, constitutes the so-called (total) transfer function "of the loop."

The control has the following two basic goals:

- to reduce as much as possible the effects of the disturbance d on y;
- to realize a suitable dependence of y on r, defined by a "desired" transfer function $F_{\text{des}}(s)$.

If r and d were sinusoidal at frequency ν, their effect on y would be determined (under the hypothesis of asymptotic stability) by the quantities $F(\tilde{j}\nu)$ and $F_d(\tilde{j}\nu)$; by imposing $F_d(\tilde{j}\nu) \to 0$, $F(\tilde{j}\nu) \to F_{\text{des}}(\tilde{j}\nu)$, it would result:

$$\left.\begin{aligned} G_1 G_2'(\tilde{j}\nu) &\longrightarrow \infty \\ H(\tilde{j}\nu) &\longrightarrow \frac{1}{F_{\text{des}}(\tilde{j}\nu)} \end{aligned}\right\} \qquad [A3.4.2]$$

recalling that $G_2 = G_2' G_2''$, $G_d = G_d' G_2''$ (see Fig. A.3b), and obviously assuming $G_2''(\tilde{j}\nu) \neq 0$, $G_d'(\tilde{j}\nu) \neq \infty$.

Generally, if the behaviors of r and d are defined in "spectral" terms, the conditions [A3.4.2] would be satisfied at least for the frequency ranges that are more involved. The matching of these conditions may be, in practice, contrasted by the stability requirement (as seen below), further than by different constraints, of technological or economical type, etc. However, the convenience of achieving a loop transfer function ($G_1 G_2 H(s)$) with a "gain" possibly high, by particularly acting on $G_1(s)$ (e.g., through the use of an amplifier), is evident, whereas the choice of $H(s)$ essentially must be adapted to the desired function $F_{\text{des}}(s)$.

A3.5. CLOSED-LOOP DYNAMIC BEHAVIOR

Based on Equations [A3.4.1], the characteristic roots of the closed-loop system are constituted — apart from cancellations in the transfer functions, which will not be considered for simplicity — by:

- the poles of $G'_d(s)$ that can be intended as known (and which will be assumed to have a negative real part);
- the solutions of the equation:

$$0 = 1 + \mathcal{G}(s) \qquad \text{[A3.5.1]}$$

where $\mathcal{G}(s) \triangleq G_1 G_2 H(s)$ is the loop transfer function.

To check the closed-loop asymptotic stability without solving Equation [A3.5.1], it is possible to apply the *Nyquist criterion*, which is based on properties (that may be assumed as known) of the function $\mathcal{G}(s)$.

More precisely, according to this criterion it is necessary and sufficient that the Nyquist diagram of $\mathcal{G}(\jmath\nu)$:

- does not pass for the point $(-1 + \jmath 0)$;
- describes around the point $(-1 + \jmath 0)$, for increasing ν from $-\infty$ to $+\infty$, a number N_{ao} of anticlockwise rotations equal to:

$$N_{ao} = P_{(+)} \qquad \text{[A3.5.2]}$$

where $P_{(+)}$ is the number of poles of $\mathcal{G}(s)$ having a positive real part[(6)].

In particular, as it is $P_{(+)} \geq 0$, the criterion is certainly not satisfied if $N_{ao} < 0$, i.e., if there are clockwise rotations.

The Nyquist criterion appears often more preferable than the Routh-Hurwitz criterion (already described) for various reasons, among which:

- it requires the knowledge of the function $\mathcal{G}(\jmath\nu)$, i.e., a knowledge that may be useful for other purposes (analysis of frequency response);
- if the elements of the loop (or some of them) are asymptotically stable, it is not necessary to know the analytical expression of $\mathcal{G}(s)$ (as it instead is for the Routh-Hurwitz criterion), as the function $\mathcal{G}(\jmath\nu)$ (or a part of it) can be derived experimentally, by frequency response tests in open-loop conditions;
- the number N_{ao} also can be derived starting from the Bode diagrams of $\mathcal{G}(\jmath\nu)$ (instead that from the Nyquist diagram), which are much easier to be traced;
- the criterion application usually permits several simplifications, based — as described in the following — on the consideration of certain parameters

(6) Actually, if $\mathcal{G}(s)$ exhibits one or more than one purely imaginary poles of the type $\jmath\nu_k$, the Nyquist diagram opens itself to the infinite for each value $\nu = \nu_k$. For the criterion application, it is necessary to intend that, for each of the considered poles, the diagram closes itself, between $\nu = \nu_k^-$ and $\nu = \nu_k^+$, by means of a semicircle of infinite radius described in the clockwise sense.

(cutoff frequency, phase margin, etc.) particularly significant also for the synthesis of the controlling system.

In most of the practical cases it is $P_{(+)} = 0$ (otherwise the open-loop system, defined by $\mathcal{G}(s)$, would be unstable); for the closed-loop asymptotic stability it is therefore necessary and sufficient that the Nyquist diagram of $\mathcal{G}(\tilde{\jmath}\nu)$ does not pass for $(-1 + \tilde{\jmath}0)$, nor does it describe any rotation around that point.

Furthermore, usually:

- it exists a single (positive) frequency ν_t, called *"cutoff" frequency*, for which $|\mathcal{G}(\tilde{\jmath}\nu_t)| = 1$, whereas $|\mathcal{G}(\tilde{\jmath}\nu)| \gtrless 1$, respectively, according to $\nu \lessgtr \nu_t$;
- the Nyquist criterion leads to the simple condition $\angle \mathcal{G}(\tilde{\jmath}\nu_t) > -180°$, i.e.,

$$\gamma > 0$$

where the angle $\gamma \triangleq \angle \mathcal{G}(\tilde{\jmath}\nu_t) + 180°$ is called *"phase margin"*: see Figure A.4[7].

For the first of Equations [A3.4.1], the "zeros" of the function $F(s)$ are simply constituted by the zeros of the functions $G_1G_2(s)$ and $(1/H)(s)$, whereas the "poles" are the solutions of Equation [A3.5.1].

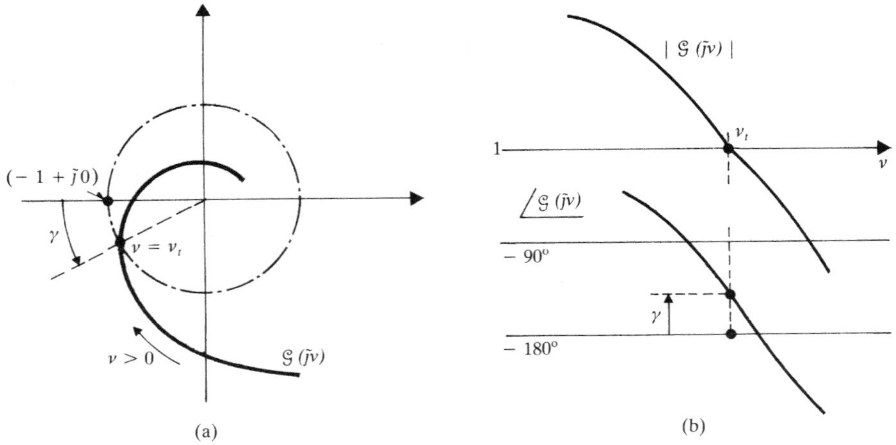

Figure A.4. Graphic definition of the cutoff frequency ν_t and the phase margin γ, with reference to: (a) the Nyquist diagram; (b) the Bode diagrams.

[7] Often, the phase $\angle \mathcal{G}(\tilde{\jmath}\nu_t)$ can be estimated based on the slope by which the (magnitude) asymptotic diagram cuts the unit gain axis (at an "asymptotic" cutoff frequency ν_t', close to ν_t). It is then possible to derive to the so-called *Bode criterion* (which is extremely simple to be applied), for which the above-mentioned slope must be -1 or even, possibly, -2.

Furthermore, the function $F(\tilde{j}\nu)$ can be written in the form:

$$F(\tilde{j}\nu) = \frac{G_1 G_2 \dfrac{1}{H}}{G_1 G_2 + \dfrac{1}{H}}(\tilde{j}\nu)$$

so that, at each frequency ν for which the magnitudes of $G_1G_2(\tilde{j}\nu)$ and $(1/H)(\tilde{j}\nu)$ are considerably different from each other, the behavior of $F(\tilde{j}\nu)$ is approximately equal to that of the function (of the two above mentioned) which has the smaller magnitude. Similar considerations can be applied to the function F_d, based on the second of Equations [A3.4.1].

In practice, there is usually only a single cutoff frequency ν_t, as said, and it is possible to define the following ranges of frequency:

- low-frequency range, with $\nu \ll \nu_t$ and $|G_1G_2(\tilde{j}\nu)| \gg |(1/H)(\tilde{j}\nu)|$ (recall the convenience of having high $|G_1G_2H(\tilde{j}\nu)|$), and thus:

$$F(\tilde{j}\nu) \cong \frac{1}{H}(\tilde{j}\nu)$$

- high-frequency range, with $\nu \gg \nu_t$ and $|G_1G_2(\tilde{j}\nu)| \ll |(1/H)(\tilde{j}\nu)|$ (because of the unavoidable response delays of the loop elements), and thus:

$$F(\tilde{j}\nu) \cong G_1G_2(\tilde{j}\nu)$$

separated by a suitable medium-frequency range within which—passing from one approximation to the other—the asymptotic diagram of $|F(\tilde{j}\nu)|$ undergoes a relatively small variation in its slope, e.g., equal to -1 or -2 (recall the Bode criterion, footnote(7), and consider that the slope of $|G_1G_2H|$ is equal to the relative slope of $|G_1G_2|$ with respect to $|1/H|$).

On the other hand, the knowledge of $F(\tilde{j}\nu)$ can allow (through the "critical" frequencies, etc.) estimation of the poles of $F(s)$, whereas the zeros are already known.

As a first approximation (using the denomination "at low frequency" for the poles and zeros the critical frequencies of which lie within the low-frequency range, and so on), it is possible to state that:

- at low frequency, the poles of $F(s)$ are close to the poles of $(1/H)(s)$ (and to the zeros of $G_1G_2(s)$, by which, however, they are practically cancelled);
- at high frequency, on the contrary, the poles of $F(s)$ are close to the poles of $G_1G_2(s)$ (and to the zeros of $(1/H)(s)$, by which, however, they are practically cancelled);

whereas the remaining poles of $F(s)$, in a small number (e.g., one or two), lie in the medium-frequency range and therefore have a critical frequency equal to or close to the cutoff frequency[8].

A3.6. ELEMENTARY CRITERIA FOR SYNTHESIS

Based on previous information, it can be generically concluded that:

- at low frequency, the control goals can be well achieved by assuming $|G_1 G_2'|$ quite high (to properly reduce the disturbance effects) and moreover $H \cong 1/F_{des}$ (to have $F \cong 1/H \cong F_{des}$);
- the response delays that more greatly limit the control efficiency are, then, those associated to the medium-frequency poles (strictly dependent on the values of the cutoff frequency and of the phase margin: see also footnote[8]).

The synthesis criteria can be generally carried back to few fundamental specifications, such as:

- a high value of $|G_1 G_2'|$ at low frequency, and particularly a high static gain $G_1 G_2'(0)$, to be realized by acting on G_1 (if $G_1 G_2'(s)$ has a pole at the origin, such a static gain is infinite; recall the first of Equations [A3.4.2], with $\nu = 0$);
- a high value of the cutoff frequency ν_t, provided that the stability and more strictly the damping characteristics are not compromised (and thus the phase margin does not become modest); on the other hand, the larger is ν_t, the smaller the phase margin usually is, so that it is necessary to adopt a compromise solution; typically, ν_t is chosen to have $\gamma \cong 30° - 60°$ (usually, it then holds what is said in footnote[8], with a damping factor of 0.25–0.50).

In most of the practical cases, the synthesis procedure can be defined as follows (with the functions G_2', G_d', G_2'' known):

- $H(s)$ is chosen to have $H(\tilde{j}\nu) \cong (1/F_{des})(\tilde{j}\nu)$ in a frequency range extended at its most;
- as a first attempt, it is assumed $G_1(s) = K_1$ (constant), and K_1 is chosen to achieve an acceptable value ($\gamma_{(1)}$) of the phase margin;

---

[8] As a result of control system synthesis, at medium frequency there are often two complex conjugate poles, with:

- resonance frequency close to the asymptotic cutoff frequency ν_t',
- damping factor close to half of the value, in radians, of the phase margin.

- in $G_1(s)$ a *"low-pass"* effect is introduced, defined by a factor of the type:

$$\frac{T_b}{T_b'}\frac{1+sT_b'}{1+sT_b}$$

  with $T_b > T_b' > 0$, and T_b and T_b' are chosen to adequately increase the low-frequency gain, without practically worsening the phase margin ($1/T_b'$ must be smaller enough than the first-attempt cutoff frequency $\nu_{t(1)}$);
- a possible *"high-pass"* effect is introduced into $G_1(s)$, defined by a factor of the type:

$$\frac{1+sT_a'}{1+sT_a}$$

  with $T_a' > T_a > 0$, and T_a' and T_a (with $1/T_a'$ in the order of $\nu_{t(1)}$) are chosen to increase the cutoff frequency ($\nu_t > \nu_{t(1)}$), still keeping the phase margin γ at an acceptable value[(9)].

A typical example of the procedure application is reported in Figure A.5.

Whenever the measures described up to now (or similar ones) are not sufficient, it is suitable to consider modifications in the control scheme, which, for example, may include (Fig. A.6):

- a *"feed-forward" action*, defined by the transfer function $G_a(s)$, starting from the reference r;
- a *"compensation" of the disturbance d*, defined by the transfer function $G_c(s)$ (it is, however, necessary that the disturbance is measurable in some way).

Instead of Equations [A3.4.1], it is then derived:

$$\begin{cases} F(s) = \dfrac{(G_1+G_a)G_2}{1+G_1G_2H}(s) \\ F_d(s) = \dfrac{G_d - G_cG_2}{1+G_1G_2H}(s) \end{cases}$$

(9) In particular, if a low-pass effect is introduced with $T_b = \infty$, it follows:

$$G_1(s) = K_p + \frac{K_i}{s}$$

with $K_p = K_1$, $K_i = K_1/T_b'$, i.e., a controlling system of the *proportional-integral* type (PI).

Similarly, if a high-pass effect is added with $T_a \to 0$, it follows:

$$G_1(s) \longrightarrow K_p + \frac{K_i}{s} + K_d s$$

with $K_p = K_1(1 + T_a'/T_b')$, $K_i = K_1/T_b'$, $K_d = K_1 T_a'$, i.e., a controlling system of the *proportional-integral-derivative* type (PID).

In the previous expressions, K_p, K_i, and K_d, respectively, represent the "proportional," "integral," and "derivative" gains.

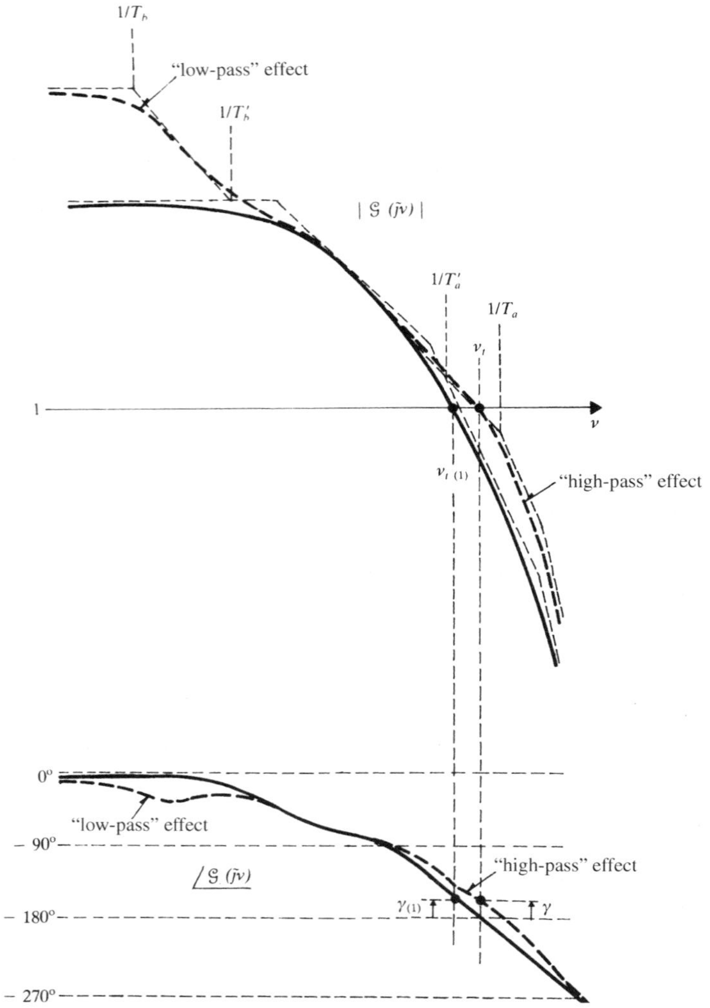

Figure A.5. Typical example of application of the synthesis procedure.

from which:

$$\begin{cases} F(s) \longrightarrow F_{\text{des}}(s) \\ F_d(s) \longrightarrow 0 \end{cases}$$

by respectively assuming:

$$\begin{cases} G_a \longrightarrow F_{\text{des}}/\text{G}_2, \quad H \longrightarrow 1/F_{\text{des}} \\ G_c \longrightarrow G_d/G_2 \end{cases}$$

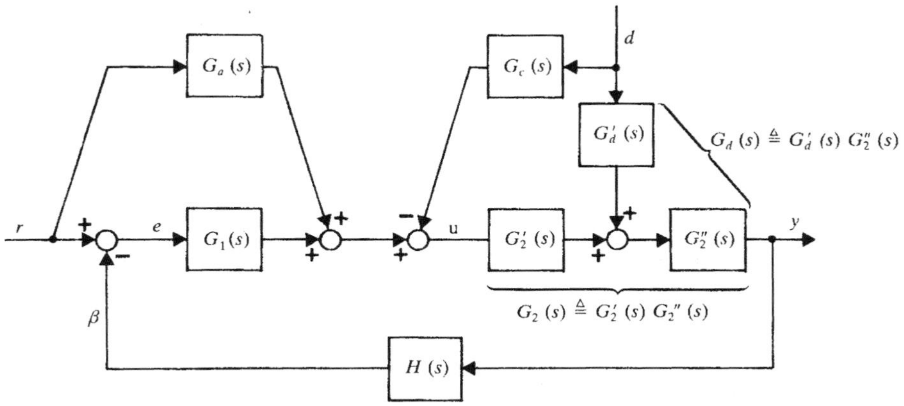

Figure A.6. Example of a general control system.

These last conditions are often nonachievable, but they can be usefully approximated (for $s = \tilde{j}\nu$), at least for the frequency ranges that are mostly interested, respectively, by the spectra of r and d.

Other measures may consist in the preliminary realization of one or more *"auxiliary" feedbacks* around the block G_2 (and possibly G_1) or a part of it, to obtain a better system (from the point of view of the control synthesis) with respect to the original system.

In fact, similar to information already discussed for the control loop, the closure of auxiliary loops can improve the dynamic response to the variables used for the control, reduce the effect of disturbances or even, in some cases, "stabilize" the system (recall the Nyquist criterion, with $P_{(+)} > 0$).

If the transfer function of the generic loop is expressed in the form $\mu g(s)$ (Fig. A.7), the choice of the gain μ must consider the effect of μ on the closed-loop characteristic roots, which are the solutions of the equation:

$$0 = 1 + \mu g(s)$$

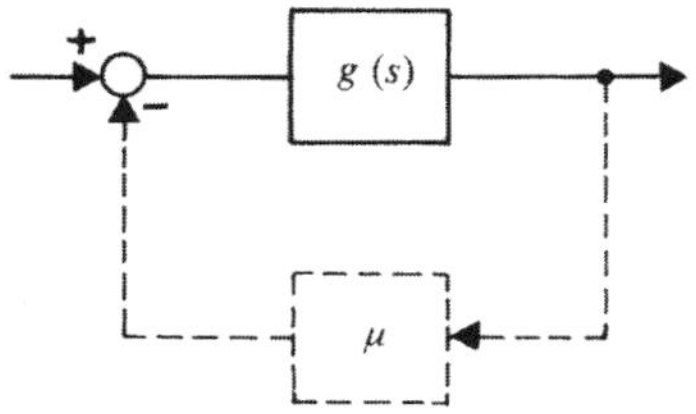

Figure A.7. Closure of a loop through a block with gain μ.

In this concern, some useful indications can be obtained by analyzing the case $\mu \to 0$; if p_h is the generic pole of $g(s)$, it can be found that:

- for each simple pole p_h there is, correspondingly, a (closed-loop) characteristic root r_h which is approximable by:

$$r_h \cong p_h - \mu C_h$$

  where C_h is the "residual" of $g(s)$ at p_h, i.e.,

$$C_h \triangleq ((s - p_h)g(s))_{s=p_h}$$

- for each multiple pole p_h, having a multiplicity m, there are, correspondingly, m (closed-loop) characteristic roots which are approximable by the m solutions:

$$p_h + (-\mu C_{hm})^{1/m}$$

  where:

$$C_{hm} \triangleq ((s - p_h)^m g(s))_{s=p_h}$$

The previous relationships allow definition of the *"sensitiveness" of the characteristic roots* with respect to μ (for $\mu \to 0$); in particular, the sensitiveness in correspondence of a simple pole is equal to:

$$\frac{dr_h}{d\mu} = -C_h \qquad [A3.6.1]$$

i.e., equal to the residual with inverted sign. This last result is particularly important, as each single residual can be evaluated experimentally, by tests on the open-loop system, without previous knowledge of the analytical expression of $g(s)$.

If there are *more output variables to be controlled* ($y_1, \ldots, y_q$), it is possible to reference the scheme in Figure A.8a, where (under the hypothesis of linearity and stationarity) G_1, G_2, H, G_d are suitable transfer matrices, functions of s.

Generally, each of the "controlling" variables $u_1, \ldots, u_m$ may influence all output variables $y_1, \ldots, y_q$, and this circumstance must be considered in the synthesis of the controlling system.

Assuming for simplicity that $H(s)$ is diagonal, each "error" $e_i = r_i - \beta_i$ is sensitive, through the feedback variable β_i, to the only output y_i. A rather spontaneous synthesis criterion consists into imposing (if possible) the *"noninteraction"* of the controls, i.e., each reference r_i ($i = 1, \ldots, q$) influences only the corresponding output y_i, without any effect on the y_j's, $j \neq i$. Thus operating, in fact, r_i acquires a precise meaning in terms of y_i, and by varying r_i no disturbance is caused to the control of the other outputs. Moreover, the control becomes q

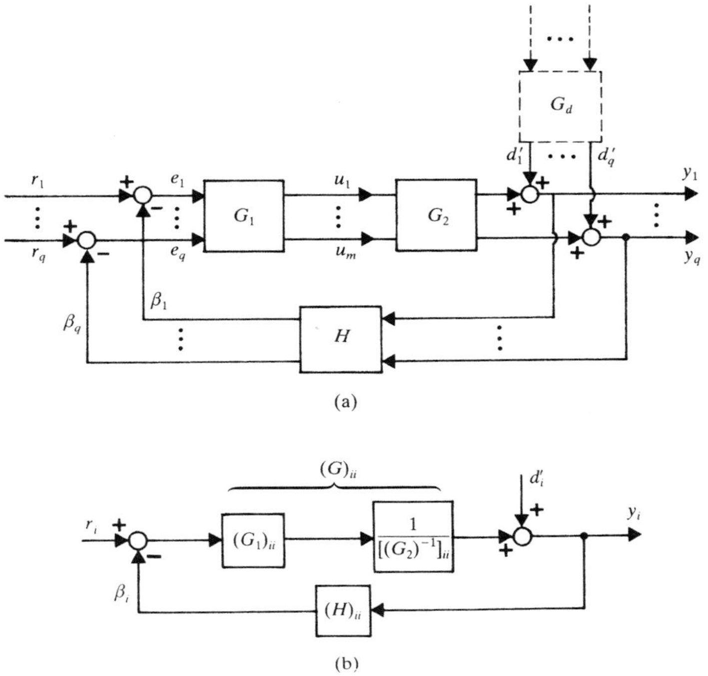

Figure A.8. Control system in the multivariable case: (a) transfer matrices of the different parts; (b) transfer functions corresponding, for $m = q$, to the generic control loop ($i = 1, \ldots, q$) in case of noninteraction.

noninteracting loops, for each one of which it is possible to apply the synthesis criteria already seen, relative to the control of a single variable.

The noninteraction criterion is satisfied if, and only if, the matrix $G_2G_1(s)$ is diagonal, i.e.,:

$$G_2G_1(s) = G(s)$$

where $G(s)$ is a suitable (q, q) diagonal matrix.

On the other hand, $G(s)$ must be nonsingular (otherwise, some of the $(G)_{ii}$ elements would be null, and the corresponding y_i's could not be controlled), and thus $G_2(s)$ must have a rank q (and this implies, in particular, $m \geq q$).

If such an hypothesis is satisfied, it is possible to derive the matrix $G_1(s)$ and to define, under feasibility conditions, the whole controlling system. In particular, for $m = q$, it can be derived (omitting for simplicity the indication of the variable s) $G_1 = (G_2)^{-1}G$, and thus:

$$(G_1)_{ii} = [(G_2)^{-1}]_{ii}(G)_{ii}$$

Therefore, the block diagram in Figure A8.b can be derived, in everything similar to that in Figure A3.b, further than the following conditions:

$$(G_1)_{ji} = [(G_2)^{-1}]_{ji}(G)_{ii} = \frac{[(G_2)^{-1}]_{ji}}{[(G_2)^{-1}]_{ii}}(G_1)_{ii} \quad (j \neq i)$$

relative to the nondiagonal elements of the matrix G_1 $(i,\ j = 1, \ldots, q)$.

ANNOTATED REFERENCES

The following works are particularly indicated: 8, 18.

REFERENCES

The subjects treated in the text are the object of many works, predominantly oriented to specific fields. The specialist aspect is obviously more marked in papers, the very high number of which also constitutes a testimony to the continuous developing — not few times fragmented, or even disorganic — of the subject matter. The following list is limited to some significant works, in a chronological order. The specific recalls are reported — with some comments — at the end of each chapter and of the appendices, with reference to the respective subjects.

BOOKS

1. KRON G., *Tensor analysis of networks*, Wiley, 1939.
2. STEINBERG M. J., SMITH T. H., *Economy loading of power plants and electric systems*, Wiley, 1943.
3. EVANGELISTI G., *La regolazione delle turbine idrauliche*, Zanichelli, 1947.
4. CRARY S. B., *Power system stability*, 2 vols, Wiley, 1947.
5. *Electrical transmission and distribution reference book*, Westinghouse, 1950.
6. RUDENBERG R., *Transient performance of electric power systems*, McGraw-Hill, 1950.
7. CONCORDIA C., *Synchronous machines*, Wiley, 1951.
8. TRUXAL J. G., *Automatic feedback control system synthesis*, McGraw-Hill, 1955.
9. VOIPIO E., *The influence of magnetic saturation on transients and voltage regulating properties of synchronous alternators, with special reference to large capacitive loads*, Trans. of the Royal Institute of Technology, Stockholm, 1955.
10. KIMBARK E. W., *Power system stability*, 3 vols, Wiley, 1956.
11. CLARKE E., *Circuit analysis of ac power systems*, 2 vols, Wiley, 1958.
12. KIRCHMAYER L. K., *Economic operation of power systems*, Wiley, 1958.

13. KIRCHMAYER L. K., *Economic control of interconnected systems*, Wiley, 1959.
14. COHN N., *Control of generation and power flow on interconnected systems*, Wiley, 1959.
15. WHITE D. C., WOODSON H. H., *Electromechanical energy conversion*, Wiley, 1959.
16. FITZGERALD A. E., KINGSLEY C. Jr., *Electric machinery*, McGraw-Hill, 1961.
17. ADKINS B., *The general theory of electrical machines*, Chapman-Hall, 1962.
18. ZADEH C. A., DESOER L. A., *Linear system theory*, McGraw-Hill, 1963.
19. VENIKOV V. A., *Transient phenomena in electrical power systems*, Pergamon Press, 1965.
20. EDELMAN H., *Théorie et calcul des réseaux de transport d'énergie électrique*, Dunod, 1966.
21. HANO I., *Operating characteristics of electric power systems*, Denki Shoin (Tokyo), 1967.
22. STAGG G. W., EL-ABIAD A. H., *Computer methods in power system analysis*, McGraw-Hill, 1968.
23. GUILE A. E., PATERSON W., *Electrical power systems*, 2 vols, Oliver-Boyd, 1969.
24. KIMBARK E. W., *Direct current transmission*, Vol.I, Wiley, 1971.
25. ELGERD O., *Electric energy systems theory: an introduction*, McGraw-Hill, 1971.
26. GREENWOOD A., *Electrical transients in power systems*, Wiley-Interscience, 1971.
27. RIBBENS PAVELLA M. (editor), *Stability of large-scale power systems*, Western Periodical Co. (California), 1972.
28. HANDSCHIN E. (editor), *Real-time control of electric power systems*, Elsevier, 1972.
29. ROEPER R., *Short-circuit currents in three-phase networks*, Siemens-Pitman, 1972.
30. MARIANI E., *Esercizio di un sistema di produzione e trasmissione dell'energia elettrica*, La Goliardica, 1973.
31. QUAZZA G., *Controllo dei processi*, CLUP, 1973.
32. BYERLY R. T., KIMBARK E. W., *Stability of large electric power systems*, IEEE Press, 1974.
33. HANCOCK N. N., *Matrix analysis of electrical machinery*, Pergamon Press, 1974.
34. ADKINS B., HARLEY R. G., *The general theory of alternating current machines: application to practical problems*, Chapman-Hall, 1975.
35. DIMO P., *Nodal analysis of power systems*, Abacus Press, 1975.
36. *Analysis and control of subsynchronous resonance*, IEEE PWR, 1976.
37. *Training courses on Power system analysis*, ENEL-CESI, 1977.
38. VENIKOV V. A., *Transient processes in electrical power systems*, Mir Publishers, 1977.
39. ANDERSON P. M., FOUAD A. A., *Power system control and stability*, Vol.I, Iowa State Univ. Press, 1977.
40. STERLING M. J. H., *Power system control*, Peregrinus, IEE, 1978.
41. EL-HAWARY M. E., CHRISTENSEN G. S., *Optimal economic operation of electric power systems*, Academic Press, 1979.
42. WEEDY B. M., *Electric power systems*, Wiley, 1979.
43. ERISMAN A. M., NOVES K. W., DWARAKANATH M. H. (editors), *Electric power problems: the mathematical challenge*, SIAM, 1980.
44. HAPP H. H., *Piecewise methods and applications to power systems*, Wiley, 1980.
45. MILLER T. J. E., *Reactive power control in electric power systems*, Wiley, 1982.
46. STEVENSON W. D., *Elements of power system analysis*, McGraw-Hill, 1982.
47. ARRILLAGA J., ARNOLD C. P., HARKER B. J., *Computer modelling of electrical power systems*, Wiley, 1983.

48. YAO-NAN YU, *Electric power system dynamics*, Academic Press, 1983.
49. WOOD A. J., WOLLENBERG B. F., *Power generation, operation and control*, Wiley, 1984.
50. MARCONATO R., *Sistemi elettrici di potenza*, 2 vols, CLUP, 1985.
51. FERRARI E., *Regolazione della frequenza e controllo dell'eccitazione di gruppi generatori*, CLUP, 1986.
52. PAI M. A., *Energy function analysis for power system stability*, Kluwer Academic Publisher, 1989.
53. SACCOMANNO F., *Sistemi elettrici per l'energia. Analisi e controllo*, UTET, 1992.
54. KUNDUR P., *Power system stability and control*, McGraw-Hill, 1994.
55. TAYLOR C. W., *Power system voltage stability*, McGraw-Hill, 1994.
56. PAVELLA M., MURTHY P. G., *Transient stability of power systems. Theory and practice*, Wiley, 1994.
57. MARIANI E., MURTHY S. S., *Control of modern integrated power systems*, Springer, 1997.
58. ILIC M. D., GALIANA F., FINK L., *Power system restructuring*, Kluwer Academic Publishers, 1998.
59. VAN CUTSEM T., VOURNAS C., *Voltage stability of electric power systems*, Kluwer Academic Publishers, 1998.
60. ANDERSON P. M., *Power system protection*, IEEE Press, 1999.
61. PAVELLA M., ERNST D., RUIZ VEGA D., *Transient stability of power systems. An unified approach to assessment and control*, Kluwer's Power Electronics and Power Systems Series, 2000.

PAPERS

62. THOMA D., *Zur theorie des wasserschlosses bei selbsttatig geregelten turbinenanlangen*, Munchen, Oldenbourg, 1910.
63. BAUM F. G., *Voltage regulation and insulation for large power long distance transmission systems*, *Journ. AIEE* 40, 1921.
64. FORTESCUE C. L., WAGNER C. F., *Some theoretical considerations of power transmission*, *Journ. AIEE*, 1924.
65. PARK R. H., *Two-reaction theory of synchronous machines*. Pt.I. Pt.II, *AIEE Trans.*, July 1929 and June 1933.
66. DARRIEUS G., *Réglage de la fréquence et de la puissance dans les réseaux interconnectés*, *Bull. ASE*, Oct. 1937.
67. CONCORDIA C. et al., *Effects of prime mover speed governor characteristics in power system frequency variations and tie-line power swings*, *AIEE Trans.*, 1941.
68. CONCORDIA C., *Steady-state stability of synchronous machines as affected by voltage-regulator characteristics*, *AIEE Trans.*, May 1944.
69. MAGNUSSON P. C., *The transient-energy method of calculating stability*, *AIEE Trans.*, 1947.
70. WARD J. B., *Equivalent circuits for power-flow studies*, *AIEE Trans.*, 1949.
71. BLACK P. M., LISCHER L. F., *The application of a series capacitor to a synchronous condenser for reducing voltage flicker*, *AIEE Trans.*, 1951.
72. VOIPIO E., *The voltage stability of a synchronous alternator with capacitive loading*, *ASEA Journ.* 1952.
73. CONCORDIA C., KIRCHMAYER L. K., *Tie-line power and frequency control of electric power systems*, *AIEE Trans. on PAS*, June 1953–April 1954.

74. Concordia C., *Voltage dip and synchronous-condenser swings caused by arc-furnaces loads*, *AIEE Trans.*, Oct. 1955.
75. Early E. D., Watson R. E., *A new method of determining constants for the general transmission loss equation*, *AIEE Trans. on PAS*, Febr. 1956.
76. Ward J. B., Hale H. W., *Digital computer solution of power-flow problems*, *AIEE Trans. on PAS*, June 1956.
77. Venikov V. A., Litkens I. V., *Etude expérimentale et analytique de stabilité de réseaux électriques avec excitation de générateurs réglée automatiquement*, CIGRE, 1956.
78. Cahen F., Robert R., Favez B., *La détermination expérimentale du temps de lancer d'un réseau de production et de distribution d'énergie électrique*, R.G.E., Oct. 1956.
79. Concordia C., Levoy L. G., Thomas C. H., *Selection of buffer reactors and synchronous condensers on power systems supplying arc-furnace loads*, *AIEE Trans.*, July 1957.
80. Aylett P. D., *The energy-integral criterion of transient stability limits of power systems*, *IEEE Trans.*, July 1958.
81. Aldred A. S., Shackshaft G., *The effect of a voltage regulator on the steady-state and transient stability of a synchronous generator*, *IEE*, August 1958.
82. *Rapport de l'UCPTE sur les essais effectués pour la détermination de l'énergie réglante et du statisme de l'ensemple des réseaux interconnectés de l'Europe occidentale*, CIGRE, 1958.
83. Priori L., Reggiani F., Velcich A., *Determinazione dei parametri caratteristici della rete del Gruppo Edison ai fini della regolazione frequenza-potenza*, *L'energia elettrica*, No. 3, 1959.
84. Baldwin C. J., Dale K. M., Dittrich R. F., *A study of economic shutdown of generating units in daily dispatch*, *AIEE Trans. on PAS*, Dec. 1959.
85. Quazza G., Saccomanno F., *Considerazioni sul proporzionamento dei regolatori di frequenza per turbine idrauliche in relazione alle caratteristiche della rete*, AEI Annual Meeting No. 160, 1960.
86. Bernholz B., Graham L. J., *Hydrothermal economic scheduling*, Pt.I/V, *AIEE Trans.*, Dec. 1960, Jan. 1962–Febr. 1962, Febr. 1962–June 1963.
87. Carpentier J. *Contribution à l'étude du dispatching économique*, Societé française des électriciens, August 1962.
88. Goodwin J., *Voltage regulator requirements for steady state stability of water-wheel generators connected to a system through long transmission lines*, CIGRE, 1962.
89. Dimo P., Groza L., Moraite G., *Optimisation des conditions d'essais pour la détermination des paramètres d'un réseau*, CIGRE, 1962.
90. Sokkappa B. G., *Optimum scheduling of hydrothermal systems. A generalized approach*, *IEE Trans.*, April 1963.
91. Bernard P. J., Dopazo J. F., Stagg G. W., *A method for economic scheduling of a combined pumped hydro and steam generating system*, *IEEE Trans. on PAS*, Jan. 1964.
92. Kirchmayer L. K., Ringlee R. J., *Optimal control of thermal hydro system operation*, *IFAC Proc.*, 1964.
93. Gless G. E., *Direct method of Liapunov applied to transient power system stability*, *IEEE Trans. on PAS*, No. 2, 1966.
94. Bainbridge E. S., Mc Namee J. M., Robinson D. J., Nevison R. D., *Hydrothermal dispatch with pumped storage*, *IEEE Trans. on PAS*, May 1966.

95. Ellis H. M., Hardy I. E., Blythe A. L., Skooglund J. W., *Dynamic stability of the Peace River transmission system*, *IEEE Trans. on PAS*, No. 6, 1966.
96. Quazza G., *Noninteracting controls of interconnected electric power systems*, *IEEE Trans. on PAS*, No. 7, 1966.
97. Ross C. W., *Error adaptive control computer for interconnected power systems*, *IEEE Trans. on PAS*, July 1966.
98. Reggiani F., Saccomanno F., *Considerations upon stability margins with special regard to the effect of generator, excitation system and voltage regulation characteristics*, CIGRE, 1966.
99. Saccomanno F., *Considerazioni sul comportamento dinamico degli elementi passivi di una rete elettrica*, *L'energia elettrica*, No. 9, 1966.
100. Ferrara E., *Applicazione delle condizioni di Kuhn e Tucker al dispatching economico di reti elettriche a generazione termica*, AEI Annual Meeting 1966, No. II-44.
101. Saccomanno F., *Alcune considerazioni sulla stabilità della regolazione delle centrali idroelettriche*, *L'energia elettrica*, No. 11, 1966.
102. Saccomanno F., *Identificazione di una rete elettrica attraverso brusche perturbazioni di struttura*, *L'elettrotecnica*, Dec. 1966.
103. Schleif F. R., Martin G. E., Angell R. R., *Damping of system oscillations with a hydrogenerating unit*, *IEEE Trans. on PAS*, No. 4, 1967.
104. Dy Liacco T. E., *The adaptive reliability control system*, *IEEE Trans. on PAS*, May 1967.
105. Zaslavskaya T. B., Putilova A. T., Tagirov M. A., *Liapunov function as synchronous transient stability criterion*, *Electrichestvo*, No. 6, 1967.
106. Kattelus J., *Effect of excitation control on the transmission capacity of power lines*, Sakho-Electricity in Finland, 1967.
107. Tinney W. F., Hart C. E., *Power flow solution by Newton's method*, *IEEE Trans. on PAS*, Nov. 1967.
108. Yu Y., Vongsuriya K., *Nonlinear power system stability study by Liapunov function and Zubov's methods*, *IEEE Trans. on PAS*, Dec. 1967.
109. Dandeno P. L., Karas A. N., Mc Clymont K. R., Watson W., *Effect of high-speed rectifier excitation systems on generator stability limits*, *IEEE Trans. on PAS*, No. 1, Jan. 1968.
110. Saccomanno F., *A new transmission loss formula for electric power systems*, IEEE Winter meeting, 1968.
111. Glavitsch H., *Moglichkeiten der verbesserung der stabilitat und der spannungsregelung von synchronmaschinen mit hilfe der gleichrichtererregung*, *Elektrotechnik u. maschinenbau (Austria)*, No. 2, 1968.
112. Ramsden V. S., Zorbas N., Booth R. R., *Prediction of induction-motor dynamic performance in power systems*, *Proc. IEE*, No. 4, 1968.
113. Hanson O. W., Goodwin C. J., Dandeno P. L., *Influence of excitation and speed control parameters in stabilizing intersystem oscillations*, *IEEE Trans. on PAS*, No. 5, 1968.
114. *Computer representation of excitation systems*, *IEEE Trans. on PAS*, No. 6, 1968.
115. Schleif F. R., Hunkins H. D., Martin G. E., Hattan E. E., *Excitation control to improve powerline stability*, *IEEE Trans. on PAS*, No. 6, 1968.
116. Chiabrera A., Curtarelli F., *Comportamento in regime variabile di alternatori a rotore massiccio*, *L'energia elettrica*, No. 10, 1968.

117. Di Caprio U., Saccomanno F., *Analysis of multimachine power systems stability by Liapunov's direct method.* Pt.I-II, International Meeting on Automation and Instrumentation, Nov. 1968.
118. Willems J. L., *Improved Liapunov function for transient power system stability, Proc. IEE*, Sept. 1968.
119. Rao N. D., *Generation of Liapunov functions for the transient stability problem, Trans. of the Eng. Inst. of Canada*, Oct. 1968.
120. Dommel H. W., Tinney W. F., *Optimal power flow solutions, IEEE Trans. on PAS*, Oct. 1968.
121. Muckstadt J. A., Wilson R. C., *An application of mixed-integer programming duality to scheduling thermal generating systems, IEEE Trans. on PAS*, Dec. 1968.
122. Saccomanno F., *Equivalente, visto da un nodo, di una rete elettrica comprendente macchine sincrone meccanicamente solidali*, ENEL-CRA Report, 1968–1969.
123. Canay I. M., *Causes of discrepancies on calculation of rotor quantities and exact equivalent diagrams of the sunchronous machines*, IEEE Winter meeting, Jan. 1969.
124. Brown H. E., Shipley R. B., Coleman D., Nied R. E. Jr., *A study of stability equivalents, IEEE Trans. on PAS*, March 1969.
125. De Mello F. P., Concordia C., *Concepts of synchronous machine stability as affected by excitation control, IEEE Trans. on PAS*, No. 4, 1969.
126. *Proposed excitation system definitions for synchronous machines, IEEE Trans. on PAS*, August 1969.
127. Saccomanno F., *Analisi completa della stabilità per grandi variazioni per una classe di reti elettriche*, AEI Annual Meeting No. 1.1.08, 1969.
128. Grohmann D., Tolomeo N., *Scelta ottima delle unità termoelettriche da tenere in servizio in un sistema elettrico per la copertura del carico*, AEI Annual Meeting No. 4.1.23, 1969.
129. Ariatti F., Grohmann D., Venturini D., *Ripartizione economica della produzione per il sistema primario ENEL 380–220 kV*, AEI Annual Meeting, No. 4.1.30, 1969.
130. Hano I., Tamura Y., Narita S., Matsumoto K., *Real time control of system voltage and reactive power, IEEE Trans. on PAS*, Oct. 1969.
131. Schweppe F. C., Wildes J., Rom D. P., *Power system static-state estimation.* Pt.I: *Exact model.* Pt.II: *Approximate model.* Pt.III: *Implementation, IEEE Trans. on PAS*, No. 1, 1970.
132. Larson R. E., Tinney W. F. et al., *State estimation in power systems.* Pt.I: *Theory and feasibility.* Pt.II: *Implementation and applications, IEEE Trans. on PAS*, March 1970.
133. Elgerd O., Fosha C. E., *Optimum megawatt-frequency control of multi-area electric energy systems, IEEE Trans. on PAS*, No. 4, 1970.
134. Willems J. L., Willems J. C., *The application of Liapunov methods to the computation of transient stability regions for multimachine power systems, IEEE Trans. on PAS*, May-June 1970.
135. Brown P. G., De Mello F. P., Lenfest E. H., Mills R. J., *The effects of excitation, turbine energy control, and transmission on transient stability, IEEE Trans. on PAS*, No. 6, 1970.
136. *Definitions and terminology for automatic generation control on electric power systems, IEEE Trans. on PAS*, July-August 1970.
137. Baughman M., Schweppe F. C., *Contingency evaluation: real power flows from a linear model*, IEEE Summer meeting, 1970.

138. CUENOD M., KNIGHT V. G., PERSOZ H., QUAZZA G., RENCHON R., *Tendances actuelles dans la surveillance et la maitrise automatique de la sécurité d'exploitation des réseaux électriques*, CIGRE 32–12, 1970.
139. GALLI F., SACCOMANNO F., SCHIAVI A., VALTORTA M., *Prospettive offerte dai collegamenti in corrente continua nel controllo delle reti elettriche interconnesse*, AEI Annual Meeting No. 2.4.01, 1970.
140. Di CAPRIO U., SACCOMANNO F., *Non-linear stability analysis of multimachine electric power systems*, *Ricerche di Automatica*, vol. **1**, No. 1, 1970.
141. MARTIN P., MERLIN A., *Méthode combinatoire de répartition à court terme d'un ensemble de moyens de production thermique et hydraulique*, Revue générale de l'électricité, Oct. 1970.
142. BARBIER C., COUSTÈRE A., GOUGEUIL J. C., PIOGER G., *Point de quelques questions concernant la stabilité dynamique des réseaux etc.*, Societé française des électriciens, Jan. 1971.
143. HAPP H. H., *The interarea matrix: a tie line flow model for power pools*, *IEEE Trans. on PAS*, Jan.-Febr. 1971.
144. KUPPURAJULU A., ELANGOVAN S., *Simplified power system models for dynamic stability studies*, *IEEE Trans. on PAS*, Jan.-Febr. 1971.
145. RIBBENS PAVELLA M., *Transient stability of multimachine power systems by Liapunov's direct method*, IEEE Winter meeting, Jan.-Febr. 1971.
146. KLOPFENSTEIN A., *Experience with system stabilizing excitation controls on the generation of the Southern California Edison Company*, *IEEE Trans. on PAS*, No. 2, 1971.
147. CONCORDIA C., BROWN P. G., *Effects of trends in large steam turbine driven generator parameters on power system stability*, *IEEE Trans. on PAS*, No. 5, 1971.
148. PESCHON J., BREE D. W., HAJDU L. P., *Optimal solutions involving system security*, PICA Conf., 1971.
149. HAPP H. H., JOHNSON P. C., WRIGHT W. J., *Large scale hydro-thermal unit commitment. Method and results*, *IEEE Trans. on PAS*, May-June 1971.
150. DI PERNA A., MARIANI E., *Programmazione giornaliera delle centrali idroelettriche a bacino e a serbatoio in un sistema di produzione misto*, L'energia elettrica, July 1971.
151. SACCOMANNO F., *Un modello matematico per la macchina sincrona. Criteri di messa a punto ed esperienze di impiego*, Symposium on Electrical Machines and Frequency Static Converters, Sorrento, 1971.
152. *Present practices in the economic operation of power systems*, *IEEE Trans. on PAS*, July-August 1971.
153. ARCIDIACONO V., FERRARI E., INSERILLO M., MARZIO L., *Prove per l'identificazione sperimentale dei carichi di una rete elettrica*, AEI Annual Meeting, No. 6.24, 1971.
154. ARCIDIACONO V., FASANI A., FERRARI E., SACCOMANNO F., *Prove sperimentali di identificazione di reti elettriche*, AEI Annual Meeting No. 6.25, 1971.
155. UNDRILL J. M., TURNER A. E., *Construction of power system electromechanical equivalents by modal analysis*, *IEEE Trans. on PAS*, Sept.-Oct. 1971.
156. UNDRILL J. M., CASAZZA J. A., GULACHESKI E. M., KIRCHMAYER L. K., *Electromechanical equivalents for use in power system stability studies*, *IEEE Trans. on PAS*, Sept.-Oct. 1971.
157. GLOVER J. D., SCHWEPPE F. C., *Advanced load frequency control*, *IEEE Trans. on PAS*, No. 5, 1972.

158. SACCOMANNO F., *Development and evaluation of simplified dynamical models for multimachine electric power systems*, PSCC, 1972.
159. DE MELLO F. P., MILLS R. J., B'RELLS W. F., *Automatic generation control.* Pt.I: *Process modeling*. Pt.II: *Digital control techniques*, IEEE PES Summer meeting, July 1972.
160. DEVILLE T. G., SCHWEPPE F. C., *On-line identification of interconnected network equivalents from operating data*, IEEE PES Summer meeting, July 1972.
161. CUSHING E. W., DRECHSLER G. E., KILLGOAR W. P., MARSHALL H. G., STEWART H. R., F*ast valving as an aid to power system transient stability and prompt resynchronization and rapid reload after full load rejection*, *IEEE Trans. on PAS*, July-August 1972.
162. *Dynamic models for steam and hydro turbines in power systems studies*, IEEE Winter meeting, Jan.-Febr. 1973.
163. MARIANI E., *Esercizio giornaliero di centrali idroelettriche modulabili*, *L'energia elettrica*, No. 1, 1973.
164. *System load dynamics. Simulation effects and determination of load constants*, *IEEE Trans. on PAS*, March-April 1973.
165. SASSON A. M., EHRMAN S. T., LYNCH P., VAN SLYCK L. S., *Automatic power system network topology determination*, *IEEE Trans. on PAS*, March-April 1973.
166. FRIEDMAN P. G., *Power dispatch strategies for emission and environmental control*, Proc. of the Instr. Soc. of America, 1973.
167. DY LIACCO T. E., RAMARAO K., WEINER A., *Network status analysis for real time systems*, PICA Conf., 1973.
168. CARPENTIER J. W., *Differential injections model, a general method for secure and optimal load flows*, PICA Conf., 1973.
169. DOPAZO J. F., KLITIN O. A., SASSON A. M., *State estimation for power systems: detection and identification of gross measurement errors*, PICA Conf., 1973.
170. ITAKURA S., YOKOYAMA R., TAMURA Y., *Construction of dynamic equivalents and determination of suboptimal control based on aggregation*, CIGRE S.C. 32, 1973.
171. SACCOMANNO F., *Global methods of stability assessment*, UMIST Symp., Manchester, Sept. 1973.
172. LEE S. T. Y., SCHWEPPE F. C., *Distance measures and coherency recognitions transient stability equivalents*, *IEEE Trans. on PAS*, Sept.-Oct. 1973.
173. REITAN D. K., RAMA RAO N., *A method of improving transient stability by bang-bang control of tie-line reactance*, *IEEE Trans. on PAS*, No. 1, 1974.
174. CAMERON D. E., KOEHLER J. E., RINGLEE R. J., *A study mode multi-area economic dispatch program*, IEEE PES Winter meeting, 1974.
175. *Long term power system dynamics*, EPRI-EL-367 final report, June 1974.
176. IONESCU et al., *Note rumaine sur les equivalents du réseau utilisés dans le calcul de la stabilité dynamique*, CIGRE Task force on Dynamic equivalents, 1974.
177. ALSAC O., STOTT B., *Optimal load flow with steady-state security*, *IEEE Trans. on PAS*, May-June 1974.
178. STOTT B., ALSAC O., *Fast decoupled load flow*, *IEEE Trans. on PAS*, May-June 1974.
179. HAPP H. H., *Optimal power dispatch*, *IEEE Trans. on PAS*, May-June 1974.
180. SACCOMANNO F., *Dynamic modelling of multimachine electric power systems*, Formator Symposium, Praga, June 1974.
181. *Report on the tests of parallel operation between Yugoslav and Italian power systems made in April 1974 (from 17th to 19th): damping of electromechanical oscillations*

in Yugoslav network by means of additional signals in Djerdap station excitation control, JUGEL-ENEL Report, SUDEL meeting, 1974.

182. SCHWEPPE F. C., HANDSCHIN E., *Static state estimation in power systems*, *IEEE Proc.*, July 1974.
183. SACCOMANNO F., *Contributo allo studio degli smorzamenti elettromeccanici nelle reti elettriche*, *L'energia elettrica*, No. 8, 1974.
184. DANESI G., GIORGI A., *Alcune considerazioni sulla regolazione della tensione in fase e in quadratura nelle reti di trasmissione*, AEI Annual Meeting, No. A.31, 1974.
185. INNORTA M., MARANNINO P., MOCENIGO M., *Programmazione economica delle generazioni attive e reattive nell'ambito di vincoli di sicurezza e qualità di servizio*, AEI Annual Meeting, No. A.46, 1974.
186. ALBANI G., ANDERLONI A., DURAZZO M., *Una apparecchiatura per l'equilibratura dei carichi nell'esercizio della rete elettrica in condizioni di emergenza*, AEI Annual Meeting, No. A.66, 1974.
187. ARIATTI F., BRASIOLI C., DI PERNA A., GALLI F., QUAZZA G., *Orientamenti per un nuovo sistema di controllo della produzione e trasmissione ENEL*, AEI Annual Meeting, No. A.67, 1974.
188. ILICETO F., *Considerazioni sulla possibilità di trasmissione dell'energia con linee compensate in corrente alternata a distanze superiori a 1500 km*, AEI Annual Meeting No. A.78, 1974.
189. ARIATTI F., MARZIO L., *Stima dello stato di una rete di trasmissione in regime permanente*, AEI Annual Meeting, No. A.80, 1974.
190. DI CAPRIO U., MARCONATO R., MARIANI E., *Studio di alcuni piani per il controllo in emergenza di una rete elettrica a mezzo di alleggerimento automatico del carico*, AEI Annual Meeting, No. A.89, 1974.
191. ARCIDIACONO V., FERRARI E., SACCOMANNO F., *Recenti studi ed esperienze sullo smorzamento di oscillazioni in reti elettriche*. Part I: *Schematizzazioni del problema e risultati analitici*, AEI Annual Meeting, No. A.92, 1974.
192. ARCIDIACONO V., FERRARI E., TAGLIABUE G., (*Idem.*) Part II: *Studi mediante simulazioni e risultati sperimentali di stabilizzazione tramite controllo di generatori*, AEI Annual Meeting, No. A.93, 1974.
193. CLERICI A., MAZZOLA L., RUCKSTUHL G., *Sistemi di trasporto di energia elettrica a grande distanza ed altissima tensione: esperienza CESI in studi di stabilità e regolazione di tensione e della potenza reattiva*, AEI Annual Meeting, No. A.104, 1974.
194. GAGLIARDI F., *Sulla opportunità dell'utilizzazione dei trasformatori a rapporto variabile sotto carico nelle reti a maglie B.T.*, AEI Annual Meeting, No. A.105, 1974.
195. FERRARI E., *Report on the answers to the questionnaire on characteristics and performances of turbine speed governors*, Electra, Oct. 1974.
196. WOODFORD D. A., TARNAWECKY M. Z., *Compensation of long distance ac transmission lines by shunt connected reactance controllers*, *IEEE Trans. on PAS*, March-April 1975.
197. HANDSCHIN E., SCHWEPPE F. C., KOHLAR J., FIECHTER A., *Bad data analysis for power system state estimation*, *IEEE Trans. on PAS*, March-April 1975.
198. BAUER D. L. et al., *Simulation of low frequency undamped oscillations in large power systems*, *IEEE Trans. on PAS*, March-April 1975.
199. *A status report on methods used for system preservation during under-frequency conditions*, *IEEE Trans. on PAS*, March-April 1975.
200. DENZEL D., GRAF F. R., VERSTEGE J. F., *Practical use of equivalents for unobservable networks in on-line security monitoring*, PSCC, 1975.

201. Saccomanno F., *Sensitivity analysis of the characteristic roots of a linear time-invariant dynamic system: application to the synthesis of damping actions in electric power systems*, IFAC Congress, August 1975.
202. Arcidiacono V., Ferrari E., Saccomanno F., *Studies on damping of electromechanical oscillations in multimachine systems with longitudinal structure*, IEEE Summer meeting, 1975.
203. Arcidiacono V., Ferrari E., Marconato R., Brkic T., Niksic M., Kajari M., *Studies and experimental results on electromechanical oscillation damping in Yugoslav power system*, IEEE PES Summer meeting, 1975.
204. Burns R. M., Gibson C. A., *Optimization of priority lists for a unit commitment program*, IEEE PES Summer meeting, 1975.
205. Di Perna A., Mariani E., *Medium term production schedules in a hydro-thermal electric power system*, PSCC, Sept. 1975.
206. Spalding B. D., Goudie D. B., Yee H., Evans F. J., *Use of dynamic equivalents and reduction techniques in power system transient stability computation*, PSCC, 1975.
207. Knight U. G., *A survey of control strategies in emergency*, Electra, Jan. 1976.
208. Kilgore L. A., Ramey D. G., Hall M. C., *Simplified transmission and generation system analysis procedures for subsynchronous resonance problems*, IEEE Winter meeting, 1976.
209. Bowler C. E. J., *Understanding subsynchronous resonance*, IEEE Symp., 1976.
210. Converti V., Gelopulos D. P., Housley M., Steinbrenner G., *Long-term stability solution of interconnected power system, IEEE Trans. on PAS*, 1976.
211. Davison E. J., Tripathi N. K., *The optimal decentralized control of a large power system: load and frequency control*, Proc. CDC, 1976.
212. Fenwick D. R. et al., *Review of trends in excitation systems and possible future developments, Proc. IEE*, No. 5, 1976.
213. Reddy K. R., Dave M. P., *Non-interacting LFC of interconnected power systems by state feedback*, IEEE Summer meeting, July 1976.
214. Arcidiacono V., Ferrari E., Marconato R., Saccomanno F., *Analysis of factors affecting the damping of low-frequency oscillations in multimachine systems*, CIGRE, 1976.
215. Arcidiacono V., Brkic T., Epitropakis E., Ferrari E., Marconato R., Saccomanno F., *Results of some recent measurements of low-frequency oscillations in a european power system with longitudinal structure*, CIGRE, 1976.
216. *Report on tests of parallel operation between Greece and Yugoslavia in autumn 1975*, JUGEL-PPC Report, SUDEL meeting, 1976.
217. Esclangon P. E., Pioger G., *Voltage regulation on the french EHV network*, IFAC Symposium, Melbourne, 1977.
218. Calvaer A. J., *Diffusion of reactive power perturbations and some related problems*, IFAC Symposium, Melbourne, 1977.
219. Elgerd O., *Automatic generation control*, IFAC Symposium, Melbourne, 1977.
220. Najaf-zadeh K., Nikolas J. T., Anderson S. W., *Optimal power system operation analysis techniques*, Proc. Am. Power Conf., 1977.
221. Germond A., Podmore A. R., *Development of dynamic equivalents for transient stability studies*, EPRI project RP763, May 1977.
222. Happ H. H., *Optimal power dispatch. A comprehensive survey, IEEE Trans. on PAS*, May-June 1977.

223. KIMBARK E. W., *How to improve system stability without risking subsynchronous resonance*, *IEEE Trans. on PAS*, 1977.
224. ILICETO F., CINIERI E., *Comparative analysis of series and shunt compensation schemes for ac transmission systems*, *IEEE Trans. on PAS*, 1977.
225. ARCIDIACONO V., CORSI S., GARZILLO A., MOCENIGO M., *Studies on area voltage and reactive power control at ENEL*, CIGRE, Dortmund, 1977.
226. BERKOVITCH M. A. et al., *The emergency automation of the USSR united power system*, Sov. power eng., Dec. 1977.
227. BARBIER C., CARPENTIER L., SACCOMANNO F., *Tentative classification and terminologies relating to stability problems of power systems*, Electra, Jan. 1978.
228. ELSLIGER R. et al., *Optimization of Hydro-Québec's 735-kV dynamic-shunt-compensated system using static compensators on a large scale*, IEEE Winter meeting, 1978.
229. LACHS W. R., *Voltage collapse in EHV power systems*, IEEE Winter meeting, 1978.
230. WALKER D. N., BOWLER C. E. J., BAKER D. H., *Torsional dynamics of closely coupled turbine generators*, *IEEE Trans. on PAS*, 1978.
231. FINK L. H., CARLSEN K. C., *Operating under stress and strain*, IEEE Spectrum, March 1978.
232. SEKI A., NISHIDAI J., MUROTANI K., *Suppression of flicker due to arc furnaces by a thyristor-controlled VAr compensator*, IEEE Summer meeting, July 1978.
233. PRICE W. W., GULACHENSKI E. M., KUNDUR P., LANGE F. J., LOEHR G. C., ROTH B. A., SILVA R. F., *Testing of the modal dynamic equivalents technique*, *IEEE Trans. on PAS*, July-August 1978.
234. DY LIACCO T. E., *An overview of practices and trends in modern power system control centres*, IFAC Congress, 1978.
235. MARIANI E., MARZIO L., RICCI P. et al., *Miglioramenti all'esercizio apportabili tramite il nuovo sistema di controllo centralizzato dell'ENEL*. Part I: *Criteri generali di progetto del sistema di controllo*. Part II: *L'on-line dispatching delle centrali termoelettriche*, AEI Annual Meeting, Nos. I.43 and I.59, 1978.
236. DY LIACCO T. E., SAVULESCU S. C., RAMARAO K. A., *An online topological equivalent of a power system*, *IEEE Trans. on PAS*, Sept.-Oct. 1978.
237. DILLON T. S., EDWIN K. W., KOCHS H. D., TAUD R. J., *Integer programming approach to the problem of optimal unit commitment with probabilistic reserve determination*, *IEEE Trans. on PAS*, Nov.-Dec. 1978.
238. SACCOMANNO F., *Analysis of electromechanical phenomena in multimachine electric power systems: some new extensions and generalizations*, *Ricerche di automatica*, No. 2, 1978.
239. KAPOOR S. C., *Dynamic stability of long transmission systems with static compensators and synchronous machines*, *IEEE Trans. on PAS*, No. 1, 1979.
240. WIRGAN K. A., *Reactive power dispatching*, *Electric forum*, No. 1, 1979.
241. CASTRO-LEON E. G., EL-ABIAD A. H., *Bibliography on power system dynamic equivalents and related topics*, IEEE Winter meeting, 1979.
242. BOIDIN M., DROUIN G., *Performance dynamiques des compensateurs statiques à thyristors et principles de regulation*, *Rev. gen. de Electr.*, 1979.
243. SHACKSHAFT G., HENSER P. B., *Model of generator saturation for use in power system studies*, *Proc. IEE*, No. 8, 1979.
244. ASHMOLE P. H., *Technical summary of reactive compensation in power systems industrial applications*, *IEE Digest*, 1979.
245. HOSONO I. et al., *Suppression and measurement of arc furnace flicker with a large static VAr compensator*, *IEEE Trans. on PAS*, 1979.

246. Schiavi A., *L'automazione ed il controllo centralizzato degli impianti elettrici di produzione e trasmissione dell'ENEL*, AEI-Anipla Annual Meeting, No. 54, 1979.
247. Busi T., Maffezzoni C., Quatela S., *Caratteristiche dei sistemi di controllo dei gruppi termoelettrici e metodi moderni per il loro progetto*, AEI-Anipla Annual Meeting, No. 110, 1979.
248. Arcidiacono V., *Metodi e risultati sperimentali di identificazione del modello matematico di grandi alternatori*, AEI-Anipla Annual Meeting, No. 151, 1979.
249. Dopazo J. F., Irisarri G., Sasson A. M., *Real-time external system equivalent for on-line contingency analysis*, *IEEE Trans. on PAS*, Nov. 1979.
250. Knight U. G., *Aids for the emergency control of power systems*, Electra, Dec. 1979.
251. Kokotovic P. V., Allemong J. J., Winkelman J. R., Chow J. H., *Singular perturbations and iterative separation of time scales*, Automatica, 1980.
252. Capasso A., Mariani E., Sabelli C., *On the objective functions for reactive power optimization*, IEEE Winter meeting, 1980.
253. Stott B., Alsac O., Marinho J. L., *The optimal power flow problem*, SIAM Int. Conf., March 1980.
254. Arcidiacono V., Ferrari E., Marconato R., Dos Ghali J., Grandez D., *Evaluation and improvement of electromechanical oscillation damping by means of eigenvalue-eigenvector analysis. Practical results in the central Peru power system*, *IEEE Trans. on PAS*, March-April 1980.
255. Arnold C. P., Turner K. S., Arrillaga J., *Modelling rectifier loads for a multimachine transient stability programme*, *IEEE Trans., PAS-99*, 1980.
256. Avramovic B., Kokotovic P. V., Winkelman J. R., Chow J. H., *Area decomposition for electromechanical models of power systems*, IFAC Symposium, Toulouse, June 1980.
257. Krumpholz G. R., Clements K. A., Davis P. W., *Power system observability: a practical algorithm using network topology*, *IEEE Trans. on PAS*, July-August 1980.
258. Calvaer A. J., Boulanger P., *Application of the external equivalents in the case of machine outages in the study system*, IFAC Symposium, Pretoria, 1980.
259. Barbier C., Barret J. P., *An analysis of phenomena of voltage collapse on a transmission system*, *Rev. Gen. Electr.*, Oct. 1980.
260. Deckmann S., Pizzolante A., Monticelli A., Stott B., Alsac O., *Studies on power system load flow equivalencing*, *IEEE Trans. on PAS*, Nov. 1980.
261. *Description and bibliography of major economy-security functions*. Pt.I/III, *IEEE Trans. on PAS*, Jan. 1981.
262. Meisel J., Sen A., Gilles M. L., *Alleviation of a transient stability crisis using shunt braking resistors and series capacitors*, Electrical power and energy systems, Jan. 1981.
263. *Proposed terms and definitions for power system stability*, IEEE paper, Febr. 1981.
264. Arnold C. P., Duke R. M., Arrillaga J., *Transient stability improvement using thiristor controlled quadrature voltage injection*, *IEEE Trans. on PAS*, March 1981.
265. Kalnitsky K. C., Kwatny H. G., *A first principles model for steam turbine control analysis*, *Journ. of Dynamic systems measurement and control*, March 1981.
266. Aldrich J. F., Fernandes R. A., Happ H. H., Wirgau K. A., *The benefits of voltage scheduling in system operations*, *IEEE Trans. on PAS*, March 1981.
267. Savulescu S. C., *Equivalents for security analysis of power systems*, *IEEE Trans. on PAS*, May 1981.
268. Burchett R. C., Happ H. H., Vierath D. R., Wirgau K. A., *Developments in optimal power dispatch*, IEEE PICA Conf., May 1981.

269. Happ H. H., Wirgau K. A., *A review of the optimal power flow*, *J.Franklin Inst.*, 1981.
270. Kakimoto N., Hayashi M., *Transient stability analysis of multimachine power system with automatic voltage regulator via Lyapunov's direct method*, PSCC, July 1981.
271. Franchi L., Innorta M., Marannino P., Mariani E., Sabelli C., Sapora D., *Optimal short-term reactive scheduling for a large power system*, PSCC, July 1981.
272. Price W. W., Roth B. A., *Large-scale implementation of modal dynamic equivalents*, *IEEE Trans. on PAS*, August 1981.
273. Ribbens Pavella M., Evans F. J., *Direct methods in the study of the dynamics of large-scale electric power systems. A survey*, IFAC Congress, August 1981.
274. Olwegard A., Walve K., Waglund G., Frank H., Torseng S., *Improvement of transmission capacity by thyristor controlled reactive power*, *IEEE Trans. on PAS*, No. 8, 1981.
275. Venikov V. A., Stroev V. A., Tawfik M. A. H., *Optimal control of electrical power systems containing controlled reactors*. Pt.I: *Effect of controlled reactors on the transient stability limit*. Pt.II: *Optimal control problem solution*, *IEEE Trans. on PAS*, No. 9, 1981.
276. Housos E., Irisarri G., *Real-time results with online network equivalents for control centre applications*, *IEEE Trans. on PAS*, Dec. 1981.
277. Lawler J., Schlueter R. A., *An algorithm for computing modal coherent equivalents*, *IEEE Trans. on PAS*, 1982.
278. Allemong J. J., Radu L., Sasson A. M., *A fast and reliable state estimation algorithm for AEP's new control center*, *IEEE Trans. on PAS*, April 1982.
279. Ishikawa T., Akita S., Tanaka T., Kaminosono H., Masuda M., Shintomi T., *Power system stabilization by superconducting magnetic energy storage system*, CRIEPI Report, April 1982.
280. Mamandur K. R. C., *Emergency adjustments to VAr control variables to alleviate over-voltages, under-voltages, and generator VAr limit violations*, *IEEE Trans. on PAS*, No. 5, 1982.
281. Van Amerongen R. A. M., Van Meeteren H. P., *A generalized Ward equivalent for security analysis*, *IEEE Trans. on PAS*, June 1982.
282. Schultz R. D., Smith R. A., Stevens R. A., *Incorporation of linearized sensitivities in the Ward method of equivalencing*, IEEE Summer meeting, 1982.
283. Van De Meulebroeke F., Debelle J., *Fast controlled valving: a new philosophy for turbine control*, CIGRE, 1982.
284. Marconato R., Vergelli L., *Problems concerning the design of automatic load shedding plans*, CIGRE, 1982.
285. Burri P., Zajc G., *A quality index for evaluating network equivalents*, CIGRE, 1982.
286. Abe S., Fukunaga Y., Isono A., Kondo B., *Power system voltage stability*, *IEEE Trans. on PAS*, 1982.
287. Baker R., Guth G., Egli W., Eglin P., *Control algorithm for a static phase shifting transformer to enhance transient and dynamic stability of large power systems*, *IEEE Trans. on PAS*, No. 9, 1982.
288. Byerly R. T., Taylor E. R., *Static reactive compensation for power transmission systems*, *IEEE Trans. on PAS*, No. 10, 1982.
289. Ferrari E., *Block diagrams and typical transfer functions of turbine speed regulations systems*, CIGRE WG 31/32-O3, Nov. 1982.

290. Barbier C., Di Caprio U., Humphreys P., *The techniques and applications of power system dynamics equivalents at CEGB, EDF and ENEL, L'energia elettrica*, Dec. 1982.
291. *Selective modal analysis in power systems*, EPRI Report, Jan. 1983.
292. Sharaf A. M., Doraiswami R., *Stabilizing an ac link by using static phase shifters, IEEE Trans. on PAS*, n.4, 1983.
293. Hiroshi Suzuki et al., *Effective application of static VAr compensators to damp oscillations*, IEEE Summer meeting, 1983.
294. Arcidiacono V., *Automatic voltage and reactive power control in transmission systems*, CIGRE-IFAC Symposium, Florence, Sept. 1983.
295. Craven R. H., Michael M. R., *Load characteristic measurement and the representation of loads in the dynamic simulation of the Queensland power system*, CIGRE-IFAC Symposium, Florence, Sept. 1983.
296. He Ren-mu, Germond A., *Identification de modeles de charges en fonction de la tension et de la fréquence*, CIGRE-IFAC Symposium, Florence, Sept. 1983.
297. Handschin E., Reissing T., *Theory and practice of load modelling for power system dynamics*, CIGRE-IFAC Symposium, Florence, Sept. 1983.
298. *Bibliography of literature on steam turbine-generator control systems, IEEE Trans. on PAS*, Sept. 1983.
299. Wu F. F., Monticelli A., *Critical review of external network modelling for on-line security analysis*, Electrical power and energy systems, Oct. 1983.
300. *Coordinated decentralized emergency operating state control*, Report of the U.S. Department of Energy, Nov. 1983.
301. Newell R. J., Stuhem D. L., *Coherency based dynamic equivalents*, Minnesota P.S. Conf., Nov. 1983.
302. Grebe E., *Entwurf der turbinenregelung under berucksichtigung der mittelzeitdynamik von kraftwerk und netz*, ETZ Archiv, 1984.
303. Biancoli L., *Tecnologia dei sistemi di dispacciamento elettrico*, Automazione e strumentazione, June 1984.
304. Arcidiacono V., Pelizzari G., *Problemi di stabilità del ciclo di corrente del raddrizzatore corso nel funzionamento biterminale con una delle due stazioni italiane*, ENEL Report, July 1984.
305. Bose A., *Modeling of external networks for on-line security analysis, IEEE Trans. on PAS*, August 1984.
306. Paris L., Zini G., Valtorta M., Manzoni G., Invernizzi A., De Franco N., Vian A., *Present limits of very long transmission systems*, CIGRE 1984.
307. Hein F., Denzel D., Graf F. R., Schwarz J., *On-line security analysis using equivalents for neighbouring networks*, CIGRE 1984.
308. *Systems research*, Program report, U.S. Department of Energy, Dec. 1984.
309. Newell R. J., Rao K. S., *Utility experience with dynamic equivalents of very large systems*, IEEE PES Winter meeting, 1985.
310. Chow J. H., Kokotovic P. V., *Time-scale modeling of sparse dynamic networks, IEEE Trans. on AC*, 1985.
311. Brown J. E., Hallenius K. E., Vas P., *Computer simulation of saturated cylindrical rotor synchronous machines*, IMACS Congress, 1985.
312. Mariani E., *Programmazione a lungo, medio e breve termine dell'esercizio di una rete elettrica a generazione idraulica e termica, L'energia elettrica*, No. 5, 1985.
313. *IEEE guide for Protection, interlocking and control of fossil-fueled unit-connected steam stations*, IEEE Std 502, June 1985.

314. Mocenigo M., *Controllo in tempo reale di una grande rete elettrica*, Automazione e strumentazione, Nov. 1985.

315. *Turbine fast valving to aid system stability: benefits and other considerations*, *IEEE Trans. on PS*, No. 1, 1986.

316. Machowski J., Cichy A., Gubina F., Omaken O., *Modified algorithm for coherency recognition in large electrical power systems*, IFAC Symposium, Beijing, 1986.

317. Geeves S., *A modal-coherency technique for deriving dynamic equivalents*, IEEE Winter meeting, Febr. 1987.

318. Nozari F., Kankam M. D., Price W. W., *Aggregation of induction motors for transient stability load modelling*, *IEEE Trans.*, PWRS-2, 1987.

319. Iliceto F., Cinieri E., *Analysis of half-wave length transmission lines with simulation of corona losses*, IEEE, 1987.

320. Fusco G., Marconato R., Menditto V., Natale A., Scarpellini P., *Azioni automatiche di controllo del sistema di produzione e trasmissione dell'ENEL in condizioni di emergenza*, AEI Annual Meeting, No. 1.c.7, 1987.

321. Demartini G., Franchi L., Mocenigo M., Sapora D., Scapeccia D., *Affidabilità delle informazioni nel controllo in tempo reale: esperienze operative della stima dello stato*, AEI Annual Meeting, No. 1.c.3, 1987.

322. Canepa A., Delfino B., Invernizzi M., Pinceti P., *Voltage regulation via automatic load tap changing transformers: evaluation of voltage stability conditions*, Electric power systems research, 1987.

323. Hingorani N. G., *Power electronics in electric utilities: role of power electronics in future power systems*, IEEE Proceedings, Apr. 1988.

324. Granelli G., Innorta M., Marannino P., Montagna M., Silvestri A., *Security constrained dynamic dispatch of real power for thermal groups*, *IEEE Trans. on PAS*, May 1988.

325. Giglioli R., Paris L., Zini G., Saccomanno F., Morrissy C. A., Vian A., *Reactive power balance optimization to improve the energy transfer through ac transmission systems over very long distance*, CIGRE, 1988.

326. Arcidiacono V., Corsi S., Natale A., Raffaelli C., *New developments in the applications of ENEL transmission system automatic voltage and reactive power control*, CIGRE, 1990.

327. Germond A. J., Clerfeuille J. P., Geeves S., Graf F. R., Saccomanno F., Van Amerongen R. A. M., *Steady-state and dynamic external equivalents*. Pt.I: *State-of-the-art report*. Pt.II: *Results of a questionnaire*, Electra, February and April 1991.

328. Delfino B., Invernizzi M., Marconato R., Morini A., Scarpellini P., *Considerations on modelling and behaviour of some typical industrial loads following large variations of frequency and voltage*, SEPOPE Conference, May 1992.

329. Jaleeli N., Van Slyck L. S., Ewart D. N., Fink L. H., Hoffmann A. G., *Understanding Automatic Generation Control*, IEEE Trans. on Power Systems August 1992.

330. Gyugyi L., Schauder C. D., Williams S. L., Rietman T. R., Torgerson D. R., Edris A. D., *Unified power controller: a new approach to power transmission control*, IEEE Trans. on Power Delivery, April 1995.

331. Corsi S., Marannino P., Losignore N., Moreschini G., Piccini G., *Coordination between the reactive power scheduling function and the hierarchical voltage control at the EHV ENEL System*, IEEE Trans. on Power Systems, May 1995.

332. Concordia C., Fink L. H., Poullikkas G., *Load shedding on an isolated system*, IEEE Trans. on Power Systems, August 1995.
333. Bagnasco A., Delfino B., Denegri G. B., Massucco S., *Management and dynamic performances of combined cycle power plants during parallel and islanding operation*, IEEE Trans. on Energy Conversion, June 1998.
334. Aresi R., Delfino B., Denegri G. B., Invernizzi M., Massucco S., *Dynamic equivalents of tap-changing-under-load transformers in a transmission system*, IEE Proceedings On Generation Transmission Distribution, January 1999.
335. Dy Liacco T., *Enabling technologies for operation and real-time control in a liberalized environment*, EPRI, November 1999.
336. Cigre Task Force 38.03.12, *Advanced angle stability controls*, December 1999.
337. Special issue on the *Technology of power system competition*, IEEE Proceedings, February 2000.
338. Delfino B., Fornari F., Massucco S., Mocenigo M., Sforna M., *Evaluating different load-frequency control schemes for restructured power systems*, CIGRE 2000.
339. Delfino B., Massucco S., Morini A., Scalera P., Silvestro F., *Implementation and comparison of different under frequency load-shedding schemes*, IEEE Summer Meeting, July 2001.

INDEX

ABOUT THE AUTHOR

Fabio Saccomanno was born in Genoa, Italy in 1933. He received the Laurea degree in electrical engineering at the University of Genoa in 1957. Professor Saccomanno has acquired his vast experience in the field power systems through a variety of sources. He began his career in industry as designer of control devices and analyst of control problems. Then called by ENEL (the Italian government agency for electrical power), he headed the System Analysis and Hybrid Computation offices and was a research leader for several international projects in the field of analysis and control of electric power systems. His original contributions cover almost all the subjects included in the book.

Currently, Dr. Saccomanno is Full Professor of Automatic Controls in the Electrical Engineering Department at the University of Genoa. His teaching activities also include a second lecture of Automation of Power Systems. He has been active on the international level (CIGRE, IFAC, IEEE, PSCC, PICA, etc.) as a member of committees and working groups, chairman at congresses, and lecturer and coordinator of advanced courses for technicians and researchers. His past positions include Chairman of the IEEE North-Italy Section.

TECHNICAL EDITORS

Stefano Massucco was born in Genoa, Italy in 1954. He received the Laurea degree in electrical engineering at the University of Genoa in 1979. After eight years at the Electric Research Center of the Italian Electricity Board (ENEL) and at Ansaldo, he became Associate Professor at the University of Pavia and then joined the Electrical Engineering Department of the University of Genoa, where he is currently Full Professor of Power System Automation. His main research interests are in the field of power systems design and control, power systems analysis and simulation, and intelligent systems application to power systems. Professor Massucco is a member of the IEEE PES-Power Engineering Society.

Prabha Kundur holds a Ph.D. from the University of Toronto and has more than 30 years of experience in the electric power industry. He is currently President and CEO of Powertech Labs Inc., the research and technology subsidiary of BC Hydro. Prior to joining Powertech Labs in 1993, he worked at Ontario Hydro for 25 years and was involved in the planning and design of power systems. He has also served as an adjunct professor at the University of Toronto and the University British Columbia. He is the author of the book *Power System Stability and Control* (McGraw-Hill, 1994). He has chaired numerous IEEE and CIGRE committees and working groups. He is the recipient of the 1997 IEEE Nikola Tesla Award and the 1999 CIGRE Technical Committee Award.